D1716166

# CRYSTAL STRUCTURES

## I. Patterns and Symmetry

M. O'KEEFFE
Department of Chemistry
Arizona State University
Tempe AZ 85287

B. G. HYDE
Research School of Chemistry
The Australian National University
Canberra ACT 0200

MINERALOGICAL SOCIETY OF AMERICA
WASHINGTON, D.C.

# CRYSTAL STRUCTURES

## I. Patterns and Symmetry

Michael O'Keeffe and Bruce G. Hyde



Printed by
BookCrafters, Inc., Chelsea, Michigan, U.S.A.

ISBN 0-939950-40-5

Published by
**MINERALOGICAL SOCIETY OF AMERICA**
1015 Eighteenth Street, NW, Suite 601
Washington, DC 20036 USA

Cover: the pattern that results when the twenty faces of an icosahedron are capped by truncated tetrahedra. This occurs as an 84-atom unit of the structure of $\beta$-rhombohedral boron. The same pattern is conspicuous in the structure of some intermetallic compounds (see p. 160-163).

MINERALOGICAL SOCIETY OF AMERICA

*MONOGRAPH SERIES*

## FOREWORD

In 1993 the Mineralogical Society of America (MSA) began publishing its Monograph Series with a major work by Frank S. Spear, entitled *Metamorphic Phase Equilibria and Pressure-Temperature-Time Paths*. MSA's second monograph was a reprinting of a slightly revised version of the 1971 book, *Crystallography and Crystal Chemistry*, by F. Donald Bloss.

This volume by Michael O'Keeffe and Bruce G. Hyde is the third contribution. As Series Editor, I was responsible for obtaining reviewers. Mike O'Keeffe prepared the entire camera-ready text and the figures which were all drawn specially for the book. As noted in the Preface below, a second volume, *Crystal Structures II. Inorganic Materials* is anticipated in the near future.

Paul H. Ribbe
Blacksburg, Virginia

## PREFACE

This book (the first of two volumes) is devoted to the topic of the description of structures, especially periodic structures, and their symmetries. Much of the material is a prerequisite for serious students of solid state chemistry and related sciences (e.g. mineralogy, materials science and solid state physics). Earlier drafts served as part of the lecture notes used for some years for a course in solid state chemistry at Arizona State University; the order of presentation of topics, occasional repetitiveness and the sometimes hectoring tone to some extent reflect this origin.

From a chemist's point of view, probably the most fundamental piece of information about a chemical compound is the way its atoms are arranged in space. For small molecules this information can be fairly readily assimilated, but the task becomes increasingly difficult for macromolecules, as any student of biochemistry can attest. For crystalline solids (which are really "infinite" molecules) the difficulty is equally great, and requires learning the methods appropriate to describing infinite periodic objects. These methods are generally unfamiliar to those who are not professional crystallographers (a fact that greatly hinders the development of solid state chemistry) and one of the primary aims of this book is to provide a usable introduction to them. This leads inexorably to a discussion of symmetry (translation is a symmetry operation) so we devote the first three chapters to an introduction to point and space symmetry groups using crystallographic conventions.

These three chapters are intended to introduce the language of space groups, and are neither rigorous nor complete. The goal is to enable the reader to be in a position to be able to extract and understand useful information from the crystallographic literature that is the primary source of information about the structure of solids and which contains enormous quantities of buried treasure.

Chapter 4 treats crystal geometry and includes a discussion of transformations of coordinate systems (unit cell transformations). This topic is essential to aspects of crystal chemistry such as describing structural relationships. The chapter also contains a compendium of formulas useful for crystallographic calculations. They are given without proof, but the derivations are, for the most part, elementary and to be found in standard crystallographic texts.

Chapters 5-7 are devoted to the description of simple geometrical patterns that underlie crystal structures. The development proceeds from simple to more complex: polyhedra, clusters of polyhedra, plane patterns, layers of polyhedra, sphere packings, cylinder packings, nets and infinite polyhedra. Many examples are given of the occurrence of these patterns in crystal structures. Most of the more common binary and ternary crystal structures are introduced along the way and, to enhance the value of the book as a self-contained text, we summarize them and their occurrence in chemical compounds in Appendix 5 at the end. (The second volume of this series, which is longer than this one, is devoted to a more complete account of crystal structures.)

Each chapter ends with sections entitled "Notes" and "Exercises." The former are often extended footnotes, somewhat peripheral to the main theme, but considered useful or amusing (or both); they often contain useful reference material or refer to some more-advanced topics. The Exercises are "homework", designed to provide a diagnosis of the understanding of the material in the chapter and also often contain useful results.

The first four appendices are "Notes" that became too long for inclusion in individual chapters. We hope that they will whet the appetites of more-adventuresome readers. The relegation of some material to Notes and Appendices necessitates some cross referencing. This seems to us to be a small price to pay for having a book that can serve both as a textbook for courses and as a general reference. A book with these aspirations has of necessity a split personality. In Chapters 5-7, some sections could well be omitted in a formal course at the instructor's discretion. The same is true of the material in the Notes.

No novelty (except perhaps for the errors) is claimed for the first four chapters, although we hope that even some experienced readers might find one or two items that provide food for thought. The later chapters do (we believe) contain some material of interest that has not been published before.

Although the book has a distinct crystallographic flavor, it is definitely not a textbook on crystallography and we do not discuss, except peripherally, topics such as quasicrystals and incommensurate crystals that are currently of active concern to inorganic crystallographers and sometimes even the occasion for controversy.

We have spent some years on the study of structures, a delightful pursuit that would not have been possible without the generous support, for which we are indebted, of our respective institutions. M. O'K. also gratefully acknowledges the support of the U.S. National Science Foundation which has funded a program of research in crystal chemistry for a number of years. We came to the study of crystal structures indirectly and in complete ignorance. Over the years we slowly collected a body of information needed to pursue research in the field. We often wished that someone had provided us with this material, and the present text attempts to provide such a package for students who may be in the position we were in 40 years ago.

The text has benefited particularly from the sure touch of Paul Ribbe, who read it in its entirety, identified errors, solecisms, and infelicities, and also made valuable suggestions for improving the presentation. We owe him special thanks.

The book could not have been written without the seemingly infinite patience and forbearance of our wives, Lita O'Keeffe and Marie Hyde, and we dedicate it to them.

Michael O'Keeffe
Tempe

Bruce Hyde
Canberra

## *A Note to the Reader*

Matrices are represented by bold capitals thus: **A**; and often written row-by-row on one line as ($a_{11}$ $a_{12}$ $a_{13}$ / $a_{21}$ $a_{22}$ $a_{23}$ / $a_{31}$ $a_{32}$ $a_{33}$). $\mathbf{A^t}$ is the transpose of **A**. Column vectors are lower case as: **a**. The corresponding row vector is $\mathbf{a^t}$. The magnitude of **a** is *a*. **E** is a unit matrix (with elements $e_{ii} = 1$, $e_{ij} = 0$).

Bold face is also generally used for names of structures. Thus **bcc** refers to the arrangement of points on a body-centered cubic lattice and **NaCl** refers to the structure of NaCl and all iso-structural compounds.

Braces around an atom symbol indicate that reference is being made to its coordination polyhedron. In **quartz** $GeO_2$, the coordination of Ge is a $\{Ge\}O_4$ tetrahedron, i.e., a Ge-centered $O_4$ tetrahedron.

Prefixed lower case Roman numeral superscripts in chemical formulas refer to coordination numbers as in $^{iv}Si_2{}^{iii}N_2{}^{ii}O$. Upper case Roman numerals in parentheses after chemical symbols refer to oxidation states as in $Ag(I)Ag(III)O_2$. Arabic numerals in parentheses refer to an arbitrary numbering to distinguish crystallographically-distinct, but chemically-identical, atoms as in $B_2O(1)_2O(2)$.

The notation ($N\times$) after interatomic distances or angles means that there are *N* equivalent (symmetry-related) distances or angles of the same magnitude (read "*N* times"). Thus the lengths of the six Ti-O bonds formed by a Ti atom in **rutile** $TiO_2$ are 1.948 ($4\times$) and 1.980 ($2\times$) Å. (1 Å = $10^{-10}$ m.)

We sometimes want to distinguish between compounds with (a) bonds formed between electropositive and electronegative elements (such as metal oxides and halides) (b) bonds between electronegative elements (such as in diamond) and (c) bonds between electropositive elements (as in brass, CuZn). For want of better terms we call these (a) ionic, (b) covalent, and (c) intermetallic. The use of these terms should not be construed to mean that we think that any particular theory of bonding is or is not applicable. Indeed they are merely labels of convenience, and in this book we go to some pains to describe structures and to resist, as far as is humanly possible, the temptation to interpret them. The reader is free from such constraints, of course.

We draw attention to the fact that conventions in crystallography are occasionally subject to change so that readers of the older literature can be misled. For example the short symbols for some cubic point groups and space groups were altered some years ago (see § 3.3.6). The method of describing unit cell transformations requires some care (see § 4.7.5). Our own view on conventions is that it is not so important that they should be logical (which, of course, is desirable), than that their users be consistent, and that changes should be made only for very compelling reasons.

Students of solid state chemistry would be well advised to obtain (or write!) a computer program that calculates distances and angles in crystal structures and another that assists in making drawings. On occasion, structures of lesser interest are simply presented as lists of coordinates—it is often tedious to convert these to drawings or models entirely by hand (although we have done just that for many years). Several good programs that do one or another of these things are now available commercially. At ASU and ANU we use a program, EUTAX, that will, among other things, do most of the numerical exercises presented herein.[1]

Many molecular graphics programs can be "tricked" into accepting fragments of crystal structures as molecules. Usually all coordinates will have to be entered explicitly so we generally give all the coordinates of symmetry-related atoms in crystal structures. For this purpose we use a condensed notation explained in § 3.4. For some comments on drawing structures see § 4.6. All the drawings in this book were made using a Macintosh® computer and Cricket Draw® with, in the case of structure drawings, the help of Cartesian coordinates generated by EUTAX.

At several places we give advice on the construction of models. Our experience is that the best way to learn a structure is to make a model of it (failing that, at least make a drawing or two). One should then look at it repeatedly, especially down principle symmetry axes, until it can be visualized clearly with one's eyes closed. Contrary to a common belief, learning three-dimensional structures is a skill not difficult to acquire but, like that of riding a bicycle, requires some initially frustrating practice. The reward is that, not only will a knowledge of crystal structures be acquired, but also one will learn to appreciate the sometimes stunning beauty of three-dimensional structures.

We entreat the reader working alone not to be discouraged if some material is difficult to understand at first reading. The book is not "bedside" reading but is intended to be read with pencil and paper at hand to verify the statements made. The level of difficulty is not uniform and we include some material (especially in the Notes) in condensed form for future reference if needed. Our advice (we follow it ourselves constantly) to those confronted with a topic that appears incomprehensible, is simply to read it, to read on, and then to return to the topic later. Be aware also that virtually every book (including this one) has *some* errors (M. O'K. would be very grateful if informed of these). If all else fails, get another book! Readers new to the subject who do not have such difficulties have our unstinted admiration.

The Exercises in many instances illustrate important aspects of crystal chemistry. (We call them Exercises to emphasize that a solid state scientist who aspires to obtain some virtuosity needs to exercise constantly, just as does a violinist or a tennis player.) Some of them are cast as statements; one should interpret these as a requirement to demonstrate the validity of the statements made. Yet others simply present interesting structures for the reader to explore. The lazy reader who is unwilling to do this may still find some of the results useful and should at least read them. Some Exercises will be found simple, some are more challenging. Sometimes we give hints and partial answers and some are answered later in the text. We emphasize though, that it is not sufficient to do a problem and obtain the correct answer. One must do a problem, get the correct answer, and *know* that it is the correct answer. After all, this is the situation faced by a practicing scientist (and by us in setting the Exercises).

[1]A Macintosh® version of this program is available from EMLab Software, 16203 S. 26th Place, Phoenix AZ 85048. A Windows® version is planned.

The book contains crystallographic data for two kinds of structure. Data of the first kind are found mainly in Chapters 6 and 7 in which we describe sphere packings and nets. In these cases the coordinates and unit cell parameters given are such as to have unit distance between sphere centers, or between the vertices of nets. These data can easily be recognized because the unit cell parameters are dimensionless. In most instances they have been calculated specially by us and published herein for the first time. On the other hand crystallographic data for real compounds are given throughout the text (specially in Exercises and in Appendix 5). In this second case unit cell edges are invariably given in Å and the data refer to structures published in the open literature. All these data come from one or other of the data bases listed in Section D of the Book List (p. 446); these (or similar sources) should be consulted for references to the original literature. The formula index contains a complete listing of compounds for which crystallographic data are given.

# TABLE OF CONTENTS

## 4 LATTICE GEOMETRY



## 5 POLYHEDRA AND TILINGS

## 6 SPHERE AND CYLINDER PACKINGS

## APPENDICES

### A1 More infinite symmetry groups



### A2 A glimpse into higher dimensions



### A3 The topology of polyhedra, nets and minimal surfaces

CHAPTER 1

# SYMMETRY IN TWO DIMENSIONS

## 1.1 Point symmetry operations in two dimensions

The concepts of *symmetry, symmetry operations* and *symmetry elements* play essential roles in the description of the structure and properties of matter, and it is very difficult to understand the crystallographic literature without some knowledge of symmetry theory and its jargon. In this and the next two chapters a brief account is given of the symmetries of objects with particular emphasis on the approach and symbolism that has been adopted as standard by crystallographers. No attempt is made to be either rigorous or complete—that would take us too far afield.

For pedagogic reasons we begin with a discussion of the symmetries of two-dimensional objects. In this case it is not too difficult to give an account of the various symmetries that are possible in a way that is plausibly complete. We can then proceed to a discussion of three-dimensional symmetry using the analogy to two dimensions as a prop to understanding. Two-dimensional symmetries are often encountered in practice (the surface of a crystal and the image of a crystal structure observed in an electron microscope are two-dimensional, for example).

A square confined to a plane looks exactly the same if it is rotated 90° around the point at its center—we say that this operation, which leaves the object unchanged in aspect, is a *symmetry operation*. Thus consider the shaded square shown below (Fig. 1.1). Rotating the square about its center point (the small black square) by 90°, 180° (= 2 × 90°) or 270° (= 3 × 90°) leaves it unchanged. Note that the fourfold repetition of rotation by 90° is equivalent to rotation by 360°. Any object is left unchanged after a rotation of 360° about *any* point and this operation is called the *identity* operation. We generally refer to a symmetry element that entails rotation by 360°/$N$ ($N$ is an integer) as $N$-fold.

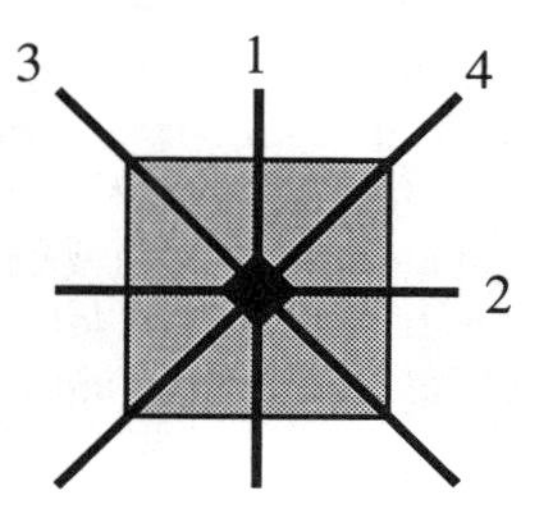


**Fig. 1.1**. Symmetries of a square.

The point at the center of the square is the location of a *4-fold rotation point* and this one of the *symmetry elements* of the square. Distinguish between the symmetry *element* and the symmetry *operations* (in this case repeated rotations by one-fourth of a circle, by convention[1] anticlockwise, about a point). Note also that the application of any symmetry operation on a finite object will always leave at least one point unmoved; for this reason such symmetries are referred to as point operations (or point *isometries*).

The square has other symmetry elements. Reflection of the square in the vertical line (labeled "1" in the figure) going through the center will again leave the square unchanged. The vertical line is a *mirror line*. There is another mirror line (labeled "2") at right angles to this (i.e. horizontal) as the square has 4-fold rotation symmetry.

There is also a second set of mirror lines among the symmetry elements of the square; these (labeled "3" and "4" in the figure) are at 45° to the first set. The reader might try to show that these are *generated* by combinations of 4-fold rotations and one or other of the first set of mirrors. (We will find a simple way to show this later.)[2]

Repeated application of any point symmetry operation will always eventually result in the identity operation (thus: two successive reflections in the same mirror line or four successive rotations of 90° about the same point). For mathematical reasons it is important to always consider the identity as one of the symmetry elements of any object.

We have identified the entire set of operations of the symmetry elements of the square. It forms a representation of a mathematical group[3], and is therefore called the symmetry group—or in this case the *point symmetry group* because at least one point is left invariant by all the symmetry operations.

It is easy to show (we do it below) that in a periodic object (such as a crystal) there is a restriction to point groups that include only 1-, 2-, 3-, 4- or 6-fold rotation symmetry elements. There are ten of these groups in two dimensions; they are called the *crystallographic* point groups. We enumerate them and show patterns with each of the symmetry groups in Fig. 1.2. The patterns are generated by the action of all the symmetry operations of the group on an asymmetric object (in this case a 7).

There are five groups with only rotation points (we include the group that contains only the "one-fold" rotation = identity). A rotation point is symbolized by a number equal to the order of the rotation. For the groups without mirrors, these numbers become also the symbols for the groups which are therefore 1, 2, 3, 4 and 6. These are called the *pure rotation groups*, as the symmetry operations are rotations only. A pattern with pure rotation symmetry will be different from its mirror image and so can be said to have a *hand*. (If the object is left-handed, its mirror image is right-handed and *vice versa*)

Additional groups are obtained by adding mirror symmetry to the pure rotation groups. The simplest such group is just that obtained by addition of a mirror line (symbol $m$) to group 1 to give group $1m$. This is the symmetry group of a plane object that has but one mirror line. An example is an isosceles triangle or the letter V. Note that group $1m$ consists of two operations: reflection in the mirror line and the identity 1.

[1]This is the convention in crystallography. Many computer graphics programs consider positive rotations to be clockwise.

[2]For example rotation by 90° anticlockwise followed by reflection in the horizontal mirror is equivalent to reflection in the mirror line running from top left to bottom right.

[3]The reader unfamiliar with group theory may find the Notes (§ 1.10.2) for this chapter useful.

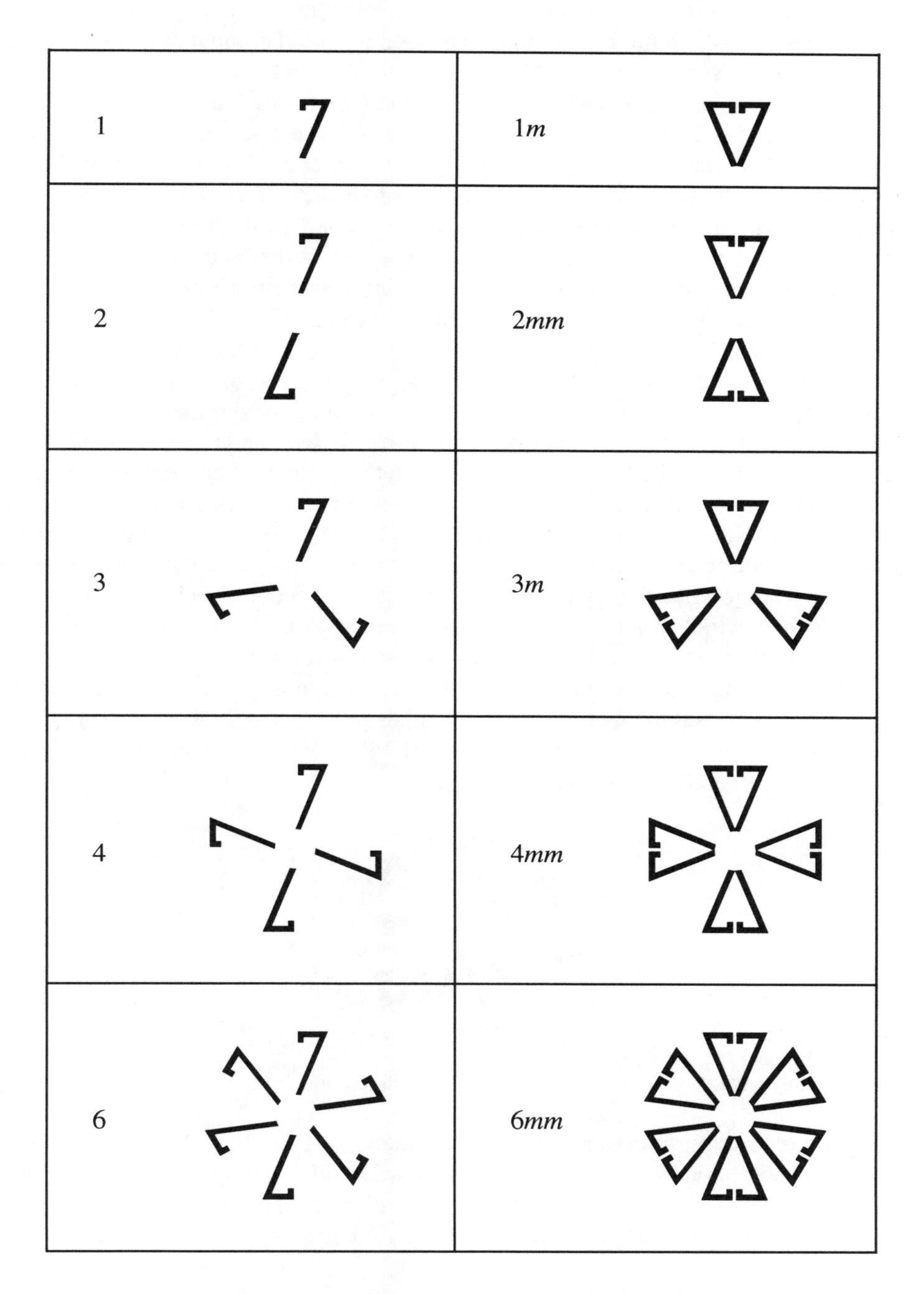


**Fig 1.2**. Illustrating the 10 two-dimensional crystallographic point groups.

If the symmetry group of an object contains mirror lines, the object is the same as its mirror image and thus will not have a hand. The reader should compare the patterns on the left of Fig. 1.2, which have a hand, to those on the right, which do not.

Next we combine a mirror with a 2-fold rotation (see Fig. 1.2 again). We find in fact that combination of a 2-fold rotation and a mirror results in a second mirror line at right angles to the first, so the group is symbolized $2mm$ to indicate the presence of the two sets of mirror lines (each set consists of one line). Conversely, if we had started with two mirror lines at right angles we would find that we generated a 2-fold rotation point at their intersection.[1] This is the symmetry group of a rectangle or of the letter H.

Proceeding in this way we next generate group $3m$ from a mirror and a 3-fold rotation. This is the symmetry of (for example) an equilateral triangle. Combining a mirror with a 4-fold rotation produces the symmetry group of the square which we have already discussed. The symbol is $4mm$ with two $m$'s to signify the presence of two sets (of two each) of mirrors. In each set the two mirror lines are related by symmetry. Thus (see Fig. 1.1) a quarter turn takes mirror 1 into mirror 2; mirror lines 3 and 4 are similarly related. On the other hand there is no symmetry operation in the group that converts either mirror 1 or 2 to mirrors 3 or 4—this is why we say there are two *sets* of mirror lines.

Finally we combine a mirror with a 6-fold rotation. Again we get in the symmetry group two sets (now of three each) of mirrors. Accordingly the symbol of the group is $6mm$. A regular plane hexagon has this symmetry. The location of the mirror lines is indicated in Fig. 1.3 below in which the two sets of mirrors are shown as dashed and dotted lines respectively. As in $4mm$, we have a symmetry operation (now rotation by a multiple of a sixth turn) to relate mirror lines in a set, but there is no symmetry operation in the group relating mirrors in one set to any of those in the other set. (Mirror lines of one set are at 30° to those of the other set, i.e. related by a rotation of 360°/12; but rotation by 1/12 of a circle is not a symmetry operation of the group).

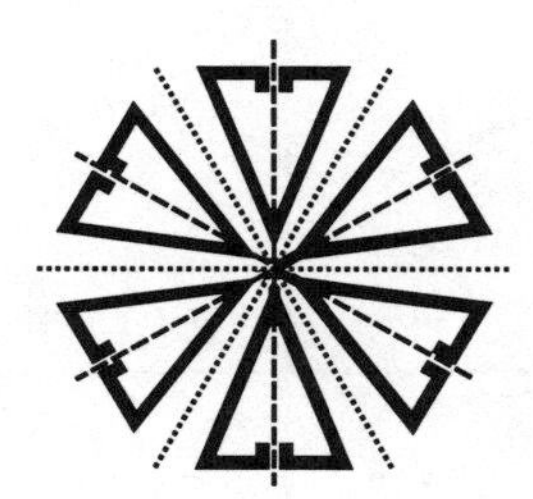

**Fig 1.3**. Illustrating $6mm$. Dotted and broken lines indicate the location of the two sets of mirrors.

Lifting the restriction to crystallographic symmetry (1-, 2-, 3-, 4-, or 6-fold rotations) we get non-crystallographic point groups. These are rotation groups $N$ with $N = 5$ or $> 6$, and the point groups $Nm$ with $N > 3$ and odd (a regular pentagon has symmetry $5m$), and $Nmm$ with $N > 6$ and even (a regular octagon has symmetry $8mm$).

[1]The three-dimensional occurrence of this phenomenon is familiar to those who have looked at themselves in two mirrors at right angles. If you have never done this please try!

## 1.2 Coordinate systems and symmetry operations

Let us consider a Cartesian coordinate system with the $x$ axis pointing down the page and the $y$ axis horizontal as in Fig. 1.4. Further, let the origin be at the 4-fold rotation point of $4mm$. Three of the symmetry elements of that group are indicated: the 4-fold rotation point by a square and two mirror lines by heavy lines.

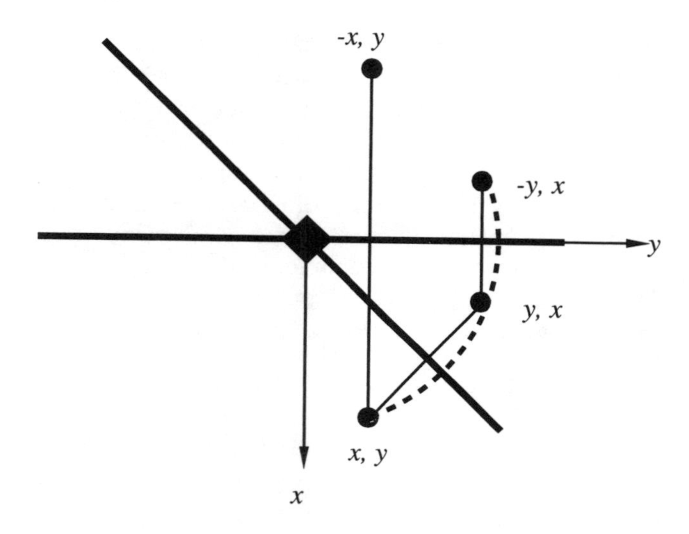


**Fig. 1.4.** Illustrating the effects of rotations and reflections on a point (see text).

Imagine now the operation of a quarter turn (90° rotation anti-clockwise about the origin) on the point $x,y$. The point will be translated to a position $-y,x$. This means that the new $x$ coordinate is equal to minus the old $y$ coordinate and the new $y$ coordinate is equal to the old $x$ coordinate; thus if the original point had coordinates 0.2,0.1 the new point would have coordinates –0.1,0.2. The transformation can be represented by multiplying the column vector ($x$ /$y$) (representing the coordinates $x,y$) by a matrix to give new coordinates $x',y'$:

$$\begin{pmatrix} x' \\ y' \end{pmatrix} = \begin{pmatrix} 0 & -1 \\ 1 & 0 \end{pmatrix} \begin{pmatrix} x \\ y \end{pmatrix} = \begin{pmatrix} \bar{y} \\ x \end{pmatrix}$$

The matrix can be written on one line as $(0\ \bar{1}\ /\ 1\ 0)$, with $\bar{1}$ representing –1, and this is a convenient way of representing the symmetry operation. The matrix corresponding to reflection in the diagonal mirror is (0 1 / 1 0).[1] This transforms the point $x,y$ to $y,x$ as

[1] The reader who is not entirely familiar with matrices and their multiplication is well advised to work out the examples given here and subsequently. See the Notes at the end of the chapter (§ 1.10.4) for help with manipulating matrices.

shown in the figure. Finally the operation of reflecting in the horizontal mirror transforms $x,y$ to $\bar{x},y$ with the corresponding matrix $(\bar{1}\ 0\ /\ 0\ 1)$. We may now verify by matrix multiplication that rotation by 90° followed by reflection in the horizontal mirror is equivalent to reflection in the diagonal mirror shown in Fig. 1.4:[1]

$$(\bar{1}\ 0\ /\ 0\ 1)(0\ \bar{1}\ /\ 1\ 0)\ (x\ /\ y) = (0\ 1\ /\ 1\ 0)(x\ /\ y) = (y\ /\ x)$$

as should be obvious from an inspection of the figure.

Instead of using the matrices to represent the symmetry operations, we could use the transformed coordinates themselves. The resulting coordinates are sometimes known as the Jones symbol. We list in Table 1.1 the matrices and coordinates for all the operations of 4*mm*. Reference to Fig 1.2 shows that the original asymmetric object is transformed to eight such objects, so there are eight symmetry operations (eight is the *order* of the group). It is common to symbolize an anticlockwise rotation by a quarter of a circle by $4^1$ or $4^+$, rotation by two quarters of a circle (= one half circle) by $4^2$ (= $2^1$) and by three quarters of a circle by $4^3$ or $4^-$. The last case emphasizes that rotation anticlockwise (the positive sense) by 3/4 of a circle is equivalent to rotation clockwise (the negative sense) by 1/4 of a circle.

**Table 1.1.** Symmetry operations of group 4*mm*.

| operation | matrix | coordinates |
|---|---|---|
| identity = 1 | $(1\ 0\ /\ 0\ 1)$ | $x,y$ |
| 90° rotation = $4^1$ | $(0\ \bar{1}\ /\ 1\ 0)$ | $\bar{y},x$ |
| 180° rotation = $4^2 = 2^1$ | $(\bar{1}\ 0\ /\ 0\ \bar{1})$ | $\bar{x},\bar{y}$ |
| 270° rotation = $4^3$ | $(0\ 1\ /\ \bar{1}\ 0)$ | $y,\bar{x}$ |
| reflection in horizontal mirror | $(\bar{1}\ 0\ /\ 0\ 1)$ | $\bar{x},y$ |
| reflection in vertical mirror | $(1\ 0\ /\ 0\ \bar{1})$ | $x,\bar{y}$ |
| reflection in mirror 3 (Fig. 1.1) | $(0\ 1\ /\ 1\ 0)$ | $y,x$ |
| reflection in mirror 4 (Fig. 1.1) | $(0\ \bar{1}\ /\ \bar{1}\ 0)$ | $\bar{y},\bar{x}$ |

In crystallography we does not always use coordinate axes at right angles. In the case of 3-fold or 6-fold symmetry, it is more convenient to take axes inclined at 120° to each other as shown in Fig. 1.5. The figure illustrates the effect of 3-fold rotations on a point $x,y$ from the rotation point taken as origin. The long and short dashed lines have lengths equal to the magnitudes of $x$ and $y$ respectively. Note that to get to the point $x,y$ from the origin, we translate by a distance $x$ parallel to the $x$ axis and then by a distance $y$ parallel to the $y$ axis.

It should be obvious from the figure that after rotation by one-third of a circle (120°) about the origin, the new $x$ coordinate is equal to $-y$, and that the new $y$ coordinate is equal to $x-y$. We therefore write the coordinates of the new point $\bar{y},x-y$. If the original point were at $x = 0.4$ and $y = 0.1$, the coordinates of the point generated by the rotation of 1/3 of

[1]Note that the order of matrix multiplication is important. Reversing the order of the two matrices is equivalent to reversing the order of carrying out the two symmetry operations. (What symmetry operation does one get then?)

a circle (anticlockwise) would be $x = -0.1$, $y = 0.3$. Repeating this operation takes $\bar{y},x-y$ to $y-x,\bar{x}$ (i.e. $-0.1,0.3$ to $-0.3,-0.4$).

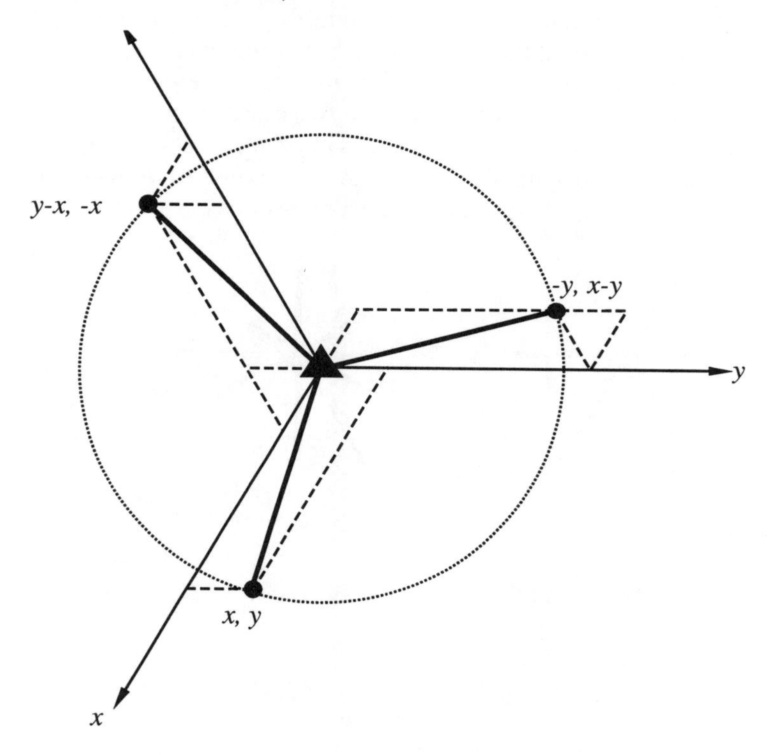


**Fig. 1.5**. Illustrating a coordinate system with axes ($x$ and $y$) at 120°.

The symmetry operations of point group 3 can hence be represented as shown in Table 1.2. As there are three symmetry operations, the order of the group is three.

**Table 1.2.** Representations of the symmetry operations of group 3.

| operation | matrix | coordinates |
|---|---|---|
| identity = 1 | (1 0 / 0 1) | $x,y$ |
| 120° rotation = $3^1$ = $3^+$ | (0 $\bar{1}$ / 1 $\bar{1}$) | $\bar{y},x-y$ |
| 240° rotation = $3^2$ = $3^-$ | ($\bar{1}$ 1 / $\bar{1}$ 0) | $y-x,\bar{x}$ |

It is important to note that the matrices and (transformed) coordinates only have the particular form given for the particular *basis* (coordinate system) chosen. A recommended exercise (see Exercise 7 at the end of the chapter) is to find the corresponding matrices and coordinates for a 120° rotation using a Cartesian basis (axes at right angles).

Even with axes at 120° we have to be careful when we consider the point group $3m$.

There is more than one way we can orient the axes with respect to the mirror lines.[1] There are two ways that prove to be useful in practice as shown in Fig. 1.6. On the left there are mirror lines parallel to the $x$ and $y$ axes; on the right they are perpendicular to the axes. Now it is important to realize that the crystallographer's symbols for symmetry groups not only signify the symmetry elements present, but *also* their orientation with respect to a coordinate system. Accordingly the symbols in the two cases above are different *viz.* $31m$ and $3m1$. (The full significance of the symbols will be made clear later—for now it is important to note that the two symbols refer to the *same* point group with the coordinate axes in two different orientations with respect to the mirror lines.)

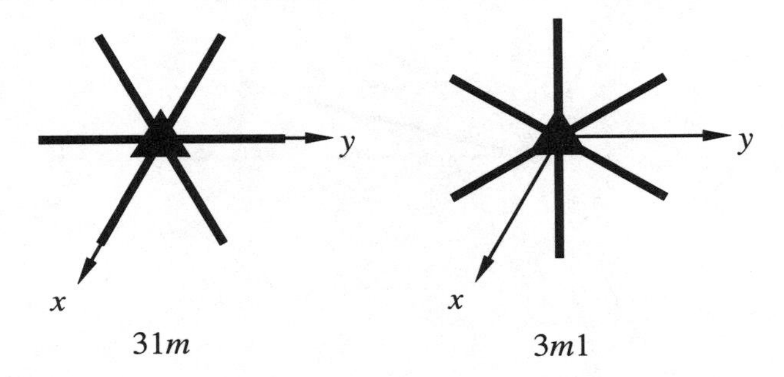


**Fig 1.6**. Illustrating two orientations of the $x$ and $y$ axes with respect to the symmetry elements (mirror lines) of $3m$.

For the record note that the three mirror reflections operating on $x,y$ give

$$\text{for } 31m: \quad \bar{x},y-x \; ; \quad x-y,\bar{y} \; ; \quad y,x$$
$$\text{for } 3m1: \quad x,x-y \; ; \quad y-x,y \; ; \quad \bar{y},\bar{x}$$

These, combined with the three points produced by the 3-fold rotation point (given in Table 1.2) produce a total of six points, so the order of the group is six.

## 1.3 Translational symmetry: lattices and unit cells

We now turn our attention from finite objects to infinite objects with *translational* symmetry. The first important concept that we need is that of a *lattice*. A lattice is an infinite array of points each of which is identically surrounded. In two dimensions a lattice is generated by repeated translations of a point by two non-collinear vectors **a** and **b**. Thus starting from an arbitrary origin we generate a lattice as the infinite set of points at the end of the vectors $m\mathbf{a} + n\mathbf{b}$ where $m, n$ are positive, negative or zero integers. The lattice is completely determined by the magnitudes, $a$ and $b$, of the vectors **a** and **b** and the angle $\gamma$

[1]In a periodic object (crystal) there is *translational* symmetry and we use axes parallel to translation vectors, so the orientation of the axes is determined by the crystal lattice.

between them.[1]

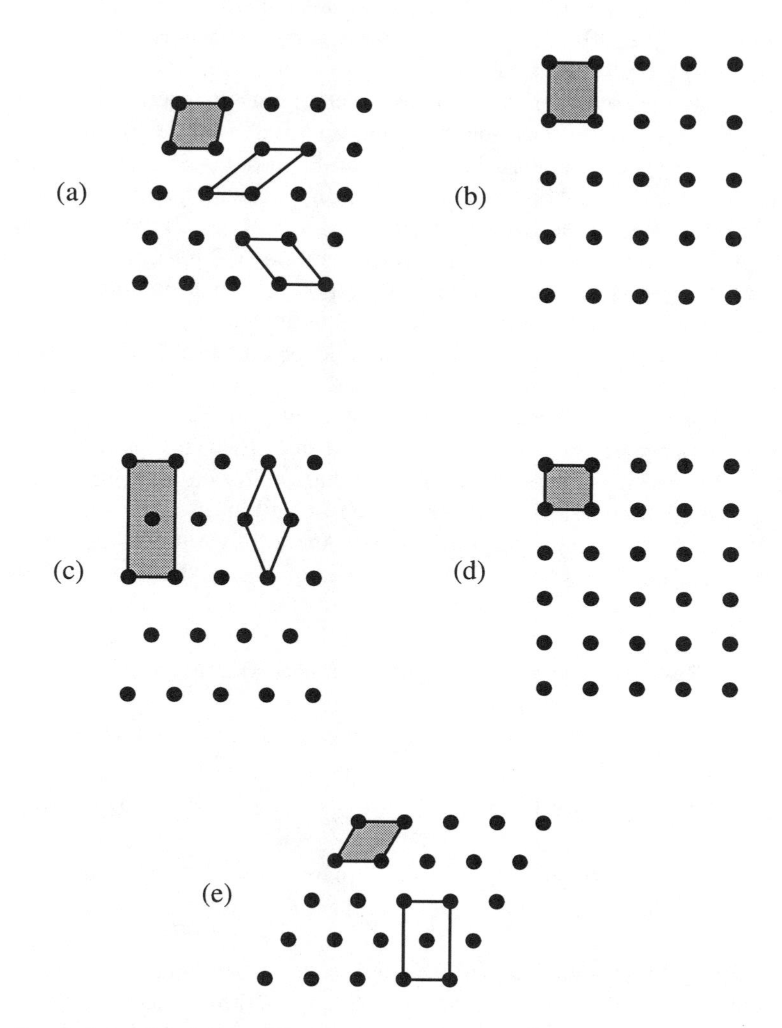


**Fig 1.7**. The five 2-dimensional Bravais lattices.

The parallelepiped defined by **a** and **b** is the basic repeat unit of the lattice and is known as the *unit cell*. As illustrated in Fig. 1.7(a), more than one unit cell (with corresponding lattice vectors **a** and **b**) can be chosen with lattice points only at the corners. Such unit cells

[1]The term "lattice" is widely abused. Crystallographers, in particular, become very upset when *lattice* is used to mean *structure* as in "the diamond lattice" or "the wurtzite lattice" (in neither case do all the atoms fall on lattice points). Such usage is considered a gross solecism. However terms such as *lattice dynamics* and *lattice defects* are in such common usage that one cannot avoid them. The word "lattice" is also used in a quite different sense by mathematicians.

contain one lattice point per cell (each point at each of the four corners is shared by four cells), so no matter how they are chosen each cell has the same area. Usually the shortest pair of vectors with $\gamma \geq 90^\circ$ are chosen as shown by the shaded parallelogram. The quantities $a$, $b$ and $\gamma$ are referred to as the *unit cell parameters*.

Lattices are classified by their symmetries. Every point in a general two-dimensional lattice is a site of 2-fold rotation symmetry (remember that a lattice extends to $\pm\infty$ in both directions). For some special relationships of the unit cell parameters, the lattice points are at points of higher symmetry. This is illustrated in Fig. 1.7 which shows the different two-dimensional lattices (known as *Bravais lattices*). The lattices shown in (b) and (c) also have mirror symmetry (the points are at sites of symmetry $2mm$), that shown in (d) has 4-fold rotational symmetry and that shown in (e) has 6-fold rotational symmetry. After we have discussed space group symmetry we will see that the lattices (b) and (c) have different space groups ($p2mm$ and $c2mm$ respectively), and hence are classified as different Bravais lattices. Every two-dimensional lattice has one of only five possible space group symmetries and thus is one of the Bravais lattice types illustrated.

The point symmetry operations that leave a lattice invariant (and one lattice point unmoved) form the point symmetry group of the lattice. As lattice translations leave the lattice invariant (the lattice extends infinitely) the translations are also symmetry elements and the full symmetry group (space group) includes the infinite number of translations.

The *constraints* on $a$, $b$ and $\gamma$ in the five lattices, and names descriptive of the shape of the unit cells are listed in Table 1.3. The table also lists the symmetry at a lattice point.

**Table 1.3.** Properties of the two-dimensional Bravais lattices.

| | constraints | system | point symmetry |
|---|---|---|---|
| (a) | none | oblique | 2 |
| (b) | $\gamma = 90^\circ$ | rectangular | $2mm$ |
| (c) | $a = b$ | rectangular | $2mm$ |
| (d) | $a = b$, $\gamma = 90^\circ$ | square | $4mm$ |
| (e) | $a = b$, $\gamma = 120^\circ$ | hexagonal | $6mm$ |

It may be objected that the unshaded unit cell in (c) is not rectangular, and that is true, but we can take an alternative unit cell that *is* rectangular (shown shaded in the figure) with lattice points at the corners *and* one at the center. Most commonly this is done and the constraint for lattice (c) above could have been stated as $\gamma = 90^\circ$ for a cell with lattice points at the corner and in the center.

Note that the unit cells with only one lattice point per unit cell are *primitive*. On the other hand the rectangular unit cell shown in (c) contains two lattice points (one corresponding to the corners, and the one at the center) and is called *centered*. The unit cell parameters for the centered cell in (c) have $a \neq b$, $\gamma = 90^\circ$ as in (b) but, as noted above, the two lattices have different space group symmetries. Two-dimensional lattices represented by a primitive cell are symbolized by $p$ and that conventionally represented by a centered cell is symbolized by $c$.

The terms *oblique*, *rectangular*, *square*, and *hexagonal* refer to the crystal *system*. Clearly there are four crystal systems in two dimensions.

## 1.4 Allowed rotational symmetries of lattices

We pause now to prove the assertion that the only rotational symmetries allowed in periodic structures are 1-, 2-, 3-, 4- and 6-fold rotations. Certainly one or more of the lattices (a)-(e) of the previous section have one or more of these symmetries (note that e.g. a 6-fold symmetry includes 1-, 2- and 3-fold).

A proof, which is very old, proceeds as follows. We take an arbitrary lattice and identify the shortest lattice vector **a**. Now consider two lattice points $A$ and $B$ separated by $a$ (see Fig. 1.8). Let there be $N$-fold rotation points at $A$ and $B$. Let $B'$ be the image of $B$ after rotation by $360°/N$ about $A$, and $A'$ the image of $A$ after rotation by $-360°/N$ about $B$. $A'$ and $B'$ must also be lattice points. The distance $B'A'$ is equal to $a[1\text{-}2\cos(360°/N)]$ and must be equal to 0, $a$ or $>a$ (if $B'A' < a$ then **a** was not the shortest lattice vector). We consider each of these possibilities in turn:

| | | |
|---|---|---|
| $B'A' = 0$: | $1 - 2\cos(360°/N) = 0$ | $N = 6$ |
| $B'A' = a$: | $1 - 2\cos(360°/N) = 1$ | $N = 4$ |
| $B'A' > a$: | $1 - 2\cos(360°/N) > 1$ | $N < 4$ (i.e. 3, 2 or 1) |

Thus the only possible values of $N$ are 1, 2, 3, 4 or 6. *Q.E.D.*[1]

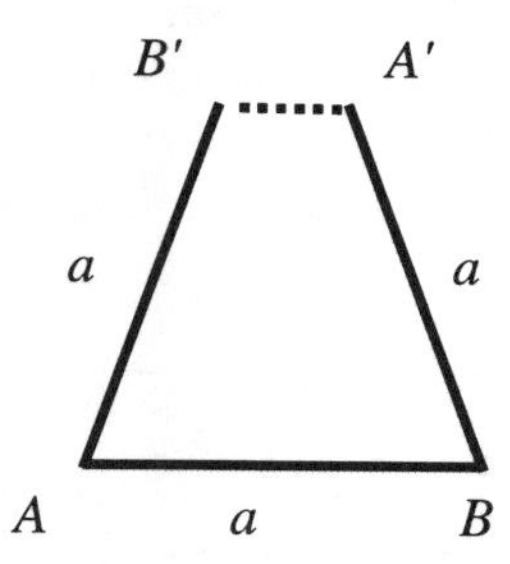


**Fig. 1.8**. Illustrating the proof that only 1-, 2-, 3-, 4- or 6-fold rotations are allowed rotational symmetries of lattices.

## 1.5 Unit cell coordinates: describing structures

For a periodic structure we now know how to specify the unit cell. To completely define the structure the next thing we have to do is to specify what is in the unit cell.

[1]The unconvinced reader is invited to consider the case of $N = 5$.

The reference coordinate system is always taken with the $x$ axis parallel to **a** and the $y$ axis parallel to **b**. Usually the $y$ axis is drawn horizontal and the $x$ axis down the page so that the origin of coordinates is at the top left corner of the unit cell. The scale of $x$ is in units of $a$ (the magnitude of the vector **a**) and the scale of $y$ is in units of $b$. In these terms the coordinates of the unit cell corners are 0,0 ; 0,1 ; 1,0 and 1,1. Note that $x$ and $y$ are dimensionless; for example the distance from the origin at 0,0 to a point $x$,0 is $xa$ and $a$ has dimensions of length.

Any point in the unit cell will be specified by $x,y$ where $x$ and $y$ will be in the range 0 to <1. It is important to note that coordinates equal to, or greater than, 1 are not used. This is because (for example) the existence of a point at 0,$y$ implies a point at 1,$y$. Both these points are shared with two unit cells so we would be counting points twice if we gave the coordinates of both points. Likewise $x$,0 implies $x$,1 and 0,0 implies 0,1 and 1,0 and 1,1.

Sometimes crystallographers use negative coordinates such as $\bar{x}$ or $\bar{y}$. These should be interpreted as 1–$x$ and 1–$y$ respectively. It should be clear that adding (or subtracting) an integer to either $x$ or $y$ will always produce an identical point in another unit cell (remember that we have translational symmetry). The point in the reference cell will always have $0 \leq x,y < 1$. It should be noted, however, that when *illustrating* unit cells of structures it is conventional to show points (if there are any) on all edges and corners.

We now cosider some examples: A certain pattern has a rectangular unit cell with $a$ = 1.73 (actually √3) and $b$ = 2.0. There are three points per unit cell with coordinates A: 0,0.5 ; B: 0.5,0.25 ; C: 0.5,0.75. Fig. 1.9 shows (on the left) a unit cell with the points plotted as circles (note that the point at 1,0.5 corresponding to A is also shown). In the center is part of the pattern obtained by repeating the unit cell. To make the pattern clearer, on the right the points nearest to each other have been joined together and the outline of the unit cell omitted. This pattern is observed in layers of atoms in some crystal structures. This example illustrates what is generally the case—although the pattern is completely specified by the shape and content of the unit cell, it can only be fully appreciated by drawing a number of unit cells.

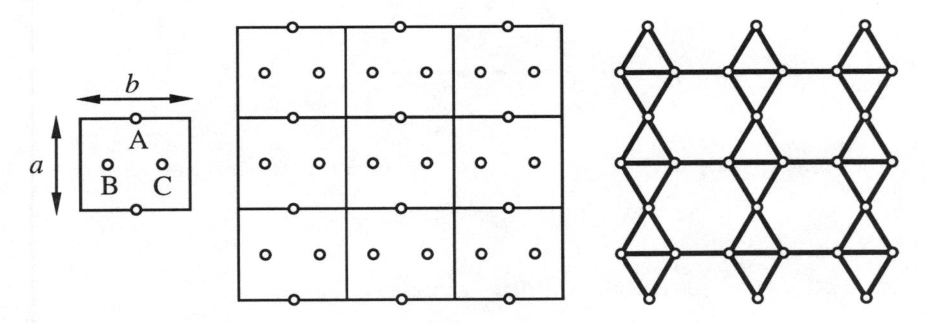


**Fig. 1.9**. Left: a unit cell of a pattern. Middle: nine unit cells. Right: the same pattern with nearest neighbors joined by lines.

A second example involves a hexagonal cell. Now $b = a$ and $\gamma = 120°$. Points are at 0,0.5 ; 0.5,0 and 0.5,0.5. Fig 1.10 shows (on the left) a unit cell, in the center nine unit cells and on the right, the pattern outlined by joining nearest neighbors in the same way as in Fig. 1.9. This pattern is one that is encountered repeatedly in crystal chemistry. It is often called the *kagome* pattern after the Japanese word for bamboo weaves used in basket making. It is also very frequently encountered in ornamental designs.

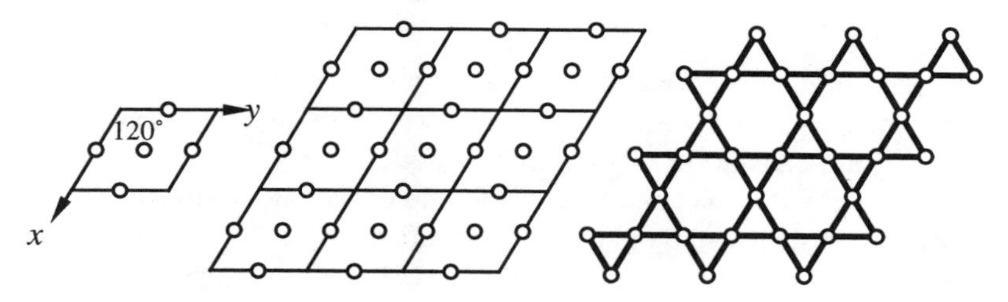


**Fig. 1.10**. Left: a (hexagonal) unit cell of a pattern. Middle: nine unit cells. Right: the kagome pattern 3.6.3.6.

Note that in the kagome pattern every point has the same surroundings in the sense that the polygons *in sequence* meeting at the vertex are hexagon, triangle, hexagon, triangle. By contrast in the pattern of Fig.1.9 there are two kinds of point; at one kind (A) the sequence is the same as in kagome (hexagon, triangle, hexagon, triangle) but at the other kind (B and C) it is hexagon, hexagon, triangle, triangle. Both patterns are examples of *tilings* or *tessellations* of the plane by a combination of regular polygons.

A digression on tilings: If the plane is covered completely by just one kind of regular polygon, we have a *regular* tiling. It is not difficult to prove that there are just the three of these. The first, in which six triangles (3-gons) meet at a point, is often symbolized $3^6$; the second, in which four squares (4-gons) meet at a point, is symbolized $4^4$; the third in which three hexagons (6-gons) meet at a point is symbolized $6^3$. (This last is also known as the *honeycomb* pattern, for an obvious reason.) The symbols $3^6$, $4^4$ and $6^6$ are known as *Schläfli* symbols.[1]

If the plane is similarly covered by more than one kind of regular polygon, and all points at which three or more polygons meet are of the same kind (related by symmetry), we have the *semiregular* or *Archimedean* tilings of the plane. The kagome pattern (Fig. 1.10) is of this type and is symbolized 3.6.3.6 as the polygons in cyclic order around each point are 3-gon, 6-gon, 3-gon, 6-gon. There are eight different semi-regular tilings; we describe them later (Chapter 5). It is also convenient to refer to them by their Schläfli symbols. End of digression.

We have noted that the pattern in Fig. 1.9 contains two kinds of point. The kagome pattern in Fig. 1.10 contains only one kind of point, as in this pattern the points are related by symmetry. Thus starting from one point we can generate two more by rotating about a 3-fold point (there are 3-fold rotation points in the centers of all the triangles) and get an equilateral triangle of points. Repeating the pattern of three points by the translation vectors

[1]After the great Swiss mathematician, Ludwig Schläfli (1814-1895). Mathematicians write $3^6$ (for example) as (3,6) as this allows easier extension to higher dimensions (as in Schläfli's pioneering work).

**a** and **b** recovers the pattern of points. This array of points however is *not* a lattice (this is an important point)—a lattice is generated by translations alone.[1]

## 1.6 Glide symmetry

We have identified three kinds of symmetry operation: rotation, reflection and translation. It transpires that we can combine two of these operations to give a new symmetry operation: translation + reflection = *glide*. The translation direction and the mirror line must be parallel to each other and to one of the lattice translation vectors. The symmetry operation is the compound one of first reflection in a line (a *glide line*) and then translation parallel to the line by a distance $d$.[2] Repetition of this operation will produce a pattern that is periodic with a period $2d$. The vector of length $2d$ must therefore be a lattice vector. Fig 1.11 illustrates nine successive glides of a triangle initially on the left: the glide line is shown in the conventional representation as a dashed line. The symbol for a glide line is $g$ (compare with $m$ for a mirror line). If you walk in a straight line, and with a uniform step, along a beach, your footprints will be related by glide symmetry.

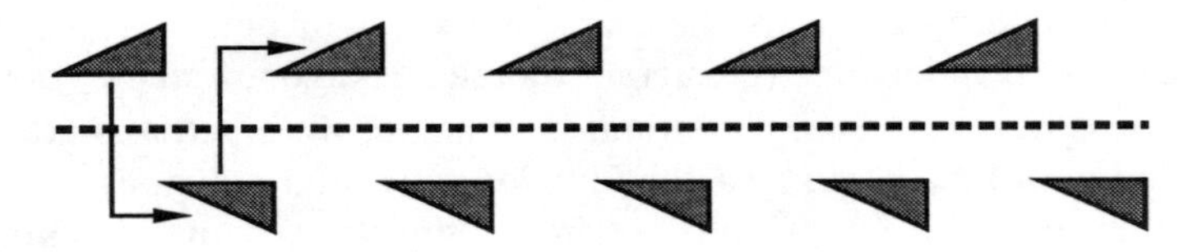

**Fig. 1.11**. Illustrating glide. The glide line is shown as a broken line.

## 1.7 Two-dimensional space groups

We are now in a position to enumerate the symmetry groups that are obtained in two dimensions by combining the point operations with those that involve translation. We will find that there are 17 such two-dimensional space groups. We first combine the point group operations of rotation and reflection with translation and then consider the cases where reflection lines are replaced by glide lines. Groups obtained by the combination of point operations with translations are called *symmorphic*. Additional *non-symmorphic* groups are obtained by replacing (where appropriate) mirror lines by glide lines. The diligent reader will work through each example by letting the symmetry operations act on an asymmetric object; Exercise 8 shows how to proceed. If this is done, patterns similar to those shown in the different parts of Fig. 1.12 (below) will be obtained.

[1]Note that points (or atoms *etc.*) related by lattice translations are often referred to as *identical* (a lattice is an infinite array of *identical* points). Points that are related by symmetry operations other than pure translations are sometimes called *equivalent*.

[2]Actually in this case the order in which these operations is carried out is unimportant; one could translate and *then* reflect.

### *1.7.1 Oblique system*

We start with the simplest case in which there is translational symmetry only with an oblique lattice. This is the symmetry of the pattern obtained from periodic translations of an asymmetric object in two dimensions. The symbol for the symmetry group is $p1$. The $p$ symbolizes the lattice (primitive) and the 1 the associated point group.

As already noted, an oblique lattice by itself has 2-fold rotation symmetry at the lattice points, so it is also compatible with the pattern obtained by repeating an object (which may be *e.g.* a collection of points) with 2-fold rotation symmetry by oblique lattice vectors. The symmetry group consisting of the combination of the translations with 2-fold rotations is symbolized $p2$.

No other symmetries are possible with an oblique lattice so we have found the two possible space groups in the oblique system: $p1$ and $p2$. Patterns with these symmetries are shown in Fig. 1.12 (*a*). Note that in the parts of Fig. 1.12, the patterns are generated by the symmetry operations operating on a single asymmetric object (a scalene triangle) and the number of such triangles per primitive unit cell is the same as the order of the point group from which the space group is derived.[1]

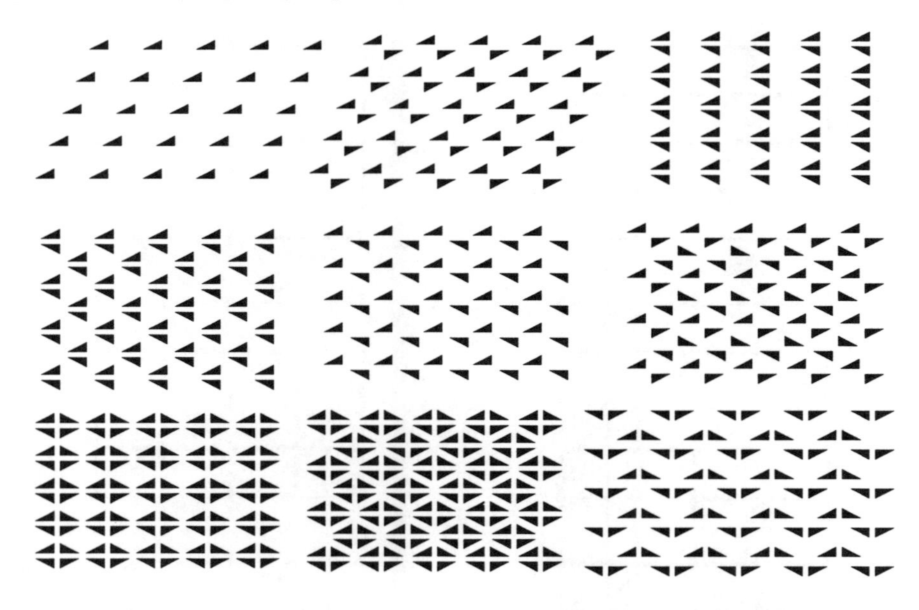

**Fig. 1.12 (*a*).** Illustrating the oblique and rectangular two-dimensional space groups. Each pattern has different symmetry (which should be identified).

[1]The device of using a scalene triangle as an asymmetric object is old [see *e.g.* L. Weber, *Zeits. Kristallogr.* **70**, 309 (1929)] and patterns similar to those in the various parts of Fig. 1.12 are to be found for example in *Elementary Crystallography* by M. J. Buerger [Wiley, New York (1963)]. The space groups are deliberately left unidentified in the figure. The reader should provide the labels. This is an easy exercise, but it should be done.

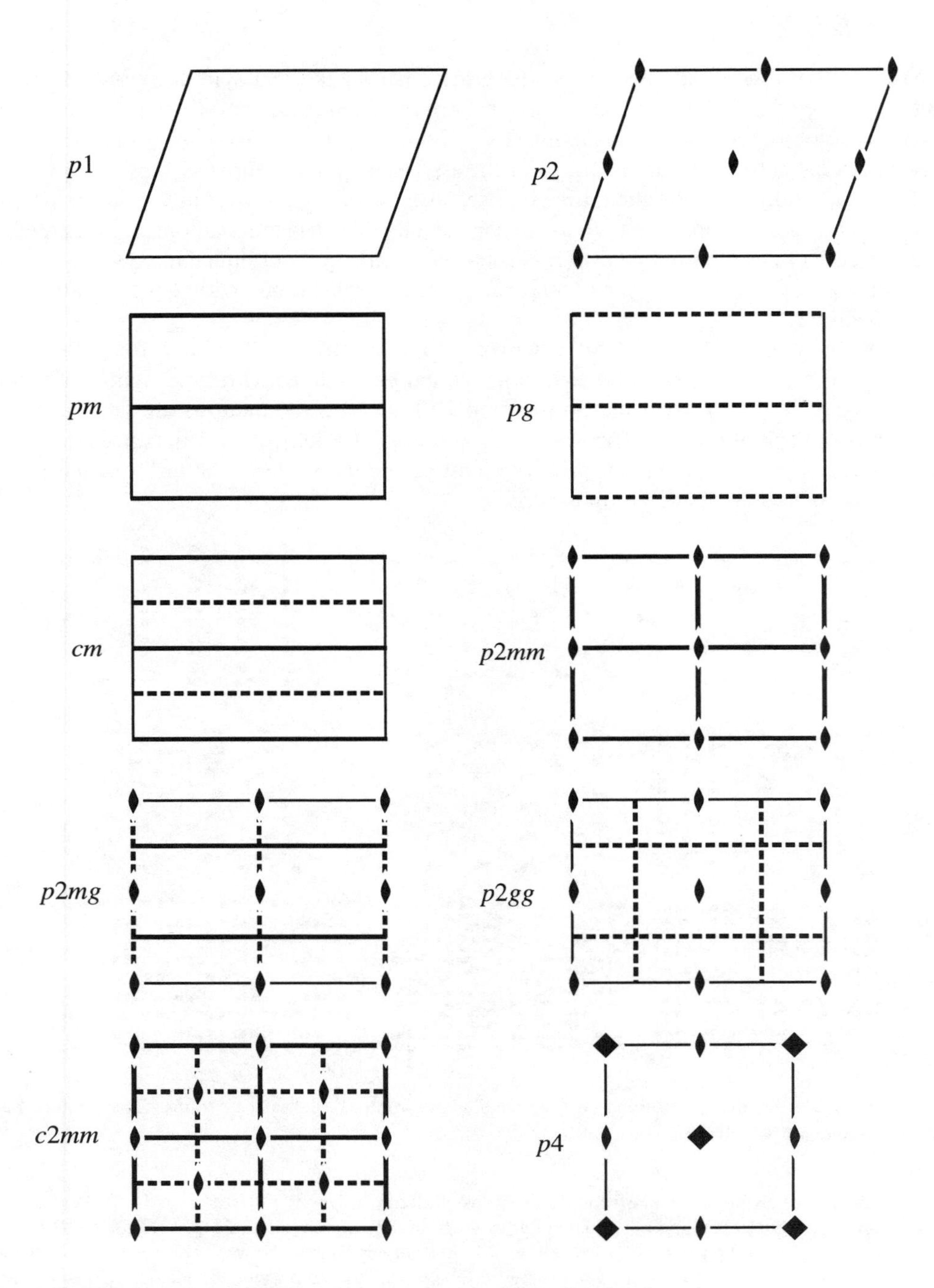


Fig. 1.13. This figure is continued on the next page (see the caption there).

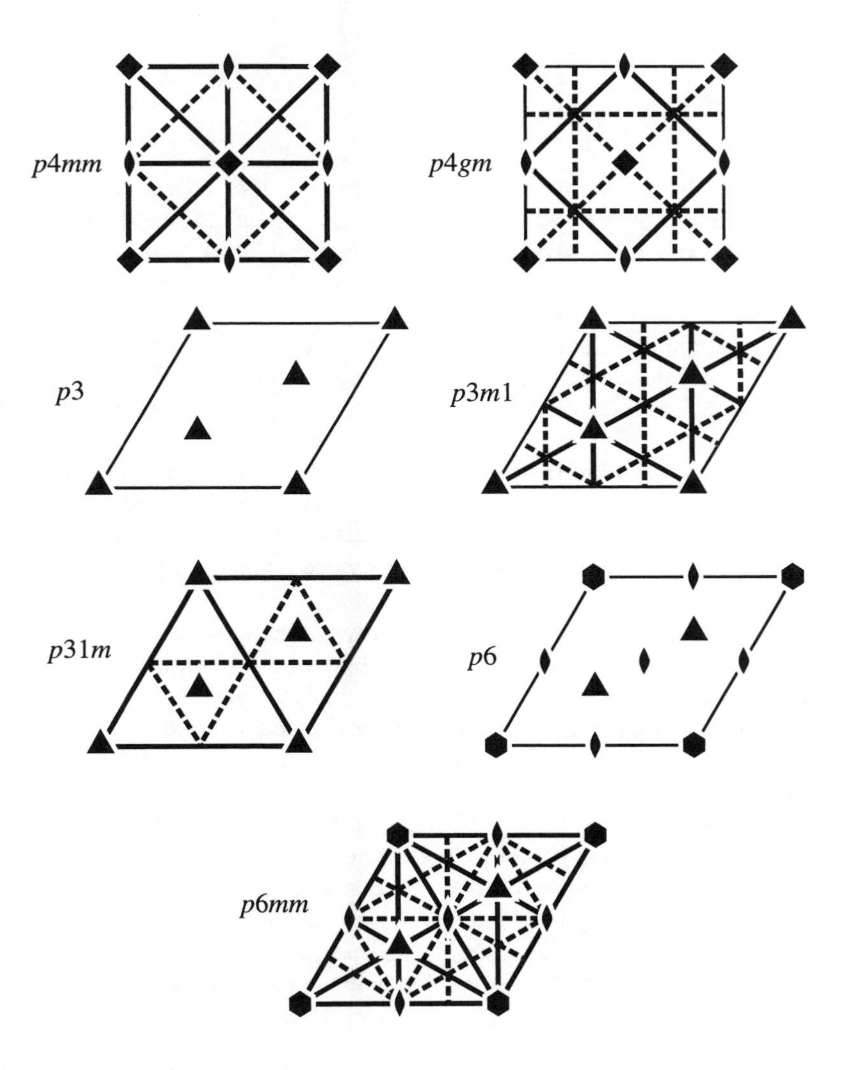


**Fig. 1.13**. The symmetry elements of the two-dimensional space groups. Light lines outline a unit cell. Heavy solid lines indicate the location of mirror lines and broken lines indicate the positions of glide lines. 2-, 3-, 4- and 6-fold rotation points are represented by "lenses", triangles, squares and hexagons respectively.

Figure 1.13 shows the location of the symmetry elements in one unit cell for the space groups. In $p1$, there are only translations, but in $p2$, in addition to the 2-fold rotation points at the corners of the cell, there are also 2-fold rotation points (whose locations are indicated by the two-pointed symbols) in the middle of the edges and the center of the unit cell. Note that *all* the symmetry elements of $p2$ are obtained by combining one rotation point with translations.[1]

### *1.7.2 Rectangular system*

The rectangular lattice $p$ has mirror symmetry and the lattice translations can be combined with point group $1m$ to give a symmetry group $p1m1$ (often abbreviated to $pm$).[2] The symmetry elements of this group consist of orthogonal lattice translations and a family of mirror lines parallel to one of the axes. The spacing between the mirror lines is one-half of the lattice repeat.

We can also combine the elements of $1m$ with the translations of the centered rectangular lattice $c$ to get space group $c1m1$ (abbreviated symbol $cm$). Note in Fig. 1.13 that the space group also contains glide lines interleaved with the mirror lines. These should be identified in the appropriate part of Fig. 1.12.

Point symmetry group $2mm$ is also compatible with a rectangular lattice, so we have space groups $p2mm$ and $c2mm$ (often abbreviated to $pmm$ and $cmm$ respectively). These are the space groups of the primitive and centered rectangular lattices respectively. Note that they are different symmetries and in particular $c2mm$ contains glide lines that are absent in $p2mm$ (Fig. 1.13 again).

Now that we have mirror lines in the space groups, we must consider the possibilities that arise when mirror lines are replaced by glide lines. Thus the mirror line in the symmorphic group $p1m1$ can be replaced by a glide line to give the non-symmorphic group $p1g1$ (often abbreviated $pg$). Again the symmetry elements should be identified in Figs. 1.12 (*a*) and 1.13.

We remarked that the symmorphic group $c1m1$ already contains glide lines so there is *not* a new group "$c1g1$" to be obtained by replacing the mirrors of $c1m1$ by glide.[3]

From the symmorphic group $p2mm$ we get the non-symmorphic groups $p2mg$ and $p2gg$ (often abbreviated $pmg$ and $pgg$) as illustrated in Fig. 1.13.

As explained below, the symbol $p2mg$ refers to the space group in which there are mirror lines perpendicular to **a** and glide lines perpendicular to **b**. The group $p2gm$ is the same symmetry group but now the axis perpendicular to the mirror lines is **b** and the axis perpendicular to the glide lines is **a**. Thus $p2mg$ and $p2gm$ are two different *settings* of the

[1]A reminder that here, and throughout the rest of § 1.7, the reader who is unfamiliar with the material will: (a) Verify that the patterns in Fig. 1.12 are indeed generated by the basic point group operations plus translations. (b) Verify the existence of the symmetry elements shown in Fig. 1.13 from the patterns of Fig. 1.12.

[2]The significance of the "1"s in the symbol is explained below. The novice may well wish to reread this section after reading § 1.8.

[3]The skeptical reader should nevertheless do the experiment. It will be found that replacing the mirror lines in $c1m1$ by glide lines will produce mirror lines where the old glide lines were.

same space group with the first arbitrarily chosen as the *standard setting*. This illustrates that symbols for space groups can change with relabeling of the axes. This is not much of a problem in two dimensions, but we will need to devote careful attention to the analogous question in three dimensions.

It is important to recall that glide lines *already* exist in *cm* and *c2mm* so there are no new groups *cg*, *c2mg* or *c2gg*. We have therefore identified all the rectangular space groups.

### *1.7.3 Square system*

In the square system, we can combine the translations of the square lattice with the point symmetries that include a 4-fold rotation, i.e. 4 and 4*mm*, getting space groups *p*4 and *p*4*mm* (short symbol *p*4*m*). It should be seen in Fig. 1.13 that the combination of lattice translations and a 4-fold point at the origin generates a second 4-fold point at the center of the cell and 2-fold points in the middle of the cell edges. The pattern of symmetry *p*4*mm* should be identified in Fig. 1.12 (*b*) and the glide lines parallel to the unit cell diagonals (shown in Fig. 1.13) identified.

The existence of one set of glide lines parallel to the second set of mirrors (symbolized by the second "*m*") in *p*4*mm* means that although there is a new space group *p*4*gm* (short symbol *p*4*g*) there are not new groups *p*4*mg* or *p*4*gg*. Again identify the space groups in Fig. 1.12 (*b*) and locate the symmetry elements. We have now identified all the square space groups.

**Fig. 1.12 (*b*).** Illustrating the square two-dimensional space groups. Each pattern has different symmetry.

### *1.7.4 Hexagonal system*

In the hexagonal system we can combine the translations of the hexagonal lattice with the point symmetries 3, 3*m*, 6, and 6*mm*.

As already explained (see Fig. 1.6) 3*m* can be oriented in two ways with respect to the coordinate axes (which are parallel to the lattice translations) to give two *distinct* space groups *p*3*m*1 and *p*31*m*. Patterns with these two symmetries are shown in Fig. 1.12 (*c*). To determine which is which (a) outline a unit cell, (b) identify the mirror lines. *p*3*m*1 is the one that has mirror lines normal to the cell edges (compare Fig. 1.13).

The other symmorphic space groups are *p*3, *p*6 and *p*6*mm* (sometimes abbreviated to *p*6*m*). Note again in Fig. 1.13 the presence of additional symmetry elements, in particular in the hexagonal system we always have 3-fold points at 1/3,2/3 and 2/3,1/3.

The groups $p3m1$ and $p31m$ both already have glide lines so there are no new groups to be derived by replacing mirrors by glides in these instances. Finally there are glide lines parallel to both sets of mirrors in $p6mm$ so again there are no new groups to be obtained by replacing the mirrors by glides.[1] We have therefore concluded the enumeration all the two-dimensional space groups.

**Fig. 1.12** (***c***). Illustrating the hexagonal two-dimensional space groups.

[1]In the centered rectangular lattice mirror lines and glide lines are interleaved throughout space. The hexagonal lattice may be considered as a special case of a centered rectangular lattice (see Table 1.3, p. 10 and Exercise 2, p. 26).

### *1.7.5 Synopsis of the two-dimensional space groups*

Table 1.4 below summarizes the two-dimensional space groups. The final column lists first the symmorphic groups and secondly the non-symmorphic groups obtained when mirror lines are replaced by glide lines. Note that the *system* is determined by the shape of the conventional unit cell, or equivalently by the point symmetry at a point of the lattice (Table 1.3). The point group from which a space group is derived is the *class*.

**Table 1.4.** Synopsis of the two-dimensional space groups (short symbols in parentheses)

| system | point group (class) | space group symmorphic | $m \rightarrow g$ |
|---|---|---|---|
| oblique | 1 | *p*1 | |
| | 2 | *p*2 | |
| rectangular | 1*m* | *p*1*m*1 (*pm*) | *p*1*g*1 (*pg*) |
| | | *c*1*m*1 (*cm*) | |
| | 2*mm* | *p*2*mm* (*pmm*) | *p*2*mg* (*pmg*) |
| | | | *p*2*gg* (*pgg*) |
| | | *c*2*mm* (*cmm*) | |
| square | 4 | *p*4 | |
| | 4*mm* | *p*4*mm* (*p*4*m*) | *p*4*gm* (*p*4*g*) |
| hexagonal | 3 | *p*3 | |
| | 3*m* | *p*3*m*1 | |
| | | *p*31*m* | |
| | 6 | *p*6 | |
| | 6*mm* | *p*6*mm* (*p*6*m*) | |

## 1.8 Construction and interpretation of space group symbols

We now summarize the rules for constructing and interpreting space group symbols. It is emphasized that the space group symbol both identifies the space group and specifies the orientation of the reference axes with respect to the symmetry elements. In specifying the direction of mirror or glide lines we actually specify the directions of the normals to them (for reasons that will become apparent in Chapter 3). Note that in interpreting the space group symbol, we have first to identify the system (oblique, rectangular, square or hexagonal).

The first letter of a space group symbol refers to the lattice type and must be "*p*" or "*c*."

In the **oblique** system we have either *p*1 or *p*2 as space group symbol.

In the **rectangular** system there are (in the long, or full, symbol) three places after the lattice symbol. The first of these three places has a symbol which refers to the nature of the rotation point and is either "1" or "2." The next symbol refers to the nature of the symmetry line ("*m*" or "*g*") normal to **a** (the normal to the mirror or glide line is parallel to **a**). The last symbol refers to the nature of the symmetry element normal to **b** and is either "1", "*m*" or "*g*." Note the use of "1" as a *place marker* to avoid ambiguity when no symmetry element is present. The standard setting for *pm* is with the mirror planes normal to **a** hence the long symbol *p*1*m*1; if for some reason we wished to label the axis normal to the mirror plane **b**, the space group symbol would now be written *p*11*m*.

In the **square** system the first position of the space group symbol is for the lattice (must be *p*). The second is for the rotational symmetry (must be 4) the third (if present) is for symmetry elements along *x* and *y* (i.e. parallel to **a** and **b**—because of the 4-fold symmetry these symmetry elements must be the same) and the fourth position (in the full symbol) signifies the presence of symmetry elements at 45° to *x* and *y* (i.e. along the directions **a**±**b**).

In the **hexagonal** system the first position of the space group symbol is again *p*. The second indicates the rotational symmetry (3 or 6). The third position signifies symmetry elements along the *x* and *y* directions (which are at 120°), i.e. parallel to **a** and **b** and to the third equivalent direction which is parallel to –(**a**+**b**). The fourth positions refer to symmetry elements at 90° to *x* and *y*. Recall that the orientations of mirror lines are specified by the directions of their normals and note again the use of 1 as a place marker in *p*3*m*1 and *p*31*m*. Accordingly in *p*3*m*1 the normals to the mirrors are parallel to **a** and **b** [the mirror lines are normal to **a** and **b** and –(**a**+**b**)]; in *p*31*m* the normal to the mirrors are at right angles to **a** and **b** [the mirror lines are parallel to **a** and **b** and –(**a**+**b**)].

## 1.9 Using the *International Tables*

Volume *A* of the *International Tables for Crystallography* (see the Book List) is the standard (and indispensable) source of information about space groups.[1] Fig. 1.13 closely follows the diagrams in that book which we often refer to just as the *International Tables*.

The *International Tables* also gives the coordinates of the points in one unit cell obtained by the operation of the symmetry operations of the space group on an arbitrary point *x*,*y*. These are the coordinates of the *general* positions. The number of general positions will equal the order of the point group multiplied by the number of lattice points per unit cell (1 for *p* and 2 for *c*).

[1]Two points to note here about the *Tables*: (a) The two-dimensional space groups are there called "plane groups". (b) The short symbols (which are widely used elsewhere) are not used for these groups (see p. 16 of the *Tables* for more on this point).

There are also *special* positions in the unit cell. These correspond to points that are either at rotation points or on mirror lines. Thus consider the group *p6mm* (see Fig. 1.13, p. 17). The point at the origin is on a 6-fold rotation point and operation of the 6-fold rotation will not generate any new points. Likewise the same point is also on the mirror planes so the mirror planes produce no new points. In fact the symmetry operations produce no new points at all. This may be verified by substituting $x = 0$ and $y = 0$ in the coordinates of the general positions listed below.

Other special positions of *p6mm* are on the 3-fold points at 1/3,2/3 and 2/3,1/3. The symmetry operations acting on one of these points will produce only the other as may be verified by substituting $x = 1/3$ and $y = 2/3$ in the coordinates of the general positions.[1]

In Table 1.5 below we give for *p6mm* (short symbol *p6m*): first the number of positions of a given kind, then the *Wyckoff* notation (*a*, *b*, etc.) and then the coordinates of the positions. Refer to Fig. 1.13 to see that the positions 6 *e* correspond to points constrained to lie on one set of mirror lines and thus at sites of symmetry *m*. The positions 6 *d* are likewise arbitrary points on the second set of mirror lines and again at sites of symmetry *m*. Positions 3 *c* are points on 2-fold rotation points and at the intersection of two mirrors, and so are at sites with point symmetry 2*mm*. Positions 2 *b* are on 3-fold rotation points at the intersection of three mirror lines so they are at sites with symmetry 3*m*. Point 1 *a* (at the origin) has symmetry 6*mm*. The letters *a*, *b*, etc. have no special significance except to serve as (universally accepted!) labels. Each space group will have special and general positions labeled from *a* (for the highest symmetry points), alphabetically to whatever letter is necessary for the general positions. The last entry in the table below is the point symmetry at each kind of site.

**Table 1.5.** Special and general positions of *p6mm*

| wyckoff notation | coordinates | symmetry |
|---|---|---|
| 12 *f* | $x,y$ ; $\bar{y},x-y$ ; $y-x,\bar{x}$ ; $y,x$ ; $\bar{x},y-x$ ; $x-y,\bar{y}$<br>$\bar{x},\bar{y}$ ; $y,y-x$; $x-y,x$ ; $\bar{y},\bar{x}$ ; $x,x-y$ ; $y-x,y$ | 1 |
| 6 *e* | $x,\bar{x}$ ; $x,2x$ ; $2\bar{x},\bar{x}$ ; $\bar{x},x$ ; $\bar{x},2\bar{x}$ ; $2x,x$ | *m* |
| 6 *d* | $x,0$ ; $0,x$ ; $\bar{x},\bar{x}$ ; $\bar{x},0$ ; $0,\bar{x}$ ; $x,x$ | *m* |
| 3 *c* | 1/2,0 ; 0,1/2 ; 1/2,1/2 | 2*mm* |
| 2 *b* | 1/3,2/3 ; 2/3,1/3 | 3*m* |
| 1 *a* | 0,0 | 6*mm* |

Having available the coordinates of the general and special positions allows a considerable economy in specifying structure. Thus only the coordinates of one of a set of general positions need to be given. In some cases no coordinates need be given explicitly. Thus to completely specify the positions of the points in the kagome pattern (3.6.3.6) described above, it is sufficient to give the space group (*p6mm*), the magnitude of the unit cell parameter (*a*) and to state that points are at positions *c*. The points of the honeycomb pattern ($6^3$) are at *b* (make a quick sketch). Another example using this space group is

[1]Remember that $x = -1/3$ is equivalent to $x = 2/3$ etc.

given in Exercise 3 at the end of the chapter.

In a space group with a centered lattice the existence of a point $x,y$ implies the existence of an identical point at $1/2+x,1/2+y$ because translation by 1/2,1/2 is a symmetry operation. In the *International Tables* the positions of *cm* are given as shown in table 1.6.

**Table 1.6.** Special and general positions of *cm*.

| | | (0,0 ; 1/2,1/2) + |
|---|---|---|
| 4 *b* | $x,y$ ; $\bar{x},y$ | |
| 2 *a* | $0,y$ | |

Thus the coordinates of the general positions are $x,y$ ; $\bar{x},y$ ; $1/2+x,1/2+y$ ; $1/2-x,1/2+y$ and the coordinates of the special positions *a* are $0,y$ ; $1/2,1/2+y$ (i.e. on the mirror lines at $x = 0$ and $x = 1/2$).

## 1.10 Notes

### *1.10.1 Symmetry operations*

The symmetry operations we have discussed are special cases of general geometrical *transformations* or *mappings*. They are those which preserve distances and angles and hence the term *isometry*. There are many mappings in which there is a one-to-one correspondence between the object before and after the transformation. A *similarity* transformation is one in which angles, but not distances are conserved. It may be considered as the result of an isometry followed by an expansion or contraction in scale. Another transformation of interest is inversion or reflection in a circle. For examples of the remarkable patterns that can arise from repeated reflections in a system of circles see R. Courant & H. Robbins, *What is Mathematics?*, Oxford University Press, Oxford (1941) *p.* 162.

### *1.10.2 Groups*

Introductions to symmetry group theory[1] are F. A. Cotton, *Chemical Applications of Group Theory*, 3rd Edition, Wiley, New York (1990) and Boisen & Gibbs (Book List).

A group is a set of (at least one) things with the following properties.

•There is defined a *binary operation* symbolized by * such that $\alpha*\beta = \gamma$, where $\alpha$, $\beta$ and $\gamma$ are all members of the group. In the case of symmetry operations, $\alpha*\beta$ means operation $\beta$ followed by operation $\alpha$, and $\gamma$ is also a symmetry operation. Note that in general $\alpha*\beta$ and $\beta*\alpha$ are not equal (but may be). If for every pair of elements in the group $\alpha*\beta = \beta*\alpha$ the group is said to be *commutative* or *abelian*.

[1]Note that in this chapter we haven't done any group *theory* (nor shall we in subsequent chapters). All we have done is to *enumerate* certain kinds of symmetry group.

• $\alpha*(\beta*\gamma) = (\alpha*\beta)*\gamma$ (this is the *associative rule*).

• There exists an element $\varepsilon$ in the group such that $\varepsilon*\alpha = \alpha*\varepsilon = \alpha$ for $\alpha$ equal to any member of the group. In symmetry groups $\varepsilon$ corresponds to the identity operation.

• For any element $\alpha$ in the group there is another $\beta$ (the *inverse* of $\alpha$) such that $\alpha*\beta = \beta*\alpha = \varepsilon$. $\alpha$ and $\beta$ may be the same (for example two reflections in the same mirror result in the identity) or different [rotation by one-third of a circle ($3^1 = 3^+$) followed by rotation by two-thirds of a circle ($3^2 = 3^-$), or *vice versa*, again results in the identity ($3^1*3^2 = 3^2*3^1 = 1$)].

The number of distinct elements in the group is the *order* of the group. The simplest possible group (of order 1) contains only $\varepsilon$.

An example of a finite group (of order three) is the numbers 0, 1 and 2 if the group operation $*$ is defined as "addition modulo 3", i.e. $\alpha*\beta \equiv (\alpha+\beta)\ \mathrm{mod}(3)$. Thus we have:

$$
\begin{array}{lll}
0*0 = 0 & 1*0 = 1 & 2*0 = 2 \\
0*1 = 1 & 1*1 = 2 & 2*1 = 0 \\
0*2 = 2 & 1*2 = 0 & 2*2 = 1
\end{array}
$$

Here 0 corresponds to $\varepsilon$, and 1 and 2 are the inverses of each other. We can write out the above relationships as a "multiplication" table. "Multiplication" refers to the group operation, in this case "addition modulo 3."

|   | 0 | 1 | 2 |
|---|---|---|---|
| 0 | 0 | 1 | 2 |
| 1 | 1 | 2 | 0 |
| 2 | 2 | 0 | 1 |

This group has the same structure as (is *isomorphic* with) the point symmetry group 3 whose elements are the identity and rotations by one-third and by two-thirds of a circle. These two sets of things with their appropriate group operation ($*$) form two different representations of the same abstract group.

The matrices that we derived for symmetry operations (see *e.g.* the case of $4mm$ Table 1.1) form a representation of the same abstract group as do the symmetry operations themselves if the group operation now is identified as matrix multiplication.

### *1.10.3 Two-dimensional space groups and decoration*

The occurrence of periodic patterns in decoration is familiar and dates from antiquity. A most striking source of periodic patterns is the XIV century designs in the Alhambra of Granada in Spain. More recently, the Dutch artist M. C. Escher was inspired by the Alhambra to produce some remarkable periodic designs which have been used as an aid to teaching crystallography [C. H. MacGillavry, *Symmetry Aspects of M. C. Escher's Periodic Drawings*, Oosthoek, Utrecht (1965)]. Escher's art also provides some beautiful examples of repeated similarity transforms.

The Moors did not use all the two-dimensional space groups in the decoration of the Alhambra. It is interesting that the explicit enumeration of the 17 two-dimensional groups was only done *after* the 230 three-dimensional space groups were enumerated. The symmetry groups of three-dimensional objects with translations in only two dimensions are the 80 *layer groups* which are sometimes of use in crystal chemistry. We refer to these again in Chapter 3 (see also Appendix A1).

### *1.10.4 Matrix manipulations*

The reader entirely unfamiliar with matrix manipulations will have to consult one of the very many elementary mathematics texts that deal with such matters. In this chapter we use only $2 \times 2$ matrices such as $\mathbf{A} = (a_{11}\ a_{12}\ /\ a_{21}\ a_{22})$ and $2 \times 1$ matrices (2 rows and 1 column; also known as column vectors) such as $\mathbf{x} = (x_1\ /\ x_2)$.

Multiplication of a $2 \times 1$ matrix by a $2 \times 2$ matrix give a new $2 \times 1$ matrix as in $\mathbf{Ax} = \mathbf{y}$, where $\mathbf{y} = (a_{11}x_1 + a_{12}x_2\ /\ a_{21}x_1 + a_{22}x_2)$.

Multiplication of a $2 \times 2$ matrix by a $2 \times 2$ matrix gives a new $2 \times 2$ matrix as in $\mathbf{AB} = \mathbf{C}$, where $\mathbf{C} = (a_{11}b_{11} + a_{12}b_{21}\ \ a_{11}b_{12} + a_{12}b_{22}\ /\ a_{21}b_{11} + a_{22}b_{21}\ \ a_{21}b_{12} + a_{22}b_{22})$. Note that the order of multiplication is important and in general $\mathbf{AB} \neq \mathbf{BA}$ (recall that the order in which symmetry operations are carried out *may* be important).

## 1.11 Exercises

1. Circles of unit diameter are packed as closely as possible on a plane. A fraction $\pi/\sqrt{12}$ of the area of the plane is covered. The symmetry is *p6mm* and the centers of the circles are on the points of a hexagonal lattice. (Do this by packing coins of one kind on a table top.)

2. A hexagonal lattice can be described by a centered rectangular cell with $b/a = \sqrt{3}$. (The centered cell is sometimes called *orthohexagonal*).

3. A pattern often encountered in crystal chemistry and in ornament is the Archimedean tiling 3.4.6.4 which has symmetry *p6mm*. If the edges of the polygons are equal to 1, the unit cell parameter is $a = 1+\sqrt{3}$ and the points are in 6 *e* (Table 1.5) with $x = 1/(3+\sqrt{3})$. Verify by plotting the pattern. [Hint: if you have difficulty plotting with hexagonal axes see Fig. 4.10, p. 121.]

4. The symmetry of the face of a conventional brick wall is *c2mm*.

5. A pair of parallel mirror lines separated by a distance *d* produces an infinite set of mirror lines with spacing *d* and translational symmetry with period 2*d* (the barber shop phenomenon).

6. Use the matrices given in Table 1.1 to construct a multiplication table for *4mm*.

7. Referred to Cartesian coordinates (axes at 90°), rotation by 120° about the origin of a point $x,y$ gives a new point at $-x/2-\sqrt{3}y/2$, $\sqrt{3}x/2-y/2$. The trace (sum of diagonal terms) of the matrix generating the new coordinates from the old is $-1$, and is independent of the choice of coordinate system (compare the matrix given in Table 1.2 for axes at 120°).

8. In Fig. 1.14 (below) the broken lines represent two glide lines at right angles; the light rectangle represents a unit cell of edges $a$ and $b$. Start with an asymmetric object (7) in the upper left corner of the cell and use the glide reflections and translations (**a** and **b**) to generate images of the original object. This process has been begun in the figure in which straight arrows represent translations and the bent ones represent glide reflections. In this way we will generate all the symmetry elements of *pgg* (*p*2*gg*). Identify all the symmetry elements in the unit cell (objects of every pair must be related by a symmetry operation). Compare with Fig. 1.13 (in which the unit cell origin is taken at a different point).

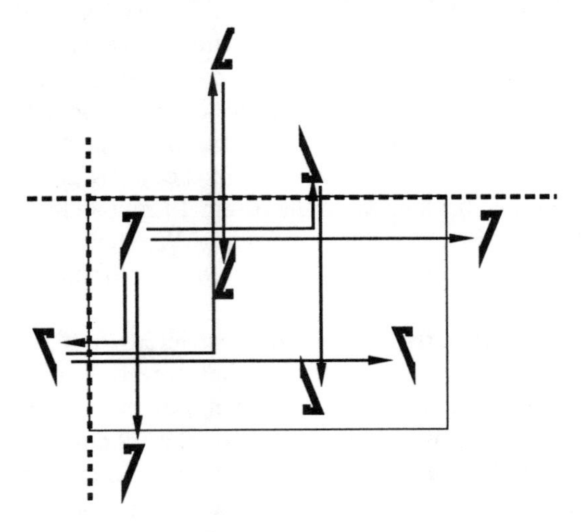

**Fig. 1.14**. See Exercise 8.

9. The orders of the point symmetry groups can be found by counting the numbers of replicas of the asymmetric object that are in the patterns of Fig. 1.2. The number of asymmetric objects (scalene triangles) in a unit cell of the patterns of Fig. 1.12 will be the same as the order of the point group of the space group if the cell is primitive, and twice that if the cell is centered. Verify for each space group.

10. Many examples of two-dimensional patterns (tilings) appear in Chapter 5. Identify the symmetry elements and verify their space groups.

Then there is the story of the crystallographer who was asked for the next letter in the sequence A,B,C,D,E... His response naturally was K as this is the next letter in the alphabet with symmetry $m$ (not *exactly* in the font used here!). What are the symmetries of F, G, H, I, J, S, +, #, ✳, ✯, 916 ? (Answers: 1, 1, 2*mm*, 2*mm*, 1, 2, 4*mm*, 2, 8*mm*, 5*m*, 2.)

CHAPTER 2

# THREE-DIMENSIONAL POINT GROUPS

## 2.1 Point symmetry operations in three dimensions

A major difference between point operations in two and three dimensions is that in three dimensions there are rotation *axes* which leave lines invariant (instead of rotation points that leave points invariant in two dimensions) and mirror *planes* which leave planes invariant (instead of mirror lines which leave lines invariant).[1] Accordingly we expect a new symmetry operation in three dimensions that leaves a point invariant.

This new symmetry operation is *inversion* in a point (discussed below). It is convenient to consider the point symmetry operations of three-dimensional space as (a) rotation and (b) rotation *combined* with inversion. The corresponding symmetry elements are often referred to as *proper* and *improper* axes respectively. A proper operation acting on an asymmetric object merely moves it through space whereas an improper operation also converts it into its mirror image. Two successive improper operations convert an asymmetric object first into its mirror image and then back again to the original form, so that a *combination* of two improper operations is equivalent to a proper operation. We shall see that a mirror reflection in a plane (an improper operation) can be considered as combination of a 2-fold rotation plus inversion.

As in the case of two dimensions, we are restricted to 1-, 2-, 3-, 4-, and 6-fold symmetry axes in crystal symmetries. There are therefore five proper and five improper axes for a total of ten different kinds of point symmetry element. We will see that there are 32 distinct crystallographic point symmetry groups that contain combinations of these symmetry elements. We will not actually demonstrate that the number of groups is exactly 32. We will describe 32 groups and ask the reader to accept that our enumeration is complete. The dissatisfied reader is directed to § 2.5.3.

Those trained in molecular chemistry will be familiar with point group symmetry, and their task will be mainly to learn new symbols for these groups, although some point group symmetries do not have simple molecular examples. It will prove a rewarding exercise to verify the symmetries of the polyhedra discussed in Appendix 4 (some of which occur as carbon molecules). Another useful exercise is to check off the point groups listed in Tables 2.1 (§ 2.2.6) and 2.2 (§ 2.2.8) as they are described. In our experience the only way to acquire a useful working knowledge of point group symmetries is to constantly practice identifying the symmetries of objects such as molecules and polyhedra.[2] Only when the groups are well known is it possible to appreciate a rigorous mathematical treatment (which we will not develop).

[1] In one dimension the only point symmetry element is a mirror *point*.

[2] We use several examples of polyhedra in this chapter. The reader unfamiliar with such objects will find them described in § 5.1.

The discussion of point symmetries in terms of pure rotation and roto-inversion axes is the approach that proves most useful in crystallography. An alternative approach, using rotation-reflection axes, is embodied in the Schoenflies system that is favored by molecular chemists; as it leads to a more elegant mathematical treatment the latter is also preferred by mathematicians. The crystallographic system with its associated symbolism (Hermann-Mauguin symbols) is distinctly better for space groups and is now in universal use by crystallographers and solid state chemists. A concordance between the Hermann-Mauguin symbols and the Schoenflies symbols for the crystallographic point groups is given in the tables at the end of the book, and for non-crystallographic groups (those involving rotations of order different from 1, 2, 3, 4 or 6) in § 2.5.6.

The five proper axes should provide little difficulty. If coordinate axes are chosen so that the $z$ direction is along the rotation axis, the $x$ and $y$ coordinates will transform as in the two dimensional case of rotation about a point, and $z$ will remain unchanged. The positive sense of rotation is anticlockwise when viewed along the $+z$ to $-z$ direction. We will discuss the improper operations individually.

### *2.1.1 Inversion in a center*

Inversion in a center[1] at the origin 0,0,0 will convert a point at $x,y,z$ to a point at $\bar{x},\bar{y},\bar{z}$ regardless of the coordinate system used. Fig 2.1 illustrates the inversion of an asymmetric object through a center at the origin of coordinates (we always take the origin at the center if it is present). Inversion is symbolized $\bar{1}$ (1-fold rotation plus inversion). This would also be the symbol for the point group if the inversion center were the only symmetry element (other than the identity) of the object. Objects that include an inversion center in their symmetry are said to be *centrosymmetric*. Those that do not are *acentric.*[2]

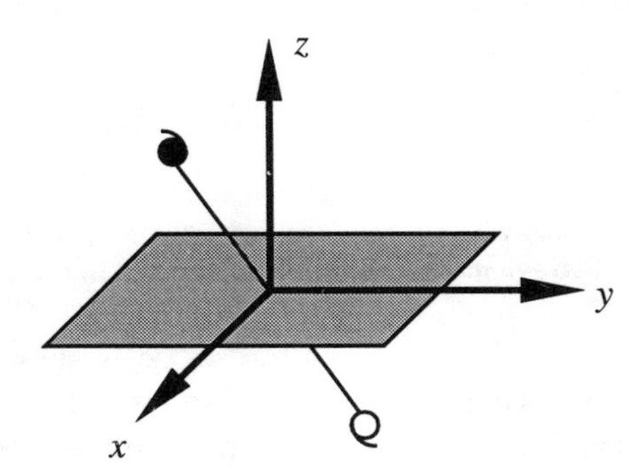


**Fig. 2.1**. Illustrating the inversion operation.

[1]The term *center* is synonymous with *inversion center* in this context.

[2]Objects (including molecules) other than crystals belonging to symmetry group $\bar{1}$ are rather rare (try to identify some). A pair of shoes (or gloves) can be arranged so that the assembly has $\bar{1}$ symmetry.

Figure 2.1 also illustrates the choice of a *right handed* set of coordinate axes (this is the "standard" choice). Interchanging any two axes will change the hand of the coordinate system unless the *direction* of one of the axes is also reversed. Cyclic permutations such as $x \rightarrow y \rightarrow z \rightarrow x$ or the reverse will not change the hand of the coordinate system.

### 2.1.2 Rotation plus inversion: axes $\overline{N}$

The combination of a 2-fold rotation with inversion is equivalent to reflection in the plane normal to the 2-fold axis and containing the inversion point. The combined operation is illustrated in Fig. 2.2. A 2-fold rotation axis through the origin and parallel to $z$ converts a point at $x,y,z$ to $\bar{x},\bar{y},z$. Inversion through a center at 0,0,0 will convert $\bar{x},\bar{y},z$ to $x,y,\bar{z}$. The net result is to transform $x,y,z$ to $x,y,\bar{z}$—it should be clear that this is equivalent to reflection in the plane $z = 0$. This symmetry element which could be symbolized $\overline{2}$ is in fact symbolized $m$ (for *mirror*). In the figure it may be seen that rotation by 180° around the axis shown, followed by inversion, is also the same as inversion followed by rotation about the axis, and in either case equivalent to reflection in the plane normal to the 2-fold axis. The orientation of a mirror plane is always specified by giving the direction of the normal to the plane.

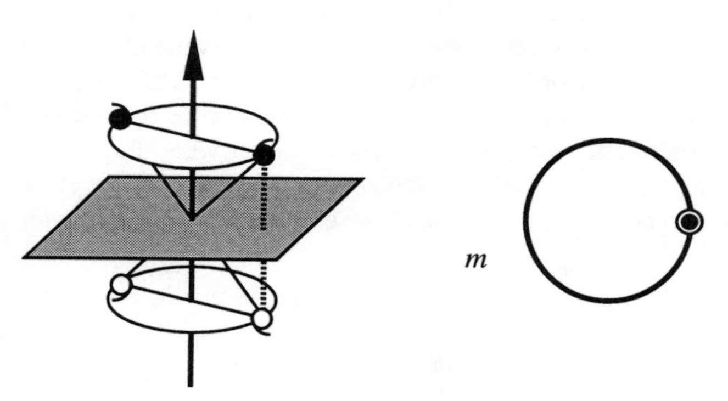


**Fig 2.2.** (Left) Illustrating that $\overline{2}$ is equivalent to reflection. An object (white) is rotated about the two-fold axis (arrow) to give a similar object which is the inverted through the center to give a mirror-image object (black). (Right) showing the two symmetry-related points in projection along $z$.

On the right in Fig 2.2 a stylized view down the $z$ axis is shown where the original object above the plane $z = 0$ is shown as an open circle and its reflection below the plane is shown as a smaller filled circle. Such diagrams are very useful to show the effects of other symmetry elements and combinations of symmetry elements. It is important to recognize that the circles in such diagrams really represent a general asymmetric objects.

It should be clear that an object (such as a chair) that has only one mirror plane (and the identity) for symmetry elements, has neither a 2-fold axis nor a center of symmetry. In general an $N$-fold inversion axis only contains a separate $N$-fold rotation axis and a separate inversion center for $N$ odd. Indeed, as explained in the next paragraphs, $\overline{3}$ is the only crystallographic symmetry operation (other than $\bar{1}$ itself) that contains a separate $\bar{1}$.

Fig 2.3 shows the effects of $\bar{3}$, $\bar{4}$ and $\bar{6}$ axes on a point. The $\bar{3}$ operation (rotation by one-third of a circle followed by inversion) has to be carried out six times before the original point is returned to. The reader should verify the *separate* existence of a $\bar{1}$ and a 3 axis.[1] In the diagrams in Fig. 2.3, the symbol in the center is the symbol for the symmetry element. The small circles represent a general asymmetric object and thus in these examples, filled and open circles should really be replaced by asymmetric objects related one to the other by inversion or reflection (i.e. of opposite hand).

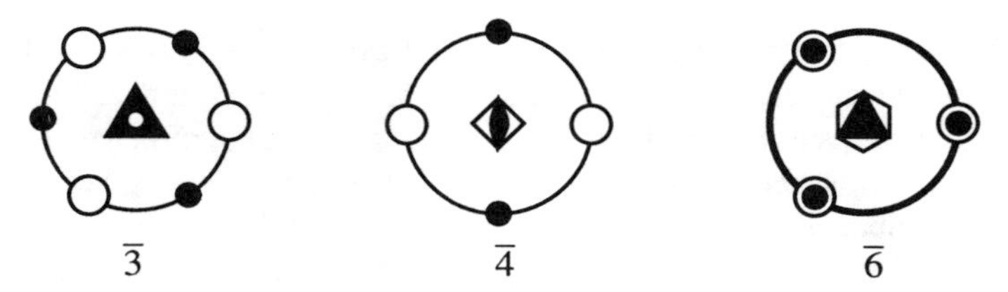


**Fig. 2.3**. Illustrating the effects of $\bar{3}$, $\bar{4}$ or $\bar{6}$ axes (see text). Filled and empty circles are points on opposite sides of the horizontal plane (the plane of the paper). In $\bar{6}$ such points are superimposed in projection and the heavier outline of the circle indicates the presence of a horizontal mirror plane.

A $\bar{4}$ axis (Fig. 2.3) generates four distinct points on repeated application; two above and two below the plane. It should be clear that an object with only $\bar{4}$ symmetry is not centrosymmetric (but does have a 2-fold rotation axis).

A $\bar{6}$ axis generates six points (Fig. 2.3 again). Note again the absence of an independent inversion center. This symmetry element includes a separate 3 axis and a mirror normal to that axis (indicated in the figure by a heavier outline for the circle).[2]

It should be obvious that in order to completely specify the position of a $\bar{3}$, $\bar{4}$ or $\bar{6}$ axis we have to specify the direction of the axis *and* the location of the inversion point even though $\bar{4}$ and $\bar{6}$ do not contain a *separate* inversion center.

A quick sketch should convince the reader that whether one rotates and then inverts, or first inverts and then rotates, is immaterial. The combined operation is the same; i.e. the component operations *commute.*

A word on terminology is in order. We refer to an *N* axis as *N*-fold as in "2-fold," "3-fold," etc. More commonly perhaps (as in the *International Tables*) the usage "twofold," "threefold," etc. is seen. We find our usage clearer in such phrases as "four 3-fold axes." Other authors refer to 1-, 2-, 3-, 4- and 6-fold rotation axes as "monads," "diads," "triads," "tetrads," and "hexads" respectively. A difficulty arises in the case of $\bar{3}$ and $\bar{6}$. As can be seen from Fig. 2.3 both these operations require *six* repetitions to produce the identity yet we refer to the former as a "3-fold inversion axis" and the latter as a

[1]The reader is urged to verify that (a) applying the rotation+inversion operation three times is equivalent to inversion alone and (b) applying the combined operation four times is equivalent to rotation alone. (Remember that rotations are by convention anticlockwise when viewed from the +*z* direction.)

[2]An *N*-fold rotation axis with a mirror normal to it is written *N*/*m* so $\bar{6}$ is sometimes written as 3/*m*. However this obscures the 6-fold nature of the axis and this notation should be avoided.

"6-fold inversion axis."[1] Crystals with parallel 3-fold axes (including $\overline{3}$) but without 6-fold axes are classed as *trigonal*. Crystals with 6-fold axes (including $\overline{6}$) are classified as *hexagonal*. $\overline{3}$ is also referred to as an "inversion triad" and $\overline{6}$ as an "inversion hexad." The difficulty is compounded by the fact that in the Schoenflies system (see Exercise 17) $\overline{3}$ is labeled $S_6$ and $\overline{6}$ is labeled $S_3$.

## 2.2 Enumeration of the point groups

There are 10 point groups corresponding to the presence of just one of the symmetry elements we have described. These are 1, 2, 3, 4, 6, $\overline{1}$, $m$, $\overline{3}$, $\overline{4}$ and $\overline{6}$. There are 22 more (crystallographic) groups to be obtained by combining several symmetry elements. We will not derive these groups very systematically, but it is worth seeing why there is only a relatively small number of them. The serious student of solid state science will get to know them all intimately. The material in this section is rather condensed—the reader interested in fully appreciating it would be well advised to have a pencil and paper at hand to make sketches to verify some of the statements. We find sketches of the sort shown in Fig. 2.3 to be particularly helpful.

### *2.2.1 Pure rotation groups: dihedral groups N2(2)*

We start by considering just those cases in which we have proper rotation axes only. The groups in this case are the *pure rotation* groups. We will have to consider two intersecting rotation axes inclined to each other. Combination of two such rotations always produces a rotation about a new axis through the point of intersection of the first two axes.[2] (Remember we are considering point group symmetry so there must be one point at which all symmetry elements intersect). It is important to recognize that we consider the rotation axes to be fixed with respect to the coordinate system and only the rotated object to move.

Fig. 2.4 shows two 2-fold axes (labeled 1 and 2) lying in a plane and inclined at an angle $\phi$. Rotation of the point $a$ above the plane about axis 1 will produce point $b$ below the plane. Rotation of $b$ about axis 2 will now produce point $c$ above the plane. It should be clear that the transition $a \rightarrow c$ is equivalent to a rotation by $2\phi$ about an axis normal to 1 and 2 and passing through their point of intersection (small shaded circle in the diagram).

In general we consider rotation about an axis $X_1$ by an amount $\rho_1$ followed by rotation by an amount $\rho_2$ about an axis $X_2$ inclined to $X_1$ by an angle $\phi_3$. This is equivalent to

[1]Curiously, the *International Tables* (which usually is our arbiter in such matters) names other symmetry axes (p. 9) but avoids the issue for inversion axes as in "Inversion axis: '3 bar'" for $\overline{3}$. The fraction of the world population that cares, appears to be approximately equally divided into those who use "bar three" and those who use "three bar" in speech for $\overline{3}$. We prefer the former, as "bar" is to be considered as standing for "minus".

[2]As rotations are proper operations they will not change the hand of an asymmetric object. The operation corresponding to two rotations about axes intersecting in a point must be another proper operation leaving that point invariant and thus can only be a rotation. The result of combining rotations about axes without a common point will also include a component of translation (also a proper operation).

rotation by an amount $\rho_3$ about a third axis $X_3$ inclined to axes $X_1$ and $X_2$ by angles of $\phi_2$ and $\phi_1$ respectively. These angles are related by an equation due to Rodrigues:[1]

$$\cos(\rho_3/2) = \cos\phi_3 \sin(\rho_1/2)\sin(\rho_2/2) - \cos(\rho_1/2)\cos(\rho_2/2) \qquad (2.1)$$

Two additional equations are obtained by cyclic permutation of indices $1 \to 2 \to 3 \to 1$. This gives us three equations for the unknowns $\rho_3$, $\phi_2$ and $\phi_1$.

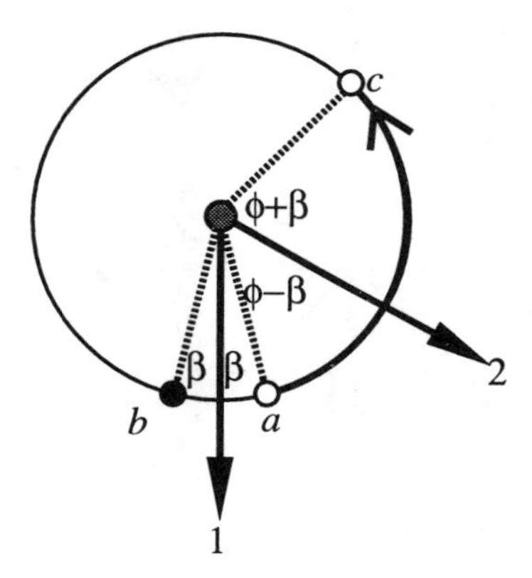


**Fig. 2.4.** Two 2-fold rotations about axes at an angle of $\phi$ to each other (about 1 to take $a$ to $b$, then about 2 to take $b$ to $c$) are equivalent to rotation (by $2\phi$) about an axis orthogonal to the 2-fold axes ($a$ to $c$).

Equation 2.1 is general, but if the rotations are limited (as they will be from now on) to integral fractions of a circle so that $360°/\rho_i = p_i$ (an integer) then it is not difficult to show (see § 2.5.1) that all possible pure rotation groups are generated[2] by rotations such that:

$$\sum_{i=1}^{3} \frac{1}{p_i} > 1 \qquad (2.2)$$

Further analysis shows that the possible solutions of Eq. 2.2 fall into two classes: (a) the dihedral groups in which there is an $N$-fold axis with 2-fold axes at right angles to that axis ($p_3 = N$, $p_1 = p_2 = 2$ which gives in turn from Eq. 2.1, $\phi_3 = 360°/2N$). (b) the *cubic* and *icosahedral* groups in which there are more than one rotation axis of order greater than two.

Consider category (a) first. When $\phi = 360°/N$ with $N$ an integer, the two 2-fold axes generate a finite number of symmetry elements as shown in Fig. 2.5 for the case $\phi_3 = 45°$.

[1]This equation is often ascribed to Euler. For a nice account both of the historical importance of the equation and of Rodrigues see *Icons and Symmetries* by S. L. Altmann (Oxford, 1992). For a derivation of Eqs. 2.1 and 2.2 see the Note § 2.5.1 at the end of this chapter.

[2]Strictly speaking only two rotation axes are needed to generate all the rotations of the group: the third rotation entering into the sum in Eq. 2.2 being generated by the other two.

The 2-fold axes 1 and 2 in that figure generate the 4-fold axis which in turn generates the other 2-fold axes shown (note that co-linear axes are really just one axis). The set of symmetry elements form a group symbolized 422. The first position is for the principle axis and the second and third positions in the point group symbol are for the two sets of 2-fold rotation axes.[1]

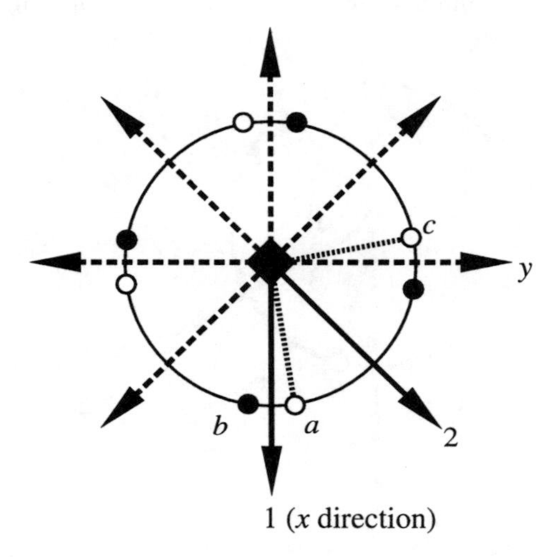


**Fig. 2.5**. Group 422. Arrows represent 2-fold rotation axes. See also Fig. 2.6.

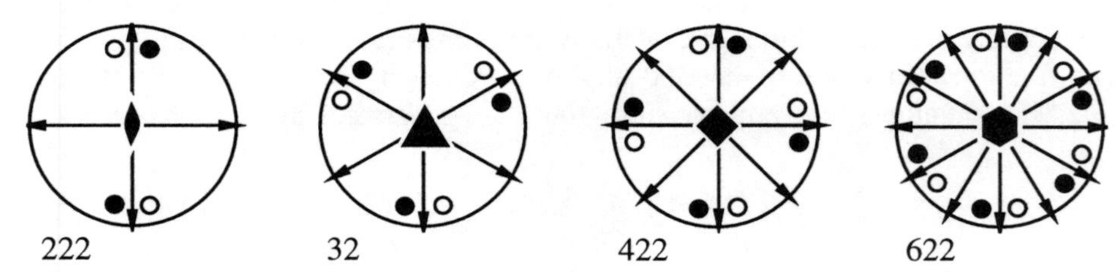


**Fig. 2.6**. Illustrating the symmetry elements of groups 222, 32, 422 and 622. Lines terminating in arrow heads represent 2-fold rotation axes in the plane of the paper. Small circles are sets of points generated by the group operations acting on an arbitrary point. Filled and open circles are on opposite sides of the plane of the two-fold axes. The symbol in the center of the large circle is that for the *N*-fold axis (also a 2-fold axis for 222) normal to the plane of the 2-fold axes.

Likewise there are groups 222, 32 and 622 generated by 2-fold axes at 90°, 60° and 30°. Note that there is only one distinct set of 2-fold axes in 32. This is because 2-fold axes inclined to each other by 60° are also inclined to each other at 120°. These groups (together with 422) are called the *dihedral* groups. It should be obvious that pure rotation groups contain only proper operations and so do not contain a center of symmetry. In Fig. 2.6 the

[1]Cf. the discussion in § 1.1 of the sets of mirror lines in two-dimensional point groups such as 4*mm*.

location of the symmetry elements of these groups is shown. A set of points generated by the symmetry operations acting on an arbitrary point is also shown. Notice that (in contrast to the case of Fig. 2.3) the filled and open circles now represent asymmetric objects of the *same* hand.

### 2.2.2 Groups *Nm(m)*

In Fig. 2.7, the two 2-fold axes of Fig. 2.4 are replaced with vertical mirror planes. It should be clear that two reflections in mirrors with normals in the same plane and at an angle $\phi$ also produce a rotation $2\phi$. (The point *b*, which was a filled circle in Fig. 2.4, is now an open circle in Fig. 2.7).

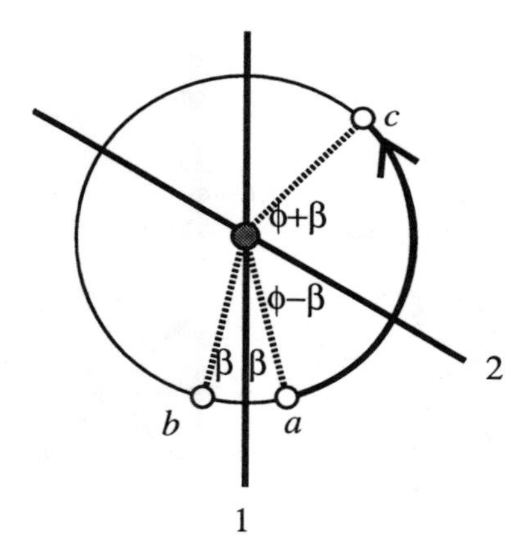


**Fig. 2.7.** Two reflections (in mirror 1 to take *a* to *b*, then in mirror 2 to take *b* to *c*) are equivalent to rotation about an axis along the line of intersection of the mirrors (*a* to *c*). Compare with Fig. 2.4 above.

The result of having two mirrors at 45° is to generate group 4*mm* shown in Fig. 2.8 (which should be compared with Fig 2.5).

Analogously we can generate the groups 2*mm* (usually written *mm*2) from two mirrors at 90°; 3*m* with mirrors at 60° (now just one *set* of mirrors related by 120° rotations); and 6*mm* from mirrors at 30° (giving two sets each of three mirrors analogous to the two sets of 2-fold rotation axes in 622).[1] Fig 2.9 illustrates these groups in a way similar to Fig. 2.6.

Symmetry *mm*2 is commonly encountered; it is for example, the symmetry of the water molecule. If the writing were to be erased, a book would have symmetry *mm*2 also. The (non-planar) $NH_3$ molecule has symmetry 3*m*. More generally *Nm*(*m*) is the symmetry of a pyramid with a regular *N*-gon as a base. The groups *Nm*(*m*) do not contain a center of symmetry (think of the pyramids).

[1]It can be rewarding to experiment with two small rectangular mirrors hinged together at one edge with sticky tape and inclined at different angles to generate these rotations.

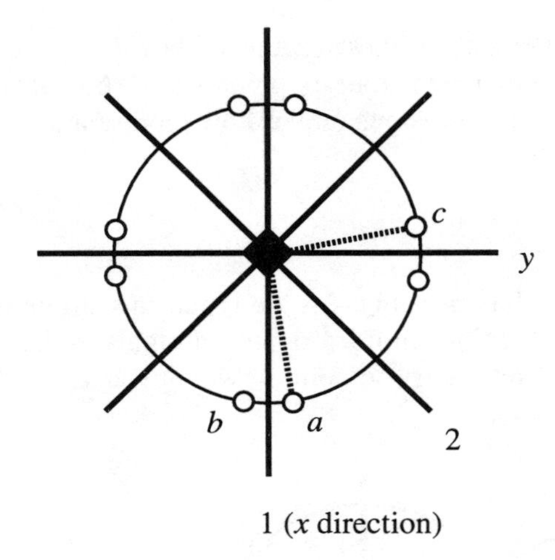


**Fig. 2.8**. Group 4*mm*. *b* is the image of *a* obtained by reflection in line 1 and *c* is the image of *b* obtained by reflection in line 2. $a \to c$ corresponds to rotation by a fourth of a circle about the center point.

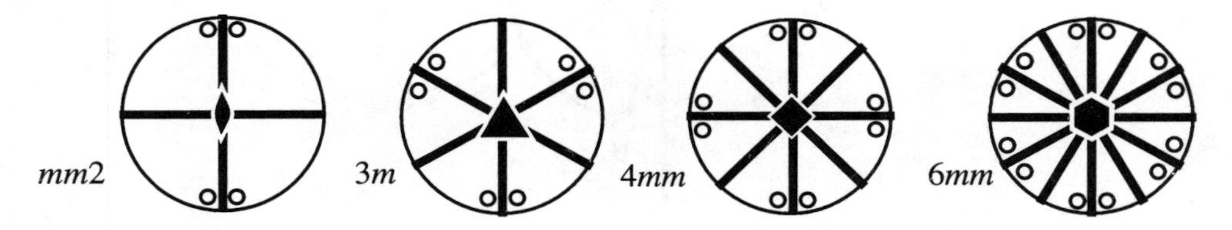


**Fig. 2.9**. Illustrating the symmetry elements of groups *mm*2, 3*m*, 4*mm* and 6*mm*. The heavy lines are the traces of mirror planes normal to the paper. Compare with Fig. 2.6.

### *2.2.3 Groups N/m*

In the next set of point symmetry groups, mirrors are added normal to the rotation axes of groups *N* giving groups *N*/*m*. As we will see later, it is convenient to consider *N*/*m* to be one symbol (because the *N*-fold rotation axis and the normal to the mirror are in the same direction).

1/*m* is the same as *m* so that is not a new group. 3/*m* is the same as $\bar{6}$, so that is not a new group either. Thus the new groups are 2/*m*, 4/*m* and 6/*m*. Note that the combination of an even order rotation axis (which contains a 2 axis) with a mirror plane normal to it generates an inversion center at the point of intersection (refer back to Fig. 2.1) so that 2/*m*, 4/*m* and 6/*m* all contain such a center. In particular 2/*m* is of order 4; the symmetry operations being the mirror reflection, the 2-fold rotation, the inversion and the identity. Fig. 2.10 illustrates these groups in the same way as used in Figs. 2.6 and 2.9. It should be apparent from the figure (compare with Fig. 2.3) that 4/*m* includes a $\bar{4}$ axis, and that 6/*m* includes $\bar{3}$ and $\bar{6}$ axes.

A useful mental exercise is to imagine a two-dimensional object (such as a letter S) with symmetry 2. Now give it some thickness in the third dimension by translation normal to the plane—the three-dimensional symmetry is $2/m$. Repeat for plane objects with symmetry 4 and 6. Although $2/m$ is an unusual symmetry for molecules, it is one of the most common point symmetries of crystals (the crystal class).

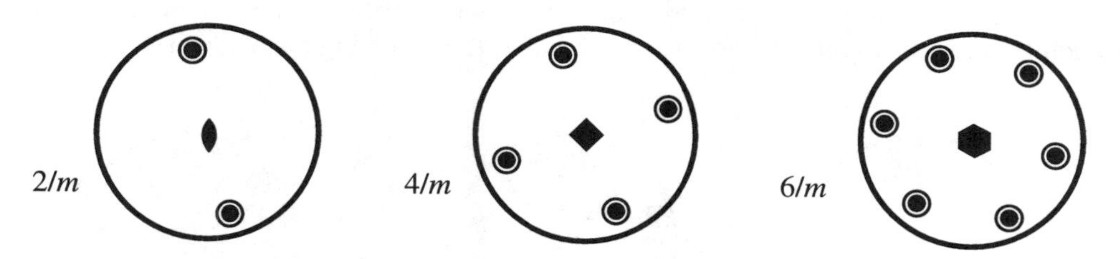


**Fig. 2.10.** Illustrating, from left to right, symmetry groups $2/m$, $4/m$ and $6/m$. Symbols have the same significance as in Figs. 2.6 and 2.9. Note that in each case there is a horizontal mirror plane whose presence is indicated by the heavy outline of the large circle (contrast e.g. Fig. 2.9).

### 2.2.4 Groups $N/m\,2/m\,2/m$

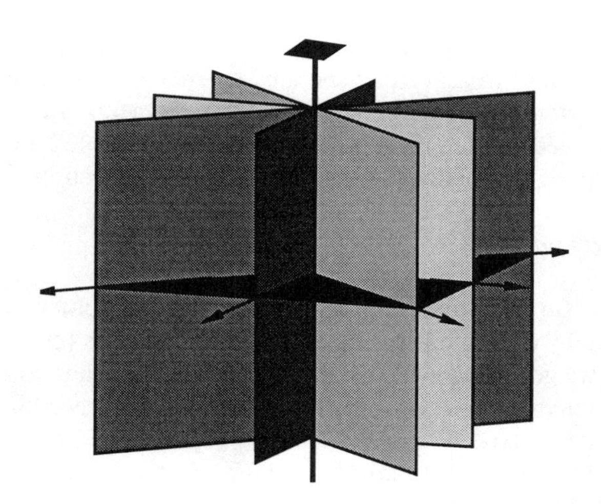

**Fig. 2.11.** Group $4/mmm$. The 4-fold rotation axis is vertical, arrows show the location of 2-fold rotation axes and mirror planes are shaded.

The next set of groups is obtained by adding mirror planes normal to rotation axes in $N22$ giving $N/m\,2/m\,2/m$. Again it is helpful to consider $N/m$ as one symbol corresponding to a single axis. The possibilities here are $2/m\,2/m\,2/m$, $4/m\,2/m\,2/m$ and $6/m\,2/m\,2/m$. The symbols for these groups are often abbreviated $mmm$, $4/mmm$ and $6/mmm$ respectively. $mmm$ is the symmetry of a brick with three different edge lengths. $4/mmm$ is the symmetry of a right square prism (a brick with a square cross section). The arrangement of symmetry elements of this group is illustrated in Fig. 2.11 in which 2-fold axes are shown as arrows and mirrors as planes. *6/mmm* is the symmetry of a right hexagonal prism; a diagram representing its symmetry elements would be similar to Fig. 2.11 but there would be six mirror planes intersecting in the 6-fold axis and six 2-fold rotation axes in those planes and normal to the 6-fold axis.

Fig. 2.12 illustrates the symmetry elements of these groups in a projection down the principle axis in a way that should now be familiar (compare Figs. 2.6, 2.9 and 2.10).

**Fig. 2.12**. Illustrating symmetry groups *mmm*, *4/mmm* and *6/mmm*. Note the presence of a mirror plane in the plane of the paper.

As a 2-fold axis normal to a mirror generates a center of inversion, these groups all include a center of symmetry among their symmetry elements. Indeed the groups may alternatively be generated by adding an inversion center to 222, 422 and 622. If a center is added to 32 the group $\bar{3}m$ is generated as discussed in the next section.

### *2.2.5 Groups* $\bar{3}m$, $\bar{4}2m$ *and* $\bar{6}m2$

There are no new (in the sense of not being already encountered above) symmetry groups to be obtained by adding mirrors normal to inversion axes (do Exercise 11 to verify this statement). But we get new groups by adding mirrors with their normals perpendicular to the $\bar{N}$ axis so that the mirror planes contain the $\bar{N}$ axis. This procedure also generates 2-fold axes normal to the $\bar{N}$ axis.

The case of $N = 2$ corresponds to mirror planes at right angles (recall that $\bar{2} \equiv m$) and generates *mm*2, which is not new.

The other possibilities ($N = 3$, 4 or 6) generate $\bar{3}2/m$ (often abbreviated to $\bar{3}m$), $\bar{4}2m$ and $\bar{6}m2$ which are new.[1] The reader is urged to demonstrate this by starting with a $\bar{N}$ axis ($N = 3$, 4, or 6) and a mirror plane containing this axis and to allow the symmetry operations of these symmetry elements to operate on an arbitrary point (i.e. one not on the mirror plane) and identify the generated symmetry elements. If the roto-inversion axis and the mirror plane are vertical you should generate diagrams like those in Fig. 2.13, in which the

[1]The significance of the order of the symbols in the last two groups is explained below. For the moment note that in $\bar{4}2m$ and $\bar{6}m2$, the 2 and the *m* respectively refer to 2-fold axes and to mirrors with normals not parallel to the 2 axes as illustrated in Fig. 2.13.

symmetry elements of these groups are shown in projection. Note also the symbols used to represent $\bar{3}$, $\bar{4}$ and $\bar{6}$ axes (compare Fig. 2.3).

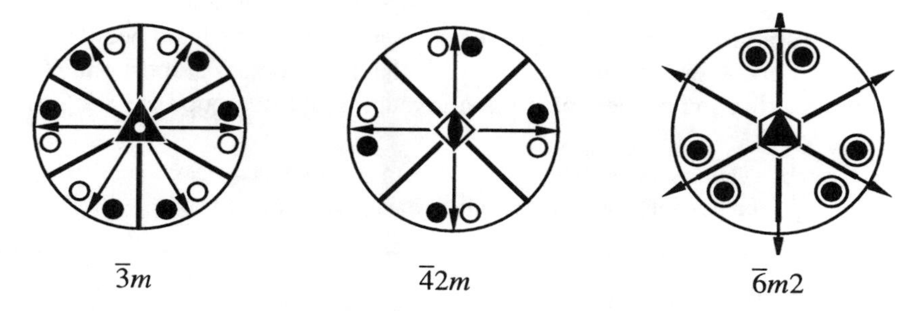


**Fig 2.13**. Illustrating groups $\bar{3}m$, $\bar{4}2m$ and $\bar{6}m2$. Mirror planes are shown as heavy lines and 2-fold axes as lighter lines terminating with arrow heads. See also the legends of Figs. 2.6 and 2.9.

An example of $\bar{3}m$ symmetry is the ethane ($C_2H_6$) molecule in its staggered conformation (Fig. 2.14, right). A right triangular prism and eclipsed ethane (Fig. 2.14, left) have symmetry $\bar{6}m2$. A tetrahedron with only one pair of opposite edges at right angles[1] (see Fig. 2.15) has symmetry $\bar{4}2m$. A baseball or tennis ball (taking into account the seams, but not any other markings that may be on it) also has symmetry $\bar{4}2m$.

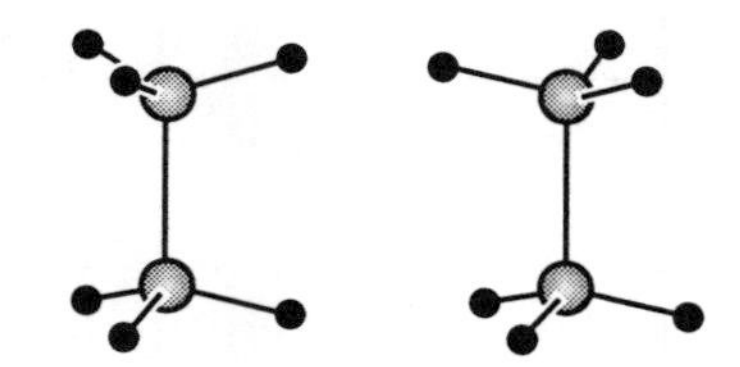

**Fig. 2.14**. Illustrating ethane in its eclipsed (left) and staggered (right) forms.

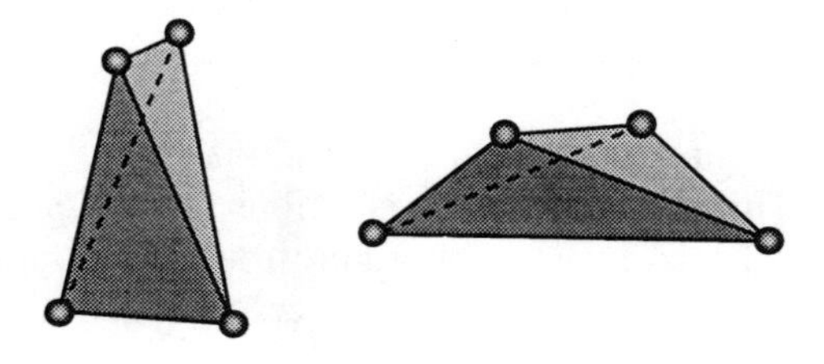

**Fig. 2.15**. Tetrahedra with symmetry $\bar{4}2m$. *The* $\bar{4}$ axis runs up the page.

$\bar{3}$ already contains a center of symmetry, therefore $\bar{3}m$ does also. It should be clear from the pattern of points in Fig. 2.13 as well as Figs. 2.14 and 2.15 that $\bar{4}2m$ and $\bar{6}m2$ do not include a center of symmetry.

[1]This is a necessary, but not a sufficient, condition for a tetrahedron to have $\bar{4}2m$ symmetry.

### *2.2.6 Summary of the non-cubic crystallographic point groups*

Table 2.1 lists the 27 crystallographic point groups we have enumerated so far. The columns in the Table correspond to the order in which the groups have been described in the indicated sections. A feature of these groups is that they contain at most one $N$-fold axis for $N > 2$. For an extension of the table to non-crystallographic groups see § 2.5.6.

**Table 2.1.** The non-cubic crystallographic point groups. Symbols in parentheses are short symbols corresponding to the long symbols immediately above them.

| $N$ § 2.1 | $\bar{N}$ § 2.1 | $N2(2)$ § 2.2.1 | $Nm(m)$ § 2.2.2 | $N/m$ § 2.2.3 | $N/m2/m2/m$ § 2.2.4 | $\bar{N}+m$ § 2.2.5 |
|---|---|---|---|---|---|---|
| 1 | $\bar{1}$ | | | | | |
| 2 | $m$ | 222 | $mm2$ | $2/m$ | $2/m2/m2/m$ $(mmm)$ | |
| 3 | $\bar{3}$ | 32 | $3m$ | | | $\bar{3}2/m$ $(\bar{3}m)$ |
| 4 | $\bar{4}$ | 422 | $4mm$ | $4/m$ | $4/m2/m2/m$ $(4/mmm)$ | $\bar{4}2m$ |
| 6 | $\bar{6}$ | 622 | $6mm$ | $6/m$ | $6/m2/m2/m$ $(6/mmm)$ | $\bar{6}m2$ |

### *2.2.7 Cubic and icosahedral rotation groups*

The dihedral rotation groups result from the solutions of Eq. 2.2 (p. 33) with $p_1, p_2$ and $p_3$ equal to 2, 2 and $N$. Three other possible solutions for $p_1, p_2, p_3$ are 2, 3, 3 ; 2, 3, 4 and 2, 3, 5 and these will lead to three new pure rotation groups. The last possibility containing 5-fold rotations will not give rise to a crystallographic point group, but is of sufficient interest to detain us briefly.

Consider the case $p_1 = 2, p_2 = 3$ and $p_3 = 3$, i.e. $\rho_1 = 180°$, $\rho_2 = \rho_3 = 120°$. Equation 2.1 shows that $\phi_1 = \cos^{-1}(1/3) = 70.53°$ and that $\phi_2 = \phi_3 = \cos^{-1}(1/\sqrt{3}) = 54.74°$. It is remarkable that starting with two of these three rotation axes (at the angles indicated!) we generate a finite group with four 3-fold and three 2-fold axes. Their orientations can be visualized with reference to a cube (Fig. 2.16). The 3-fold axes are parallel to the body diagonals and the 2-fold axes parallel to the cube edges. Of course all the rotation axes have a point in common. This symmetry group, symbolized 23, is in fact one of the five *cubic* point groups.[1]

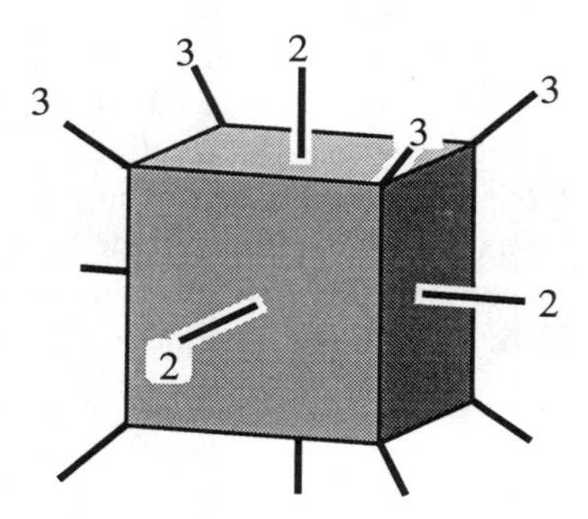


**Fig 2.16**. Group 23. 3-fold rotation axes are symbolized by "3" and 2-fold by "2."

Now consider the case $p_1 = 2$, $p_2 = 3$ and $p_3 = 4$, i.e. $\rho_1 = 180°$, $\rho_2 = 120°$ and $\rho_3 = 90°$. Equation 2.1 shows now that $\phi_1 = \cos^{-1}(1/\sqrt{3}) = 54.74°$, $\phi_2 = 45°$ and $\phi_3 = \cos^{-1}(\sqrt{2}/\sqrt{3}) = 35.26°$. Again we get a finite group, this time containing three 4-fold axes, four 3-fold axes and six 2-fold axes, and again the axes are oriented along principle directions of a cube. The 4-fold axes are parallel to the cube edges (i.e. as the 2-fold axes in group 23, Fig. 2.16), the 3-fold axes are parallel to the body diagonals (as in 23) and the 2-fold axes are parallel to the face diagonals as shown in Fig. 2.17. This second cubic group is symbolized 432. As discussed below, the order of the numbers (such as 23 or 432) in the group symbol indicates the orientation of the rotation axes.

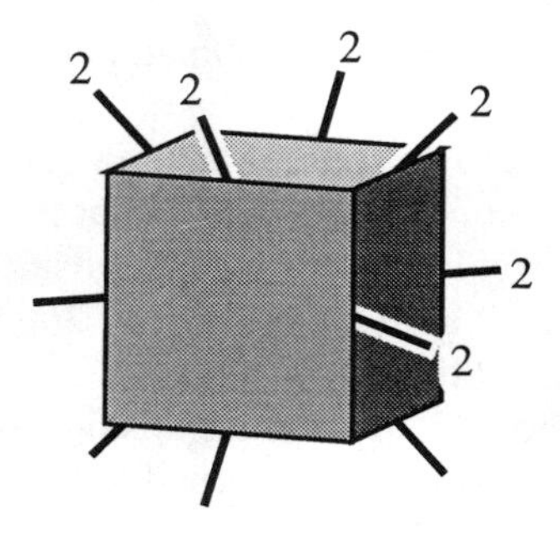


**Fig 2.17.** The 2-fold axes of group 432 (compare Fig. 2.16).

The last pure rotation group has $p_1 = 2$, $p_2 = 3$ and $p_3 = 5$. The generated group has six 5-fold, ten 3-fold and fifteen 2-fold axes. Their orientations can be related to a regular icosahedron (see Fig. 2.18). The 5-fold axes are along the directions joining the six pairs of opposite vertices, the 3-fold axes are along lines joining the centers of the ten pairs of

[1]A point group is *cubic* if it contains exactly four 3-fold axes among the symmetry elements.

opposite faces and the 2-fold axes are along lines joining the midpoints of the fifteen pairs of opposite edges. The smallest angle between 5-fold axes is 63.43°, the smallest angle between 3-fold axes is 41.81°, and the smallest angle between 2-fold axes is 36°. The smallest angle between a 2-fold and a 3-fold axis is 20.90° and between a 2-fold and a 5-fold axis it is 37.38°. The group is the *icosahedral* rotation group, often symbolized *I*; we will also use the symbol 235. Note that although, for simplicity, we use an icosahedron to illustrate the orientation of the axes of 235, the icosahedron has additional symmetry elements (mirror planes and a center). For more on this group (including the angles between axes) see Exercises 15 & 16, and see Appendix 4 for some examples of objects with symmetry 235.

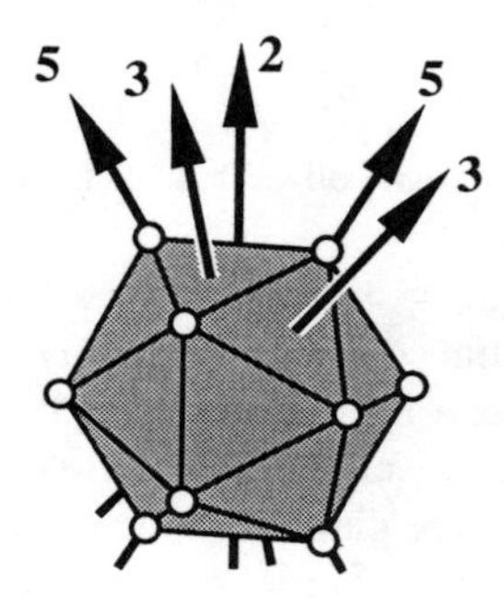


**Fig. 2.18**. The location of some of the symmetry axes of group 235 shown with respect to an icosahedron. For a regular icosahedron the axes marked "2" are $2/m$, those marked "3" are $\bar{3}$, and those marked "5" are $\bar{5}$.

### 2.2.8 *Cubic and icosahedral groups* $m\bar{3}$, $m\bar{3}m$, $\bar{4}3m$ and $m\bar{3}\bar{5}$

The remaining symmetry groups to be considered are obtained by adding mirrors to the icosahedral and the two cubic rotation groups in a way that is suggested in Table 2.2 below in which the last two groups are icosahedral. The results (with short symbols[1] in parentheses) are: $4/m\bar{3}\,2/m$ ($m\bar{3}m$), $2/m\bar{3}$ ($m\bar{3}$) $\bar{4}3m$, and $2/m\bar{3}\bar{5}$ ($m\bar{3}\bar{5}$). Underneath each Hermann-Mauguin symbol is the Schoenflies symbol.

We generate $2/m\bar{3}$ (short symbol $m\bar{3}$) by adding a center to 23. The combination of a center and a 2-fold axis generates mirror planes normal to the 2-fold axes and converts 3 to $\bar{3}$.

Similarly $4/m\bar{3}\,2/m$ (short symbol $m\bar{3}m$) is generated from 432. In this case we generate mirror planes normal to the 4-fold and 2-fold axes of 432 and again convert 3 to $\bar{3}$.

The final cubic group $\bar{4}3m$ is obtained as a subgroup of $m\bar{3}m$ and is not centrosymmetric.

The group $2/m\bar{3}\bar{5}$ is similarly obtained by adding a center to the icosahedral rotation

[1]The short symbols follow the usage in Volume A of the *International Tables* (1983). Previously, the bar was removed over the $\bar{3}$ in $m\bar{3}$ and $m\bar{3}m$ so in the older literature the short symbols were written as $m3$ and $m3m$ respectively.

group 235. The 2 axes become $2/m$ and the 3 and 5 axes become $\bar{3}$ and $\bar{5}$ respectively. This is the group of all the symmetries of a regular icosahedron (Figs. 2.18 and 2.25) and is also symbolized $I_h$.

**Table 2.2.** The cubic and icosahedral point groups.

| rotation group | plus center | other group |
|---|---|---|
| 23<br>$T$ | $2/m\bar{3}$ ($m\bar{3}$)<br>$T_h$ | |
| 432<br>$O$ | $4/m\,\bar{3}\,2/m$ ($m\bar{3}m$)<br>$O_h$ | $\bar{4}3m$<br>$T_d$ |
| 532<br>$I$ | $2/m\bar{3}\bar{5}$ ($m\bar{3}\bar{5}$)<br>$I_h$ | |

We now adduce examples of familiar objects with these cubic symmetries.

$\bar{4}3m$ is the symmetry of a *regular* tetrahedron or of the molecule $CH_4$. The tetrahedron has three $\bar{4}$ axes along the lines joining the centers of opposite edges and four 3 axes along the lines joining the vertices to the centers of opposite faces. There are also six mirror planes, each of which contains an edge and the center of the opposite edge. These symmetry elements should be identifiable in Fig. 2.19 which shows (from left to right) a clinographic projection of a regular tetrahedron, a projection down a $\bar{4}$ axis, a projection along a 3 axis and a projection normal to a mirror plane.[1] Objects with $\bar{4}3m$ symmetry are often said to have *tetrahedral* symmetry. Note the absence of an inversion center and the fact that $\bar{4}$ includes a 2 axis.

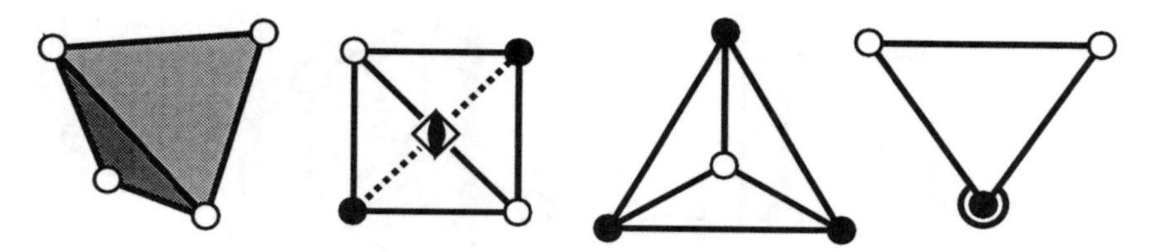

**Fig. 2.19**. Different views of a tetrahedron.

Fig 2.20 shows how the six mirror planes of $\bar{4}3m$ are arranged with respect to the framework of a cube. The $\bar{4}$ axes are parallel to the cube edges and the 3 axes are parallel to

[1]There is really no substitute for holding a model of a polyhedron and identifying its symmetry elements. The reader who finds cubic symmetry difficult is urged to make models of a tetrahedron, an octahedron and a cube by taping or gluing together equilateral triangles or squares of light cardboard.

the body diagonals. The same set of mirror planes also occurs (together with others parallel to the cube faces) in $m\bar{3}m$.

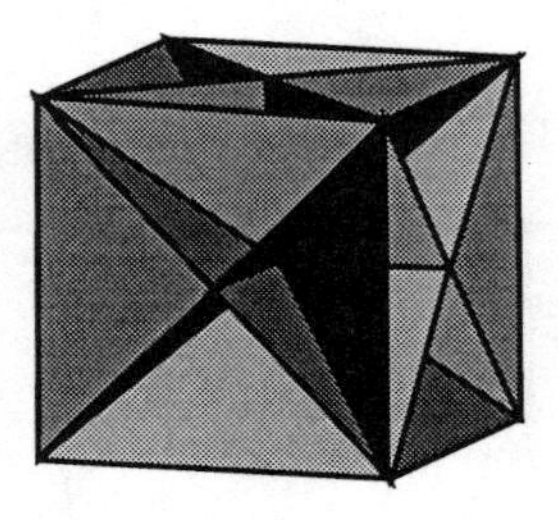

**Fig. 2.20**. The mirror planes of $\bar{4}3m$.

$4/m\,\bar{3}\,2/m$ (abbreviated to $m\bar{3}m$) is the symmetry of a cube itself. A regular octahedron and an octahedral molecule such as $SF_6$ also have symmetry $m\bar{3}m$. Objects with this symmetry are said to have *octahedral* symmetry. Fig 2.21 shows different views of an octahedron similar to those of the tetrahedron in Fig. 2.19. Second from the left is a view down a $4/m$ axis, third from the left is a view down a $\bar{3}$ axis and on the right is a view down a $2/m$ axis.

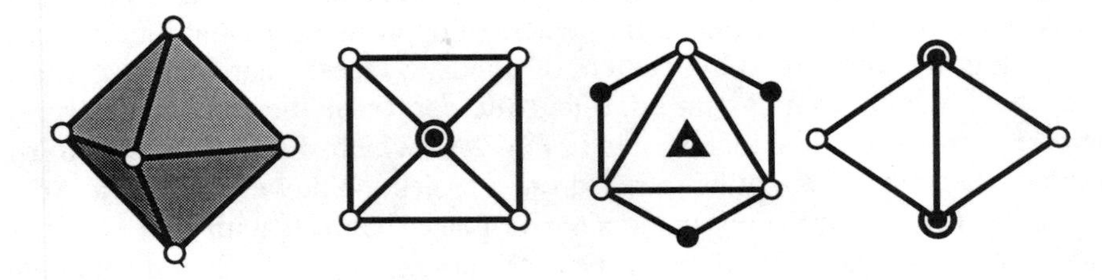

**Fig 2.21**. Different views of an octahedron.

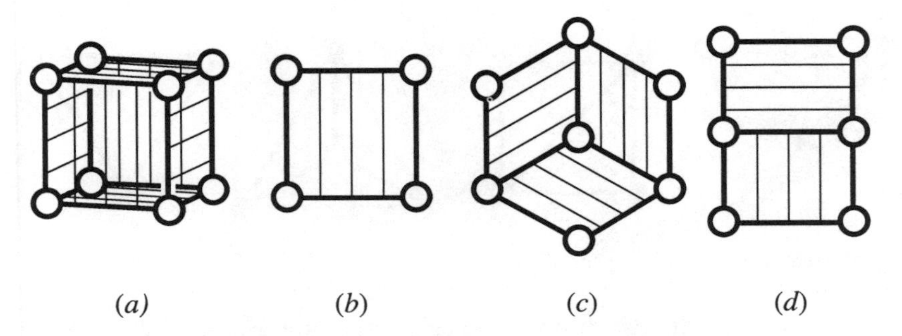


**Fig. 2.22.** (*a*) A cube with parallel markings (light lines) on each face (opposite faces are marked in the same direction) to produce an object with $m\bar{3}$ symmetry. (*b*) A projection on a cube face (down a $2/m$ axis of the marked cube). (*c*) A view down a body diagonal ($\bar{3}$ axis). (*d*) A projection down a face diagonal (note the absence of a 2-fold axis normal to the paper in this projection).

To summarize the symmetry elements of $m\bar{3}m$ (the most complex of the crystallographic point groups—the order is 48): there are three $4/m$ axes (by this is meant a 4 axis with a mirror plane normal to it) parallel to the edges of a reference cube, four $\bar{3}$ axes parallel to the body diagonals of the cube, and six $2/m$ axes parallel to the face diagonals of the cube.

$m\bar{3}$ is quite a common symmetry in crystals but rare for molecules. Crystals of pyrite ("fool's gold" = $FeS_2$) often crystallize as spectacular cubes but if examined closely, striations will be noticed on the faces. Fig 2.22 shows schematically how these markings remove the 4-fold axes of the cube and also eliminate the symmetry elements parallel to the face diagonals. The pyritohedron described on p. 195 (Fig. 5.68) has this symmetry.

## 2.3 Point groups by system

In the next chapter we will discuss three-dimensional lattices and unit cells. We will identify crystal systems just as in two dimensions, and find seven of them (see Chapter 3). For reference the point groups are listed by system in the tables at the end of the book (p. 440). Also given in the list is the Schoenflies symbol for each group.

A crystal symmetry is obtained by combining translations with point symmetries. The point group of the crystal is its *class*. If the crystal point group contains an inversion center, the crystal will be centrosymmetric. The table lists the space group numbers corresponding to each class and also indicates whether that class is centrosymmetric.

## 2.4 Coordinate systems and the order of symbols

The symbols for the point groups assume a reference coordinate system which may differ from one crystal system to another. In a crystal, the axes are always chosen parallel to lattice vectors and this determines the reference coordinate system used. This in turn determines the symbols of the derived space groups, so it is very well worth the little effort it requires to memorize the system. Remember that the orientation of a mirror plane is specified by the direction of its normal.

In the **triclinic** system there is at most an inversion center which is at the origin of coordinates. Triclinic point groups are 1 and $\bar{1}$.

In the **monoclinic** system there is a unique 2-fold axis. The point groups are 2, $m$ and $2/m$. Coordinates are usually chosen so the $y$ axis is parallel to the 2-fold axis (normal to the mirror in $m$). Occasionally other choices are made: then symbols for the symmetry elements parallel to the $x$, $y$, and $z$ axes are used, as illustrated for $2/m$. (1 means no symmetry parallel to that axis and acts as a place marker).

| | |
|---|---|
| $2/m$ parallel to $x$ | $2/m11$ |
| $2/m$ parallel to $y$ | $12/m1$ (or just $2/m$) |
| $2/m$ parallel to $z$ | $112/m$ |

In the **orthorhombic** system there are three mutually perpendicular 2-fold axes. The axes (parallel to the symmetry axes) are also mutually at right angles.The point groups are 222, *mm*2 and *mmm*. The first position in the symbol for the group refers to a symmetry element parallel to *x*, the second parallel to *y* and the third parallel to *z*. It should be clear that 2*mm, m*2*m* and *mm*2 refer to the same point group but with the direction of the 2-fold rotation axis labeled *x*, *y* and *z* respectively.

In the **tetragonal** system there is a unique 4-fold axis and the *z* axis is always chosen to coincide with it. The first position of the point group symbol is the symbol for this axis (4 or $\bar{4}$). The *x* and *y* axes (at right angles to each other and to *z*) are equivalent by symmetry and the second position of the point group symbol is the symbol for symmetry elements (if present) along *x and y*. The third position refers to symmetry elements at 45° to *x* and *y*. Note in particular that $\bar{4}2m$ can also (with a 45° rotation of the coordinate system about *z*) be written $\bar{4}m2$, so there are two *different* symbols for the *same* point group.

Tetragonal point groups are 4, $\bar{4}$, 4/*m*, 422, 4*mm*, $\bar{4}2m$, and 4/*mmm*.

In the **trigonal** and **hexagonal** systems there is a unique 3-fold or 6-fold axis and the *z* axis is always chosen to coincide with it. The first position of the point group symbol is occupied by the symbol (3, $\bar{3}$, 6 or $\bar{6}$ ) for this axis. The *x* and *y* axes are chosen at right angles to *z* and at 120° to each other, so that the *x* and *y* axes are equivalent by symmetry. The second position of the group symbol is then taken by the symbol for symmetry elements parallel to *x* and *y* [and to the third equivalent direction at 120° to both *x* and *y*, i.e. $-(x+y)$]. The third position is reserved for the symbol for symmetry elements at right angles to *x* or *y*. Fig. 2.23 should make clear the directions referred to in the second and third positions.

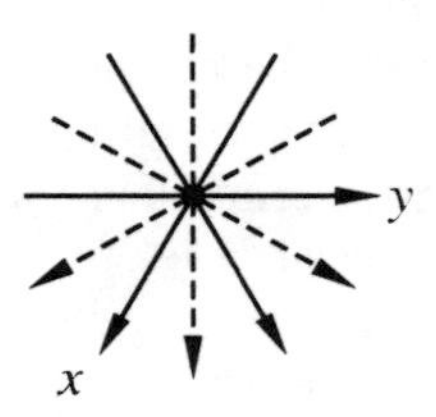


**Fig. 2.23**. The directions corresponding to the second (solid lines) and third (broken lines) positions in the symbols for the trigonal and hexagonal symmetry groups.

Note that 3*m* can also be written as 3*m*1 and (with a 30° rotation of the coordinate system about *z*) as 31*m* (see Fig. 1.6). Likewise 32 can be 321 or 312 and $\bar{6}m2$ can also be $\bar{6}2m$.

Some trigonal crystals can be referred to a **rhombohedral** unit cell with equi-inclined axes. We defer a discussion of that case until later (Chapter 3).

Trigonal point groups are 3, $\bar{3}$, 32, 3*m* and $\bar{3}m$.

Hexagonal point groups are 6, $\bar{6}$, 6/*m*, 622, 6*mm*, $\bar{6}m2$ and 6/*mmm*.

In the **cubic** system we always use axes at right angles to each other. Imagine these axes imbedded in a cube with the origin as the center and $x$, $y$ and $z$ parallel to the cube edges. The first position in the point group symbol refers to symmetry elements parallel to $x$, $y$ and $z$. The second position refers to symmetry elements parallel to the four body diagonals (so the second symbol will always be 3 or $\bar{3}$—this is diagnostic of a cubic group). The third position refers to symmetry elements (either 2, $m$ or $2/m$) parallel to the six face diagonals if they are present.

Cubic point groups are 23, 432, $m\bar{3}$, $\bar{4}3m$ and $m\bar{3}m$.

## 2.5 Notes

### *2.5.1 Rotations*

Eq. 2.1 is derived by Boisen & Gibbs (see Book List). A useful expression in this regard is that for the *Cartesian rotation matrix* which determines how a point $x,y,z$ is transformed by rotation about an axis. Let **i**, **j** and **k** be unit vectors in the $x$, $y$ and $z$ directions respectively. A unit vector from the origin is given by $\mathbf{r} = l\mathbf{i} + m\mathbf{j} + n\mathbf{k}$ where $l^2 + m^2 + n^2 = 1$. ($l$, $m$ and $n$ are the direction cosines of **r**.) Consider a rotation by an angle $\rho$ about this axis; the new coordinates $x'$, $y'$ and $z'$ are given by:

$$\begin{pmatrix} x' \\ y' \\ z' \end{pmatrix} = \begin{pmatrix} ll(1-c)+c & lm(1-c)-ns & ln(1-c)+ms \\ ml(1-c)+ns & mm(1-c)+c & mn(1-c)-ls \\ nl(1-c)-ms & nm(1-c)+ls & nn(1-c)+c \end{pmatrix} \begin{pmatrix} x \\ y \\ z \end{pmatrix} \tag{2.3}$$

Here $c = \cos\rho$ and $s = \sin\rho$ and, as a mnemonic aid, $l^2$ is written as $ll$ etc. For the special case of rotation about the $z$ axis ($l = m = 0$, $n = 1$), the matrix greatly simplifies to $(c\ -s\ \ 0\ /\ s\ \ c\ \ 0\ /\ 0\ \ 0\ \ 1)$. Note that the inverse of a rotation matrix (corresponding to rotation in the opposite sense) is the transpose of the original matrix.

For a roto-inversion, change the sign of all the matrix elements. For reflection ($\bar{2}$), $c = -1$ and $s = 0$ and remember that $l$, $m$, and $n$ are the direction cosines of the normal to the mirror. Thus for reflection in the plane $z = 0$, the matrix is $(1\ 0\ 0\ /\ 0\ 1\ 0\ /\ 0\ 0\ \bar{1})$. [Compare with $(\bar{1}\ 0\ 0\ /\ 0\ \bar{1}\ 0\ /\ 0\ 0\ 1)$ for a 2-fold rotation about $z$.]

If we imagine the three rotation axes ($X_1$, $X_2$ and $X_3$; § 2.2.1) to intersect a unit sphere and the points to be connected by arcs of great circles, then the surface of the sphere is divided into congruent spherical triangles. This is illustrated in Fig. 2.24 which shows two 2-fold rotation axes separated by an angle of $\rho/2$ in a horizontal plane. The generated rotation axis is a $360°/\rho$–fold axis. It should be clear that the angles of the spherical triangle (heavy outline) are $180°/2$ (twice) and $\rho/2$. If $360°/\rho$ is an integer, the sphere can be exactly covered by triangles congruent to the one shown.

More generally, the spherical triangles on a unit sphere corresponding to any rotation group will have sides equal to $\phi_1$, $\phi_2$ and $\phi_3$ (the angles between the axes) and angles

$\rho_1/2$, $\rho_2/2$ and $\rho_3/2$ (using the same symbols as in Eq. 2.1). As $\rho_1/2 = 180°/p_1$ and so on, Eq. 2.2 then follows from the fact that the sum of the angles of a spherical triangle must be greater than 180° (i.e. $\rho_1/2 + \rho_2/2 + \rho_3/2 > 180°$).

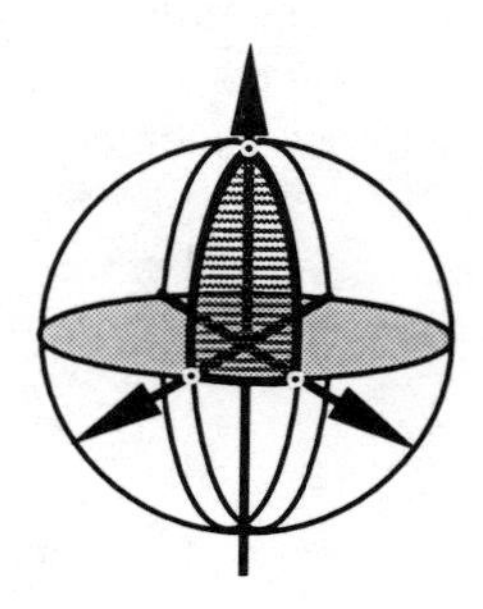

**Fig. 2.24**. See text.

Equation 2.1 refers to rotation axes fixed with respect to the coordinate system. Usually when tilting something, such as a crystal on a goniometer stage, by sequential rotations, the rotation axes move with the crystal. The net rotation is usually described in terms of rotations by Euler angles about such moving axes. A useful text is *Mathematical Methods for Physicists* [3rd ed. Academic Press, New York (1985)] by G. Arfken who warns that "There are almost as many definitions of Euler angles as there are authors."

### *2.5.2 Groups of symmetry operations and their orders*

A group of symmetry operations consists of all the operations associated with the symmetry elements. Thus the group 4 consists of four members: a quarter-turn ($4^1 = 4^+$), two quarter-turns ($4^2 = 2^1$), three quarter-turns ($4^3 = 4^-$) and four quarter-turns ($4^4 = 1$). Any combination of these will produce another. As there are four symmetry operations the order of the group is four.

The order of $\bar{1}$ is two (the two symmetry operations are the identity and the inversion). Reference to Fig. 2.3 (p. 31) should make it apparent that the order of $\bar{3}$ is six, the order of $\bar{4}$ is four and of $\bar{6}$ is six.

The group $2/m$ consists of four elements: a half-turn ($2^1$), two half-turns (1), reflection ($m$) and also an inversion ($\bar{1}$) which is the result of combining the rotation with reflection (and of course a reflection is the result of a half-turn combined with inversion).

The group $4/m$ requires a little more thought. The 4-fold axis along $z$ will generate four points with the same value of $z$. Reflection in the plane $z = 0$ will generate four more for a total of eight so the order of the group is eight. The symmetry operations are the four rotations $4^1$, $4^2$ (= $2^1$), $4^3$, $4^4$ (= 1) and the result of combining these with the mirror reflection which are: ($4^1$ then $m$) = $\bar{4}^3$, ($4^2$ then $m$) = $\bar{1}$, ($4^3$ then $m$) = $\bar{4}^1$, ($4^4$ then $m$) = $m$.

Now consider the group 23. There are three 2-fold axes and four 3-fold axes. In enumerating the different symmetry operations, we agree not to count the identity for a moment. The symmetry operations are $2^1$ in three different directions and $3^1$ and $3^2$ in four directions for a total of $3 + 4 \times 2 = 11$ different rotations. Counting also the identity we see that the order of the group is 12.

The largest crystallographic point group is $m\bar{3}m$ for which the order is 48. The full symbol is $4/m\bar{3}2/m$. There are three $4/m$ axes, four $\bar{3}$ axes and six $2/m$ axes. We agree now not to count the identity and the point of inversion until we have counted the other operations. Besides the identity and inversion each $4/m$ contains six operations (see above) so we get $3 \times 6 = 18$ distinct operations from the three of them. A $\bar{3}$ axis contains six operations which again include the identity and inversion ($\bar{3}^4 = \bar{1}$) which we agreed not to count for the moment, so from the four $\bar{3}$ axes we get $4 \times 4 = 16$ new operations. From the six $2/m$ we get (counting only the $m$ and $2^1$) $6 \times 2 = 12$ new operations. Adding in the identity and inversion we find $18 + 16 + 12 + 1 + 1 = 48$ for the order of the group.

The order of 432, $6/mmm$, $\bar{4}3m$ and $m\bar{3}$ is 24 in each case; all other crystallographic groups are smaller (their order is a divisor of 48). The order of 235 ($I$) is 60 and the order of $m\bar{3}\bar{5}$ ($I_h$) is 120.

It is worth noting that 432 and $\bar{4}3m$ are isomorphic to each other (as are several other sets of groups) so they do not represent different abstract groups.

### *2.5.3 Derivation of the point groups*

The enumeration of the crystallographic point groups can be done starting from the eleven pure rotation groups. The eleven centrosymmetric groups are then obtained by adding an inversion center (in mathematical terms this corresponds to group multiplication of the rotation groups by the group $\bar{1}$). Ten subgroups of the centrosymmetric groups that do not contain a center, but that do contain elements other than pure rotations, can then be found. This scheme is outlined in Table 2.3.

**Table 2.3.** The crystallographic point groups as pure rotation groups, centrosymmetric groups and other groups.

| rotation group | centrosymmetric group | other groups |
|---|---|---|
| 1 | $\bar{1}$ | |
| 2 | $2/m$ | $m$ |
| 3 | $\bar{3}$ | |
| 4 | $4/m$ | $\bar{4}$ |
| 6 | $6/m$ | $\bar{6}$ |
| 222 | $mmm$ | $mm2$ |
| 32 | $\bar{3}m$ | $3m$ |
| 422 | $4/mmm$ | $4mm$ $\bar{4}2m$ |
| 622 | $6/mmm$ | $6mm$ $\bar{6}m2$ |
| 23 | $m\bar{3}$ | |
| 432 | $m\bar{3}m$ | $\bar{4}3m$ |

A good account of the derivation of the groups in this way is given by M. B. Boisen & G. V. Gibbs, *Amer. Mineral.* **61**, 145-165 (1976). For the more mathematically inclined *Geometry and Symmetry* by P. B. Yale [Dover, New York 1988] is recommended.

### *2.5.4 Curie's law, Friedel's law, Laue classes, optical activity and polarity*

*Curie's law* states that an effect cannot have lower symmetry than its cause so that any asymmetry of an effect must be found in its cause. Thus the result of an experiment can give information about symmetry, but symmetry arguments should not be used to predict *a priori* the result of an experiment.[1] In X-ray diffraction it is often found that the three-dimensional diffraction pattern is of higher symmetry than that of the crystal (but never of lower symmetry). In the absence of anomalous dispersion the diffraction pattern is in fact always centro-symmetric (*Friedel's law*). The point group of the diffraction pattern is therefore that obtained by adding a center of symmetry to the point group of the crystal and the apparent crystal class is that of one of the centrosymmetric groups. The *Laue Classes* are comprised of those groups that result in the same centrosymmetric group when a center is added. In Table 2.3 (p. 49), each Laue class consists of a centrosymmetric group and the non-centrosymmetric (*acentric*) groups on the same line (thus one of the eleven Laue classes consists of groups 422, 4*mm*, $\bar{4}2m$ and 4/*mmm*).

The *enantiomorphous* groups consist of those in the first column of Table 2.3. Crystals belonging to these classes will have left- and right-handed forms (that cannot be superimposed on their mirror images). They will also be optically active (rotate the plane of polarized light). Contrary to a belief popular among chemists, enantiomorphism is not a necessary condition for optical activity, which may also be found in crystals of classes *m*, *mm*2, $\bar{4}$, and $\bar{4}2m$. In these latter cases, both left- and right-handed rotations will occur.

The acentric crystal classes are often referred to as *polar* by crystallographers, but this term is correctly given a more restricted meaning: those classes in which a spontaneous electric polarization is possible.[2] In this more restricted sense (which we use subsequently) the *polar* classes are 1, 2, *m*, *mm*2, 4, 4*mm*, 3, 3*m*, 6 and 6*mm*. In all but 1 and *m*, there is a definite polar axis: **b** in class 2, **c** in the rest.

In piezoelectric crystals (quartz is a notable example) a polarization can be induced by stress; piezoelectricity is possible in all acentric classes except 432.

Good references to symmetry constraints on crystal properties are *Physical Properties of Crystals* by J. F. Nye (Oxford, 1955) and *Tensors and Group Theory for the Physical Properties of Crystals* by W. A. Wooster (Oxford, 1973).

[1]Crystallographers, who are otherwise admirable people, sometimes put the cart before the horse, and say that a certain structural feature (such as an 180° bond angle) is *required* by symmetry. The structure, and its symmetry, is determined by the often inscrutable interplay of interatomic forces, and if these dictate a certain symmetry, so be it.

[2]See the discussion in the *International Tables A*, p. 782. The polarization is defined as dipole moment per unit volume.

### *2.5.5 Cubic and icosahedral groups; generators*

Objects with symmetry $\bar{4}3m$ and $m\bar{3}m$ are commonly found. Four points at the vertices of a regular tetrahedron have symmetry $\bar{4}3m$ and six points at the vertices of an octahedron have symmetry $m\bar{3}m$. By contrast a minimum of twelve points is required make an arrangement with symmetry 23 or $m\bar{3}$ so it is not surprising that this group is not often encountered in molecular chemistry; as a minimum a molecule of the form $A_{12}$ or $AB_{12}$ is needed—a possible candidate is neopentane, $C(CH_3)_4$. A minimum of 24 points is needed to make an arrangement with symmetry 432—the vertices of a *snub cube* ($3^4.4$) provide the simplest example (see Fig. 2.26). Very few examples of crystals in this class are known ($\beta$-Mn is one). Twelve points at the vertices of a regular icosahedron have symmetry $m\bar{3}\bar{5}$ ($I_h$) but a minimum of 60 points is required to generate a pattern with symmetry 235 ($I$) so examples of molecules with this symmetry are also hard to find.[1] The snub dodecahedron, $3^4.5$ (§ 5.1.3, p. 136) with 60 vertices is the simplest object with this symmetry. Further information (and other useful information about subgroup relations) is to be found in a classic paper by H. A. Jahn & E. Teller, *Proc. Roy. Soc. (London)*, A**261**, 220 (1937).

The symmetry elements of 432, $\bar{4}3m$ and $m\bar{3}$ are all to be found in $m\bar{3}m$ so they are all *subgroups* of $m\bar{3}m$. The symmetry elements of 23 are contained in all the other cubic groups so they are all *supergroups* of 23. The cubic subgroups of $m\bar{3}\bar{5}$ are $m\bar{3}$ and 23 and 23 is also a subgroup of 235. The group hierarchy is therefore:

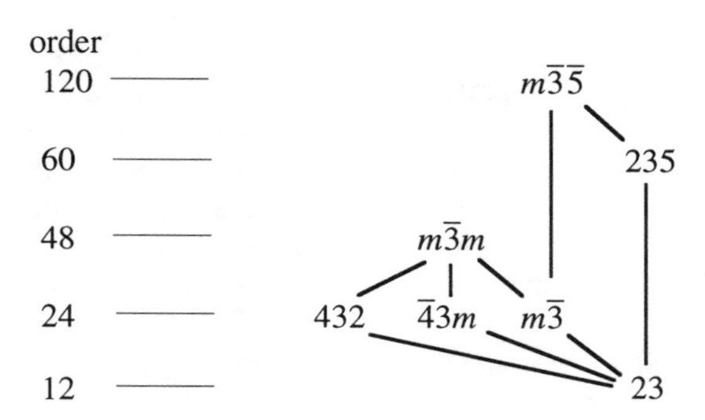


We mentioned in § 2.2.1 that rotations about two axes at an angle and through a common point would generate all the pure rotation groups. The two rotations are *generators* of the group. We can specify the orientation of an axis through the origin 0,0,0 of a Cartesian coordinate system by giving the coordinates of another point. Thus in group 23 we can take as generators a 2-fold rotation about an axis passing through 1,0,0 and a 3-fold rotation about an axis passing through 1,1,1. We can label positive rotations about the these axes as $2^+(100)$ and $3^+(111)$ respectively. The ten other operations of the group are

[1]In Appendix 4 we mention a possible "fullerene" molecule $C_{140}$ with symmetry $I$. However, theoretical studies indicate that this molecule will undergo a Jahn-Teller distortion to lower symmetry.

then generated as combinations of these two. Thus a rotation first about the 3-fold axis and then about the 2-fold axis is equivalent to a positive rotation about a 3-fold axis passing through –1,1,–1. We can symbolize this as $2^+(100)*3^+(111) = 3^+(\bar{1}1\bar{1})$. Other examples are:

$$2^+(111)*2^+(111) = 1$$
$$3^+(111)*3^+(111) = 3^-(111)$$
$$3^-(111)*3^+(\bar{1}1\bar{1}) = 3^+(111)*3^+(111)*2^+(100)*3^+(111) = 2^+(001)$$

The other cubic and icosahedral groups are similarly generated starting from two generators. Particularly convenient sets involving a 3-fold and a 2-fold axis are given below. Note that the orientation of mirror planes through the origin are specified by a point on the normal to the plane from the origin and that $q = (3-\sqrt{5})/2$.

| | | |
|---|---|---|
| $m\bar{3}\bar{5}$ | $\bar{3}^+(1q0)$ | $m(100)$ |
| 235 | $3^+(1q0)$ | 2(100) |
| $m\bar{3}m$ | $\bar{3}^+(111)$ | $m(100)$ |
| 432 | $3^+(111)$ | 2(110) |
| $\bar{4}3m$ | $3^+(111)$ | $m(110)$ |
| 23 | $3^+(111)$ | 2(100) |

*2.5.6 Non-crystallographic point groups*

Molecules with a non-crystallographic symmetry are common and their symmetries are almost invariably described by the Schoenflies symbol. Right prisms with a regular $N$-gonal base have symmetries $D_{Nh}$ in the Schoenflies notation. $C_{70}$ (see Appendix 4) is an example of a molecule with $D_{5h}$ symmetry. A special case of interest is a cylinder for which $N = \infty$ and for which the symmetry is $D_{\infty h}$. Linear molecules with a center of symmetry such as $O_2$ or $CO_2$ have this symmetry. So does a cricket ball (a ball with an equatorial seam).

Pyramids with a regular $N$-gon base have symmetry $C_{Nv}$ in the Schoenflies notation. A cone is the special case with $N = \infty$ and has symmetry $C_{\infty v}$. A linear molecule without a center such as CO also has this symmetry.

Other group symbols worth knowing about include $K_h$ for the symmetry of a sphere. $C_{3i}$ ($\bar{3}$) is sometimes labeled $S_6$ and $D_2$ is sometimes labeled $V$.

The symmetry of an antiprism with a regular $N$-gon base is $D_{Nd}$ in the Schoenflies notation. The symmetry of a regular square antiprism (§ 5.1.4, p. 139) contains a $\bar{8}$ axis and may be written $\bar{8}2m$ ($D_{4d}$ in Schoenflies notation). Thus the only *regular* antiprism that can occur in crystal structures is the triangular antiprism (symmetry $\bar{3}2/m = D_{3d}$) although figures approximating square antiprisms are rather common. Ferrocene, $Fe(C_5H_5)_2$, should be familiar (to chemists at least) as an example of a molecule with the symmetry of a pentagonal antiprism ($D_{5d}$).

To generalize Table 2.1 to axes of arbitrary order $N$ and to show the correspondence to the Schoenflies notation we have to consider three cases (here $n$ is an integer): (a) The order of the axis is $4n$. (b) The order of the axis is $4n+2$. (c) The order of the axis is $2n+1$. The first two cases ($N$ even) result in identical Hermann-Maugin symbols, but require different Schoenflies symbols when there is a $\bar{N}$ axis (see exercise 17). Table 2.4 below gives the Hermann-Maugin symbol with the corresponding Schoenflies symbol directly underneath under the same headings as in Table 2.1. Table 2.4 combined with Table 2.2 (p. 43) gives a complete listing of all the finite point symmetry groups in three dimensions.

What happens as $N$ goes to infinity? See Appendix A.1 (§ A1.5) for the surprising answer.

**Table 2.4.** Point symmetry groups other than cubic or icosahedral.

| $N =$ | $N$ | $\bar{N}$ | $N2(2)$ | $Nm(m)$ | $N/m$ | $N/m2/m2/m$ | $\bar{N}+m$ |
|---|---|---|---|---|---|---|---|
| $4n$ | $N$<br>$C_n$ | $\bar{N}$<br>$S_N$ | $N22$<br>$D_N$ | $Nmm$<br>$C_{Nv}$ | $N/m$<br>$C_{Nh}$ | $N/m2/m2/m$<br>$D_{Nh}$ | $\bar{N}2m$<br>$D_{N/2d}$ |
| $4n+2$ | $N$<br>$C_n$ | $\bar{N}$<br>$C_{N/2h}$ | $N22$<br>$D_N$ | $Nmm$<br>$C_{Nv}$ | $N/m$<br>$C_{Nh}$ | $N/m2/m2/m$<br>$D_{Nh}$ | $\bar{N}2m$<br>$D_{N/2h}$ |
| $2n+1$ | $N$<br>$C_n$ | $\bar{N}$<br>$C_{Ni}$ | $N2$<br>$D_N$ | $Nm$<br>$C_{Nv}$ | | | $\bar{N}2m$<br>$D_{Nd}$ |

### *2.5.7 Symmetry, and relations between polyhedra*

We take axes oriented as described in § 2.4 for cubic symmetry. The operations of $m\bar{3}$ on an arbitrary point $x,y,z$ will produce a pattern of 24 points with symmetry $m\bar{3}$. If the point is on a mirror plane (e.g. $0,y,z$) only 12 points are produced. For special values of $y$ and $z$ the symmetry may be higher. Thus if the point is 0,1,1 the vertices of a *cuboctahedron* (symbol 3.4.3.4) are produced with symmetry $m\bar{3}m$. If the point is $0,1/\tau,1$ [$\tau$ is the golden ratio $(1 + \sqrt{5})/2 = 1.6180$] the vertices of a regular icosahedron ($3^5$) are produced with symmetry $m\bar{3}\bar{5}$. This illustrates that $m\bar{3}$ is a subgroup of both $m\bar{3}m$ (the symmetry of the cuboctahedron) and $m\bar{3}\bar{5}$ (the symmetry of the icosahedron).

Fig. 2.25 shows on the left a cuboctahedron and in the center a regular icosahedron. The $\{Si\}Cr_{12}$ icosahedron in the $Cr_3Si$ structure (§ 6.6.4) is obtained from a point at 0,1/2,1 and is illustrated on the right in the figure. The darker-shaded triangles are normal to 3-fold axes in each case.

The operations of 432 applied to a point $x,y,1$ will produce a snub cube ($3^4.4$, see Fig. 2.26) if $x$ is the solution of $x^3 + x^2 + 3x = 1$ and $y = (1 - x)/(1 + x)$. [The solution of the first equation is $x = q - 8/9q - 1/3$, where $q = (26/27 + \sqrt{44}/\sqrt{27})^{1/3}$ giving $x = 0.2956$, $y$

= 0.5437.] Interchanging $x$ and $y$ produces the mirror image polyhedron.

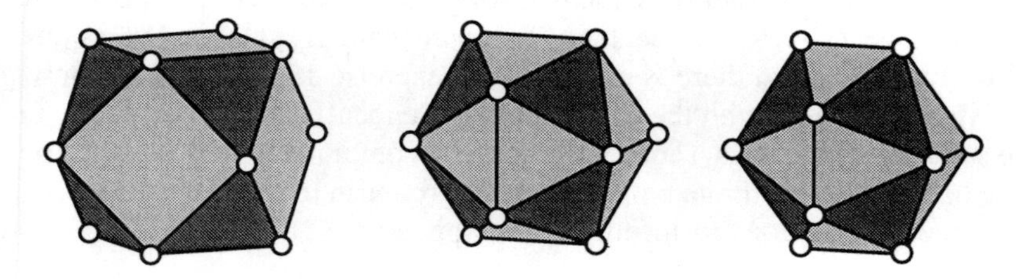

**Fig. 2.25**. Relationship between a cuboctahedron (left) and an icosahedron (see text).

Fig 2.26 shows on the left a snub cube and on the right its enantiomorph. In the center is an intermediate case with $x = y = \sqrt{2} - 1 = 0.414$. This polyhedron is a *rhombicuboctahedron* (symbol $3.4^3$) and it is centrosymmetric (symmetry $m\bar{3}m$). This illustrates that 432 is a subgroup of $m\bar{3}m$. In the diagram the triangular faces normal to 3-fold axes are darker shaded.

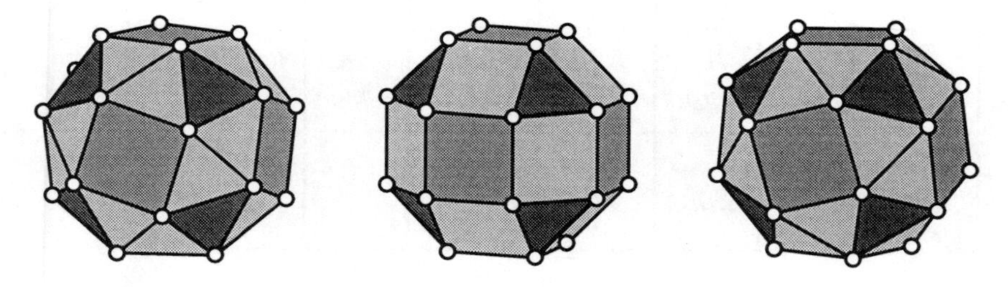

**Fig. 2.26**. Relationship between a rhombicuboctahedron (center) and a snub cube (see text).

### *2.5.8 Antisymmetry: magnetic or black-and-white groups*

We have been discussing transformations of a point whose position is described by three coordinates ($x$, $y$ and $z$). Students of quantum mechanics will know that in addition to positional coordinates an electron has a fourth (spin) coordinate that can take one of two values (commonly signified $\alpha$ and $\beta$ or $\downarrow$ and $\uparrow$). We could consider the set of symmetry operations that change not only coordinates, but which also change $\alpha$ to $\beta$ and *vice versa*. Such an operator is called an antisymmetry operator. The discussion is often in terms of *black-and-white* symmetry groups in which the antisymmetry operation changes black to white or *magnetic* symmetry groups in which the antisymmetry operation reverses the direction of magnetization.

Let us signify an antisymmetry operation by underlining; so that for example an antimirror is $\underline{m}$ and an anti-2-fold rotation is $\underline{2}$. An antimirror reflects in a plane and changes black to white and *vice versa*. The symmetry operation $\underline{3}$ cannot occur as it will repeatedly change a black point at a given place to a white point at the same place and then back to black and a given point must be either black or white (but not black *and* white). In fact there is an antisymmetry operation corresponding to all the crystallographic point operations except 1 and 3 (the only ones of odd period).

Some simple binary crystal structures $AB$ have antisymmetry in the sense that interchanging $A$ and $B$ produces the same structure. Examples are the structures of CsCl, NaCl and the polytypes of SiC.

As well as the classical group $2/m$ there are the black and white groups $\underline{2}/m$, $2/\underline{m}$ and $\underline{2}/\underline{m}$. In all there are 58 crystallographic antisymmetry groups for a total of 90 (= 58 + 32) crystallographic black-and-white (or magnetic) point groups.

If the fourth coordinate can have a finite number (>2) of values the *polychromatic* symmetry groups are obtained. If the fourth coordinate can have *any* value we have of course reached four dimensions.

A good place to start reading about such groups is Shubnikov & Kopstik (Book List). Magnetic space groups are obviously of interest in the description of ordered magnetic structures in solids.

## 2.6 Exercises

1. A right triangular prism with equilateral faces has symmetry $\bar{6}m2$. Locate the symmetry elements.

2. The square antiprism (see § 5.1.4, p.139) has symmetry $D_{4d}$ . It has a $\bar{8}$ axis. What are its other symmetry elements?

3. Another common 8-coordination figure is one with atoms at the vertices of a bis-disphenoid (see § 5.1.6, p. 141). The symmetry of this figure is $\bar{4}2m$ and there are two sets of bonds with bonds of one set unrelated by symmetry to those of the other set. Locate the 2-fold axes in this polyhedron.

4. Show that adding an inversion center to 622 will produce $6/mmm$.

5. For 2-fold rotations about the $x$, $y$ and $z$ directions respectively the matrix in Eq. 2.3 becomes:

$$(1\,0\,0\,/\,0\,\bar{1}\,0\,/\,0\,0\,\bar{1}),\quad (\bar{1}\,0\,0\,/\,0\,1\,0\,/\,0\,0\,\bar{1}),\quad (\bar{1}\,0\,0\,/\,0\,\bar{1}\,0\,/\,0\,0\,1)$$

Thus the transformed coordinates for 222 are $x,y,z$ ; $x,\bar{y},\bar{z}$ ; $\bar{x},y,\bar{z}$ ; $\bar{x},\bar{y},z$

6. For a rotation by one-third of a circle about a body diagonal in the $+x$, $+y$, $+z$ direction the matrix in Eq. 2.3 reduces to (0 0 1 / 1 0 0 / 0 1 0). For a two-thirds rotation the matrix becomes (0 1 0 / 0 0 1 / 1 0 0). Operation of these matrices corresponds to cyclic permutation of the coordinates.

7. The transformed coordinates for the operations of group 23 are now easily derived from the matrices (given above) for the identity and rotations about the three 2-fold axes followed by rotations about a 3-fold axis. They are:

$$x,y,z\ ;\ z,x,y\ ;\ y,z,x\ ; \qquad x,\bar{y},\bar{z}\ ;\ \bar{z},x,\bar{y}\ ;\ \bar{y},\bar{z},x$$
$$\bar{x},y,\bar{z}\ ;\ \bar{z},\bar{x},y\ ;\ y,\bar{z},\bar{x}\ ; \qquad \bar{x},\bar{y},z\ ;\ z,\bar{x},\bar{y}\ ;\ \bar{y},z,\bar{x}$$

8. For each of the symbols given in Exercise 7 identify the symmetry operation that generates it from $x,y,z$. Incidentally, we have confirmed that the order of 23 is twelve.

9. Multiplying the matrices corresponding to the symbols in Exercise 7 by the identity: ( 1 0 0 / 0 1 0 / 0 0 1) and by the inversion: ($\bar{1}$ 0 0 / 0 $\bar{1}$ 0 / 0 0 $\bar{1}$) will now produce the symbols for $m\bar{3}$. They are simply the twelve given above plus the twelve obtained by changing the signs of all coordinates.

10. Still with Cartesian coordinates, a mirror in the plane $x = y$ will interchange $x$ and $y$ coordinates of a point and leave $z$ unchanged (cf. Fig. 1.4) so the matrix representation of this symmetry operation is (0 1 0 / 1 0 0 / 0 0 1). Multiplying the coordinates of Exercise 9 by this matrix (interchanging the first two coordinates in each triplet) and by the identity will now produce the transformed coordinates for $m\bar{3}m$. They are all the 48 permutations of $\pm x,\pm y,\pm z$.

11. Verify the assertion that adding mirrors normal to inversion axes will not produce new groups. In particular "$\bar{3}/m$" ≡ $6/m$ and "$\bar{4}/m$" ≡ $4/m$. What is "$\bar{6}/m$"? A simple way to do this is to construct diagrams like those shown in Fig. 2.3 and 2.10. (Hence $4/m$ includes $\bar{4}$, and $6/m$ includes $\bar{3}$.)

12. Show that successive reflections in mirror planes normal to Cartesian $x$, $y$ and $z$ axes is equivalent to inversion through the point in common to the three mirror planes. (Three improper operations combine to produce another improper operation). In optics such a configuration of mirrors is known as a corner cube.

13. We have seen that combinations of rotation by a 1/2 circle ($\rho_1 = 180°$) and by 1/3 circle ($\rho_2 = 120°$) can result in rotation ($\rho_3$) equal to $360°/N$, where $N = 2, 3, 4$ or 5. Find the angles between the three rotation axes in each case. Hint: from Eq. 2.1, $\cos\phi_3 = (2/\sqrt{3})\cos(\rho_3/2)$.

14. If you don't have one, make or borrow a model of a regular icosahedron (which has symmetry $I_h$). Convince yourself that the lines joining opposite vertices are $\bar{5}$ axes, the lines joining the centers of opposite faces are $\bar{3}$ axes and the lines joining the mid-points of opposite edges are $2/m$ axes. We therefore write $I_h$ as $2/m\bar{3}\bar{5}$ (short symbol $m\bar{3}\bar{5}$).

15. The smallest angle between two 5-fold axes in $I = 235$ (or $I_h = m\bar{3}\bar{5}$ is 63.435° =

$2\tan^{-1}(1/\tau)$ where $\tau = (1 + \sqrt{5})/2$. The combination of a fifth turn about each these two axes is a third turn.

16. With three of the 2-fold axes of 235 aligned along Cartesian $x$, $y$ and $z$ axes (as for 23), a 5-fold axis is in the $yz$ plane at $\tan^{-1}(1/\tau) = 31.717°$ from $z$. The Cartesian rotation matrix (Eq. 2.3) for a fifth turn about this axis is $\mathbf{R} = (c_1 {-c_2}\ 1/2\ /\ c_2\ 1/2\ c_1\ /{-1/2}\ c_1\ \ c_2)$ where $c_1 = \cos(\pi/5) = (\tau\text{-}1)/2$ and $c_2 = \cos(\pi/10) = \tau/2$. Transforming the 12 points of 23 (Exercise 7) by multiplying by powers of $\mathbf{R}$ ($\mathbf{R}$, $\mathbf{R}^2$, $\mathbf{R}^3$, $\mathbf{R}^4$, $\mathbf{R}^5 = \mathbf{E}$) will produce the 60 transformed coordinates for 235 for this orientation of Cartesian axes. Adding an inversion (reversing the signs of all coordinates) will result in the 120 symbols for $m\bar{3}\bar{5}$. [Hint: see the drawing of an icosahedron in Fig. 2.18.]

17. Instead of using rotation + inversion axes $\bar{N}$, the point groups can be generated using rotation + reflection axes $S_N$ which involve $N$-fold rotation followed by reflection in a plane normal to the rotation axis (this is the Schoenflies system). The correspondence between the $S_N$ symmetry elements and the $\bar{N}$ symmetry elements is:

$$S_1 \leftrightarrow m\ (=\bar{2})\ ;\ S_2 \leftrightarrow \bar{1}\ ;\ S_3 \leftrightarrow \bar{6}\ ;\ S_4 \leftrightarrow \bar{4}\ ;\ S_6 \leftrightarrow \bar{3}$$

Generalize for $S_N$. (Hint: there are three cases to consider, $N = 4n$, $N = 4n+2$ and $N = 2n+1$, where $n$ is an integer).

18. What is the result of combining 1/6 turns about intersecting axes at right angles?

## CHAPTER 3

# THREE-DIMENSIONAL SPACE GROUPS

We now proceed to generate three-dimensional space symmetry groups by a procedure analogous to that followed in the case of two dimensions. Three-dimensional lattices are described and then the symmetry operations (glide and screw) that combine point operations with translation are discussed. Finally it is shown how the three-dimensional space groups arise. No effort is made to be systematic or complete (there are 230 three-dimensional space groups), however the ideas involved should be clear to the those who have read and understood Chapters 1 and 2.

## 3.1 Three-dimensional lattices

A unit cell of a three-dimensional lattice has edges that are three non-coplanar vectors **a**, **b** and **c** with magnitudes $a$, $b$ and $c$. The angle between **a** and **b** is $\gamma$, that between **b** and **c** is $\alpha$ and that between **c** and **a** is $\beta$. The unit cell of the lattice is specified by the parameters $a$, $b$, $c$, $\alpha$, $\beta$ and $\gamma$. There are 14 three-dimensional Bravais lattices each having a different space group symmetry (compare five in two dimensions).

As in the two-dimensional case, centered cells are sometimes chosen for convenience[1] (see Fig. 3.1 below). Symbols are given to the lattices according to the kind of centering (recall the symbols $p$ and $c$ for two-dimensional lattices). These are given in Table 3.1.

**Table 3.1.** Symbols for three-dimensional lattices. $n$ is the number of lattice points per unit cell

| symbol | name | description | $n$ |
|---|---|---|---|
| $P$ | primitive | lattice points at corners only | 1 |
| $R$ | rhombohedral | lattice points at corners only | 1 |
| $A$ | A-centered | lattice points at corners and centers of **b**,**c** faces | 2 |
| $B$ | B-centered | lattice points at corners and centers of **a**,**c** faces | 2 |
| $C$ | C-centered | lattice points at corners and centers of **a**,**b** faces | 2 |
| $I$ | body-centered | lattice points at corners and body center | 2 |
| $F$ | face-centered | lattice points at corners and all face centers | 4 |

The symbol $R$ is reserved for the lattice having a primitive cell of a particular shape ($a = b = c$, $\alpha = \beta = \gamma$) as explained below.

The 14 Bravais lattices are divided into seven crystal systems according to the

[1]We emphasize that a primitive cell can *always* be used for any lattice. The advantage of using centered cells is that it allows the use of orthogonal axes where they would not otherwise be possible.

constraints on the unit cell parameters imposed by symmetry. These are summarized in Table 3.2. The parameters are considered to be able to take any value within the constraints imposed. Note that there are 15 entries in the table. The reason is that the *lattice* for a crystal with trigonal symmetry and $\alpha = \beta = 90^\circ$, $\gamma = 120^\circ$, $a = b$ is the same as that for a hexagonal crystal. The other lattice ($R$) listed as trigonal is often referred to a centered hexagonal unit cell.[1]

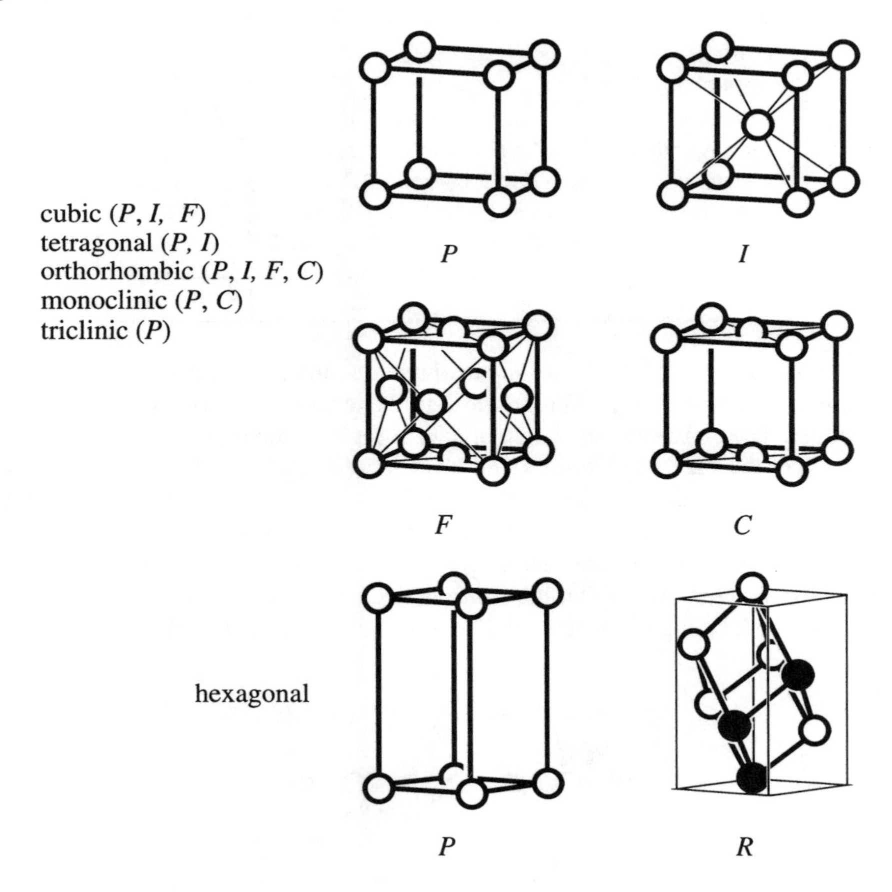


**Fig. 3.1**. Primitive and centered unit cells for lattices (see text).

Conventional unit cells for lattices are shown in Fig. 3.1 in which the *shape* of the unit

[1]Different definitions of crystal system are found. If the classification is by the symmetry of the lattice, then the hexagonal and trigonal symmetries with a primitive hexagonal lattice ($P$) belong to the same system, but there is a separate system for symmetries with a rhombohedral lattice ($R$). In the *International Tables* the classification (adopted here) is by space group symmetry. Trigonal symmetries (including rhombohedral) possess only 3-fold axes (all parallel)—hexagonal symmetries contain 6-fold axes.

cell is not necessarily cubic; for example a cell with just one face centered cannot have cubic symmetry. The data for the space groups in the *International Tables* refer to these cells.

**Table 3.2.** The three-dimensional Bravais lattices.

| system | constraints | lattices |
|---|---|---|
| triclinic | *none* | $P$ |
| monoclinic | $\alpha = \gamma = 90°$ | $P, C$ |
| orthorhombic | $\alpha = \beta = \gamma = 90°$ | $P, C, I, F$ |
| tetragonal | $\alpha = \beta = \gamma = 90°,\ a = b$ | $P, I$ |
| [trigonal | $\alpha = \beta = 90°,\ \gamma = 120°,\ a = b$ | $P$] |
| trigonal | $a = b = c,\ \alpha = \beta = \gamma$ | $R$ |
| hexagonal | $\alpha = \beta = 90°,\ \gamma = 120°,\ a = b$ | $P$ |
| cubic | $\alpha = \beta = \gamma = 90°,\ a = b = c$ | $P, I, F$ |

A monoclinic lattice has 2-fold axes in one direction only and the standard choice of axes for monoclinic cells is with **b** parallel to the 2-fold axes. However other choices are found in the literature. The conventional choice of axes for a centered monoclinic cell is such that the **a**,**b** face (i.e. the face containing **a** and **b**) is centered, so that the lattice symbol is $C$. However interchanging the names of **a** and **c** (and reversing the direction of **b** to maintain a right-handed coordinate system) will result in the **b**,**c** face being centered and the lattice symbol now becomes $A$. Yet another choice of axes will give a body-centered cell as shown in Fig. 3.2. This means that the *same* lattice can have *different* symbols ($A$, $C$, or $I$) according to the labeling of the axes and/or choice of unit cell vectors.

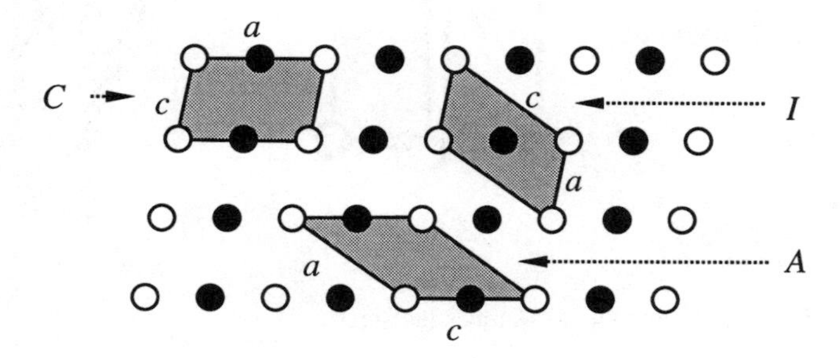


**Fig 3.2**. Three choices of unit cell for a centered monoclinic lattice. **b** is normal to the plane of the paper. Open circles are at $y = 0$ and filled circles are at $y = 1/2$. On the top left is a $C$-centered cell, on the top right is a body-centered ($I$) cell and below is an $A$-centered cell.

The coordinate axes for conventional choices of cell for orthorhombic lattices are mutually perpendicular. The one-face-centered orthorhombic lattice is normally labeled $C$; this means that the **c** direction is normal to the centered face. It should be clear that relabeling the axes can again result in the lattice being *B* or *A*.

Rhombohedral crystals are often described using a centered hexagonal cell (i.e. one with $\alpha = \beta = 90°$, $\gamma = 120°$, $a = b$) with lattice points at 0,0,0 (the unit cell corners) and at 1/3,2/3,2/3 and at 2/3,1/3,1/3. We will discuss this again, but note now that the *R* lattice does not have 6-fold symmetry (the symmetry at the points of the lattice is $\bar{3}m$) even though the cell is referred to as "hexagonal." In the bottom of Fig. 3.1 we show on the left a primitive hexagonal cell and on the right we show (heavy lines) a primitive rhombohedral cell with a centered hexagonal cell lightly outlined (for more detail see Figs. 4.4 and 4.5); the hexagonal cell contains three lattice points (filled circles).

The cubic *F* lattice can be described using a primitive rhombohedral cell with $\alpha = 60°$, and the cubic *I* lattice by a primitive rhombohedral cell with $\alpha = \cos^{-1}(-1/3) = 109.47°$.

**Table 3.3.** Names and extended symbols for Bravais lattices.

| | symbol | point symmetry | name |
|---|---|---|---|
| 1. | *aP* | $\bar{1}$ | primitive triclinic (anorthic) |
| 2. | *mP* | $2/m$ | primitive monoclinic |
| 3. | *mC* | $2/m$ | one-face-centered monoclinic |
| 4. | *oP* | $mmm$ | primitive orthorhombic |
| 5. | *oC* | $mmm$ | one-face-centered orthorhombic |
| 6. | *oI* | $mmm$ | body-centered orthorhombic |
| 7. | *oF* | $mmm$ | (all) face-centered orthorhombic |
| 8. | *tP* | $4/mmm$ | primitive tetragonal |
| 9. | *tI* | $4/mmm$ | body-centered tetragonal |
| 10. | *hP* | $6/mmm$ | primitive hexagonal |
| 11. | *hR* | $\bar{3}m$ | rhombohedral [using a hexagonal cell] |
| 12. | *cP* | $m\bar{3}m$ | primitive cubic |
| 13. | *cI* | $m\bar{3}m$ | body-centered cubic |
| 14. | *cF* | $m\bar{3}m$ | (all) face-centered cubic |

Lattices are sometimes given extended symbols that consist of first (in lower case) a letter that indicates the unit cell shape and then a symbol (upper case) that indicates the centering. Using this system[1] the symbols for lattices are given in Table 3.3 which also lists the point symmetry at a lattice point. The space group symmetry of the lattice (see Exercise 3) is simply found by combining the lattice symbol with the point group symbol (so that for example the face-centered cubic lattice has symmetry $Fm\bar{3}m$). Note that every *lattice* is centrosymmetric.

[1]An example of the use of this system is in *Pearson's Handbook of Crystallographic Data for Intermetallic Phases* (see Book List). In this work structures are classified by a *Pearson symbol* which is one of the lattice symbols followed by the number of atoms in the unit cell. Thus the rutile structure (§ 3.4) is found under the heading *tP*6. The *Inorganic Crystal Structure Database* (Book List) can also be searched by Pearson symbol.

Note that in Table 3.3 above, trigonal is subsumed under hexagonal. In practice there is usually no problem with distinguishing trigonal and hexagonal symmetries, but when there is a wish to avoid ambiguity the following are useful:

| | |
|---|---|
| hexagonal (*sensu lato*) | includes trigonal symmetry |
| hexagonal (*sensu stricto*) | only symmetry groups with a 6-fold axis |

## 3.2 Glide and screw axes

Just as in two dimensions, before obtaining space groups by combining translational symmetry with point symmetry operations, we have to consider the possibility of compound symmetry operations that are combinations of point symmetry operations with translation. Two cases are recognized. These are: *glide*, which we met in two dimensions, and which is a combination of *reflection* and *translation*; and a new operation, that of *screw*, which is a combination of *rotation* and *translation.*

### *3.2.1 Glide*

Glide combines reflection in a plane (the *glide plane*) with translation. The translation must be parallel to the glide plane and in a direction parallel to a lattice vector. The magnitude of the glide translation must be one-half that of the corresponding lattice vector so that action of the glide operation twice will transform a point to an identical point related to the first by a lattice vector. Figure 3.3 illustrates the glide symmetry operation (compare Fig. 1.11, p. 14 and also Fig. 3.8, p. 67).

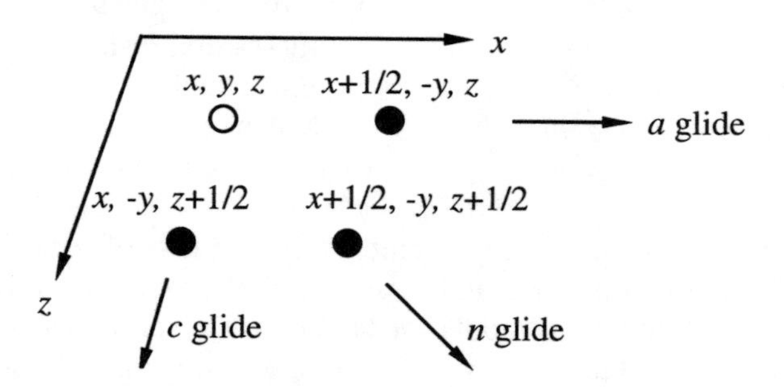


**Fig. 3.3**. Illustrating glide in the monoclinic system (**b** is normal to the page). The glide plane is in the plane of the paper and the coordinates of points produced by one operation of each of $a$, $c$, and $n$ glide are shown (open circle to a filled circle). The origin has been chosen so that the glide plane is at $y = 0$.

If the glide is parallel to **a**, **b**, or **c** the symbol for the glide plane that appears in a space group symbol is $a$, $b$ or $c$ respectively. This case is called *axial* glide. A further possibility is that with (*e.g.*) a glide plane parallel to the **a**,**b** plane, the glide is along the direction $\mathbf{a} \pm \mathbf{b}$ (also a lattice vector!). The glide translation is then $(\mathbf{a} \pm \mathbf{b})/2$ and the symbol is $n$.

Likewise if the glide plane is parallel to the **a**,**c** plane, the glide translation may also be (**a** ± **c**)/2 and if the glide plane is parallel to the **b**,**c** plane, the glide translation may be (**b** ± **c**)/2. The symbol is $n$ in each of these cases also as the direction of the glide translation (now called *diagonal* glide) is clear from the orientation of the glide plane.

In lattices with centered cells there are primitive lattice vectors shorter than those defining the unit cell edges. The glide direction may be along a primitive cell vector such as (**a** ± **b**)/2 etc. (face-centered cells) or (**a** ± **b** ± **c**)/2 (body-centered cells). In these instances the magnitude of the glide vector is one-half that of the primitive-lattice vector and the symbol is $d$. Glide of this sort is called *diamond* glide as it is one of the symmetry elements of the diamond structure (which is face-centered cubic).

### *3.2.2 Screw*

Screw axes are a combination of proper rotations with translation. The translation must be along the rotation axis (why?). Let the axis be **c**; the combined operation is then a counterclockwise rotation about **c** followed by a translation **t** along the +**c** direction.[1] Now consider an $N$-fold screw axis. Repeating the screw operation $N$ times must result in the transformation of a point to an identical point separated from the original one by $n\mathbf{c}$, where $n$ is an integer less than $N$.[2] Thus we have at once that $N\mathbf{t} = n\mathbf{c}$ or $t = (n/N)c$. The symmetry element corresponding to these symmetry operations is called an $N_n$ screw axis. Figure 3.4 illustrates the case of a $3_1$ axis ($N = 3$, $n = 1$).

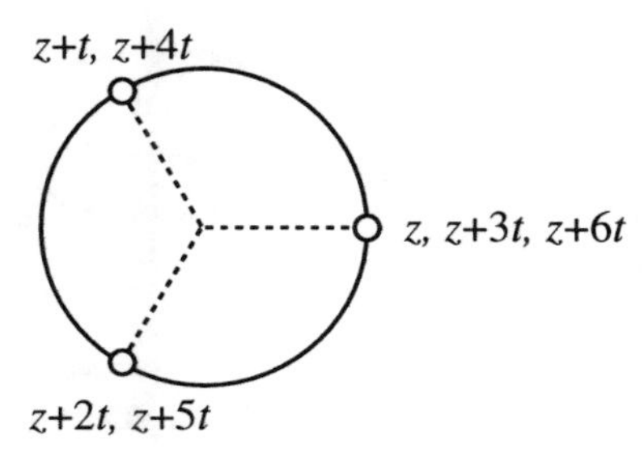


**Fig. 3.4.** Illustrating the effect of six applications of a $3_1$ screw axis with translation $t$ on a point originally with height $z$. The $z$ axis is normal to the page with the $+z$ direction up.

In general then we have screw axes $N_n$ with $1 \le n < N$ and $N$ = 2, 3, 4, 6. The

[1]As in the case of glide, the order of carrying out the components of the combined operation is unimportant.

[2]This should be obvious from the definition of an $N$-fold axis. For a point symmetry element, carrying out the $N$-fold operation $N$ times is equivalent to the identity operation. For the corresponding screw axis, carrying out the operation $N$ times must be equivalent to a lattice translation, as we consider points separated by a lattice translation (which may be a multiple of a primitive translation) also to be *identical*. The case of $n = N$ is the same as a pure rotation followed separately by a lattice translation.

possibilities are:

$$\begin{array}{ll} 2 \rightarrow & 2_1 \\ 3 \rightarrow & 3_1, 3_2 \\ 4 \rightarrow & 4_1, 4_2, 4_3 \\ 6 \rightarrow & 6_1, 6_2, 6_3, 6_4, 6_5 \end{array}$$

Recall that the translation is $n/N$ of the lattice repeat vector along the axis.

Let's examine $3_1$ and $3_2$ in more detail. Referred to axes with $z$ along the screw direction and with $x$ and $y$ at right angles to $z$, and at 120° to each other (compare Fig. 1.5, p. 7), successive applications of $3_1$ along the axis $z = 0$ will send a point at $x,y,z$ to $\bar{y},x-y,z+1/3$ ; $y-x,\bar{x},z+2/3$ ; $x,y,z+1$ (note that, as usual, we are measuring $z$ in units of the repeat distance $c$). The fourth point is identical to the first as it is just a lattice translation **c** away.

$3_2$ acting on $x,y,z$ will give $\bar{y},x-y,z+2/3$ ; $y-x,\bar{x},z+4/3$ ; $x,y,z+2$. But note that we can always add or subtract an integer from the $z$ coordinate so we could equally express the new coordinates as $\bar{y},x-y,z+2/3$ ; $y-x,\bar{x},z+1/3$ ; $x,y,z+1$.

Fig. 3.5 shows schematically a plot of these points. If the **c** axis were to be ascended by using successively higher points of $3_1$ as steps, the path would be counterclockwise and a right-handed screw. Conversely, ascending $3_2$ in the same way would result in clockwise motion along a left-handed screw.

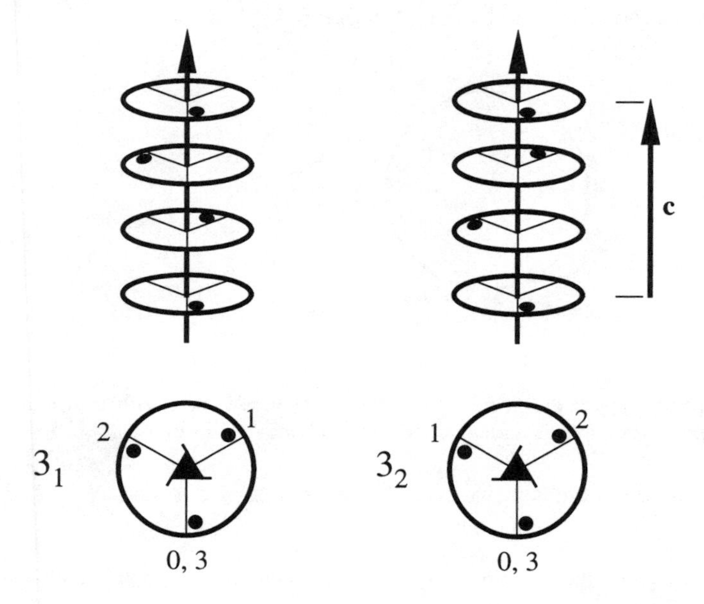


**Fig. 3.5**. Illustrating the operations of $3_1$ (left) and $3_2$ (right). The bottom portion is a projection down **c**. Numbers indicate heights in multiples of $c/3$. Notice the symbols for 3-fold screw axes.

The same discussion would hold for $4_1$ and $4_3$ (Fig. 3.6) except that now we use a quarter turn and it takes four steps up the staircase to go up a height *c*. $4_1$ is right-handed and $4_3$ is left-handed. What about $4_2$? Reference to the figure shows that at each point there are higher steps at the same height on the left and on the right so the ascent could be either clockwise or anticlockwise and $4_2$ does not have a hand.

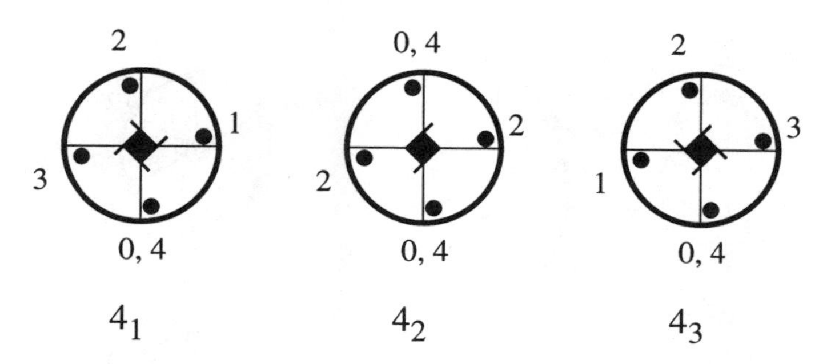


**Fig. 3.6.** Illustrating 4-fold screw axes. The numbers are elevations in multiples of $c/4$. Note that $4_1$ and $4_3$ include a $2_1$ and that $4_2$ includes a 2 axis. Notice also the symbols for 4-fold screw axes.

For 6-fold screws the possibilities are $6_1$, $6_2$, $6_3$, $6_4$, $6_5$ (Fig. 3.7). $6_1$ and $6_5$ give right- and left-handed stairs related as mirror images (analogous to the pairs $3_1$, $3_2$ and $4_1$, $4_3$) but now there are six steps for each revolution. $6_3$ is neutral (analogous to $2_1$ and $4_2$). $6_2$ and $6_4$ show a new feature: the points fall on two intertwined circular helices (*double helices*) as shown in Fig. 3.7. $6_2$ is right-handed and $6_4$ is left-handed.[1]

Some properties of screw axes (which also should be verified from the figures) are:

$4_1$, $4_3$, $6_1$, $6_3$ and $6_5$ axes (only) include a $2_1$ axis.
$4_2$, $6_2$ and $6_4$ axes (only) include a 2 axis.
$6_2$ and $6_5$ axes include a $3_2$ axis.
$6_1$ and $6_4$ axes include a $3_1$ axis.
A $6_3$ axis includes a 3 axis.
Points generated by $6_2$ or $6_4$ lie on a double helix.

Also to be noted is that there can be a mirror normal to $2_1$, $4_2$, and $6_3$ but not to $3_1$, $3_2$, $4_1$, $4_3$, $6_1$, $6_2$, $6_4$, and $6_5$ as the latter group have a hand.

It is important to realize that our labels "left-handed" and "right-handed" are arbitrary, just as they are for right- and left-handed nuts and bolts and coordinate systems. We could have used terms such as "positive" and "negative" (as in electricity) or even "north" and "south" (as in magnetism). Indeed, when talking about pure rotations, we used the terms "positive" and "negative." Thus we saw that the "right-handed" screw $6_2$ contains a "left-

[1]A "spiral" (better "helical") staircase is a familiar example of an object with a screw axis. If the staircase continued indefinitely the axis would be $N_1$ (right-handed) or $N_{N-1}$ (left-handed). The famous "miraculous" staircase in the Loretto Chapel in Santa Fe, New Mexico is $16_{15}$. There is a nice double helix stairway in King's Park, Perth (Western Australia) that is $56_2$.

handed" screw $3_2$. Our stair climber could have ascended the $3_2$ screw in an anti-clockwise ("right handed") sense if he took not the next highest step (up $c/3$) but the step in the other direction (up $2c/3$—see Fig. 3.19, p. 87).

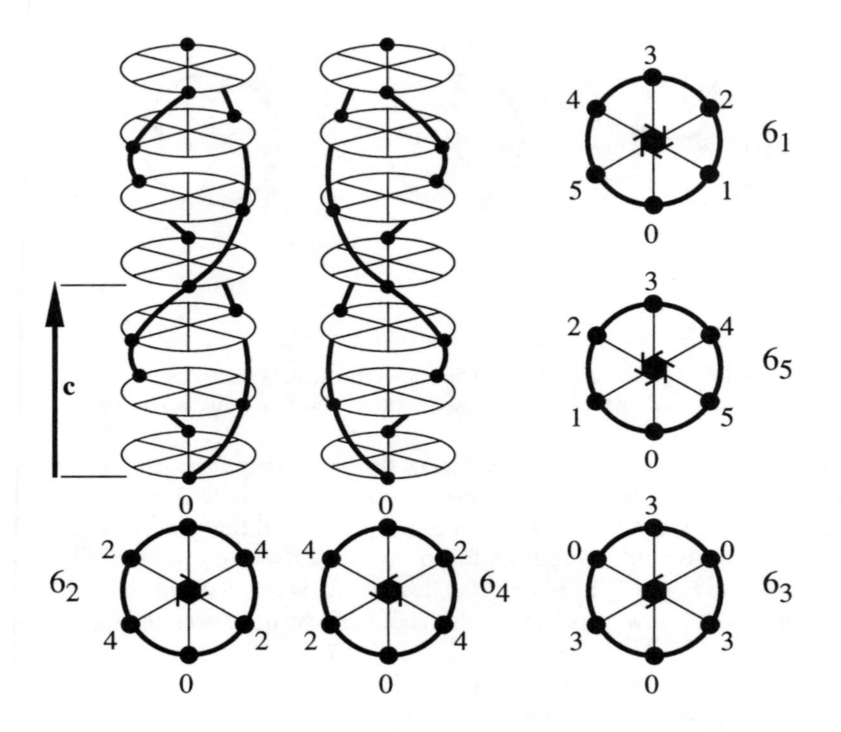


**Fig. 3.7**. Illustrating six-fold screw axes. The numbers are elevations in multiples of $c/6$. The double helices in $6_2$ and $6_4$ are suggested by heavy lines in the sketches in the top left (which correspond to the projections immediately below them). **c** is the repeat vector along the screw axis.

In this connection we quote from a famous lecture by Weyl[1]: "the inner structure of space does not permit us, except by arbitrary choice, to distinguish a left from a right screw...on [this fundamental concept] depends the entire theory of relativity...." If we were to pursue the subject here we would soon find ourselves in deep water.

Notice also that an asymmetric periodic object may well have screw axes of opposite hand as symmetry elements. An example is the cylinder packing labelled **β-Mn** in § 6.7.3 (p. 265). $I4_1$ is discussed as an example of a space group with $4_1$ and $4_3$ axes in § 3.3.4 (p. 74). For more on screw axes and the "hand" (left or right) of *crystals* see § 3.6.

[1]Chapter 1 in *Symmetry* by H. Weyl [Princeton University Press (1952)]. This was written before the "non-conservation of parity" was discovered. On the latter topic in connection with hand, a readable account, that discusses Weyl's lecture, is Chapter 3 in *Elementary Particles* by C. N. Yang [Princeton University Press (1962)].

### *3.2.3 Comparison of screw and glide*

In our illustrations of the effect of symmetry operations, for simplicity we show their effects on a point (a small circle). The reader should mentally replace the point with an asymmetric object. In Fig. 3.8 we contrast the effect of glide (arbitrarily labeled *a*) with the effect of a $2_1$ axis on an asymmetric object (a scalene triangle that is black on one side and white on the other).

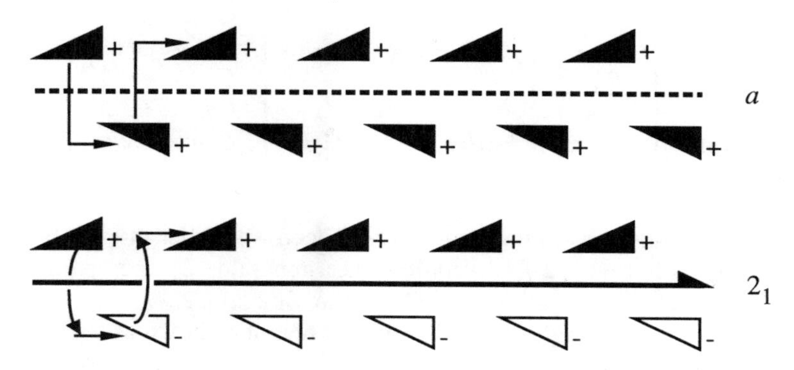


**Fig. 3.8.** Contrasting glide with $2_1$. Figures marked "+" are above the plane of the paper, and those marked "-" are below.

## 3.3 Three-dimensional space groups

We now consider how the space groups arise. We will consider just some of the simpler possibilities—our aim is to suggest how to proceed rather than to be rigorous. We emphasize that the immediate goal is that of being able to interpret (and use) space group symbols. The reader will find it helpful to work through the examples provided. It would also be very useful to have *International Tables A* at hand.

### *3.3.1 Triclinic space groups*

In the triclinic system we have just a primitive lattice (*P*) to combine with the point groups with 1-fold axes. The only possibilities are therefore $P1$ and $P\bar{1}$.

### *3.3.2 Monoclinic space groups*

In the monoclinic system we have to combine the *P* and *C* lattices with the point groups that have just one 2-fold axis (2, *m* and 2/*m*). We then get the symmorphic groups *P*2, *Pm*, *P*2/*m*, *C*2, *Cm* and *C*2/*m*.

The next thing to do is to consider the possibilities that arise when 2 axes are changed to $2_1$ axes and/or when mirrors are changed to glide. In the latter case we note that the mirror planes are necessarily normal to **b** so that the glide planes are either *a*, *n* or *c*. However the labeling of the axes normal to **b** is arbitrary, so we adopt the convention (not universally adhered to) that the glide, if present, is *c* (see Fig. 3.12 below for the choice of axes that converts *Pc* to *Pn* or *Pa*). The distinct cases are listed in Table 3.4.

**Table 3.4.** The monoclinic space groups.

| class | *P* lattice | | | | *C* lattice | |
|---|---|---|---|---|---|---|
| 2 | *P*2 | $P2_1$ | | | *C*2 | |
| *m* | *Pm* | *Pc* | | | *Cm* | *Cc* |
| 2/*m* | *P*2/*m* | $P2_1/m$ | *P*2/*c* | $P2_1/c$ | *C*2/*m* | *C*2/*c* |

The reader interested in confirming that this is indeed a full list might note that the combination of a *C* lattice with 2-fold rotation axes generates $2_1$ axes parallel to and interlaced with the 2 axes. Thus (see Fig. 3.9) $C2_1$ is the same as *C*2.

In Fig. 3.9, the two black triangles at the top left (at height *y* shown as "+") are related by a 2-fold rotation axis. The pair to the right (at 1/2+*y*) are generated by *C* centering [(1/2,1/2,0)+]. On the right the figure shows the symmetry elements generated by the combination of 2-fold rotations (symbolized by ellipses) and primitive unit cell translations [(**a**±**b**)/2 and **c**]. Note the $2_1$ axes (symbolized by ellipses with two arms)—the reader should verify their existence in the pattern on the left in Fig. 3.9. The same pattern and symmetry elements would have been generated by starting with a $2_1$ axis and primitive unit cell translations.

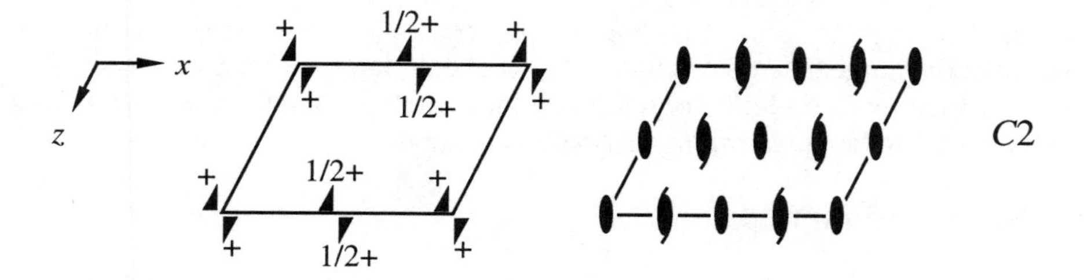


**Fig. 3.9**. Illustrating a unit cell of space group *C*2. **b** is the unique axis normal to the paper, **a** is horizontal and **c** runs down the page. Left: The pattern generated by 2-fold rotations, *C* centering and unit cell translations. Right: The generated symmetry elements.

We now consider space group *C*2/*m* (a rather common symmetry for crystals). Fig. 3.10 illustrates this space group in much the same way as Fig. 3.9 illustrated *C*2. A major difference is that a mirror plane at $y = 0$ generates pairs of triangles above and below the plane and superimposed in the projection down **b** (and shown as gray triangles). This situation is symbolized by giving elevations as ±. The *C* centering operation produces corresponding triangles at elevations 1/2±. Note that there is also a mirror plane at $y = 1/2$. On the right the mirror plane is symbolized by a heavy bent line with arms parallel to the cell edges; by convention no elevation is shown for mirror planes at heights 0 and 1/2. Also on the right the generated centers are shown as small open circles. Those with no elevations marked are at $y = 0$ and $y = 1/2$ and the site symmetry at these points is $2/m$. The other centers with elevations marked as "1/4" are at $y = 1/4$ and 3/4 and the symmetry at these points is $\bar{1}$. In a centrosymmetric crystal there are always eight centers per primitive cell (so in this case there are sixteen in the centered cell of twice the volume). It should be seen that $C2/m$ also contains glide planes normal to **b**. These are symbolized in the same way as the mirror planes but with an arrow head pointing in the glide direction which can be seen to be **a**. These planes are at 1/4 and 3/4 (again by convention the elevation is just given as 1/4). The reader should verify the presence of the *a* glide operation noting that reflection in a plane at $y = 1/4$ will transform an elevation "+" to "1/2–" and so on. When we discuss subgroups of space groups, it will be seen that knowledge of the existence of the extra symmetry elements ($2_1$ axes and *a* glide in this example) is very useful.

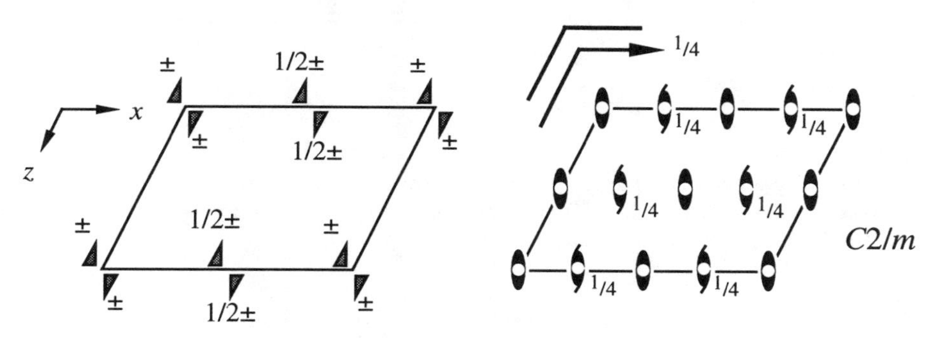


**Fig. 3.10**. Illustrating a unit cell of space group $C2/m$. **b** is the unique axis normal to the paper, **a** is horizontal and **c** runs down the page. Left: The pattern generated by the $2/m$ axis, $C$ centering and unit cell translations. Right: The generated symmetry elements.

Although $C2/m$ contains glide planes, they are *a* glide, and $C2/c$ is a distinct space group. Fig. 3.11 illustrates $C2/c$ in the same way that Fig. 3.10 illustrated $C2/m$. Now triangles do not superimpose in projection with their mirror images, which are generated by the *c* glide plane at $y = 0$. The mirror images of black triangles are shown as white triangles and *vice versa*. The reader should work through this example as for $C2/m$ discussed in the previous paragraph. Note now the existence of *n* glide planes (as indicated by the direction of the arrow) at $y = 1/4$ and $y = 3/4$. In contrast to $C2/m$, in $C2/c$ the 2-fold axes do not intersect centers and there are no mirror planes (only glide planes).

Even with **b** unique the three possibilities of choosing **a** and **c** (see *e.g.* Fig. 3.2) result in a variety of possible symbols for monoclinic space groups that are centered and/or have glide planes. Thus *C* can become *A* or *I*. In a similar way a glide *c* can become *a* or *n* as illustrated in Fig. 3.12.

Because there can be ambiguity about which axis is chosen as the unique axis it is a common practice to use extended symbols with 1's as place markers (cf. § 2.4, p. 45) and the symbol is to be interpreted the same way as that for an orthorhombic symmetry group. Thus with **b** unique $P2$ becomes $P121$ (2-fold axis parallel to $y$) and with **c** unique it becomes $P112$ (2-fold axis parallel to $z$). $Pc$ is written $P1c1$ or $P11a$ and so on. Fortunately the *International Tables* (vol. A) considers all these possibilities. The various symbols encountered for monoclinic space groups are listed in the tables at the end of this book. The different choices of axes are referred to as different *settings* of the space group.

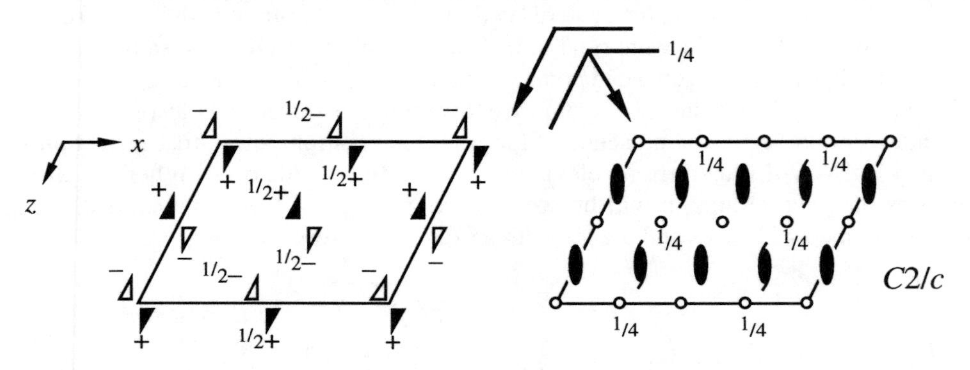


**Fig. 3.11.** Illustrating a unit cell of space group $C2/c$. **b** is the unique axis normal to the paper, **a** is horizontal and **c** runs down the page. Left: The pattern generated by the 2-axis, $c$ glide, $C$ centering and unit cell translations. Right: The generated symmetry elements.

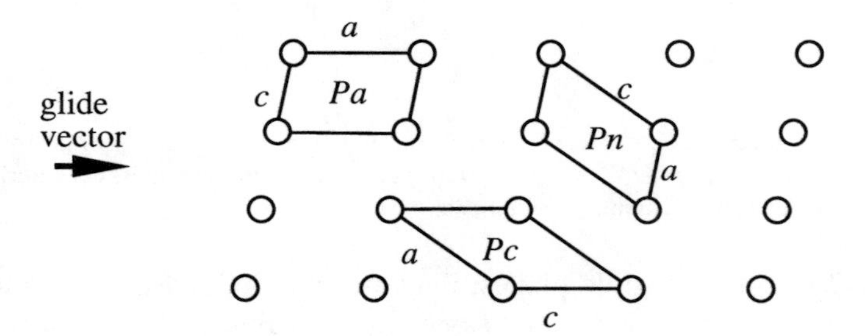


**Fig. 3.12.** Illustrating three choices of unit cell for $Pc$. **b** is the unique axis normal to the plane of the paper and the glide plane is in that plane. The glide vector is **c**/2 ($Pc$) = **a**/2 ($Pa$) = (**a** + **c**)/2 ($Pn$).

### *3.3.3 Orthorhombic space groups*

In the orthorhombic system there are three axes at right angles with 2-fold symmetry elements along each axis. The point groups to consider are therefore 222, $mm2$ and $mmm$. There are also four types of lattice ($P$, $C$, $F$ and $I$) to consider, so the number of possibilities becomes much greater (it turns out that there are 69 distinct orthorhombic space groups). Thus in contrast to the monoclinic pair $P2$ and $P2_1$, we have $P222$, $P222_1$, $P2_12_12$ and $P2_12_12_1$.[1] Mirrors can become glide ($a$, $b$, $c$, $n$, or $d$). Because of this complexity it would take us much too long to systematically generate all the orthorhombic space groups in the manner suggested in the monoclinic case.

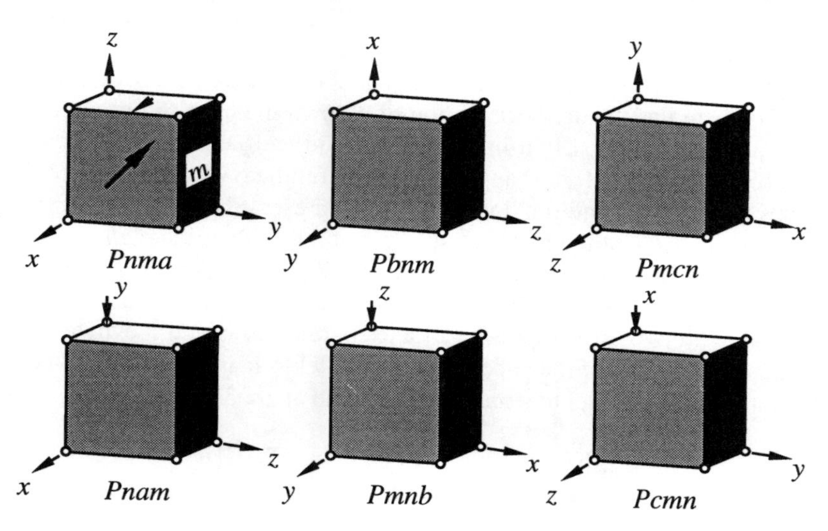


**Fig 3.13**. Different settings of the space group *Pnma*. The diagrams show the orientations of the *n* glide plane (shaded, front face), the axial glide plane (unshaded, top face) and the mirror, *m* (black face) with respect to the axes. The arrows show the directions of the axes for a right-handed system.

Again we have to adopt some conventions for the labeling of axes. These are outlined in the *International Tables* and will not be given here. The conventions adopted there we call the standard setting; it is important to recognize that other settings are often chosen.[2] The significance of the order of symbols in a space group symbol is the same as that given in § 2.4 for point groups except that the first symbol represents the lattice type. Thus in the symbol *Pnma* the "*P*" tells us that the lattice is primitive. The next three symbols indicate that there are respectively: an *n* glide plane normal to $x$, a mirror plane normal to $y$ and then an *a* glide plane normal to $z$.[3] As there are six possible permutations of the $x$, $y$ and $z$ axes, this space group can have six different symbols: *Pnma*, *Pbnm*, *Pmcn*, *Pnam*, *Pmnb* and *Pcmn*. Fortunately a concordance of symbols is given in the *International Tables* (see also the Tables at the end of the book). Figure 3.13 illustrates the arrangement of the

[1]Note that when there is an "odd man out" it is labeled the $c$ axis in the standard setting. Thus $P222_1$ not $P2_122$ *etc.* similarly we have *Pmm*2 and not *Pm*2*m* etc. as the standard setting.

[2]Mineralogists (and some others) often use the convention that $c < a < b$. Another recommendation, which we prefer, is to use standard settings, and to use $a < b < c$ when there is flexibility of choice of all three axes (as in *Pmmm* etc.) and $a < b$ in cases such as *Pmm*2 where two axes can be chosen arbitrarily.

[3]The orientation of glide planes (as that of mirror planes) is given by the directions of their normals. Thus the direction of the normal to the *n* glide plane is parallel to the $x$ direction (the *n* glide plane is normal to the $x$ direction as stated).

symmetry elements with respect to the axes in this particular case.[1]

The full symbol for *Pnma* is $P2_1/n2_1/m2_1/a$. If the inversion center is removed one of the acentric space groups $P2_12_12_1$, $Pnm2_1$, $Pn2_1a$ or $P2_1ma$ (see § 3.5) is obtained. It is not uncommon for materials with closely related structures to have one or other of these symmetries. Note that the standard settings for the last three groups (which are polar) result in the symbols $Pnm2_1$, $Pna2_1$ and $Pmc2_1$.

Fig. 3.14 shows how the symmetry elements of *Pnma* are illustrated in the *International Tables* in the standard orientation of **a** down the page, **b** horizontally on the page and **c** normal to the page. Light lines outline a unit cell and small circles show the locations of $\bar{1}$ points (centers) at $z = 0$ and $z = 1/2$. The full heavy lines represent the traces of mirror planes normal to **b** and the dot-dash lines represent the traces of *n* glide planes normal to **a**. The bent arrow at the top right shows that there is a glide plane at $z = 1/4$ (i.e. normal to **c**) with the glide direction along the arrow (i.e. *a*). As the spacing between mirror or glide planes is half the translation normal to them, there is also an *a* glide plane at $z = 3/4$. The ovals with two arms represent $2_1$ axes parallel to **c**, and the arrows with half heads are $2_1$ axes in the plane of the paper. Those parallel to **b** are at $z = 0$ (and necessarily also at $z = 1/2$), and those parallel to **a** are at $z = 1/4$ (and at $z = 3/4$).

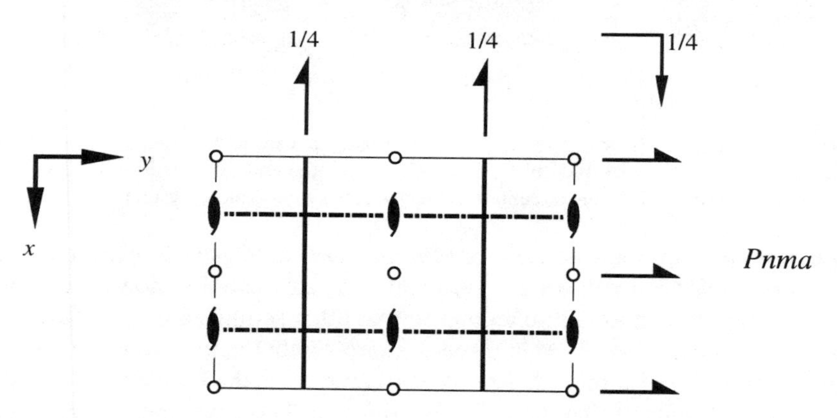


**Fig. 3.14**. The symmetry elements of *Pnma*. For an interpretation see text.

Recall that monoclinic space groups with centered cells have additional symmetry elements that do not appear in the space group symbol. Thus *C*2 also has $2_1$ axes, and *C*2/*m* also has $2_1$ axes and *a* glide planes. Similarly orthorhombic space groups with centered cells have additional symmetry elements that do not appear in the space group symbol. In the *International Tables* a useful table (Table 4.3.1) lists these extra symmetry elements as in the following examples:

| *C* | *m* | *c* | *m* | | *I* | *m* | *m* | *m* |
|---|---|---|---|---|---|---|---|---|
| | *b* | *n* | *n* | | | *n* | *n* | *n* |

This means that in *Cmcm*, normal to **a** there are also *b* glide planes, and normal to **b** and

[1]Symmetry *Pnma* is particularly common, occurring for more than 8% of inorganic structures. For more on the occurrence of particular symmetry groups see § 3.7.7 (p. 94).

**c** there are also *n* glide planes. Likewise in *Immm* there are *n* glide planes normal to all three axes. We *could* write *Cmcm* as *Cbnn* or even *Cmcn*, etc. and write *Immm* as *Innn*. Fortunately practical, if arbitrary, rules were long ago decided on for preference of symbols (*e.g.* *m* has preference over glide). Fortunately also, in this instance, the rules are universally obeyed by crystallographers.[1] But note that with permutation of axes the symbols of the other symmetry elements may also change as in (for example) two different settings of *Cmcm*:

$$\begin{matrix} C & m & c & m \\ & b & n & n \end{matrix} \qquad \begin{matrix} A & m & m & a \\ & n & c & n \end{matrix}$$

### 3.3.4 Tetragonal space groups

In the tetragonal system there is normally no ambiguity about the choice of axes. The $z$ axis is always parallel to the 4-fold axis, and $x$ and $y$ are normal to $z$ and to each other. The symbol for the space group is again a symbol for the lattice type followed by a three-position symbol derived from the point group symbol (see § 2.4). Thus with a body-centered lattice there are in the class $4/mmm$ (among others) the space groups $I4/mmm$ and $I4_1/amd$.

In the class $\bar{4}2m$ there are two distinct space groups $P\bar{4}m2$ and $P\bar{4}2m$. In the first of these the mirror planes are normal to $x$ and $y$, and 2-fold axes at 45° to $x$ and $y$; in the second the mirrors are normal to directions at 45° to $x$ and $y$, and the 2-fold axes along $x$ and $y$. (Compare the positions of "2" and "$m$" in the space group symbols.) Recall that the orientation of the axes is determined by the lattice translations.

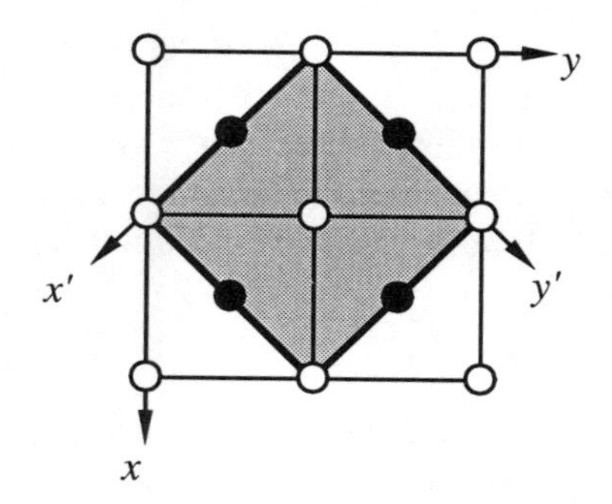


**Fig 3.15**. A body-centered tetragonal lattice (projected down **c**) with four unit cells indicated with light lines and a face-centered cell (heavier lines, shaded). Open circles are lattice points at $z = 0$ and filled circles are lattice points at $z = 1/2$.

Occasionally a body-centered tetragonal crystal is described in terms of a face-centered tetragonal cell of twice the volume. As shown in Fig. 3.15, the $x$ and $y$ axes are rotated by 45° when the cell changes. The last two symbols in the space group symbol have then to be interchanged. Thus $I\bar{4}2m$ becomes $F\bar{4}m2$ and $I\bar{4}m2$ becomes $F\bar{4}2m$. Sometimes the

[1]But see P. M. de Wolff *et al.*, *Acta Crystallographica* A**48**, 727 (1992) for proposed changes to five orthorhombic space group symbols. Although logical, we hope these suggestions are not adopted.

change is more subtle: $I4_1/amd$ becomes $F4_1/ddm$.[1]

There are primitive tetragonal space groups $P4$, $P4_1$, $P4_2$ and $P4_3$, but only two corresponding body-centered tetragonal groups $I4$ (which contains an equal density of $4_2$ axes) and $I4_1$ (which contains an equal density of $4_3$ axes). This means that an acentric structure that has symmetry $I4_1$ (and hence has distinct right- and left-handed forms) will have both $4_1$ and $4_3$ axes, but the arrangement around the two axes will not be related by mirror symmetry. Contrast the situation in the centrosymmetric space group $I4_1/a$ where there are again $4_1$ and $4_3$ axes but they are now related by the *a* glide operation.

### *3.3.5 Trigonal and hexagonal space groups*

In the trigonal and hexagonal systems one setting is usually adopted. Except as noted below for crystals with a rhombohedral lattice, the *z* axis is taken parallel to the 3- or 6-fold axis. The *x* and *y* axes are perpendicular to *z* and at 120° to each other. Again the significance of the last two positions of the space group symbol is the same as for point groups (§ 2.4). Thus in $P31m$ the normals to the mirrors are at 90° to the *x* and *y* axes (i.e. the *x* and *y* axes are parallel to the mirrors) whereas in $P3m1$ the normals to the mirrors are parallel to the *x* and *y* axes (compare $p31m$ and $p3m1$ in Fig. 1.13). Note also the two distinct space groups $P6_3/mmc$ and $P6_3/mcm$.

In the case of rhombohedral crystals the 3-fold axis is along a body diagonal of the primitive unit cell—parallel to $\mathbf{a} + \mathbf{b} + \mathbf{c}$. The lattice symbol is now always *R*. Thus we have the space groups $R3$, $R\bar{3}$, $R32$, $R3m$, $R3c$, $R\bar{3}m$ and $R\bar{3}c$. It is worth noting that the *R* lattice already contains both $3_1$ and $3_2$ axes.[2]

Rhombohedral crystals are very often described in terms of a hexagonal unit cell with three times the volume and with **c** parallel to the 3-fold axis as described in § 3.1 and in more detail later (§ 4.4.2, p. 104). In this case *the space group symbol is unchanged.* The nature of the cell chosen is always clear from the parameters given (*a* and $\alpha$ for a primitive cell; *a* and *c* for a hexagonal cell). It never hurts to be explicit however.

### *3.3.6 Cubic space groups*

In the cubic system there is a universal choice of axes. In the space group symbol, after the symbol for the lattice, the significance of the positions is the same as given for the point groups (§ 2.4, p. 47). Thus in the class $m\bar{3}m$ there are (for example) $Fd\bar{3}m$ and $Ia\bar{3}d$. It might be noted that in the older literature these symbols are $Fd3m$ and $Ia3d$ respectively (i.e. the bar over the 3 is dropped just as in the short symbols for the point groups). The full symbols for these space groups are $F4_1/d\bar{3}2/m$ and $I4_1/a\bar{3}2/d$. Other points to notice include the fact that there are four separate groups $P432$, $P4_132$, $P4_232$ and $P4_332$ but only groups $I432$ and $F432$ (both of which already contain $4_2$ axes) and $I4_132$ and $F4_132$

[1]The *a* glide planes normal to the 4-fold axis have translations alternately **a**/2 and **b**/2 in the *I* cell. These directions become $(\mathbf{a} \pm \mathbf{b})/4$ in the *F* cell so now the glide is *d* instead of *a*.

[2]Thus there are not separate space groups $R3_1$ or $R3_2$. Contrast the three separate space groups $P3$, $P3_1$ and $P3_2$.

(both of which already contain $4_3$ axes)—cf. the discussion of tetragonal groups above.

The group *I*23 also contains $2_1$ axes but there is nevertheless a separate group $I2_13$ which also contains 2 axes. The difference between them (as discussed below) is that in *I*23 the 2 axes all intersect (as do the $2_1$ axes) in $I2_13$ they do not. In this case the symbols for the two space groups have to be assigned arbitrarily.

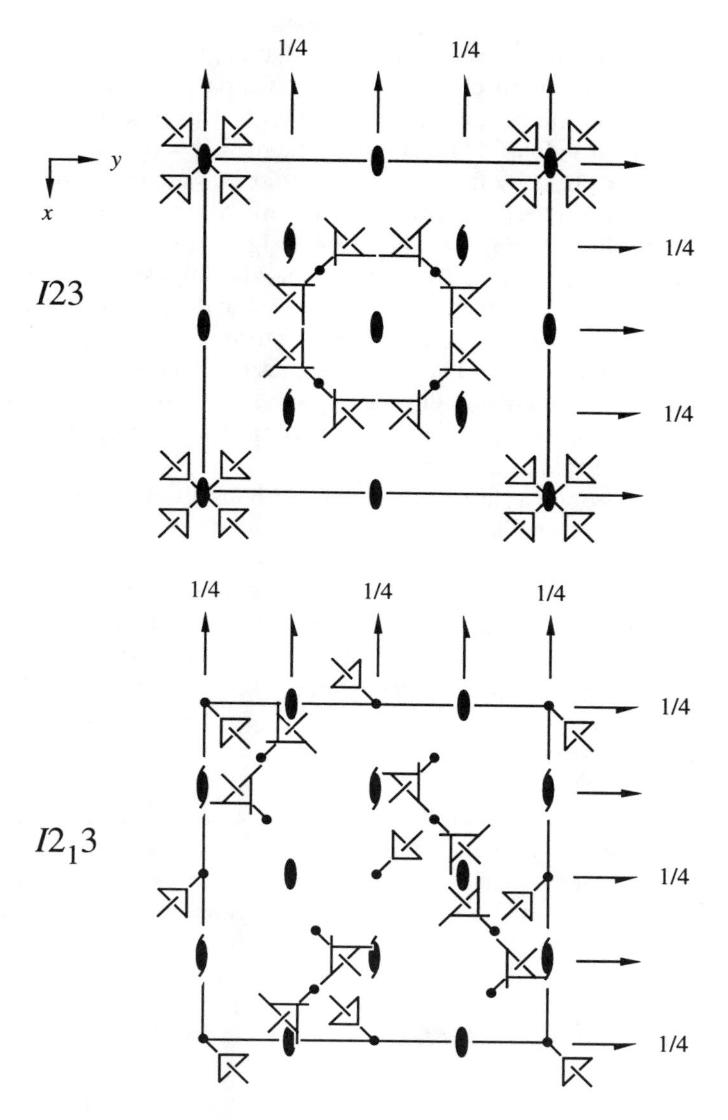


**Fig 3.16**. The symmetry elements of *I*23 and $I2_13$. The meaning of the symbols is explained in the text (§ 3.3.6).

In addition to the 3 or $\bar{3}$ axes in four different directions, the cubic space groups have $3_1$ and $3_2$ axes each in four different directions. Their location will depend on the space group. There are two different cases to consider. In the first case the 3 or $\bar{3}$ axes intersect (at the origin) and the $3_1$ and $3_2$ axes intersect in pairs. In the second case the 3 or $\bar{3}$ axes do not intersect, and in fact there is no intersection of any of the three-fold axes.

Figure 3.16 illustrates this point for *I*23 (first case) and *I*$2_1$3 (second case). To interpret the figure, note that $x$ is down the page, $y$ is horizontal (left to right) and $z$ extends up out of the plane. 3-fold axes (shown as short lines through triangles) intersect the plane $z = 0$ at the points shown as small filled circles and go upwards from there parallel to body diagonals. Triangles represent 3 axes and triangles with arms represent screw ($3_1$ or $3_2$) axes. Thus in *I*$2_1$3, $3_1$ axes along[1] [111] intersect the plane $z = 0$ at 1/3,2/3,0 and $3_2$ axes along [111] intersect the plane $z = 0$ at 2/3,1/3,0. 2-fold axes are symbolized as described above (§ 3.3.3, p. 70 and Fig. 3.14); those parallel to the plane of the paper are either at $z = 0$ and 1/2 (no height shown) or at $z = 1/4$ and 3/4 (height shown as 1/4).

Note also that in *I*$2_1$3, the 2 axes do not intersect with themselves or with any of the 3-fold axes. The same is true of the $2_1$ axes, but the 2 and $2_1$ axes do intersect. Cubic crystal structures with space groups that have non-intersecting symmetry axes are often rather hard to understand and depict, but can sometimes be described as based on packings of cylinders whose axes are along non-intersecting symmetry axes (§ 6.7). In space groups with non-intersecting 3-fold axes there is no site of cubic point symmetry (which can only be at a point where four 3-fold axes intersect).

The locations of the symmetry elements in cubic groups are shown in a similar way in the *International Tables.* It is rewarding to learn the symbolism and to practice reading the diagrams.

Two choices of origin are made for centrosymmetric space groups in which the intersection of 3 or $\bar{3}$ axes is not at an inversion center. The first choice is at the intersection of the 3-fold axes and is therefore at a site of cubic symmetry ($\bar{4}3m$, 432 or 23). The second choice is at a site of inversion symmetry; see § 3.7.4 (p. 91) for details.

### *3.3.7. Space group and crystal class*

It should be obvious that to obtain the crystal class (the point group from which the space group was derived) from the space group symbol, one should (a) drop the lattice symbol, (b) drop all subscripts, and (c) change all glide symbols to "*m*." Thus *I*$4_1$*/amd* → *4/mmm*, *P*$2_1$*/c* → *2/m* etc. If the point group is centrosymmetric then the space group is also (do Exercise 4).

## 3.4 Using the *International Tables*

The *International Tables* provide a wealth of information about the space groups including the nature and location of all symmetry elements in the unit cell. The tables

[1]We explain notation such as [111] in the next chapter. [111] refers to the direction from 0,0,0 to 1,1,1.

should also be consulted for the symbols for symmetry elements. All that are needed for our immediate purposes are the coordinates of equivalent points in a unit cell. Our first example is for $P2_1/c$; the special and general positions are given with Wyckoff notation (*a*, *b*, etc.) in Table 3.4. Note that the origin of coordinates is taken on a center of symmetry:

**Table 3.4.** Special and general positions of $P2_1/c$.

| | | |
|---|---|---|
| general: | 4 *e* | $x,y,z$ ; $\bar{x},\bar{y},\bar{z}$ ; $\bar{x},1/2+y,1/2-z$ ; $x,1/2-y,1/2+z$ |
| special | 2 *d* | 1/2,0,1/2 ; 1/2,1/2,0 |
| | 2 *c* | 0,0,1/2 ; 0,1/2,0 |
| | 2 *b* | 1/2,0,0 ; 1/2,1/2,1/2 |
| | 2 *a* | 0,0,0 ; 0,1/2,1/2 |

It is important to recognize that in order to derive the general positions (4 *e* in this example) the location of the symmetry elements in the unit cell must be known. Fortunately this has been done for every space group in the *International Tables*. The symmetry operations are in the present case:

| | |
|---|---|
| inversion through the origin | $x,y,z \rightarrow \bar{x},\bar{y},\bar{z}$ |
| *c* glide about plane at $y = 1/4$ [1] | $x,y,z \rightarrow x,1/2-y,1/2+z$ |
| rotation about $2_1$ along $x = 0$, $z = 1/4$ | $x,y,z \rightarrow \bar{x},1/2+y,1/2-z$ |

The structure of AgO has symmetry $P2_1/c$ and a crystallographic description of AgO is:

AgO $P2_1/c$, $a = 5.859$, $b = 3.484$, $c = 5.500$ Å, $\beta = 107.51°$
Ag(1) in 2 *a*; Ag(2) in 2 *d*; O in 4 *e*, $x = 0.296$, $y = 0.345$, $z = 0.222$

It may be seen that there are two kinds of Ag atom in the structure. When atoms of a given element appear on sites that are not related by symmetry, we say that they are *crystallographically* distinct and distinguish them by numbering. We translate the above description as follows: There is one kind of Ag atom at positions 0,0,0 and 0,1/2,1/2 in the unit cell and a second kind of Ag which is at 1/2,0,1/2 and 1/2,1/2,0. In AgO it turns out that the compound is really $Ag(I)Ag(III)O_2$ and that Ag(1) is Ag(I) and Ag(2) is Ag(III).[2]

The O positions in the unit cell are obtained as:

| | | |
|---|---|---|
| 0.296, 0.345, 0.222 | ≡ | 0.296, 0.345, 0.222 |
| −0.296,−0.345,−0.222 | ≡ | 0.704, 0.655, 0.778 |
| −0.296, 0.845, 0.278 | ≡ | 0.704, 0.845, 0.278 |
| 0.296, 0.155, 0.722 | ≡ | 0.296, 0.155, 0.722 |

[1]Compare this case with Fig. 3.3 where the glide plane is at $y = 0$.

[2]Be sure to note that Arabic numerals (1 and 2 in this instance) are used as arbitrary identification numbers, but Roman numerals (I and III in this case) refer to oxidation states which are *inferred* from structural details (see Exercise 4.7.3).

In the first column the values of $x$, $y$ and $z$ have been substituted into the expressions for the 4 $e$ positions; in the second column 1.0 has been added to any negative coordinates to bring the atoms all into the same unit cell (i.e. so that $0 \le x < 1, 0 \le y < 1, 0 \le z < 1$).

As a second example we take a crystallographic description of the rutile form of $TiO_2$.

| | |
|---|---|
| $TiO_2$ | $P4_2/mnm$, $a = 4.594$, $c = 2.958$ Å |
| | Ti in 2 $a$ ; O in 4 $f$, $x = 0.305$ |

We recognize the symmetry to be tetragonal. Turning to the *International Tables* we find for this space group that 2 $a$ correspond to 0,0,0 and 1/2,1/2,1/2, so this is where the Ti atoms are located in the unit cell. The 4 $f$ positions are given as $x,x,0$ ; $\bar{x},\bar{x},0$ ; $1/2+x,1/2-x,1/2$ ; $1/2-x,1/2+x,1/2$. Proceeding as for the O atoms in AgO we determine the coordinates of the four O atoms in the unit cell of rutile are:

0.305, 0.305, 0 ; 0.695, 0.695, 0 ; 0.805, 0.195, 0.5 ; 0.195, 0.805, 0.5

Sometimes there is a remarkable economy in this type of description. For example the structure of spinel, $MgAl_2O_4$ has 56 atoms in the unit cell, yet it is completely specified by symmetry information and just two numbers ($a$ and $x$):

| | |
|---|---|
| $MgAl_2O_4$ | $Fd\bar{3}m$, $a = 8.080$ Å |
| | Mg in 8 $a$. Al in 16 $d$. O in 32 $e$, $x = 0.262$ |

From the *International Tables* we find that the coordinates of 32 $e$ are:

(0,0,0 ; 0,1/2,1/2 ; 1/2,0,1/2 ; 1/2,1/2,0) +
$x,x,x$ ; $x,1/4-x,1/4-x$ ; $1/4-x,x,1/4-x$ ; $1/4-x,1/4-x,x$
$\bar{x},\bar{x},\bar{x}$ ; $\bar{x},3/4+x,3/4+x$ ; $3/4+x,\bar{x},3/4+x$ ; $3/4+x,3/4+x,\bar{x}$

The interpretation of this is that to the coordinates in the second and third lines above, we must add in turn 0,0,0 ; 0,1/2,1/2 ; 1/2,0,1/2 and 1/2,1/2,0. The last three quantities are in fact primitive lattice translations for a face-centered cell ($Fd\bar{3}m$ is face-centered cubic). The coordinates of O in the unit cell are then (please verify, noting that $\bar{x} = 1-x = 0.738$; $1/4-x = -0.012 \equiv 0.988$; $3/4+x = 1.012 \equiv 0.012$; etc.):

| | | | |
|---|---|---|---|
| 0.262, 0.262, 0.262 | 0.262, 0.762, 0.762 | 0.762, 0.262, 0.762 | 0.762, 0.762, 0.262 |
| 0.262, 0.988, 0.988 | 0.262, 0.488, 0.488 | 0.762, 0.988, 0.488 | 0.762, 0.488, 0.988 |
| 0.988, 0.262, 0.988 | 0.988, 0.762, 0.488 | 0.488, 0.262, 0.488 | 0.488, 0.762, 0.988 |
| 0.988, 0.988, 0.262 | 0.988, 0.488, 0.762 | 0.488, 0.988, 0.762 | 0.488, 0.488, 0.262 |
| 0.738, 0.738, 0.738 | 0.738, 0.238, 0.238 | 0.238, 0.738, 0.238 | 0.238, 0.238, 0.738 |
| 0.738, 0.012, 0.012 | 0.738, 0.512, 0.512 | 0.238, 0.012, 0.512 | 0.238, 0.512, 0.012 |
| 0.012, 0.738, 0.012 | 0.012, 0.238, 0.512 | 0.512, 0.738, 0.512 | 0.512, 0.238, 0.012 |
| 0.012, 0.012, 0.738 | 0.012, 0.512, 0.238 | 0.512, 0.012, 0.238 | 0.512, 0.512, 0.738 |

The metal atom positions are found from the *International Tables* by obtaining the coordinates for positions 8 $a$ and 16 $d$. These are:

| | (0,0,0 ; 0,1/2,1/2 ; 1/2,0,1/2 ; 1/2,1/2,0) + |
|---|---|
| 8 *a* | 1/8,1/8,1/8 ; 7/8,7/8,7/8 |
| 16 *d* | 1/2,1/2,1/2 ; 1/2,1/4,1/4 ; 1/4,1/2,1/4 ; 1/4,1/4,1/2 |

We now know where all the atoms are, but are not really much wiser about the structure. In fact the situation of the would-be crystal chemist at this point may be likened to that of a student of architecture who is given coordinates of bricks, rather than an architectural drawing. Accordingly, the next steps are to draw the structure (or make a model), to find nearest neighbors and coordination numbers and to calculate bond lengths and angles.

Because we realize that it would be tedious for the reader to have to reach for the *International Tables* every time that he/she wants to draw or do calculations on a structure (which we hope is often), we usually give explicitly the coordinates of special and general positions of a space group when reporting a structure. Especially for cubic groups it is desirable to have some concise way of doing this. We use the following conventions.

1. The origin is always taken at a center of symmetry if present.

2. ± means plus *and* minus
   $\pm(x,y,z)$ means $x,y,z$ and $\bar{x},\bar{y},\bar{z}$
   $\pm x,y,z$ means $x,y,z$ and $\bar{x},y,z$
   $\pm(\pm x,y,z)$ means $x,y,z$ ; $\bar{x},y,z$ ; $\bar{x},\bar{y},\bar{z}$; $x,\bar{y},\bar{z}$

3. Centering is expressed as a letter followed by + or ±
   *I* refers to (0,0,0 ; 1/2,1/2,/12)
   *F* refers to (0,0,0 ; 0,1/2,1/2; 1/2,0,1/2; 1/2,1/2,0)
   *A* refers to (0,0,0 ; 0,1/2,1/2)
   *B* refers to (0,0,0 ; 1/2,0,1/2)
   *C* refers to (0,0,0 ; 1/2,1/2,0)
   *R* refers to (0,0,0 ; 1/3,2/3,2/3 ; 2/3,1/3,1/3)

4. Cyclic permutation is expressed as $(\ldots)\kappa$
   $(x,y,z)\kappa$ means $x,y,z$ ; $z,x,y$ ; $y,z,x$
   (This corresponds to the operation of a threefold axis through the origin and along [111].)

We now give examples (crystal structure information is usually given in this format subsequently): [1]

| | |
|---|---|
| $MgAl_2O_4$ | $Fd\bar{3}m$, $a$ = 8.080 Å |
| | Mg in 8 *a*: $F \pm (1/8,1/8,1/8)$ |
| | Al in 16 *d*: $F + (1/2,1/2,1/2; (1/2,1/4,1/4)\kappa)$ |
| | O in 32 *e*: $F \pm (x,x,x; (x,1/4-x,1/4-x)\kappa)$, $x = 0.262$ |

[1] The first line always has the space group and lattice parameter(s).

$Ca_3Al_2Si_3O_{12}$ $Ia\bar{3}d$, $a$ = 11.846 Å
Ca in 24 $c$: $I\pm$ (1/8,0,1/4 ; 5/8,0,1/4)κ
Al in 16 $a$: $I+$ (0,0,0 ; 1/4,1/4,1/4 ; (0,1/2,1/2 ; 1/4,3/4,3/4)κ))
Si in 24 $d$: $I\pm$ (3/8,0,1/4 ; 7/8,0,1/4)κ
O in 96 $h$: $I\pm$ ($x,y,z$ ; 1/2–$x$,1/2+$y$ ,$z$ ; $x$,1/2–$y$,1/2+$z$ ; 1/2+$x$,$y$ 1/2–$z$
1/4+$x$,1/4+$z$,1/4+$y$ ; 3/4+$x$,1/4–$z$,3/4–$y$ ; 3/4–$x$,3/4+$z$,1/4–$y$ ;
1/4–$x$,3/4–$z$,3/4+$y$)κ, $x$ = –0.0381, $y$ = 0.0449, $z$ = 0.1514

$Ca_3Al_2Si_3O_{12}$ is one of a large group of natural and synthetic materials known as *garnets*. Note that although this is a complex structure with 160 atoms in the unit cell, most of the structural information is contained in the symmetry and only four *numbers* ($a$, $x$, $y$, $z$) are necessary to specify the structure. For the O atoms in the structure we get 96 coordinate triplets. Any one of these could be given as the coordinates of the typical atom. Which to choose? Often the choice is that which corresponds to a minimum of $x^2+y^2+z^2$. In a cubic crystal this is an atom closest to the origin. In the case of $Ca_3Al_2Si_3O_{12}$ all six O atoms closest to an Al atom at the origin (Al has six equidistant O neighbors) have at least one negative coordinate, and the coordinates given above for O are the coordinates of one of them. It might be verified that $x$, $y$, $z$ = 0.0381, 0.0449, 0.6514 are also coordinates of one of the O atoms obtained from the first by $I+$ (1/2–$x$,1/2+$y$,$z$) and could equally have been given (and often are: some authors avoid negative coordinates and use the smallest set of positive coordinates) as the O coordinates to generate the other 95. Remember that you can always add an integer to, or subtract an integer from, any coordinate.

## 3.5 Sub- and super-groups of space groups

A *subgroup* **H** of a group **G** is one that contains some, but not all, of the elements of **G**. Conversely **G** is a *supergroup* of **H**. The number of elements in the supergroup is always an integer ($n$) times that of the subgroup. In the jargon, $n$ is the index of **H** in **G**. If $n$ is prime then **H** is a *maximal* subgroup of **G**.

The *International Tables* give information on subgroups and supergroups of the space groups that is of interest in many contexts. For example a small distortion of a symmetrical crystal structure may result in a lower symmetry which is a subgroup of the parent structure. Knowing the possible symmetries can be an invaluable aid to determining the structure.

The full symbol for *Pnma* is $P2_1/n2_1/m2_1/a$ and the point group is $2/m2/m2/m$ with order 8. Systematically removing half of the symmetry elements will result in a space group with point group of order 4 that is a subgroup of $2/m2/m2/m$ (i.e. $mm2$, 222 or $2/m$). There are eight possibilities which are given in the *Tables*. We give first the symbol for the axes labeled as for *Pnma* and in parentheses the symbol in the standard setting.

These are the maximal subgroups of *Pnma*. In this instance each point group has an order (4) half that of the parent so the index is 2. Lower order subgroups are subgroups of the maximal subgroups and so on until finally $P1$, which is a subgroup of every space group, is reached.

| | |
|---|---|
| $Pnm2_1$ | $(Pmn2_1)$ |
| $Pn2_1a$ | $(Pna2_1)$ |
| $P2_1ma$ | $(Pmc2_1)$ |
| $P2_12_12_1$ | $(P2_12_12_1)$ |
| $P112_1/a$ | $(P2_1/c)$ |
| $P12_1/m1$ | $(P2_1/m)$ |
| $P2_1/n11$ | $(P2_1/c)$ |

In the above example, the subgroups retain the same lattice translation symmetry; such subgroups are called *translationengleiche* ("translationally equivalent"—*gleich* means "equal" or "same" in German) or *t* subgroups. $t_n$ is used to denote a translationengleiche subgroup of index *n*.

Another kind of subgroup is possible for space groups with centered lattices in which the centering is lost but the point group (crystal class) remains the same—these subgroups are called *klassengleiche* ("same class") or *k* subgroups. Thus for *Cmcm* we find listed as maximal subgroups, eight *t* subgroups with a *C* lattice and eight *k* subgroups with a *P* lattice. An example of the latter is *Pbnm* (standard setting *Pnma*). That *Pbnm* is a subgroup of *Cmcm* is by no means obvious unless it is known that *Cmcm* has *b* glide planes normal to **a**, and *n* glide planes normal to **b** (and so could also be written *Cbnm* as explained in § 3.3.3, p. 72). Providing such information is yet another invaluable service provided by the *International Tables*. $k_n$ is used to denote a klassengleiche subgroup of index *n*.

The reader is urged to use Fig. 3.10 (p. 69) to see that the $t_2$ subgroups of *C*2/*m* are *C*2, *Cm* and $C\bar{1}$. In the last case the symmetry is triclinic and with the conventional (primitive) cell the symbol is $P\bar{1}$.

Similarly it should be apparent that the $k_2$ subgroups of *C*2/*m* are *P*2/*m*, $P2_1/m$ (*C*2/*m* contains $2_1$ axes), *P*2/*a* (*C*2/*m* contains *a* glide), and $P2_1/a$. In the last two cases interchanging the labels of the **a** and **c** axes gives the "standard" symbols *P*2/*c* and $P2_1/c$ respectively.[1]

A special kind of *k* subgroup is an *isomorphic* subgroup (or *i* subgroup for short). This is one that has the same space group as its parent, but because some of the translational symmetry has been lost, the unit cell is enlarged. An example is given below.

In Chapter 6 we discuss how symmetrical, low-density sphere packings can be smoothly distorted into denser, lower-symmetry forms. The space group of the latter will be a subgroup of that of the low-density parent. An example is the distortion of an 11-coordinated sphere packing (described in § 6.3.1) with symmetry $P4_2/mnm$ (full symbol $P4_2/m2_1/n2/m$) to a denser version with symmetry *Pnnm*. The *International Tables* lists as a *t* subgroup of $P4_2/m2_1/n2/m$ the non-standard space group $P2/m2_1/n1$. This symbol is to be interpreted *as if* it referred to tetragonal symmetry (the symbol is called the "tetragonal version"): thus parallel to *z* we have 2/*m*, and parallel to *x* and *y* we have $2_1/n$. We see that in fact there are 2-fold axes along *x*, *y* and *z* and the symmetry is

[1]For a detailed discussion of the subgroups of *C*2/*m* (including isomorphic subgroups with a doubled **c** axis) see § 7.13 of *Geometrical and Structural Crystallography* by J. V. Smith (Book List).

actually orthorhombic. The "orthorhombic version" of this space group symbol is $P2_1/n2_1/n2/m$ (short symbol *Pnnm*). The symmetry of the distorted structure is thus shown to be indeed a subgroup of the parent structure.

The symmetries of the structures of rutile ($TiO_2$) and $CaCl_2$ are the same as in the above example, and indeed the anion arrangements are related in much the same way as the two sphere packings.

Another structure related to that of rutile is the *trirutile* structure such as that of tapiolite, $Ta_2FeO_6$. In this compound the Ta and Fe atoms order in layers along **c** in a sequence FeTaTa..., compared with TiTiTi... in rutile, thus tripling $c$. The space group remains the same so this is an example of ordering producing an isomorphic subgroup. The space group of trirutile is a subgroup of that of rutile because 2/3 of the translations (along **c**) have been lost; accordingly the index of the subgroup is 3 (it is an $i_3$ subgroup).

Another example of a parent structure being reduced in symmetry is provided by the structure of sodalite, $Na_4Al_3Si_3O_{12}Cl$. We consider just the $(Al,Si)O_2$ part of the structure which consists of a framework of $\{Al,Si\}O_4$ tetrahedra sharing corners, and for the moment we consider the (Al,Si) atoms to be disordered in a random manner. In its most symmetrical arrangement the O packing is as described in § 6.8.5 with symmetry $Im\bar{3}m$, but as discussed in that section, can distort to a denser arrangement with symmetry $I\bar{4}3m$ which is a $t_2$ subgroup of $Im\bar{3}m$. In both cases the tetrahedron centers (disordered Si,Al) are at the same positions: $I \pm (1/4,0,1/2)\kappa$. Ordering of the Si and Al atoms into two sets $\pm(1/4,0,1/2)\kappa$ for one atom and $\pm(1/4,1/2,0)\kappa$ for the other removes the body centering and reduces the symmetry further to $P\bar{4}3n$: a $k_2$ subgroup of $I\bar{4}3m$.

A phase transition in $KH_2PO_4$ has been intensively studied.[1] The room temperature phase (with disordered H atoms) has symmetry $I\bar{4}2d$ and the low-temperature phase (in which the H atoms are ordered) has symmetry *Fdd*2. If the high temperature phase is described by a doubled cell as shown in Fig. 3.15, the space group symbol becomes $F\bar{4}d2$ and the unit cells of both phases have very nearly equal dimensions.[2] The H ordering reduces the $\bar{4}$ axis to a 2 axis and removes the 2 axes normal to **c**. The space group symbol written as if the system were still tetragonal (it is actually orthorhombic) is now *F*2*d*1 ("tetragonal version") which becomes *Fdd*2 ("orthorhombic version"). Note that both the last two symbols imply that the 2-fold axis is parallel to **c**. Note also that the translations of $I\bar{4}2d$ are retained in *Fdd*2 so the latter is a translationengleiche subgroup of the former. For comments on coordinates in these structures see § 3.7.4.

There is something apparently paradoxical in the observation that ordering reduces symmetry and disordering increases it. One's intuitive idea is surely that high symmetry is associated with "orderly" patterns. The solution to the paradox lies in recognizing that the

[1]The interest in this material stems in part from the fact that at low temperatures it has a spontaneous polarization which can be reversed by an electric field (*i.e.* it is *ferroelectric*). The elucidation of the role of ordering of H atoms in producing the ferroelectric phase was one of the early big successes of neutron crystallography [G. E. Bacon & R. S. Pease, *Proc. Roy. Soc. (London), Ser A* **220**, 397 (1953) and **230**, 359 (1955)].

[2]For the room temperature phase with the $F\bar{4}d2$ cell $a$ = 10.54, $c$ = 6.98 Å. At 77 K the *Fdd*2 structure has $a$ = 10.54, $b$ = 10.46, $c$ = 6.92 Å.

symmetry we talk about in crystals is the *average* symmetry as revealed, for example, by diffraction of radiation from a large volume (many unit cells) of the crystal. Thus in disordered sodalite the tetrahedral sites are occupied *on average* by one half Al and one half Si and thus are considered equivalent and related by symmetry.[1] In the ordered version there are two sets of sites, one occupied by Al and the other by Si, and there cannot be a symmetry operation relating the two sets.

There is some interest in knowing which cubic groups have non-intersecting 3-fold axes. These will be the symmetry groups of cubic structures constructed of non-intersecting rods of atoms aligned parallel to the cube body diagonals (§ 6.7.3). In the table below the groups in each column are subgroups of the ones above them. In addition the groups in the center column are subgroups of the ones in the left and right columns of the row above.

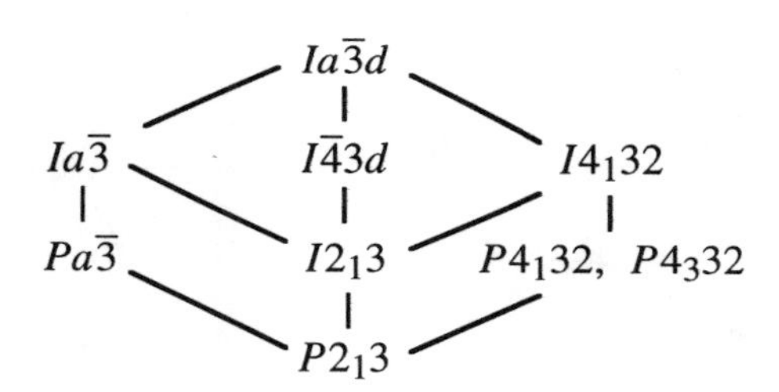


## 3.6 Symmetry and the quartz structure—a case study

The $\alpha$-quartz form of $SiO_2$ is an important mineral (comprising about 15% of the continental crust of the earth), it also has valuable physical properties (for example it is piezoelectric) that are a consequence of its symmetry. The structure is based on a framework of corner-sharing $\{Si\}O_4$ tetrahedra and has occasioned a great deal of discussion, here we just focus on the symmetry aspects. The discussion illustrates several topics: the use of non-standard origins, subgroup-supergroup relationships, twinning, and the relationship between the hand of enantiomorphs and symmetry.

We start with a description of the structure of the high-temperature or $\beta$-quartz form.

$\beta$-quartz — $P6_222$, $a = 5.01$, $c = 5.52$ Å
Si in 3 $c$: (1/2,0,0 ; 0,1/2,2/3 ; 1/2,1/2,1/3)
O in 6 $j$: ($x$,2$x$,1/2 ; 2$\bar{x}$,$\bar{x}$,1/6 ; $x$,$\bar{x}$,5/6 ; $\bar{x}$,2$\bar{x}$,1/2 ; 2$x$,$x$,1/6 ; $\bar{x}$,$x$,5/6), $x = 0.208$

We recognize the symmetry to be hexagonal and the crystal class (622) to be one of the enantiomorphous classes so that quartz exists as left- and right-handed forms (in essentially

[1] Only crystallographers really believe in fractional atoms. In a disordered crystal there may be no symmetry at all on an atomic level (*i.e.* on a *specific* as opposed to *average* site). In particular the microscopic translational symmetry is lost (if a translation is from a Si atom to an Al atom it is not a symmetry operation).

equal amounts as the $\alpha$- form in nature). As discussed below, the form we have described is called right-handed.[1] The left-handed enantiomorph has symmetry $P6_422$. The description of that structure is:

| | |
|---|---|
| $\beta$-quartz | $P6_422$, $a = 5.01$, $c = 5.52$ Å |
| | Si in 3 *c*: (1/2,0,0 ; 0,1/2,1/3 ; 1/2,1/2,2/3) |
| | O in 6 *j*: ($x$,2$x$,1/2 ; 2$\bar{x}$,$\bar{x}$,5/6 ; $x$,$\bar{x}$,1/6 ; $\bar{x}$,2$\bar{x}$,1/2 ; 2$x$,$x$,5/6 ; $\bar{x}$,$x$,1/6), $x = 0.208$ |

The only difference from the right-handed form is that we have replaced all the $z$ coordinates by 1-$z$. This corresponds to reflection in an imaginary mirror plane at $z = 1/2$, so that the two enantiomorphs are related as mirror images as expected.

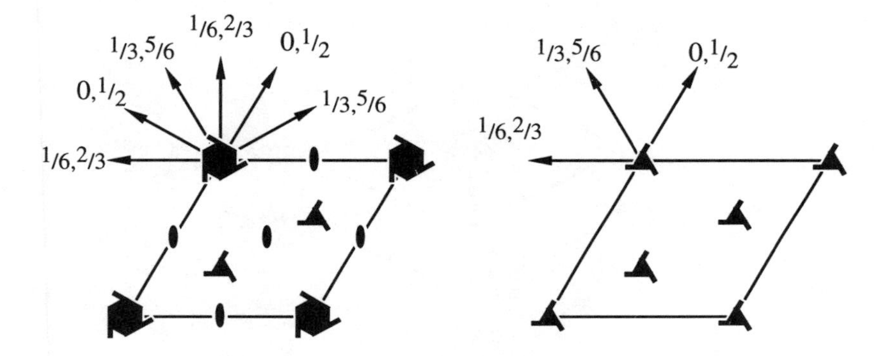


**Fig. 3.17**. The location of some of the symmetry elements of $P6_222$ and $P3_221$. **c** is vertical out of the plane of the paper, and numbers represent elevations in fractions of $c$. The locations of $2_1$ axes perpendicular to **c** are not shown. Note that the origin chosen for $P3_221$ in *not* the same as that in the *International Tables*.

As well as the $6_2$ axis (represented by a hexagon with three arms) at 0,0,$z$, $P6_222$ contains $3_2$ axes (represented by triangles with three arms) at 1/3,2/3,$z$ and 2/3,1/3,$z$ as shown in Fig. 3.17. In the structure of right-handed quartz the {Si}$O_4$ tetrahedra spiral around the $3_2$ axes as shown in Fig. 3.18. We earlier described a $3_2$ axis as "left-handed" and indeed the $P6_222$ (right-handed) enantiomorph of quartz is sometimes described as "structurally left-handed." On the other hand, the $6_2$ axis was described as "right-handed" and if we focus instead on the helix of tetrahedra around that axis we would say that we have a right-handed structural unit; so it is not appropriate to refer to a structural "hand." Why then do we call the $P6_222$ form of $\beta$-quartz "right-handed"? The answer lies in the fact that the hand of a crystal is determined from its optical activity (rotation of the plane of polarization of light); clockwise rotation (when viewed looking towards the light source, i.e. light coming to the viewer) being called right handed (this is the Biot convention) or

[1]It is sometimes said that X-ray diffraction cannot distinguish between enantiomorphs of a crystal. This is not true when anomalous dispersion (absorption) is taken into account, and the absolute configuration of a number of enantiomorphic crystals has been determined by this technique. For an evaluation of the literature see A. M. Glazer & K. Stadnicka, *J. Appl. Crystallogr.* **19**, 108 (1986).

dextrorotatory (+). This is the sense of the rotation of the plane of polarization of light when passing through a slice of $P6_222$ quartz in a direction parallel to **c**. (Recall though that in a right-handed *screw* the rotation is anticlockwise when the translation is in a direction towards the viewer.) The opposite of "dextrorotatory" is "levorotatory" (–).[1]

Below about 573 °C, the $\beta$ form transforms to $\alpha$-quartz, the main difference in the low-temperature structure (Fig. 3.18) being that the Si-O-Si angle is decreased by rotating the {Si}$O_4$ tetrahedra in a concerted way about 2-fold axes perpendicular to **c**. In the $P6_222$ enantiomorph this destroys the 2-fold rotation axis contained in the $6_2$-axis at 0,0,$z$ and the latter is degraded to a $3_2$ axis. The symmetry of the crystal becomes $P3_221$. Reference to Fig. 3.17 shows that the $3_2$ axes at 1/3,2/3,$z$ and 2/3,1/3,$z$ remain. $P3_221$ is a $t_2$ subgroup of $P6_222$. A description of right-handed $\alpha$-quartz often seen is (note that we do not give the Wyckoff symbols for the positions for reasons to become apparent later):

$\alpha$-quartz

$P3_221$, $a = 4.92$, $c = 5.41$ Å
Si at ($x$,0,0 ; 0,$x$,2/3 ; $\bar{x}$,$\bar{x}$,1/3), $x = 0.470$
O at ($x,y,z$ ; $y-x,\bar{x}$,1/3+$z$ ; $\bar{y},x-y$,2/3+$z$ ; $x-y,\bar{y},\bar{z}$ ; $y,x$,2/3−$z$ ; $\bar{x},y-x$,1/3−$z$)
$x = 0.415$, $y = 0.266$, $z = 0.119$

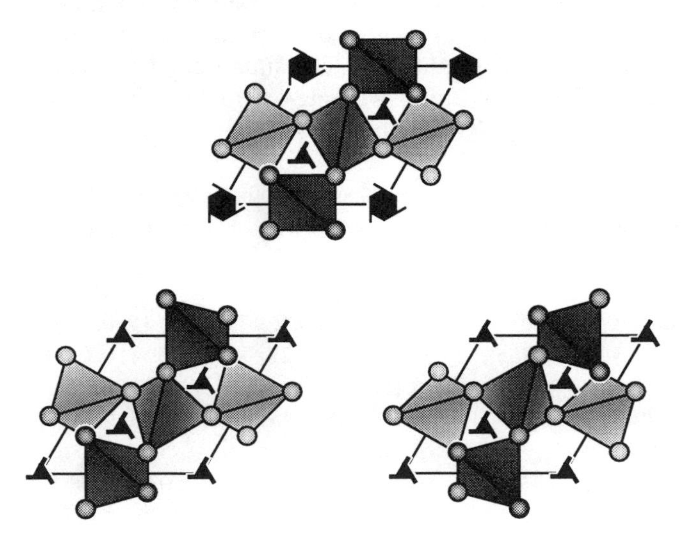

**Fig. 3.18**. Top: the oxygen atom positions in $\beta$-quartz ($P6_222$). Bottom: on the left oxygen atom positions in $\alpha_1$ and on the right $\alpha_2$ twins of $\alpha$-quartz ($P3_221$). The projection is along **c** and $O_4$ tetrahedra are outlined. Silicon atoms (not shown) center the tetrahedra. In order of increasing depth of shading the tetrahedron centers are at $z$ = 0, 1/3 and 2/3.

[1]See the discussion by Glazer & Stadnika (*op. cit.* previous footnote). The sign of the optical activity associated with a helix of polarizable atoms depends on the orientation of the axis of maximum polarizability with respect to the helix. If the polarizability of the atoms were to be isotropic there would be no optical activity. For an account of the confusing history of the description of the quartz structure see P. J. Heaney in *Silica* [*Reviews in Mineralogy* **29** (1994), p. 1-40]. According to Heaney dextrorotary quartz is morphologically right-handed (*i.e.* a macroscopic crystal of dextrorotary quartz is right handed).

It should be apparent that if $x_{Si}$ = 0.5 in $\alpha$-quartz, the silicon atoms in the $\alpha$- and $\beta$-forms would then be in the same positions. In Table 3.5 below, the oxygen atom positions (rounded off for simplicity) in $\beta$-quartz are listed in the first column, and those given above for $\alpha$-quartz are listed in the second column. These differ only slightly. The oxygen positions in the two structures would be exactly the same if $x,y,z$ in the $P3_221$ form were 0.416,0.208,0.167.

Rotating the $\alpha$ structure (by 180°) around a 2-fold axis (now nonexistent in the crystal) along 0,0,$z$ will change the sign of the $x$ and $y$ coordinates to give the coordinates in the third column of the table. They correspond to substituting $x$ = 0.585, $y$ = 0.784, $z$ = 0.119 for the oxygen positions given above for $P3_221$. We have of course exactly the same structure, but in a different orientation.

As $\beta$-quartz is cooled to transform it to the $\alpha$ form, these two orientations (which we call $\alpha_1$ and $\alpha_2$) nucleate at random points in the crystal. In the region of the interface where domains of $\alpha_1$ and $\alpha_2$ eventually meet, presumably there is an average structure, which is just a stranded element of $\beta$. When a crystal is composed of two parts related by a symmetry operation that is not a part of the structure (in this case a 2-fold rotation axis along 0,0,$z$), we say that it is composed of *twins* and the phenomenon is known as *twinning*. Quartz can be twinned in a number of different ways, the example we have just discussed is known as Dauphiné twinning. The situation where left- and right-handed forms are intergrown is known as Brazil twinning—here the twins are related by reflection (which again is not a symmetry element of the structure).

**Table 3.5.** Oxygen atom positions ($x,y,z$) in right-handed quartz.
Replacing all the $z$ coordinates by $1-z$ will give coordinates for left-handed quartz.

| $\beta$ | $\alpha_1$ | $\alpha_2$ |
|---|---|---|
| 0.21, 0.42, 0.50 | 0.27, 0.42, 0.55 | 0.15, 0.42, 0.45 |
| 0.58, 0.79, 0.17 | 0.58, 0.85, 0.21 | 0.58, 0.73, 0.12 |
| 0.21, 0.79, 0.83 | 0.15, 0.73, 0.88 | 0.27, 0.85, 0.79 |
| 0.79, 0.58, 0.50 | 0.85, 0.58, 0.45 | 0.73, 0.58, 0.55 |
| 0.42, 0.21, 0.17 | 0.42, 0.27, 0.12 | 0.42, 0.15, 0.21 |
| 0.79, 0.21, 0.83 | 0.73, 0.15, 0.79 | 0.85, 0.27, 0.88 |

We haven't finished our story. For inscrutable reasons the origin of coordinates in $P3_221$ is chosen differently in the *International Tables*—it is displaced by 0,0,1/3 from the position used above to describe $\alpha$-quartz. Thus the 2-fold axis in the $x$ direction runs along $x$,0,0 in $P6_222$ but, with the origin chosen for $P3_221$ in the *International Tables* the same axis runs along $x$,0,-1/3. The "non-standard" origin is almost invariably chosen by quartz aficionados (who are legion) as it makes the relation between the $\alpha$- and $\beta$-forms transparent. Unfortunately they do not always make this clear to outsiders.

In the *International Tables* the relevant positions in $P3_221$ are given as:

Si in 3 *a*: ($x$,0,2/3 ; 0,$x$,1/3 ; $\bar{x},\bar{x}$,0)
O in 6 *c*: ($x,y,z$ ; $\bar{y},x-y,z$+2/3 ; $y-x,\bar{x},z$+1/3 ; $y,x,\bar{z}$ ; $x-y,\bar{y}$,1/3$-z$ ; $\bar{x},y-x$,2/3$-z$)

To describe right-handed $\alpha$-quartz using this origin we can use the same values of $x$ and $y$ and subtract 1/3 from (or add 2/3 to) the value of $z$ given above (0.119). Specifically for $\alpha_1$ we have for Si: $x = 0.470$ and for O: $x = 0.415$, $y = 0.266$, $z = 0.786$. For $\alpha_2$ we have for Si: $x = 0.530$ and for O: $x = 0.585$, $y = 0.734$, $z = 0.786$.

To describe left-handed $\alpha$-quartz using the origin given in the *International Tables* for $P3_121$ use the coordinates in the previous paragraph but with $z$ replaced by $1-z$.

Many framework silicates have analogs in which 2Si are replaced, in an ordered fashion, by Al and P. The analog of quartz is the berlinite form of $AlPO_4$ in which Al and P alternate along the three-fold helices of the quartz structure. If the helices in the parent structure were $3_2$ the symmetry along the helix of ordered atoms becomes $3_1$ (with a doubled $c$) as illustrated in Fig. 3.19.

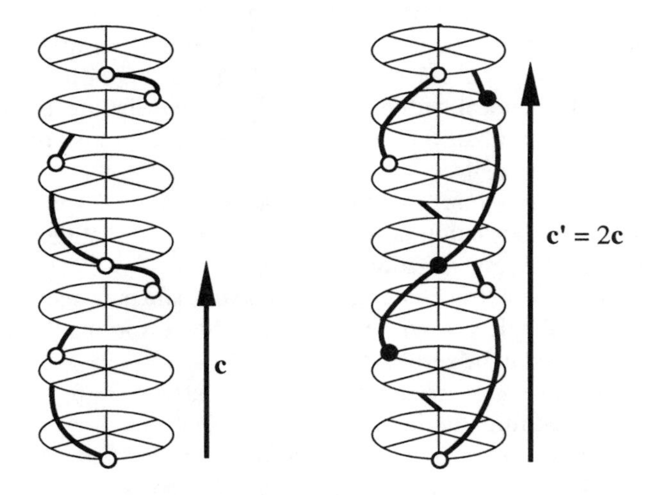


**Fig. 3.19**. Ordering on a $3_2$ helix (shown on the left) produces a structure with a $3_1$ axis (on the right) and with doubled repeat distance.

In right-handed $\alpha$-quartz the silicon atoms are in positions 3 $a$ of $P3_221$: $\bar{x},\bar{x},0$; $0,x,1/3$; $x,0,2/3$ with $x = 0.47$. If the **c** axis repeat is doubled these positions become:

| $\bar{x},\bar{x},0$ | $0,x,1/6$ | $x,0,1/3$ | $\bar{x},\bar{x},1/2$ | $0,x,2/3$ | $x,0,5/6$ |
|---|---|---|---|---|---|
| Si | Si | Si | Si | Si | Si |
| Al | P | Al | P | Al | P |

The Al positions are in fact 3 $a$ of $P3_121$ and the P positions are 3 $b$ of $P3_121$. In $AlPO_4$ $x_{Al} \approx x_P \approx 0.46$.

$P3_121$ is seen to be an isomorphic subgroup of index 2 of $P3_221$ (and *vice versa*). Note that as $P3_221$ quartz is optically right handed, $P3_121$ berlinite is also right handed.

One of the major uses of quartz is as a piezoelectric oscillator in watches and other timing devices. It would be ironic if digital clocks using quartz oscillators entirely replaced traditional clocks with their analog display of time, so that future generations lost sight of the meaning of the term "clockwise."[1]

## 3.7 Notes

### *3.7.1 Additional symmetry elements in the unit cell*

Let there be an inversion center at $x_0,y_0,z_0$. A point at $x,y,z$ will be transformed to $2x_0-x,2y_0-y,2z_0-z$ by the inversion operation. Inversion centers with coordinates obtained by adding 1/2 to $x_0$ and/or $y_0$ and/or $z_0$ will produce points with the same coordinates with 1 added on to the new values of $x$ and/or $y$ and/or $z$ (i.e. in adjacent unit cells). In particular if there is an inversion center at 0,0,0 there will also be centers at 1/2,0,0 ; 0,1/2,0 ; 0,0,1/2 ; 0,1/2,1/2 ; 1/2,0,1/2 ; 1/2,1/2,0 ; 1/2,1/2,1/2. These operating on a given point will always produce points identical with respect to a lattice translation. Thus the spacing of inversion centers is always one-half that of the *primitive* lattice translations and there are always exactly 8 per *primitive* cell.

Similar reasoning shows that mirror and glide planes are always spaced at intervals of one half the shortest lattice translation normal to the mirror and that the spacing between 2-fold axes is one-half of a lattice translation vector.

When elevations are given for centers, 2-fold axes, and symmetry planes in the *International Tables*, 1/2 should always be added to these elevations. Note that when no elevation is given it is implied to be zero and 1/2.

Trigonal and hexagonal space groups will always have a 3- or 6-fold axis along $1/3,2/3,z$ and $2/3,1/3,z$ in addition to the one at $0,0,z$.

In tetragonal space groups there are always 4-fold axes of the same kind (4, $4_1$, $\bar{4}$ etc.) with axes displaced by 1/2,1/2,0. Be aware that the origin is not always chosen on a 4-fold axis in tetragonal space groups.

### *3.7.2 General and special positions and lattice complexes*

General positions are points at which there is no symmetry. The coordinates of general positions completely specify the space group. If $\Gamma$ is a point group then the general positions of the space group $P\Gamma$ also provide the coordinate symbols for the point group operations. The number of general positions in the primitive cell is the order of the point group. The maximum number of general positions will therefore be $4 \times 48 = 192$ for a

[1]Peizoelectricity is the production of electric polarization by an applied mechanical stress. It can occur for all non-centrosymmetric crystal classes other than 432. In class 321 (that of $\alpha$-quartz) the **a** and **b** directions are polar and reversing their directions reverses the sign of the piezoelectric coefficents. Recall that the forms that we have called $\alpha_1$ and $\alpha_2$ are related in just this way (replacing $x$ and $y$ in one by $\bar{x}$ and $\bar{y}$ in the other) so that a crystal composed of equal amounts of each form would not be piezoelectric. To exploit the piezoelectric properties of quartz one must therefore avoid having both forms present (i.e. Dauphiné twinning—often called electrical twinning in this context). By convention the piezoelectric coefficients are referred to the $\alpha_1$ form.

face-centered cubic cell in class $m\bar{3}m$.

Points lying on mirror planes, but otherwise unrestricted, will have two degrees of freedom (will be *bivariant*) and will thus have coordinates of the sort $x,y,z_0$, where $z_0$ is 0 or a simple fraction (the mirror plane is $z = z_0$), or $x,x,z$ (the mirror plane is $x = y$), etc. The site symmetry at such a point will be $m$ and there will be only one-half as many such points as general positions.

Points lying on rotation axes will have only one degree of freedom (*univariant*) and will thus have coordinates such as $x$,0,0 or $x,x,x$, etc. There will be at most one-half as many as there are general positions, but there may be less if the axis also coincides with the intersections of mirror planes and/or the order of the rotation axis is greater than 2. The site symmetry will now be $N$ or $Nm(m)$. As univariant points lie on a rotation axis the coordinates can be construed as specifying the line of intersection of two planes. Thus $x$,0,0 is the line of intersection of the planes $y = 0$ and $z = 0$; likewise $x,x,x$ represents the intersection of the planes $x = y$ and $y = z$.

Points lying on centers, or at the intersections of rotation axes, or the intersection of a rotation axis and a mirror, will have no free parameters (*invariant*). The site symmetry at such points will be one of the crystallographic point groups (it must of course be the same as, or a subgroup of, the crystal point group).

The fact that a point lies on a glide plane or a screw axis does not on the other hand reduce the number of points as the translational component will always take a given point to a new position. Thus (to take a random example) there are no special positions for $Pca2_1$ because it contains no point operations (those that leave at least one point invariant).

Sets of symmetry-related points are sometimes called lattice complexes (but the term is sometimes used differently, so beware).

There is special interest in the cubic invariant lattice complexes which crop up in many different contexts, and they have been given symbols in the *International Tables* although these have not yet achieved wide currency (they are given in § 6.8.7). As an example, the positions 8 $a$ of $Fd\bar{3}m$ (symbol $D$) are the positions of the atoms in the **diamond** form of carbon, silicon etc. as well as in a large number of compounds (Mg in $MgAl_2O_4$ cited above). The positions 16 $d$ of the same space group (symbol $T$), which are the Al positions in $MgAl_2O_4$, are also the Cu positions in the important structure type $MgCu_2$ (§ 6.6.3). Many other examples of the recurrence of common motifs will be adduced in later chapters.

### *3.7.3 Matrix representations of symmetry operations*[1]

A symmetry operation acting on a point $x,y,z$ produces another point at $x',y',z'$. Using the conventional coordinate systems, the set of new points in a unit cell resulting from all the symmetry operations in a space group are given as the general positions in the *International Tables*. Let **x** be the column vector $(x\,/\,y\,/\,z)$ and **x'** be the column vector $(x'\,/\,y'/\,z')$, then **x'** = **Ax** + **t** where **A** is the matrix corresponding to a point symmetry operation about the origin and **t** is another column vector $(t_x\,/\,t_y\,/\,t_z)$. i.e.:

[1]This note is not for those who are uneasy with matrix manipulation; as we do not use the results later it may be omitted. See § 4.6.5 for some help with matrices.

$$\begin{pmatrix} x' \\ y' \\ z' \end{pmatrix} = \begin{pmatrix} a_{11} & a_{12} & a_{13} \\ a_{21} & a_{22} & a_{23} \\ a_{31} & a_{32} & a_{33} \end{pmatrix} \begin{pmatrix} x \\ y \\ z \end{pmatrix} + \begin{pmatrix} t_x \\ t_y \\ t_z \end{pmatrix}$$

This equation is sometimes expressed in terms of a four-dimensional matrix:

$$\begin{pmatrix} x' \\ y' \\ z' \\ 1 \end{pmatrix} = \begin{pmatrix} a_{11} & a_{12} & a_{13} & t_x \\ a_{21} & a_{22} & a_{23} & t_y \\ a_{31} & a_{32} & a_{33} & t_z \\ 0 & 0 & 0 & 1 \end{pmatrix} \begin{pmatrix} x \\ y \\ z \\ 1 \end{pmatrix}$$

A compact symbol for the above matrix is $\{\mathbf{A}|\mathbf{t}\}$ such that $\mathbf{x}' = \{\mathbf{A}|\mathbf{t}\}\mathbf{x}$. $\{\mathbf{A}|\mathbf{t}\}$ is known as the *Seitz Operator*.

With conventional choices of bases all the elements of **A** are ±1 or 0. The nature of the symmetry operation can be found as follows [H. Wondratschek & J. Neubüser, *Acta Crystallogr.* **23**, 349-352 (1967)]: Let $\tau$ be the trace (sum of the diagonal elements) of **A** and $\Delta$ the determinant of **A**, then the point symmetry element is determined as shown below:

| $\tau\Delta$ = | 3 | −1 | 0 | 1 | 2 |
|---|---|---|---|---|---|
| $\Delta = 1$ | 1 | 2 | 3 | 4 | 6 |
| $\Delta = -1$ | $\bar{1}$ | $m$ | $\bar{3}$ | $\bar{4}$ | $\bar{6}$ |

The space group operation is identified as follows:

If $\Delta = 1$ there is an $N$-fold rotation or screw axis along axis **q** given by the solutions of

$$(\mathbf{A} - \mathbf{E})\mathbf{q} = 0 \quad (\mathbf{E} \text{ is the identity matrix})$$

with translation **d** (parallel to **q**) given by

$$\mathbf{d} = (1/N)\mathbf{B}\mathbf{t} \quad (\mathbf{B} = \mathbf{A}^{N-1} + \mathbf{A}^{N-2} + \ldots + \mathbf{E})$$

points **x** on the axis are the solutions of

$$(\mathbf{A} - \mathbf{E})\mathbf{x} = (\mathbf{B}/N - \mathbf{E})\mathbf{t}$$

If $\Delta = -1$ then there is:

(a) $\tau\Delta = 3$ an inversion center.

(b) $\tau\Delta = -1$ a mirror or glide with translation **d** given by

$$\mathbf{d} = (1/2)(\mathbf{A} - \mathbf{E})\mathbf{t}$$

and the points **x** on the plane are the solutions of

$$(\mathbf{A} - \mathbf{E})\mathbf{x} = (1/2)(\mathbf{A} - \mathbf{E})\mathbf{t}$$

(c) $\tau\Delta = 0$, 1 or 2 there is a $\bar{3}$, $\bar{4}$ or $\bar{6}$ axis along **q** given by

$$(\mathbf{A} + \mathbf{E})\mathbf{q} = 0$$

and the inversion point **x** is given by the solution of:

$$(\mathbf{A} - \mathbf{E})\mathbf{x} = -\mathbf{t}$$

It will be found that working through one or two examples (see *e.g.* Exercise 2) will make the above clear.

The inverse problem of finding the matrix corresponding to a given space group operation is easily solved from the above equations. For a clear and thorough exposition see M. B. Boisen & G. V. Gibbs, *Canadian Mineralogist*, **16**, 293 (1978).

### *3.7.4 Alternative origins and unit cells*

There are advantages to taking the origin at a center of symmetry in space groups that have them. Computer programs that generate space group coordinates often do it for a centric space group by adding a center to an acentric space group by also reversing the sign of all the coordinates (in getting *Pmmm* from *P*222 for example). They then require that you use such an origin. Properties such as the electron density distribution [$\rho(x,y,z)$] of a centrosymmetric crystal become an even function [$\rho(x,y,z) = \rho(-x,-y,-z)$] if the origin is taken at a center. Expression of $\rho(x,y,z)$ as a Fourier series requires only cosine terms in this case.

Why then ever take an alternative origin? Consider the diamond structure. The symmetry is $Fd\bar{3}m$ and with origin at the center (half way between two neighboring C atoms, site symmetry $\bar{3}m$) the C atom positions are 8 *a*: $F \pm (1/8,1/8,1/8)$; however if the origin is taken on an atom (with site symmetry $\bar{4}3m$) the coordinates are (again 8 *a*) $F$ + (0,0,0 ; 1/4,1/4,1/4). Now consider the cubic form of SiC. The structure is derived from that of diamond by replacing half the C atoms by Si destroying the inversion center and the symmetry is now $F\bar{4}3m$ (an acentric subgroup of $Fd\bar{3}m$). The point that was the old inversion center now has symmetry $3m$ and the natural origin is the site of $\bar{4}3m$ symmetry where the atoms are located, and which is chosen as the standard origin for $F\bar{4}3m$. The description of the structure is: Si in 4 *a*: $F$ + (0,0,0) and C in 4 *c*: $F$ + (1/4,1/4,1/4). The relationships between the two structures are more readily apparent if the origin is chosen at the site with $\bar{4}3m$ symmetry in both cases.

When alternative origins are given, the first choice used in the *International Tables* is at $x_0,y_0,z_0$ from an inversion center (the second choice is on the center). To convert coordinates appropriate for the first choice to those appropriate for the second choice,

$x_0,y_0,z_0$ must be added to the first set. Values of $x_0,y_0,z_0$ are given in Table 3.6 below.
See § 3.6 (p. 86) for choice of origin in $P3_121$ and $P3_221$.

**Table 3.6**. Coordinates $x_0,y_0,z_0$ to add to convert from origin "choice 1" to "choice 2" in the *International Tables*

| | |
|---|---|
| *Pnnn* | 1/4,1/4,1/4 |
| *Pban, Pmmn* | 1/4,1/4,0 |
| *Ccca* | 0,1/4,1/4 |
| *Fddd*, $Fd\bar{3}$, $Fd\bar{3}m$ | –1/8,–1/8,–1/8 |
| *P4/n, P4/nmm, P4/ncc* | –1/4,1/4,0 |
| $P4_2/n$, *P4/nnc*, $Pn\bar{3}$, $Pn\bar{3}n$, $Pn\bar{3}m$ | –1/4,–1/4,–1/4 |
| $I4_1/a$ | 0,–1/4,–1/8 |
| $I4_1/amd$, $I4_1/acd$ | 0,1/4,–1/8 |
| *P4/nbm* | –1/4,–1/4,0 |
| $P4_2/nbc$, $P4_2/nnm$, $P4_2/nmc$, $P4_2/ncm$ | –1/4,1/4,–1/4 |
| $Fd\bar{3}c$ | –3/8,–3/8,–3/8 |

A more subtle (and maddening) cause of confusion can arise in some polar space groups (i.e. those belonging to one of the polar crystal classes—see § 2.5.4). The general positions of *Fdd*2 (class *mm*2) are given in the *International Tables* as:

$$F + (x,y,z\ ;\ \bar{x},\bar{y},z\ ;\ 1/4+x,1/4-y,1/4+z\ ;\ 1/4-x,1/4+y,1/4+z)$$

The second position is generated from the first by rotation about the 2-fold axis parallel to **c** and passing through the origin. The next two positions are generated from the first two by diamond glide: reflection in the plane $y = 1/8$ followed by translation by **a**/4+**c**/4.

One of the most studied of all crystals with this symmetry is the low-temperature phase of $KH_2PO_4$. For reasons of their own, people who work with this material reverse the **c** component of the glide (so that it is **a**/4 – **c**/4) and the general positions become (with –1/4 replaced by 3/4):

$$F + (x,y,z\ ;\ \bar{x},\bar{y},z\ ;\ 1/4+x,1/4-y,3/4+z\ ;\ 1/4-x,1/4+y,3/4+z)$$

To use their coordinates with the positions given in the *International Tables* the sign of all values of $z$ must be reversed This is simple enough to do if you are told, but compilations of structural data often neglect to make clear the choice of direction of $z$. The same problem can arise in $I4_1md$ and $I4_1cd$ which are polar supergroups of *Fdd*2.

The high-temperature phase of $KH_2PO_4$ is often described as having space group $F\bar{4}d2$ which is $I\bar{4}2d$ using with a doubled cell (see Fig. 3.15).[1] Your computer program quite possibly does not recognize $F\bar{4}d2$ either! We explain in Chapter 4 how to transform coordinates when a unit cell is transformed. To transform coordinates $x,y,z$ for a face-centered tetragonal cell to $x',y',z'$ for a body-centered tetragonal cell use $x' = x-y$, $y' =$

[1]The reason for this is that the unit cells and coordinates of both phases are then very nearly the same. See § 3.5.

$x+y$, $z' = z$. You may still have to be careful about choice of origin!

Occasionally a primitive tetragonal crystal is described by a similar doubled cell which is now *C* centered. The same rule applies.

### *3.7.5 Standardized description of crystal structures*

Frequently crystal structures are determined and not recognized as the same as one previously determined for a different compound (for an example see Exercise 11). Among the reasons for this are choices of different settings of axes and choice of different origins as described in § 3.7.4, or (less frequently) assignment of a wrong space group.

Even if the "standard" settings of the *International Tables* are chosen there can still be freedom of choice of axes; for example space group *Pmmm* has the same symbol for any permutation of the axes. There can also be freedom of choice in assigning atoms to Wyckoff positions: for example in the structure of PtS with symmetry $P4_2/mmc$, the Pt atoms occupy the positions 2 *c*. The S atoms can be placed in either 2 *e* or 2 *f* resulting in identical structures but described with different origins (in both cases on a center).

With an increasing number of structures being published and entered into computer databases, it is important that they be reported in as "standard" as possible a way. Sensible proposals for this and a computer program to implement them have been described [E. Parthé & L. M. Gelato, *Acta Crystallogr.* A**40**, 169, (1984); L. M. Gelato & E. Parthé, *J. Appl. Crystallogr.* **20**, 139 (1987)]. Unfortunately relationships between structures of different symmetries are sometimes obscured if standard descriptions are used. In those cases it would seem appropriate to use two descriptions.

To illustrate that this is not a trivial problem for non-crystallographers we recount an anecdote about the crystal structure of zircon ($ZrSiO_4$). Two structure determinations in essential agreement were reported, but with different choices of origin (see Exercise 10). A structure simulation was subsequently carried out (by a third party) to determine which was correct! The reader is urged to do Exercise 11 (taken from Parthé & Gelato) which provides a more subtle example.

### *3.7.6 Other symmetry groups with translations*

The enumeration of the classical space groups was performed 100 years ago. Most of the credit goes to E. S. Fedorov (Russia) and A. M. Schoenflies (Germany) who worked independently, but it should be noted that the final correct tally was only achieved by cross-checking each other's work. We should also mention in this connection the remarkable work at the same time of W. Barlow (England) who a (possibly apocryphal) tale has arranging hundreds of gloves (asymmetric objects!) on racks to simulate the space groups.[1]

The symmetry groups of three-dimensional objects with only two-dimensional

[1]Barlow (who has been called "one of the last great amateurs of science") also contributed greatly to our knowledge of sphere packings and invented some simple crystal structures, such as those of NaCl and CsCl. This last work was just what W. H. & W. L. Bragg needed for their early investigations of crystal structures (that of NaCl was the first determined). The proof in § 1.4 is attributed to Barlow.

periodicity (*layers*) or one-dimensional periodicity (*rods*) are sometimes of interest. The restriction to 1-, 2-, 3-, 4- or 6-fold rotations does not apply with one-dimensional periodicity, so the rod groups subject to that restriction are called *crystallographic* (as for the point groups). As the crystallographic one-dimensional and two-dimensional groups are also subgroups of space groups, information about them is implicit in the *International Tables*. Appendix 1 lists these groups and gives some hints for using the *Tables* to obtain the general and special positions for these groups and to locate their symmetry elements. Layer groups are of special interest to electron microscopy and we use the rod groups in the discussion of cylinder packings in Chapter 6.

### 3.7.7 The occurrence of symmetry groups

Experienced crystallographers often have a knack of knowing what kind of structure to expect once they know the symmetry and unit cell parameters. It is useful therefore to have some ideas about the factors determining the occurrence of symmetry groups.

Organic and many inorganic *molecules* are often assembled with great skill piecewise and are generally of low symmetry (and metastable). The crystals they form have symmetries suitable for efficient packing of such molecules; generally this requires the absence of pure rotation axes and mirror planes, but allows a low density of screw axes and glide planes. Thus in a survey[1] of about 40,000 organic crystal structures it was observed that about one third have symmetry $P2_1/c$ but there was not a single example of a structure with symmetry $P2/m$. Only 64 examples of cubic symmetry were found and almost a third of these had symmetry $Pa\bar{3}$ which is a favorable symmetry for packing of molecules [like $CO_2$ and "congressane" (§ 5.1.10)] with large quadrupole moments. The same symmetry is found for pyrite, $FeS_2$, which contains $S_2$ groups.

The inorganic crystals with which we are largely concerned in this book are very often the stable configurations (at a given temperature and pressure) of combinations of atoms in predetermined ratios and generally have higher symmetries. *Pearson's Handbook* (see Book List) contains about 50,000 inorganic structures and about 2500 structure *types*. The distribution of symmetries among the latter and the more common space groups (in percent) are:

| | | | | | |
|---|---|---|---|---|---|
| triclinic | 2.8 | | | | |
| monoclinic | 20.1 | $C2/m$ | 6.1 | $P2_1/c$ | 5.5 |
| orthorhombic | 29.7 | $Pnma$ | 6.1 | $Cmcm$ | 3.1 |
| tetragonal | 15.1 | | | | |
| trigonal | 10.7 | $R\bar{3}m$ | 3.7 | | |
| hexagonal | 12.1 | $P6_3/mmc$ | 4.3 | | |
| cubic | 9.5 | | | | |

About three-quarters of the structure types are centrosymmetric. Of course some structure types have hundreds of representatives (they are usually of high symmetry) and many others have just one. With reference to the above table, note that *Cmcm* is a maximal

[1]See the article by A. J. C. Wilson in *International Tables C*, p. 792.

subgroup of index 3 of $P6_3/mmc$ and that *Pnma* and *C2/m* are maximal subgroups of order 2 of *Cmcm*. Likewise $P2_1/c$ is a maximal subgroup of index 2 of *Pnma* and *C2/m* is a maximal subgroup of index 3 of $R\bar{3}m$. The most common cubic symmetries are $Fm\bar{3}m$ and $Fd\bar{3}m$ (both supergroups of $R\bar{3}m$).

The presence of high symmetry means that apparently complex structures (such as **spinel** and **garnet**; see § 3.4) can often be described by just a few parameters and appreciation of possible reasons for adoption of a particular symmetry can be a great aid to understanding such structures. For this reason we spend some time on the symmetries of sphere and cylinder packings in Chapter 6 (see especially § 6.8.1 and 6.8.10). For the symmetries of certain types of layer structure see § 5.6.14.

### *3.7.8 Incommensurate (modulated) crystals, quasicrystals and non-classical symmetries*[1]

We should mention that in this book we are only concerned with structures that have symmetries described by one of the classical symmetry groups, in particular for "infinite" solids the symmetry is one of the 230 space groups. Corresponding to the lattice of these structures there is a reciprocal lattice (§ 4.5.1) in the space (*reciprocal* space or *Fourier* space) of the diffraction pattern of the structure. The diffraction pattern is conveniently indexed in terms of three non-coplanar reciprocal lattice vectors. The reader should know however, that the diffraction patterns of many solid materials (incommensurate crystals and quasicrystals) cannot be so indexed, but require more than three (sometimes as many as six) vectors to be indexed.

A simple kind of incommensurate crystal structure can arise as in the following example. Consider a structure of $Na_2CO_3$ composed of Na atoms and $CO_3$ groups in well-defined positions in a unit cell and described by a conventional space symmetry group. Now imagine the structure to be modulated by rotations of the $CO_3$ groups by an angle whose magnitude depends on distance along an axis and with a periodicity different from, and incommensurate with, the basic lattice periodicity. In addition to the basic unit cell and its contents, we have also to specify the modulation function to describe the structure.[2] The diffraction pattern will have main diffraction spots at positions specified in terms of three reciprocal lattice vectors corresponding to the underlying lattice and satellite reflections whose positions relative to the main reflections are specified in terms of a fourth vector, and four numbers will be required to index the spots of the diffraction pattern. In *this* sense it may be considered "four-dimensional."

We mention quasicrystals briefly in § 5.5. The diffraction patterns of these have "non-crystallographic" symmetries. The first to be discovered (and many others subsequently) have icosahedral diffraction patterns that require six digits for indexing.

For crystallographic purposes it is sometimes convenient to consider such structures as projections onto three dimensions of higher dimensional structures with symmetries of

[1]Some acquaintance with X-ray diffraction is required to fully appreciate the content of this Note.

[2]$Na_2CO_3$ was one of the first incommensurately-modulated crystals discovered. The situation in the real crystal is a little more complicated than we have described.

higher dimensional space groups (see Appendix 2).

A recent interesting development is the recognition that the symmetries of diffraction patterns are not restricted to "crystallographic" symmetries (1-, 2-, 3-, 4- and 6-fold) axes. That restriction in real space came about (§ 1.4) because we restricted ourselves to lattices in which lattice points could not come arbitrarily close to each other—this is an eminently sensible restriction in real space, as atoms (which must be related to identical atoms by a lattice translation) can not approach arbitrarily close to each other. In reciprocal space, however, there is no physical reason for such a restriction and if it is lifted, axial symmetry groups with an $N$-fold axis with arbitrary $N$, and icosahedral groups are allowed. It turns out that the derivation of symmetry groups starting from reciprocal space is in many ways much more simple and elegant than the traditional way hinted at in this chapter, and it is quite possible that future texts will adopt the reciprocal space approach when the details are fully worked out. Key references are D. A. Rabson *et al.*, *Rev. Mod. Phys.* **63**, 699 (1991) and N. D. Mermin, *Rev. Mod. Phys.* **64**, 3 (1992). In these papers the 230 space groups are derived very elegantly, as are the 11 icosahedral space groups of reciprocal space. Note that there are no real space lattices corresponding to the icosahedral cases, so the term "reciprocal" lattice is unfortunate.

Despite the above remarks, we emphasize that in this book we are interested primarily in real space *structures* (arrangements of atoms in space) and that we use the language of space groups simply for a convenient, and succinct, method of describing such structures; and for this purpose a real space description is most intuitive. We also recognize that, unless one is completely familiar with Fourier transforms and their properties, working in reciprocal space can be somewhat daunting. Finally we observe that although the occurrence of incommensurate and quasi-crystals is being found to be quite common, at present structural details, on the level available for conventional crystals, are available in only a few cases.

## 3.8 Exercises

1. By choosing new axes $P2_1/c$ (§ 3.4, p. 77) becomes $P2_1/n$. The coordinates of general positions now are (verify from the *International Tables*):

$$x,y,z\ ;\ \bar{x},\bar{y},\bar{z}\ ;\ 1/2+x,1/2-y,1/2+z\ ;\ 1/2-x,1/2+y,1/2-z$$

2. The general positions of $Pnma$ (full symbol $P2_1/n2_1/m2_1/a$) are

$$\pm\ (x,y,z\ ;\ 1/2+x,1/2-y,1/2-z\ ;\ \bar{x},1/2+y,\bar{z}\ ;\ 1/2-x,\bar{y},1/2+z)$$

From these the locations of the $2_1$ axes parallel to $x$,$y$ and $z$ and the locations of the mirror and glide planes can be found. For example, $x,y,z \rightarrow \bar{x},1/2+y,\bar{z}$ corresponds to operation of a $2_1$ axis passing through the origin and parallel to $y$.

The special positions can be generated by appropriate substitutions for the general positions. Thus $y = 1/4$ (or $y = 3/4$) will generate positions 4 $c$ on the mirror planes. There are eight inversion centers in the cell (see Fig. 3.14, p. 72); substitute their coordinates for the general positions to generate 4 $a$ and 4 $b$.

3. The symmetries of the Bravais lattices are (identify each lattice):

$P\bar{1}$, $P2/m$, $C2/m$, $Pmmm$, $Cmmm$, $Immm$, $Fmmm$, $P4/mmm$, $I4/mmm$, $R\bar{3}m$, $P6/mmm$, $Pm\bar{3}m$, $Im\bar{3}m$, $Fm\bar{3}m$.

4. For the following space groups determine ($a$) the Bravais lattice, ($b$) the crystal system, ($c$) the crystal class, ($d$) whether there is a center of symmetry. (Use the *International Tables* to check your answers).

$Ia$, $Ccca$, $Amm2$, $I4_1/a$, $P4nc$, $I4_1/acd$, $R3c$, $P31c$, $P6_3mc$, $Ia\bar{3}$, $Fd\bar{3}c$

5. Draw a projection down **c** of the rutile structure (§ 3.4, p. 78). Show that Ti has six near O atoms and O has three near Ti atoms. Shade in the $\{Ti\}O_6$ polyhedra to see how they are connected. You should draw *at least* 4 unit cells. {Hint: see Chapter 4 for advice on drawing structures.]

6. Show that the following sets of points in a cubic unit cell all describe a face-centered lattice with a different choice of origin:

(a) 0,0,0 ; 0,1/2,1/2 ; 1/2,0,1/2 ; 1/2,1/2,0
(b) 0,0,1/2 ; 0,1/2,0 ; 1/2,0,0 ; 1/2,1/2,1/2
(c) 1/4,1/4,1/4 ; 1/4,3/4,3/4 ; 3/4,1/4,3/4 ; 3/4,3/4,1/4
(d) 3/4,3/4,3/4 ; 3/4,1/4,1/4 ; 1/4,3/4,1/4 ; 1/4,1/4,3/4

7. Show that the Al atoms in the structure of $Ca_3Al_2Si_3O_{12}$ structure (§ 3.4, p. 80) lie on points of a body-centered lattice described by a cell with doubled edge (so the cell contains 16 lattice points rather than 2).

8. The crystal structure of $Cu_2O$ (cuprite) is cubic with:

Cu at 0,0,0 ; 0,1/2,1/2 ; 1/2,0,1/2 ; 1/2,1/2,0
O at 1/4,1/4,1/4 ; 3/4,3/4,3/4

The Cu atoms fall on the points of a face-centered lattice, and the O atoms fall on the points of a body-centered lattice. The lattice of the structure is primitive (in fact the space group is $Pn\bar{3}m$). What is the coordination of Cu by O and of O by Cu (i.e. how many nearest neighbors and how are they arranged)?

9. The crystal structure of cuprite was one of the first determined. From the unit cell

parameter ($a$ = 4.26 Å) and the density ($\rho$ = 6.1 g cm$^{-3}$) show that there must be 4 Cu and 2 O atoms in the unit cell. Now get a copy of the *International Tables* and show, by elimination, that the only possible cubic symmetry is $Pn\bar{3}m$ (in which case the structure *must* be that given in Exercise 8) or $P\bar{4}3m$.[1] (Why can $P23$, $P4_232$ and $Pn\bar{3}$ be ruled out?). [Hint: begin by eliminating all space groups in which positions $a$ (which have the lowest multiplicity) consist of more than 2 points].

10. The structure of zircon, $ZrSiO_4$, has been described as:

$I4_1/amd$, $a$ = 6.607, $c$ = 5.982 Å. Origin at $\bar{4}2m$
Zr in 4 $a$: $I$ + (0,0,0 ; 0,1/2,1/4). Si in 4 $b$: $I$ + (0,0,1/2 ; 0,1/2,3/4)
O in 16 $h$: $I$ + (0,$\pm x$,$z$ ; $\pm x$,0,$\bar{z}$ 0,1/2$\pm x$,1/4$-z$ ; $\pm x$,1/2,1/4+$z$), $x$ = 0.1839, $z$ = 0.3203

Transforming to an origin at a center (see § 3.7.4, p.91):

Zr in 4 $a$: $I$ ± (0,3/4,1/8). Si in 4 $b$: $I$ ± (0,1/4,3/8)
O in 16 $h$: $I$ ± (0,$x$,$z$ ; 0,1/2$-x$,$z$ ; 1/4+$x$, 1/4,3/4+$z$ ; 3/4$-x$,1/4,3/4+$z$), $x$ = 0.0661, $z$ = 0.1953

11. An important structure type is that of $CeCu_2$; $KHg_2$ has been described as having the same symmetry:

| | |
|---|---|
| $CeCu_2$ | $Imma$, $a$ = 4.43, $b$ = 7.06, $c$ = 7.48 Å<br>Ce in 4 $c$: $I$ ± (0,1/4,$z$), $z$ = 0.538<br>Cu in 8 $h$: $I$ ± (0,$y$,$z$ ; 0,1/2$-y$,$z$), $y$ = 0.051, $z$ = 0.165 |
| $KHg_2$ | $Imma$, $a$ = 8.10, $b$ = 5.16, $c$ = 8.77 Å<br>K in 4 $c$: $I$ ± (0,1/4,$z$), $z$ = 0.703<br>Hg in 8 $i$: $I$ ± ($\pm x$,1/4,$z$), $x$ = 0.190, $z$ = 0.087 |

Plot both structures in projection down the *shortest* axis and thus show that they are essentially the same (except for a change of scale).[2] An exactly equivalent description of the structure of $KHg_2$ is (compare with that given for $CeCu_2$): $a$ = 5.16, $b$ = 8.10, $c$ = 8.77 Å. K in 4 $a$, $z$ = 0.457; Hg in 8 $h$, $y$ = 0.060, $z$ = 0.163. [Hints: for some help with drawing structures see Chapter 4. These structures are nicely illustrated if the four shortest Cu-Cu or Hg-Hg bonds are drawn in (two of these superimpose in projection)—see § 7.3.5]

12. Use the *International Tables A* to find the maximal *non-polar t* subgroups (**H**) of **G** = $Pm\bar{3}m$. [Hint: they will be found as maximal subgroups of maximal subgroups; the two highest symmetries have index 6 and 8 in **G** respectively. For the polar crystal classes, see p. 50.] These are the possible symmetries of the polar structures formed when a cubic **perovskite** (such as $BaTiO_3$ or $SrTiO_3$) deforms to polar structures at low temperature.

[1]Systematic absences in the powder pattern immediately point to $Pn\bar{3}m$. Specifically, reflections with indices $hk0$ are absent for $h + k$ odd (100, 210 etc.).

[2]This example is taken from *Elements of Inorganic Structural Chemistry* by E. Parthé (available from K. Sutter Parthé, 49 Chemin du Gué, CH-1213 Petit-Lancy, Switzerland).

CHAPTER 4

# LATTICE GEOMETRY

In this chapter we explain how the directions of lines and the orientations of planes in a crystal are specified. The description necessarily involves reference to the coordinate system and if alternative unit cell descriptions are used the description of directions and so on. will change, so we next discuss unit cell transformations. Next we give a compendium of formulas useful for crystallographic calculations. Finally we present some comments and hints on drawing crystal structures.

## 4.1 Directions in a crystal

### 4.1.1 General

A direction in a crystal is specified by three integers $u,v,w$ and is written $[uvw]$. The direction is then that of the vector $u\mathbf{a} + v\mathbf{b} + w\mathbf{c}$ where **a**, **b** and **c** are the unit cell (lattice) vectors. Normally $u$, $v$ and $w$ are integers that have no common factors other than 1 (i.e. they are *coprime*). Thus the direction of the $x$-axis is [100], that of the $y$-axis is [010] and that of the $z$-axis is [001]. Note that $[\bar{1}00]$ refers to the $-x$ direction. Fig. 4.1 illustrates these and some other principal directions. In the figure the shape of the unit cell is to be considered quite general (i.e. not necessarily a cube) and the heavy lines meet at the origin; the filled circle is a point at 0,1/2,1. In crystallography the symbol $[uvw]$ is often said to represent a *zone axis* and all planes parallel to $[uvw]$ are said to fall in that zone.

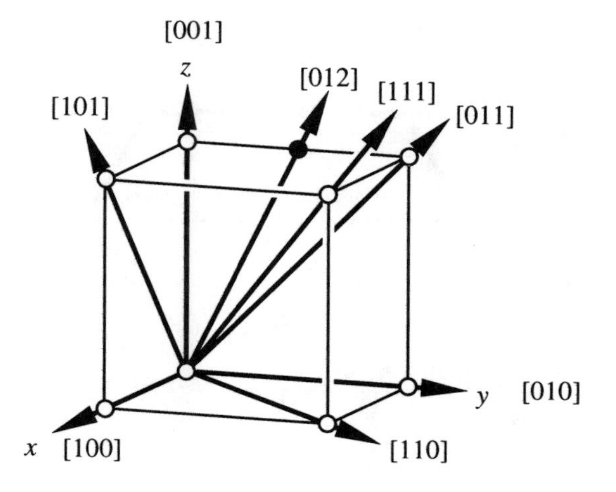


**Fig. 4.1** Directions in a crystal.

The set of equivalent (symmetry-related) directions in a crystal is represented <*uvw*>. Thus in a centrosymmetric *cubic* crystal <111> represents the family of specific directions along the body diagonals:

$$[111], [\bar{1}11], [1\bar{1}1], [11\bar{1}], [\bar{1}\bar{1}1], [\bar{1}1\bar{1}], [1\bar{1}\bar{1}] \text{ and } [\bar{1}\bar{1}\bar{1}]$$

For the same symmetry, <100> is similarly the set:

$$[100], [010], [001], [\bar{1}00], [0\bar{1}0] \text{ and } [00\bar{1}].$$

In the jargon <*uvw*> is a *form* of zone axes.

Often $[uvw]$ is used to represent not only a direction but is considered to have a magnitude associated with it. Usually the context makes this clear. Thus a vector of magnitude $a/2$ along the $x$ direction is written $1/2[100]$, and $1/2[11\bar{1}]$ refers to $(\mathbf{a} + \mathbf{b} - \mathbf{c})/2$. Vector sums can be expressed as (e.g.) $1/6[\bar{1}12] + 1/6[\bar{1}10] = 1/3[\bar{1}11]$ which is shorthand for $(-\mathbf{a} + \mathbf{b} + 2\mathbf{c})/6 + (-\mathbf{a} + \mathbf{b})/6 = (-\mathbf{a} + \mathbf{b} + \mathbf{c})/3$.

### *4.1.2 Directions in hexagonal crystals*

In the hexagonal system two systems are used for specifying directions. The first uses a four index symbol which we write $[UVJW]$. The corresponding direction is that of the vector $U\mathbf{a}_1 + V\mathbf{a}_2 + J\mathbf{a}_3 + W\mathbf{c}$ where $\mathbf{a}_1$ is the $x$ direction, $\mathbf{a}_2$ is the $y$ direction (at 120° to each other) and $\mathbf{a}_3 = -(\mathbf{a}_1 + \mathbf{a}_2)$ is the equivalent direction at 120° to both $\mathbf{a}_1$ and $\mathbf{a}_2$ (and at 90° to $\mathbf{c}$) as shown in Fig. 4.2. An important point is that as $\mathbf{a}_1$, $\mathbf{a}_2$ and $\mathbf{a}_3$ are not independent there are many choices of $U$, $V$ and $J$ that correspond to the same direction. *The choice made is such that* $J = -(U + V)$. Although, as we will see, there are advantages to this system, it is not intuitive. Thus (see again Fig. 4.2) the $y$ direction (parallel to $\mathbf{a}_2$) is $[\bar{1}2\bar{1}0]$ and the vector $\mathbf{a}_2 = 1/3[\bar{1}2\bar{1}0]$.

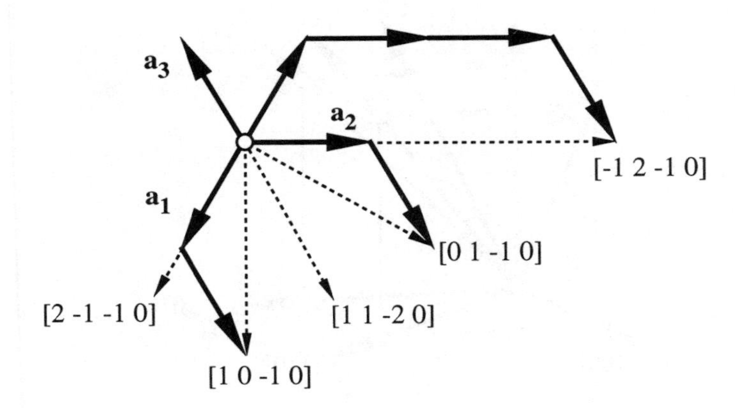


**Fig. 4.2.** Directions in hexagonal crystals.

Three-index symbols $[uvw]$ can also be used for directions in hexagonal crystals. These are also sometimes written as $[uv.w]$ or $[uv0w]$ which can be confusing and definitely should be avoided. The four-index symbols are easily obtained from three-index symbols based on a cell with dimensions $a_1$, $a_2$, $c$ as follows:

$$U = (2u - v)/3 \ ; \ V = (2v - u)/3 \ ; \ J = -(U + V) \ ; \ W = w \tag{4.1}$$

Note that $U$, $V$, $J$ and $W$ should be converted to the smallest possible integers by multiplying or dividing by a constant factor if necessary. Our recommendation is to use three-index symbols for all calculations (see e.g. sections 4.3 and 4.4) and to convert at the end to four-index symbols (to avoid ambiguity) for communication.

## 4.2 Planes and Miller indices

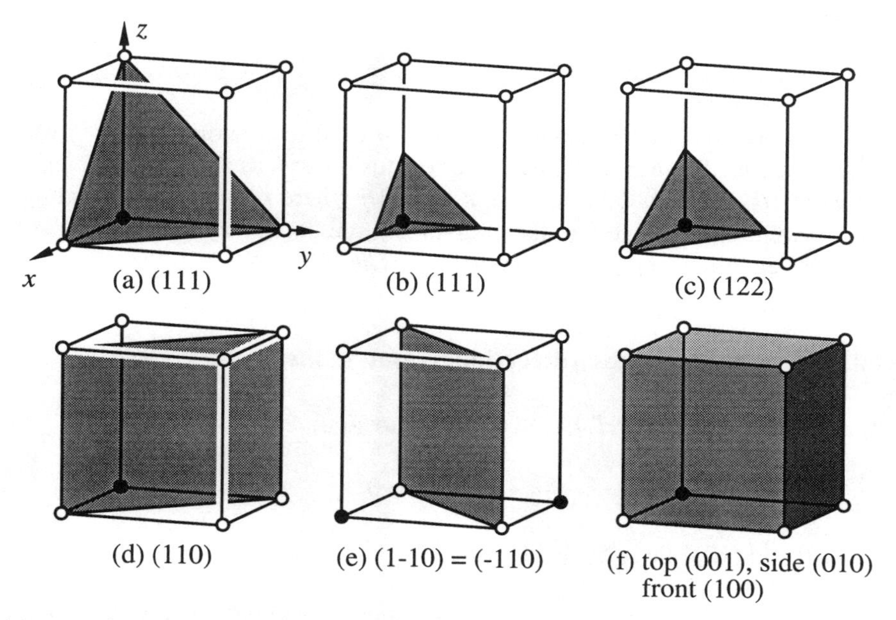


**Fig. 4.3.** Miller indices for planes.

The orientation of planes relative to a coordinate system defined by translation vectors **a**, **b** and **c** is given by three integers $h$, $k$, $l$ (known as Miller indices) and written $(hkl)$. The significance of these numbers is that the intercepts of the plane with the reference axes (in units of $a$, $b$ and $c$ respectively) are in the ratio $1/h : 1/k : 1/l$. Again (as for the indices representing directions) $h$, $k$ and $l$ are integers with no factor common to all three. Note that $h$, $k$ and $l$ specify only the *orientation* of the plane, not its location. Thus all parallel planes have the same Miller indices.[1] Planes which intersect lattice points are referred to as *lattice planes* or *net planes.* The equation of the plane in the lattice coordinate system is $hx + ky + lz$ = constant. $(hkl)$ specifies the *orientation* of the plane and the constant specifies its *position* relative to the origin.

Figure 4.3 provides a number of examples of low-index planes with respect to a unit cell which may be triclinic. In (a) the plane intercepts the axes at 1/1, 1/1, 1/1 so its indices are (111). In (b) the intercepts are 1/2, 1/2, 1/2 so after factoring out a common 2, the indices are again (111). In (c) the intercepts are 1/1, 1/2, 1/2 so now the indices are (122). In (d) the plane is parallel to **c** so the intercept with that axis is at infinity ($1/\infty = 0$). The plane shown is in fact (110). Part (e) of the figure shows a plane with intercepts either at 1/1, 1/-1, 1/0 or 1/-1, 1/1, 1/0 according to whether the origin is taken at one or other of the two filled circles shown. These planes are of course the same and thus should have the same indices which we could write either as $(1\bar{1}0)$ or $(\bar{1}10)$. Part (f) shows planes (100), (010) and (001); they are parallel to the faces of the unit cell.

A form of planes (a set of planes related by symmetry) is represented by braces as in $\{hkl\}$; thus the faces of a cubic unit cell are {100}.

*Hexagonal crystals*

As for directions, four-index symbols are usually used for the orientation of planes in the hexagonal system. There is little chance for confusion now as a plane that intercepts $\mathbf{a_1}$ at $1/h$ and $\mathbf{a_2}$ at $1/k$ will inevitably intercept $\mathbf{a_3}$ at $1/i$ where $i = -(h+k)$. Accordingly the three-index symbol $(hkl)$ becomes the four-index symbol $(hkil)$ with $i = -(h+k)$. The superfluous $i$ is sometimes replaced by a point as in $(hk.l)$.

## 4.3 Relations between zones (directions) and planes

(a) The *zone law*: A plane $(hkl)$ lies in the zone $[uvw]$ if:

$$hu + kv + lw = 0 \tag{4.2}$$

thus both $(1\bar{1}0)$ and $(10\bar{1})$ lie in the [111] zone.

(b) The zone $[uvw]$ corresponding to the intersection of planes $(h_1k_1l_1)$ and $(h_2k_2l_2)$ is given by:

$$u = k_1l_2 - k_2l_1\ ;\ v = l_1h_2 - l_2h_1\ ;\ w = h_1k_2 - h_2k_1 \tag{4.3}$$

thus the intersection of $(1\bar{1}0)$ and $(10\bar{1})$ is [111] [compare (a) above].

[1]Authors of elementary texts often seem to us to get this point wrong, confusing Miller indices with Bragg indices which are discussed below (§ 4.7.2). Explicitly for example: no distinction should be made between a (222) and a (111) plane.

(c) Some perpendicularity conditions in different systems:

**orthorhombic**: [100], [010] and [001] are perpendicular to (100), (010) and (001) respectively.

**tetragonal**: [001] is perpendicular to (001) and [*pq*0] is perpendicular to (*pq*0).

**hexagonal**: [0001] is perpendicular to (0001) and [*pqi*0] is perpendicular to (*pqi*0).[1] In the three-index system one has [2*p*+*q* *p*+2*q* 0] perpendicular to (*pq*0) or equivalently, [*pq*0] perpendicular to (2*p*–*q* 2*q*–*p* 0).
More generally (four-index system) [$p\ q\ i\ (3a^2/2c^2)l$] is normal to (*pqil*).

**rhombohedral**: [111] is perpendicular to (111) (but if a hexagonal cell is used, see above).

**cubic**: [*pqr*] is perpendicular to (*pqr*).

## 4.4 Unit cell transformations

### *4.4.1 General*

There are many reasons why it would be desirable to describe a crystal structure with a unit cell other than the one provided. For example the original investigator may have chosen a "non-standard" setting of the axes, or a rhombohedral crystal might be more conveniently described using a hexagonal cell rather than a rhombohedral one. Sometimes structural relationships are made more evident using unconventional cells (such as a hexagonal cell for a cubic crystal).

Let a "new" unit cell defined by translation vectors **a'**, **b'**, **c'** be derived for a structure with an "old" cell defined by **a, b, c**. The relationship between the new and old cells is given by (using the notation of the 1965 *International Tables*, Vol I):

$$\begin{aligned} \mathbf{a}' &= s_{11}\mathbf{a} + s_{12}\mathbf{b} + s_{13}\mathbf{c} & \qquad \mathbf{a} &= t_{11}\mathbf{a}' + t_{12}\mathbf{b}' + t_{13}\mathbf{c}' \\ \mathbf{b}' &= s_{21}\mathbf{a} + s_{22}\mathbf{b} + s_{23}\mathbf{c} & \qquad \mathbf{b} &= t_{21}\mathbf{a}' + t_{22}\mathbf{b}' + t_{23}\mathbf{c}' \\ \mathbf{c}' &= s_{31}\mathbf{a} + s_{32}\mathbf{b} + s_{33}\mathbf{c} & \qquad \mathbf{c} &= t_{31}\mathbf{a}' + t_{32}\mathbf{b}' + t_{33}\mathbf{c}' \end{aligned} \qquad (4.4)$$

Let **S** be the matrix of coefficients $s_{ij}$ and **T** the matrix $t_{ij}$.[2] Then **S** and **T** are the

[1]This is one of the advantages of the 4-index system.

[2]Note that Boisen & Gibbs (*Mathematical Crystallography* in Book List) use **S** and **T** matrices that are the transposes of ours. The newer *International Tables* (volume *A*) uses **P** and **Q** that are also the transposes of our **S** and **T**. These authors consider (more logically) the set of lattice vectors to be a 1 × 3 (row) vector rather than as a 3 × 1 (column) vector. The system and notation we use appears to us to be simpler and is also to be found in most texts. This point is addressed in § 4.7.5 (p. 128).

inverses of each other (i.e. **ST** = **TS** = **E**). As the new and old cells are both unit cells the elements of the matrices will be rational numbers.

The ratio of the volumes of the new and old unit cells is det(**S**) and likewise the ratio of the volumes of the old and new unit cells is det(**T**). Should the determinants be negative, the hand of the coordinate system has been changed in the transformation.

When the coordinate system is changed the indices representing the orientation of planes and directions will also transform, as will the coordinates of points in the unit cell. The rules for the transformations are as given below ($\mathbf{T}^t$, $\mathbf{S}^t$ are the transposes of **T** and **S** respectively):[1]

| | new from old | old from new |
|---|---|---|
| **a, b, c** | **S** | **T** |
| (*hkl*) | **S** | **T** |
| [*uvw*] | $\mathbf{T}^t$ | $\mathbf{S}^t$ |
| *x*,*y*,*z* | $\mathbf{T}^t$ | $\mathbf{S}^t$ |
| **a*,b*,c*** | $\mathbf{T}^t$ | $\mathbf{S}^t$ |

The table shows that axes and indices of planes transform the same way as each other, and that coordinates of points and indices for lines also transform the same way as each other but using the transposed reciprocal matrix. The bottom row shows for completeness how reciprocal lattice vectors (discussed below) transform.

In calculating coordinates of points (e.g. atom positions), it must be taken into account that if the new and old unit cells have different volumes, the number of points in the cell will change accordingly. How this is handled should be clear from the examples below. To calculate the length of the new axes, note that e.g. $\mathbf{a}' = s_{11}\mathbf{a} + s_{12}\mathbf{b} + s_{13}\mathbf{c}$ so that $a'$ is the distance from 0,0,0 to $s_{11},s_{12},s_{13}$ in the *old* coordinate system. Likewise $\alpha'$ is the angle between $[s_{21}\ s_{22}\ s_{23}]$ and $[s_{31}\ s_{32}\ s_{33}]$ (again in the *old* coordinate system). Calculating distances and angles is discussed in § 4.5.2 and § 4.5.3 and a general expression for unit cell parameters of transformed cells is given in § 4.5.4 (p. 112).

### *4.4.2 Rhombohedral to hexagonal and vice versa*

Rhombohedral crystals are often referred to hexagonal axes and the hexagonal cell is three times the volume of the rhombohedral one. Fig. 4.4 shows the relationship of the rhombohedral cell to the hexagonal one. The filled circles in the figure are at 0,0,0 ; 1,0,0 ; 1,1,0 and 0,0,1 in the rhombohedral cell and at cell corners (0,0,0 ; 1,0,0 ; etc.) ; 2/3,1/3,1/3 ; 1/3,2/3,2/3 in the hexagonal cell (see also Fig. 4.5). The open circles are the remaining corners of the rhombohedral cell that are in adjacent hexagonal cells. It may be seen that if the rhombohedral cell is rotated by 60° about $c_h$ the hexagonal cell contains

[1] A simple derivation (using the same notation) of these rules is given by A. Kelly & G. Groves *Crystallography and Crystal Defects*, Addison-Wesley (1970).

points at 1/3,2/3,1/3 and 2/3,1/3,2/3; this is the so-called *reverse* setting which is not used. The setting used here (Fig. 4.4) is the "standard" one and is called the *obverse* setting.

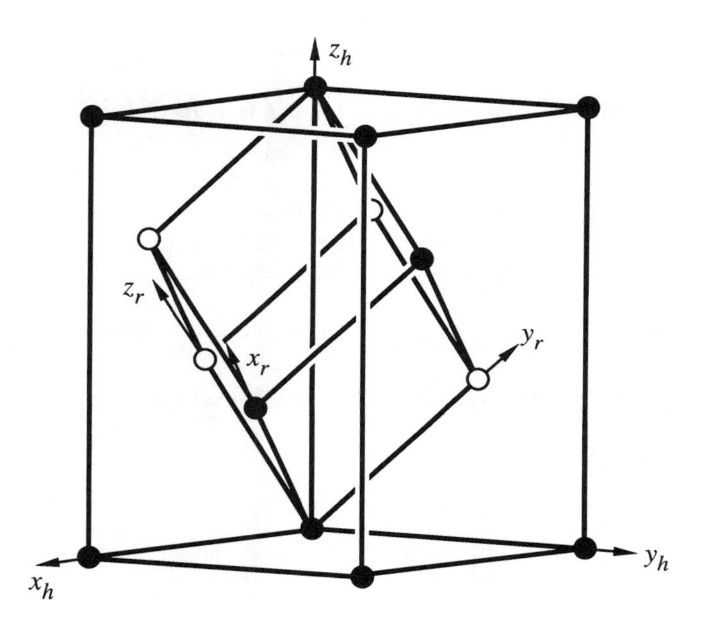


**Fig. 4.4**. The relationship between centered and primitive rhombohedral cells.

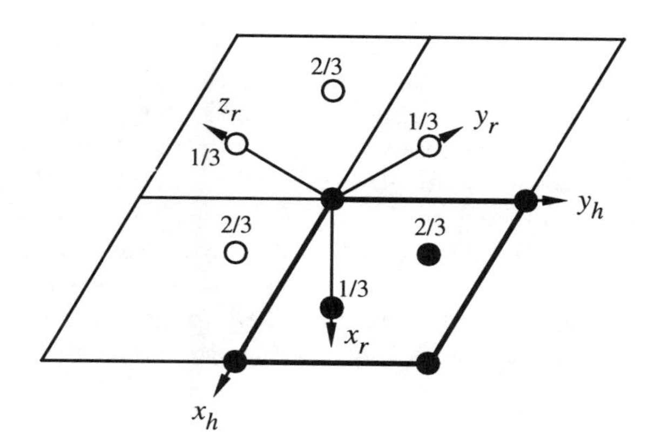


**Fig. 4.5**. The points of Fig. 4.4 projected down $\mathbf{c}_h$ (the $z_h$ direction).

Figure 4.5 shows the same rhombohedral and hexagonal unit cells in projection along the 3-fold axis. It should be observed that:

$$\begin{aligned} \mathbf{a}_h &= \mathbf{a}_r - \mathbf{b}_r \\ \mathbf{b}_h &= \quad\;\; \mathbf{b}_r - \mathbf{c}_r \\ \mathbf{c}_h &= \mathbf{a}_r + \mathbf{b}_r + \mathbf{c}_r \end{aligned}$$

The transformation matrix is therefore $\mathbf{S} = (1\ \bar{1}\ 0\ /\ 0\ 1\ \bar{1}\ /\ 1\ 1\ 1)$ (rhombohedral → hexagonal). Likewise:

$$\begin{aligned} \mathbf{a}_r &= 2\mathbf{a}_h/3 + \mathbf{b}_h/3 + \mathbf{c}_h/3 \\ \mathbf{b}_r &= -\mathbf{a}_h/3 + \mathbf{b}_h/3 + \mathbf{c}_h/3 \\ \mathbf{c}_r &= -\mathbf{a}_h/3 - 2\mathbf{b}_h/3 + \mathbf{c}_h/3 \end{aligned}$$

so $\mathbf{T} = (2/3\ 1/3\ 1/3\ /\ -1/3\ 1/3\ 1/3\ /\ -1/3\ -2/3\ 1/3)$ is the corresponding inverse matrix.

The relations between the unit cell parameters in the two cells are:

$$a_r = \left(\tfrac{1}{3}\right)\sqrt{(3a_h^2 + c_h^2)}$$

$$\alpha = 2\sin^{-1}\left(\frac{3}{2\sqrt{3 + (c_h^2 / a_h^2)}}\right)$$

$$a_h = 2a_r \sin(\alpha/2)$$

$$c_h = a_r\sqrt{(3 + 6\cos\alpha)}$$

As we use $\mathbf{T}^t$ to transform coordinates, a point with coordinates $x,y,z$ in the rhombohedral cell will have coordinates $(2x-y-z)/3$, $(x+y-2z)/3$, $(x+y+z)/3$ in the hexagonal cell. Note that 1,0,0 transforms to 2/3,1/3,1/3; 0,1,0 transforms to −1/3,1/3,1/3 ≡ 2/3,1/3,1/3 and 0,0,1 transforms to −1/3,−2/3,1/3 ≡ 2/3,1/3,1/3. 1,1,0 transforms to 1/3,2/3,2/3 and so on. Examination of all the possibilities shows that the rhombohedral lattice points correspond to 0,0,0 or 2/3,1/3,1/3 or 1/3,2/3,2/3 in the hexagonal cell. Accordingly, these quantities must be added to the new coordinates of each point calculated from the original $x,y,z$ (recall that the hexagonal cell volume is three times that of the rhombohedral cell). The transformations may be summarized (note that we use three-index symbols for hexagonal):

**rhombohedral → hexagonal**

$$x,y,z \quad \rightarrow \quad \begin{aligned} &(0,0,0\ ;\ 2/3,1/3,1/3\ ;\ 1/3,2/3,2/3)\ + \\ &(2x-y-z)/3,\ (x+y-2z)/3,\ (x+y+z)/3 \end{aligned}$$

$[uvw] \rightarrow \quad (1/3)[2u{-}v{-}w \;\; u{+}v{-}2w \;\; u{+}v{+}w]$
(thus $[111]_r$ becomes $1/3[003]_h = [001]_h$)

$(hkl) \rightarrow \quad (h{-}k \;\; k{-}l \;\; h{+}k{+}l)$
[thus $(111)_r$ becomes $(003)_h \equiv (001)_h$ etc.]

**hexagonal → rhombohedral**

$x,y,z \rightarrow \quad x+z,\ -x+y+z,\ -y+z.$

$[uvw] \rightarrow \quad [u{+}w \;\; v{+}w{-}u \;\; w{-}v]$
(thus $[001]_h$ becomes $[111]_r$)

$(hkl) \rightarrow \quad (1/3)(2h{+}k{+}l \;\; {-}h{+}k{+}l \;\; {-}h{-}2k{+}l)$
[thus $(001)_h$ becomes $(1/3\ 1/3\ 1/3)_r \equiv (111)_r$ etc.]

When transforming from hexagonal cell to a rhombohedral one, and the new coordinates are expressed modulo 1, it will be found that sets of three points in the hexagonal cell reduce to one point in the rhombohedral cell.

### *4.4.3 Cubic to hexagonal*

Cubic crystals may be considered as special cases of rhombohedral symmetry, the *P*, *I*, and *F* cells having primitive rhombohedral cells with $\alpha = 90°$, $109.47°$, and $60°$ respectively. They can also be transformed to hexagonal cells (among other things this is very useful for drawing cubic structures projected down a 3-fold axis).

For the primitive cubic cell the transformation is exactly the same as outlined above. The centered cells should first be transformed to primitive cells before being converted to hexagonal ones. It should be clear that the final hexagonal cell will contain 3/2 times as many atoms as a body-centered cell, and 3/4 times as many atoms as the face-centered cell. For convenience the **S** and **T** matrices are:

face-centered → primitive $\quad \mathbf{S} = (0\ 1/2\ 1/2\ /\ 1/2\ 0\ 1/2\ /\ 1/2\ 1/2\ 0)$
$\mathbf{T} = (\bar{1}\ 1\ 1\ /\ 1\ \bar{1}\ 1\ /\ 1\ 1\ \bar{1})$

face-centered → hexagonal $\quad \mathbf{S} = (-1/2\ 1/2\ 0\ /\ 0\ -1/2\ 1/2\ /\ 1\ 1\ 1)$
$\mathbf{T} = (-4/3\ -2/3\ 1/3\ /\ 2/3\ -2/3\ 1/3\ /\ 2/3\ 4/3\ 1/3)$

body-centered → primitive $\quad \mathbf{S} = (-1/2\ 1/2\ 1/2\ /\ 1/2\ -1/2\ 1/2\ /\ 1/2\ 1/2\ -1/2)$
$\mathbf{T} = (0\ 1\ 1\ /\ 1\ 0\ 1\ /\ 1\ 1\ 0)$

body-centered → hexagonal $\quad \mathbf{S} = (\bar{1}\ 1\ 0\ /\ 0\ \bar{1}\ 1\ /\ 1/2\ 1/2\ 1/2)$
$\mathbf{T} = (-2/3\ -1/3\ 2/3\ /\ 1/3\ -1/3\ 2/3\ /\ 1/3\ 2/3\ 2/3)$

In the hexagonal description of cubic cells the axes are (with subscript $c$ for cubic and $h$ for hexagonal):

| | | |
|---|---|---|
| primitive | $a_h = \sqrt{2}a_c$ | $c_h/a_h = \sqrt{(3/2)} = 1.225$ |
| body-centered | $a_h = \sqrt{2}a_c$ | $c_h/a_h = \sqrt{(3/8)} = 0.614$ |
| face-centered | $a_h = a_c/\sqrt{2}$ | $c_h/a_h = \sqrt{6} = 2.449$ |

*4.4.4 Hexagonal to orthohexagonal*

Hexagonal crystal structures are sometimes conveniently illustrated in projection normal to **c**. A good way to do this is to transform to an *orthohexagonal* cell as shown in Fig. 4.6 (old cell in light lines, new cell in heavier lines). As **a' = a + b**, **b' = –a + b**, **c' = c**, the transformation matrices are **S** = (1 1 0 / $\bar{1}$ 1 0 / 0 0 1), **T** = (1/2 –1/2 0 / 1/2 1/2 0 / 0 0 1). The unit cell parameters are $a' = a$, $b' = \sqrt{3}b$ and $c' = c$. Notice that the new cell is $C$ centered (Fig. 4.6) so that 0,0,0 and 1/2,1/2,0 must be added to each $x',y',z'$.

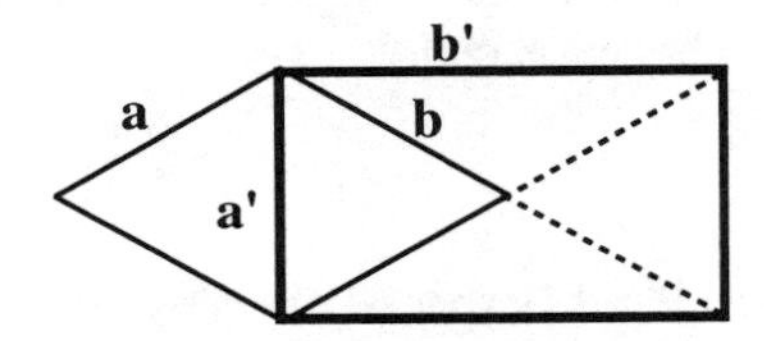


**Fig. 4.6.** Illustrating derivation of an orthohexagonal cell (heavy lines) from a hexagonal cell.

To plot normal to **c** usually project down **a'** (the short axis) of the orthohexagonal cell [this is on (11$\bar{2}$0) of the hexagonal cell]. Note that the same transformation is useful for plotting a cubic or tetragonal structure on (110) (i.e. along [110]).

If the original cell were rhombohedral, it should first be converted to a hexagonal cell and then to an orthohexagonal cell. The combined operation rhombohedral → orthohexagonal is described by **S** = (1 0 $\bar{1}$ / $\bar{1}$ 2 $\bar{1}$ / 1 1 1), **T** = (1/2 –1/6 1/3 / 0 1/3 1/3 / –1/2 –1/6 1/3). Now the new unit cell is six times larger and (0,0,0 ; 1/2,1/2,0 ; 0,1/3,1/3 ; 1/2,5/6,1/3 ; 0,2/3,2/3 ; 1/2,1/6,2/3) must be added to the new coordinates.

See § 4.6.3 for some worked examples and for a transformation from face-centered cubic to orthohexagonal.

## 4.5 Crystallographic calculations

*4.5.1 Unit cell volume and reciprocal lattice unit cell parameters*

The volume of a unit cell is given by

$$V = abc\sqrt{(1 + 2\cos\alpha\cos\beta\cos\gamma - \cos^2\alpha - \cos^2\beta - \cos^2\gamma)} \quad (4.5)$$

Special cases are:

| | |
|---|---|
| Monoclinic | $V = abc\sin\beta$ |
| Orthorhombic | $V = abc$ |
| Hexagonal | $V = \sqrt{3}a^2c/2$ |

For many reasons it is convenient to define a lattice *reciprocal* to the lattice of a crystal. Let a lattice be defined by vectors **a**, **b** and **c**; then we define reciprocal lattice vectors **a***, **b*** and **c*** such that **a*** is normal to the *bc* plane (i.e. the plane containing **b** and **c**), **b*** is perpendicular to the *ac* plane and **c*** is perpendicular to the *ab* plane.[1] The defining relationships are:

$$\begin{array}{lll} \mathbf{a.a^*} = 1 & \mathbf{b.a^*} = 0 & \mathbf{c.a^*} = 0 \\ \mathbf{a.b^*} = 0 & \mathbf{b.b^*} = 1 & \mathbf{c.b^*} = 0 \\ \mathbf{a.c^*} = 0 & \mathbf{b.c^*} = 0 & \mathbf{c.c^*} = 1 \end{array} \quad (4.6)$$

From these equations it follows that the dimensions of the reciprocal lattice vectors are 1/distance (units e.g. $Å^{-1}$).

The volume of the reciprocal lattice unit cell is $V^* = 1/V$.

The magnitudes of the reciprocal lattice vectors and the angles between them ($\alpha^*$, $\beta^*$ and $\gamma^*$) are

$$\begin{aligned} a^* &= bc\sin\alpha/V \\ b^* &= ac\sin\beta/V \\ c^* &= ab\sin\gamma/V \\ \cos\alpha^* &= (\cos\beta\cos\gamma - \cos\alpha)/\sin\beta\sin\gamma \\ \cos\beta^* &= (\cos\alpha\cos\gamma - \cos\beta)/\sin\alpha\sin\gamma \\ \cos\gamma^* &= (\cos\alpha\cos\beta - \cos\gamma)/\sin\alpha\sin\beta \end{aligned} \quad (4.7)$$

There is a set of equations analogous to Eq. 4.7 obtained by interchanging starred and unstarred quantities. Note the simplification if $\alpha = \beta = \gamma = 90°$; when $a^* = 1/a$, $b^* = 1/b$, $c^* = 1/c$, $\alpha^* = \beta^* = \gamma^* = 90°$. Other useful special cases are:

Monoclinic: $a^* = 1/(a\sin\beta)$, $b^* = 1/b$, $c^* = 1/(c\sin\beta)$, $\beta^* = 180° - \beta$

Hexagonal: $a^* = 2/(\sqrt{3}a)$, $c^* = 1/c$, $\gamma^* = 60°$

The lattice reciprocal to *bcc* is *fcc* and *vice versa*. This may readily be confirmed by finding the reciprocal unit cell parameters of the primitive cells.[2]

[1]It is virtually impossible to do practical crystallography without reference to the reciprocal lattice. The direct lattice and reciprocal lattice are related as Fourier transforms.

[2]Thus find the reciprocal lattice parameters if $a = b = c$, $\alpha = \beta = \gamma = 60°$.

### *4.5.2 Interatomic distances and the G matrix*

We often need to calculate the distance between points $x_1,y_1,z_1$ and $x_2,y_2,z_2$ (e.g. to determine bond lengths in a structure). Let $\delta x = (x_2-x_1)$, $\delta y = (y_2-y_1)$ and $\delta z = (z_2-z_1)$ then the distance $d$ between the points is given by:

$$d^2 = (\delta x)^2a^2 + (\delta y)^2b^2 + (\delta z)^2c^2 + 2(\delta x)(\delta y)ab\cos\gamma + 2(\delta y)(\delta z)bc\cos\alpha + 2(\delta z)(\delta x)ca\cos\beta \qquad (4.8)$$

Remember in using Eq. 4.8 to keep the *signs* of $\delta x$, $\delta y$ and $\delta z$. For hand calculations it is worth taking into account the simplifications that arise for more symmetrical unit cells.

cubic $\quad d^2 = a^2[(\delta x)^2 + (\delta y)^2 + (\delta z)^2]$

tetragonal $\quad d^2 = a^2[(\delta x)^2 + (\delta y)^2] + c^2(\delta z)^2$

orthorhombic $\quad d^2 = a^2(\delta x)^2 + b^2(\delta y)^2 + c^2(\delta z)^2$

hexagonal $\quad d^2 = a^2[(\delta x)^2 + (\delta y)^2 - (\delta x)(\delta y)] + c^2(\delta z)^2$

rhombohedral $\quad d^2 = a^2[(\delta x)^2 + (\delta y)^2 + (\delta z)^2 + 2\{(\delta x)(\delta y) + (\delta y)(\delta z) + (\delta z)(\delta x)\}\cos\alpha]$

monoclinic $\quad d^2 = (\delta x)^2a^2 + (\delta y)^2b^2 + (\delta z)^2c^2 + 2(\delta z)(\delta x)ca\cos\beta$

Equation 4.8 is conveniently expressed in matrix notation as follows. With

$$\mathbf{G} = \begin{pmatrix} a^2 & ab\cos\gamma & ac\cos\beta \\ ab\cos\gamma & b^2 & bc\cos\alpha \\ ac\cos\beta & bc\cos\alpha & c^2 \end{pmatrix} \qquad (4.9)$$

If $\boldsymbol{\delta}$ is the column vector $(\delta x / \delta y / \delta z)$ and $\boldsymbol{\delta}^t$ the corresponding row vector (i.e. the transpose of $\boldsymbol{\delta}$) then Eq. 4.8 may be written:

$$d^2 = \boldsymbol{\delta}^t\mathbf{G}\boldsymbol{\delta} \qquad (4.8a)$$

The matrix $\mathbf{G}^* = \mathbf{G}^{-1}$ (the inverse of $\mathbf{G}$) is obtained by replacing the parameters in Eq. 4.9 by reciprocal unit cell parameters. **G** and **G*** are known as the *metric tensor* and *reciprocal space metric tensor* respectively (**G** is also called the Niggli matrix).

$$\mathbf{G}^* = \begin{pmatrix} a^{*2} & a^*b^*\cos\gamma^* & a^*c^*\cos\beta^* \\ a^*b^*\cos\gamma^* & b^{*2} & b^*c^*\cos\alpha^* \\ a^*c^*\cos\beta^* & b^*c^*\cos\alpha^* & c^{*2} \end{pmatrix}$$

Note that the unit cell volume is simply given as $V^2 = \det(\mathbf{G})$ and that $V^{*2} = \det(\mathbf{G}^*)$.

### *4.5.3 Angles*

The simplest way to calculate angles (especially if distances have already been calculated) is to use the cosine formula for the angle $A$ between sides $b$ and $c$ of a triangle with opposite side $a$:

$$\cos A = (b^2 + c^2 - a^2)/2bc \tag{4.10}$$

A simple, if inelegant, way to calculate the dihedral angle $ABCD$ (i.e. the angle between planes $ABC$ and $BCD$ intersecting in $BC$) is to find the perpendicular equations of the planes $ABC$ and $BCD$ in Cartesian coordinates from which the angle between them is easily obtained (see below). Often the interest in dihedral angles lies in molecular chemistry when Cartesian coordinates are used. A familiar example is that of hydrogen peroxide where interest focuses on the H-O-O-H dihedral angle.

If the Miller indices of the planes are given and if **g** is the column vector $(h_1 / k_1 / l_1)$ and **h** is the column vector $(h_2 / k_2 / l_2)$, the angle between the planes $(h_1k_1l_1)$ and $(h_2k_2l_2)$ is $\phi$ given by

$$\cos\phi = \mathbf{g}^t\mathbf{G}^*\mathbf{h}/\sqrt{[(\mathbf{g}^t\mathbf{G}^*\mathbf{g})\cdot(\mathbf{h}^t\mathbf{G}^*\mathbf{h})]} \tag{4.11}$$

In the case of a cubic crystal the above formula simplifies to

$$\cos\phi = (h_1h_2 + k_1k_2 + l_1l_2)/\sqrt{[(h_1{}^2 + k_1{}^2 + l_1{}^2)(h_2{}^2 + k_2{}^2 + l_2{}^2)]} \tag{4.12}$$

Likewise if the indices of directions are given[1] and if **u** is the column vector $(u_1 / v_1 / w_1)$ and **v** is the column vector $(u_2 / v_2 / w_2)$, the angle between the lines $[u_1v_1w_1]$ and $[u_2v_2w_2]$ is $\phi$ given by:

$$\cos\phi = \mathbf{u}^t\mathbf{G}\mathbf{v}/\sqrt{[(\mathbf{u}^t\mathbf{G}\mathbf{u}) - (\mathbf{v}^t\mathbf{G}\mathbf{v})]} \tag{4.13}$$

In the case of a cubic crystal the above formula simplifies to

$$\cos\phi = (u_1u_2 + v_1v_2 + w_1w_2)/\sqrt{[(u_1{}^2 + v_1{}^2 + w_1{}^2)(u_2{}^2 + v_2{}^2 + w_2{}^2)]} \tag{4.14}$$

If the solid angle at a vertex $D$ of a tetrahedron $ABCD$ is wanted, it is simplest to calculate the angles $\alpha = \angle CDB$, $\beta = \angle ADC$ and $\gamma = \angle ADB$. The solid angle at $D$ is then $\varpi$ given by:

$$\tan(\varpi/4) = [\tan\sigma\tan(\sigma - \alpha/2)\tan(\sigma - \beta/2)\tan(\sigma - \gamma/2)]^{1/2} \quad \text{where } \sigma = (\alpha + \beta + \gamma)/4 \tag{4.15}$$

[1]Note that $u_1,v_1,w_1$, $u_2,v_2,w_2$ do not have to be integers.

### *4.5.4 The **G** matrix and unit cell parameters for transformed cells*

Let a new unit cell (primed quantities) be obtained from an old one using the transformation matrix **S** as described in § 4.4. The new **G'** matrix is given in terms of the old by (see p. 128 for a derivation):

$$\mathbf{G'} = \mathbf{SGS^t} \tag{4.16}$$

As $g_{11}' = a'^2$, $g_{22}' = b'^2$, $g_{33}' = c'^2$, and $g_{21}' = a'b'\cos\gamma'$, $g_{31}' = a'c'\cos\beta'$, $g_{23}' = b'c'\cos\alpha'$ the new unit cell parameters are readily obtained from the elements of the **G'** matrix. The elements of **G'** are:

$$g'_{ij} = \sum_{k=1}^{3}\sum_{l=1}^{3} s_{ik}s_{jl}g_{kl} \tag{4.16a}$$

### *4.5.5 Cartesian coordinates*

Sometimes it is convenient to transform from crystal coordinates to Cartesian coordinates; for example, for drawing structures, for calculating volumes of polyhedra, etc. We take the Cartesian axis $x_c$ to be along the crystal $x$ direction (i.e. parallel to **a**). The $y_c$ axis (perpendicular to $x_c$) is in the $ab$ plane and the $z_c$ axis normal to $x_c$ and $y_c$. Then:

$$\begin{pmatrix} x_c \\ y_c \\ z_c \end{pmatrix} = \begin{pmatrix} a & b\cos\gamma & c\cos\beta \\ 0 & b\sin\gamma & -c\cos\alpha^*\sin\beta \\ 0 & 0 & 1/c^* \end{pmatrix} \begin{pmatrix} x \\ y \\ z \end{pmatrix} \tag{4.17}$$

A useful special case is for a monoclinic crystal for which $\alpha = \gamma = 90°$. The matrix then becomes ($a$ 0 $c\cos\beta$ / 0 $b$ 0 / 0 0 $c\sin\beta$). Writing out Eq. 4.17 for this case:

$$x_c = ax + cz\cos\beta\ ;\ y_c = by\ ;\ z_c = cz\sin\beta \tag{4.18}$$

It may be convenient to change the relative orientations of the Cartesian and crystal axes according to the task at hand as described in § 4.6.1. The following table gives six choices that we find useful. To use the table, notice that each quantity on the right in Eq. 4.17 appears on the top row of the table. For the indicated orientation of Cartesian axes replace that quantity by the corresponding entry underneath in the appropriate row.

| | | | | | | | | | |
|---|---|---|---|---|---|---|---|---|---|
| $x_c$ along **a**, $y_c$ in **ab** plane | $a$ | $b$ | $c$, $c^*$ | $\alpha^*$ | $\beta$ | $\gamma$ | $x$ | $y$ | $z$ |
| $x_c$ along **a**, $y_c$ in **ac** plane | $a$ | $c$ | $b$, $b^*$ | $\alpha^*$ | $\gamma$ | $\beta$ | $x$ | $z$ | $-y$ |
| $x_c$ along **b**, $y_c$ in **bc** plane | $b$ | $c$ | $a$, $a^*$ | $\beta^*$ | $\gamma$ | $\alpha$ | $y$ | $z$ | $x$ |
| $x_c$ along **b**, $y_c$ in **ba** plane | $b$ | $a$ | $c$, $c^*$ | $\beta^*$ | $\alpha$ | $\gamma$ | $y$ | $x$ | $-z$ |
| $x_c$ along **c**, $y_c$ in **ca** plane | $c$ | $a$ | $b$, $b^*$ | $\gamma^*$ | $\alpha$ | $\beta$ | $z$ | $x$ | $y$ |
| $x_c$ along **c**, $y_c$ in **cb** plane | $c$ | $b$ | $a$, $a^*$ | $\gamma^*$ | $\beta$ | $\alpha$ | $z$ | $y$ | $-x$ |

Some expressions valid in Cartesian coordinates follow.
The plane through three points $x_1,y_1,z_1$ ; $x_2,y_2,z_2$ ; $x_3,y_3,z_3$ is given by:

$$\begin{vmatrix} x & y & z & 1 \\ x_1 & y_1 & z_1 & 1 \\ x_2 & y_2 & z_2 & 1 \\ x_3 & y_3 & z_3 & 1 \end{vmatrix} = 0 \tag{4.19}$$

This can be written

$$Ax + By + Cz = D \tag{4.20}$$

with

$$A = \begin{vmatrix} y_1 & z_1 & 1 \\ y_2 & z_2 & 1 \\ y_3 & z_3 & 1 \end{vmatrix}, B = -\begin{vmatrix} x_1 & z_1 & 1 \\ x_2 & z_2 & 1 \\ x_3 & z_3 & 1 \end{vmatrix}, C = \begin{vmatrix} x_1 & y_1 & 1 \\ x_2 & y_2 & 1 \\ x_3 & y_3 & 1 \end{vmatrix}, D = \begin{vmatrix} x_1 & y_1 & z_1 \\ x_2 & y_2 & z_2 \\ x_3 & y_3 & z_3 \end{vmatrix} \tag{4.21}$$

Equation 4.20 can be written in the form (the perpendicular equation):

$$lx + my + nz = p \tag{4.22}$$

Here $p$ is the length of the perpendicular from the plane to the origin and $l$, $m$ and $n$ are the cosines of the angles of this line with the $x$, $y$ and $z$ axes respectively. If $s$ is the sign ($\pm 1$) of $D$ in Eq. 4.21 and $Q$ is $(A^2 + B^2 + C^2)^{-1/2}$ then:

$$l = sAQ \; ; \; m = sBQ \; ; \; n = sCQ \; ; \; p = sDQ \tag{4.23}$$

The dihedral angle between two planes $l_1x + m_1y + n_1z = 0$ and $l_2x + m_2y + n_2z = 0$ is

$$\phi = \cos^{-1}(l_1l_2 + m_1m_2 + n_1n_2) \tag{4.24}$$

The volume of a tetrahedron with vertices $x_i,y_i,z_i$ ($i$ = 1,2,3,4) is:

$$V = \frac{1}{6}\left\| \begin{matrix} x_1 & y_1 & z_1 & 1 \\ x_2 & y_2 & z_2 & 1 \\ x_3 & y_3 & z_3 & 1 \\ x_4 & y_4 & z_4 & 1 \end{matrix} \right\| \tag{4.25}$$

In Eq. 4.25 the double lines indicate that the absolute value of the determinant is to be taken. Note that Eq. 4.19 follows from Eq. 4.25.

### *4.5.6 Distances between planes*

Let **h** be the column vector $(h\ /\ k\ /\ l)$ where $1/h$, $1/k$ and $1/l$ are the intercepts of a plane with the crystal axes in units of $a$, $b$ and $c$ respectively. The equation of the plane in the lattice coordinate system is $hx + ky + lz = 1$. The perpendicular distance from the origin to the plane is then $d_{hkl}$ which is given by

$$d_{hkl} = 1/\sqrt{(\mathbf{h}^t\mathbf{G}^*\mathbf{h})} \tag{4.26}$$

This also gives the interplanar spacing between *families* of planes with *Bragg* indices *hkl* (see § 4.7.2). As such it is useful to have the simplified forms for the more symmetrical unit cells:

| | |
|---|---|
| cubic | $d_{hkl} = a/\sqrt{(h^2 + k^2 + l^2)}$ |
| tetragonal | $d_{hkl} = 1/\sqrt{[(h^2 + k^2)/a^2 + l^2/c^2]}$ |
| orthorhombic | $d_{hkl} = 1/\sqrt{(h^2/a^2 + k^2/b^2 + l^2/c^2)}$ |
| hexagonal | $d_{hkl} = 1/\sqrt{[4(h^2 + hk + k^2)/3a^2 + l^2/c^2]}$ |

## 4.6 Drawing crystal structures and using cell transformations

### *4.6.1 Orthographic and clinographic projections*

By far the best way to "learn" a structure is by building a model; failing that it is good to draw it oneself. Many complex crystal structures can only be appreciated if a good model or drawing is available.

If a structure is to be communicated, at least one good drawing is essential, and this requires that a three-dimensional structure be projected onto the two dimensions of a page. Finding the best way to do this is a major problem. We have no patience at all with the crystallographer who determines a beautiful structure and then presents it as a projection of a unit cell with circles of different sizes representing the positions of different kinds of atoms. For some simple low-coordination structures "ball-and-stick" drawings may be satisfactory, but it is always almost always better to outline coordination polyhedra and to shade them in. It is a rare structure that does not profit from being drawn in more than one way; for example many "ionic" structures are usefully illustrated separately in terms of both anion coordination and cation coordination polyhedra. It is better to draw too many unit cells than too few.

Computer programs for drawing structures automatically are available. These will satisfy many people, but we have yet to find one that does all that we want. Particularly if it is wanted to present a structure in a novel way it will be found useful to be able to draw structures oneself using a drawing program, and the rest of this section is devoted to providing some guidance in this respect. It will be found essential to use a program that allows translation of duplicated objects by precise amounts to simulate plotting.[1]

Many complex structures are most readily (and often best) shown as a projection down a principle axis (orthographic projection). Usually if there is more than one such axis (as e.g. for an orthorhombic crystal), the shortest axis is most appropriate. Sometimes all atoms are on a set of mirror planes so a projection normal to those planes is appropriate—the atoms will be at one of two heights separated by half the repeat distance along that axis. Some hexagonal structures with large $c/a$ are best drawn in projection normal to **c**. In that case transform to an orthohexagonal cell using the transformation for axes described in § 4.4.4 (p. 108), and project along the new **a** axis (equal in length to the old $a$). Other unit cell transformations that are useful in preparing drawings are described in § 4.6.3.

Sometimes, particularly when dealing with cubic structures, it may be found that projections along a principal axis are confusing, as atoms in a unit cell are superimposed. Then it is best to make a clinographic projection in which the view point is tilted away from a principal axis. The way to do this is first to obtain Cartesian coordinates as in Eq. 4.17. This involves just a simple scaling if the crystal is cubic, tetragonal or orthorhombic.

Now imagine the Cartesian coordinates so that $x_c$ is horizontal on the paper (the $H$ direction) and $y_c$ is vertical on the paper (the $V$ direction); the $z_c$ axis is coming out of the paper (the $O$ direction). The view is now along $z_c$; we want to tilt away from this. First tilt the coordinate system by an angle $\theta$ clockwise about the $y_c$ axis and then by an amount $\phi$ anticlockwise about an axis normal to $y_c$ but in the plane of the paper (Fig. 4.7). Good choices of tilt angles are $\theta = 20°$ and $\phi = 10°$.[2] The new coordinates for plotting the structure on paper are now $H$ horizontal and $V$ vertical on the paper. The points projected on the paper are really a distance $O$ above that plane. $H$, $V$ and $O$ are given by:

$$\begin{aligned} H &= x_c\cos\theta - z_c\sin\theta \\ V &= -x_c\sin\theta\sin\phi + y_c\cos\phi - z_c\cos\theta\sin\phi \\ O &= x_c\sin\theta\cos\phi + y_c\sin\phi + z_c\cos\theta\cos\phi \end{aligned} \tag{4.27}$$

[1]Most modern computer graphics programs have far too many "bells and whistles" to be useful in this context. All that is needed is the capability to draw (and duplicate) in black and white, simple objects such as lines, circles and polygons and to add shading and to control which objects are in front of which. All the drawings in this book were made using the original (1988) Cricket Draw® on a Macintosh® computer using coordinates generated by EUTAX. The drawing instructions that follow assume that the reader has such a program available.

[2]Imagine that we are looking at the coordinate axes in a three-dimensional world with the view along $z_c$ and with $y_c$ vertical. The rotations are equivalent to moving our point of view to the right and up. Long ago many authors chose $\theta = \tan^{-1}(1/3) \approx 18.5°$ and $\phi = \tan^{-1}(1/6) \approx 9.5°$ as a "standard." For a discussion of drawing macroscopic crystals (still an important aspect of mineralogy) see Smith (Book List). Another good discussion is given by de Jong (Book List). The brain soon gets used to clinographic projections at a certain angle. You have reached this stage when it is found that such illustrations look almost incomprehensible when viewed upside down.

Some authors prefer a perspective drawing with parallel lines not parallel on the page; except for very simple structures or stereo pairs this tends to make the illustration less clear (in contrast to e.g. architectural drawings) and, except in stereo pairs, we use drawings in which parallel lines in the structure are parallel on the page ("view from infinity"). Notice although parallel edges are parallel on the paper in the projection of a cube (Fig. 4.7) none of the angles between edges is 90°; most drawings in the older literature incorrectly show the front face of a cube seen in clinographic projection as a rectangle, indeed commonly as a square.[1]

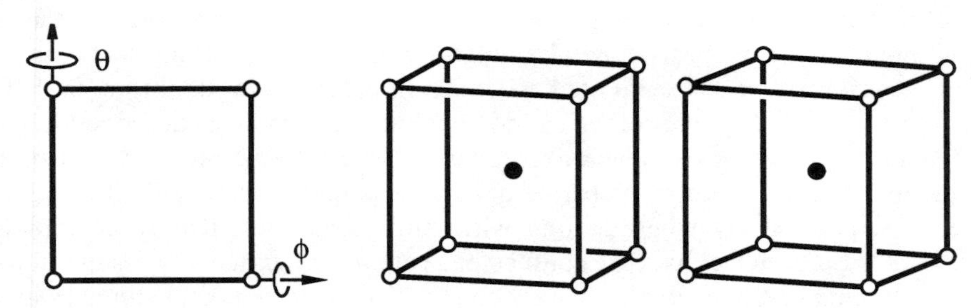


**Fig 4.7**. Obtaining a clinographic projection of a cube by rotation from a symmetry axis. Middle: $\phi = 10°$, $\theta = 17°$ ; Right: $\phi = 10°$, $\theta = 23°$. The filled circle is at the cube body center.

Approximations to stereo pairs can be made by using two different values of $\theta$ (the pair of cubes in Fig. 4.7 have $\theta = 17°$ on the left and $\theta = 23°$ on the right), but many people have difficulty merging the images.[2] In our experience, stereo pairs are most effective for stick models. Some stereo pairs of this sort are presented in § 7.11.8; if only one of each pair shown there is viewed it will be found to be rather uninformative.

For non-cubic crystals, it is necessary first to decide the point of view. Thus it must be decided which crystal plane will be parallel to the plane of the paper and which axis will be horizontal in that plane before tilting by $\theta$ and $\phi$. In practice this means choosing one of the six orientations of Cartesian axes with respect to the crystal axes discussed in § 4.5.5.

### *4.6.2 Examples of clinographic drawings* (ZnS and $CaF_2$)

We now go through the steps of drawing a simple structure, that of sphalerite ZnS. The procedure is more complicated to describe than to implement, as readers who work through the example will find. The structure is cubic with symmetry $F\bar{4}3m$, $a$ = 5.436 Å. Zn atoms are in 4 $c$: $F$ + (1/4,1/4,1/4) and S atoms in 4 $a$: $F$ + (0,0,0). In this case (a cubic structure)

[1]We have erred several times in the past in this respect. But we are in *very* distinguished company.

[2]Making a true stereo pair is rather subtle. The reader is invited to merge the two images on the right of Fig. 4.7 (holding a piece of opaque paper perpendicular to the page so that each eye can only view one image helps). Most viewers will see the back face and circles of the cube as much larger than the front ones even though they are drawn exactly the same size—the brain is adding perspective that is absent in the drawings. For references to drawing true stereo images see C. K. Johnson in *Crystallographic Computing* (F. R. Ahmed, ed.), Munksgaard, Copenhagen (1970).

to get Cartesian coordinates we simply have to multiply all the atomic coordinates by $a$. To make a clinographic projection we will take **a** horizontal and **b** vertical and then rotate by $\theta = 20°$ and $\phi = 10°$. Coordinates ($H$, $V$, and $O$) for plotting are then calculated as described in the previous section (Eq. 4.27).[1] For simplicity we take a scale of 1 cm = 1 Å for plotting (the drawings shown below have been subsequently shrunk to fit the page).

The first thing to do is to calculate the unit cell dimensions on the scale of the plot. The point at 1,0,0 has Cartesian coordinates (in Å) 5.409,0,0. We then calculate $H = 5.409\cos20° = 5.083$ cm, $V = -5.409\sin20°\sin10° = -0.321$ cm. Although not needed for plotting it is useful to record the "out" coordinate $O = 5.409\sin20°\cos10° = 1.822$ to keep track of what is in front of what in the drawing. The coordinates of this point and the points 0,1,0 and 0,0,1 are listed below. It should be clear that adding the three sets of coordinates will give the plotting coordinates for 1,1,1 and so on. The calculation of S atom plotting coordinates from those (1/2,1/2,0 etc.) in the unit cell proceeds similarly and they are also listed.

| $x$ | $y$ | $z$ | $x_c$ | $y_c$ | $z_c$ | $H$ | $V$ | $O$ |
|---|---|---|---|---|---|---|---|---|
| 1 | 0 | 0 | 5.409 | 0.0 | 0.0 | 5.083 | –0.321 | 1.822 |
| 0 | 1 | 0 | 0.0 | 5.409 | 0.0 | 0.0 | 5.327 | 0.939 |
| 0 | 0 | 1 | 0.0 | 0.0 | 5.409 | –1.850 | –0.883 | 5.006 |
| 1 | 1 | 1 | 5.409 | 5.409 | 5.409 | 3.233 | 4.123 | 7.767 |
| 0 | 0 | 0 | 0.0 | 0.0 | 0.0 | 0.0 | 0.0 | 0.0 |
| 1/2 | 1/2 | 0 | 2.705 | 2.705 | 0.0 | 2.541 | 2.503 | 1.381 |
| 1/2 | 0 | 1/2 | 2.705 | 0.0 | 2.705 | 1.616 | –0.602 | 3.414 |
| 0 | 1/2 | 1/2 | 0.0 | 2.705 | 2.705 | –0.925 | 2.222 | 2.972 |

The top row in Fig. 4.8 shows some stages in the drawing of the structure. On the left, the S and Zn atoms are shown as open or filled circles plotted with the appropriate values of $H$ and $V$ relative to the origin of coordinates. Notice that S atoms are at the corners and face centers of the cell so we actually show 14 atoms. In the middle, Zn-S bonds are drawn in to make a ball and stick drawing. Notice the use of broken lines as depth cues.[2] On the right the same pattern is shown as {Zn}$S_4$ tetrahedra sharing corners. The "tetrahedra" of the drawing are actually two triangles corresponding to the two visible (front) faces of each tetrahedron. It is generally best to consider coordination polyhedra to be opaque so the centering Zn atom and the back faces are omitted. We often (as here) shade the visible faces of polyhedra with a density that varies from one side to the other as this allows overlapping polyhedra to be more easily differentiated.

Notice that the structure is the same if the Zn and S positions are interchanged (this just corresponds to a shift of origin by 1/4,1/4,1/4) so the figure on the right could equally be illustrating {S}$Zn_4$ tetrahedra.

[1]EUTAX will calculate $H$, $V$ and $O$ for a specified orientation and scale. However this program recognizes that most computer drawing programs consider the positive direction of vertical coordinates as down the page, and reports $V$ coordinates as the negative of those given here.

[2]In the drawings shown , we actually have slightly thicker opaque white lines immediately behind black lines, so the white lines occlude black lines that are in turn behind them. All computer drawing programs that we have examined keep track of which objects are behind which, and usually allow changing the order.

The structure of fluorite $CaF_2$ can be illustrated similarly. The structure is cubic: $Fm\bar{3}m$, $a$ = 5.463 Å. Ca atoms are in 4 $a$: $F$ + (0,0,0) and F in 8 $c$: $F$ ± (1/4,1/4,1/4). Again we show metal atoms as filled circles. A ball and stick drawing on the bottom left of Fig. 4.8 shows that F atoms are in {F}$Ca_4$ tetrahedra. These polyhedra are shown in the center where it may be seen that tetrahedra now share edges.

As there are twice as many F atoms as Ca atoms, the latter must be in eight coordination. This is not immediately apparent from the ball and stick drawing, so parts of adjacent unit cells need to be added as shown in the figure on the bottom right, which shows that the Ca atoms are in {Ca}$F_8$ cubes (shown here as three differently shaded quadrilaterals). The question of how much to include is always tricky—too little does not show all aspects of the structure, and too much leads to too great an overlap of the component parts.

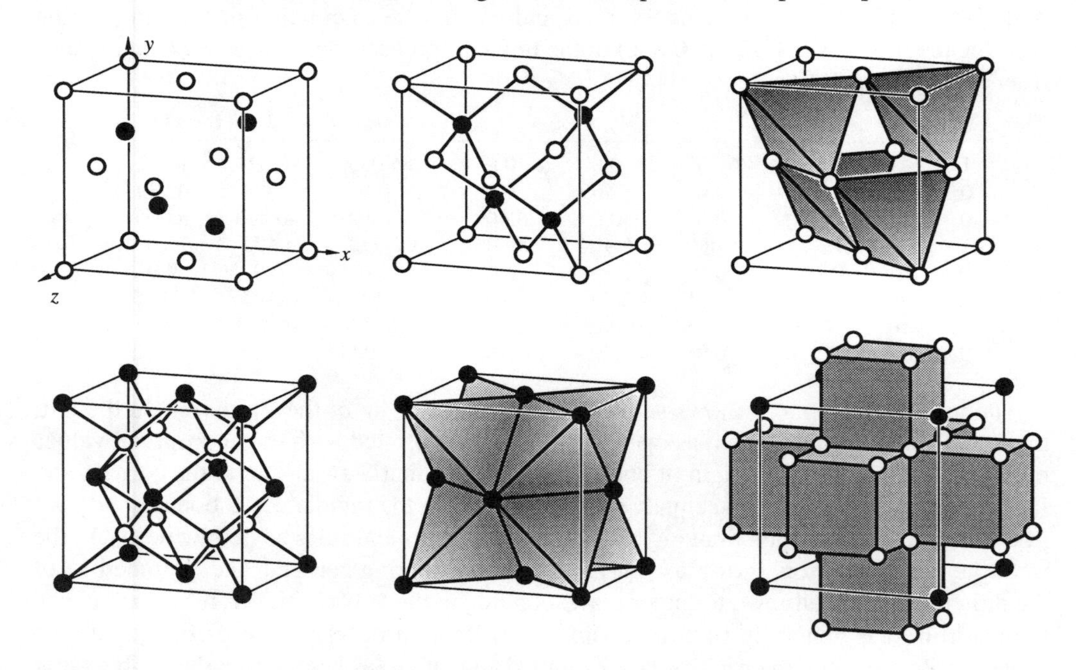


**Fig. 4.8**. Aspects of the structures of sphalerite, ZnS (top) and fluorite, $CaF_2$ (bottom). See the text for discussion.

### *4.6.3. Projections of some structures with octahedral coordination:* **NaCl, NiAs** *and* **TiP**. *Using orthohexagonal cells for cubic and hexagonal structures*

In this section we consider three structure types of composition $AX$ in which metal atoms are {$A$}$X_6$ octahedra. For the purpose of illustration we use idealized versions of **NaCl, NiAs** *and* **TiP** (recall that bold face formulas refer to structures not compounds) with regular octahedra of unit edge. The structures then are:

**NaCl** $Fm\bar{3}m$, $a$ = 1.414 ; **Na** in 4 $b$: $F$ + (0,0,1/2) ; **Cl** in 4 $a$: $F$ + (0,0,0)

**NiAs** $P6_3/mmc$, $a$ = 1.0, $c$ = 1.633 ; **Ni** in 2 $a$: (0,0,0 ; 0,0,1/2) ; **As** in 2 $d$: ±(1/3,2/3,1/4)

**TiP** $P6_3/mmc$, $a$ = 1.0, $c$ = 3.266 ; **Ti** in 4 $f$: ±(1/3,2/3,$z$ ; 1/3,2/3,1/2–$z$), $z$ = 0.125
**P(1)** in 2 $a$: (0,0,0 ; 0,0,1/2) ; **P(2)** in 2 $d$: ±(1/3,2/3,1/4)

In these structures there are layers of octahedra sharing edges. In **NaCl** the layers are normal to [111], in **NiAs** and **TiP** they are normal to **c**. To compare the three structures we will project them in a similar way in each case. For the hexagonal structures we project normal to **c**; specifically along **a** of the orthohexagonal cell described in § 4.4.4.

For **NaCl** we convert from *fcc* to hexagonal by $\mathbf{S_1}$ = (–1/2 1/2 0 / 0 –1/2 1/2 / 1 1 1) (p. 107) and then from hexagonal to orthohexagonal by $\mathbf{S_2}$ = (1 1 0 / $\bar{1}$ 1 0 / 0 0 1). In fact we do not need to do each transformation individually, but instead transform by $\mathbf{S} = \mathbf{S_2 S_1}$ = (–1/2 0 1/2 / 1/2 $\bar{1}$ 1/2 / 1 1 1); the inverse matrix, **T** = ($\bar{1}$ 1/3,1/3 / 0 –2/3 1/3 / 1 1/3 1/3). The final cell has 3/2 the volume of the original and contains 6 Na and 6 Cl.

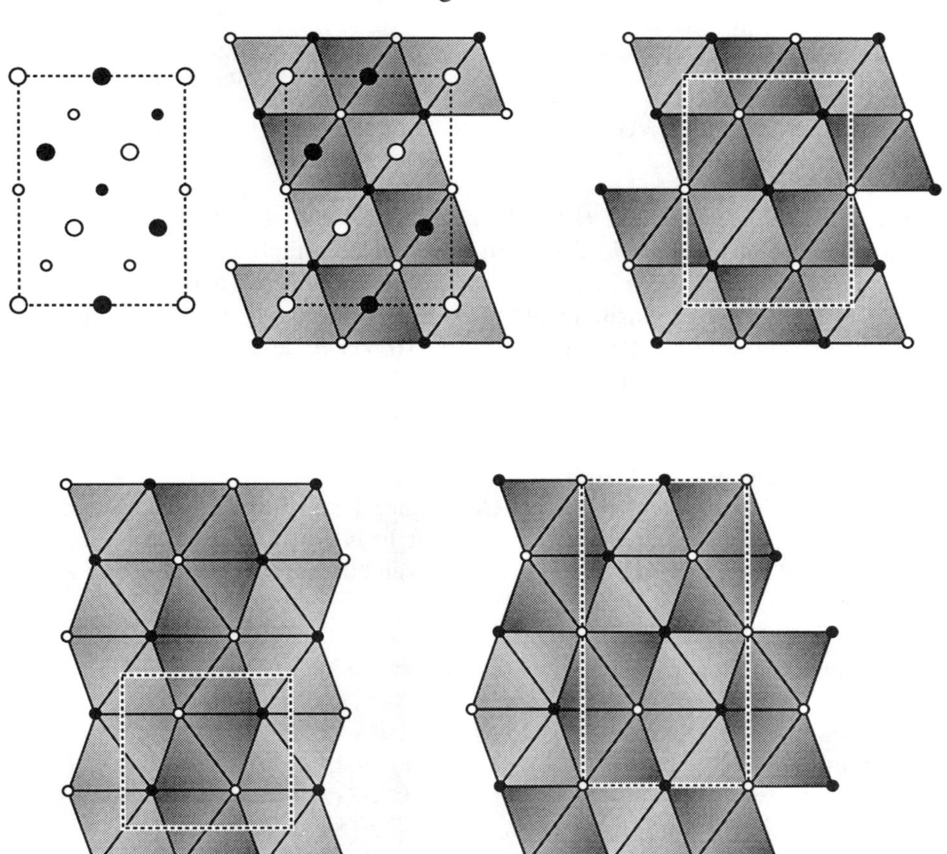

**Fig. 4.9**. Top; projection of **NaCl** normal to [111] (vertical on the page) showing {**Na**}$\mathbf{Cl_6}$ octahedra. Bottom: **NiAs** (left) and **TiP** (right) as {**Ni**}$\mathbf{As_6}$ or {**Ti**}$\mathbf{P_6}$ octahedra projected on (11$\bar{2}$0). See text.

The hexagonal cell has $a_h = a/\sqrt{2} = 1.000$, and $c_h = \sqrt{6}a_h = 2.450$; and the orthohexagonal cell has $a_o = a_h = 1.000$, $b_o = \sqrt{3}a_h = 1.732$ and $c_o = c_h = 2.450$ (see pp 107-108). We get the Cl coordinates using $(x' / y' / z') = \mathbf{T^t}(x / y / z)$. (specifically, in this case, $x' = -x+z$, $y' = x/3-2y/3+z/3$, $z' = x/3+y/3+z/3$.) The set (1/2,0,0 ; 0,1/2,0 ; 0,0,1/2 ; 1/2,1/2,1/2) in the cubic cell transform to (1/2,1/6,1/6 ; 0,2/3,1/6 ; 1/2,1/6,1/6 ; 0,0,1/2) with due allowance for the equivalence of −1/2 and 1/2, −1/3 and 2/3 etc. Thus we have three distinct new coordinate triplets. Recalling that the orthohexagonal cell is *C* centered, we find the other three Cl atoms in the cell by adding (1/2,1/2,0) to the coordinates of the first three. These, and the transformed Na coordinates, are listed below (notice that the new Na and Cl coordinates differ by 0,0,1/2). Also listed are plotting coordinates using the scale $b = 5$ cm; these are simply obtained as $H = 5y$, $V = 5zc/b = 7.071z$.

| | | Na | | | | | Cl | | |
|---|---|---|---|---|---|---|---|---|---|
| $x$ | $y$ | $z$ | $H$ | $V$ | $x$ | $y$ | $z$ | $H$ | $V$ |
| 0 | 0 | 0 | 0 | 0 | 0 | 0 | 1/2 | 0 | 3.536 |
| 1/2 | 1/2 | 0 | 2.5 | 0 | 1/2 | 1/2 | 1/2 | 2.5 | 3.536 |
| 0 | 1/3 | 1/3 | 1.667 | 2.357 | 0 | 1/3 | 5/6 | 1.667 | 5.892 |
| 1/2 | 5/6 | 1/3 | 4.167 | 2.357 | 1/2 | 5/6 | 5/6 | 4.167 | 5.892 |
| 0 | 2/3 | 2/3 | 3.333 | 4.714 | 0 | 2/3 | 1/6 | 3.333 | 1.179 |
| 1/2 | 1/6 | 2/3 | 0.833 | 4.714 | 1/2 | 1/6 | 1/6 | 0.833 | 1.179 |

Transformed coordinates and *H* and *V* for the other two structures are obtained in a similar way using first a transformation from hexagonal to orthohexagonal.

In Fig. 4.9 we show, at top left, a transformed unit cell of **NaCl** with **b** horizontal and **c** vertical. Also shown are the positions of the **Na** atoms (large circles) and **Cl** atoms (small circles). The elevations of all atoms are either $x = 0$ (open circles) or $x = 1/2$ (filled circles). On the right of that we show a little more of the structure with the **{Na}**$\mathbf{Cl_6}$ octahedra outlined. Those with centers at $x = 0$ are lighter shaded than those at $x = 1/2$. On the top right just the **{Na}**$\mathbf{Cl_6}$ octahedra are shown. In the bottom half of the figure **NiAs** and **TiP** are similarly depicted.

Notice that **NaCl** is the same if **Na** and **Cl** are interchanged, so Fig. 4.9 also depicts the arrangement of **{Cl}**$\mathbf{Na_6}$ octahedra. In contrast in **NiAs** there are **{As}**$\mathbf{Ni_6}$ trigonal prisms and in **TiP** there are both **{P}**$\mathbf{Ti_6}$ octahedra and triangular prisms. These aspects of these structures are met again in Chapter 6.

### *4.6.4. Further example of projections of crystal structures.* ZnS *again*

ZnS occurs in two forms known as sphalerite and wurtzite. Sphalerite was discussed and the structure given in § 4.6.2. Wurtzite is hexagonal, space group $P6_3mc$, $a = 3.823$, $c = 6.261$ Å. Zn atoms are in 2 *b*: (1/3,2/3,$z$ ; 2/3,1/3,1/2+$z$) with $z = 0.0$ ; S atoms are in 2 *b* with $z = 0.375$.[1] In both structures there are $\{Zn\}S_4$ tetrahedra linked by sharing corners, but the topology of linkage is different. We will illustrate the difference by

[1]Notice that this is an example of a polar structure. In particular $z$ only enters as $+z$ so the position of the origin on the $z$ axis is arbitrary and it has been chosen so that the Zn atoms are at $z = 0$ and $z = 1/2$.

projecting wurtzite down **c** and sphalerite down [111] of the cubic cell.

To calculate coordinates for plotting a hexagonal structure down **c** we take a scale for $a$ (say $a = s$ cm). Then plotting coordinates are horizontal $H = s(y - x/2)$ and vertical $V = \sqrt{3}sx/2$ (see Fig. 4.10). In wurtzite, Zn and S atoms with the same $x$ and $y$ coordinates differ in elevation by $\delta zc = 0.375c = 2.35$ Å (Zn below S) so we just plot the S atoms and make a mental note that the Zn atoms are underneath them. The plotting coordinates for the unit cell corners and S atoms are given below.

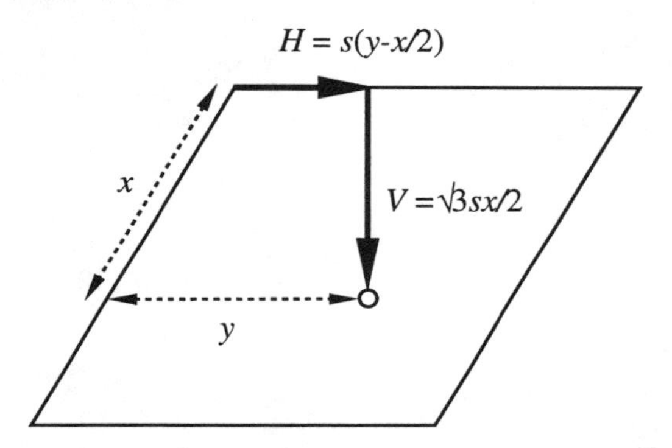


**Fig. 4.10**. Plotting coordinates $H$ and $V$ for a point $(x,y,z)$ in a hexagonal cell with scale $a$ = s.

Now the sphalerite structure is transformed to a hexagonal cell (§ 4.4.3, p. 107) for which it is found that $a = 3.825$, $c = 9.369$ Å. The atom coordinates in this new cell are for Zn: $R$ + (0,0,1/4) and for S: $R$ + (0,0,0). Notice that Zn and S atoms with the same $x$ and $y$ coordinates differ in elevation by $\delta z = c/4 = 2.34$ Å but now with S below Zn. In order to compare with wurtzite we would like to have Zn below S (cf. the previous paragraph) accordingly we reverse the direction of $z$ and, to avoid changing hand of the axes, interchange $x$ and $y$. The S coordinates are now (0,0,0 ; 2/3,1/3,2/3 ; 1/3,2/3, 1/3) with Zn underneath by $\delta z$ = 1/4) The S coordinates are also given in the table below.

The plotting coordinates ($H$ and $V$) in the two structures are (with $s$ = 5 cm) are:

| wurtzite | | | | | | sphalerite | | | | | |
|---|---|---|---|---|---|---|---|---|---|---|---|
| atom | $x$ | $y$ | $z$ | $H$ | $V$ | atom | $x$ | $y$ | $z$ | $H$ | $V$ |
| | 0 | 1 | | –2.500 | 4.330 | | 0 | 1 | | –2.500 | 4.330 |
| | 1 | 0 | | 5.000 | 0.000 | | 1 | 0 | | 5.000 | 0.000 |
| | 1 | 1 | | 2.500 | 4.330 | | 1 | 1 | | 2.500 | 4.330 |
| S | 1/3 | 2/3 | 0.375 | 2.500 | 1.433 | S | 0 | 0 | 0 | 0.000 | 0.000 |
| S | 2/3 | 1/3 | 0.875 | 0.000 | 2.887 | S | 1/3 | 2/3 | 1/3 | 2.500 | 1.433 |
| | | | | | | S | 2/3 | 1/3 | 2/3 | 0.000 | 2.887 |

When the wurtzite structure is first plotted (Fig. 4.11, top left) the unit cell is outlined and the positions of two S atoms in the cell are plotted with a filled circle for the atom with $z = 0.875$ and an open circle for the atom at $z = 0.375$. The Zn atoms below these two S atoms are in tetrahedra pointing up (see e.g. Fig. 2.19) and their other three neighbors form a triangle at a lower elevation. The two tetrahedra are drawn in the figure. Notice that the darker tetrahedron around the Zn atom with $z = 0$ has three S atoms at the base with elevation $0.875–1 = –0.125$. More of the structure is drawn on a smaller scale at the bottom left of the figure.

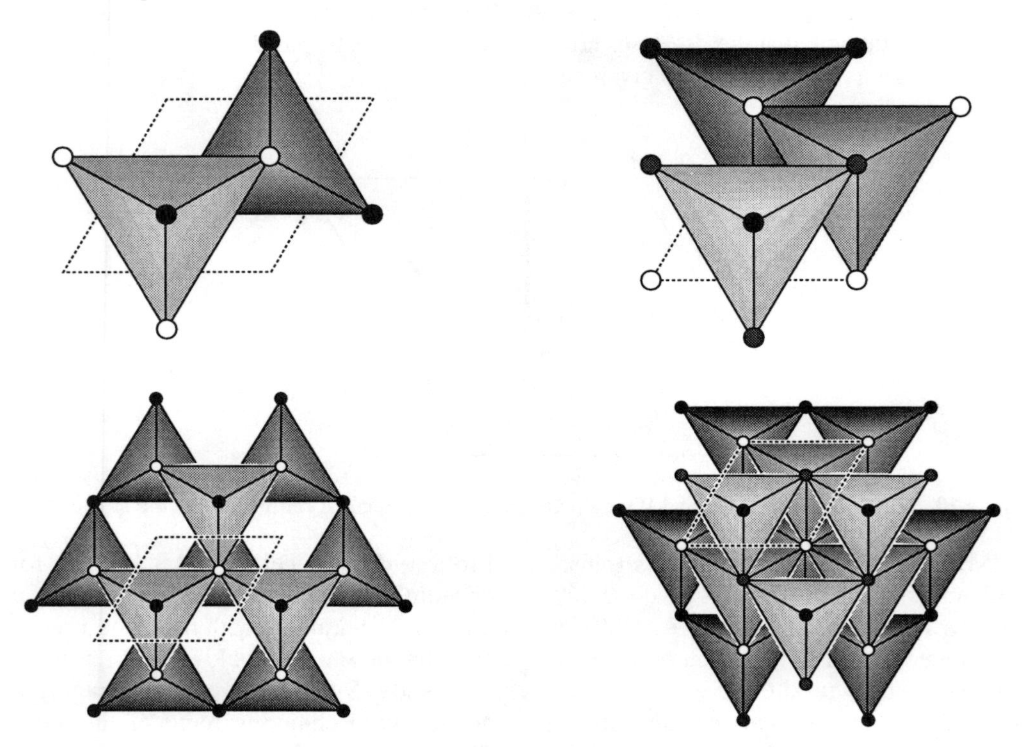

**Fig. 4.11**. The wurtzite (left) and sphalerite (right) forms of ZnS as $\{Zn\}S_4$ tetrahedra.

The drawing of sphalerite proceeds similarly. Now there are three S atoms in the cell at $z = 0$, $1/3$ and $2/3$ and these are shown as open, shaded and filled circles respectively in the top right of Fig 4.11. The tetrahedra about the three Zn atoms in the unit cell are also drawn. A larger portion of the structure is shown on a reduced scale underneath.

We will discuss these structures again in Chapters 6 and 7.

### *4.6.5 Drawing monoclinic structures*

Often monoclinic structures are drawn in projection down **b** (which is normal to **a** and **c**) —see Exercises 6 and 7 at the end of this chapter. Cartesian coordinates are given in equation 4.18 (p. 112). All that is necessary to do is to scale by a suitable amount: Thus if the scale chosen is $a = s$, the plotting coordinates for **a** horizontal are $H = (s/a)x_c$ and $V = (s/a)z_c$. If it is wanted to have **c** horizontal on the page (Fig. 4.12) then the roles of $a$ and $c$ and of $x$ and $z$ must be interchanged in the formulas for $H$ and $V$. Explicitly:

**a** horizontal: $H = sx + (szc/a)\cos\beta$ $\quad V = (szc/a)\sin\beta$
**c** horizontal: $H = (szc/a) + sx\cos\beta$ $\quad V = sx\sin\beta$

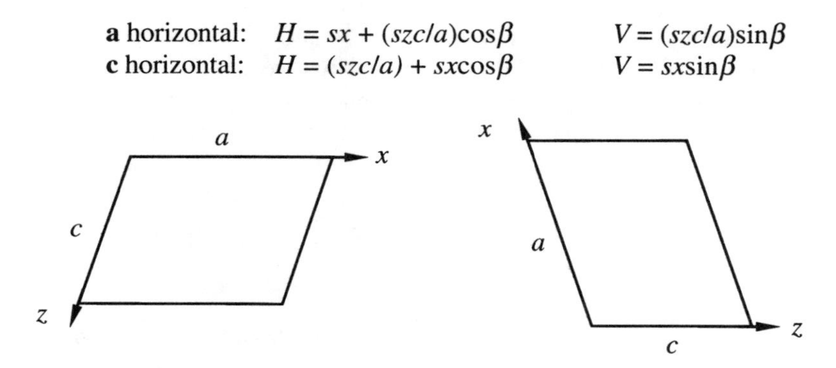


**Fig. 4.12**. Two orientations of a monoclinic cell projected down **b** (the positive sense of **b** is up out of the page in both cases). In the orientation on the left the coordinates $H$ and $V$ are to the right and down the page. In the orientation on the right the coordinates $H$ and $V$ are to the right and up the page.

For a projection of a monoclinic structure on (100) or (001), Cartesian coordinates should be calculated, but with the tilts $\theta$ and $\phi$ set equal to 0 in calculating $H$ and $V$. An example of such a projection is to be found in Fig. 7.28 (p. 314).

Occasionally a projection down **a** or **c** [note: this is *not* the same as projection on (100) or (001)] is required. Such projections should be carried out just like clinographic projections. The cell in projection is rectangular: for a projection down **a** it has projected dimensions (i.e. on the page) $b$ and $c\sin\beta$ and for a projection down **c** it has projected dimensions $b$ and $a\sin\beta$. The following table of tilts ($\theta$ and $\phi$) may be useful (as always we assume that **a**, **b** and **c** form a right-handed coordinate system):

| | | | |
|---|---|---|---|
| Projection down **a** | **b** horizontal | $\theta = 0$ | $\phi = 90° - \beta$ |
| | **b** vertical | $\theta = 90° - \beta$ | $\phi = 0$ |
| Projection down **c** | **b** horizontal | $\theta = 0$ | $\phi = \beta - 90°$ |
| | **b** vertical | $\theta = 90° - \beta$ | $\phi = 0$ |

To illustrate some of these points we use the structure of $MnB_4$ which was reported as:

$MnB_4$ $\quad$ $C2/m$, $a = 5.503$, $b = 5.367$, $c = 2.949$ Å, $\beta = 122.71°$
Mn in 2 $a$: $C + (0,0,0)$ ; B in 8 $j$: $C \pm (x,\pm y,z)$, $x = 0.200$, $y = 0.343$, $z = 0.197$

First we project down **b**, i.e. on (010), with **c** horizontal and a scale $a = 5$ cm. We calculate plotting coordinates $H$ and $V$ using the expressions given above. Thus for Mn at 0,0,0 we have $H = 0.000$, $V = 0.000$ and for Mn at 1/2,1/2,0 $H = -1.350$, $V = 2.104$. For B at 0.200,0.343,0.197 we find $H = -0.013$, $V = 0.841$ and so on (the reader may wish to verify these numbers and continue the calculation). In Fig 4.13 we have plotted the atoms in the unit cell and some in adjacent unit cells. For the atoms in the unit cell, elevations in multiples of $b/100$ are indicated (notice that pairs of B atoms superimpose in projection). The B atoms form a network in which each is bonded to four neighbors; three of the bonds are drawn in, the fourth joins pairs of atoms at $y = 0.34$ and $y = 0.66$ (open circles labeled ±34) or pairs of atoms at $y = \pm 0.16$ (filled circles). Thus we see that the B atoms are on puckered $6^3$ 2-dimensional nets that are linked by bonds "up" or "down" to similar nets to form the three-dimensional network. Mn atoms (shown as larger circles) are in the cavities of the B framework.

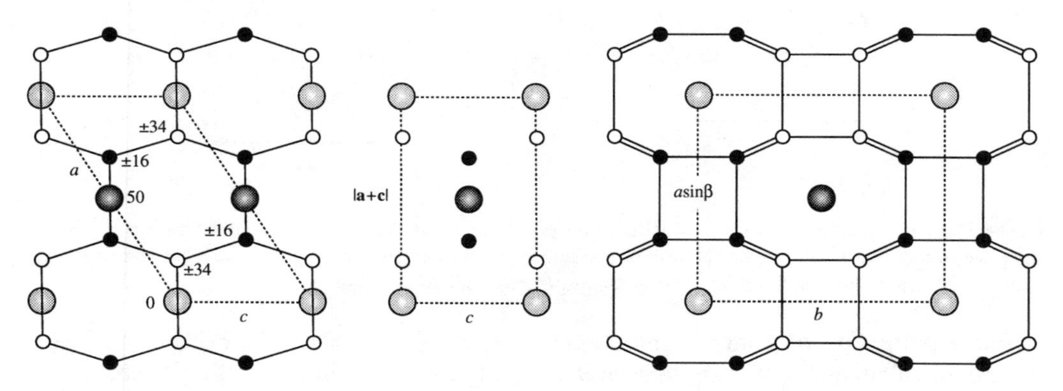


**Fig. 4.13**. The structure of $MnB_4$ (large circles are Mn). Left projected down **b**. Middle: showing a body-centered (*I*2/*m*) cell. Right: projected down **c**. See the text for further details.

Just a quick glance at Fig. 4.13 shows that the structure appears to be close to orthorhombic symmetry. In the middle of the figure a body-centered cell obtained by the transformation **S** = (1 0 1 / 0 1 0 / 0 0 1) is indicated. **T** = (1 0 $\bar{1}$ / 0 1 0 / 0 0 1) is the inverse matrix. The description of the structure using this cell is:

$MnB_4$    *I*2/*m*, $a' = 4.630$, $b' = 5.367$, $c' = 2.949$ Å, $\beta' = 90.31°$
Mn in 2 *a*: $I$ + (0,0,0) ; B in 8 *j*: $I \pm (x',\pm y',z')$, $x' = 0.200$, $y' = 0.343$, $z' = -0.003$

Actually the atoms are less than 0.01 Å away from positions with symmetry *Immm*; all that is needed is for $\beta'$ to be 90° and $z'$ to be 0. The description would then be:

$MnB_4$    *Immm*, $a' = 4.630$, $b' = 5.367$, $c' = 2.949$ Å
Mn in 2 *a*: $I$ + (0,0,0) ; B in 8 *n*: $I \pm (x',\pm y',0)$, $x' = 0.200$, $y' = 0.343$

With this description of the structure, it is seen that the short axis (**c'**) is a symmetry axis and a projection of the real structure down the corresponding **c'** axis should be rewarding. This is made on the same scale ($a = 5$ cm) as the original drawing. The unit cell edges in the drawing are $b = 4.876$ cm and $a\sin\beta = 4.207$ cm. The height of $c$ out of the page is 2.679 cm. B atoms (in the original unit cell) have elevations –0.013, –0.010, 1.341 and –1.363 cm. These correspond closely to elevations of $z = 0$ and ±1/2 (±1.34 cm), and they are shown as open and filled circles in the drawing on the right side of the figure.

The B arrangement is a 4-connected net that we discuss in Chapter 7 (see § 7.3.3). In its most symmetrical form the net is tetragonal; it is named for $CrB_4$ which is reported to be orthorhombic, and it does seem likely that $MnB_4$ is really orthorhombic also.[1]

## 4.7 Notes

### *4.7.1 Orientation of direct and reciprocal lattices*

For crystals with orthogonal axes (cubic, tetragonal and orthorhombic) the reciprocal lattice vectors **a***, **b*** and **c*** are parallel to the direct lattice vectors **a**, **b** and **c**. For hexagonal crystals **c*** is parallel to **c** and now **a*** and **b*** are not parallel to **a** and **b** (although these last four vectors are in the same plane) as shown in Fig. 4.14. Note that **a*** is perpendicular to **b** and that **b*** is perpendicular to **a**. In monoclinic crystals **b*** is parallel to **b** and normal to **a** and **c**.

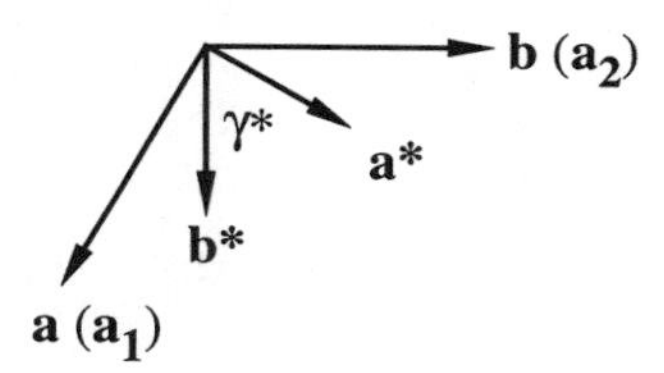


**Fig. 4.14**. The relative orientations of direct and reciprocal lattice vectors for a hexagonal lattice. $\gamma^* = 60°$.

### *4.7.2 Bragg and Miller indices*

Many text books do not distinguish between Miller indices of a plane and Bragg indices of a family of planes. In the Bragg condition for diffraction, which we write ($\lambda$ is the wavelength and $2\theta$ is the diffraction angle):

$$(2\sin\theta)/\lambda = 1/d_{hkl}$$

$d_{hkl}$ is the spacing between a family of parallel planes. It is the perpendicular distance from the origin to a plane that intercepts the axes at $1/h$, $1/k$, $1/l$, where now $h$, $k$ and $l$ may have a common divisor and are written without parentheses as in 222. $1/d_{hkl}$ is better thought of as the distance from the origin in reciprocal space to a reciprocal lattice point $h\mathbf{a}^*+k\mathbf{b}^*+l\mathbf{c}^*$ (i.e. the length of a reciprocal lattice vector). Thus $d_{111} = 2d_{222}$. On the other hand Miller indices only give the *orientation* of a plane, and $(222) \equiv (111)$.

[1]It appears that many more structures than might be expected are reported with the wrong symmetry. For some typical corrections and a discussion of the importance of knowing correct space groups see R. E. Marsh & I. Bernal, *Acta Crystallogr.* **B51**, 300 (1995). This paper should be required reading for non-crystallographers who wish to interpret the details (bond lengths, angles, etc.) of crystal structures.

### 4.7.3 More equations for triangles and tetrahedra

Some occasionally useful results for triangles and tetrahedra include:

The *medians* of a triangle join the vertices to the midpoints of the opposite edges. They all intersect at a point, the *centroid*, which divides them in the ratio 2:1.

For a triangle with edges $a$, $b$ and $c$, and opposite angles $A$, $B$ and $C$, and with $s = (a+b+c)/2$:

The diameter of the circumscribed circle is:

$$a/\sin A = b/\sin B = c/\sin C$$

The perpendicular from $C$ to $AB$ is:

$$h = c/(\cot A + \cot B) = 2\sqrt{[s(s-a)(s-b)(s-c)]}/c$$

The area is given by:

$$\text{area} = hc/2 = \sqrt{[s(s-a)(s-b)(s-c)]}$$

The medians of a tetrahedron join the vertices and the centroids of opposite faces. They all intersect at a point, also called the centroid, which divides them in the ratio 3:1. The *bimedians* join the midpoints of opposite edges. They also pass through the centroid which bisects them.

For a tetrahedron of sides $a$, $b$, $c$, $d$, $e$ and $f$ with $a$, $b$, and $c$ meeting at a common vertex $P$ the length $l$ of the median from $P$ is given by:

$$l^2 = (a^2+b^2+c^2)/3 + (d^2+e^2+f^2)/9$$

A *circumscriptible* tetrahedron is one defined by the centers of four spheres, each in contact with the other three. If the radii of the spheres are $1/p$, $1/q$, $1/r$ and $1/s$, the volume is:

$$V = [(p+q+r+s)^2 - 2(p^2+q^2+r^2+s^2)]^{1/2}/3pqrs$$

Consider a tetrahedral unit $AX_1X_2X_3X_4$ (i.e. $\{A\}X_4$), There are six angles $\theta_{ij}$ between the vectors $AX_i$ and $AX_j$ $(i \neq j = 1,2,3,4)$. These are related by:

$$\begin{vmatrix} 1 & \cos\theta_{12} & \cos\theta_{13} & \cos\theta_{14} \\ \cos\theta_{12} & 1 & \cos\theta_{23} & \cos\theta_{24} \\ \cos\theta_{13} & \cos\theta_{23} & 1 & \cos\theta_{34} \\ \cos\theta_{14} & \cos\theta_{24} & \cos\theta_{34} & 1 \end{vmatrix} = 0$$

Special cases that often arise (with symmetry) are:

(a) $\bar{4}2m$ ($D_{2d}$) $\theta_{12} = \theta_{34} = \beta$ ; $\theta_{13} = \theta_{14} = \theta_{23} = \theta_{24} = \alpha$. $\cos\beta = -(1 + 2\cos\alpha)$

(b) $3m$ ($C_{3v}$) $\theta_{12} = \theta_{13} = \theta_{14} = \alpha$ ; $\theta_{23} = \theta_{24} = \theta_{34} = \beta$ $\sin\alpha = (2/\sqrt{3})\sin(\beta/2)$

For a square-pyramidal group $AX_1X_2X_3X_4X_5$ with $X_1$ at the apex and with 4-fold symmetry $4mm$ ($C_{4v}$) so that $\theta_{12} = \theta_{13} = \theta_{14} = \theta_{15} = \alpha$ and $\theta_{23} = \theta_{34} = \theta_{45} = \theta_{25} = \beta$; $\sin\alpha = \sqrt{2}\sin(\beta/2)$.

*4.7.4. Some matrix expressions written out explicitly*

Our experience is that many people (such as ourselves) like to have matrix expressions written out explicitly in a handy place. Here are some involving the 3 × 3 matrices used in this chapter.

1. **Ax** gives another vector [here **x** is the column $(x_1 / x_2 / x_3)$]:

$$\begin{pmatrix} a_{11} & a_{12} & a_{13} \\ a_{21} & a_{22} & a_{23} \\ a_{31} & a_{32} & a_{33} \end{pmatrix} \begin{pmatrix} x_1 \\ x_2 \\ x_3 \end{pmatrix} = \begin{pmatrix} a_{11}x_1 + a_{12}x_2 + a_{13}x_3 \\ a_{21}x_1 + a_{22}x_2 + a_{23}x_3 \\ a_{31}x_1 + a_{32}x_2 + a_{33}x_3 \end{pmatrix}$$

2. det **A** (the determinant of **A** which is scalar):

$$\begin{vmatrix} a_{11} & a_{12} & a_{13} \\ a_{21} & a_{22} & a_{23} \\ a_{31} & a_{32} & a_{33} \end{vmatrix} = a_{11}a_{22}a_{33} + a_{12}a_{23}a_{31} + a_{13}a_{21}a_{32} - a_{11}a_{23}a_{32} - a_{12}a_{21}a_{33} - a_{13}a_{22}a_{31}$$

3. $\mathbf{A}^{-1}$ (the inverse of **A**, another 3 × 3 matrix):

$$\mathbf{A}^{-1} = \frac{1}{\det \mathbf{A}} \begin{pmatrix} a_{22}a_{33} - a_{23}a_{32} & a_{13}a_{32} - a_{12}a_{33} & a_{12}a_{23} - a_{13}a_{22} \\ a_{23}a_{31} - a_{21}a_{33} & a_{11}a_{33} - a_{13}a_{31} & a_{13}a_{21} - a_{11}a_{23} \\ a_{21}a_{32} - a_{22}a_{31} & a_{12}a_{31} - a_{11}a_{32} & a_{11}a_{22} - a_{12}a_{21} \end{pmatrix}$$

4. $\mathbf{A}^{\mathbf{t}}$ (the transpose of **A**): interchange rows and columns. In particular the column **x** = $(x_1 / x_2 / x_3)$ becomes the row $\mathbf{x}^{\mathbf{t}} = (x_1\ x_2\ x_3)$.

5. $\mathbf{x}^{\mathbf{t}}\mathbf{G}\mathbf{y}$ (a scalar; here **G** is a symmetric matrix with $g_{ij} = g_{ji}$):

$$\mathbf{x}^{\mathbf{t}}\mathbf{G}\mathbf{y} = g_{11}x_1y_1 + g_{22}x_2y_2 + g_{33}x_3y_3 + g_{12}(x_1y_2 + x_2y_1) + g_{13}(x_1y_3 + x_3y_1) + g_{23}(x_2y_3 + x_3y_2)$$

6. **AB** = **C** (multiplication of two matrices gives a third). The elements $c_{ij}$ of **C** are:

$$c_{ij} = a_{i1}b_{1j} + a_{i2}b_{2j} + a_{i3}b_{3j}.$$

7. $(\mathbf{AB})^t = \mathbf{B}^t\mathbf{A}^t$.

To derive Eq. 4.16 relating the **G'** matrix for a new cell to the **G** matrix for an old cell, we start with Eq. 4.8a for a distance between two points:

$$d^2 = \boldsymbol{\delta}^t\mathbf{G}\boldsymbol{\delta} \tag{4.8a}$$

Now to express old coordinates in terms of new we use $\mathbf{S}^t$ so that $\boldsymbol{\delta} = \mathbf{S}^t\boldsymbol{\delta}'$ and, using 7 above, $\boldsymbol{\delta}^t = \boldsymbol{\delta}'^t\mathbf{S}$. Substituting in Eq. 4.8a we get:

$$d^2 = \boldsymbol{\delta}'^t\mathbf{S}\mathbf{G}\mathbf{S}^t\boldsymbol{\delta}' \tag{4.8b}$$

but as the distance between two points is independent of the coordinate system, we must have:

$$d^2 = \boldsymbol{\delta}'^t\mathbf{G}'\boldsymbol{\delta}' \tag{4.8c}$$

Comparing Eqs. 4.8b and 4.8c, we see that $\mathbf{G}' = \mathbf{S}\mathbf{G}\mathbf{S}^t$ which is Eq. 4.16. The elements of **G'** were given in Eq. 4.16a (p. 112).

*4.7.5 Unit cell transformations in International Tables Vol. A*

For good reasons that we won't go into, in the *International Tables* volume *A*, the triplet of lattice vectors is written as a $1 \times 3$ row matrix (**a b c**) and the transformation to a new cell expressed as (**a' b' c'**) = (**a b c**)**P**. This means explicitly (with $p_{ij}$ the elements of **P**) that $\mathbf{a}' = p_{11}\mathbf{a}+p_{21}\mathbf{b}+p_{31}\mathbf{c}$, $\mathbf{b}' = p_{12}\mathbf{a}+p_{22}\mathbf{b}+p_{32}\mathbf{c}$, $\mathbf{c}' = p_{13}\mathbf{a}+p_{23}\mathbf{b}+p_{33}\mathbf{c}$. In Eq. 4.4 we write (**a' / b' / c'**) = **S**(**a / b / c**) so that $\mathbf{a}' = s_{11}\mathbf{a}+s_{12}\mathbf{b}+s_{13}\mathbf{c}$, $\mathbf{b}' = s_{21}\mathbf{a}+s_{22}\mathbf{b}+s_{23}\mathbf{c}$, $\mathbf{c}' = s_{31}\mathbf{a}+s_{32}\mathbf{b}+s_{33}\mathbf{c}$. It should be clear that the matrix **P** is the transpose of **S**. With this convention $\mathbf{G}' = \mathbf{P}^t\mathbf{G}\mathbf{P}$. The inverse transformation is written as (**a b c**) = (**a' b' c'**)**Q** and likewise **Q** is the transpose of our matrix **T**.

As there are two systems current, it is advisable, when specifying transformation matrices, always to specify also whether the transformation is effected by multiplying the column of lattice vectors by the matrix (i.e. using **S**), or whether it is effected by multiplying the matrix by a row of lattice vectors (i.e. using **P**). In this book, as in virtually all the literature up to the present time, we use the first system.

## 4.8 Exercises

1. A good way to check computer programs for crystallographic calculations is to take a simple cubic structure and transform to a triclinic cell.

The transformation **S** = (2 $\bar{1}$ 0 / 1 3 1 / $\bar{1}$ $\bar{1}$ 1) applied to a cubic cell with $a = a_c$ will produce a ten-times larger unit cell with $a = \sqrt{5}a_c$, $b = \sqrt{11}a_c$, $c = \sqrt{3}a_c$, $\alpha = \cos^{-1}[-\sqrt{(3/11)}]$, $\beta = \cos^{-1}(-1/\sqrt{15})$ and $\gamma = \cos^{-1}(-1/\sqrt{55})$. Interatomic distances, angles etc. in the new cell can be calculated and compared with the known values in the cubic cell.

[The inverse matrix is **T** = (0.4 0.1 –0.1 / –0.2 0.2 –0.2 / 0.2 0.3 0.7).]

Show that the coordinates of *original* lattice points 0,0,0 ; 0,0,1 ; etc. (referred to the *old* cell) are in the *new* cell (there must be ten such points):

0,0,0 ; 0.4,0.1,0.9 ; 0.2,0.3,0.7 ; 0,0.5,0.5 ; 0.8,0.2,0.8
0.6,0.4,0.6 ; 0.4,0.6,0.4 ; 0.2,0.8,0.2 ; 0.8,0.7,0.3 ; 0.6,0.9,0.1

Show that the *old* [111] and (111) become [232] and (15$\bar{1}$) respectively in the *new* cell.

2. The general and special positions of space group $Cmc2_1$ are

| | |
|---|---|
| 8 *b* | $C + (x,y,z\ ;\ \bar{x},y,z\ ;\ \bar{x},\bar{y},1/2+z\ ;\ x,\bar{y},1/2+z)$ |
| 4 *a* | $C + (0,y,z\ ;\ 0,\bar{y},1/2+z)$ |

As the $z$ coordinate appears only as $+z$ or as $1/2+z$ in each case the position of the origin on the $z$ axis is arbitrary.

$Si_2N_2O$ has this space group:

$Si_2N_2O$ — $Cmc2_1$, $a = 8.872$, $b = 5.491$, $c = 4.850$ Å.
Si 8 in *b*, $x = 0.1767$, $y = 0.1511$, $z = 0.2815$
N 8 in *b*, $x = 0.2191$, $y = 0.1228$, $z = 0.6267$ ; O 4 in *a*, $y = 0.2127$, $z = 0.230$

Plot the structure as a projection on (001). It is made up of vertex-sharing $\{Si\}N_3O$ tetrahedra which should be sketched in (cf. Fig. 4.11).

Calculate the Si-O and Si-N bond lengths. What are the coordinations of N and O by Si?

The high-pressure phase of $B_2O_3$ was reported to have space group $Ccm2_1$:

$B_2O_3$ (HP) — $Ccm2_1$, $a = 4.613$, $b = 7.803$ and $c = 4.129$ Å.
B in 8 *b*, $x = 0.161$, $y = 0.165$, $z = 0.436$
O(1) in 4 *a*, $x = 0.248$, $z = 0.5$. O(2) in 8 *b*, $x = 0.370$, $y = 0.291$, $z = 0.580$

By appropriate transformation of axes (and coordinates) *and* shift of the origin show that these two structures are essentially the same. [Transform from $Ccm2_1$ first.]

3. The structure of AgO is given in § 3.4 (p. 77). Show that if the cell is transformed according to **S** = (1/2 0 –1/2 / 1/2 –1 1/2 / –1/2 –1 –1/2) the new cell is "almost" metrically cubic with $a = 4.58$ Å, $b = c = 4.84$ Å, $\alpha = 87.9°$, $\beta = 92.6°$ and $\gamma = 87.4°$. The Ag atoms are arranged in a face-centered array if no distinction is made between the two kinds of Ag atom, but the new cell is not a true unit cell as it does not contain the same kind of Ag atom at every corner of the unit cell.

For comparison note that in $Ag_2O$ (which has the $Cu_2O$ structure Exercise 3.6) Ag is exactly on the points of a face-centered cubic lattice with $a = 4.74$ Å. The change in volume in adding two O atoms to the unit cell in going from $Ag_2O \rightarrow AgO$ is only 0.6 Å$^3$.

Show that in AgO, Ag(1) is bonded to two O atoms in a linear arrangement and that Ag(2) is bonded to four O atoms in an approximately square planar arrangement. (Calculate Ag-O bond lengths and O-Ag-O angles). [It is helpful to recognize that the Ag atoms are on inversion centers].

4. The zircon ($ZrSiO_4$) structure was given in Chapter 3, Exercise 8, and that of rutile ($TiO_2$) was given in § 3.4.

Zr is 8-coordinated in zircon. Draw the coordination polyhedron in clinographic projection.

Bragg, Claringbull & Taylor (Book List) state: "The structure of zircon...was first wrongly interpreted...as being similar to [that of] rutile" and "there is no relation between the two structures." Plot the cation positions in rutile in a projection on (110) and compare with a projection on (100) (on the same scale) of the cation positions in zircon. Be sure to draw several unit cells.

When you have finished read Hyde & Andersson (Book List) p. 285-288.

5. Data for $MgCl_2$ are:

| | |
|---|---|
| $MgCl_2$ | $R\bar{3}m$, $a_r = 6.252$ Å, $\alpha = 33.81°$ |
| | Mg in 1 *a*: 0,0,0 ; Cl in 2 *c*: ± (*x*,*x*,*x*), $x = 0.2578$ |

Show that Mg has six Cl neighbors at the vertices of an octahedron. Calculate all the edge lengths of the octahedron.

Transform to a hexagonal cell and plot the structure projected on $(11\bar{2}0)$. The $MgCl_6$ octahedra share edges to form layers normal to **c**. Outline these layers (cf. NaCl, p. 119).

6. Here are two monoclinic structures to draw in projection down **b**.

$\beta$-$Ga_2O_3$ — $C2/m$, $a = 12.23$, $b = 3.04$, $c = 5.80$ Å, $\beta = 103.7°$
All atoms are in 4 *i*: $C \pm (x,0,z)$ with:

| | Ga(1) | Ga(2) | O(1) | O(2) | O(3) |
|---|---|---|---|---|---|
| $x =$ | 0.0904 | 0.3414 | 0.1674 | 0.4957 | 0.8279 |
| $z =$ | 0.7948 | 0.6857 | 0.1011 | 0.2553 | 0.4365 |

$NaCuO_2$ — $C2/m$, $a = 6.351$, $b = 2.747$, $c = 6.103$ Å, $\beta = 120.77°$
Na in 2 *d*: $C$ + (0,1/2,1/2) ; Cu in 2 *a*: $C$ + (0,0,0) ; O in 4 *i* : 0.333,0,0.777

Determine the coordination of *all* the atoms and illustrate as cation-centered polyhedra. [Hints: See § 4.6.5 for a discussion of how to draw monoclinic structures. In $Ga_2O_3$ one Ga is in octahedral coordination and the other is in tetrahedral coordination by oxygen (i.e. $\{Ga\}O_6$ and $\{Ga\}O_4$). In $NaCuO_2$, Cu is in square planar $\{Cu\}O_4$ groups and Na is in $\{Na\}O_6$ octahedra.]

These structures are examples of an interesting phenomenon: surprisingly many monoclinic crystals can be described by an almost (metrically) orthorhombic cell. Transform the $\beta$-gallia cell by **S** = (2 0 1 / 0 1 0 / 0 0 1) and find the new unit cell parameters [you should find $\alpha' = \gamma = 90°$ and $\beta' = 89.98°$ (= 90° within the precision of the unit cell parameters given)]. Your drawing should also convince you, however, that the structure is *not* orthorhombic. If it were orthorhombic there would be mirror/glide planes and/or 2-fold axes parallel to **a** and **c**. (Contrast the situation in $MnB_4$, § 4.6.5.)

Repeat with the appropriate **S** for $NaCuO_2$. Compare your structure drawing with that for $MgCl_2$ (Exercise 5). You should have similar layers of $\{Mg\}Cl_6$ and $\{Na\}O_6$ octahedra sharing edges.

7. Another monoclinic structure that is nice to plot in projection down **b** is that of $Li_4Al_9$. All atoms lie at $y = 0$ and $y = 1/2$. A good way to illustrate the structure is to use large and small circles for Li and Al and to use open and filled circles for $y = 0$ and $y = 1/2$.

$Li_4Al_9$ — $C2/m$, $a = 19.155$, $b = 4.499$, $c = 5.429$ Å, $\beta = 107.67°$
Li(1) in 2 $a$ $C$ + (0,0,0)
Li(2) in 4 $i$: $C \pm (x,0,z)$, $x = 0.0863$, $z = 0.531$
Li(3) in 4 $i$, $x = 0.2326$, $z = 0.622$ ; Li(4) in 4 $i$, $x = 0.3080$, $z = 0.144$
Li(5) in 4 $i$, $x = 0.4564$, $z = 0.239$
Al(1) in 4 $i$, $x = 0.1505$, $z = 0.087$ ; Al(2) in 4 $i$, $x = 0.3853$, $z = 0.706$

The $\beta$-brass structure of alloys $AB$ has symmetry $Pm\bar{3}m$ with $A$ at 0,0,0 and $B$ at 1/2,1/2,1/2. Transform this structure by **S** = (4 4 1 / $\bar{1}$ 1 0 / –0.5 –0.5 3/2) using $a = 3.25$ Å for the cubic cell and plot down the new **b** axis using a similar convention to illustrate atoms. You should discover that the $Li_4Al_9$ structure is closely related to that of $\beta$-brass.

8. A triclinic structure to test computer programs for transformations is that of kyanite, $Al_2SiO_5$, which has symmetry $P\bar{1}$ with $a = 7.126$, $b = 7.852$, $c = 5.572$ Å, $\alpha = 89.99°$, $\beta = 101.11°$, $\gamma = 106.03°$. The unit cell volume is 293.56 Å$^3$.

It is claimed that the structure is based on a cubic packing of O atoms.

Show that the transformation **S** = (0.4 0.1 –0.4 / 0 0.5 0 / 0.4 0.1 0.6) indeed gives a subcell (not a unit cell!) of 1/5 the volume with $a' = 3.8627$, $b' = 3.9260$, $c' = 3.8744$ Å, $\alpha' = 90.02°$, $\beta' = 92.14°$, $\gamma' = 90.03°$ that is "nearly cubic."

The inverse matrix $\mathbf{S}^{-1}$ is **T** = (1.5 –0.5 1 / 0 2 0 / $\bar{1}$ 0 1).

9. Cartesian coordinates for a regular tetrahedron are 1,1,1 (–1,–1,1)κ. Calculate:

(a) the edge length
(b) the volume
(c) the dihedral angle
(d) the solid angle at a vertex
(e) the distance from a vertex to the centroid (at 0,0,0)
(f) the angle subtended at the centroid by an edge.

Repeat for an octahedron. Cartesian coordinates for the vertices are: (±1,0,0)κ.

## CHAPTER 5

# POLYHEDRA AND TILINGS

Chapters 5-7 are devoted to a description of simple geometrical structures that are frequently important components of crystal structures. They often arise in a purely mathematical context and we generally consider them from this point of view, deferring most of the crystallographic aspects until later. We believe that "knowing" these basic structures is essential to understanding crystal structures and structural relationships.

We start with a description of some simple polyhedra, most of which are encountered in crystal chemistry. Finite polyhedra are closed figures with polygonal faces. The polygons are in general not regular and in fact only a very small subset of all polyhedra can be made with regular polygonal faces. In crystal chemistry the interest in polyhedra arises mainly (but not entirely) because their vertices represent the positions of atoms in the coordination sphere of a central atom. For this reason we generally place small spheres (which appear as circles in the drawings) at the vertices for emphasis. We use the terms "large" and "small" polyhedra in a relative sense to mean those having many or few vertices respectively.

There is a sense in which a tiling of the plane by polygons is a special case (with an infinite number of vertices) of a polyhedron so we consider such patterns in this chapter also. The vertices of such tilings often correspond to the positions of atoms in a plane of a crystal structure.

## 5.1 Polyhedra

A number of polyhedra have been met already in Chapter 2. A more systematic listing of common convex polyhedra is now given. Convex polyhedra are those for which all dihedral angles (angles between faces) are less than 180° when viewed from inside. The discussion is by no means complete; other (coordination) polyhedra will be met subsequently. Some large polyhedra have attracted interest as molecular forms of carbon ("fullerenes") and as the shapes of viruses; these polyhedra are discussed in Appendix 4.

### *5.1.1 Regular polyhedra*

Most important and well known are the *regular* polyhedra. These are polyhedra with all vertices related by symmetry and with all faces congruent regular polygons.[1] It is trivial to show that there are at most five of these and that all possibilities can be realized. The faces must have fewer than six edges, as three regular hexagons meeting at a point must all lie in the same plane. In fact if $n$ regular $N$-gons meet at a point the angles of the polygons must be less than $360°/n$. The possibilities are given in Table 5.1. In the table the symbol $N^n$ is

[1] We restrict the discussion to convex polyhedra and thus exclude the four beautiful Kepler-Poinsot polyhedra which are constructed with intersecting faces.

the *Schläfli symbol* for the polyhedron. The regular polyhedra are also known as the Platonic solids.

**Table 5.1** The regular convex polyhedra
$n$ is the number of faces meeting at a vertex and $V$, $E$ and $F$ are the numbers of vertices, edges and faces.

| faces | $n$ | name | symbol | $V$ | $E$ | $F$ | symmetry |
|---|---|---|---|---|---|---|---|
| triangles | 3 | tetrahedron | $3^3$ | 4 | 6 | 4 | $\bar{4}3m$ |
| triangles | 4 | octahedron | $3^4$ | 6 | 12 | 8 | $m\bar{3}m$ |
| triangles | 5 | icosahedron | $3^5$ | 12 | 30 | 20 | $m\bar{3}\bar{5}$ |
| squares | 3 | cube | $4^3$ | 8 | 12 | 6 | $m\bar{3}m$ |
| pentagons | 3 | dodecahedron | $5^3$ | 20 | 30 | 12 | $m\bar{3}\bar{5}$ |

We have already met the tetrahedron, cube and octahedron but they are illustrated again in Fig. 5.1. In the figure a cube is shown on the left. Next an octahedron is shown inscribed in a cube; the vertices of the octahedron center the faces of the cube. A second polyhedron obtained by centering the faces of a polyhedron is said to be the *dual* of the first so the octahedron is the dual of the cube. It should be apparent that the polyhedron with vertices centering the faces of the octahedron is a cube (now smaller) so the cube ($4^3$) and octahedron ($3^4$) are the duals of each other. In general $p^q$ is the dual of $q^p$. The right half of Fig. 5.1 shows a tetrahedron inscribed in a cube in two different ways. It should be apparent that the dual of the tetrahedron on the left is a (smaller) tetrahedron with the same orientation as the one on the right so that the tetrahedron ($3^3$) is *self dual.*

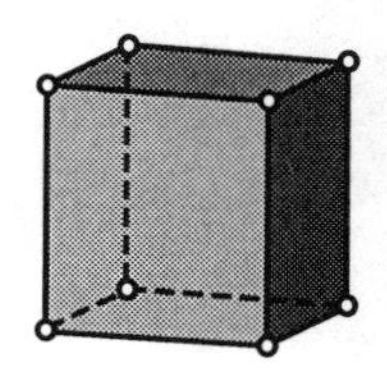
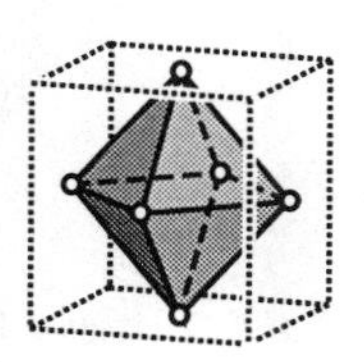
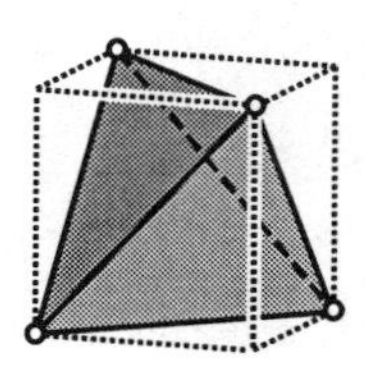
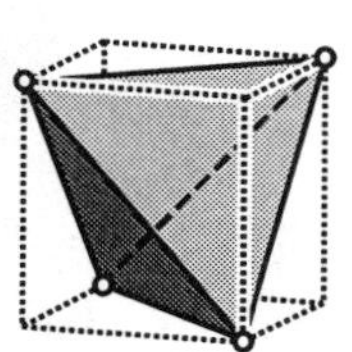

**Fig. 5.1**. From left to right: a cube, an octahedron and tetrahedra in two different orientations. Broken lines are edges obscured by the front faces, and dotted lines outline a cube.

The relationship to a cube allows an easy determination of some useful metrical properties of the octahedron and tetrahedron. Recall first that for a cube of unit edge, the length of a face diagonal is $\sqrt{2}$ and the length of a body diagonal is $\sqrt{3}$. The reader should verify that for a tetrahedron of unit edge, the distance from the center to a vertex is $\sqrt{(3/8)}$. The distance from a vertex to the center is 3/4 of the distance from a vertex to the center of the opposite face, so the perpendicular distance from a vertex to its opposite face (the "height" of the tetrahedron) is $\sqrt{(2/3)} = 0.8165$. This last result is sufficiently useful that it (including the numerical value) is worth committing to memory. The volume of a tetrahedron of unit edge is $1/(6\sqrt{2})$.

For an octahedron of unit edge, the distance from the center to a vertex is $1/\sqrt{2}$, and the distance between opposite vertices is twice that, i.e. $\sqrt{2}$. Less obvious is that the perpendicular distance between opposite faces is the same as the height of a tetrahedron with unit edge which, we repeat, is $\sqrt{(2/3)} = 0.8165$. The octahedron volume is $\sqrt{2}/3$, four times that of a tetrahedron with the same (unit) edge length.

The angle between the faces of a tetrahedron (the dihedral angle) is $\cos^{-1}(1/3) = 70.53°$ and that between adjacent faces of an octahedron is the supplementary angle $\cos^{-1}(-1/3) = 109.47°$. Opposite faces of an octahedron are parallel.

An icosahedron and a dodecahedron are illustrated in Fig. 5.2. Both have the same symmetry: $m\bar{3}\bar{5} = I_h$. We already remarked that $m\bar{3}$ is a cubic subgroup of $m\bar{3}\bar{5}$ and illustrated that point using the icosahedron in Fig. 2.25. In the illustration of the dodecahedron, eight of the twenty vertices are shown as filled circles—these lie on the vertices of a cube. Of course all the vertices are equivalent, and four other sets of eight vertices forming a cube could have been picked out. Note an important distinction. The operations of $m\bar{3}$ relate all the vertices of an icosahedron to each other, but only sets of eight of the twenty vertices of the dodecahedron are so related. This is is because the vertices of the icosahedron are on 5-fold axes, but in the dodecahedron the 5-fold axes run through the centers of pairs of opposite faces and vertices on a given face are related by a 5-fold axis which does not exist in $m\bar{3}$. In the dodecahedron the vertices are on $\bar{3}$ axes.

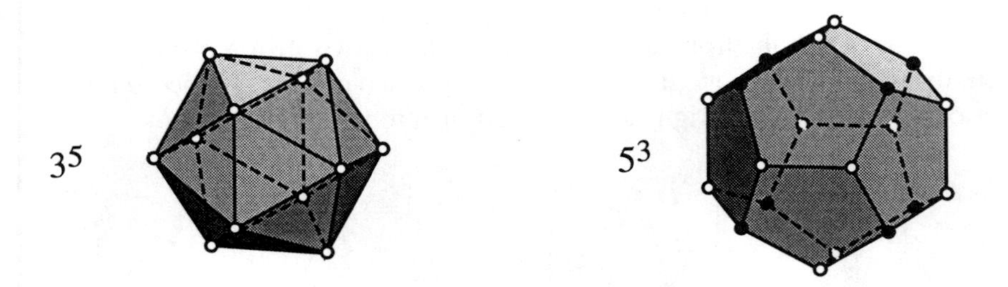


**Fig. 5.2.** An icosahedron, $3^5$ (left), and a dodecahedron, $5^3$ (right). Broken lines are edges obscured by the front faces.

Figure 5.2 should also make it apparent that the icosahedron ($3^5$) is the dual of the dodecahedron ($5^3$). Likewise the dual of the icosahedron is a (smaller) dodecahedron. For an icosahedron of unit edge, the distance from the center to a vertex is $5^{1/4}\sqrt{[(1 + \sqrt{5})/8]} = 0.9511$; i.e. slightly less than the edge length.

A note on terminology: The term "tetrahedron" refers to any polyhedron with four faces (which are triangles but not necessarily *equilateral* ones). Sometimes the term is used to refer specifically to a regular tetrahedron; this is particularly the case in such usage as "tetrahedral symmetry" which usually refers to the point group $\bar{4}3m$ ($T_d$) which is the symmetry of a regular tetrahedron, but strictly refers also to 23 ($T$) and $m\overline{3}$ ($T_h$).[1]

Similarly, the term "octahedron" strictly refers to any solid with eight faces, of which there are 257 (!) distinct examples, but it is often used only for the polyhedron with all triangular faces.[2] "Octahedral symmetry" likewise usually refers to $m\overline{3}m$ ($O_h$) which is the symmetry of a regular octahedron, but strictly should refer also to 432 ($O$). We will sometimes use terms such as "distorted octahedron" when we wish to emphasize the departure from regularity of the polyhedron with all triangular faces ($3^4$).

For many chemists the term dodecahedron refers, not to the regular dodecahedron, but to a polyhedron with twelve triangular faces (for which we prefer the term "bisdisphenoid").[3] The dodecahedron with twelve rhombic faces appears conspicuously later (we call it the rhombic dodecahedron). A regular dodecahedron is often called the pentagonal dodecahedron to distinguish it from the millions of other dodecahedra.

### *5.1.2 Combinations of octahedra and tetrahedra*

Capping two opposite faces of a regular octahedron with two tetrahedra with equal edge lengths will produce a rhombohedron as shown in Fig. 5.3. The angles of the faces are 60° and 120° and three angles of 60° meet at two of the vertices. We note that a rhombohedron has symmetry $\overline{3}m$ and that the two opposite vertices where three equal angles $\alpha$ meet are on the $\overline{3}$ axis. If $\alpha < 90°$ the rhombohedron is called *acute* (or *prolate*), and if $\alpha > 90°$ the rhombohedron is termed *obtuse* (or *oblate*). If $\alpha = 90°$ the rhombohedron is a cube of course. The rhombohedron with $\alpha = 60°$ will be met many times; it is the primitive cell of the face-centered cubic lattice.

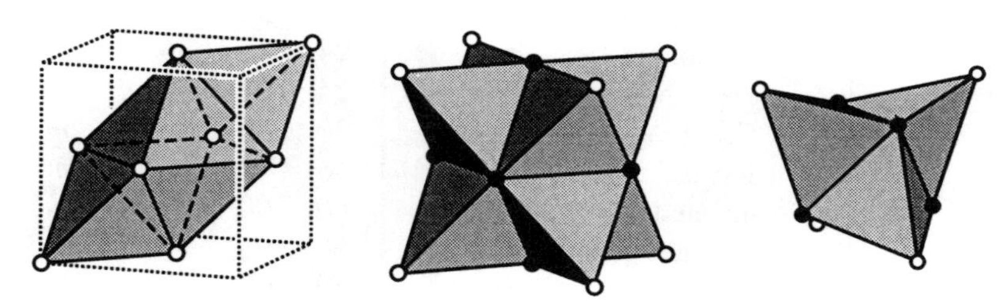

**Fig. 5.3.** Left: a 60° rhombohedron produced by capping two opposite faces of an octahedron with tetrahedra (compare with Fig. 5.1). Middle: a stella octangula produced by capping all the faces of an octahedron with tetrahedra—filled circles are the octahedron vertices. Right: a stella quadrangula obtained by capping all the faces of a tetrahedron—filled circles are the vertices of the central tetrahedron.

[1] Please note that $T_d$ refers to a symmetry group and is *not* shorthand for the word "tetrahedral". Thus an atom at a site with $T_d$ symmetry may, or may not, be four-coordinated; and conversely an atom with four neighbors at the vertices of a tetrahedron may, or may not, be at a site of $T_d$ symmetry.

[2] Some readers will be familiar with "Dürer's octahedron" which is a conspicuous feature of the celebrated engraving *Melencolia I* (1514). This polyhedron is an acute rhombohedron with the two acute vertices truncated and has two triangular faces and six pentagonal ones. Truncated tetrahedra and hexagonal prisms, discussed below, are other familiar examples of octahedra in this general sense.

[3] Another polyhedron with twelve triangular faces is the hexagonal bipyramid (see § 5.1.5).

Similarly capping *all* the faces of an octahedron with tetrahedra will produce the polyhedron known as a *stella octangula* shown in the middle of Fig. 5.3.[1] As the shading in the figure suggests, the stella octangula can also be considered as two interpenetrating larger tetrahedra.

Also shown in Fig. 5.3 is the polyhedron obtained by capping the four faces of a tetrahedron with tetrahedra. By analogy it is called a *stella quadrangula*; it is of some interest in crystal chemistry.[2] See § 5.2.4 and § 5.6.8 for some examples of its occurrence.

Other combinations of tetrahedra and/or octahedra will be met subsequently (§ 5.2). Of great importance in crystal chemistry is the fact that regular octahedra and tetrahedra with equal edges and in the number ratio of 1:2 can be packed together to fill space; this is a topic covered in Chapter 6.

### *5.1.3 Archimedean polyhedra*

**Table 5.2.** The Archimedean polyhedra other than prisms and antiprisms. *V*, *E* and *F* are the numbers of vertices, faces and edges. The numbers *N* in the Schläfli symbol represent in cyclic order the polygons (*N*-gons) meeting at a vertex.

| Schläfli symbol | name | *V* | *E* | *F* | symmetry |
|---|---|---|---|---|---|
| $3.6^2$ | truncated tetrahedron | 12 | 18 | 8 | $\bar{4}3m$ |
| 3.4.3.4 | cuboctahedron | 12 | 24 | 14 | $m\bar{3}m$ |
| $4.6^2$ | truncated octahedron | 24 | 36 | 14 | $m\bar{3}m$ |
| $3.8^2$ | truncated cube | 24 | 36 | 14 | $m\bar{3}m$ |
| $3.4^3$ | rhombicuboctahedron | 24 | 48 | 26 | $m\bar{3}m$ |
| $3^4.4$ | snub cube | 24 | 60 | 38 | 432 |
| 3.5.3.5 | icosidodecahedron | 30 | 60 | 32 | $m\bar{3}\bar{5}$ |
| 4.6.8 | truncated cuboctahedron | 48 | 72 | 26 | $m\bar{3}m$ |
| $3.10^2$ | truncated dodecahedron | 60 | 90 | 32 | $m\bar{3}\bar{5}$ |
| $5.6^2$ | truncated icosahedron | 60 | 90 | 32 | $m\bar{3}\bar{5}$ |
| 3.4.5.4 | rhombicosidodecahedron | 60 | 120 | 62 | $m\bar{3}\bar{5}$ |
| $3^4.5$ | snub dodecahedron | 60 | 150 | 92 | 235 |
| 4.6.10 | truncated icosidodecahedron | 120 | 180 | 62 | $m\bar{3}\bar{5}$ |

Polyhedra with equivalent (i.e. symmetry-related) vertices but with more than one kind of regular polygonal face are referred to as *semi-regular* or *Archimedean*. These consist of the infinite families of regular prisms and antiprisms together with thirteen others. Prisms and antiprisms are discussed in the next section; the remaining polyhedra, all of which have either cubic or icosahedral symmetry, are listed in Table 5.2, together with the number of vertices (*V*), edges (*E*) and faces (*F*). These quantities are related by the Euler condition for

[1]So christened by the great geometer, astronomer and mystic, Johannes Kepler (1571-1630).
[2]See Hyde & Andersson (Book List) p. 342.

finite polyhedra $V - E + F = 2$. Of the polyhedra in the table, the cubic ones are the most important in crystal chemistry, but some of the others are also met. Some are conveniently derived by cutting off (truncating) vertices of simpler polyhedra in a symmetrical way and hence their names.

The rhombicuboctahedron ($3.4^3$) is sometimes called the *small* rhombicuboctahedron and the truncated cuboctahedron (4.6.8) is sometimes called the *great* rhombicuboctahedron. We find it easier to remember, and to recall mentally, these polyhedra from their Schläfli symbols rather than from their often cumbersome names.[1]

The cuboctahedron (3.4.3.4) was introduced in Fig. 2.25 (p. 54) where we saw that it was simply related to the regular icosahedron. It is illustrated again in Fig. 5.4 together with its dual, the rhombic dodecahedron, which we will describe in § 5.1.5 below. The cuboctahedron and its relative the "twinned cuboctahedron" are discussed further when we consider sphere packings in Chapter 6. An important property of the cuboctahedron is that distance from the center to a vertex is equal to the edge length. It is also an example of a *quasiregular* polyhedron, which is one in which all edges and vertices (but not faces) are equivalent. (3.5.3.5 is the other quasiregular polyhedron.) As the edges are all equivalent, there is only one dihedral angle which is $\cos^{-1}(-1/\sqrt{3}) = 126.26°$.

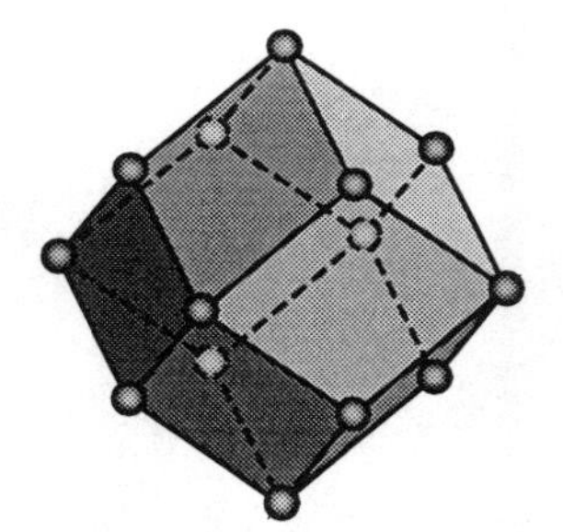
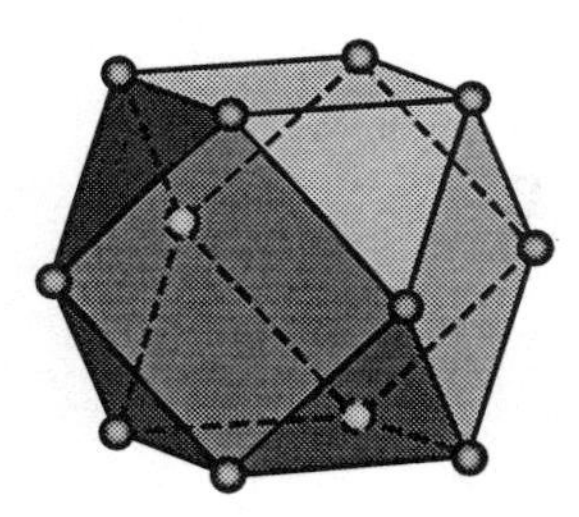

**Fig. 5.4**. Right: a cuboctahedron (3.4.3.4). Left: a rhombic dodecahedron, the dual of the cuboctahedron. Broken lines are edges obscured by the front faces.

The remaining cubic Archimedean polyhedra are illustrated in Fig. 5.5. They are sufficiently important that it is worth learning their names and shapes.

The truncated tetrahedron ($3.6^2$) is an important 12-coordination polyhedron. It occurs particularly in structures of intermetallic compounds (see § 5.2.5). The reader with polyhedra at hand may like to verify that four regular octahedra and seven regular tetrahedra (all with equal edge length) can be combined to make a truncated tetrahedron.

The snub cube ($3^4.4$) occurs in left and right-handed forms; Fig. 2.26 (p. 54) illustrates both and their relationship to the rhombicuboctahedron ($3.4^3$).

Truncated octahedra ($4.6^2$) have the interesting property that they can be packed together

[1]A reminder that the numbers $N$ in the Schläfli symbol represent in cyclic order the polygons ($N$-gons) meeting at a vertex. Schläfli symbols are most useful for polyhedra with just one kind of vertex and we do not use them for polyhedra such as the rhombic dodecahedron (Fig. 5.4) with two different kinds of vertex ($4^3$ and $4^4$).

to fill space as discussed in Chapter 7. In such an array the centers of the polyhedra fall on a body-centered cubic lattice and the surface of a polyhedron encloses all points in space closer to the lattice point at its center than to any other lattice point, so it is the *Voronoi polyhedron* associated with the lattice point. In a terminology used by physicists, it is the *Wigner-Seitz cell* of the body-centered cubic lattice. The dihedral angles are $\cos^{-1}(-1/3) = 109.47°$ between square and hexagonal faces and $\cos^{-1}(-1/\sqrt{3}) = 125.26°$ between hexagonal faces.

We meet the truncated cuboctahedron (4.6.8) later also in connection with polyhedron packings and in zeolite structures. It is the largest polyhedron with equivalent vertices and cubic symmetry (the number of vertices is 48—equal to the order of group $m\bar{3}m$).

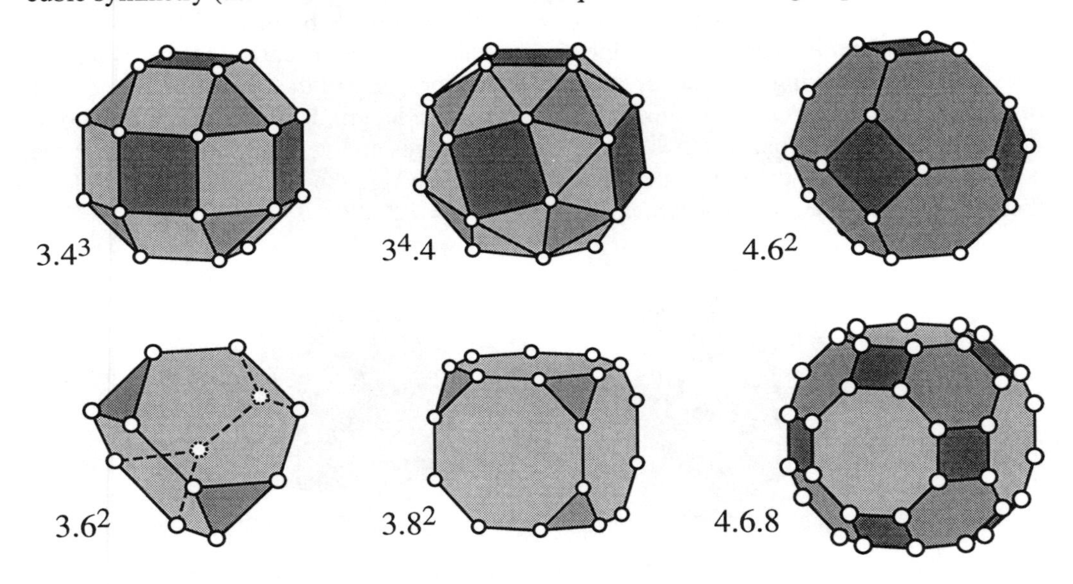


**Fig. 5.5.** Six of the cubic Archimedean polyhedra. Top left: rhombicuboctahedron ($3.4^3$). Top middle: snub cube ($3^4.4$). Top right: truncated octahedron ($4.6^2$). Bottom left: truncated tetrahedron ($3.6^2$). Bottom middle: truncated cube ($3.8^2$). Bottom right: truncated cuboctahedron (4.6.8). In each polyhedron faces related by symmetry have the same depth of shading. For $3.6^2$ (only) edges obscured by the front faces are shown as broken lines.

The six beautiful icosahedral Archimedean polyhedra are illustrated in Fig. 5.6. It will be found that models of them are useful aids to appreciating icosahedral symmetry. Most familiar will be the truncated icosahedron ($5.6^2$) which represents the structure of $C_{60}$;[1] it may also be familiar as the structure of a soccer ball (balls based on 3.5.3.5 are also sometimes seen). It should be apparent that $3^4.5$ is related to 3.4.5.4 in the same way as $3^4.4$ is related to $3.4^3$. Note that $3^4.5$ with symmetry 235 (= *I*) is enantiomorphic.

[1]Note that there two distinct edges in the truncated icosahedron; one (66) that is common to hexagons and a second (65) common to a hexagon and a pentagon. In $C_{60}$ the 66 bond is significantly shorter than the 65 bond. The symmetry remains icosahedral of course.

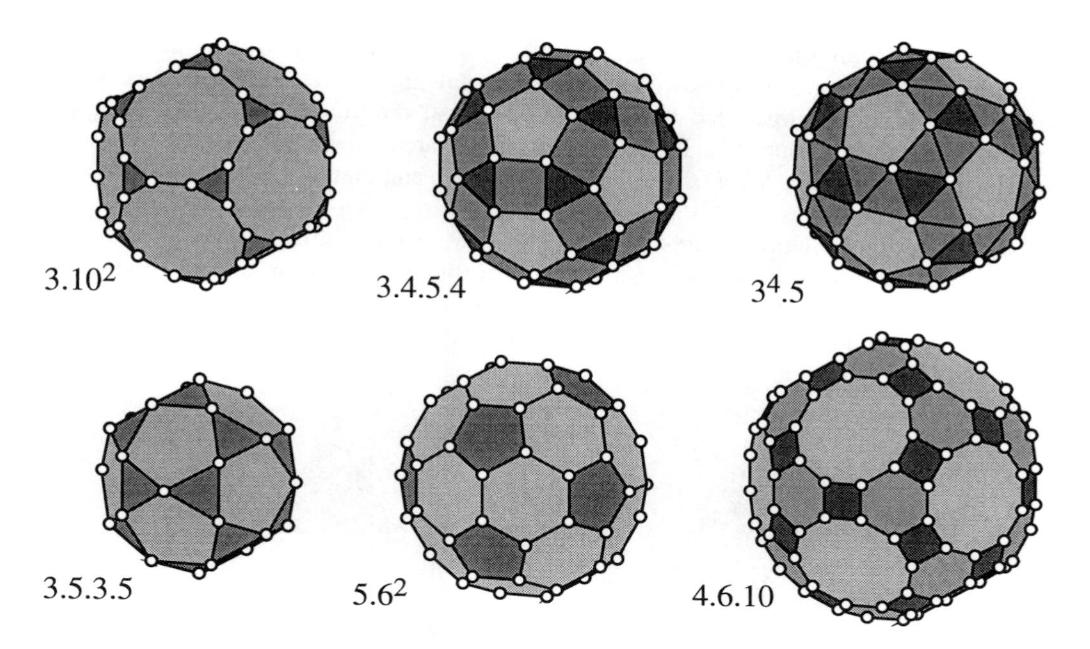


**Fig. 5.6**. The icosahedral Archimedean polyhedra. Top left: truncated dodecahedron ($3.10^2$). Top middle: rhombicosidodecahedron (3.4.5.4) Top right: snub dodecahedron ($3^4.5$). Bottom left: icosidodecahedron (3.5.3.5). Bottom middle: truncated icosahedron ($5.6^2$). Bottom right: truncated icosidodecahedron (4.6.10). In each polyhedron, faces related by symmetry have the same depth of shading.

### *5.1.4 Prisms, antiprisms and capped prisms*

The remaining semiregular solids are prisms and antiprisms. The semiregular prisms are right prisms with two faces that are $N$-gons and $N$ square faces (symbol $4^2.N$). The triangular (or *trigonal*) prism should be familiar. It has Schläfli symbol $3.4^2$. Note that the triangular prisms encountered as coordination figures in solids usually have rectangular faces that are not square (see the Note § 5.6.2 at the end of this chapter). A cube is a square prism. The term "prism" is almost invariably used by crystal chemists to mean "right prism" (i.e. that *all* the quadrangular faces are rectangles).

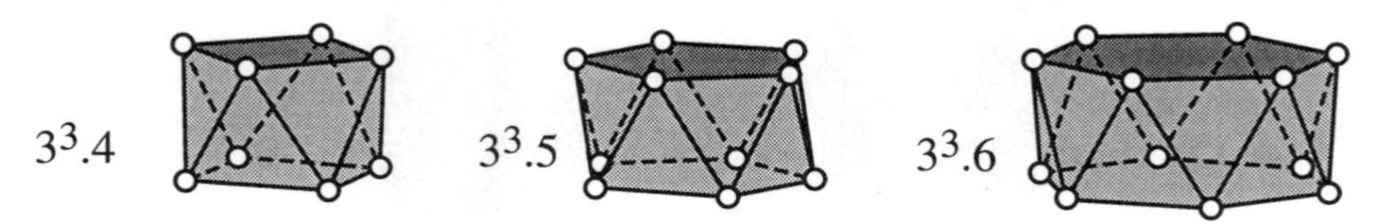


**Fig. 5.7**. From left to right: a square antiprism ($3^3.4$), a pentagonal antiprism ($3^3.5$) and a hexagonal antiprism ($3^3.6$). Edges obscured by the front faces are shown as broken lines.

Antiprisms conclude the enumeration of the semiregular solids. They have two $N$-gon faces and $2N$ triangular ones (symbol $3^3.N$). The regular octahedron is a triangular antiprism. The square antiprism ($3^3.4$), pentagonal antiprism ($3^3.5$) and hexagonal antiprism ($3^3.6$) are illustrated in Fig. 5.7. Note that they have non-crystallographic symmetries (see § 2.5.6, p. 52). The symmetry of a square antiprism is $\bar{8}m2 = D_{4d}$, that of a pentagonal antiprism is $\bar{5}2/m$ (short symbol $\bar{5}m$) $= D_{5d}$ and that of a hexagonal antiprism is $\overline{12}m2 = D_{6d}$.

Sometimes coordination figures (particularly for 6- or 8-coordination) are encountered that are intermediate between prisms and antiprisms. We refer to such solids as *metaprisms.* Fig. 5.8 illustrates the case of 6-coordination. Square metaprisms are also common coordination figures. The symmetry of metaprisms with $2N$ vertices is $N2$ for $N$ odd and $N22$ for $N$ even.

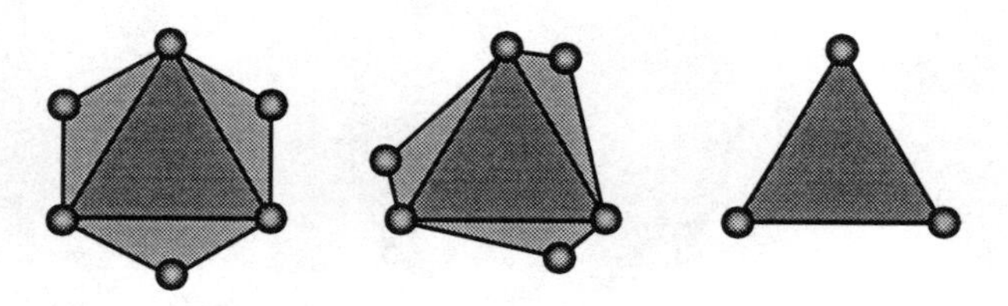

**Fig. 5.8**. Left: a trigonal antiprism (symmetry $\bar{3}m$) with the bottom face the same as the top one but rotated by 60°. Middle: a trigonal metaprism (symmetry 32) with bottom face the same as the top one but rotated by 30°. Right a trigonal prism (symmetry $\bar{6}m2$).

The process of adding an extra vertex outside a polyhedron is usually known as *capping*. Examples of particular interest in chemistry are coordination figures obtained by capping the rectangular faces of trigonal prisms. According to whether there are one, two or three such extra vertices, the polyhedra are referred to as monocapped, bicapped or tricapped trigonal prisms respectively. A tricapped trigonal prism is illustrated in Fig. 5.9 in two ways; first as the full polyhedron and secondly with the "capping" vertices treated separately, as is often done for clarity in crystal structure drawings. As already remarked, the rectangular faces are generally not square, but elongated along the direction of the $\bar{6}$ axis. The symmetry is $\bar{6}m2$.

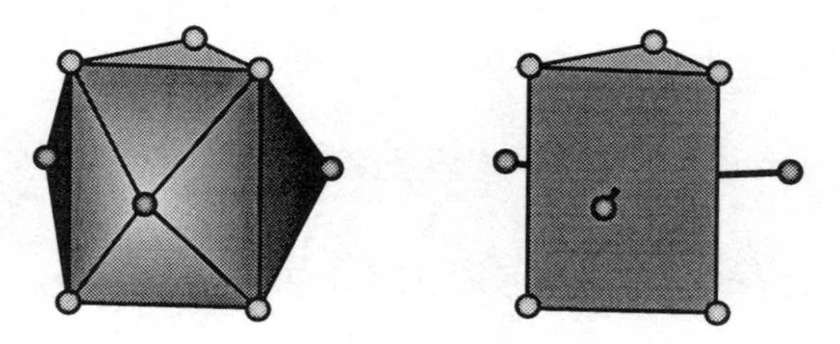

**Fig. 5.9**. Two illustrations of a tricapped trigonal prism with all distances from the vertices to the center equal and with all edges not parallel to the $\bar{6}$ axis equal.

Capping the two pentagonal faces of a pentagonal antiprism results in an icosahedron.

### *5.1.5 Catalan polyhedra: the rhombic dodecahedron, bipyramids and pyramids*

The *Catalan* solids (named after the Belgian mathematician) are the duals of the Archimedean polyhedra. The Archimedean polyhedra have all *vertices* equivalent, so the Catalan polyhedra have all *faces* equivalent. The Archimedean polyhedra are generally of interest as symmetrical coordination polyhedra, whereas the Catalan polyhedra are more relevant to the external shapes of symmetrical crystals. As our interest is mainly in the internal structure of crystals we only mention here some Catalan polyhedra of special interest.[1]

The most interesting to us is the dual of the cuboctahedron. This is the rhombic dodecahedron (Fig. 5.4, p. 137) which has twelve faces that are rhombuses with alternating angles of $\cos^{-1}(-1/3) = 109.47°$ and $\cos^{-1}(1/3) = 70.53°$. The cuboctahedron has 14 faces and 12 vertices whereas the rhombic dodecahedron has 12 faces and 14 vertices. Both polyhedra have 24 edges that are all equivalent in each case. The dihedral angles of the rhombic dodecahedron are thus all the same (and equal to 120°), and it is another a space-filling polyhedron. When packed together to fill space, the polyhedron centers fall on the points of a face-centered cubic lattice. It is the Voronoi polyhedron (Wigner-Seitz cell) of the face-centered cubic lattice.

The duals of the right prisms are the bipyramids. A cube is a special case of a square prism and an octahedron is a square bipyramid. A pyramid is of course half of a bipyramid. Pyramids are self dual (a tetrahedron is a triangular pyramid). The two distinct polyhedra with five vertices are the trigonal bipyramid and the square pyramid; Fig. 5.10.

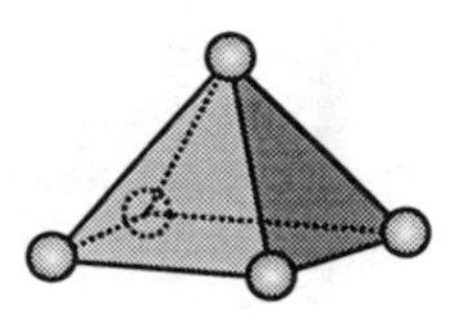

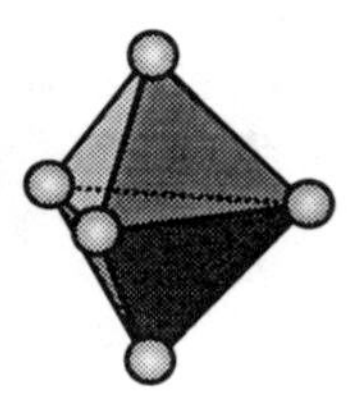

**Fig. 5.10**. The two polyhedra with five vertices. Left: a square pyramid. Right: a triangular bipyramid.

### *5.1.6 Deltahedra and the bisdisphenoid*

Another class of polyhedra that we consider in this chapter is that of *deltahedra*. These are convex polyhedra in which all faces are equilateral triangles. Some have been met before. The maximum number of triangles meeting at a vertex is five and the minimum is three. The minimum number of vertices is four (in the regular tetrahedron) and the maximum number is twelve (in the icosahedron). In Table 5.3, $V_3$, $V_4$ and $V_5$ are the numbers of vertices at which 3, 4 or 5 triangles meet.

[1]The Catalan polyhedra are described in detail by Cundy & Rollett (Book List). A good account in the context of crystal shapes is given by J. V. Smith (Book List).

**Table 5.3.** The eight convex deltahedra

| vertices | $V_3$ | $V_4$ | $V_5$ | faces | edges | name |
|---|---|---|---|---|---|---|
| 4 | 4 | 0 | 0 | 4 | 6 | tetrahedron |
| 5 | 2 | 3 | 0 | 6 | 9 | trigonal bipyramid |
| 6 | 0 | 6 | 0 | 8 | 12 | octahedron |
| 7 | 0 | 5 | 2 | 10 | 15 | pentagonal bipyramid |
| 8 | 0 | 4 | 4 | 12 | 18 | bisdisphenoid |
| 9 | 0 | 3 | 6 | 14 | 21 | tricapped trigonal prism |
| 10 | 0 | 2 | 8 | 16 | 24 | bicapped square antiprism |
| [*11?* | *0* | *1?* | *10?* | *18?* | *27?* | *impossible!*] |
| 12 | 0 | 0 | 12 | 20 | 30 | icosahedron |

The reader should observe the regularity of the numbers of vertices, edges and faces, and is invited to prove the impossibility of the deltahedron with eleven faces (not an entirely trivial task). The polyhedron with ten vertices is obtained by capping each of the two square faces of a square antiprism with a square pyramid.

The boron atom skeletons of the *closo* borane anions $B_nH_n^{2-}$ are close to ideal deltahedra for $6 \geq n \geq 12$, except for $n = 11$; however $B_{11}H_{11}^{2-}$ *does* exist, as a polyhedron with irregular triangular faces in which $V_4 = 2$, $V_5 = 8$, and $V_6 = 1$. The isoelectronic carboranes $B_{n-2}C_2H_n$ are isostructural.

The *bisdisphenoid* is an interesting polyhedron with symmetry $\bar{4}2m$ that is illustrated in Fig. 5.11. We will use this term for dodecahedra (= 12 faces!) with this symmetry even when the faces are not *equilateral* triangles. It is a rather common 8-coordination figure.

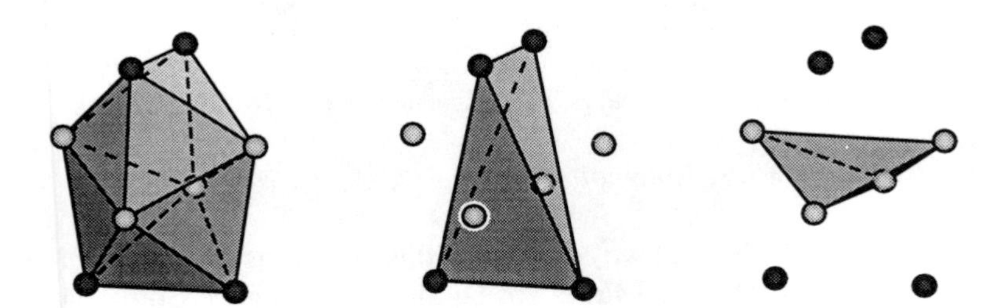

**Fig 5.11**. Left: A bisdisphenoid. The $\bar{4}$ axis is vertical and the faces are equilateral triangles. In the center and on the right the two sets of four vertices on "elongated" and "squashed" tetrahedra are indicated.

The term sphenoid (which comes from the Greek word for a wedge) refers to a figure with inclined planes meeting at an edge. A tetrahedron can be decomposed into two pairs of such sphenoids and is sometimes called a disphenoid. The origin of the term *bisdisphenoid* come from the fact that a bisdisphenoid can be decomposed into two tetrahedra, one "elongated" and one "squashed" as indicated in Fig. 5.11. It should be apparent that the distance from the vertices of the squashed tetrahedron to the center is less than the distance from the vertices of the elongated tetrahedron to the center. In the bisdisphenoids occurring in crystal structures, the distances from the vertices to the center are often (approximately) equal; the faces can no longer be equilateral triangles in this case.

If we remove the restriction to *convex* polyhedra, there are many other possibilities for deltahedra; one of them is the polyhedron constructed from two inter-penetrating icosahedra (shown in Fig. 5.12) which has six triangles meeting at some of the vertices.

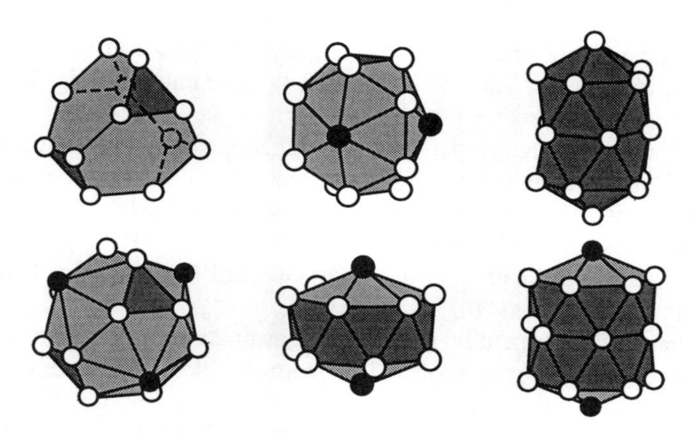

**Fig. 5.12.** Frank-Kasper and related polyhedra. On the left is shown on the top a truncated tetrahedron and on the bottom the Friauf polyhedron ($V_6 = 4$) obtained by capping the hexagonal faces of the truncated tetrahedron. In the middle on top is shown the 15 vertex polyhedron ($V_6 = 3$) and underneath it is the 14 vertex polyhedron ($V_6 = 2$) obtained by capping the hexagonal faces of a hexagonal antiprism. In these four drawings, 6-coordinated vertices are shown as filled circles and equilateral triangles are darker shaded. On the right are shown top: intergrown icosahedra (capped pentagonal antiprisms) and bottom: intergrown capped hexagonal antiprisms.

### *5.1.7 Frank-Kasper polyhedra and intergrown polyhedra*

If the restriction to equilateral triangles is lifted, there is an infinite number of polyhedra with triangular faces. These are often referred to as *simplicial* polyhedra. Of special interest are those in which either five or six triangles meet at a vertex. Let the number of such vertices be $V_5$ and $V_6$. It is easy to show that $V_5 = 12$ in every case. "Geodesic" domes are parts of such polyhedra. Their duals have faces that are either hexagons or pentagons, and three meet at every vertex; for these latter polyhedra there must be exactly twelve pentagonal faces—see Exercise 6 and Appendix 4.

The *Frank-Kasper* polyhedra (Fig. 5.12) are the four simplest simplicial polyhedra with $V_5 = 12$; they are found as coordination figures in dense intermetallic structures. $V_6 = 0$ corresponds to the icosahedron (12 vertices). As a periodic structure cannot have 5-fold symmetry axes, icosahedra in crystal structures will not be strictly regular, but they often come surprisingly close. The next case, $V_6 = 1$ cannot be realized. The next three possibilities, $V_6 = 2$, 3, and 4, can be realized and have 14, 15, and 16 vertices respectively. They are the only possibilities not having contiguous 6-coordinated vertices. The edges meeting at a 6-coordinated vertex must be longer than the other edges if the polyhedron is to be convex.

The polyhedron with 14 vertices is made from a hexagonal antiprism by capping the hexagonal faces. The polyhedron with 16 vertices is similarly derived from a truncated tetrahedron ($3.6^2$) and is often called a Friauf polyhedron (it is a conspicuous feature of the structures of the intermetallic phases known as Friauf-Laves phases—see § 6.6.3). The polyhedron with 15 vertices (see Fig. 5.12) has $\bar{6}m2$ symmetry (the same as that of a trigonal prism) and is sometimes called the $\mu$-phase polyhedron (after the structure in which it was first identified). The 16-, 15-, and 14-coordinated polyhedra are sometimes symbolized $P$, $Q$, and $R$ respectively.

"Ionic" crystal structures are typified by the fact that generally the coordination of cations is entirely by anions and *vice versa*. These structures can be described in terms of catenated (vertex-, edge-, or face-sharing) coordination polyhedra. It may be recalled, for example, that we described the quartz structure in terms of corner-sharing $\{Si\}O_4$ polyhedra in § 3.6. These structures are characterized by generally low coordination numbers (average for all atoms $< 8$).

On the other hand, in intermetallic structures, coordination numbers are generally higher ($\geq 12$) and usually include coordination of like atoms by like. If they are to be described in terms of coordination polyhedra such as the polyhedra of this section, it will be found that the polyhedra must interpenetrate each other. Fig. 5.12 illustrates intergrowths of two icosahedra and also of two bicapped hexagonal antiprisms. The vertices in the interior of the intergrowths are (approximately) at the centers of icosahedra and bicapped hexagonal antiprisms respectively, and thus serve as both the *center* of one polyhedron and as the *vertex* of another. In intermetallic crystal structures there is a continuous intergrowth (rather than just the pairs discussed here) which makes such structures particularly difficult to illustrate satisfactorily. In $\mathbf{MgCu_2}$, for example (§ 6.6.3), there are intergrown $\{\mathbf{Mg}\}\mathbf{Cu_{12}Mg_4}$ Friauf polyhedra which produce, *mirabile dictu*, for the **Cu** coordination $\{\mathbf{Cu}\}\mathbf{Mg_6Cu_6}$ icosahedra.

Simplicial polyhedra can be considered as made up of (irregular) tetrahedra with one vertex at the center and the other three on a face of the polyhedron. In this way, the icosahedron can be decomposed into 20 tetrahedra and the Friauf polyhedron similarly decomposed into 28 tetrahedra. Structures which can be described as a packing of irregular tetrahedra are often described as "topologically close-packed."

### *5.1.8 Relationships between polyhedra with eight vertices*

We mentioned earlier the fact that there are 257 topologically distinct octahedra. Their duals will be the 257 distinct polyhedra with eight vertices.[1] Some other symmetrical arrangements that we have met and that often occur in crystal (and molecular) structures are

[1] A complete catalog of polyhedra with $\leq 8$ vertices is given by D. Britton & J. D. Dunitz, *Acta Crystallogr.* A**29**, 362 (1973).

the cube, square antiprism, bicapped trigonal prism and hexagonal bipyramid. Often in polyhedra representing coordination of a central atom, the distances from the vertex atoms to the central atom are approximately equal, so the vertices are approximately on the surface of a sphere. As the number of vertices increases, conversion from one polyhedron to another will generally involve smaller displacements of vertices. Because 8-coordination is rather common, and the possibilities large, the relationships between the possible coordination polyhedra have been the subject of some discussion.[1] We describe some important cases briefly here.

The relationship between a cube (square prism) and square antiprism should be obvious. The intermediate case, a square metaprism, with point symmetry 422, is commonly encountered in crystals.

The relationship between a cube and a hexagonal bipyramid is also very simple. Fig. 5.13 shows a cube viewed down a body diagonal. The six vertices not on the diagonal project as a hexagon; they are in fact on a skew hexagon with 90° angles. Converting these angles to 120° will produce a planar hexagon and the vertices will correspond to a hexagonal bipyramid. The "waist" edges of the hexagonal bipyramids must be shorter than the other edges; if the edges were all equal, the polyhedron would collapse to a two-dimensional figure. The $\{Y\}O_8$ coordination in the structure of $Y_2Ti_2O_7$ is an example of coordination intermediate between a hexagonal bipyramid and a cube.

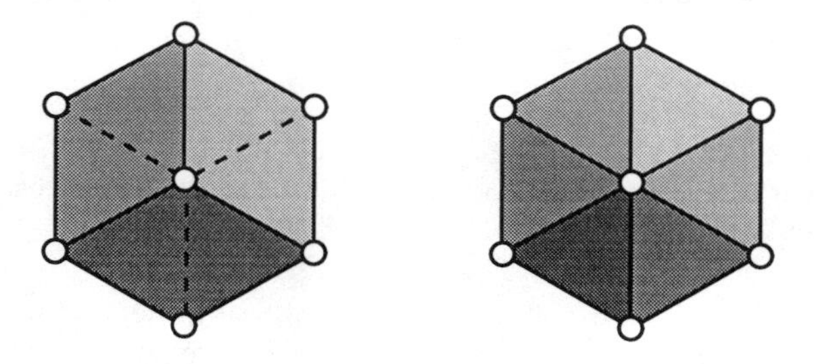

**Fig. 5.13**. The relationship between a cube and a hexagonal bipyramid.

Less obvious are the relationships between a square antiprism, a bisdisphenoid and a bicapped trigonal prism, but in fact these three polyhedra are very closely related as shown in Fig. 5.14. Converting the two square faces of the antiprism to pairs of triangles produces the topology of the bisdisphenoid and only small displacements of the vertices are necessary to produce $\bar{4}2m$ symmetry. Note that the $\bar{4}$ axis of the bisdisphenoid is parallel to one of the 2-fold axes of the antiprism. Similarly, converting just one of the faces of the square antiprism to a pair of triangles converts it into a bicapped trigonal prism.

The conversion directly from a cube to a bisdisphenoid is also important and should be obvious when it is recalled that the vertices of a cube are the same as the convex vertices of a stella quadrangula (Fig. 5.3) which can be considered as two interpenetrating regular tetrahedra. Converting one of these tetrahedra into a squashed tetrahedron and the other into

[1]See especially D. L. Kepert, *Inorganic Stereochemistry* [Springer-Verlag, Berlin (1982)].

an elongated one will produce a bisdisphenoid (compare with Fig. 5.11).

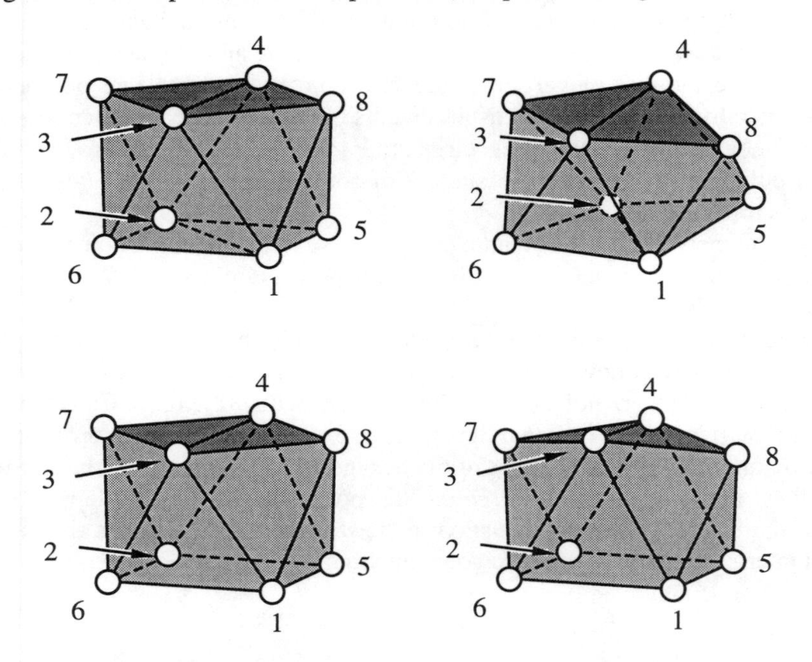


**Fig. 5.14**. Top: the relationship between a square antiprism (left) and a bisdisphenoid (right). Note that the $\bar{4}$ axis of the bisdisphenoid is horizontal. Bottom: the relationship between a square antiprism (left) and a bicapped trigonal prism (right).

It should be emphasized that, in the context of crystal or molecular structures, the use of polyhedra to describe the positions of atoms is purely for convenience. In a crystal there are only atoms and (less surely) bonds. On the other hand, the salient features of a polyhedron to the eye are its faces (hence the term *polyhedron*) which are really of little importance in the crystal structure.

In particular, to illustrate structural relationships, it is often fruitful to chose one or another polyhedron to describe a particular structure element, and even to describe the same coordination in more than one way. In the garnet structure, Ca is eight coordinated by O and has a site symmetry 222. This symmetry is a subgroup of that of a cube ($m\bar{3}m$), a square metaprism (422), a bisdisphenoid ($\bar{4}2m$) and a bicapped trigonal prism ($mm2$). It is not surprising therefore, that different authors describe the coordination figure differently [as e.g. a "twisted cube," "distorted square antiprism," or "distorted dodecahedron" (meaning bisdisphenoid)].

### 5.1.9 Tammes' problem and coordination polyhedra

In his study of pollen grains (which are approximately spherical), the biologist P. M. L.

Tammes concluded that the orifices on their surfaces are arranged so that there is a maximum number, subject to the constraint that the distance between them is not less than some minimum amount. It might be supposed that similar considerations also apply to packing coordinating atoms around a central atom. The mathematical expression of Tammes' problem is to find the arrangement of points on the surface of a sphere so that the shortest distance between pairs of points is as large as possible. Solutions of Tammes' problem, particularly its generalization to arrangements of points in space, are here referred to as *eutactic* arrangements.[1]

The general problem is difficult and most results have been obtained numerically.[2] For four points the solution is provided by the vertices of a regular tetrahedron, and for six points by the vertices of a regular octahedron.

The solution for five points is provided both by the vertices of a trigonal bipyramid and by the vertices of a square pyramid with all edges equal (i.e. half a regular octahedron). Note that in this case the shortest distance is $\sqrt{2}r$, where $r$ is the radius of the sphere; the same result is obtained for an octahedron, so there might as well be six points on the surface of the sphere rather than five.[3] It is noteworthy that 4- and 6-coordination are much more commonly found than 5-coordination of cations in crystals. When 5-coordination *is* found, the square pyramid and trigonal bipyramid occur with comparable frequencies.[4]

For eight points the solution to Tammes' problem is provided by the vertices of an Archimedean square antiprism (*not*, as is occasionally claimed, a cube). For nine points the figure obtained is that of a tricapped trigonal prism. We may note also that for twelve points the solution to Tammes' problem is provided by the vertices of a regular icosahedron and for 24 points by the vertices of a snub cube ($3^4.4$). See also the Notes (§ 5.6.2) for a discussion of related problems.

### *5.1.10 Polyhedra with divalent vertices*

In the polyhedra considered so far at least three edges meet at a vertex and each pair of contiguous edges (an *angle*) is part of only one face. However for some purposes it is convenient to generalize the concept of polyhedra to include those in which only two edges

[1]From the Greek for "well-arranged". We thought that we had invented the word, or rather that it was coined for us by Marie Hyde who knows about such things, but subsequently we found that it had been used earlier by Schläfli in a mathematical context [see H. S. M. Coxeter *Regular Polytopes*, 3rd Edition, Dover, New York (1973) p. 251] and it is in the *Oxford English Dictionary*. As the term has now gained some currency in the sense we use it, it is retained.

[2]Recent papers with extensive results are B. W. Clare & D. L. Kepert, *Proc. Roy. Soc. (London)* A**405**, 329 (1986); D. A. Kottwitz, *Acta Crystallogr.* A**47**, 158 (1991).

[3]The distances between vertices expressed as multiple of $r$ are: for the trigonal bipyramid, $\sqrt{2}$ (6×) $\sqrt{3}$ (3×) and 2; for the square pyramid, $\sqrt{2}$ (8×) and 2 (2×) and for the octahedron $\sqrt{2}$ (12×) and 2 (3×).

[4]Molecular chemists are used to considering non-bonding valence electrons as well as ligands when discussing coordination figures. Compare $PF_5$ (trigonal bipyramid, no non-bonding valence electrons) with $ClF_5$ (square pyramid with the non-bonding electron pair completing an octahedron around Cl). In crystalline PbO Pb has 4 O atoms in a square on one side of Pb with a lone pair completing a square pyramid.

meet at some vertices and the angles containing such vertices are part of two faces. We note two points about such polyhedra: (a) some of the faces are of necessity non-planar if angles of 180° are excluded, and (b) the number of vertices, edges and faces are still related by $V - E + F = 2$. We just give some simple examples here.

Polyhedra with divalent vertices can be derived in a formal way by inserting vertices in edges of conventional polyhedra. In Fig. 5.15 we show how polyhedra with six and eight vertices are derived in this way from a tetrahedron. The new polyhedra are facially-regular and, as there are four faces, they are also tetrahedra. The first (on the left in Fig. 5.15) is encountered as a building unit in the fibrous zeolites (§ 7.8.7); as it has four 4-cornered faces we call it a tetragonal tetrahedron. The second polyhedron (on the right in Fig. 5.15) has four 5-cornered faces so we call it a pentagonal tetrahedron. It occurs in the molecule $As_4S_4$ (realgar) with 3-coordinated As ($\{As\}AsS_2$) and 2-coordinated S ($\{S\}As_2$).

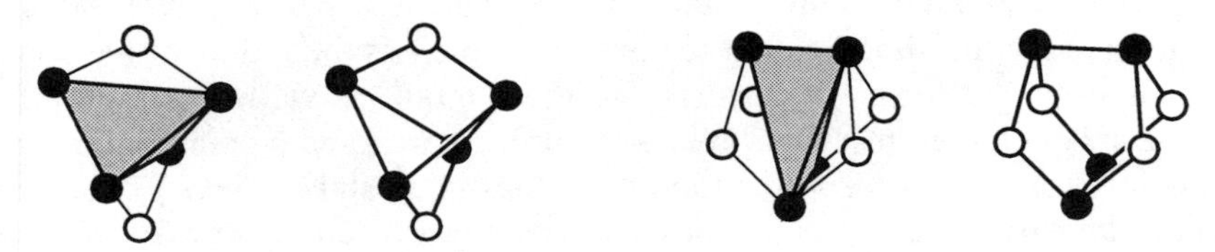

**Fig. 5.15.** On the left are two drawings of a tetragonal tetrahedron and on the right two drawings of a pentagonal tetrahedron. In each case the righ-hand member of the pair shows only the polyhedron edges. Vertices shown as open circles are divalent.

The next polyhedron in the series is a hexagonal tetrahedron with ten vertices shown in Fig. 5.16. This is an important geometry in structural chemistry. If $CH_2$ units are at the six 2-coordinated vertices, and CH units at the four 3-coordinated vertices, the hydrocarbon *adamantane*, $C_{10}H_{16}$ results. If the CH groups are replaced by the isoelectronic N, hexamethylenetetramine, $N_4(CH_2)_6$ is produced. $P_4O_6$ has the same structure.

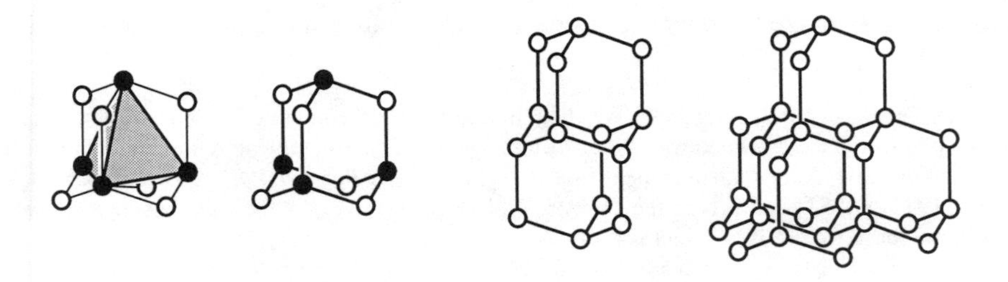

**Fig. 5.16.** From the left are: two drawings of a hexagonal tetrahedron, two such polyhedra sharing a face, a cluster of five hexagonal tetrahedra. Compare Fig. 5.15.

If a wire model of the hexagonal tetrahedron framework is dipped in a soap solution and withdrawn, the soap film will form a curved surface on the faces of the polyhedron.

Remarkably, such a polyhedron with curved faces is a space-filling solid.[1] In Fig. 5.16 two such tetrahedra are shown sharing a face and also a cluster of five tetrahedra sharing faces.[2] In fact the structure of vertices and edges obtained by filling space with hexagonal tetrahedra is just that of the atoms and bonds in the diamond form of carbon, and Fig. 5.16 should be compared with Fig. 7.10 illustrating the diamond structure. Much of the interest in polyhedra with divalent vertices is as building units of three-dimensional nets.

Another example of a tetragonal tetrahedron is shown in Fig. 5.17. Four 2-connected vertices are on a square of edge $d$, and the 4-connected vertices are a distance $d$ apart so that the edge lengths are $\sqrt{2}d/2$. *If* all vertices were 4-connected the polyhedron would be a squashed octahedron. The figure shows curved surfaces such as might be made by a soap film on the framework. Such a polyhedron is also space-filling (the packing requires three different orientations of the tetrahedra). The vertices are on the points of a body-centered cubic lattice and, as discussed in § 6.2, the centers of the polyhedra are at the "octahedral" sites of a body-centered cubic sphere packing.

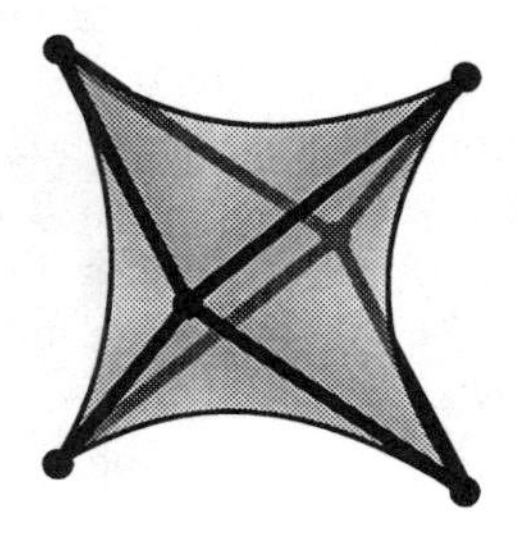

**Fig. 5.17**. A space-filling tetragonal tetrahedron with curved faces.

## 5.2 Polyhedral clusters

We are interested in polyhedra because their vertices often represent the positions of atoms coordinating a central one. In this way we can reduce a group, such as $\{M\}X_6$ to a single entity such as an octahedron or a trigonal prism. In many structures it is convenient

[1] Such polyhedra appear to have been first described by P. Pearce in Chapter 8 of *Symmetry in Nature is a Strategy for Design* [MIT Press (1978)]. A beautifully illustrated systematic account of them and their packings is given by S. T. Hyde & S. Andersson, *Zeits. Kristallogr.* **168**, 221 (1984).

[2] This topic provides an excuse for some free-association historical notes: Hexamethylenetetramine was the first organic compound to have its structure determined by X-ray diffraction [R. G. Dickenson & A. L. Raymond, *J. Amer. Chem. Soc.* **45**, 22 (1923)]. Dickenson was the first person to get a Ph.D. degree from the California Institute of Technology (in 1920); in 1922 he took on a graduate student named Linus Pauling who was to be a major force in structural chemistry for more than the next half century (Nobel prize in Chemistry, 1954). The hydrocarbon corresponding to the two face-sharing polyhedra ($C_{14}H_{20}$) is known as *congressane* as its structure served as the emblem of the XIX congress of the International Union of Pure and Applied Chemistry (1963). Its structure was solved "by inspection" from X-ray data by the famous team of I. L. Karle and J. Karle in 1965. J. Karle went on to share the Nobel prize in chemistry with H. Hauptman in 1985 (for the development of "direct methods" of X-ray diffraction analysis).

to continue the process and consider the structural units to be groups (clusters) of polyhedra (which may or may not be centered). This approach allows us to describe the structures of complex crystals in a hierarchical way: we first describe how atoms form coordination polyhedra, then we describe how the polyhedra are assembled into clusters, and finally we describe how the clusters are linked in the crystal. A similar procedure is familiar to molecular chemists who are long used to replacing functional groups (clusters!) such as, for example, $-C_6H_5$ by -Ph and $C_5H_5$ by Cp.

In this section we describe some simple examples of clusters, that are often found in crystal structures and give some examples of their occurrence.

### *5.2.1 Clusters of tetrahedra*

In § 5.1.2 (Fig. 5.3) we introduced the *stella quadrangula* which may be derived by fusing tetrahedra to the faces of a central tetrahedron, and the *stella octangula* which is similarly derived by capping the faces of a central octahedron with tetrahedra.

The composition of the stella octangula can be written $A_6B_8$ with the six vertices $A$ forming an octahedron and the eight capping vertices $B$ at the corners of a cube. This is a configuration frequently found in crystal structures. A notable example is as the $Mo_6S_8$ cluster in $PbMo_6S_8$ (one of the so-called Chevrel phases). The same cluster is often found in Nb and Mo halides; in $Nb_6I_{11}$, $Nb_6I_8$ clusters are linked by Nb-I-Nb bonds to produce overall stoichiometry $Nb_6I_8I_{6/2} = Nb_6I_{11}$.

If just four (non-adjacent) faces of an octahedron are capped with tetrahedra a "supertetrahedron" as shown in Fig. 5.18 is produced. If the small tetrahedra are centered by $T$ and the octahedron vertices are $X$ and the capping vertices $Y$, the stoichiometry is $T_4X_6Y_4$. A molecular example is provided by the molecule $Ge_4S_6Br_4$. These units are also familiar as $P_4O_{10}$ and $P_4S_{10}$ molecules, and as the complex anions $[P_4N_{10}]^{10-}$ in $Li_5P_2N_5$ and $[Si_4S_{10}]^{4-}$ in $Na_2Si_2S_5$.[1]

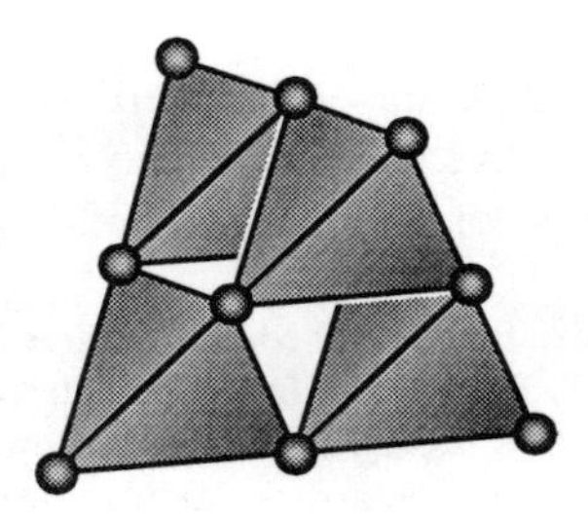

**Fig. 5.18**. A "supertetrahedron" cluster of four tetrahedra.

Joining supertetrahedra (with $Y = X$) into a three-dimensional network by sharing the four outer vertices results in stoichiometry $T_4X_6X_{4/2} = TX_2$. This is the situation in

[1]Note that $P_4O_{10}$ and $[P_4N_{10}]^{10-}$ are isoelectronic as are $P_4S_{10}$ and $[Si_4S_{10}]^{4-}$. In every case there are 80 valence electrons.

compounds such as $ZnI_2$ and $Be(NH_2)_2$.

Clusters of condensed $\{T\}X_4$ tetrahedra sharing vertices are very common in crystal structures. A pair of such tetrahedra with one common vertex has stoichiometry $T_2X_7$, and is found in oxide chemistry (especially, but by no means exclusively) with $T$ = Si, P and S (the compounds are often called pyrosilicates, pyrophosphates and pyrosulfates respectively). It is rare for the $T$-$X$-$T$ link to be linear, but an example occurs in the mineral thortveitite, $Sc_2Si_2O_7$. $SiP_2O_7$ is an elegant example of a structure made of linked $\{Si\}O_6$ octahedra and $P_2O_7$ double tetrahedra. $Cl_2O_7$ provides an example of a neutral cluster.

Larger clusters of tetrahedra are also common and we give some examples of their occurrence in silicate minerals. Rings of $n$ tetrahedra sharing vertices (Fig. 5.19) have stoichiometry $T_nX_{3n}$ ; for silicates the cluster has formal charge $[Si_nO_{3n}]^{2n-}$.

Common stoichiometries are $n$ = 3 (as in benitoite, $BaTiSi_3O_9$), 4 (as in kainosite, $Ca_2Y_2[Si_4O_{12}](CO_3){\cdot}H_2O$) and 6 (as in dioptase, $CuSiO_3{\cdot}H_2O = Cu_6Si_6O_{18}{\cdot}6H_2O$).

Rings of corner-connected tetrahedra are "flexible" and the conformation found is determined in part by the $T$-$X$-$T$ angle which usually has a characteristic value for given $T$ and $X$ (about 145° for Si-O-Si and 128° for Ge-O-Ge). Fig. 5.19 illustrates some extreme configurations of three-membered and four-membered rings and gives the $T$-$X$-$T$ angles for regular tetrahedra. Not surprisingly, three-membered rings are more common for germanates than for silicates. A special case is the linear chain ($n = \infty$) which occurs in many important minerals (such as enstatite, $MgSiO_3$).

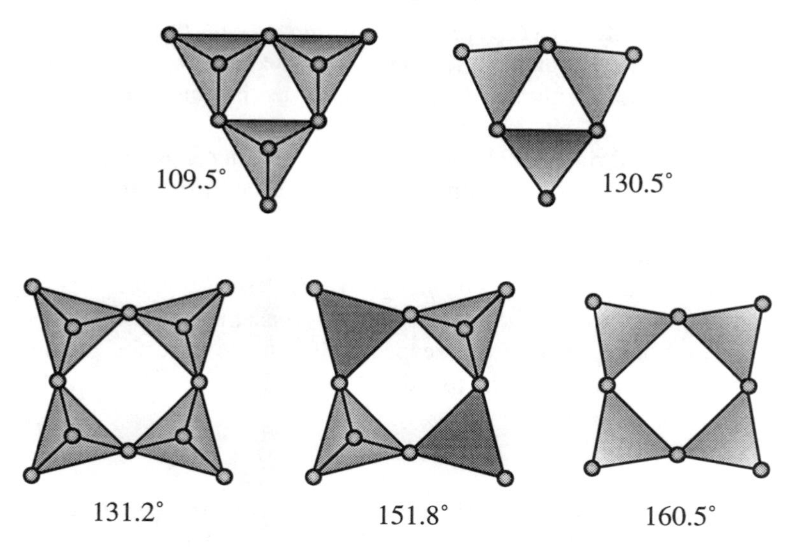


**Fig. 5.19**. Symmetrical configurations of rings of regular $\{T\}X_4$ tetrahedra. Numbers are $T$-$X$-$T$ angles.

Double rings of tetrahedra are also common in silicates. These are made by further corner sharing to produce stoichiometry $T_{2n}X_{5n}$. The $n$ = 4 member of this family is obtained by fusing tetrahedra to the eight triangular faces of a cuboctahedron (Fig. 5.20, middle) so that the $T_8$ array is a cube. For regular tetrahedra the maximum $T$-$X$-$T$ angle is 148.4 °. A nice example of the occurrence of this cluster is as $[Si_8O_{20}]^{8-}$ units in steacyite, $KNaCaThSi_8O_{20}$.

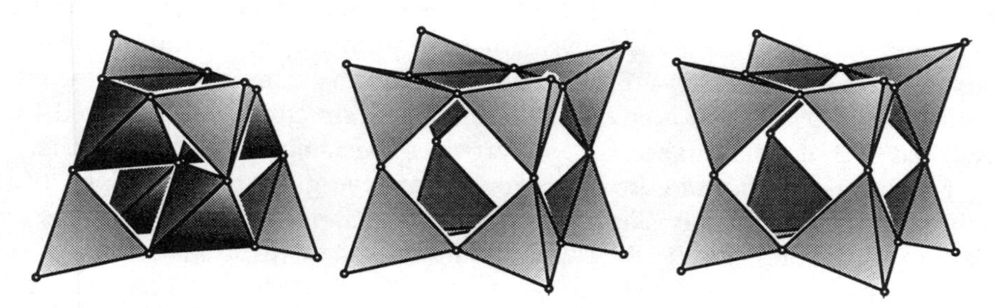

**Fig. 5.20.** Left: a $T_4T'_4X_{17}$ cluster. Middle: a $T_8X_{20}$ cluster obtained by capping the triangular faces of a cuboctahedron with tetrahedra. Right: a similar cluster with alternating tetrahedra of two different sizes.

Many compounds with vertex-sharing tetrahedra have two (or more) kinds of tetrahedra of different sizes. For example, there known many analogs of silicates with a framework such as $AlPO_4$ with alternating $\{Al\}O_4$ and $\{P\}O_4$ tetrahedra. It is interesting that if the centers alternate in a $T_4T'_4X_{20}$ cluster as shown on the right in Fig. 5.20, the maximum $T$-$X$-$T'$ angle is still 148.4° for regular tetrahedra.

A second kind of cluster ($T_4T'_4X_{17}$) of eight vertex-sharing tetrahedra occurs in $Na_{10}Be_4Si_4O_{17}$. This is shown on the left in Fig. 5.20; the $\{Be\}O_4$ tetrahedra are the inner ones (darker shaded). With regular tetrahedra (which must be congruent) the $T$-$X$-$T$ angles are all 109.5°.

With two six-membered rings of tetrahedra, the tetrahedron centers are on the vertices of a hexagonal prism, the corresponding silicate anion is $[Si_{12}O_{30}]^{12-}$ which also is found in complex minerals.

Clusters of tetrahedra commonly occur in condensed structures such as zeolite frameworks with overall stoichiometry $TX_2$. When describing the topology of such structures, it is common to omit the anions and to represent clusters such as $T_8X_{20}$ and $T_{12}X_{30}$ as cubes and hexagonal prisms respectively, and to describe the structure as a linkage of such units. Examples will be found in Chapter 7.

### 5.2.2 *Clusters of octahedra*

Discrete and linked clusters of octahedra are found in great variety in the chemistry of the oxides of the early transition elements (particularly Nb, Mo and W) and we give only a few simple examples of them here. The discrete clusters are usually anionic and referred to as *isopolyanions* if they contain just one kind of metal atom (i.e. one metallic element) and *heteropolyanions* otherwise.[1]

[1]For a review of such species see D. L. Kepert, Chapter 51 in *Comprehensive Inorganic Chemistry*, Vol. 4. [Pergamon Press, Oxford (1973)].

Face-sharing in extended clusters of centered polyhedra is rather rare, but pairs of octahedra sharing a common face are rather common as anions (e.g. $[W_2Cl_9]^{3-}$ with two $\{W\}Cl_6$ octahedra sharing a face) and such pairs linked by corners are a conspicuous feature of many structures. Usually in such a group $M_2X_9$, the $M$ atoms move "off center" to increase their mutual separation. Such displacements are only possible for pairs, so that longer chains of centered octahedra sharing faces are less common unless the cluster is stabilized by metal-metal bonding through the shared face. The "anti-structure" (with the roles of metal and nonmetal reversed) is found in $Rb_9O_2$ in which there are pairs of $\{O\}Rb_6$ octahedra with metal-metal bonding between the clusters.

The dihedral angle of a regular octahedron is $\cos^{-1}(-1/3) = 109.47°$ so that if three meet at a common edge, there will be a small gap between some of the faces. However three can meet at a common edge with three pairs of adjacent faces shared if the octahedra are distorted to make the angle between shared faces 120° (Fig. 5.21). The stoichiometry is now $M_3X_{11}$. Such clusters (linked by sharing edges) may be identified in the structure of $Nb_3Te_4$. The anti-structure cluster occurs in oxides such as $Cs_{11}O_3$ and $Rb_7Cs_{11}O_3$.

Edge-sharing pairs of $\{M\}X_6$ octahedra have stoichiometry $M_2X_{10}$ and are rather common as discrete units (e.g. $NbCl_5 = Nb_2Cl_{10}$).

With more than two octahedra joined by edges, the number of distinct topologies rapidly becomes rather large, but generally it is the most condensed (minimum value of $X/M$) and symmetrical that are of greatest interest.[1]

With three octahedra joined by edges there are three possibilities as shown in Fig. 5.21. The first two have stoichiometry $M_3X_{14}$ and the third (on the right in the figure) has stoichiometry $M_3X_{13}$. This last occurs as, for example, $[Te_3Cl_{13}]^-$ and (condensed by further edge sharing) in $Nb_3Cl_8$.

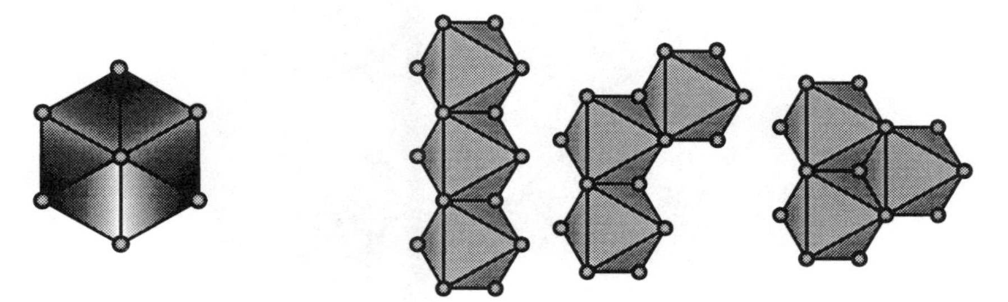

**Fig. 5.21**. Left: three distorted octahedra sharing faces and a common edge. Right: the three distinct ways of linking three octahedra by sharing edges only.

For four octahedra joined by edges, there are eight possibilities with stoichiometries ranging from $M_4X_{18}$ (five cases), through $M_4X_{17}$ (one case), to $M_4X_{16}$ (two cases). Fig. 5.22 illustrates the two $M_4X_{16}$ isomers. In the first the centers are coplanar, and fall on the vertices of a 60° rhombus, so we call it the rhombic isomer. In the second isomer the centers form a tetrahedron so we call it the tetrahedral isomer. The second cluster can also

[1]For a discussion see P. B. Moore, *Amer. Mineral.* **55**, 135 (1970).

be described as consisting of four octahedra sharing faces with a central tetrahedron. It is interesting that $Te_4Cl_{16}$ has the tetrahedral structure but $Te_4I_{16}$ has the structure of the rhombic isomer [in these clusters Te is "off center" in the octahedron to give a 3 + 3 coordination often found for Te(IV) compounds]. The tetrahedral group is also found as the anion $[W_4O_{16}]^{8-}$.

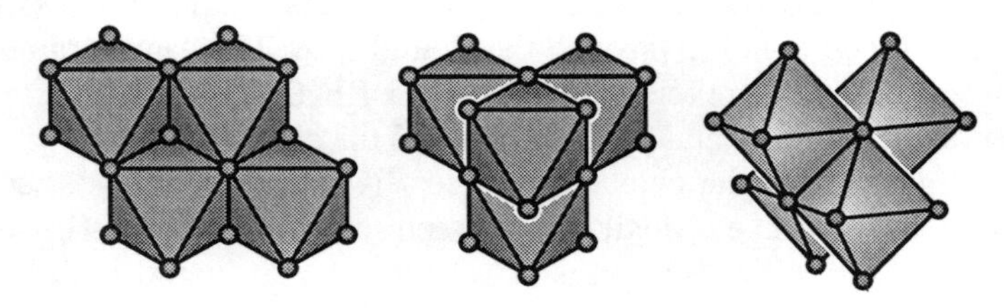

**Fig. 5.22**. The two $T_4X_{16}$ clusters of four edge-sharing octahedra. Left the rhombic isomer. Center and right: two views of the tetrahedral isomer.

The $M_4X_{16}$ unit may be linked by tetrahedra in three mutually perpendicular directions to give a cubic $M_4T_3X_{16}$ framework as shown in Fig. 5.23. Examples of compounds with structures based on this framework are $Cs_3Mo_4P_3O_{12}$ ($M$ = Mo, $T$ = P) and minerals of the pharmacosiderite family such as $BaM_4As_3O_{12}(OH)_4{\cdot}5H_2O$ ($M$ = Al or Fe, $T$ =As), and $K_3Ge_7O_{15}(OH){\cdot}4H_2O$ (with $M$ = $^{vi}$Ge and $T$ = $^{iv}$Ge).

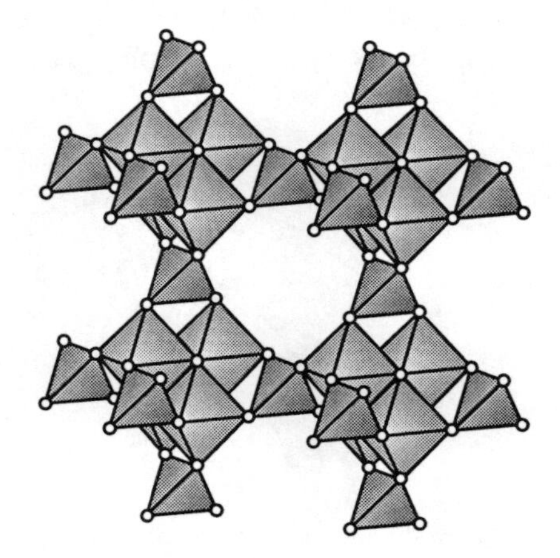

**Fig. 5.23**. A part of the $M_7X_{16}$ framework of pharmacosiderite.

In Fig. 5.24 we illustrate the most symmetrical and condensed cluster of six edge-sharing octahedra with stoichiometry $M_6X_{19}$ which we call, for reasons that should be obvious, a "superoctahedron." This is found as the anion $[Mo_6O_{19}]^{2-}$. Also shown in the figure is a five octahedron cluster $M_5X_{18}$ which is also a common anion configuration. Two of the five-octahedron clusters can condense by sharing four *vertices* to produce the configuration found in the $[W_{10}O_{32}]^{4-}$ ion. Clearly the possibilities are endless.

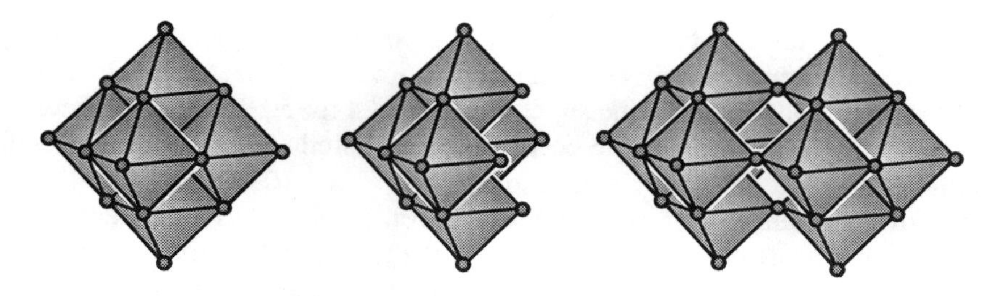

**Fig. 5.24**. Left: A "superoctahedron" $M_6X_{19}$. Center: a cluster $M_5X_{18}$. Right: a cluster $M_{10}X_{32}$.

The structure of $Nb_2F_5$ is a nice example of a three-dimensional structure built up from superoctahedra (for crystallographic data see Appendix 5). In this instance the central anion is missing so there are $\{Nb\}F_5$ square pyramids and we really have a cluster of six square pyramids (Fig. 5.25). The cluster composition is $Nb_6F_{12}F'_6$ where F' is on the outer vertices of the superoctahedron. Sharing these outer vertices produces an open three dimensional network of corner-connected clusters with composition $Nb_6F_{12}F'_{6/2}$ = $Nb_6F_{15}$. The formula is usually given as $Nb_6F_{15}$ (rather than $Nb_2F_5$) to emphasize the presence of $Nb_6$ groups in the structure

To finish the description of this beautiful structure we note that the topology of the corner-connected network of (super) octahedra is that of **$ReO_3$** (see p. 170) and in $Nb_2F_5$, two such networks interpenetrate. Fig. 5.25 shows a fragment of one network.

Three points concerning this structure might be might be noted: (a) the Nb atoms form an empty $Nb_6$ octahedron, a very common feature of the chemistry of niobium in lower oxidation states, (b) we describe the $Nb_6F_{12}$ part of the cluster as an edge-capped octahedron cluster in § 5.2.4, and (c) the F atoms and the sites (empty) at the center of the $Nb_6$ octahedra combine to form **ccp** (§ 6.1.3).

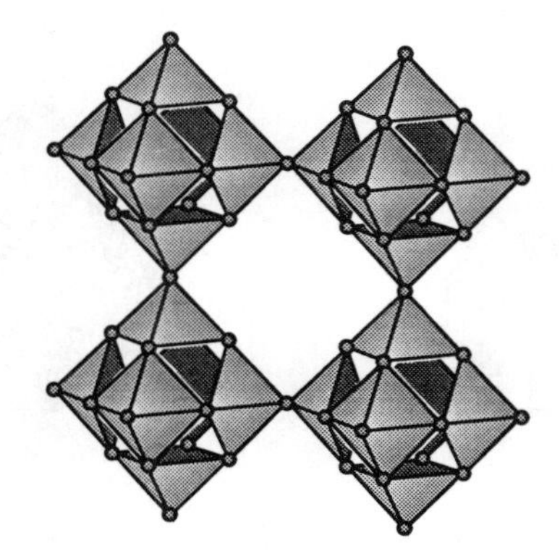

**Fig. 5.25**. Four of the corner-connected clusters of square pyramids in the $Nb_6F_{15}$ structure.

Frameworks of vertex-sharing octahedra with stoichiometry $MX_{6/2} = MX_3$ are rather important, and some will be discussed in § 5.3.4 and in Chapter 6. Here we call attention to a finite group, which may be considered as composed of four octahedra sharing non-contiguous faces of a central octahedron. In Chapter 6 we call this a "pyrochlore" unit; it is illustrated in Fig. 5.26. (Compare the supertetrahedron, Fig. 5.18.) As the dihedral angle of an octahedron is 109.48°, the angle between the unshared faces of the outer octahedra and the adjacent face of the empty octahedron is $360° - 2 \times 109.48° = 141.04°$; which is close to the dihedral angle of a regular icosahedron (138.18°). This means that if the remaining four faces of the central octahedron of the pyrochlore unit are capped with icosahedra, the assembly will fit together rather snugly. Such units are an important feature of some intermetallic structure types (with the icosahedra and octahedra very slightly distorted so that they have common vertices).

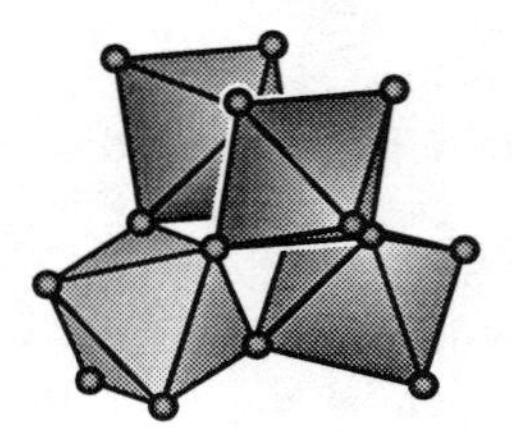

**Fig. 5.26**. The "pyrochlore unit" of four vertex-sharing octahedra.

Finally, if three regular octahedra share a common vertex, but without edge sharing, the shortest inter-octahedron distance between vertices is, of necessity, less than the octahedron edge length. Such configurations occur in the structure of compounds such as pyrite, $FeS_2$. In this structure $\{Fe\}S_6$ octahedra have all their vertices shared in this way (so the stoichiometry is $FeS_{6/3}$) and the short distances correspond to S-S bonds. The octahedron edge lengths are S...S = 3.07 and 3.32 Å and the S-S bond length is 2.18 Å. Fig. 5.27 illustrates a fragment of the structure (for crystallographic data see Appendix 5).

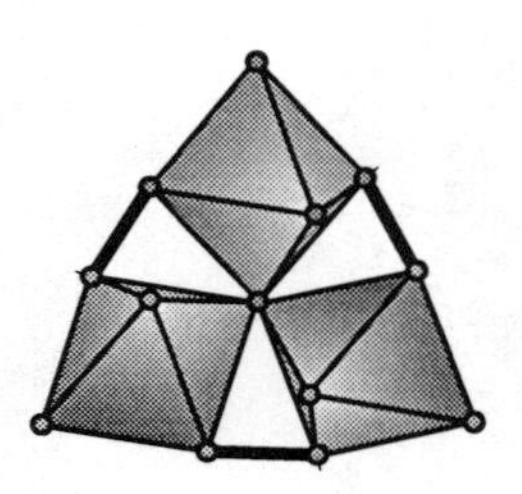

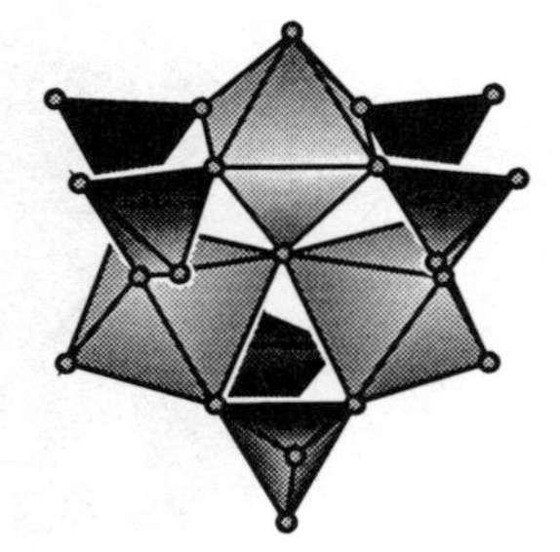

**Fig. 5.27**. Left: fragment of the pyrite, $FeS_2$, structure showing three $\{Fe\}S_6$ octahedra with a common vertex. Heavy lines represent S-S bonds. Right: the anion in Maus's salt shown as $\{Fe\}O_6$ octahedra and $\{S\}O_4$ tetrahedra.

Another example of three octahedra sharing a common vertex is provided by the complex anion $Fe_3O(SO_4)_6(H_2O)_3^{5-}$ (also shown in Fig. 5.27) which occurs in the mineral metavoltine and in the compounds known as Maus's salts.[1] The {Fe}$O_6$ octahedra are fairly regular with edge lengths of 2.73-2.90 Å. The edge lengths in the {S}$O_4$ tetrahedron range from 2.31-2.37 Å.

### *5.2.3 Octahedra plus tetrahedra: Keggin and spinel units*

There are very many niobates, molybdates and tungstates (and some other related oxides) of various dimensionalities that can be described as assemblies of {$M$}$O_4$ tetrahedra and {$M$}$O_6$ octahedra. An adequate description of all known structures would require a rather large book. Here we describe two particularly interesting clusters involving four groups each of three edge-sharing octahedra about a central tetrahedron.

First we examine the $M_3X_{13}$ cluster of Fig. 5.21 a little more closely. The cluster has symmetry $3m$ which means that the two faces normal to the 3 axis are different, and have 3-fold symmetry as illustrated in Fig. 5.28.

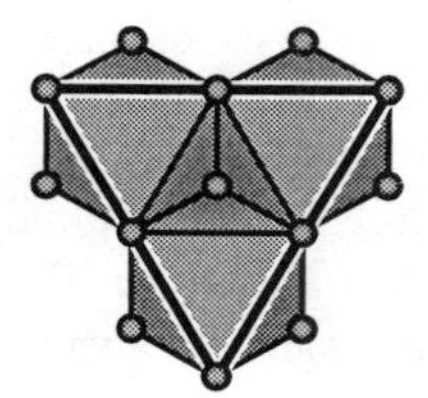
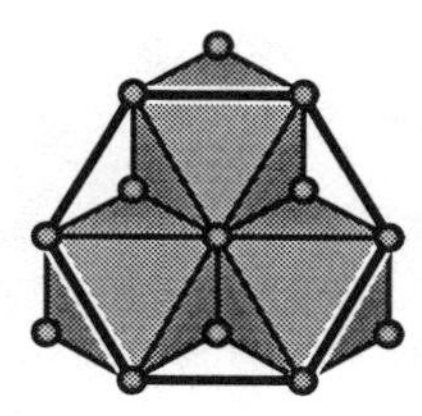

**Fig. 5.28**. Two views of the $M_3X_{13}$ cluster in projection down the 3 axis.

Examination of the figure will show (on the right) that one face of the cluster consists of a centered hexagon of vertices, so that the cluster can cap hexagonal faces of a polyhedron, but as the symmetry is only 3-fold this can be done in two different ways. The clusters we now consider involve capping of the hexagonal faces of a truncated tetrahedron.

In the first case, the clusters of three octahedra share vertices with each other, to form stoichiometry $M_{12}X_{40}$. In this structure, shown in Fig. 5.29, four central $X$ atoms form a tetrahedron which can be centered by a $T$ atom to give stoichiometry $M_{12}TX_{40}$.

This unit is usually named after J. F. Keggin who first determined the structure of phosphotungstic acid which contains a $[W_{12}PO_{40}]^{3-}$ group.[2] The same grouping has been

[1]These have the general formula $M_5Fe_3O(SO_4)_6{\cdot}nH_2O$ with $M$ = Li, Na, K, Rb, Cs, $NH_4$ and Tl. Those who object to trivial names might note that *Structure Reports* lists the anion under the heading "triaquo-μ3-oxo-hexa-μ-sulphato-triferrate(III) dihydrate."

[2]This was done by X-ray diffraction in 1934 and was a remarkable *tour de force* for the time; locating light atoms (O) in the presence of heavy atoms (W) is difficult even with the greatly improved equipment available today. The formula for phosphotungstic acid is $H_3W_{12}PO_{40}{\cdot}6H_2O$ but it is perhaps better written $(H_5O_2)_3W_{12}PO_{40}$ as it contains $H_5O_2^+$ groups. These provide a rare example of a symmetrical (linear) hydrogen bond O-H-O, linking two water molecules, with $d$(O-H) = 1.21 Å [G. M. Brown *et al. Acta Crystallogr.* **B33**, 1038 (1977)]. The record in 1995 for cluster size appears to be a remarkable toroidal oxoanion with approximate 7-fold rotation symmetry and containing 154 Mo atoms [A. Müller *et al., Angew. Chem. Int. Ed. Engl.* **34**, 2122 (1995)].

identified in numerous anions, particularly with $M$ = Mo or W and with $T$ = P, As or Si. The $M$ atoms are at the vertices of a cuboctahedron. The complex mineral zunyite contains condensed $Al_{13}O_{40}$ units with the same structure, and the $O_{40}$ arrangement also serves as structural units in some complex intermetallic structures.

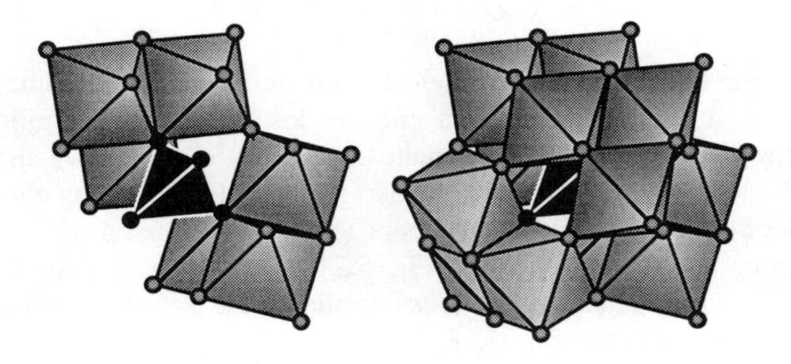

**Fig. 5.29**. The Keggin cluster, $M_{12}TX_{40}$. On the left only six of the twelve octahedra are shown. Compare Fig. 5.30.

If the three-octahedron units of the Keggin structure are rotated by 60°, they are then joined by edge sharing in the more compact cluster shown in Fig. 5.30. The $M$ atoms are now at the vertices of a truncated tetrahedron. We call this structure, which has the same stoichiometry as the Keggin structure, a "spinel unit" because the important spinel ($MgAl_2O_4$) structure can be considered as made up of a condensation of such units. Partly hydrolyzed Al salts have been found to contain $[Al_{13}O_4(OH)_{24}(H_2O)_{12}]^{7+}$ units with this structure. It is of interest that dehydration of hydrous alumina at low temperatures produces so-called "$\gamma$-alumina" (used as a polishing powder) which has a spinel-related structure.

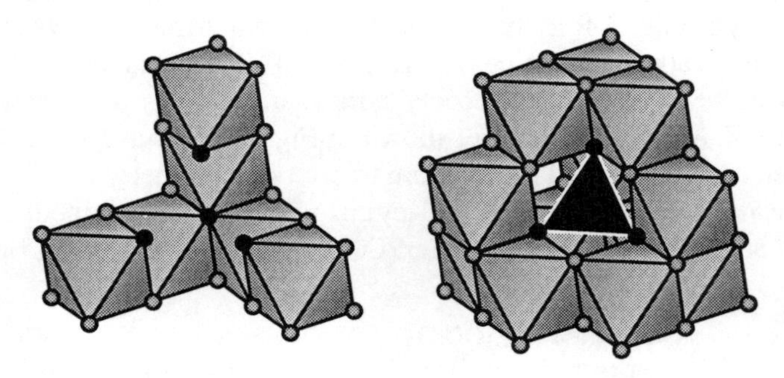

**Fig. 5.30**. The "spinel cluster." On the left only six of twelve octahedra are shown. Compare Fig. 5.29.

### *5.2.4 Edge-capped clusters*

Capping the *edges* of polyhedra with additional vertices produces what can be considered as clusters made of polyhedra and *triangles*. We can generate many useful groupings in this way. Recall that in § 5.1.10 we derived polyhedra with divalent vertices by edge-capping polyhedra (tetrahedra).

Capping the six edges of an $A_4$ tetrahedron with $B$ produces a cluster $A_4B_6$. This is shown in two ways in Fig. 5.31, first as an edge-capped tetrahedron, and second as a $B_6$ octahedron with $A$ centering four of the faces. The latter emphasizes the triangular coordination of $A$ by $B$. An example of its occurrence is as the $Cu_4S_6$ grouping in complex anions such as $[Cu_4(SPh)_6]^{2-}$.

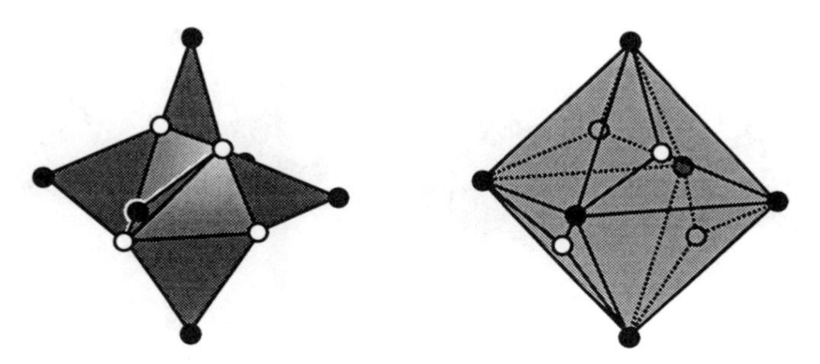

**Fig. 5.31**. Two views of an edge-capped tetrahedron.

Our second example is obtained by capping the edges of an $M_6$ octahedron with $X$ atoms to produce stoichiometry $M_6X_{12}$ as shown in Fig. 5.32. In the figure, the capping vertices have been placed so that $M$ is in square planar coordination by $X$. This grouping is very common in the halides of early transition metals, both as neutral molecules (as in $PtCl_2 = Pt_6Cl_{12}$) and as complex cations (as in $[Nb_6Cl_{12}]^{n+}$, $n = 2,3,4$). A similar cluster, but with more nearly equilateral triangles, forms the basis of an elegant description of the $Mn_5Si_3$ structure in which $Mn_6$ octahedra share opposite faces to form columns and the non-shared edges (six per octahedron) are capped by further Mn.

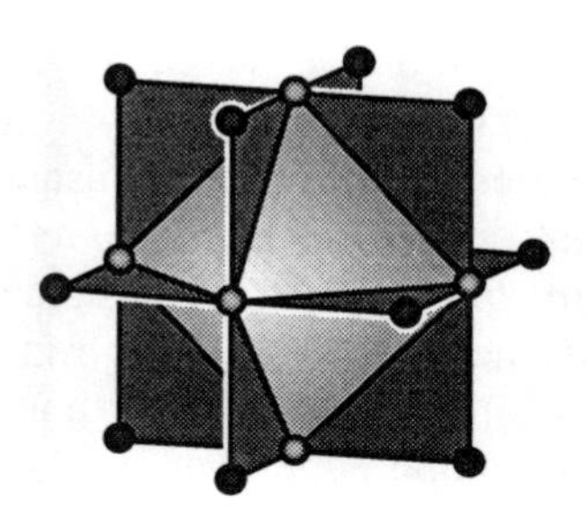

**Fig. 5.32**. An edge-capped octahedron.

As a final example of edge-capping we describe the build-up of the structural unit of $\gamma$-brass ($Cu_5Zn_8$). In Fig. 5.33 where we start (top left) with a $Zn_4$ tetrahedron and cap its faces with Cu (at the vertices of an "outer tetrahedron") to produce a $Cu_4Zn_4$ stella quadrangula. Next (top right) we cap the six edges of the $Zn_4$ tetrahedron with Cu (at the vertices of an octahedron) to produce the cluster $Cu_{10}Zn_4$. Finally (bottom left) we cap the

12 outer edges of the stella quadrangula with Zn arriving at a cluster $Cu_{10}Zn_{16}$. The crystal structure is made up of a body-centered cubic array of such discrete clusters. It might be noted that the 12 outer Zn atoms are at the vertices of a distorted cuboctahedron (bottom right in the figure) with four small and four large equilateral triangle faces; the remaining faces are rectangular instead of square. Such a distorted cuboctahedron (with tetrahedral = $\bar{4}3m$ symmetry) is met in other intermetallic structures (see e.g. Fig. 6.77, § 6.8.6).

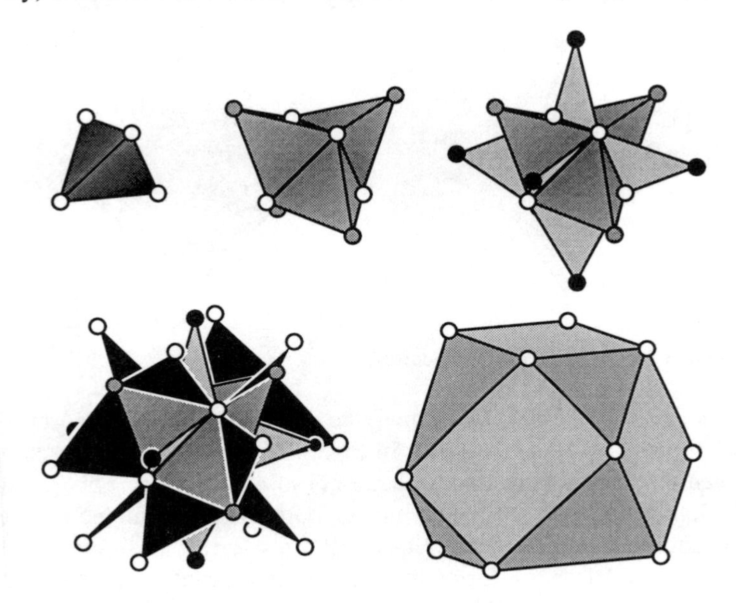

**Fig. 5.33**. Build-up of the gamma brass structure (see text).

### *5.2.5 Clusters of truncated tetrahedra and icosahedra*

Some complex crystal structures (with over a thousand atoms per unit cell in some instances) can be decomposed into clusters of truncated tetrahedra and/or icosahedra.[1] We describe one relatively simple structure to illustrate the kind of groupings found.

The dihedral angle between the hexagonal faces of a truncated tetrahedron is the same as that of a regular tetrahedron [$\cos^{-1}(1/3) = 70.53°$] so with just a minor distortion a group of five can share faces around a common edge (Fig. 5.34). We mentioned above that an icosahedron can be considered as made up of twenty tetrahedra with a common vertex (the center of the icosahedron). The tetrahedra have three edges of $d$ (the edges of the

[1]For beautifully illustrated descriptions of these structures in terms of clusters of truncated tetrahedra and icosahedra, see S. Samson in (a) *Developments in the Structural Chemistry of Alloy Phases*, B. C. Giessen, ed. Plenum Press, New York (1969); (b) *Structural Chemistry and Molecular Biology*, W. H. Freeman, San Francisco (1968). For alternative descriptions see also S. Andersson in *Structure and Bonding in Crystals*, M. O'Keeffe & A. Navrotsky, eds. Vol 2. Academic Press, New York (1981) and B. Chabot, K. Cenzual & E. Parthé, *Acta Crystallogr.* A**37**, 6 (1981).

icosahedron) and three of 0.95$d$ (the distance from the center of the icosahedron to one of its vertices). Perhaps less obvious is the fact that twenty truncated tetrahedra will fit around a central icosahedron of edge $d$ if nine of its eighteen edges are $d$ and nine are 0.95$d$. Fig. 5.35 shows an icosahedron sharing a triangular face with a truncated tetrahedron. The longer edges of the truncated tetrahedron are those of the shared face and the opposite hexagon.

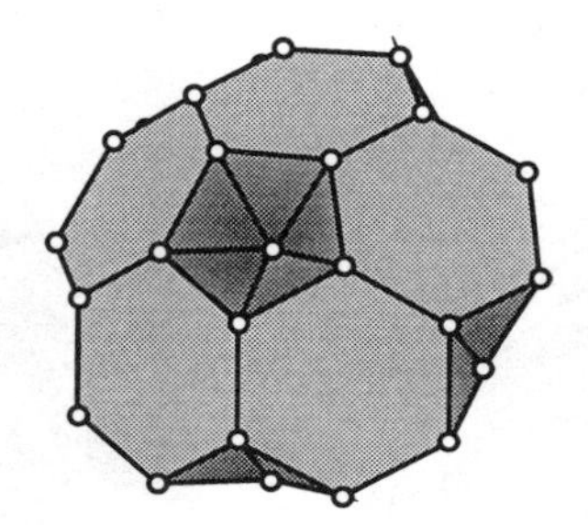

**Fig. 5.34**. A pentagonal cluster of five truncated tetrahedra

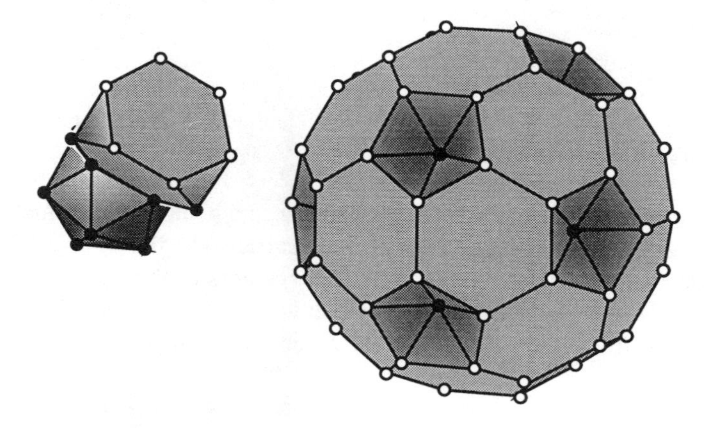

**Fig. 5.35**. Left: a truncated tetrahedron and an icosahedron sharing a face. Right: twenty truncated tetrahedra around a central icosahedron.

Figure 5.35 also shows the complete assembly of twenty truncated tetrahedra sharing faces with a central icosahedron. The twenty hexagonal faces form a regular truncated icosahedron. The pentagonal faces of the truncated icosahedron are capped on the inside producing depressions that are pentagonal pyramids. The cluster so far described contains 84 vertices (12 for the central icosahedron, 60 vertices of the truncated icosahedron and the 12 apices of the pentagonal pyramids.

This unit occurs in a number of intermetallic compounds (some approximate compositions are $Li_3Al_5Cu$, $Na_2Au_3Sn$ and $Li_2Zn_3Ga$) with the icosahedron and the 20 truncated tetrahedra centered, so the cluster now contains 105 atoms. To build up the crystal structure, the truncated icosahedron unit is put inside a truncated octahedron so that eight of the hexagonal faces of the truncated icosahedron are coplanar with the (larger) hexagonal faces of the truncated octahedron as shown on the left in Fig. 5.36. The resulting unit (shown on the right in the figure) can now be packed to fill space with their centers on a body-centered cubic lattice (see § 7.3.10 for a discussion of space filling by truncated octahedra). To count atoms in the unit cell, note that there two units per unit cell and that 48 atoms (those in the hexagonal faces of the truncated octahedron) of each 105 atom cluster are common to two units. Accordingly the number of atoms in the unit cell is $2 \times (105 - 48/2) = 162$.

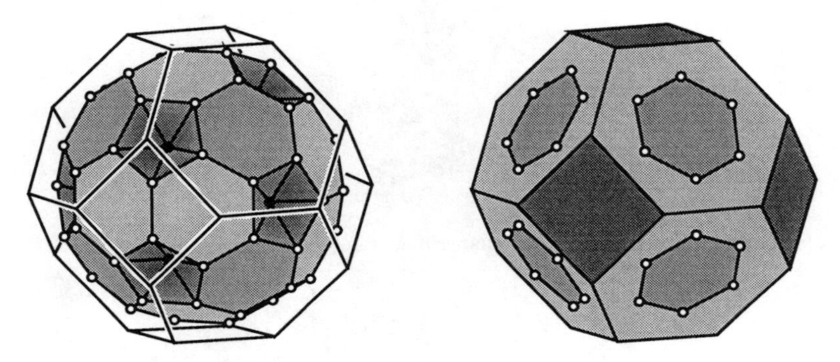

**Fig. 5.36**. Illustrating how a truncated icosahedron fits into a truncated octahedron with 48 of its vertices in the hexagonal faces of the truncated octahedron.

Note that in increasing distance from the atom at the center of the cluster: this central atom has 12 neighbors at the vertices of the central icosahedron, 20 at the vertices of a pentagonal dodecahedron (these correspond to the centers of the truncated tetrahedra), 12 more at the vertices of a larger icosahedron (the apices of the pentagonal pyramids), and 60 at the vertices of the truncated icosahedron: all fitted into a truncated octahedron which is a space filling solid.[1]

Of course in analyzing this fascinating structure more completely we would have to inquire into the coordinations of all the atoms. It transpires that the near neighbors of each atom fall at the vertices of one of the Frank-Kasper polyhedra.

The icosahedron is also a conspicuous feature of the structures of elemental boron and boron-rich compounds. The most stable form of the element is the so called "$\beta$-rhombohedral boron" which contains 105 atoms in the primitive cell. The truncated icosahedron appears in this structure also. Fig 5.37 shows a central $B_{12}$ icosahedral unit joined to twelve $B_6$ pentagonal pyramids to give a $B_{84}$ unit.

[1]How Kepler would have loved this structure! His concept of nested polyhedra, which so deluded him about the sizes of planetary orbits, perhaps comes into its own here. We can't resist quoting [A. Koestler's translation in *The Sleep Walkers*, MacMillan, New York (1959)] from Kepler's *Mysterium Cosmographicum*: "I saw one symmetrical solid after another fit in so precisely between the appropriate orbits, that if a peasant were to ask you on what kind of hook the heavens are fastened so that they don't fall down, it will be easy for you to answer him." Substitute "atoms" for "heavens."

If an icosahedron is cut in half through a plane normal to a five-fold axis, the six vertices of each half will form a pentagonal pyramid. In this sense a pentagonal pyramid is half of an icosahedron. In $\beta$-rhombohedral boron each $B_{84}$ cluster (one per primitive cell) is joined to six others by making additional icosahedra from a juxtaposition of pentagonal prisms of the neighboring $B_{84}$ units. The other six pentagonal prisms (around a waist of the cluster) are united with $B_{21}$ units (also one per primitive cell) constructed from two sets of three face-sharing pentagonal pyramids ($B_{10}$ groups—see Fig. 5.37) and one additional atom. The central icosahedron is therefore surrounded by an icosahedron of icosahedra.

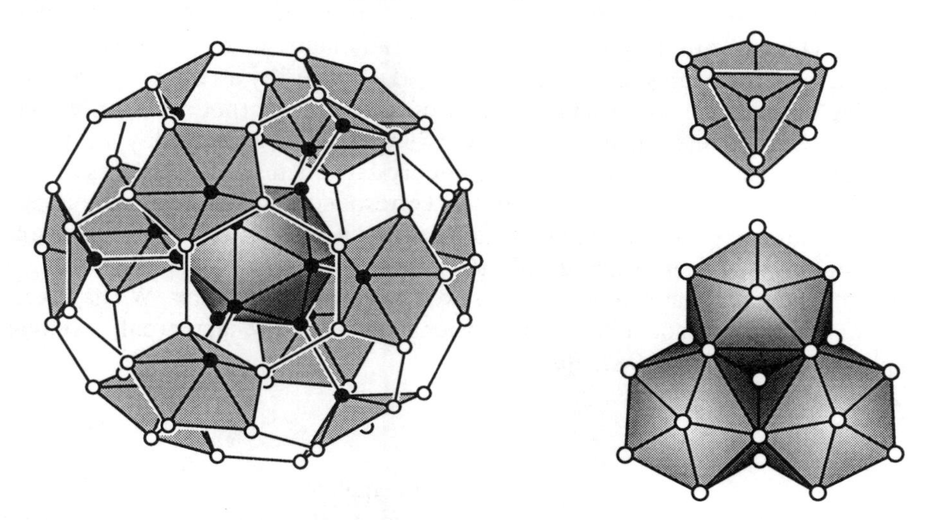

**Fig. 5.37**. Left: the $B_{84}$ cluster in $\beta$-rhombohedral boron. Top right: a $B_{10}$ group made from three face-sharing pentagonal pyramids. The $B_{21}$ group in $\beta$ rhombohedral boron is made from two $B_{10}$ groups united by an atom at a center of symmetry. Bottom right: a $B_{28}$ cluster of three face-sharing icosahedra formed when the $B_{10}$ group is capped by pentagonal pyramids (one from each of three different $B_{84}$ clusters).

Note that although the structure of $\beta$-rhombohedral boron is usually described in terms of linked and face-sharing icosahedra, the $B_{84}$ cluster also contains 20 truncated tetrahedra as shown in Fig. 5.35.

It is remarkable that so stable an element (its heat of atomization is second only to that of carbon among non-metallic elements) should have so complex a structure—there are 15 crystallographically-distinct B atoms. Boron also has a simpler structure (known as $\alpha$-rhombohedral boron) in which all the atoms are at vertices of icosahedra.

## 5.3 Circle packings and tilings of the plane

An extremum problem of interest in crystal chemistry is related to the problem of "points on a sphere" mentioned above (§ 5.1.9). It can be stated in the following way: What is the arrangement of non-overlapping equal circles on a plane such that the greatest possible fraction of the area of the plane is covered; i.e. what is the closest circle packing?

The answer is familiar to anyone who has played with arranging coins on a table top. Each circle is in contact with six others and their centers are at the points of a hexagonal lattice. Equivalently we may say that the centers lie on a $3^6$ net. If the circles have radius $r$, then their area is $\pi r^2$ per circle; the edge of the hexagonal cell is $2r$ and its area is $(\sqrt{3}/2)(2r)^2$. Thus the fraction of the plane covered is (area of circle/area of unit cell) = $\pi/\sqrt{12} = 0.907$.

In anticipation of similar questions in three dimensions we may also ask about circle packings in general. We define a *stable* circle packing to be one in which each circle has at least three neighbors, with the points of contact not all on the same semicircle. We direct attention first to arrangements of *equivalent* circles (i.e. related by a symmetry operation).

By a *tiling* of the plane we mean a covering of the plane, by (not necessarily convex) pieces or *tiles*. Our interest is mainly in periodic tilings by convex polygons (i.e. polygons with all interior angles < 180°). In the context of crystal structures, the vertices of a tiling correspond to atoms and the edges (sometimes) correspond to bonds between atoms. The pattern of vertices and edges of a tiling is often called a *net*, especially when the emphasis is on the topology of the structure. We tend to use the term "net" and "tiling" (or "tessellation") interchangeably in referring to two-dimensional structures. We have earlier[1] given a comprehensive account of two-dimensional nets; in this chapter (only) we refer in several places to that work as *OKH* for brevity.

### *5.3.1 Regular tilings*

The sequence $3^3$, $3^4$, $3^5$ of regular polyhedra with equilateral triangles as faces leads to a consideration of the pattern $3^6$. Six such triangles meeting at a vertex will lie in a plane (the sum of the vertex angles is 360°) and in fact we arrive at the tiling of the plane with equilateral triangles discussed above. Thus a tiling of a plane can be considered a special case of a polyhedron in which the sum of the angles at each vertex is 360°.

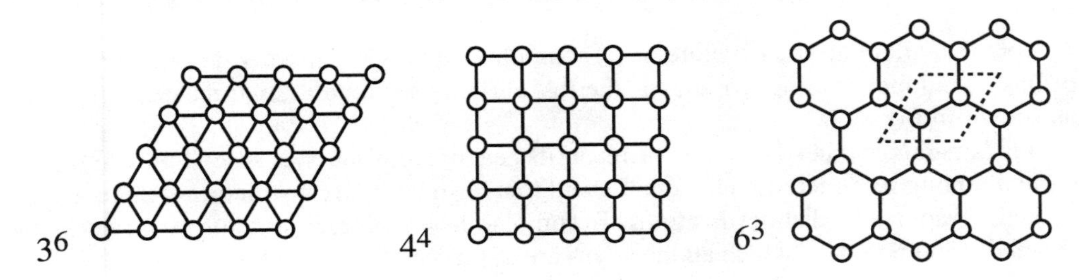


**Fig. 5.38.** The regular tilings $3^6$, $4^4$ and $6^3$. Broken lines outline a unit cell of $6^3$.

It should be apparent from a consideration of the possibilities that the only regular tilings are $3^6$, $4^4$ and $6^3$. These are illustrated in Fig. 5.38. Note that $3^6$ corresponds to a

[1]M. O'Keeffe & B. G. Hyde, *Phil. Trans. Roy. Soc. (London)* A**295**, 553 (1980).

hexagonal lattice, and that $4^4$ corresponds to a square lattice. The vertices of $6^3$, which we have called the honeycomb pattern, do not form a lattice as there are two points in a primitive cell (Fig. 5.38). The dual of $3^6$ is $6^3$ and *vice versa* and $4^4$ is self dual.

### *5.3.2 Archimedean tilings*

Tilings with one kind of vertex, but more than one kind of regular polygon as tile, are called Archimedean, or semi-regular, by analogy with the names used for finite polyhedra. The eight possibilities are listed in Table 5.4 and illustrated in Fig. 5.39. In the table $v$, $e$, and $f$ are the number of vertices, edges and faces per unit cell (it is a good exercise to verify these numbers from the drawings). These quantities are related by $v - e + f = 0$. Also listed in the table is the density (fraction of the plane covered) of the circle packing obtained by placing equal circles in contact and with their centers at the vertices (another good exercise is to verify these also).

**Table 5.4.** The Archimedean tilings. $v$, $e$ and $f$ are the numbers of vertices, edges and faces (polygons) *per unit cell.*

| symbol | symmetry | $v$ | $e$ | $f$ | density |
|---|---|---|---|---|---|
| $3^4.6$ | *p6* | 6 | 15 | 9 | 0.7773 |
| $3^3.4^2$ | *c2mm* | 4 | 10 | 6 | 0.8418 |
| $3^2.4.3.4$ | *p4gm* | 4 | 10 | 6 | 0.8418 |
| 3.4.6.4 | *p6mm* | 6 | 12 | 6 | 0.7290 |
| 3.6.3.6 | *p6mm* | 3 | 6 | 3 | 0.6802 |
| $4.8^2$ | *p4mm* | 4 | 6 | 2 | 0.5390 |
| $3.12^2$ | *p6mm* | 6 | 9 | 3 | 0.3907 |
| 4.6.12 | *p6mm* | 12 | 18 | 6 | 0.4860 |

Some properties of these patterns that are of interest are mentioned here. $3^4.6$ exists in enantiomorphic forms (recall the polyhedra $3^4.4$ and $3^4.5$ with the same property). 3.6.3.6 has all edges equivalent so it is quasiregular (recall the quasiregular polyhedra 3.4.3.4 and 3.5.3.5). The honeycomb, $6^3$, can be derived from $3^6$ by removing 1/3 of the vertices (centering the hexagons in $6^3$ recovers $3^6$). In a similar manner 3.6.3.6 can be derived from $3^6$ by removing 1/4 of the vertices, and $3^4.6$ can be derived from $3^6$ by omitting 1/7 of the vertices.

The duals of the Archimedean tilings are tilings of the plane with one kind of (irregular) tile and more than one kind of vertex.

Tilings with one kind of vertex and composed of polygons with equal edges correspond to packings of equivalent (i.e. symmetry-related) circles. As the above list (combined with the regular tilings) is complete, we can immediately answer questions such as "Which is the least dense such circle packing?" Answer: $3.12^2$. Unfortunately Nature is not so kind in simply providing the answers to similar questions (now concerning *sphere* packings) in three dimensions.

We will find $3^2.4.3.4$, 3.6.3.6, 3.4.6.4, and $4.8^2$ particularly useful in describing the patterns of layers of atoms in crystal structures and $4.8^2$ and 4.6.12 figure prominently in the derivation of three-dimensional nets in Chapter 7. We met 3.6.3.6 in Chapter 1, where it was called the kagome pattern. 3.4.6.4 is often called HTB as explained in § 5.3.4.

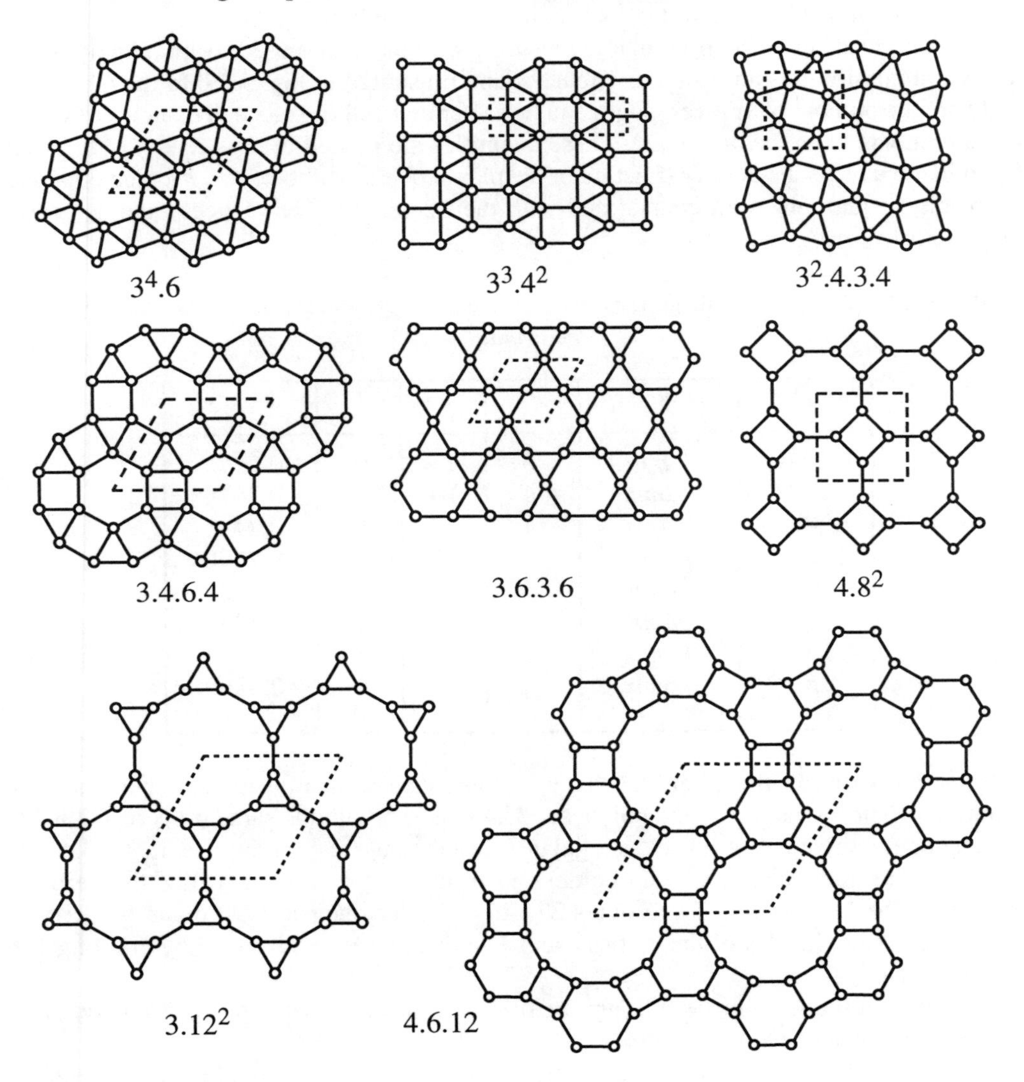


**Fig. 5.39**. The Archimedean tilings. Top row: $3^4.6$, $3^3.4^2$ and $3^2.4.3.4$. Middle row: 3.4.6.4, 3.6.3.6 and $4.8^2$. Bottom row: $3.12^2$ and 4.6.12. Unit cells are outlined with broken lines.

### *5.3.3 Relationships between tilings*

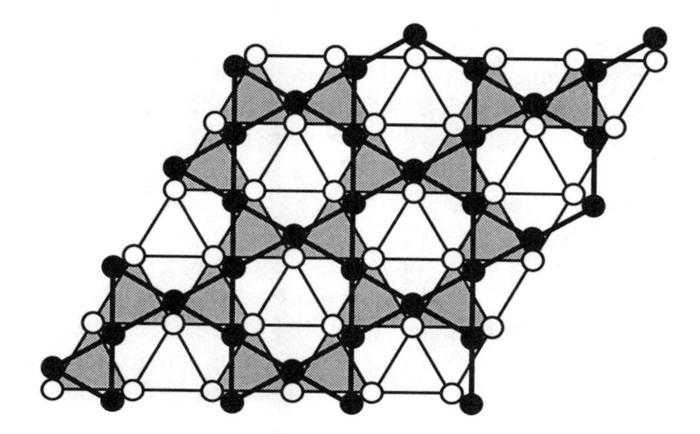

**Fig. 5.40**. Illustrating how 3.6.3.6 (filled circles) can be transformed into $3^6$ (open circles) and *vice versa*. Note how $3^6$ is obtained by rotation (by 30°) and dilation (by 15%) of the sides of the triangles of 3.6.3.6 (shown as lightly shaded in $3^6$). The numbers of open and filled circles in the diagram are the same.

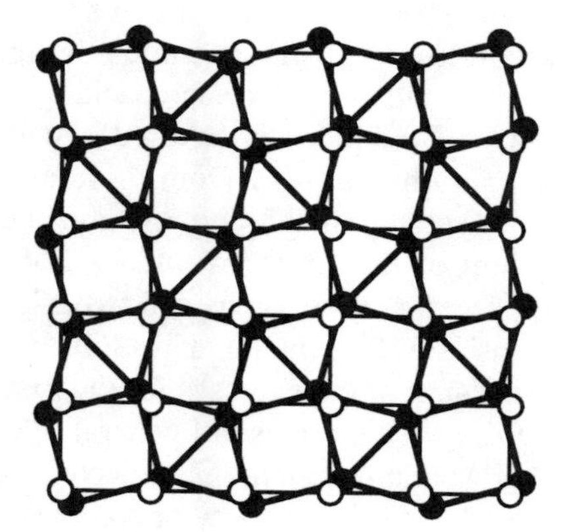

**Fig. 5.41**. Illustrating how $3^2.4.3.4$ (filled circles) can be transformed into $4^4$ (open circles) and *vice versa*. $4^4$ is obtained by rotating (by 15°) the squares of $3^2.4.3.4$, or alternatively by converting pairs of triangles (rhombuses) to squares. The numbers of open and filled circles in the diagram are the same.

It is of interest that the arrangement $3^3.4^2$ (to take an example) can be continuously deformed to an arrangement with density arbitrarily close to that of $3^6$ (which has the highest possible density) while keeping five neighbors.[1] For this reason the interesting question is usually that of finding the *least* density for a given number of neighbors (coordination number). It is a common mistake when considering the analogous problem of

[1]We will see other examples of these kinds of transformation later (Chapter 6). The present one is accomplished by shearing every second layer of $3^3.4^2$ so that the squares are transformed into rhombuses.

packing spheres in three dimensions to ask for the maximum density for a given number of neighbors, whereas, for similar reasons, it is the least-dense arrangement that is of interest.

Fig 5.40 shows, for example, how 3.6.3.6 (the least dense packing of equivalent circles with four neighbors, see Table 5.4, p. 165) which we met in Chapter 1, transforms to $3^6$ (the densest packing of equivalent circles). This transformation, which can be effected by rotating triangular groups of vertices, will be found to occur in sequences of related crystal structures.

Another transformation of importance relates $4^4$ and $3^2$.4.3.4. Now square groups of vertices are rotated as shown in Fig. 5.41.

We have seen (§ 2.5.7, p. 53) that converting a square to a pair of triangles transforms the polyhedron 3.4.4.4 (i.e. $3.4^3$) to $3^4$.4. and transforms the polyhedron 3.4.5.4. to $3^4$.5. The same operation converts the plane tiling 3.4.6.4 to $3^4$.6. (see Fig. 5.72, p. 201). These three transformations can equally be described by rotations of a polygon (square, pentagon and hexagon respectively). We return to this topic in the Notes (§ 5.6.12).

### *5.3.4 Tilings including pentagons, and "bronzes"*

The plane cannot be tiled with regular pentagons or with combinations of regular pentagons and other regular polygons. However it can *almost* be covered with regular pentagons, squares and triangles and this is good enough for nature, if not for mathematicians. Fig. 5.42 shows, at the top, two such packings that almost cover the plane (in the pattern on the left 96% of the plane is covered). At the bottom are tilings with irregular polygons derived in an obvious way from these arrangements. There are two kinds of vertex in each pattern. At the first, a 3-gon, a 4-gon and two 5-gons meet; if these were regular polygons the sum of the angles would be 366°, larger than 360° by six degrees. At the second kind of vertex, two 3-gons and two 5-gons meet and now the sum of angles for regular polygons is 336°, smaller than 360° by 24°. As the two kinds of vertex occur in the ratio 4:1, the sum of the excesses for the first vertices equals the sum of the deficits for the second vertices. This is a useful general rule that allows us to see what combination of vertices, and in what proportion, might occur in a tiling.

The tilings shown in Fig. 5.42 occur in a variety of contexts. The one on the left, which has symmetry *c2mm*, we call the **β-$U_3O_8$** net because of its occurrence as an O net in that compound (it also occurs as atom layers in intermetallic compounds). This net can serve as the "parent" of a large number of derived triangle-quadrangle-pentagon nets (see *OKH*, § 10). The tiling on the right, which has symmetry *p*4*gm*, we call the **$Mn_2Hg_5$** net as the Hg atoms in that compound fall on such a net.

These tilings have four edges meeting at each vertex; when we refer to them as nets we say that they are 4-connected. In a 4-connected net with, per unit cell, $n_3$ triangles, $n_4$ quadrangles and $n_5$ pentagons, it is easy to show (see Notes, § 5.6.11) that $n_4$ can have any value and $n_3 = n_5$. In the two nets above $n_3 = n_5 = 4$ and $n_4 = 2$.

We digress for a moment on the structure of $Mn_2Hg_5$ (for crystallographic data see Appendix 5) which serves to illustrate some instructive points. Adding translations in the third direction and mirror planes (normal to this direction) to the two-dimensional space group *p*4*gm* generates the three-dimensional group *P*4/*mbm*. In $Mn_2Hg_5$, which has this symmetry, the Hg atoms lie on the mirror plane at $z = 0$ on the **$Mn_2Hg_5$** net; this is called the *primary* net of the structure. The Mn atoms are on the mirror plane at $z = 1/2$ and lie over the centers of the Hg pentagons. The net of the Mn atoms, is called the *secondary* net, and in this instance is $3^2.4.3.4$ (which also has symmetry *p*4*gm*). Thus the $Mn_2Hg_5$ structure, and indeed many others, are succinctly described in terms of a stacking of alternating primary and secondary nets. In many instances, the secondary net is one of the regular or simpler semiregular ones. Care should be taken to distinguish a *dual* net (in which the centers of *all* the polygons of the original net correspond to vertices of the derived net) from a *secondary* net (in which the centers of only the larger polygons of the original net correspond to vertices of the derived net).

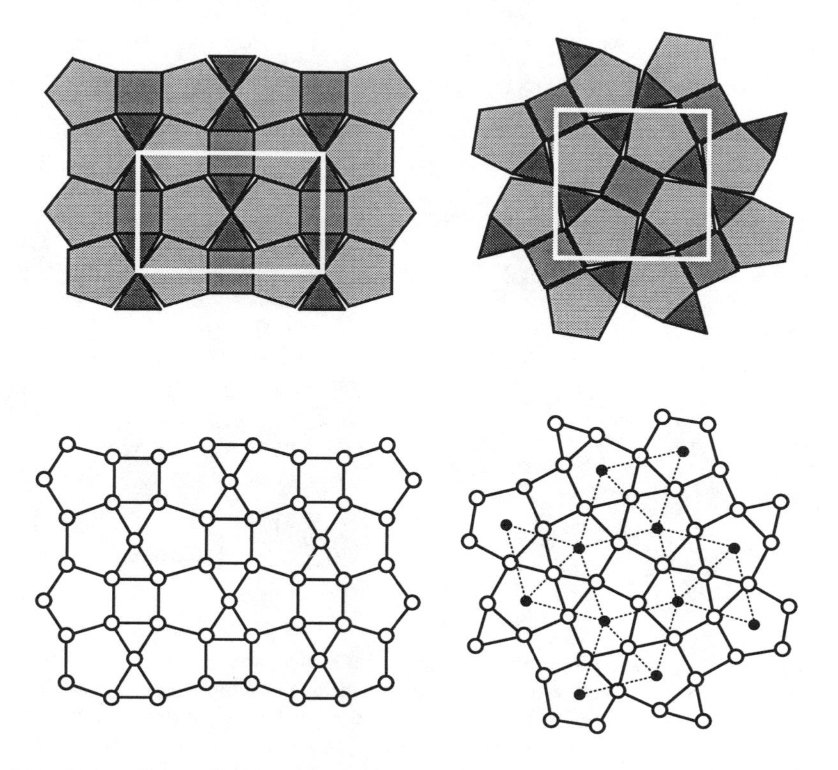

**Fig. 5.42**. Top: "almost" covering the plane with regular triangles, squares and pentagons of equal edge. Bottom: the derived tilings (nets). On the left is the **$\beta$-$U_3O_8$** net and on the right the **$Mn_2Hg_5$** net. The secondary net of the **$Mn_2Hg_5$** net is indicated with filled circles.

A new 4-connected net can be derived from a given 4-connected net by placing vertices in the middle of each edge of the old net, and joining them by edges to make a quadrangle surrounding the old vertices, which are then deleted. The reader is invited to do this with the **$Mn_2Hg_5$** net illustrated in Fig. 5.42. The new net we call the **TTB** net for a reason given below.

To most people, the term "bronze" refers to the beautiful copper-tin alloy, but solid state chemists also use the term to refer to compounds of variable composition involving (usually) alkali metal atoms combined with an early transition metal oxide. The tungsten bronzes, $M_xWO_3$ (here $M$ is an alkali metal) have been known for nearly 200 years, and are so named for their striking colors and metallic luster. Their structures also illustrate a use of plane nets in crystal chemistry, so we digress a little on this topic.

**Fig. 5.43**. Top left: a clinographic projection of a fragment of a cubic array of corner-connected octahedra. On the right of that the same structure (but with more octahedra) is shown in projection. Top right: the hexagonal tungsten bronze structure seen in projection. Bottom left: a similar view of the tetragonal tungsten bronze structure. Bottom right: The $Ba_4MgTa_{10}O_{30}$ structure (large circles Ba, Mg not shown).

The structure of $WO_3$ is based on a simple corner-connected array of $\{W\}O_6$ octahedra as shown schematically in Fig. 5.43 [this structure in its most symmetrical (cubic) form is usually called **$ReO_3$**]. When Na metal is added to $WO_3$, the metal atoms enter the cavities (actually cuboctahedra) in the structure, in the position shown as a large filled circle in the figure. The structure, seen in projection down a cube axis, is illustrated in Fig. 5.43; the shaded squares are octahedra seen in projection. Open circles are O atoms around the octahedron "waist"; let's imagine them to be in the plane of the paper. Small filled circles then represent O atoms above and below the plane completing the octahedron about W (not shown). The Na atom positions, also above and below the plane, are shown by larger filled circles. The net of the O atoms in the plane is clearly $4^4$. If every Na site were occupied, the composition would be $NaWO_3$ and the structure is called **perovskite.**[1]

If, instead of Na, Cs is used to form a tungsten bronze, a different structure is obtained (top right of Fig. 5.43). It is still based on an array of corner-connected $\{W\}O_6$ octahedra, but now the pattern of O atoms in the plane is 3.4.6.4. The large cavities in the structure are over the hexagons of this net and if they are all filled, the composition is $CsW_3O_9$. The structure is the hexagonal tungsten bronze (HTB) structure, and for this reason 3.4.6.4 is often called the **HTB** net. Note that the W atoms fall on a 3.6.3.6 net.

With K atoms another structure with composition $K_3W_5O_{15}$ is obtained.[2] Now the main oxygen atom net is made up of triangles, quadrangles and pentagons, and is in fact the net we derived from the **$Mn_2Hg_5$** net earlier. The structure is that of tetragonal tungsten bronze (TTB), so we call the net the **TTB** net (see bottom left of Fig. 5.43). Recalling how we derived this net it should be clear that the W atoms (in the octahedra, hence centering the squares of the projection) are on the **$Mn_2Hg_5$** net. Note that the tungsten compound with mixed valence [W(V) and W(VI)] is a metallic conductor, but many insulating (and colorless) compounds have been made with the same structure and they are often also referred to as "bronzes."

Another 4-connected net may be derived from the **$\beta$-$U_3O_8$** net in the same way as the **TTB** net was derived from the **$Mn_2Hg_5$** net (i.e. by putting new vertices in the middle of edges and removing the old vertices). This new net (Fig. 5.43, bottom right) is found as oxygen layers in, for example, the structure of $Ba_4MgTa_{10}O_{30}$ which is based on corner-sharing $\{Ta\}O_6$ octahedra. The Ta atoms are on the original **$\beta$-$U_3O_8$** net, the Ba atoms are in the space over the pentagons and Mg atoms fill one-half of the space over the quadrilateral tunnels.

The structure of $K_3V_5O_{14}$ is based on corner sharing $\{V\}O_6$ octahedra and $\{V\}O_5$ trigonal bipyramids; it is illustrated in the same way as the tungsten bronzes in Fig. 5.44. As before, octahedra project as centered squares, and now trigonal bipyramids project as centered triangles. Again we have a 4-connected triangle-square-pentagon net, and K atoms lie over the pentagon centers above the plane of the paper. The overall stoichiometry of a corner-sharing trigonal bipyramid is $VO_{5/2}$ and that of a corner-sharing octahedron is $VO_{6/2}$. As there are two trigonal bipyramids and three octahedra in the unit cell (together with three K atoms), the cell content is $3K + 2 \times VO_{5/2} + 3 \times VO_{6/2} = K_3V_5O_{14}$.

The net of the V atoms in $K_3V_5O_{14}$ is also shown in the figure (note that not all vertices are 4-connected in this instance). This is also the Pd net in the structure of $Th_3Pd_5$ (for crystallographic data see Appendix 5). The Th atoms of $Th_3Pd_5$ are on the secondary net of this structure (over the centers of the pentagons) which may be seen to be (slightly distorted) $3^6$. Thus the $K_3V_5$ part of $K_3V_5O_{14}$ has the same structure as $Th_3Pd_5$; indeed it is commonly found that the cation array in oxides and related materials is that of an

[1]The mineral perovskite is $CaTiO_3$ and its actual structure is a small distortion of the ideal cubic one.

[2]We have oversimplified the chemistry somewhat. Several structures occur for several different alkali atoms, and the observed structure depends on the stoichiometry as well as the size of the inserted atom.

intermetallic compound. The net of the V atoms in $K_3V_5O_{14}$ and of the Pd atoms in $Th_3Pd_5$ *also* occurs as the net of the O atoms in $\alpha$-$U_3O_8$ so we call it the **$\alpha$-$U_3O_8$** net. Although it may not be obvious from a comparison of Figs 5.42 and 5.44, the **$\alpha$-$U_3O_8$** net is closely related to the **$\beta$-$U_3O_8$** net (see § 5.6.12).

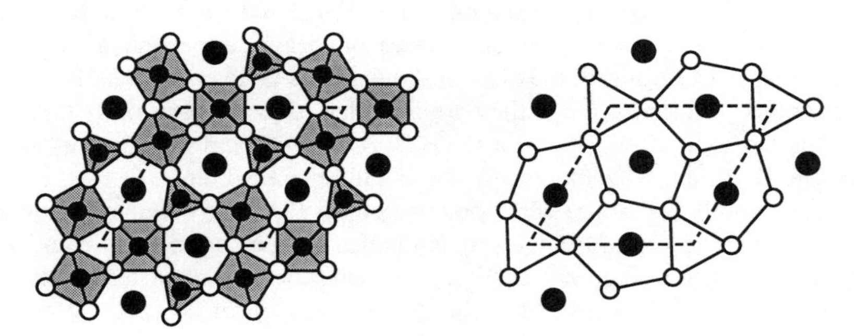

**Fig. 5.44**. Left: the structure of $K_3V_5O_{14}$ as corner-sharing octahedra and trigonal bipyramids. Larger circles are K atoms, and smaller circles are O atoms. V atoms (not shown) center the shaded octahedra and trigonal bipyramids. Right: the **$\alpha$-$U_3O_8$** net of the V atoms (open circles) in $K_3V_5O_{14}$, with the K atoms (filled circles) on the secondary net.

Many other triangle-quadrangle-pentagon nets occur in crystal chemistry; for a review of some of them, and their interrelationships, see *OKH*.

### *5.3.5 Some 3-connected boron nets:* $AlB_2$, $YCrB_4$, $ThMoB_4$ *and* $Y_2ReB_6$

Examples of 3-connected nets already met are $6^3$, $3.12^2$, $4.8^2$ and 4.6.12. A unit cell of $4.8^2$ contains one square and one octagon, so on average the polygons have six sides. The unit cell of $3.12^2$ contains two triangles and one dodecagon, so again, the average number of sides per polygon is six. It is easy to show (see Notes § 5.6.11) that this must be the case in general for 3-connected nets. As each edge belongs to two polygons, the number of edges is $e = 3f$ where $f$ is the number of polygons. In a 3-connected net $e = 3v/2$, where $v$ is the number of vertices per cell, so $f = v/2$.

In many borides, the B atoms form 3-connected plane nets with metal atoms ($M$) in interleaved layers centering the polygons. It follows that the stoichiometry is $MB_2$ and the average polygon size in the B layer is 6. The simplest case is in the structure of $AlB_2$ in which the B atoms lie on $6^3$ nets (as in the graphite form of carbon) with Al atoms over the centers of the hexagons forming larger-spacing $3^6$ nets (see Fig. 5.46 on p. 174).

In some ternary borides the B atom layer contains pentagons and heptagons (which must occur in equal numbers) as well as hexagons. Three common structure types containing such layers are those of $YCrB_4$, $ThMoB_4$ and $Y_2ReB_6$ (for crystallographic data see Appendix 5). The B nets in these compounds are illustrated in Fig. 5.45. Those on the left and the right of the figure provide nice examples of patterns with symmetry *p*2*gg*. Adding translations and mirrors perpendicular to the translations of *p*2*gg* produces the three-dimensional group *Pbam* which is the symmetry of the corresponding compounds. The pattern in the center has symmetry *cmm* and the symmetry of the crystal structure is *Cmmm*.

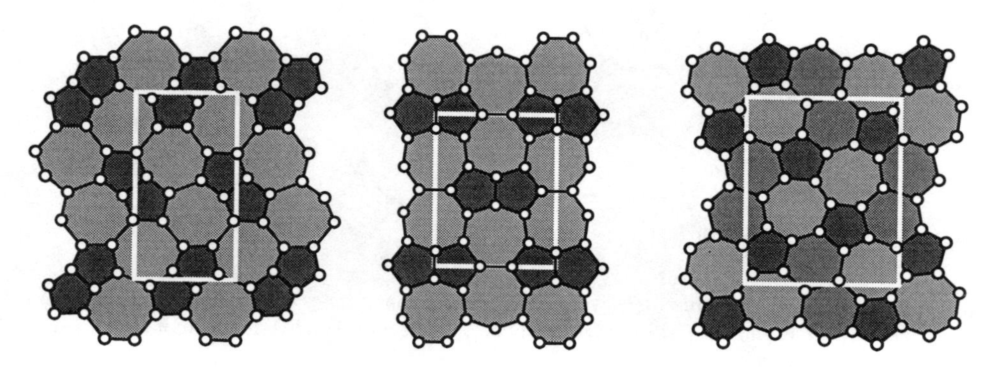

**Fig. 5.45**. The B nets in $YCrB_4$ (left), $ThMoB_4$ (left) and $Y_2ReB_6$ (right).

In each crystal structure the larger metal atoms lie over the heptagons (and the hexagons in $Y_2ReB_6$). The smaller metal atoms (Cr, Mo and Re respectively) lie over the pentagon centers and thus have pentagonal prismatic coordination by B. In $YCrB_4$ and $ThMoB_4$, the occurrence of adjacent pairs of pentagons in the net results in a very short metal-metal distances. Thus there is a Cr-Cr distance of 2.38 Å (compare the shortest distance of 2.50 Å in the metal) in $YCrB_4$; likewise there is an Mo-Mo distance of 2.56 Å (compare the shortest distance of 2.73 Å in the metal) in $ThMoB_4$.

### 5.3.6 Some boron-carbon nets

In the previous section we described some boron layers in borides; related nets are also to be found as boron-carbon layers. In these, it is often found that the C atoms are of lower connectivity than the B atoms. These nets again provide some nice exercises in recognizing symmetry.

Our first example, which occurs in the structure of $LaB_2C_2$ (for data see Appendix 5), is just a simple decoration of the $4.8^2$ net as shown in Fig. 5.46. In the crystal structure, alternate layers of the net are rotated by 90° and La atoms center the octagonal prisms, so that the coordination of La is $\{La\}B_8C_8$. The La array is close to primitive cubic. Note that although the symmetry of one *plane* is rectangular (2-dimensional group *c2mm*), the three-dimensional symmetry has $\bar{4}$ axes and the crystal is tetragonal (space group $P\bar{4}2c$).

In the $TbB_2C$ structure (for crystallographic data see Appendix 5) layers of $B_2C$ alternate with Th layers. The $B_2C$ layer is shown in Fig. 5.47. It may be considered a 3-connected net of heptagons and quadrilaterals; but, as indicated in the figure, the quadrilaterals are better considered as two triangles, and some of the B atoms are then 4-connected. The secondary net of Tb (centering the heptagons) is $3^2.4.3.4$. In the crystal structure alternate $B_2C$ layers are rotated by 90°; thus although again the symmetry of one plane is rectangular (*p2gg*), the symmetry of the crystal is tetragonal ($P4_2/mbc$).

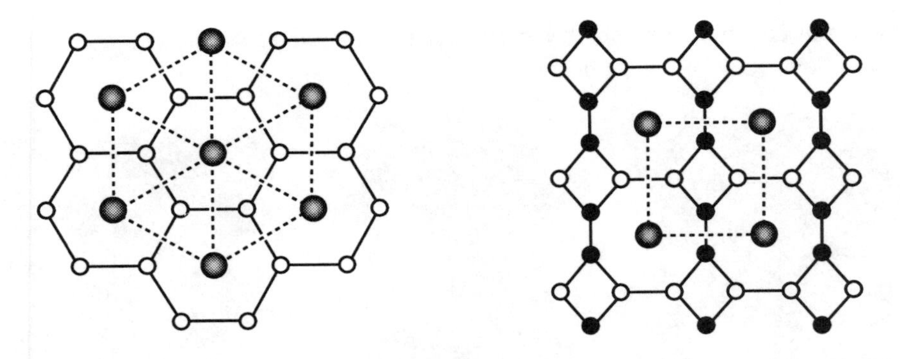

**Fig. 5.46**. Left: the $AlB_2$ structure. $6^3$ layers of B (small open cuircles) alternate with $3^6$ layers of Al (large shaded circles). Right: the The $LaB_2C_2$ structure. A $4.8^2$ net of B and C (filled circles) alternates with $4^4$ nets of La (larger shaded circles).

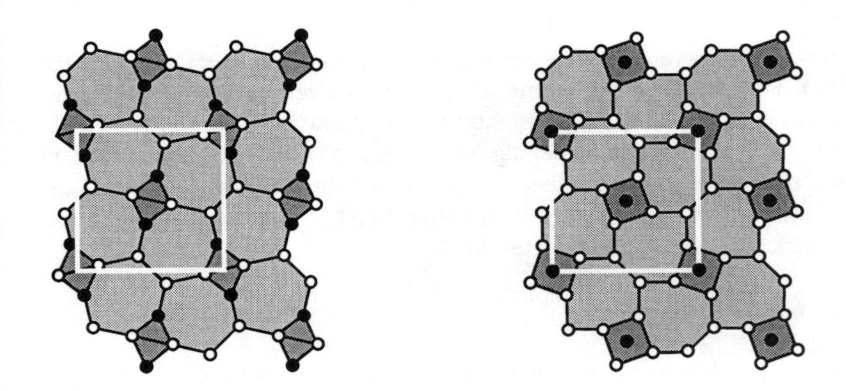

**Fig. 5.47**. Left the boron-carbon net in $TbB_2C$. Filled circles are C atoms. Right: the B net in $ThB_4$ (open circles). Now filled circles are $B_2$ groups above and below the plane of the net.

The crystal structure of $ThB_4$ (for data see Appendix 5) contains similar layers (Fig. 5.47) but now the quadrangles are squares; additional B atoms above and below the plane cap the squares to form octahedra, and strings of octahedra joined by additional edges run normal to the plane of the net. (Fig. 5.48). Thus the structure should really be considered a three-dimensional B net. We meet other example of B nets in Chapter 7.[1]

In the structure of $ThB_2C$ (also known as the $UB_2C$ structure), there are $B_2C$ layers with 3-coordinated B and 2-coordinated C as shown in Fig. 5.49. Crystallographic data are given in Appendix 5. Those who like to count electrons might care to speculate on why the

[1]Note the influence of stoichiometry on coordination number. In compounds $M_nB$ ($n \geq 2$) there are usually B atoms that are not bonded to other B atoms. In compounds *M*B (such as CrB and FeB, § 6.4.2) B is 2-connected to other B atoms, in compounds $MB_2$ (discussed in the previous section) B is 3-connected, in $ThB_4$, B is 3- and 5-connected (but 4-connected in $CrB_4$, § 7.3.3), in $CaB_6$ and $UB_{12}$ (§ 7.9) B is 5-connected, and in elemental B (§ 5.2.5) B is 6- and 7-connected.

$TbB_2C$ structure is formed by Sc, Y and lanthanides (trivalent elements) but a different structure is formed by $ThB_2C$ and $UB_2C$.

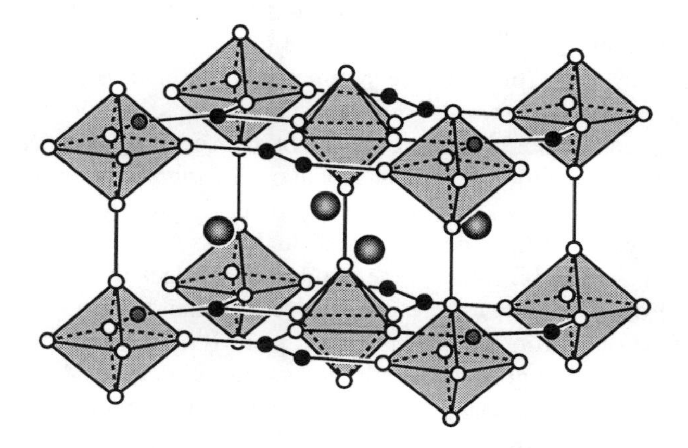

**Fig. 5.48.** The $ThB_4$ structure in clinographic projection. Small open circles are B atoms connected to five other B, small filled circles are B connected to three other B; larger circles are Th.

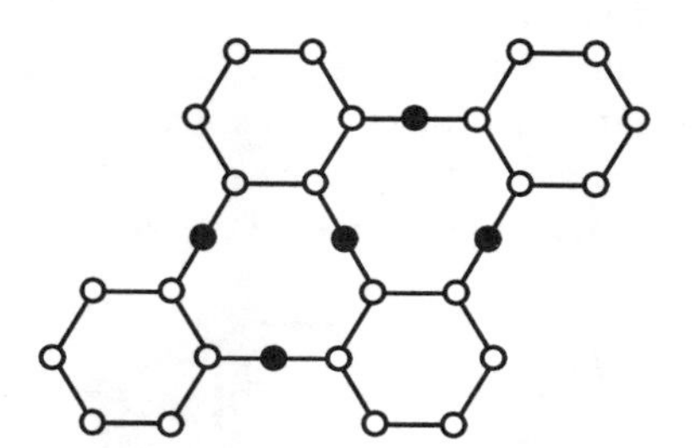

**Fig. 5.49.** The boron-carbon net in $ThB_2C$. Filled circles are carbon atoms.

### *5.3.7 Polyatomic tilings: self-dual nets*

The tilings we have considered so far generally correspond to layers of one kind of atom (or two related atoms such as B and C) in crystal structures; in this section we describe some (chemically) ternary structures in which all three kinds of atom are in the same layer. As the different atoms may be of different "sizes," in general the tilings derived from these structures will be made up of polygons with unequal edges.

Very often in structures composed of layers, all the atoms lie on mirror planes and then there are just two layers in the structure (recall that mirror planes repeat with a translation equal to half the shortest lattice vector perpendicular to them). In some instances, the two mirror planes will be related by symmetry so that there is only one distinct kind of layer. This will occur when, for example, there are axes of the sort $2_1/m$, $4_2/m$, or $6_3/m$ normal to the mirror planes (say along **c**). In each of these cases the symmetry operation has a translation component of **c**/2 taking an atom on one mirror plane into an equivalent atom on the other mirror plane. For example, in the commonly-occurring space group *Pnma*, there is a $2_1/m$ axis along 0,*y*,0 and the mirror planes are at $y = 1/4$ and $y = 3/4$ (see Fig. 3.14, p. 72). The $2_1$ axis operating on an atom at $x,1/4,z$ will take it into an equivalent atom located at $\bar{x},3/4,\bar{z}$. In fact the whole layer at $y = 1/4$ (or 3/4) is rotated by 180° about the *y* axis and translated to $y = 3/4$ (or 1/4). The symmetry of the layer is $p1g1$. To take another example, it should be clear that in body-centered space groups, mirror planes normal to the axes of the conventional (centered) cell are related by the centering translation. For more on the symmetry of structures with a two-layer repeat of one kind of layer see § 5.6.14.

It is quite common for the vertices of one net to center (in projection) the polygons of the other net and *vice versa*, so that the net is self dual. For a self-dual net, the number of polygons with *n* edges must be equal to the number of vertices that are *n*-connected. The average polygon (ring) size must therefore be equal to the average connectivity of the vertices. It follows at once from Eq. 5.4 (§ 5.6.11, p. 198) that this average is 4. It follows further that either all the polygons are quadrangles (and all vertices 4-connected), i.e. the net is $4^4$, or that some of the polygons are triangles and correspondingly that some have more than four edges. A simple example of the latter case is provided by the structure of CrB (Fig. 5.50) in which there are equal numbers of pentagons and triangles. In this structure Cr atoms in one layer center (in projection) the pentagons of layers above and below and B atoms similarly center the triangles.

An atom centering (in projection) the triangles of identical nets above and below it is in (capped) triangular-prismatic coordination, so it is not surprising that such coordination is a feature of structures that can be described in terms of a stacking of self dual nets. In Fig. 5.50 the trigonal prisms (at two elevations) in **CrB** are shaded for emphasis.

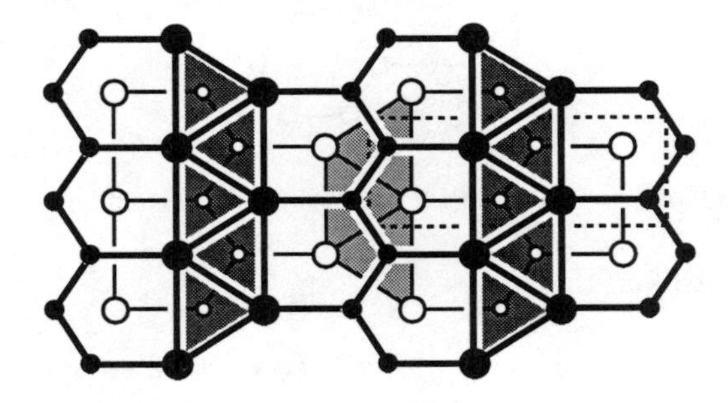

**Fig. 5.50**. The structure of CrB projected on (100). Larger circles are Cr.

As another example we illustrate (in Fig. 5.51), the structure of SrMgSi (which is one of the many compounds with the $Co_2Si$ structure type) in which all atoms lie on the mirror planes of *Pnma*. In this structure Sr centers pentagons of adjacent nets, and Mg and Si similarly center quadrangles and triangles respectively. Accordingly the net (which has symmetry *pg*) has triangles, quadrangles and pentagons in equal numbers. For crystallographic data for SrMgSi see Appendix 5.

WCoB is isostructural with SrMgSi but NbCoB and TaCoB have a related structure (Fig. 5.51) with symmetry *Pmmn* composed of self-dual nets of symmetry *pm*. Notice that in a SrMgSi layer all the triangles are $Sr_2Mg$, but that in a NbCoB layer there are triangles $Nb_3$, $Nb_2Co$ and $Co_3$.

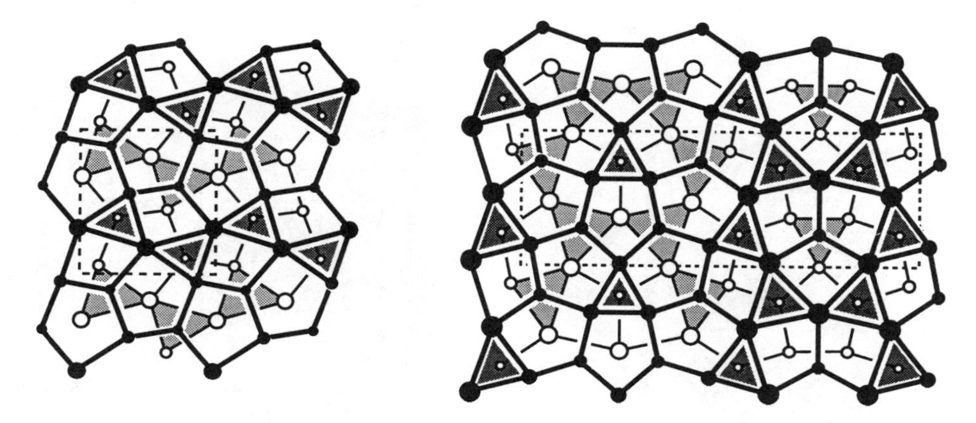

**Fig. 5.51**. Left: the structure of SrMgSi projected on (010), atoms at $y$ = 1/4 and 3/4. Large, medium and small circles are Sr, Mg and Si respectively. Right: the structure of NbCoB projected on (100), atoms at $x$ = 0 and 1/2. Large, medium and small circles are Nb, Co and B respectively. Trigonal prisms are shaded.

A second crystallographically ternary compound (the first was SrMgSi) with a structure based on self-dual nets is that $Re_3B$ (Fig. 5.52). This is better written as $Re(1)Re(2)_2B$ to emphasize that there are two kinds of Re atom, and a more typical composition for this structure type is $YAl_2Co$. The symmetry is *Cmcm* and the symmetry of the layers is *pmg*. Y centers pentagons, Al centers quadrangles and Co centers triangles. For crystallographic data for $Re_3B$ see Appendix 5.

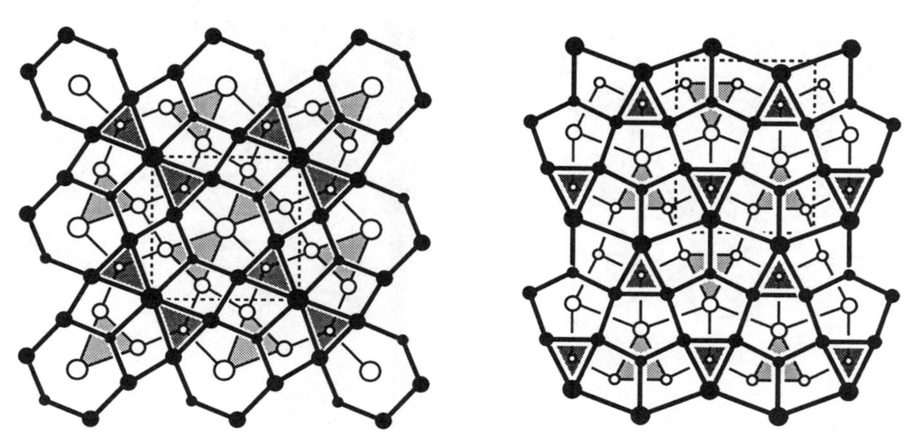

**Fig. 5.52**. Left: the structure of $ZrFe_4Si_2$ projected on (001), atoms at $z$ = 0 and 1/2. Large, medium and small circles are Zr, Fe and Si respectively. Right: the structure of $YAl_2Co$ (**$Re_3B$**) projected on (100), atoms at $x$ = 0 and 1/2. Large, medium and small circles are Y, Al and Co respectively. Trigonal prisms are shaded.

Another ternary compound with a structure based on self-dual nets is that of $ZrFe_4Si_2$ (Fig. 5.52). Now there are hexagons, quadangles and triangles (centered by Zr, Fe and Si respectively) in the ratio 1:4:2. The symmetry of the structure is $P4_2/mnm$ and that of the layers is *c2mm* (for crystallographic data see Appendix 5).

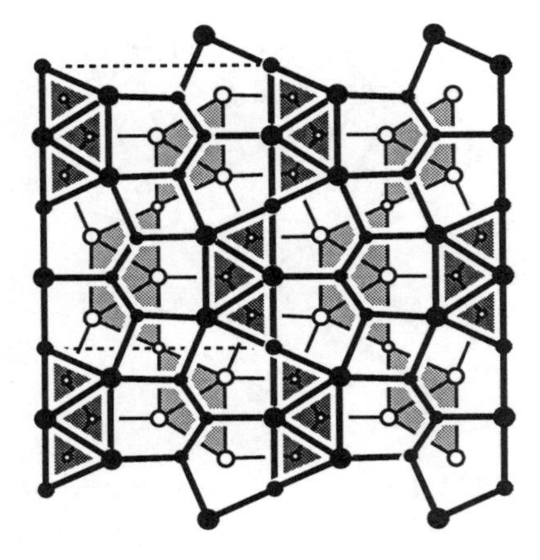

**Fig. 5.53**. The structures of $W_2CoB_2$ (left) projected on (001), atoms at $z$ = 0 and 1/2, and $W_3CoB_3$ (right) projected on (100), atoms at $x$ = 0 and 1/2. Large, medium and small circles are W, Co and B respectively. Trigonal prisms are shaded.

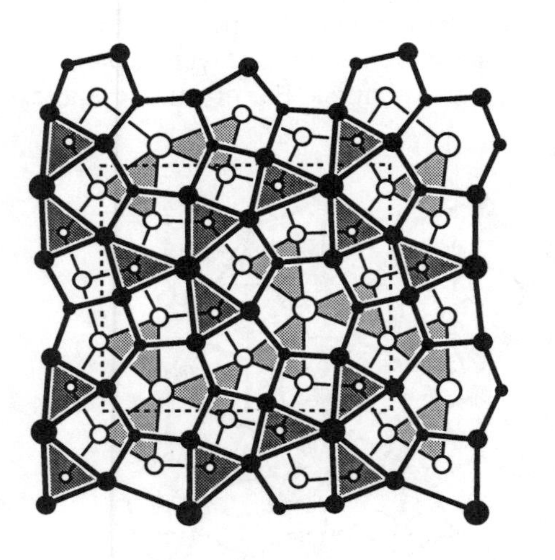

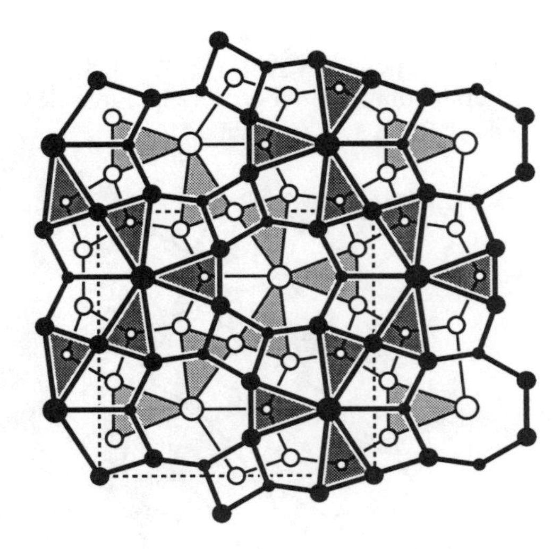

**Fig. 5.54**. Left: the structure of $YCo_5P_3$ projected on (010), atoms at $y$ = 1/4 and 3/4. Right: the structure of $LaCo_5P_3$ projected on (100), atoms at $x$ = 0 and 1/2. Large, medium and small circles are Y (La), Co and P respectively.

There is a series of compounds $W_nCoB_n$ whose structures are composed of self-dual nets in which W centers pentagons, Co centers quadrangles and B centers triangles. The case $n = \infty$ is WB which has the CrB structure (Fig. 5.50). $n = 1$ is WCoB which has the SrMgSi structure (Fig. 5.51). Fig. 5.53 illustrates the $n = 2$ and 3 cases. Note that in these structures there are groups of $n$ trigonal prisms sharing rectangular faces so that there are $B_n$ groups joined by bonds through these faces.

In $LaCo_5P_3$ (Fig. 5.54) with symmetry *Cmcm*, heptagons, quadrangles and triangles in the ratio of 1:5:3 form a self-dual net with symmetry *p2mg*. La centers the heptagons, Co the quadrangles and P the triangles. Note that the triangles are all $LaCo_2$, but the quadrangles are $LaCo_2P$ and $Co_2P_2$. This structure type is sometimes named **$LaNi_5P_3$**.

In $YCo_5P_3$ (Fig. 5.54) with symmetry *Pnma*, the Y atoms center hexagons, and P atoms center triangles so hexagons and triangles occur in the ratio 1:3. As the average ring size must be four there is also one pentagon per hexagon, and Co centers pentagons as well as quadrangles. A third structure occurs for this stoichiometry in $LaNi_5Si_3$ (Fig. 5.55) this has the same symmetry and the net contains the same number of polygons as in $YCo_5P_3$.

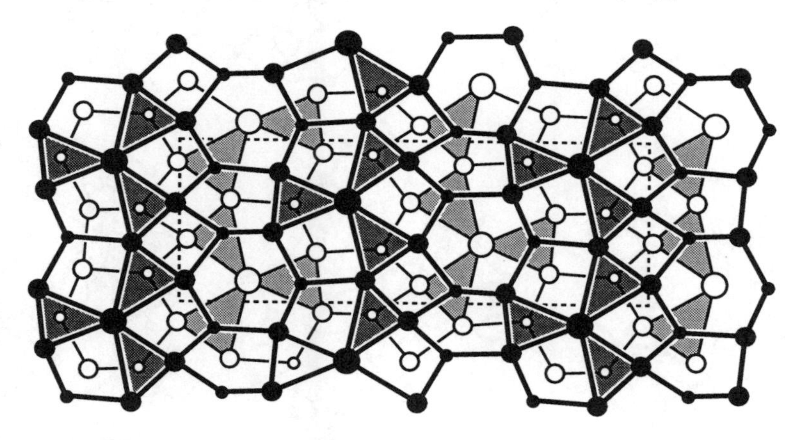

**Fig. 5.55**. The structure of $LaNi_5Si_3$ projected on (010). Large, medium and small circles are La, Ni and Si respectively at $y = 1/4$ and 3/4.

To illustrate the variety and beauty of the structures that nature has devised we illustrate a number of other self-dual nets in Fig. 5.56. Properties of these and the other nets of this section are summarized in Table 5.5 (p. 181) The chemical formula of all the compounds discussed in this section is $A_aB_bC_c$ (CrB is considered to have $b = 0$). When this formula is written in the preferred way of most electropositive element first, it is observed that $A$ centers the large polygons, $B$ centers smaller polygons (usually quadrangles) and $C$ invariably centers triangles (i.e. is in trigonal prismatic coordination). The table identifies the polygons so centered; as $A$ centers the largest polygons and $C$ the smallest, we refer to $A$, $B$ and $C$ as "large," "medium," and "small" respectively, referring, of course, to relative size. It may be verified that the average polygon (ring) size is four in every case. The structures illustrate the delicate balance between "size" and stoichiometry in determining coordination numbers although we should remark that the number of "caps" of the coordination prisms $N.4^2$ is not always $N$. Also to be noted that as $C$ is in trigonal prisms of $A$ and $B$ the structures can always be represented in terms of trigonal prisms with centers at two heights (either 0 and 1/2 or 1/4 and 3/4). The number of structures that can be so described is considerably larger than those based on self-dual nets—see for example **$Fe_2P$** (§ 5.6.15).

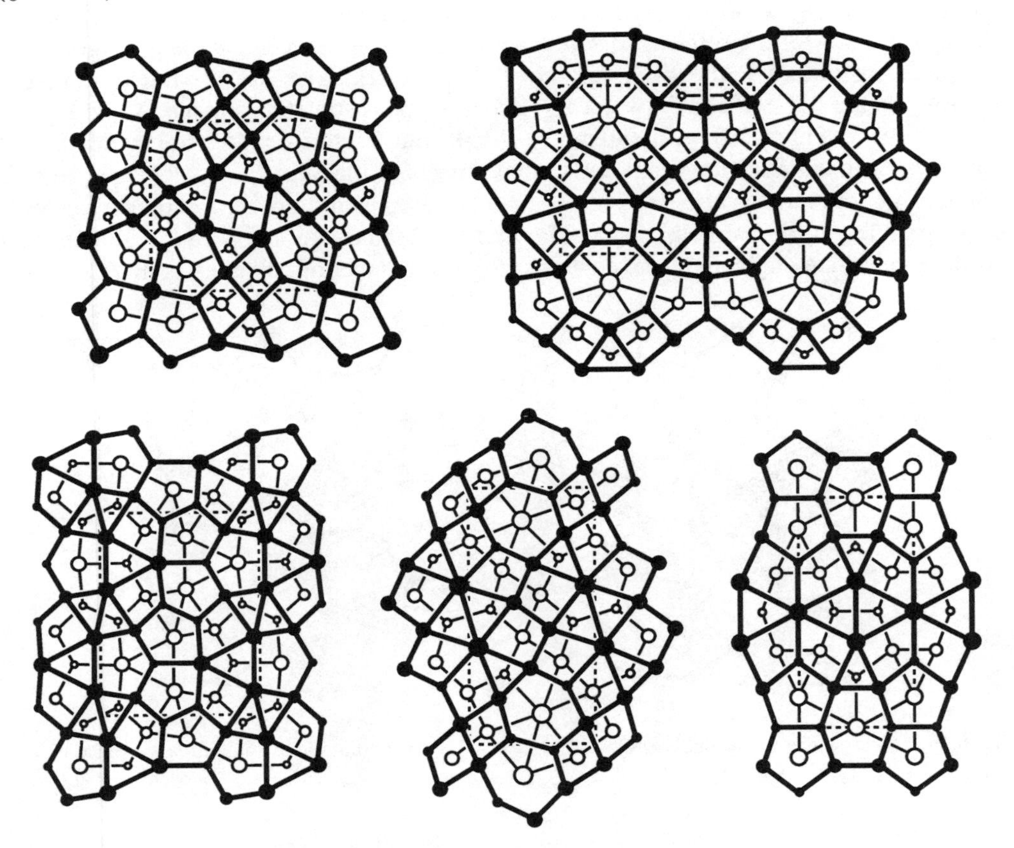

**Fig. 5.56**. Some structures based on self-dual nets. Top: left $Nb_5Cu_4Si_4$, right $EuCo_8P_5$. Bottom, from the left: $Hf_3Ni_2Si_3$, $NdRe_4Si_2$, $Zr_3Cu_4Si_4$. In all these structures of $A_aB_bC_c$ largest circles are $A$ and smallest are $C$ (the latter invariably centering triangles).

In the tetragonal structures listed in Table 5.5, atoms are on the mirror planes normal to the 4-fold axis. In some simple tetragonal structures (with small unit cell parameter $a$) the atoms are all on mirror planes parallel to the 4-fold axis (along **c**) and often the atoms on such layers form self-dual nets normal to **a**. There are of course equivalent layers normal to **b** and a description of such structures in terms of a stacking of nets is leaves something to be desired. Notable examples of such structures are **$ThCr_2Si_2$** (also named **$BaAl_4$**) and **BaMgSi** (also named **PbFCl**), perhaps the two most populous structure types (for more on these structures see § 6.4.2).

**Table 5.5**. Some structures (with symmetries) based on self-dual nets. Numbers are $N$-gons per formula unit in the net

| structure | Fig. | crystal | layer | large atom | | | | medium atom | | small atom |
|---|---|---|---|---|---|---|---|---|---|---|
| $N$ | | | | 7 | 6 | 5 | 4 | 5 | 4 | 3 |
| CrB | 5.50 | *Cmcm* | *p2mg* | | | 1 | | | | 1 |
| SrMgSi | 5.51 | *Pnma* | *pg* | | | 1 | | | 1 | 1 |
| NbCoB | 5.51 | *Pmmn* | *pm* | | | 1 | | | 1 | 1 |
| $YAl_2Co$ | 5.52 | *Cmcm* | *p2mg* | | | 1 | | | 2 | 1 |
| $ZrFe_4Si_2$ | 5.52 | $P4_2/mnm$ | *c2mm* | | 1 | | | | 4 | 2 |
| $W_2CoB_2$ | 5.53 | *Immm* | *p2mm* | | | 2 | | | 1 | 2 |
| $W_3CoB_3$ | 5.53 | *Cmcm* | *p2mg* | | | 3 | | | 1 | 3 |
| $LaCo_5P_3$ | 5.54 | *Cmcm* | *p2mg* | 1 | | | | | 5 | 3 |
| $YCo_5P_3$ | 5.54 | *Pnma* | *pg* | | 1 | | | 1 | 4 | 3 |
| $LaNi_5Si_3$ | 5.55 | *Pnma* | *pg* | | 1 | | | 1 | 4 | 3 |
| $Nb_5Cu_4Si_4$ | 5.56 | *I4/m* | *p4* | | | 4 | 1 | | 4 | 4 |
| $EuCo_8P_5$ | 5.56 | *Pmmn* | *pm* | 1 | | | | 2 | 6 | 5 |
| $Hf_3Ni_2Si_3$ | 5.56 | *Cmcm* | *p2mg* | | | 3 | | | 2 | 3 |
| $NdRe_4Si_2$ | 5.56 | *Pnnm* | *c2* | | 1 | | | | 4 | 2 |
| $Zr_3Cu_4Si_4$ | 5.56 | *Immm* | *p2mm* | | 1 | 2 | | | 4 | 4 |

## 5.4 Layers of tetrahedra and/or octahedra: sheet silicates

Many crystal structures are conveniently described as built up of slabs (multiple layers of atoms) rather than as a stacking of planes and we describe here some important layers of connected polyhedra. Specifically we describe layers of corner-connected tetrahedra and edge-connected octahedra.[1]

### *5.4.1 Layers of tetrahedra and octahedra*

Layers of corner-connected $\{T\}X_4$ tetrahedra can be constructed with each tetrahedron sharing three vertices with other tetrahedra so the stoichiometry is $TXX_{3/2} = T_2X_5$. The simplest, and most important way, of doing this is with the $T$ atoms on the vertices of a $6^3$ net as shown in Fig. 5.57.

As the figure illustrates, the layer of tetrahedra is flexible. In "expanded" form, the $T$-$X$-$T$ angle is 141.1° and the base vertices are on a 3.6.3.6 net. In "contracted" form (with the distance between vertices of neighboring tetrahedra equal to the tetrahedron edge) the $T$-$X$-$T$ angle is 109.5° and the base vertices are on a $3^6$ net (the two configurations are related exactly as shown in Fig. 5.40, p.167). In each case the $T$ atoms and the vertices above them remain on a $6^3$ net.

We note in passing that two such layers (one pointing "up," and one pointing "down") can be joined using the fourth vertex to produce stoichiometry $TX_2$. Such layers are found in one polymorph of $BaAl_2Si_2O_8$ in "expanded" form, and in one polymorph of $CaAl_2Si_2O_8$ in "contracted" form (with the top layer "anti" to the bottom layer).

[1]Slabs of edge-connected tetrahedra are discussed especially in § 6.4.2 (see Fig. 6.43, p. 249).

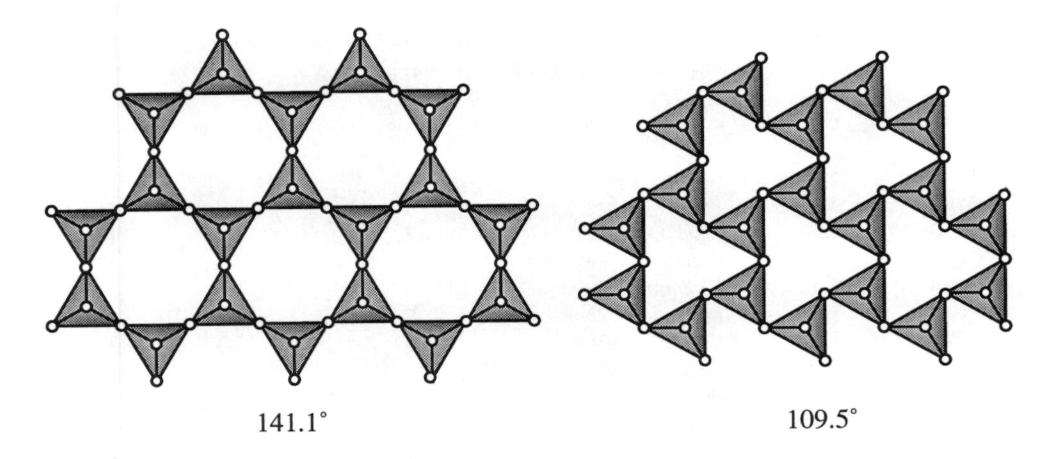


**Fig. 5.57**. Layers of corner-connected tetrahedra pointing "up." Left: in expanded form. Right: "contracted" to lower density. Numbers are the *T-X-T* angles (in degrees) for regular tetrahedra.

The *T-X-T* angle can be considerably larger (indeed up to 180°) if the tetrahedra alternate "up" and "down" as shown for two configurations in Fig. 5.58 and such alternating layers are common in silicates (in which the Si-O-Si angle is usually about 145°, or greater).

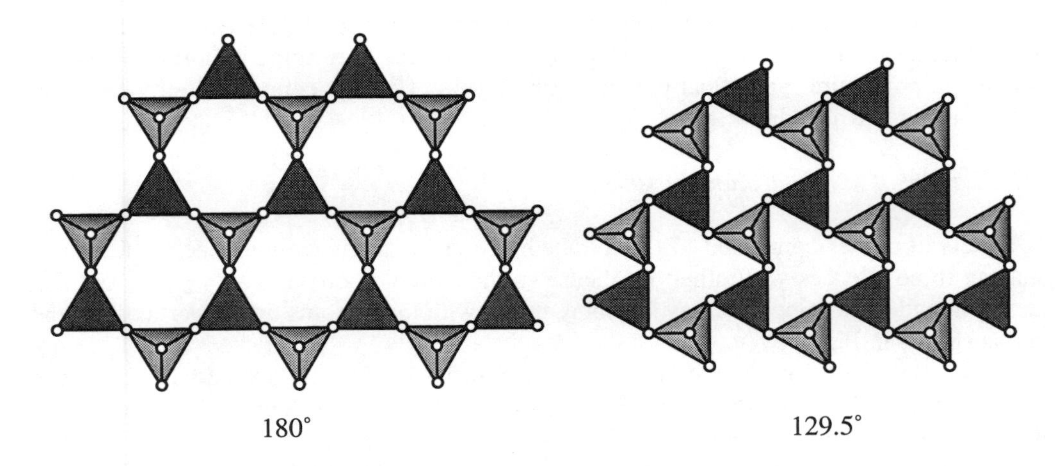


**Fig. 5.58**. Layers of corner-connected tetrahedra alternating up and down. Left: in expanded form. Right: "collapsed" to lower density. Numbers are the *T-X-T* angles (in degrees) for regular tetrahedra.

Layers of $[Si_2O_5]^{2-}$ alternate with cations in compounds such as $Na_2Si_2O_5$ and $BaSi_2O_5$. Because the apical O atoms are bonded also to cations, the layers can be considerably distorted from the high-symmetry configurations shown in Figs. 5.57 and 5.58. Fig. 5.59 illustrates the layers in $Na_2Si_2O_5$ and $BaSi_2O_5$. In the former the tetrahedra alternate up and down; in the latter *pairs* of tetrahedra alternate up and down.

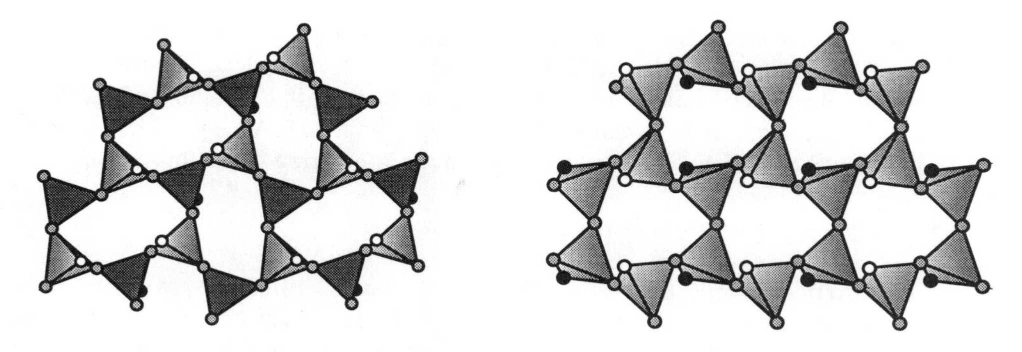

**Fig. 5.59**. The $Si_2O_5$ layers in $Na_2Si_2O_5$ (left) and $BaSi_2O_5$ (right). Filled circles are vertices pointing "down."

Layers of edge-sharing octahedra are also common in crystal structures. On the left in Fig 5.60 we show a layer of $\{M\}Y_6$ octahedra sharing six edges with stoichiometry $MY_2$. Such layers occur notably in $\mathbf{CdCl_2}$ and $\mathbf{CdI_2}$ (which differ only in the stacking of the layers—see § 6.1.5). The $MgO_2$ part of $Mg(OH)_2$ (brucite) consists of similar layers. For a reason to become apparent below, mineralogists often refer to the $MY_2$ layer as a "trioctahedral layer."

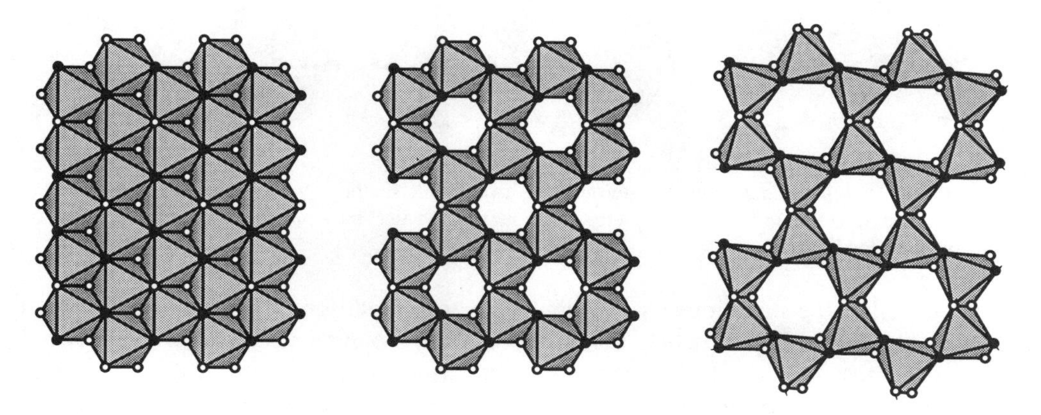

**Fig. 5.60**. Layers of edge-sharing octahedra. Left: an $MY_2$ layer. Center: an $MY_3$ layer of regular octahedra. Right: an $MY_3$ layer of metaprisms with shared edges shorter than unshared edges (see text). The filled circles are vertices that can be shared with a tetrahedral layer (see Fig. 5.57).

A second layer of edge-sharing octahedra, now with each octahedron sharing only three edges, is obtained from the first by removing one third of the $M$ atoms to produce stoichiometry $MY_3$. In this layer the $M$ atoms are at the vertices of a $6^3$ net. A notable example of its occurrence is $Al(OH)_3$. Such a layer is often referred to as "dioctahedral."

It is important to recognize that in real materials (such as oxides) the octahedra in layers are not regular. Because of metal-metal repulsions, the shared edges of octahedra are shortened and the unshared edges are lengthened. This does not show up in a drawing of a projection of the $MY_2$ layer, but in the $MY_3$ layer it does; the octahedra become metaprisms, and the structure expands as illustrated on the right in Fig. 5.60. In $Mg(OH)_2$, the lengths of the shared and unshared edges are 2.78 and 3.15 Å respectively, and in $Al(OH)_3$ they are about 2.5 and 2.8 Å (the octahedra are rather irregular in this case).

### *5.4.2 Sheet silicates (phyllosilicates)*

Many important minerals, known as "sheet" or "layer" silicates, are based on a combination of the above tetrahedral and octahedral layers. They have interesting and useful physical properties that are intimately connected to their structures. Here we outline some of the possibilities; in what follows it should be remembered that minerals typically have rather complicated compositions, and chemical formulas are often idealized.[1]

Fig. 5.61 shows how two layers can be combined. Note that if the tetrahedral layer is contracted (or partly so) two distinct configurations (known as *O* and *H*) are possible.

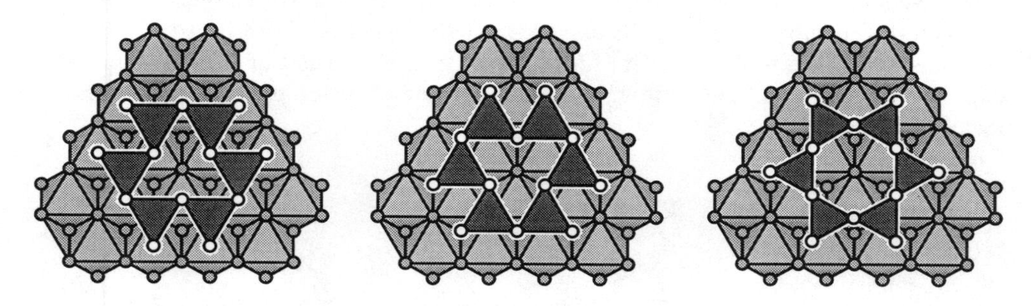

**Fig. 5.61**. Illustrating the positions of a tetrahedral layer (dark shading) with respect to an octahedral layer. Left: a contracted tetrahedral layer in the *O* orientation. Center: the same in *H* orientation. Right: an expanded tetrahedral layer.

Let's calculate the stoichiometry of a double layer formed from $MY_2$ and $T_2X_5$ layers. It should be clear from Fig. 5.61 that, per unit area, there are three $M$ atoms for every two $T$ atoms so we must combine 3 $MY_2$ (hence the term "trioctahedral") with 1 $T_2X_5$. The apices of the tetrahedra replace 2/3 of the $Y$ atoms on one side of the $MY_2$ layer, i.e. 1/3 of the total $Y$ atoms, so the double-layer stoichiometry is $M_3Y_4T_2X_5$. In layer silicates, $Y$ is usually OH and $X$ is O, so a typical composition for a neutral layer is $Mg_3(OH)_4Si_2O_5$. This is approximately the composition of antigorite and chrysotile asbestos. We return below to a further discussion of the structure of these materials.

In a double layer formed from $MY_3$ and $T_2X_5$ layers we must combine 2 $MY_3$ (hence the

[1]We can only hint at the richness of this subject. A good account, with some of the history of the subject, is in Chapter 13 of Bragg & Claringbull (book list). Good recent references are: *Micas* (S. W. Bailey, ed.), *Reviews in Mineralogy* **13** (1984) and *Hydrous Phyllosilicates* (S. W. Bailey, ed.), *Reviews in Mineralogy* **19** (1988). The term *phyllosilicate* is generally considered to include all aluminosilicates with $(Si,Al)_2O_5$ layers of corner-connected $\{Si\}O_4$ and $\{Al\}O_4$ tetrahedra.

term "dioctahedral") with 1 $T_2X_5$ and the stoichiometry is $M_2Y_4T_2X_5$; a typical composition for a neutral layer is $Al_2(OH)_4Si_2O_5$. This is approximately the composition of kaolinite ("china clay").[1] Minerals with related structures are referred to as "clay minerals"; they are important components of soils.

In silicates, the Si-O-Si angles are about 145° so the tetrahedral layer is expanded (or very nearly so). The edge of a $\{Si\}O_4$ tetrahedron is about 2.62 Å, so for such a layer to be commensurate with an octahedral layer an octahedron edge (in the layer of the plane) of about $(2/\sqrt{3}) \times 2.62 = 3.03$ Å is needed. It transpires that the $\{Al\}O_6$ layers can adapt to this length, but the larger $\{Mg\}O_6$ cannot.

Accordingly in kaolinite there are gets flat sheets, but in chrysotile the sheets curve and form small cylinders of about $r = 100$ Å radius, with the octahedral layer on the outside and the tetrahedral layer inside. The separation between the octahedral and tetrahedral layers is about $\delta r = 2.7$ Å so the spacing in the octahedral layer is increased by $\delta r/r = 2.7\%$ compared to that in the tetrahedral layer. Thus the material really has translational symmetry in only one dimension (the cylinder axis).[2]

But nature shows great ingenuity in solving the mismatch problem in a number of different ways. In antigorite the tetrahedra in alternate laths of the tetrahedral layer are "up" and "down," and the double layer is corrugated with a period of about 45 Å as shown schematically in Fig. 5.62; the radius of curvature of the sections is about 75 Å—approximately the same as in chrysotile. Still more complicated patterns are found in (for example) carlosturanite and manganpyrosmalite.[3]

**Fig. 5.62**. A schematic illustration of the layers (seen end-on) in antigorite. Octahedral layers are lightly shaded.

Triple layers consisting of an octahedral layer joined on both sides to tetrahedral layers

[1]Kaolinite was mined continuously from one location in Jiangxi province, China for about one thousand years and became known in England as "China clay"; ceramics made by firing it became known as *chinaware* or just *china*. *Kaolin* comes from the Chinese word *gaoling* ("high hill") describing the location. Currently mining of kaolinite is a $ billion industry; a major use is as a filler for paper.

[2]Actually, as well as cylinders one often finds a few turns of a cylindrical spiral (like a rolled-up carpet). The idea of cylindrical "crystals" seems to have originated with L. Pauling in 1930; subsequently electron microscopy provided dramatic direct evidence for such structures. See especially the now classic work of K. Yada, *Acta Crystallogr.* **23**, 704 (1967) and A**27**, 659 (1971).

[3]For a good review see S. Guggenheim & R. A. Eggleton in *Hydrous Phyllosilicates* (S. W. Bailey, ed.) Reviews in Mineralogy **19** (1988), p. 675-725.

cab also be formed.[1] Combining an $MY_2$ (trioctahedral) layer with two tetrahedral layers produces composition $M_3Y_2T_4X_{10}$. A neutral composition is talc (also known as soapstone), $Mg_3(OH)_2Si_4O_{10}$. Note that layers that include two tetrahedral layers cannot bend (in contrast to chrysotile) and talc normally occurs as poorly-ordered microcrystals (talcum powder).

Minerals of the vermiculite group have alternating triple layers of the talc type and $MY_2$ layers. As, per unit area, 3 $MX_2$ are combined with 1 $M_3Y_2T_4X_{10}$, the ideal composition is now $M_6Y_8T_4X_{10}$.

Combining an $MY_3$ (dioctahedral) layer with two tetrahedral layers produces a triple layer with composition $M_2Y_2T_4X_{10}$. A neutral composition is that of pyrophillite, $Al_2(OH)_2Si_4O_{10}$.

In the micas, part of the Si in the tetrahedral layer is replaced by Al and compensating cations are intercalated between the layers. Typical ideal compositions (with tetrahedral atoms in brackets) are:

(a) using trioctahedral ($MY_2$) layers: muscovite, $KAl_2[AlSi_3]O_{10}(OH)_2$

(b) using dioctahedral ($MY_3$) layers: phlogopite, $KMg_3[AlSi_3]O_{10}(OH)_2$

In constructing a three-dimensional crystal from the layers described here, attention must be paid to the way the layers stack.[2]

Clays[3] also can have some cations between the layers. Montmorillonite is made up of triple layers consisting of two tetrahedral layers and a dioctahedral layer in which some of the Al is replaced with Mg. Typical compositions are $Na_xMg_xAl_{2-x}Si_4O_{10}(OH)_2$ (bentonite) and $Ca_{x/2}Mg_xAl_{2-x}Si_4O_{10}(OH)_2$ (Fuller's earth). The intercalated cations are readily exchanged for other cations at low temperatures in aqueous suspension (ion exchange). An interesting recent development has been the intercalation of large isopolycations such as $[Al_{13}O_4(OH)_{24}(H_2O)_{12}]^{7+}$ (the "spinel cluster," see § 5.2.3, p. 157) which results in greatly increased interlayer spacing. Subsequent heat treatment removes much of the hydrogen (as water) leaving a "pillared" clay of high microporosity that has great potential for application as a catalyst.

Slurries of bentonite are thixotropic[4] and find many applications (such as use as a drilling mud) exploiting this property.

[1]Note that as the vertices on each side of an octahedral layer are not one above the other, there is an offset between the two tetrahedral layers, and the symmetry of the slab is at most rectangular.

[2]A good review and bibliography on this topic is given by J. B. Thompson in *Structure and Bonding in Crystals* Vol II [M. O'Keeffe & A. Navrotsky, eds., Academic Press, New York (1981)].

[3]The term *clay* is generally used loosely to mean any layer silicate that is (a) generally composed of very small crystallites and (b) can absorb water.

[4]A thixotropic material is a gel which becomes fluid when agitated. Platy microcrystals of bentonite have positive charges on their thin edges and negative charges on their faces and at rest orient themselves edge to face in the water producing a stiff gel. Agitation destroys this order resulting in a fluid of greatly reduced viscosity. "Non-drip" paints are thixotropic and usually contain bentonite.

## 5.5 Aperiodic tilings and quasicrystals

In recent years there has been considerable interest on the part of mathematicians and crystallographers in aperiodic tilings. These are tilings by basic tiles that cover the plane but in which there is no translational symmetry. The literature on this topic is now enormous and we can only attempt to provide some of the flavor of the subject.[1]

The interest is in those sets of tiles for which every possible tiling is aperiodic. Many have been discovered in recent years. One of the simplest and most studied is the pair of *Penrose* tiles shown in Fig. 5.63. These have edge lengths of $\tau = (1 + \sqrt{5})/2 = 1.618...$ and 1 and the angle at the top is $2\pi/5 = 72°$. If that were all, these tiles could cover the plane periodically as the two can fit together to form a rhombus. However the aperiodic property is forced by coloring the vertices of the tiles black and white as shown, and by requiring that only vertices of one color meet at a point.[2] A fragment of a tiling made from such tiles is also shown in the figure.

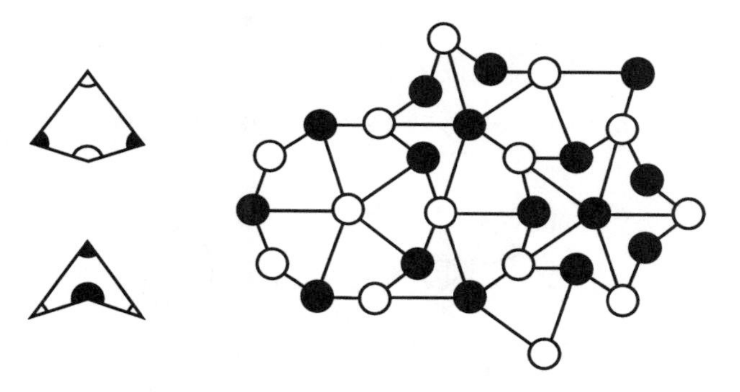

**Fig. 5.63**. On the left are shown the two Penrose tiles known as the kite (top) and dart (bottom) and on the right a fragment of an aperiodic tiling.

A closed fragment of a tiling (such as that shown in the figure) is known as a *patch*. A remarkable feature of a Penrose tiling is that it contains infinitely many patches congruent to any given patch. Even more remarkably, a local area with diameter $d$ in any Penrose tiling has an identical area not more that $(1/2 + \tau)d$ away. In every such tiling the ratio of kites to darts (see figure legend) is $\tau$.

Instead of darts and kites wo different rhombuses (with acute angles of 36° and 72° respectively) can be used with aperiodicity forced by constraints on matching vertices and edges. A pair of such tiles and a tiling by them are shown in Fig. 5.64. Note that equivalent unmarked tiles with curved edges have the same property. The two rhombuses are often referred to as *skinny* and *fat* respectively.

In three dimensions there is an analogous pair of rhombohedral tiles which can be forced

[1]A good introduction with references is in the book by Grünbaum & Shephard (Book List). A popular account is in M. Gardner *Penrose Tiles to Trapdoor Cyphers* [Freeman, New York (1989)].

[2]Aperiodicity can also be forced on monochrome tiles by curving the edges of the tiles shown.

to produce aperiodic tilings. These polyhedra[1] have faces that are fat rhombuses (angles of 72° and 108°). One rhombohedron is acute (three angles of 72° meet at a point) and the other is obtuse (three angles of 108° meet at a point). The faces have to be marked and appropriate matching rules applied to force aperiodicity.

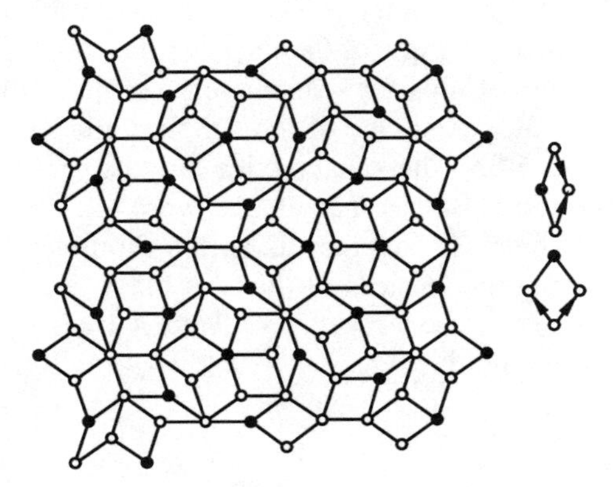

**Fig. 5.64**. A tiling (left) by rhombuses (right) with vertex colors and arrows on the edges matching.

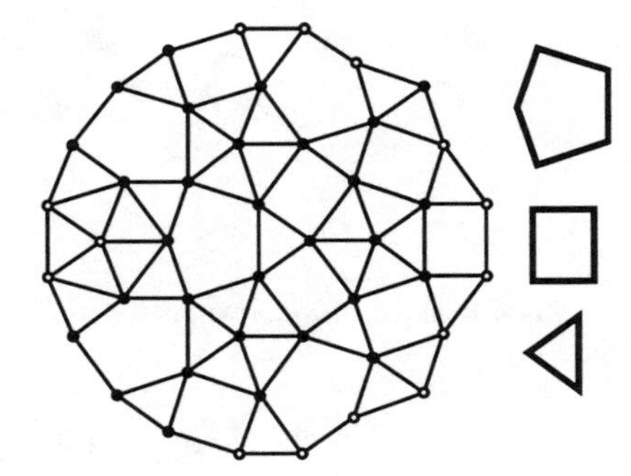

**Fig. 5.65**. The black circles are the black circles in Fig. 5.64, and open circles represent a continuation of the pattern using the tiles shown on the right and maintaining 5-coordination for each point.

The arrangement of black dots in Fig. 5.64 is suggestive. Fig. 5.65 shows that they form triangles, quadrangles and pentagons arranged to produce a five-connected net. The edge lengths of the polygons are 1 and $2\sin(\pi/10) = 1.176$. A conspicuous feature of the diagram is the groups of five triangles condensed to form regular pentagons. In three dimensions 20 tetrahedra (with edges in the ratio 1:1.05—see Exercise 3) can condense to form a regular icosahedron, and it is possible that (at least some) quasicrystal structures contain regular icosahedra and other polyhedra packed to fill space. Constraints analogous

[1]Sometimes called *golden* rhombohedra or *Ammann* rhombohedra. They appear to have been first described by Ammann in 1976.

to those on matching edges and vertices of tiles in two dimensions may well arise in crystals. Such a constraint occurs for chemical compounds (black and white circles being different kinds of atom); another constraint is coordination number (cf. Fig. 5.65).

The Fourier transform of a periodic function is discrete; by this is meant that the Fourier integral is reduced to a sum. In particular the Fourier transform of a lattice is the reciprocal lattice. There are certain aperiodic functions (known as quasiperiodic) that also have discrete Fourier transforms.[1] Penrose tilings and the three-dimensional analogs (quasicrystals in the real world) are quasiperiodic and have discrete Fourier transforms and hence sharp diffraction patterns. One of the reasons for the excitement among crystallographers is that these diffraction patterns display symmetries not allowed for strictly periodic real-space patterns. The diffraction pattern of a two-dimensional crystal based on Penrose tiling has 5-fold symmetry—note that *local* regions of 5-fold symmetry are apparent in Figs. 5.63-65. Much larger patches with 5-fold symmetry can and do appear also. Quasicrystals have been found with diffraction patterns that have $m\bar{3}\bar{5}$, $12/mmm$ and $10/mmm$ point symmetries (and also other non-crystallographic symmetries —see § 3.7.8).[2]

Readers who experiment with these tilings (as they are urged to) will find that as a tiling is being constructed, situations frequently arise where there is more than one way to proceed and that not all of these ways lead to acceptable tilings. One has therefore either to back-track frequently to eliminate "defects" in the tiling or to think ahead many moves (like a chess player). For this reason it has been argued that real quasicrystals cannot grow from such tiles, as atoms can neither think ahead nor readily undo their mistakes, although the difficulty can largely be avoided by recognizing additional rules for acceptable placing of tiles.[3] The challenges to crystallographers and crystal chemists are to (a) identify the (three-dimensional) tiles that make up quasicrystals, (b) identify the rules for their assembly and (c) determine how atoms are situated on (or *decorate*) the tiles.[4]

A final note is in order. Mathematicians are mainly interested in tilings which are *forced* to be aperiodic. Solids often form aperiodic structures (such as glasses) not because they *have* to, but for kinetic reasons—silica is a notable example. The tiles in Fig. 5.65 do not *force* aperiodicity, but they can be put together in an aperiodic way that has a high concentration of patches with 5-fold symmetry.

[1]Readers unfamiliar with Fourier transforms will have to skip this part. A simple example of a one-dimensional quasi-periodic function is $f(x) = \cos x + \cos\sqrt{2}x$.

[2]For a good introduction to quasicrystals see W. Steurer, *Zeits. Kristallogr.* **190**, 179 (1990) or *Lectures on Quasicrystals* (F. Hippert & D. Gratias, eds.) Les Edition de Physique (1994). Typical compositions are $Al_{73}Mn_{21}Si_6$, $Al_{70.3}Pd_{21.4}Mn_{8.3}$ and $Li_{30}Al_{60}Cu_{10}$—notice that Al is the main component. Many materials studied to date are rapidly cooled from the melt, but there is now some evidence that the quasicrystal state is the most stable for compositions such as $Al_{62}Fe_{12.5}Cu_{23.5}$.

[3]These rules place additional constraints on the vertex figures that are allowed, and are thus "local" see G. Y. Onoda *et al.*, *Phys. Rev. Letts.* **62**, 1210 (1990).

[4]$MnAl_3$ has a crystal structure believed to be closely related to that of a decagonal quasicrystal and relevant tiles for the quasicrystal have been described by K. Hiraga *et al.*, *Phil. Mag.* **B67**, 193 (1993). The structure of $MnAl_4$ is also believed to be relevant to quasicrystal structures; it is rather complex (574 atoms in the unit cell!)—C. B. Shoemaker *et al.*, *Acta Crystallogr.* **B45**, 13 (1989).

## 5.6 Notes

### *5.6.1 Relationships between polyhedra, and "pseudorotations"*

The transformation from a trigonal bipyramid to a square pyramid involves just small displacements of the vertices. Figure 5.66 shows how a trigonal bipyramid can be transformed to a square pyramid, and thence to a trigonal bipyramid in a different orientation and with interchange of apical and equatorial atoms. Such a transformation path is known as a *Berry pseudorotation*. A bipyramidal molecule such as $PF_5$ appears to have five equivalent F atoms in $^{19}$F NMR due to rapid changes of this sort.

Another example of a pseudorotation was given in § 5.1.8 where we showed how a square antiprism could be related to a bisdisphenoid. The different ways of converting the squares of the square antiprism into pairs of triangles will result in different orientations of the $\bar{4}$ axes of the bisdisphenoid which likewise can be converted back into antiprisms with different orientations. Analogously, small displacements can result in pseudorotations of extended structures such as layers (§ 5.6.13) or sphere packings (see e.g. § 6.3.1).

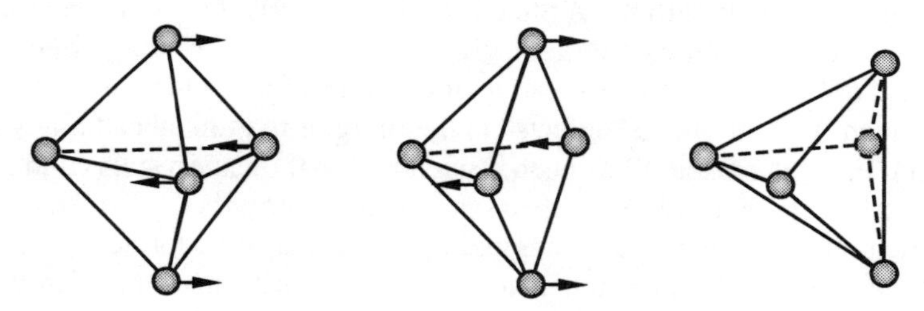

**Fig 5.66.** Transformation of a trigonal bipyramid into a square pyramid and then back to a bipyramid with the three-fold axis rotated by 90°.

### *5.6.2 Polyhedra, points on a sphere, and related topics*

The solution to Tammes' problem is appropriate for atoms in a coordination figure when there is a short range repulsion between neighboring atoms (fairly "hard" spheres). At the other extreme is the case of the long-range repulsion (Coulomb interaction) when the energy of interaction of the atoms is proportional to 1/distance. It is convenient in this case to measure the repulsion energy, $E$, in units of $q^2/4\pi\varepsilon_0 r$ where $r$ is the bond distance and $q$ the charge.[1]

For 5-coordination the repulsive energy is (note that the second value is for the most favorable type of square pyramid which has equal edges):

[1]This problem has been called the electron problem as it was the basis for early (*ca* 100 years ago) models of electronic configurations in atoms. It was first treated by L. Föppl. For recent results see J. R. Edmundson, *Acta Crystallogr.* A**48**, 60 (1992). If $q = ze$, where $e$ is the electronic charge, and $r$ is in Å, $q^2/4\pi\varepsilon_0 r = 14.4z^2/r$ eV.

| | |
|---|---|
| bipyramid: | $E = 6.475$ |
| pyramid: | $E = 6.657$ |

Turning to 6-coordination we might compare the octahedron with the trigonal prism. On a sphere of unit radius the shortest distance between points arranged as the vertices of a regular octahedron is $\sqrt{2} = 1.414$; for a regular trigonal prism it is $\sqrt{(12/7)} = 1.309$. In the case of the trigonal prism, in calculating the electrostatic repulsion, we now allow a degree of freedom which is the ratio of the height ($h$) to the length ($b$) of the triangular face; the minimum energy is for $h/b = 0.917$. In the same units as before:

| | |
|---|---|
| octahedron | $E = 9.985$ |
| prism $h/b = 1$ | $E = 10.114$ |
| prism $h/b = 0.917$ | $E = 10.096$ |

In the case of the tricapped trigonal prism (9-coordination), the solution to Tammes' problem has the distances from the "capping" atoms to their four nearest neighbors equal to the length of the base. The ratio of the height to base is now $h/b = \sqrt{(5/3)} = 1.291$. Alternatively if we again minimize the electrostatic energy, $h/b = 1.143$ for the minimum electrostatic energy, but the energy is only 1.5% lower in this second configuration. In either case we have $h/b > 1$. In general when capped trigonal prismatic coordination occurs in crystals it is observed (see Hyde & Andersson, p. 213) that as the number of capping atoms increases, the prism becomes more elongated ($h/b$ increases) as the above considerations suggest it should.

Many fascinating results, conjectures, and unsolved problems relating to polyhedra are to be found in *Unsolved Problems in Geometry* by H. T. Croft, K. J. Falconer and R. K. Guy [Springer-Verlag, Berlin (1991)].

### *5.6.3 Constructing polyhedra*

The easiest, and possibly the best, way of constructing polyhedra is to assemble them from individual polygons cut *accurately* from stiff cardboard. They can be assembled into polyhedra using masking tape, and the joints filled with white glue (such as Elmer's glue in the U.S.). Convex polyhedra are always rigid, so that when the glue has set, the masking tape can be removed and a very sturdy model will be obtained. Pictures of many remarkable models made in a similar way are to be found in A. Holden, *Space, Shapes and Symmetry* [Columbia University Press, New York (1971) also Dover reprint]. Many of the models depicted in that book are pertinent to crystal chemistry.

### *5.6.4 Schlegel diagrams and adjacency matrices*

The topology of a polyhedron (the connection of vertices by edges) is conveniently expressed in a *Schlegel diagram* which is a planar graph in which there is a one to one correspondence between edges and vertices of the polyhedron and of the graph. The graph is *planar* because it can be realized on a plane without any crossing of edges. A way to imagine such a graph is that it represents the view of the inside of the polyhedron one would have if one of its faces were transparent and the polyhedron were viewed from a point just outside that face. The Schlegel diagram of a cuboctahedron is shown in Fig. 5.67 in what appears to be two quite different graphs. (The first is a view through a square face, and the second is a view through a triangular face). They can be shown to be topologically identical by writing out the *adjacency matrix* which contains 1 as element $ij$ if vertex $i$ is connected to $j$ and 0 otherwise. The matrix below uses the vertex numberings of the figure.

| $i\backslash j$ | 1 | 2 | 3 | 4 | 5 | 6 | 7 | 8 | 9 | 10 | 11 | 12 |
|---|---|---|---|---|---|---|---|---|---|---|---|---|
| 1 | 0 | 1 | 0 | 1 | 1 | 0 | 0 | 0 | 0 | 1 | 0 | 0 |
| 2 | 1 | 0 | 1 | 0 | 0 | 1 | 0 | 0 | 0 | 1 | 0 | 0 |
| 3 | 0 | 1 | 0 | 1 | 0 | 1 | 0 | 0 | 1 | 0 | 0 | 0 |
| 4 | 1 | 0 | 1 | 0 | 1 | 0 | 0 | 0 | 1 | 0 | 0 | 0 |
| 5 | 1 | 0 | 0 | 1 | 0 | 0 | 0 | 0 | 0 | 0 | 1 | 1 |
| 6 | 0 | 1 | 1 | 0 | 0 | 0 | 1 | 1 | 0 | 0 | 0 | 0 |
| 7 | 0 | 0 | 0 | 0 | 0 | 1 | 0 | 1 | 0 | 1 | 0 | 1 |
| 8 | 0 | 0 | 0 | 0 | 0 | 1 | 1 | 0 | 1 | 0 | 1 | 0 |
| 9 | 0 | 0 | 1 | 1 | 0 | 0 | 0 | 1 | 0 | 0 | 1 | 0 |
| 10 | 1 | 1 | 0 | 0 | 0 | 0 | 1 | 0 | 0 | 0 | 0 | 1 |
| 11 | 0 | 0 | 0 | 0 | 1 | 0 | 0 | 1 | 1 | 0 | 0 | 1 |
| 12 | 0 | 0 | 0 | 0 | 1 | 0 | 1 | 0 | 0 | 1 | 1 | 0 |

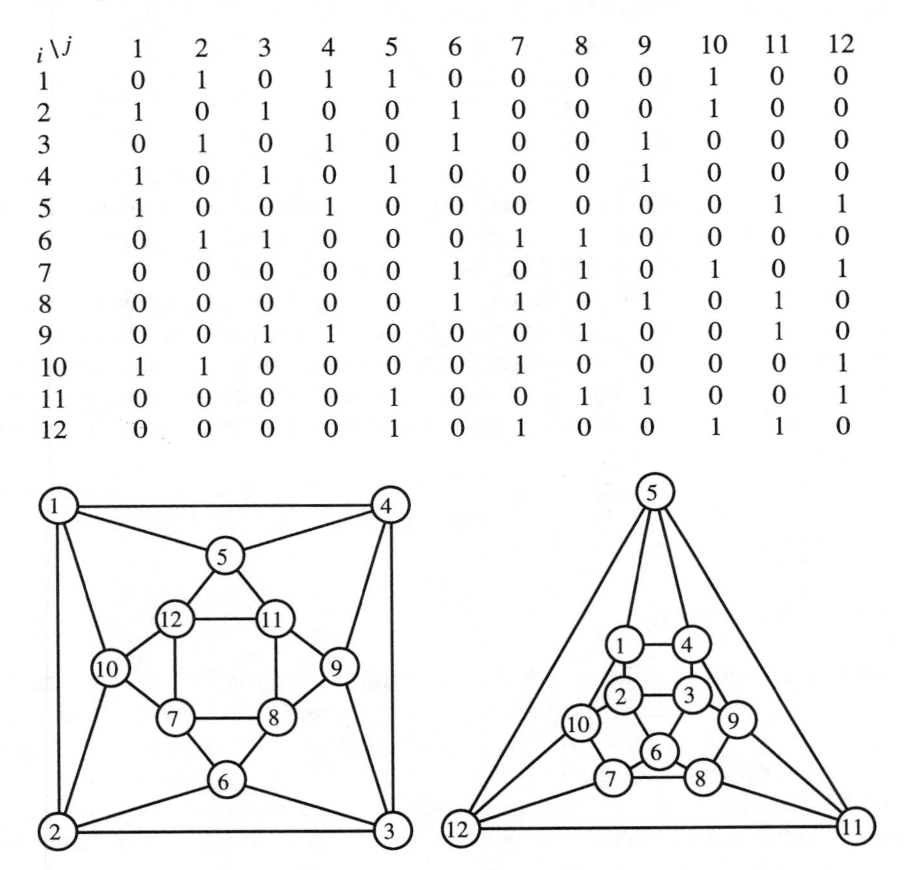


**Fig. 5.67**. Two representations of a cuboctahedron 3.4.3.4 by Schlegel diagrams. The numbering is the same as used for the adjacency matrix above.

A warning on terminology: Mathematicians refer to the graph of a polyhedron as 3-connected because *at least* three edges have to be cut to separate the graph into disjoint pieces. However, it is common usage in chemistry to call a net in which $n$ edges meet at every vertex as $n$-connected. Mathematicians refer to such vertices as $n$-valent. Schlegel diagrams are sometimes used to describe the topology of coordination figures in solids.

The problem of enumerating polyhedra with a given number of vertices is the same as that of enumerating distinct 3-connected (in the mathematical sense) planar graphs.

Adjacency matrices are utilized extensively in molecular chemistry. Thus one way to recognize molecules with identical topologies in a computer is to compare their adjacency matrices. It might be mentioned that for large molecules, this is a far from trivial task as to get identical adjacency matrices the vertices must be numbered the same in each molecule.

### *5.6.5 Coordinates for drawing polyhedra and nets*

Some important polyhedra are conveniently drawn on the framework of a cube. Let the origin be at 0,0,0 and the cube vertices ±1,±1,±1. The vertices of some other polyhedra are given below [remember κ stands for cyclic permutation and $\tau = (1 + \sqrt{5})/2$]. To get coordinates for a clinographic projection, use the method outlined in § 4.6.1.

| | |
|---|---|
| tetrahedron | 1,1,1 ; (–1,–1,1)κ or –1,–1,–1 ; (1,1,–1)κ |
| octahedron | (±1,0,0)κ |
| cuboctahedron | (0,±1,±1)κ |
| truncated octahedron | (0,±1/2,±1 ; ±1/2,0,±1)κ |
| rhombic dodecahedron | ±1,±1,±1 ; (±2,0,0)κ |
| icosahedron | (0,±τ,±1)κ |

For other cubic polyhedra the matrices for the symmetry operations of the appropriate symmetry group (see Exercises 7, 9 & 10 of Chapter 2) can be used to generate coordinates of the $N$ vertices from those of the one given below (appropriate for unit edge). To generate coordinates for polyhedra with icosahedral symmetry, it is best to first generate the matrices for the symmetry operations of $I$ and $I_h$ as described in Exercises 7 and 16 of Chapter 2. These can be then applied to the coordinates of a typical vertex (given below). In the table, $N$ is the number of vertices and the coordinates are appropriate for cartesian axes oriented as described in Exercise 16 of Chapter 2 and for unit edge length.

| polyhedron | $N$ | symmetry | coordinates |
|---|---|---|---|
| $3.6^2$ | 12 | $\bar{4}3m$ | $1/\sqrt{8},\ 1/\sqrt{8},\ 3/\sqrt{8}$ |
| 3.4.3.4 | 12 | $m\bar{3}m$ | $0,\ 1/\sqrt{2},\ 1/\sqrt{2}$ |
| $4.6^2$ | 24 | $m\bar{3}m$ | $0,\ 1/\sqrt{2},\ \sqrt{2}$ |
| $3.8^2$ | 24 | $m\bar{3}m$ | $1/2,\ y,\ y;\ y = (1+\sqrt{2})/2$ |
| $3^4.4$ | 24 | 432 | 0.3378, 0.6212, 1.1426 |
| $3.4^3$ | 24 | $m\bar{3}m$ | $1/2,\ 1/2,\ (1+\sqrt{2})/2$ |
| 4.6.8 | 48 | $m\bar{3}m$ | $1/2,\ 1/2+1/\sqrt{2},\ 1/2+\sqrt{2}$ |
| $3^5$ | 12 | $I_h$ | $0,\ 1/2,\ \tau/2$ |
| $5^3$ | 20 | $I_h$ | $\tau/2,\ \tau/2,\ \tau/2$ |
| 3.5.3.5 | 30 | $I_h$ | $0,\ 0,\ \tau$ |
| $3.10^2$ | 60 | $I_h$ | $0,\ (3\tau+1)/2,\ 1/2$ |
| $5.6^2$ | 60 | $I_h$ | $0,\ 1/2,\ 3\tau/2$ |
| 3.4.5.4 | 60 | $I_h$ | $0,\ 1+\tau/2,\ (1+\tau)/2$ |
| $3^4.5$ | 60 | $I$ | 0.3309, 0.3748, 2.0970 |
| 4.6.10 | 120 | $I_h$ | $1/2,\ 1/2,\ 2\tau+1/2$ |

Coordinates of the vertices in Archimedean tilings with unit edge are also given below. For some other two-dimensional patterns see *OKH*. Recourse to the *International Tables* will be necessary to generate equivalent positions from the ones given.

| tiling | symmetry | cell | x,y |
|---|---|---|---|
| $3^4.6$ | *p6* | $a = \sqrt{7}$ | 3/7, 1/7 (or 1/7, 3/7) |
| $3^3.4^2$ | *c2mm* | $a = 1, b = 2+\sqrt{3}$ | $0, (1+\sqrt{3})/(4+\sqrt{12})$ |
| $3^2.4.3.4$ | *p4gm* | $a = \sqrt{(2+\sqrt{3})}$ | $x, 1/2+x\ ; x = 1/\sqrt{(16+\sqrt{192})}$ |
| 3.6.3.6 | *p6mm* | $a = 2$ | 0, 1/2 |
| 3.4.6.4 | *p6mm* | $a = 1+\sqrt{3}$ | $x, \bar{x}\ ; x = 1/(3+\sqrt{3})$ |
| $4.8^2$ | *p4mm* | $a = 1+\sqrt{2}$ | $x, 1/2\ ; x = 1/(2+\sqrt{8})$ |
| $3.12^2$ | *p6mm* | $a = 2+\sqrt{3}$ | $x, \bar{x}\ ; x = 1-1/\sqrt{3}$ |
| 4.6.12 | *p6mm* | $a = 3+\sqrt{3}$ | $x, y\ ; x = 1/(3+\sqrt{27}), y = 1/3+x$ |

### *5.6.6 Names of polygons and polyhedra*

The names of polygons specify the number of angles they have, -gon comes from the Greek word for angle (a goniometer is an angle-measuring device). The names of polyhedra often specify the number of faces they have; -hedron likewise coming from the Greek for face. The prefixes come from Greek words for numbers. Some of the more commonly used are given below.

| | | | | | |
|---|---|---|---|---|---|
| 1 | mono- | 8 | octa- | 16 | hexakaideca- |
| 2 | di- | 9 | ennea- | 20 | icosa- |
| 3 | tri- | 10 | deca- | 30 | triaconta- |
| 4 | tetra- | 11 | hendeca- | 32 | icosidodeca- |
| 5 | penta- | 12 | dodeca- | 60 | hexaconta- |
| 6 | hexa- | 13 | triskaideca- | many | poly- |
| 7 | hepta- | 14 | tetrakaideca- | | |

A polygon is called *skew* if all the vertices are not all in one plane.

The words “tetrahedra,” “octahedra,” etc. are plural. The singulars are “tetrahedron,” “octahedron,” etc. The usage “tetrahedrons,” “octahedrons,” etc. for the plural is also considered acceptable. The term “polyhedron” applies *only* to three-dimensional figures. For higher dimensional analogs use “polytope” (as in Appendix 2).

### *5.6.7 The shapes of crystals*

It is worth recalling that the sciences of crystallography and solid state chemistry have their roots in mineralogy. Some symmetrical minerals crystallize in beautiful polyhedra. If a cubic crystal has a {100} habit then with planes equally spaced from a center the crystal will be a cube (look at table salt under a magnifying glass). Similarly {111} results in an octahedron in a centrosymmetric crystal and {110} results in a rhombic dodecahedron. Cuprite ($Cu_2O$) occurs with all these habits and garnets sometimes form spectacular dodecahedra. Other common shapes include cuboctahedra and truncated octahedra in which the faces are {100} (squares) and {111} (triangles). Hexagonal crystals such as beryl often occur as hexagonal prisms; the $\{11\overline{2}0\}$ faces are termed the prism faces and the top and bottom (0001) planes are referred to as pinacoid.

In class $m\overline{3}$ a form $\{hk0\}$ consists of the 12 *cyclic* permutations of ($\pm h \pm k$ 0). ($hk0$) and ($kh0$), for example, are related by a 4-fold rotation (about the $z$ axis) which is not an operation in $m\overline{3}$ [so ($hk0$) and ($kh0$) are not part of the same form]. Pyrites ($FeS_2$), which has symmetry $Pa\overline{3}$, commonly crystallizes with a {210} habit producing beautiful dodecahedra with pentagonal faces known as *pyritohedra*. Fig. 5.68 illustrates such a polyhedron which appears to be a regular pentagonal dodecahedron, but the eye is being deceived. There are two kinds of vertex: eight ($X$) at $\pm x,\pm x,\pm x$ (at the corners of a cube) shown as filled circles in the figure and twelve ($Y$) at $(\pm 3x/4,\pm 3x/2,0)\kappa$ shown as open circles. The edge lengths are $Y\text{-}Y = 3x/2$ and $X\text{-}Y = \sqrt{21}x/4$ and the face angles are $XYY = 102.6^\circ$, $XYX = 106.6^\circ$ and $YXY = 121.6^\circ$ (contrast all are 108° for a regular pentagon).

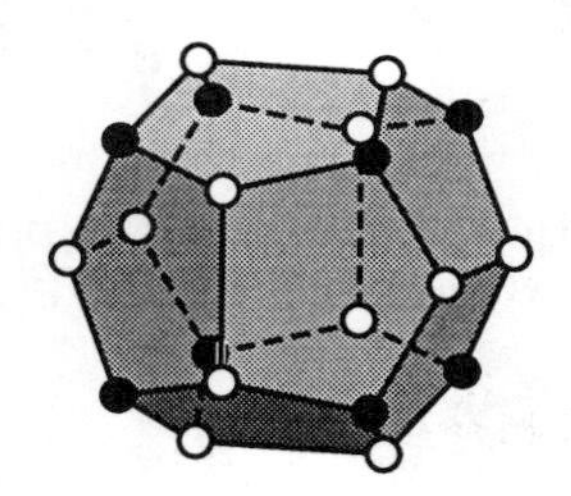

**Fig. 5.68.** The pyritohedron described in the text.

Some terms commonly used by mineralogists to describe the external form of crystals are:

| | |
|---|---|
| *euhedral* | refers to crystal completely bounded by well-formed faces |
| *anhedral* | is the opposite of euhedral |
| *acicular* | needle shaped (long thin crystals) |
| *tabular* | having two prominent parallel faces (like a tablet) |
| *micaceous* | an extreme case of tabular (occurring as thin sheets as in mica) |
| *hemimorphic* | refers to crystals with different forms at each end |
| *lamellar* | occurring as a sheaf of thin sheets like pages in a book |

### *5.6.8* $Na_3Pt_4Ge_4$: *a structure with stellae quadrangulae*

Stellae quadrangulae (§ 5.1.2, p. 135) are important in intermetallic structures. A nice example of their occurrence is in the structure of $Na_3Pt_4Ge_4$ ($Eu_3Ni_4Ga_4$ is isostructural). In this structure, a $Pt_4$ tetrahedron has each face capped to make a $Pt_4Ge_4$ stella quadrangula. Each Ge atom is then linked to a fourth Pt on a neighboring stella quadrangula to make an open body-centered cubic array (Fig. 5.69). The open space in the structure is filled by a 4-connected net of Na atoms (in § 7.3.11 we refer to this as the **NbO** net). We note in passing that $Na_3SbS_4$ has a closely-related structure (sometimes called $\mathbf{Tl_3VS_4}$) with $\{Sb\}S_4$ tetrahedra replacing the stellae quadrangulae (see Fig. 5.69).

Na is also in a distorted {Na}$S_4$ tetrahedron. Crystallographic data are in Appendix 5.

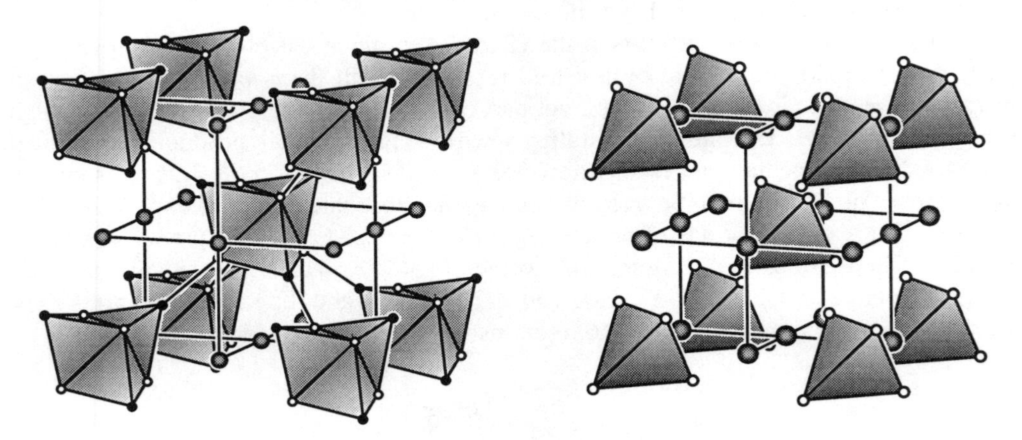

**Fig. 5.69**. Left: the structure of $Na_3Pt_4Ge_4$. Small open circles are Pt, small filled circles are Ge, and larger circles are Na. Right: the structure of $Na_3SbS_4$ showing {Sb}$S_4$ tetrahedra. Large circles are Na.

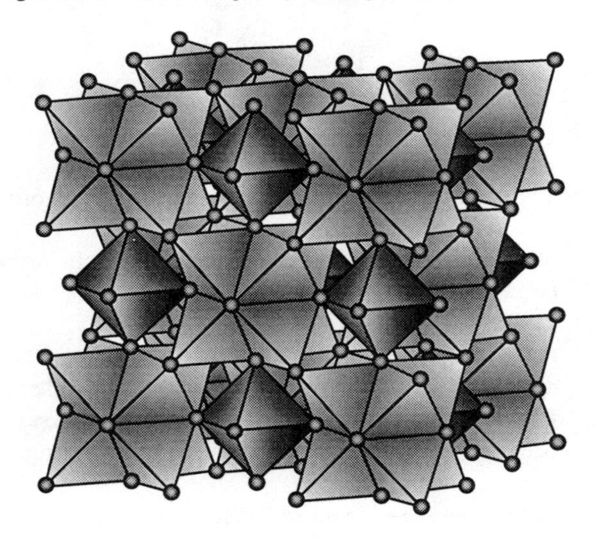

**Fig. 5.70**. The structure of $Co_9S_8$ shown as a packing of stellae octangulae and octahedra. Co atoms (not shown) are in {Co}$S_4$ tetrahedra and {Co}$S_6$ octahedra.

### *5.6.9* $Co_9S_8$: *a structure with stellae octangulae*

The structure of $Co_9S_8$ [the mineral pentlandite = $(Fe,Ni)_9S_8$ is isostructural] is a nice example of a structure in which stellae octangulae are a conspicuous feature. A stella octangula (Fig. 5.3, p. 135) has 6 inner vertices and eight outer ones. In the $Co_9S_8$ structure (Fig. 5.70) Co centers the tetrahedra of a $S_6S_8$ stella quadrangula to produce a

$Co_8S_6S_8$ unit. These units are joined together as shown in the figure; each outer vertex is shared between four units, so the stoichiometry is now $Co_8S_6S_{8/4} = Co_8S_8$. There are empty $S_6$ octahedra inside the stella quadrangulae; in the packing, new $S_6$ octahedra are generated (one per $Co_8S_8$ unit), and these are also filled with Co to make the overall stoichiometry $Co_9S_8$. The S atom arrangement is cubic closest packing (**ccp**, § 6.1.3).

Crystallographic data for $Co_9S_8$ are given in Appendix 5.

### *5.6.10 Enumeration of Archimedean polyhedra and tilings*

We remarked (§ 5.3.1) that the regular tilings were limiting cases of regular polyhedra (the idea appears to have originated with Kepler) as in the sequence: $3^3$, $3^4$, $3^5$ (polyhedra) and $3^6$ (plane tiling). In Table 5.6 below, the archimedean polyhedra and tilings are listed in families according to the number of polygons meeting at a vertex.

**Table 5.6.** Archimedean polyhedra and tilings

| polyhedra | | | | tilings | |
|---|---|---|---|---|---|
| | $3.6^2$ | $3.8^2$ | $3.10^2$ | $3.12^2$ | |
| | $4.6^2$ | | | $4.8^2$ | |
| | $5.6^2$ | | | | |
| $N.4^2$ | (prisms) | | | | |
| | | 4.6.8 | 4.6.10 | 4.6.12 | |
| $3^3.N$ | (antiprisms) | | | | |
| 3.4.3.4 | 3.5.3.5 | | | 3.6.3.6 | |
| 3.4.5.4 | | | | 3.4.6.4 | |
| $3.4^3$ | | | | | |
| $3^4.4$ | $3^4.5$ | | | $3^4.6$ | |
| | | | | $3^3.4^2$ | $3^2.4.3.4$ |

How can we be sure that the list is complete? We need first to require the sum of the angles at the vertices is $\leq 360°$ (with equality applying for plane tilings). The vertex angle of a regular plane $n$-gon is $180°–360°/n$ so we must have for the sum over the $i$ vertices:

$$\Sigma_i(1 - 2/n_i) \leq 2 \tag{5.1}$$

However this equation admits many more solutions than appear in the table. What about the polyhedron 3.4.6 which is missing? Let's calculate the number of vertices, $V$, such a polyhedron should have. The number of edges, $E$, is $3V/2$ (as three edges meet at each vertex and each edge belongs to two vertices). An $n$-gon at a vertex is shared by $n$ vertices and contributes $1/n$ of a face per vertex, so the number of faces is $F = V(1/3 + 1/4 + 1/6)$. Now solving Euler's equation $F - E + V = 2$, we find $V = 8$. Hence $F = 6$. So far, so good. But using our argument that there are $V/3$ triangular faces, $V/4$ square faces and $V/6$ hexagonal faces, we find that our polyhedron has to consist of 8/3 triangular faces, 2 square faces and $8/6 = 4/3$ hexagonal faces; clearly impossible.

Actually in this case, simple topological considerations show that 3.4.6 is impossible. To see this, draw a triangle and the third edge meeting at each of the triangle vertices. The polygons must alternately be 4- gons and 6-gons on each side of these edges; this is impossible as the number of such edges is an odd number (3).

In general, the possible polyhedra or tilings are those that satisfy Eq. 5.1 and are of the sort:

| | |
|---|---|
| $p.q.r$ | with $p,q,r$ all even |
| $p.q.q$ | with $q$ even |
| $p.p.p$ | |
| $4^4$ and $3.p.q.r$ | when $q \neq 3$, $p$ and $r \neq 3$ |
| $p.q.r.s.t$ and $3^6$ | |

It is interesting that all the topological possibilities involving one kind of vertex can be realized with regular polygons. In general this is not the case when there is more than one kind of vertex. The general problem of deciding whether a given set of polygons (regular or not) can be combined into a polyhedron is of interest in chemistry (see Appendix 4 for more on this topic), but unfortunately it is also difficult and unsolved.

*5.6.11 Euler's equation applied to plane nets*

The celebrated Euler equation relating the numbers of faces, $F$, edges, $E$ and vertices $V$ for a finite *polyhedron* is

$$V + F = E + 2 \tag{5.2}$$

For a plane *tiling*

$$V + F = E \tag{5.3}$$

As $V$, $F$ and $E$ are all infinite, Eq. 5.3 is conveniently divided through by $V$. This equation also applies to a tiling on the surface of a torus ("doughnut").[1]

Let $\phi_n$ be the fraction of the polygons in a tiling (or net) that are $n$-gons, then $<n> = \Sigma n\phi_n$ is the average size of the polygons of the pattern. Likewise let $f_i$ be the fraction of vertices at which $i$ vertices meet, then $<i> = \Sigma i f_i$ is the average connectivity of the net. Using the facts that each edge is common to two polygons and joins two vertices, it may derived, using arguments similar to those used in the previous section, from Eq. 5.3:

$$1/<n> + 1/<i> = 1/2 \tag{5.4}$$

[1]There are certain pathological kinds of tiling for which this equation does not hold: "Euler's Theorem for Tilings is of fundamental importance, but is a strong contender for one of the most frequently misquoted results in mathematics!"—B. Grünbaum & G. C. Shephard, *Tilings and Patterns* (Book List). The Swiss mathematician L. Euler (1707-83) was perhaps the most prolific in history—his bibliography requires a substantial shelf of volumes. He made significant contributions to virtually all parts of mathematics but geometry was his favorite.

An example of the application of this formula is to tilings of the plane by (not necessarily regular) triangles ($\phi_3 = 1$, $\langle n \rangle = 3$) to give $\langle i \rangle = 6$. This shows that the average number of triangles meeting at a point must be six. As $\langle n \rangle$ cannot be less than 3, $\langle i \rangle$ cannot be greater than 6.

For a pattern with three polygons meeting at a point (a 3-connected net, $\langle i \rangle = 3$) Eq. 5.4 shows that the average polygon size (ring[1] size) $\langle n \rangle = 6$. This means that there is no constraint on the number of six-rings. For the pentagon-hexagon-heptagon nets of § 5.3.5, let there be $\phi_5$ pentagons, $\phi_7$ heptagons and $1-\phi_5-\phi_7$ hexagons. Equation 5.4 becomes:

$$5\phi_5 + 6(1 - \phi_5 - \phi_7) + 7\phi_7 = 6, \quad \text{i.e. } \phi_5 = \phi_7.$$

In general for any infinite plane net in which all vertices have the same connectivity *i* (the vertices need not otherwise be the same) one has ("rings per vertex" refers to the average over all the vertices):[2]

| Connectivity, *i* | rings per vertex | average ring size |
|---|---|---|
| 6 | 2 | 3 |
| 5 | 3/2 | 10/3 |
| 4 | 1 | 4 |
| 3 | 1/2 | 6 |

### *5.6.12 Transformations between patterns: common unit cells*

It should be clear that two patterns with the same densities of points can be converted from one to the other by shuffles of the points that involve just finite displacements.

A special kind of structural relationship is one in which there is no change in shape of the unit cell. A simple example is the relationship between the kagome (3.6.3.6) and $3^6$ patterns (Fig. 5.40, p. 167). The common hexagonal cell is given in Exercise 11.

First a common cell that may be a supercell of one or both structures must be found. This supercell will, of course contain the same number of points for each of the patterns. The transformation can then be effected just by moving the points in the cell.

Consider square (two-dimensional) cells first. A primitive cell for the square lattice has sides *a*. A non-primitive cell can be chosen with translation vectors $u\mathbf{a} + v\mathbf{b}$ and $-v\mathbf{a} + u\mathbf{b}$ with area $u^2 + v^2$ and containing $u^2 + v^2$ lattice points.

The number of atoms in a unit cell can be expressed uniquely as $Ap$, where $p$ can be represented as the sum of two squares ($u^2 + v^2$, including zero as a square) and $A$ is either 1 or a number that cannot be expressed as $u^2 + v^2$. Only if patterns have the same $A$, will they have a common square supercell.

Patterns with common $A$ are described as *compatible*. Sorted by value of $A$, we have for $Ap \leq 21$:

[1]In discussing chemical compounds it is more usual to talk about "rings" (as in "benzene ring") rather than "polygons." In the present context the terms are synonymous.

[2]As the average ring size is 3 for 6-connected nets, it follows that in the absence of 2-rings (loops) all the rings in a 6-connected net are 3-rings and $3^6$ is the *only* 6-connected net.

| | |
|---|---|
| $A = 1$ | $Ap$ = 1, 2, 4, 5, 8, 9, 10, 13, 16, 17, 18, 20,... |
| $A = 3$ | $Ap$ = 3, 6, 12, 15,... |
| $A = 7$ | $Ap$ = 7, 14,... |
| $A = 11$ | $Ap$ = 11, 21,... |
| $A = 19$ | $Ap$ = 19,... |

The square lattice (the regular tiling $4^4$) has one point per cell and so belongs in the class with $A = 1$. The important tiling $3^2.4.3.4$ (Fig. 5.39) has four vertices per unit cell (§ 5.6.5) and so is compatible with $4^4$. Fig. 5.41 (p. 167) illustrated that the transformation from one structure to the other can be effected by concerted rotations of squares. This relationship is valuable in relating layers of crystal structures.

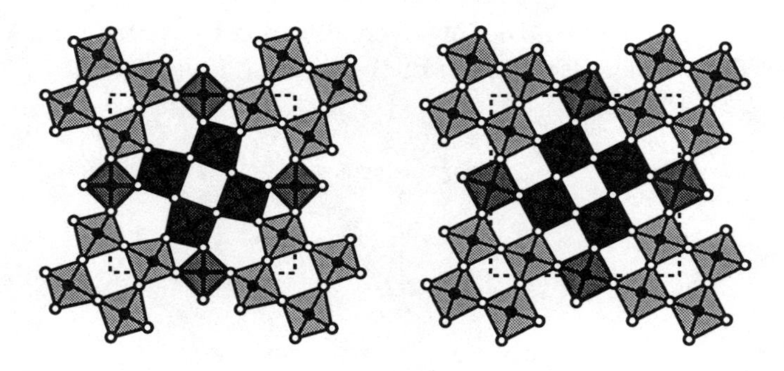

**Fig. 5.71**. The relationship between **TTB** (left) and $4^4$ (right). Corresponding parts of each diagram are equally shaded for ease of comparison.

As a second example we illustrate (Fig. 5.71) the relationship between $4^4$ and the tetragonal tungsten bronze (**TTB**) net (Fig. 5.43, p. 170) with 20 points in the unit cell (i.e. $A = 1$ again). Again the transformation between the two structures is effected by rotations of groups of vertices. For other examples see *OKH*.

In both these examples the common supercell is the larger of the two. If we wanted to inter-relate patterns with (say) 4 and 5 points per cell respectively, we would have to use a supercell with 20 points per cell to describe the transformation.

Similar considerations apply to two-dimensional hexagonal patterns. A non-primitive hexagonal cell with translation vectors $u\mathbf{a} + v\mathbf{b}$ and $-v\mathbf{a} + (u-v)\mathbf{b}$ (derived from a primitive cell defined by **a** and **b**) will have area $u^2 - uv + v^2$ containing $u^2 - uv + v^2$ lattice points. Let the number of points in a hexagonal cell be $Bq$, where $q$ can be expressed as $u^2 - uv + v^2$ and $B$ is either 1 or cannot be so expressed. Hexagonal patterns with a common $B$ will have a common super-cell and thus will be compatible.

Sorted by value of $B$ we have for $Bq \leq 20$ :

| | |
|---|---|
| $B = 1$ | $Bq$ = 1, 3, 4, 7, 9, 12, 13, 16, 19,... |
| $B = 2$ | $Bq$ = 2, 6, 8, 14, 18,... |
| $B = 5$ | $Bq$ = 5, 15, 20,... |
| $B = 10$ | $Bq$ = 10,... |
| $B = 11$ | $Bq$ = 11,... |

In Fig. 5.40 we demonstrated that $3^6$ (one vertex per cell, $B = 1$) and 3.6.3.6 (three vertices per cell, $B = 1$) were indeed compatible and illustrated their relationship.

$6^3$ (2 vertices per cell) belongs to the $B = 2$ family, so it cannot be transformed to $3^6$ (for example) without a change of shape of cell. But $6^3$ is compatible with $3^4.6$ and 3.4.6.4 (both six vertices per cell, $B = 2$). Fig 5.72 illustrates that the relationship between them is very simple and can be effected by rotating the shaded hexagonal groups of vertices.

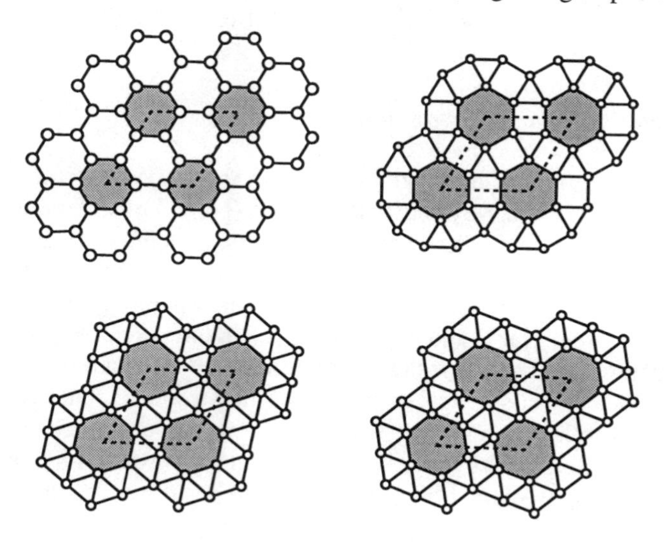

**Fig. 5.72.** Relationships between $6^3$ (top left), 3.4.6.4 (top right) and enantiomorphs of $3^4.6$ (bottom).

A slightly more subtle example involves the relationship between the **α-$U_3O_8$** and **β-$U_3O_8$** nets (see § 5.3.4). **α-$U_3O_8$** is hexagonal with five vertices per cell ($B = 5$ family). **β-$U_3O_8$** is centered rectangular with ten vertices per cell, but the primitive cell is *almost* metrically hexagonal (there are no 3- or 6-fold symmetry axes of course) and also has five vertices per cell. Fig. 5.73 shows that these two nets are simply related by rotation of pentagonal groups of vertices (shown shaded in the figure). In this example, because **β-$U_3O_8$** is not strictly hexagonal, we do have a *small* change of unit cell shape.

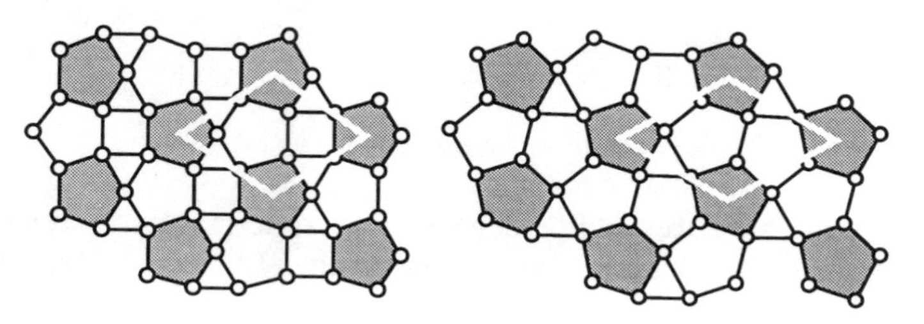

**Fig. 5.73.** Relationship between the **α-$U_3O_8$** (right) and **β-$U_3O_8$** (left) nets. A primitive cell (with an unconventional origin) is shown as white lines.

Many of the transformations between two-dimensional patterns that we have described involve rotations of polygons. We will find that in similarly relating three-dimensional patterns (such as sphere packings) we will rotate polyhedral groups of atoms.

Most of the plane patterns of interest in crystal chemistry are square with $A = 1$ or 3, or hexagonal with $B = 1$ or 2 (*OKH*).

Note that transformations between incompatible patterns (e.g. $3^6$ with $B = 1$ and $6^3$ with $B = 2$) require a change of shape of the unit cell (a metrically hexagonal cell of $3^6$ containing two points does not exist).

### *5.6.13 Pseudorotations and twinning of nets*

We remarked that two "infinite" structures with the same densities of points can be converted from one to the other by shuffles of atoms (points) by an amount less than, or at least comparable to, interatomic distances.[1] In particular structures related by rotation about an axis or by reflection can be interconverted without macroscopic displacements. Intergrowth of two otherwise identical structures in different orientations is called *twinning* and usually different structural elements are generated at the interface.

In Fig. 5.74 below we show how the orientation of a $4^4$ net can be changed merely by small rotations of individual sets of four points forming a square. The left-hand part of the figure shows open circles on a $4^4$ net. Rotating the points on the shaded squares to the positions shown by filled circles results in a fragment of $4^4$ rotated by approximately 51° (or –39°) from the original orientation. On the right the resulting structure is shown; it may be seen to be a coherent intergrowth of the two $4^4$ nets with squares replaced by triangles and pentagons in the line where the two twins meet.[2]

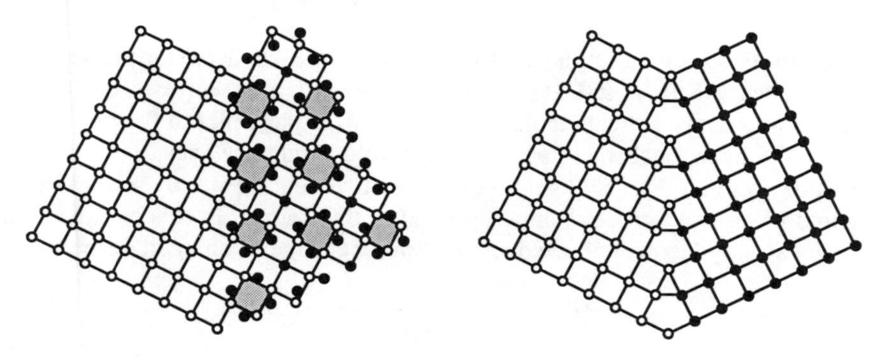

**Fig. 5.74.** Twinning of a $4^4$ net by rotation of squares. See the text for details.

Very many crystal structures can be derived from simpler ones by repeated periodic twinning (*mimetic* twinning). Numerous examples are adduced by Hyde & Andersson

[1]Transformations between lattices of the same density and the nature of the displacements are discussed by M. Duneau & C. Oguey, *Journal of Physics A* **24**, 461 (1991).

[2]In three dimensions twins meet in a plane: the *composition plane*.

[Book List, see also B. G. Hyde *et al.*, *Progr. Solid State Chem.* **12**, 273 (1979)] and it is important to realize that (at least in some cases) only small atomic displacements (as opposed to macroscopic shear) are needed to interconvert structures. An example of a three-dimensional pseudo-rotation is to be found in § 6.3.1.

### *5.6.14 Symmetries of structures derived by stacking equivalent nets*

In § 5.3.7 we described some structures consisting of a two-layer stacking of self-dual nets. In any two-layer structure, the layers are of necessity on mirror planes, say normal to **c**. If further the two layers are the same (or mirror related), there must be a symmetry operation that includes a translation **c**/2 relating the two mirror planes. In the general case all atoms are in positions $x,y,z_0$ with $z_0$ = either 0 and 1/2 or 1/4 and 3/4. Possible symmetry groups for structures with atoms in such positions are listed in Table 5.7 below, first in a setting with the mirror planes containing the nets normal to **c** and second in the "standard" setting. The symmetry of the layer is also given. Notice that there are often extra constraints on the lattice parameters of the layers. Thus in space group $P4_2/m$, layers with symmetry *p*2 are stacked one above another with alternate layers rotated by 90°; for this to be possible one must have $a = b$ for the layer.

**Table 5.7**. Possible symmetries for structures with symmetry-related layers

| space group | | layer | space group | | layer | space group | layer |
|---|---|---|---|---|---|---|---|
| *B*11*m* | *Cm* | *p*1 | *Pbcm* | *Pbcm* | *p*1*g*1 | $P4_2/m$ | *p*2 |
| $P112_1/m$ | $P2_1/m$ | *p*1 | *Pnnm* | *Pnnm* | *p*2 | *I*4/*m* | *p*4 |
| *B*112/*m* | *C*2/*m* | *p*2 | *Pmnm* | *Pmmn* | *p*1*m*1 | *P*4/*mcc* | *p*4 |
| *Pc*2*m* | *Pma*2 | *p*1 | *Pbnm* | *Pnma* | *p*1*g*1 | *P*4/*mnc* | *p*4 |
| $Pn2_1m$ | $Pmn2_1$ | *p*1 | *Amam* | *Cmcm* | *p*2*mg* | $P4_2/mmc$ | *p*2*mm* |
| *Bm*2*m* | *Cmm*2 | *p*1*m*1 | *Cmcm* | *Cmcm* | *p*2*mm* | $P4_2/mcm$ | *c*2*mm* |
| $Bb2_1m$ | $Cmc2_1$ | *p*1*g*1 | *Acam* | *Cmca* | *p*2*gg* | $P4_2/mbc$ | *p*2*gg* |
| *Am*2*m* | *Amm*2 | *p*1*m*1 | *Cmmm* | *Cmmm* | *p*2*mm* | $P4_2/mnm$ | *c*2*mm* |
| *Ac*2*m* | *Abm*2 | *p*1*g*1 | *Cccm* | *Cccm* | *c*1*m*1 | *I*4/*mmm* | *p*4*mm* |
| *Cc*2*m* | *Ama*2 | *p*1 | *Bmcm* | *Cmma* | *p*2*mm* | *I*4/*mcm* | *p*4*gm* |
| *Fm*2*m* | *Fmm*2 | *c*1*m*1 | *Bmam* | *Cmma* | *p*2*mg* | $P6_3/m$ | *p*3 |
| *Im*2*m* | *Imm*2 | *p*1*m*1 | *Fmmm* | *Fmmm* | *c*2*mm* | $P\bar{6}c2$ | *p*3 |
| *Ic*2*m* | *Ima*2 | *p*1*g*1 | *Immm* | *Immm* | *p*2*mm* | $P\bar{6}2c$ | *p*3 |
| *Pccm* | *Pccm* | *p*2 | *Ibam* | *Ibam* | *p*2*gg* | *P*6/*mcc* | *p*6 |
| *Pmcm* | *Pmma* | *p*1*m*1 | *Imcm* | *Imma* | *p*2*mg* | $P6_3/mcm$ | *p*31*m* |
| *Pncm* | *Pmna* | *p*2 | *Imam* | *Imma* | *p*2*mg* | $P6_3/mmc$ | *p*3*m*1 |

Notice too that although there are many possibilities for stacking low-symmetry layers, the number of possibilities becomes limited for more-symmetrical layers and there is only one possible stacking sequence for layers with symmetry *p*3*m*1, *p*31*m*, *p*4*mm*, *p*4*gm* and *p*6. This is because alternate layers must be related by a symmetry operation that is not a part of the symmetry of the layer. Thus in $P6_3/mcm$ adjacent layers of symmetry *p*31*m* are related to each other by reflection in mirrors normal to **a** (combined with the translation *c*/2 this corresponds to the operation of *c* glide normal to **a** as indicated in the space group symbol). In $P6_3/mmc$ adjacent layers of symmetry $p3m1$ are related to each other by reflection in mirrors parallel to **a** (operation of *c* glide plane containing **a** as indicated in the space group symbol). Layers with symmetry *p*6 can be also be stacked with alternate layers related by *c* glide but now one set of glide planes automatically generates a second set and there is again only one possibility: symmetry *P*6/*mcc*.

A square layer has 4-fold axes at the center of the cell as well as at the corners (see Fig. 1.13, p. 16) so square layers can be stacked with alternate layers displaced by 1/2,1/2,1/2. This is the only possibility for layers with symmetry *p*4*mm* and *p*4*gm*.

Finally, note that a two-layer stacking of layers of symmetry *p*6*mm* is not possible. This is because there is no three-dimensional crystallographic operation relating the layers that will not leave them unchanged, so the only possible stacking of such layers is that in which they are directly one above the other and then there is a single layer structure with symmetry *P*6/*mmm*. It follows at once that there cannot be a self-dual net with symmetry *p*6*mm*. On a little reflection (!) this result should become "obvious."

### 5.6.15 More structures with dual nets: $Fe_2P$

In § 5.3.7 we gave some examples of structures that were simply described as stackings of self-dual nets. The description is particularly attractive as once the 2-dimensional net and the spacing between layers are specified, the structure is completely determined; although to fully appreciate the structure, it should be illustrated in more than way (for example by also emphasizing the pattern of trigonal prisms).

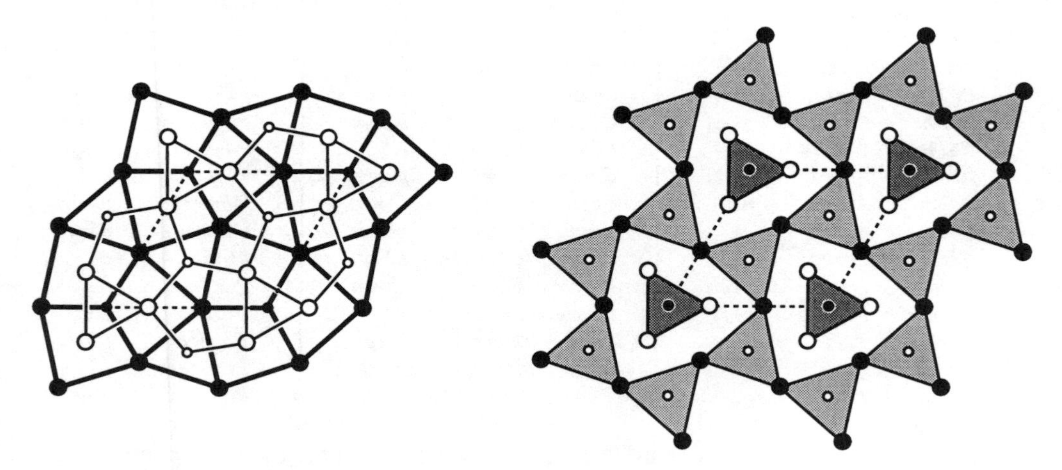

**Fig. 5.75**. The structure of $Fe_2P$ projected on (0001). Larger circles are Fe. Filled (open) circles are at $z$ = 1/2 (0). Left: showing the nets at $z$ = 1/2 (heavy lines) and at $z = 0 \equiv 1$ (lighter lines). Right: as $\{P\}Fe_6$ trigonal prisms centered at $z$ = 0 (lighter shaded) and at $z$ = 1/2.

A related group of structures can be described in terms of stackings of dual (but not self-dual) nets. These structures are generally more complicated, in the sense that there are more crystallographic kinds of atom, because atoms on the two layers are not related by symmetry. The simplest structure of this sort is $\mathbf{Fe_2P = Fe(1)_3Fe(2)_3P(1)P(2)_2}$ (for crystallographic data see Appendix 5).

Figure 5.75 shows a projection of the hexagonal ($P\bar{6}2m$) structure on (0001). It may be seen that the layers at $z = 0$ and $z = 1/2$ are different, but the nets are indeed the duals of each other.[1] Such an illustration allows a ready determination of the coordination of all the atoms, but it is hard to see the larger-scale organization of the structure. On the right in the figure $\{P\}Fe_6$ trigonal prisms centered at $z = 0$ and $z = 1/2$ are emphasized; now the linkage of these polyhedra is more apparent.

## 5.7 Exercises

1. The dual of the bisdisphenoid is an octahedron which also has $\bar{4}2m$ symmetry (of course!), and has four faces that are pentagons and four that are quadrilaterals. Can this polyhedron be made with faces that are regular polygons? (No.)

There *is* an octahedron with three pentagonal and five triangular faces that can be made with regular polygons. [Hint remove three vertices from a regular icosahedron.]

2. If polyhedra are to fill space, the solid angles at the vertices that meet at a point must sum up to $4\pi$ (720°) and the dihedral angles at a common edge must sum up to 360°. Verify that the solid angle at the vertex of a truncated octahedron is 180° (a useful formula is Eq. 4.15) and that the dihedral angles are 120°. Truncated octahedra indeed do fill space with three polyhedra meeting at an edge, and four meeting at a vertex (see § 7.3.10).

3. Regular tetrahedra alone cannot fill space. The solid angle at the vertex of a regular tetrahedron is $3\cos^{-1}(1/3) - \pi = 31.59°$ so at most 22 regular tetrahedra can have a common vertex. But at most five regular tetrahedra can have a common edge (the dihedral angle of a regular tetrahedron is $\cos^{-1}(1/3) = 70.52°$) so in fact only 20 regular tetrahedra can have a common vertex. (Do the experiment with 21 regular tetrahedra!). A regular icosahedron can be considered as made up of 20 tetrahedra with three edges equal to unity and three edges (meeting at the center) equal to $5^{1/4}\sqrt{[(1 + \sqrt{5})/8]} = 0.9511$. [These considerations are relevant to the shapes of small clusters of metal atoms].

4. A linear rod can be made of regular tetrahedra sharing faces (so that the tetrahedron centers fall on a straight line). The vertices fall on an aperiodic helix.[2] Are the positions of the vertices described by a quasiperiodic function? (Yes.)

[1] The net at $z = 0$ (open circles in the figure) should be recognized as the $\boldsymbol{\alpha}$-$\mathbf{U_3O_8}$ net (see Fig. 5.44, p. 172).

[2] A nice, but challenging, problem. See H. Nyman *et al.*, *Zeits. Kristallogr.* **196**, 39 (1991) for some help. Such a rod is known as a "Bernal Spiral". For more on this beautiful structure see also C. Zheng *et al.*, *J. Amer. Chem. Soc.* **112**, 3784 (1990).

5. It is simple to demonstrate that the volumes of regular tetrahedra, octahedra and truncated tetrahedra with the same edge length are in the ratio 1 : 4 : 23. (Hint: combine each of the larger polyhedra in turn with four tetrahedra to make larger tetrahedra.)

6. The reader might enjoy deriving results similar to Eq. 5.4 from Eq. 5.2 for finite polyhedra. For example it is not difficult to show that if a closed shell (polyhedron) of "graphite" (three-connected C atoms) is made, in addition to hexagonal rings (as in graphite), there must be 12 pentagonal rings. $C_{60}$ and $C_{70}$ are well known examples.

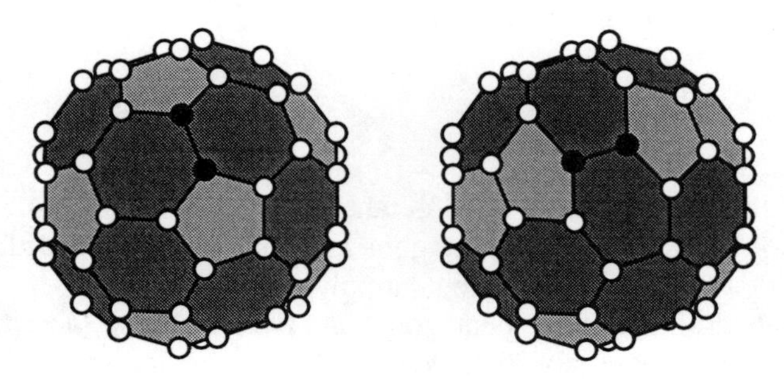

**Fig 5.76**. The polyhedron shown on the left is the truncated icosahedron ($5.6^2$). On the right is a topological isomer.

Hint: if three polygons meet at each vertex, $E = 3V/2$ [incidentally showing that the number of vertices must be even] and Eq. 5.2 becomes $2F = V + 4$. Let there be $F_5$ pentagonal faces and $F_6$ hexagonal faces so that $F = F_5 + F_6$. As each vertex belongs to three polygons there are 6/3 vertices per hexagon and 5/3 vertices per pentagon

Figure 5.76 shows, on the left, the truncated icosahedron ($5.6^2$) which has 60 vertices, 20 hexagonal faces and 12 pentagonal faces. On the right of the figure is shown a topological isomer with the same number of faces but some vertices are $5^2.6$ and $6^3$; it is derived from $5.6^2$ by rotating the two vertices shown as filled circles.[1]

7. The combination of two regular pentagons and a regular decagon meeting at a point have the sum of their angles equal to 360° but there is no tiling $5^2.10$.[2] Show that nevertheless the combination of two pentagons and a decagon per vertex satisfies Eq. 5.3.

8. Verify for the Archimedean tilings (Table 5.4, § 5.3.2) that the average ring sizes and numbers of rings per vertex have the values given in § 5.6.11.

[1]See A. J. Stone & D. J. Wales, *Chem. Phys. Letts.* **128**, 501 (1986).
[2]This problem fascinated Kepler, who produce some beautiful tilings consisting largely (but of course not entirely) of regular pentagons and decagons [see Grünbaum & Shephard (Book List)].

9. There is a simple tiling of the plane by congruent pentagons.[1] The symmetry is *p4gm* with $a = \sqrt{(4 + \sqrt{7})}$. There are two kinds of vertex: one at 2 *a*: 0,0 ; 1/2,1/2 and the second at 4 *c*: $\pm(x, 1/2+x\ ;\ 1/2-x, x)$ with $x = 0.363$. Four pentagons meet at one vertex and three at the other (Fig. 5.77). The pattern is known as *Cairo tiling* or MacMahon's net.[2]

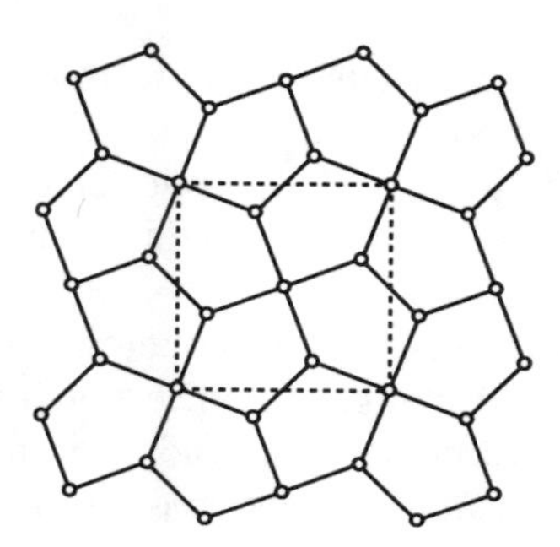

**Fig. 5.77**. MacMahon's net (Cairo tiling).

10. Show that the transformation 3.6.3.6 → $3^6$ illustrated in Fig. 5.40 can be described analytically as follows: We take a cell containing three points at positions 3 *c* of *p*31*m* ($x$,0; 0,$x$; $\bar{x},\bar{x}$) with $x = (3 - 2\sin\phi)/6$ and unit cell edge $a = d/\sqrt{(3x^2 - 3x + 1)}$ where $d$ is the shortest distance between points (the edges of the net). $\phi$ is the angle of rotation of the triangles of 3.6.3.6 so that $\phi = 0°$ corresponds to 3.6.3.6 and $\phi = 30°$ corresponds to $3^6$. The endpoints have symmetry *p6m* and the true unit cell for $3^6$ is 1/3 the area of this cell. What happens when $\phi > 30°$?

11. Describe the transformation $3^2$.4.3.4 → $4^4$ in a similar way to that in Exercise 10. (See Fig. 5.41; the symmetry in the general case in *p4gm*)

12. Consider tilings by the polygons shown in Fig. 5.65 such that every vertex is 5-connected. Show that Eq. 5.4 requires that $2/3 \leq \phi_3 \leq 5/6$, $0 \leq \phi_4 \leq 1/3$ and $0 \leq \phi_5 \leq 1/6$. Can you make a tiling with a 5-fold rotation point with these tiles? Can you make a periodic tiling using all three tiles and with all vertices 5-connected?

[1]Enumerating the tilings of the plane with congruent pentagons is a famous problem and attempts at its solution have resulted in amateurs handily beating professional mathematicians. See M. Gardner in *Time Travel and Other Mathematical Bewilderments* [Freeman, New York (1988)].

[2]In Cairo (Egypt) the tiling is common for paved sidewalks. The net is featured in P. A. MacMahon's *New Mathematical Pastimes* [Cambridge, (1921)].

CHAPTER 6

# SPHERE AND CYLINDER PACKINGS

In this chapter and the next we discuss periodic three-dimensional structures. This extends the discussion from finite polyhedra, circle packings, and two-dimensional nets, to infinite polyhedra, sphere and cylinder packings, and three-dimensional nets. Now we are approaching the real world of crystal structures and examples of them will be met more frequently.

In many crystal structures one or several kinds of atom are in positions corresponding to the centers of spheres in a sphere packing, and other atoms are in positions corresponding to interstices of that packing. It is common (especially for mathematicians) to refer to such interstices as "holes," but that word has been appropriated to have a special meaning (referring to an electronic defect) in solid state science, so we avoid it. We generally just refer to "sites." In the same vein, the word "vacancy" has the meaning of a site that *should* be occupied but at some particular point in a real crystal is not (i.e. a special kind of atomic defect). We recommend using "vacancy" *only* in the context of defect thermodynamics and kinetics (and then with great care), and at other times using a term such as "unoccupied site" instead.

It is essential at the outset to recognize that we are here only concerned with the geometry of certain patterns of points which are of common occurrence in crystal structures. It is convenient to consider such patterns as arising from packings of spheres, but as they can also arise in several other contexts, it is important not to get a mental image of crystals as assembled from a packing of hard sphere "atoms" as is sometimes seen illustrated (we do this ourselves in § 6.1 and 6.2, but nowhere else). We shall see, for example, that in several simple and familiar crystal structures of binary compounds *AB* the arrays of *both* the *A* and the *B* atoms are the same as the centers of spheres in closest packing. Unless the spheres representing *A* and those representing *B* interpenetrate substantially there cannot be simultaneous *A*-*A* and *B*-*B* contacts.

One reason for discussing sphere packings is that it is hard to read the literature of crystal chemistry without some knowledge of the subject and its associated terminology. The most compelling reason is, however, that the topic introduces patterns that are ubiquitous in crystal structures; indeed it is hard to invent a simple symmetrical sphere packing that does not occur in nature. Our organization is by coordination number, starting with the densest packing of spheres; however, this is for convenience only; we could equally have chosen one of a number of other schemes.

## 6.1 The densest packing of spheres

We consider first the classical problem of packing equal-sized spheres in space as densely as possible (closest packing). It should be apparent that this is the same problem as that of arranging points of an infinite array with given density (number per unit volume) so that the shortest distance between them is a maximum; stated this way, the problem is an extension of Tammes' problem (§ 5.1.9) which referred to arranging points (with fixed number per unit area) on the surface of a sphere such that the minimum distance between them was a maximum. Another problem with the same solution is to ask for ways of arranging points in space so that every point has twelve equidistant nearest neighbors. Yet another problem, again with the same solution, is to ask for the arrangement of vertices in space-filling packings of regular octahedra and tetrahedra with equal edges.

Thus the same geometrical arrangement arises in very different contexts, only one of which involves spheres in contact. When we want to emphasize such considerations we use the term *eutaxy* to mean "an arrangement corresponding to the centers of spheres in closest packing" and describe such arrays as *eutactic*. Generally though, we follow established usage indicated by bold face abbreviations in the next paragraph.

If we restrict ourselves to arrangements in which the points (or the centers of the spheres) fall on a lattice, there is just one solution to the above problem. The arrangement is commonly referred to as cubic closest packing (**ccp**), but we also use the term *cubic eutaxy.*[1] If the restriction to points on a lattice is lifted, we find a second arrangement of equivalent (symmetry-related) points known as "hexagonal closest packing" (**hcp**) or *hexagonal eutaxy*. There is an infinity of other arrangements with the same density, but with more than one kind (in the crystallographic sense) of point as we will see. We refer to these generically as **cp** or as eutactic. We consider only periodic patterns and state results mostly without proof.[2] The construction of models with a dozen or so polystyrene balls and toothpicks to hold them together will be found invaluable.

### *6.1.1 Stacking of close-packed layers*

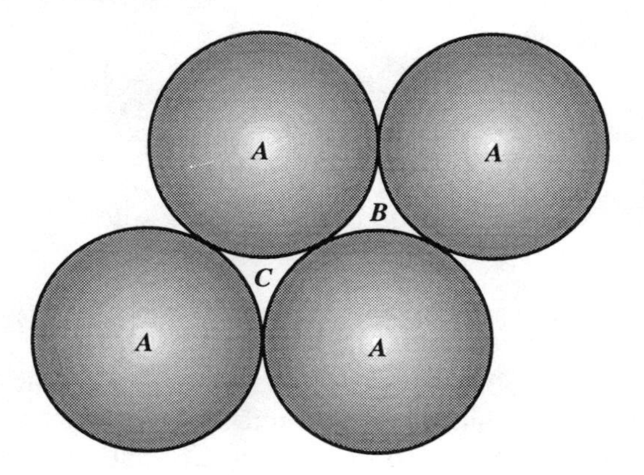


**Fig. 6.1**. Part of a layer of close-packed spheres. *A* marks the corners of a unit cell.

[1]The term is used [as is "cubic closest packing" (**ccp**)] also for arrangements that only approximate the ideal arrangement.

[2]Some "obvious" results are remarkably difficult to prove in a way acceptable to mathematicians. For comments on the proof (by W. Hsiang) that "closest packing" is just that see N. Max, *Nature* **355**, 115 (1992).

Conceptually the simplest way to generate these patterns is to start with a layer of spheres lying on a flat surface in a closest-packed way.[1] Their centers will be on the vertices of a $3^6$ net (a **cp** layer), just as in the densest circle packing. Fig. 6.1 shows four such spheres with centers at the corners of a hexagonal unit cell of $3^6$. In the figure the letters "*A*" are at these corners.

Now we add a second close-packed layer on top of the first. To have maximum density we want the spheres of the second layer to nestle in the depressions of the first, i.e. over the points marked "*B*" or "*C*." (It should be clear that *B* and *C* are too close for spheres of the second layer to be simultaneously over both of these positions). Accordingly there are two possibilities for the two-layer structure: *AB* or *AC*. These are of course identical arrangements (remember the layers are infinite in the plane). For the sake of subsequent discussion let the arrangement chosen be *AB* for the moment.

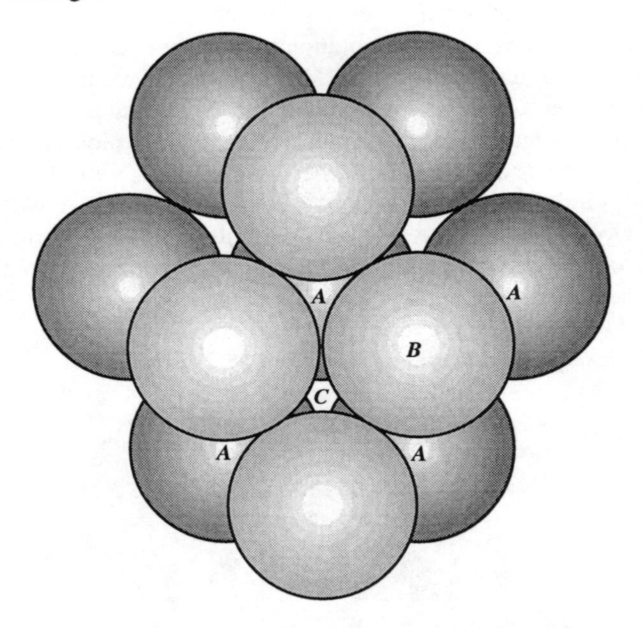


**Fig. 6.2**. Part of two layers of close-packed spheres. The letters are in the same place as in Fig. 6.1.

A small fragment of the two-layer packing is shown in Fig. 6.2. It should be clear from that figure that if we now add a third layer in a similar way, the centers of the spheres in the third layer must lie over either *A* or *C* so we have two *distinct* three-layer sequences: *ABA* and *ABC*. These differ in that in the first case the layers below and above the middle one are in same (*A*) positions, and in the second case the layers below and above the middle one are in different positions (*A* and *C*).

For unit diameter spheres in contact, the perpendicular distance between layers will be $\sqrt{(2/3)} = 0.8165$ (this is the height of a regular tetrahedron of unit edge, cf. p. 133).

[1]This is why we pack spheres rather than, for example, polyhedra.

Fig. 6.3 shows parts of the two different three-layer sequences. At the top the spheres are packed *ABC* and at the bottom the sequence is *ABA*. The two simplest infinite packings would be obtained by repeating these sequences indefinitely: *ABCABC*... and *ABAB*....

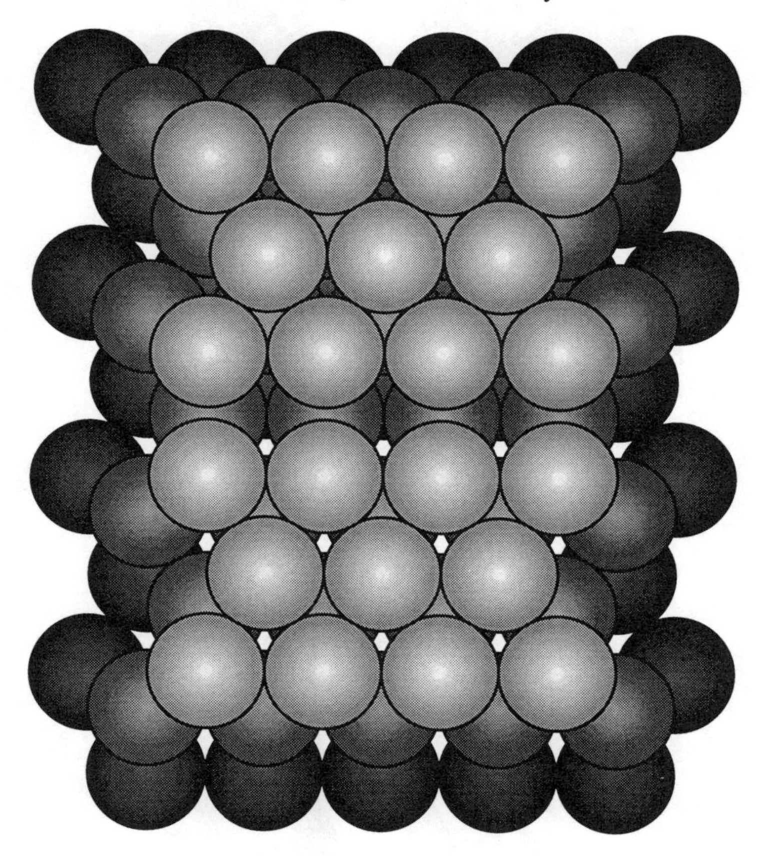

**Fig. 6.3**. Spheres packed in the sequence *ABC* (top half) and *ABA* (bottom half).

The centers of the spheres in a slab of two layers divide the slab into regular tetrahedra and octahedra. Fig. 6.4 shows how a tetrahedron and an octahedron are so defined by four and six sphere centers respectively. The eutactic arrangements of points thus also arise as the positions of the vertices when regular tetrahedra and octahedra are packed to fill space, and this is possibly the real reason for the common occurrence of eutaxy as the arrangement of cations and/or anions in so many compounds in which there is 4- and/or 6-fold coordination (see Notes § 6.8.3).

Figure 6.5 illustrates the arrangement of octahedra and tetrahedra in a two-layer slab. Each octahedron shares edges with six other octahedra. The tetrahedra can be divided into two groups (see Fig. 6.5): those "pointing down" and those "pointing up." Tetrahedra of each type share only vertices with each other, but each "up" tetrahedron shares three edges with "down" tetrahedra in the same layer and *vice versa*. Altogether space is divided into equal numbers of octahedra, "up" tetrahedra and "down" tetrahedra.

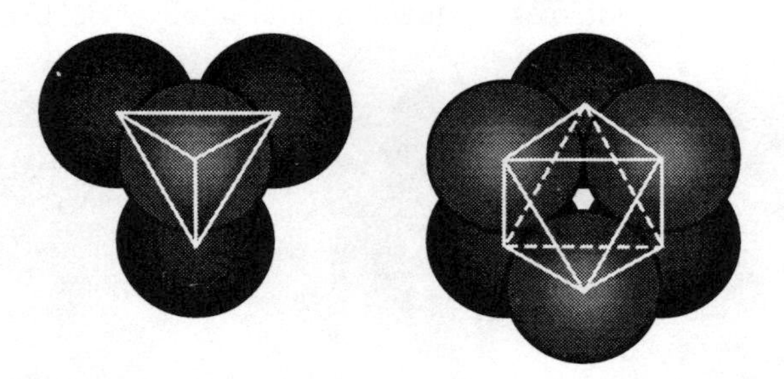

**Fig. 6.4**. A tetrahedron and an octahedron formed by spheres in contact.

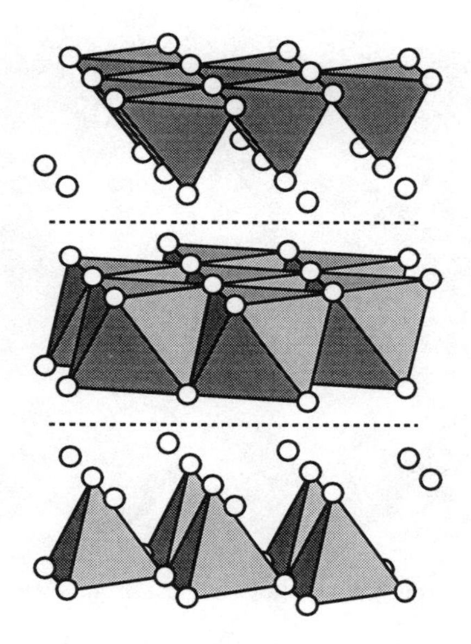

**Fig. 6.5**. Middle: the centers of 24 spheres (represented by circles) in a double layer of close packing showing the octahedral interstices. Top: the same points, but now showing the "down" tetrahedra. Bottom: the same points showing the "up" tetrahedra.

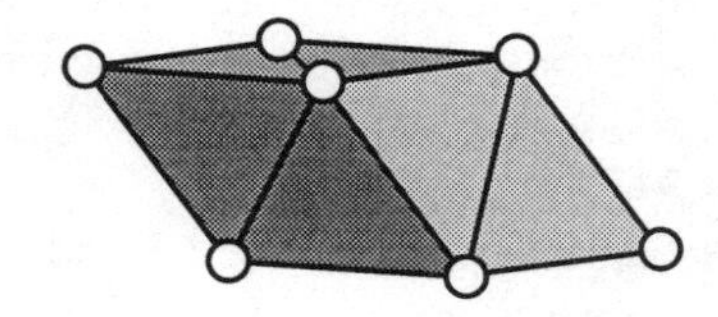

**Fig. 6.6**. A rhombohedron composed of an octahedron and two tetrahedra (compare Fig. 6.5).

The reader is urged to assemble such a layer with polyhedra. When that is done it will surely be noticed that two tetrahedra may be combined with an octahedron to form a 60° rhombohedron as shown in Fig. 6.6. Clearly this polyhedron fills space and contains one each of an "up" tetrahedron, a "down" tetrahedron and an octahedron. We will see that it is a unit cell of **ccp**. It should be clear that each two-layer slab can be divided into such rhombohedra, so all **cp** arrays are made up of octahedra, "up" tetrahedra, and "down" tetrahedra in equal amounts (one each per **cp** sphere).

### *6.1.2 Hexagonal eutaxy* (**hcp**)

We discuss the case *AB*... first. We could equally label the sequence *AC*..., *BC*..., etc. which would describe exactly the same packing but with a different choice of origin. In fact if *A* is at the origin of a hexagonal cell, it is useful now to describe the two-layer repeat as *BC*.... We can then get a convenient description in crystallographic terms of a hexagonal unit cell with sphere centers at 1/3,2/3,1/4 and 2/3,1/3,3/4. If the spheres are unit diameter, $a$ will be 1.0 (see Fig. 6.1) and $c$ will be $2\sqrt{(2/3)} = \sqrt{(8/3)} = 1.6330$. The $z$ coordinates are chosen as 1/4 and 3/4 (rather than e.g. 0 and 1/2) because we then have the origin of coordinates at a center of symmetry. The space group is $P6_3/mmc$. The arrangement of sphere center points is not a lattice, as a vector from the center of a sphere to the center of a contiguous sphere in an adjacent layer is not a lattice vector. The spheres are related by symmetry though; their centers are in the special positions 2 *c* of $P6_3/mmc$.

In **hcp**, the planes normal to **c** and containing the centers of the spheres are mirror planes. It follows therefore, that the octahedra in successive layers share common faces and form face-sharing columns (parallel to **c**). The centers of the octahedra are at 0,0,0 and 0,0,1/2 in the unit cell (2 *a* of $P6_3/mmc$). On the other hand the tetrahedra will form pairs (one "up" and one "down") with a common face. Recall that the "up" set of tetrahedra have only common vertices (are corner-sharing) as do the "down" set. The centers of the tetrahedra are in 4 *f*: $\pm(1/3,2/3,z\ ;\ 2/3,1/3,1/2+z)$ with $z = -1/8$.

### *6.1.3 Cubic eutaxy* (**ccp**)

The rhombohedron of Fig. 6.6 with points at the vertices can be considered as the unit cell of a structure with the points coinciding with a lattice. The 60° rhombohedron is, in fact, a primitive cell of the face-centered cubic lattice (see § 4.4). Fig. 6.7 shows 14 spheres with their centers at the lattice points of a face-centered cubic cell. As can be seen from the figure the centers of the spheres lie in close-packed {111} planes.

The structure we are describing is cubic eutaxy or "cubic close packing" (**ccp**). Discussed in terms of a stacking of close-packed layers the sequence is *ABC*... The simplest way to see this is to use the description of a rhombohedral lattice in terms of a centered hexagonal cell. Thus if the rhombohedral cell (Fig 6.7) has $a = 1$, $\alpha = 60°$ then the hexagonal cell has $a = 1$, $c = 3\sqrt{(2/3)} = \sqrt{6} = 2.449\ldots$ and the lattice points are at 0,0,0 (*A*); 2/3,1/3,1/3 (*B*); and 1/3,2/3,2/3 (*C*).

The face-centered cubic cell contains four lattice points (at 0,0,0 ; 1/2,1/2,0 ; 1/2,0,1/2 and 0,1/2,1/2) and must therefore contain four octahedra. The octahedron centers are at 1/2,1/2,1/2 ; 1/2,0,0 ; 0,1/2,0 and 0,0,1/2 (i.e. at the body center and in the middle of each edge). These octahedral sites also fall on the points of a face-centered cubic lattice (displaced from the first by 1/2,1/2,1/2).

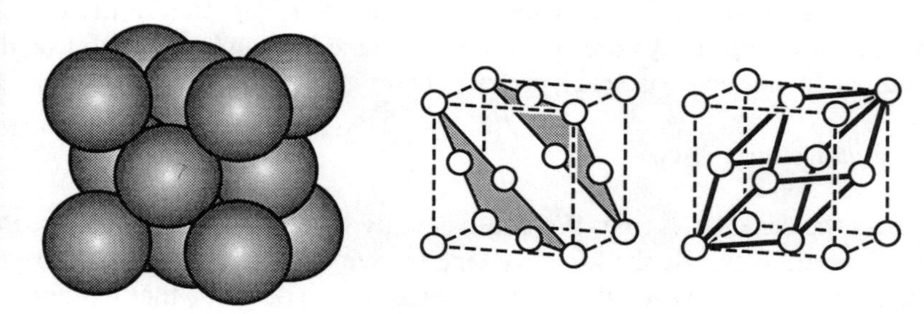

**Fig. 6.7**. Left: a face-centered unit cell with spheres centered at lattice points arranged in **ccp**. Center: the arrangement of the centers on (111) planes. Right: the primitive cell is heavily outlined.

The cubic unit cell also contains eight tetrahedral sites with centers at ±1/4,±1/4,±1/4.[1] Their centers are on a primitive cubic lattice with one-half the cell edge (see Fig. 6.11 below). The symmetry of **ccp** is $Fm\overline{3}m$ and the sphere centers are in 4 *a*. The octahedral sites are in 4 *b* and the tetrahedral sites are 8 *c*.

The centers of the spheres in the {100} planes (i.e. parallel to the faces of the cubic unit cell) are on $4^4$ nets so it can be seen that an alternative description of **ccp** is in terms of a stacking of such nets (see § 6.4.2). In close packing there is only one position for a second $4^4$ layer on top of a first one, so **ccp** is the only close packing that admits this description.

In contrast to **hcp**, the octahedra in **ccp** share only edges (i.e. not faces), and as we have seen, their centers are also in **ccp**. The tetrahedra likewise share only edges. Viewed along one of the <111> directions the tetrahedra can be considered to fall into "up" and "down" sets in each of which, as for **hcp**, they share only vertices. We shall see that the centers of each of these sets are also **ccp**.

### *6.1.4 Other eutactic* (**cp**) *arrangements*

The hexagonal and cubic arrangements *AB*... and *ABC*... are obviously just the simplest of an infinite number of possibilities of stacking close-packed layers. The next possibility is *ABAC*...

In all **cp** arrays each sphere is coordinated by 12 nearest neighbors and there are just two possibilities for the coordination figure. The first is the cuboctahedron, illustrated on the left in Fig. 6.8. The central sphere is in one of the positions *A*, *B* or *C* and the spheres in the layer above in one of the other two remaining positions and the spheres below in the third. To avoid the redundancy arising from the arbitrariness in the labels *A*, *B* or *C* it is often more convenient simply to label such a central layer *c*.

[1]Recall that −1/4 is the same as 3/4 in this context

The spheres in **hcp** have neighboring spheres at the vertices of a "twinned cuboctahedron" (also sometimes called an *anticuboctahedron*), shown on the right in Fig. 6.8. The spheres in the layers above and below the central sphere are now in the *same* positions. If the central sphere is in, for example, position *A*, those in the layers above or below are either both *B* or both *C*. Such a central layer is labeled *h*.

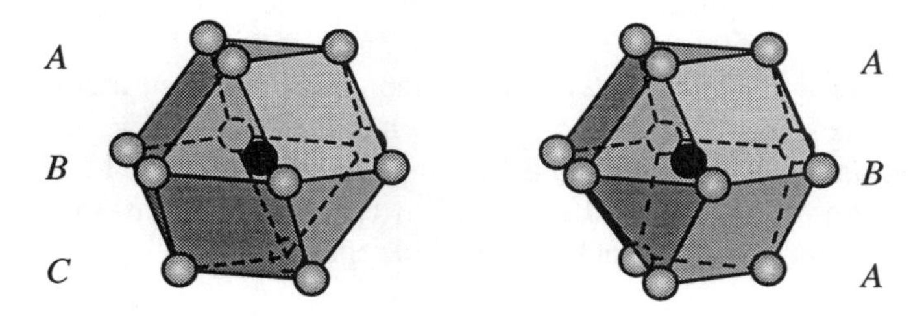


**Fig. 6.8**. The coordination of an atom (filled circle) by its neighbors in **ccp** (left) and **hcp** (right).

Any eutactic array can be described by a sequence of *c*'s and *h*'s as we now demonstrate. First write out the sequence in terms of *A*, *B* and *C* and then identify each layer as *h* or *c*. The layer will be *h* if the letters on either side are the same as each other, and will be *c* if the letters on the left and right are different from each other. Conversely a sequence of *c*'s and *h*'s can be converted to a sequence of *A*'s, *B*'s and *C*'s by starting arbitrarily with (e.g.) *AB*. A number of examples is given below (with on the right the conventional label).

| | |
|---|---|
| *h h h h* | *h* |
| *A B A B A B* | *AB* |
| *c c c c* | *c* |
| *A B C A B C* | *ABC* |
| *h c h c h c* | *hc* |
| *A B A C A B A C* | *ABAC* |
| *h c c h c c h c c h c c* | *hcc* |
| *A B A C B C A B A C B C A* | *ABACBC* |
| *h h c h h c h h c h h c h h c* | *hhc* |
| *A B A B C B C A C A B A B C B C A* | *ABABCBCAC* |

The symbolism in terms of *h* and *c* is more concise, but does not immediately reveal how many layers are in the repeat unit. Thus the repeat unit is six layers for *hcc* but nine layers for *hhc*. Many of the metallic elements crystallize as either *h* or *c* but more-complicated sequences are found. Sm for example occurs as both *hc* and *hhc* forms. We will meet **cp** arrays many times and in several different contexts later.

Another way of specifying stacking sequences is preferred by some authors. In this method, due originally to Zhdanov, a change from $A \rightarrow B$ or $B \rightarrow C$ or $C \rightarrow A$ is symbolized "+" and the reverse, i.e. $B \rightarrow A$ or $C \rightarrow B$ or $A \rightarrow C$ is symbolized "–." Thus **ccp** ($c$) is $+++\ldots$ (or $---\ldots$) and **hcp** ($h$) is $+-+-\ldots$ (or $-+-+\ldots$). The packing is specified by a sequence of numbers, each of which represents the number of repetitions of a given sign. We have then that **ccp** is $\infty$ and **hcp** is 11. *hc* is *ABAC*... i.e. $++--\ldots =$ 22. The reader should verify that *hcc* is 33 and that *hhc* is 21. The sequence *hhhc* corresponds to $++-+--+-$ i.e. 211211. A commonly encountered modification of this notation is to omit the second half of the symbol when it is a repetition of the first and to enclose the symbol in angle brackets. Thus 22 becomes <2> and 211211 becomes <211> (but 21 is <21>). The rule is that if the symbol in brackets contains an odd number of entries, the Zhdanov symbol is the bracketed symbol repeated twice.[1]

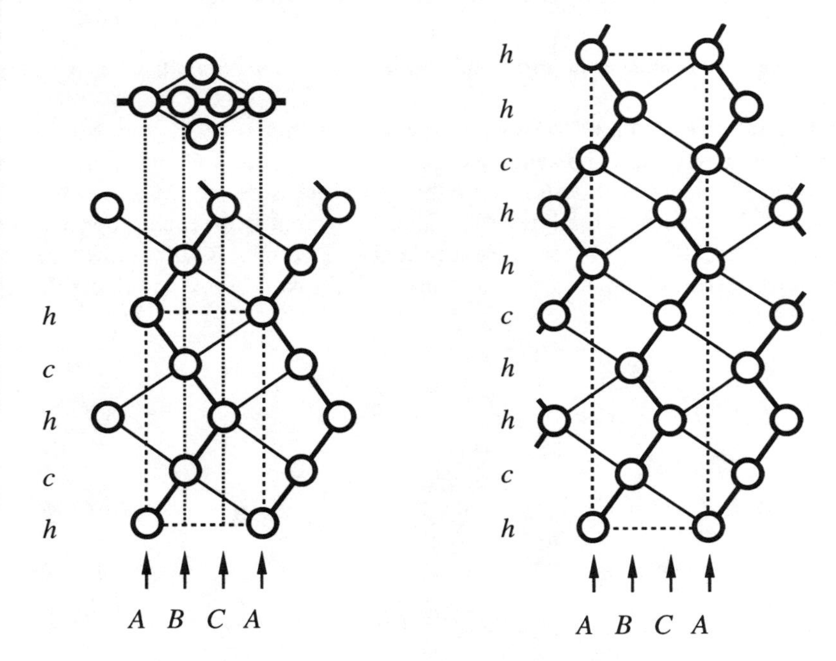


**Fig. 6.9**. $(11\bar{2}0)$ slices of *hc* (left) and *hhc* (right). **c** is vertical and a unit cell is shown with dashed lines. In the top left is shown a (0001) projection with the trace of a $(11\bar{2}0)$ plane as a heavy line.

A simple geometrical interpretation of the Zhdanov notation can be obtained from Fig. 6.9 which shows *slices* (not a projection) of two **cp** arrays. The slice is a $(11\bar{2}0)$ plane of a hexagonal cell (outlined) and heavy lines connect nearest neighbors in this plane. In *hc* = 22 (left) the line, as it ascends up in the **c** direction, goes two places right then two places left alternately. In *hhc* = 21, the line goes two places right then one left alternately. Notice

[1]It should be clear that a Zhdanov symbol (other than that for **ccp**) always has an even number of terms.

that $h$ layers are at positions of change of direction.[1]

Fig. 6.9 also shows that a **cp** array can be described as a two-layer stacking of $4^4$ nets made up of two kinds of tiles—rectangles and twinned rectangles ("kites") with edges in the ratio of $1{:}\sqrt{2}$. In **ccp** the tiles are all rectangles and in **hcp** they are all kites.

Sometimes a symbol $nX$ is used to specify a packing. $n$ is the number of layers in the repeat unit of the packing and $X$ is $H$ if the structure is hexagonal (*sensu stricto*), $R$ if the structure is rhombohedral, and $T$ if the structure is trigonal (but not rhombohedral). Thus $h$ is $2H$, and $hc$ is $4H$ and $hhc$ is $9R$. Other examples are given in Exercise 2. Unfortunately as $n$ gets large there is generally more than one packing with the same such symbol. For the use of Zhdanov symbols to determine the space group of the packing see the Notes (§ 6.8.1).

### *6.1.5 Patterns of filling interstitial sites in* **cp** *arrays*

Reference to Fig. 6.4 (p. 212) shows that the center of an octahedron between two close-packed layers $A$ and $B$ is in the $C$ position, and midway between the layers. It is a common practice to specify the positions of these interstitial sites by Greek letters: $\alpha$, $\beta$ and $\gamma$ instead of $A$, $B$ and $C$, so such an octahedral site position between $A$ and $B$ layers is labeled $\gamma$ (Fig. 6.10). Fig. 6.4 also shows that the center of an "up" tetrahedron is underneath the top sphere, so the center of an "up" tetrahedron formed by layers $AB$ with $B$ on top is in a $\beta$ position (Fig. 6.10); it is located 1/4 of the way up between the two layers. Similarly the center of a "down" tetrahedron between the same layers $AB$ is in the $\alpha$ position 3/4 of the way up.

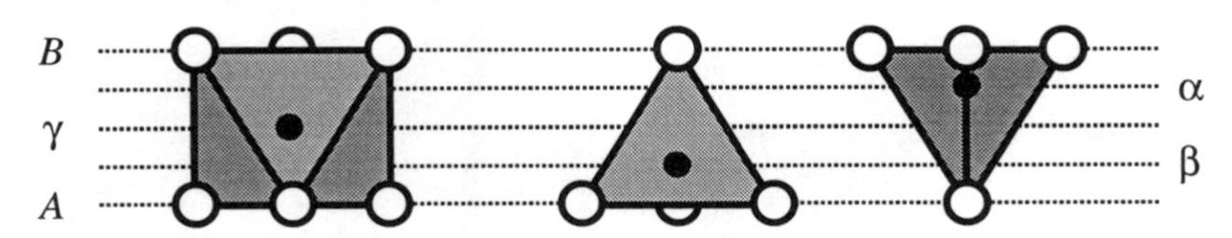


**Fig. 6.10**. Location of octahedral and tetrahedral sites in a **cp** layer *AB*. In the middle is an "up" tetrahedron; right is a "down" tetrahedron.

Suppose we fill only the "up" tetrahedral sites in **hcp**. The arrangement can be symbolized $A\beta{\cdot\cdot}B\alpha{\cdot\cdot}A\beta{\cdot\cdot}B\alpha\ldots$[2] The pattern of tetrahedral sites is $\beta{\cdots}\alpha{\cdots}\beta{\cdots}\alpha\ldots$, i.e. also **hcp**. Filling only the "down" tetrahedral sites will produce $A{\cdot\cdot}\alpha B{\cdot\cdot}\beta A{\cdot\cdot}\alpha B{\cdot\cdot}\beta\ldots$ The pattern of these tetrahedral sites is $\alpha{\cdots}\beta{\cdots}\alpha{\cdots}\beta\ldots$, i.e. again **hcp**. The structure obtained by filling either the "up" set or the "down" set of tetrahedral sites of **hcp** is that of the **wurtzite** form of ZnS which may be described *either* as an **hcp** array of Zn with S in

[1]Note also that the unit cell is chosen at an unconventional origin for clarity. Normally the origin would be taken on a $c$ atom which is at a center of symmetry (see Exercises 1 and 2).

[2]We use dots "·" as space markers for clarity, but they are not absolutely necessary and they are sometimes omitted later. Distinguish such dots from the ellipsis "..." at the end of a sequence indicating that the sequence continues indefinitely.

one-half of the tetrahedral sites (all "up" or all "down") *or* as an **hcp** array of S with Zn in one-half of the tetrahedral sites.

We repeat the exercise for **ccp**. Filling the "up" tetrahedral sites produces the sequence: $A\beta{\cdot\cdot}B\gamma{\cdot\cdot}C\alpha{\cdot\cdot}A\beta{\cdot\cdot}B\gamma{\cdot\cdot}C\alpha$... and filling the "down" sites only produces the sequence $A{\cdot\cdot}\alpha B{\cdot\cdot}\beta C{\cdot\cdot}\gamma A{\cdot\cdot}\alpha B{\cdot\cdot}\beta C{\cdot\cdot}\gamma$... The sequence of tetrahedral sites is $\beta{\cdots}\gamma{\cdots}\alpha{\cdots}\beta{\cdots}\gamma{\cdots}\alpha$... and $\alpha{\cdots}\beta{\cdots}\gamma{\cdots}\alpha{\cdots}\beta{\cdots}\gamma$... respectively; in both cases **ccp**. The structure obtained by filling either the "up" set or the "down" set of tetrahedral sites of **ccp** is that of the **sphalerite** form of ZnS which may be described *either* as a **ccp** array of Zn with S in one-half of the tetrahedral sites (either all "up" or all "down") *or* as a **ccp** array of S with Zn in one-half of the tetrahedral sites (again either all "up" or all "down"). (For the structures of ZnS see § 4.6.4, especially Fig. 4.11, for data for **wurtzite** ZnO see Appendix 5.)

Notice that both **Zn** and **S** are in 4-coordination in these structures, and we could consider the structure as a network of 4-connected atoms (a "4-connected net"). Such structures are one of the topics of the next chapter where we see (§ 7.3.1) that if all the atoms were the same (say C) then we would have the structures of the hexagonal **lonsdaleite** and cubic **diamond** forms of carbon.

Filling all the tetrahedral sites of **ccp** produces the **fluorite** structure of $CaF_2$ (with **ccp** Ca) which we can code as $A\beta{\cdot}\alpha B\gamma{\cdot}\beta C\alpha{\cdot}\gamma A\beta{\cdot}\alpha B\gamma{\cdot}\beta C\alpha$... Note that the sequence of tetrahedral sites is $\beta{\cdot}\alpha{\cdot}\gamma{\cdot}\beta{\cdot}\alpha{\cdot}\gamma{\cdot}\beta{\cdot}\alpha{\cdot}\gamma$... as in **ccp**; *but* because the spacing between layers is only one-half the distance between close-packed layers, the pattern is no longer **ccp** but is in fact primitive cubic—see Fig. 6.11. Accordingly care must be taken to ensure that the spacing between layers is appropriate before describing structures as **cp**. The spacing between each symbol (letter or "·") is (at least approximately) 1/4 of the distance between the **cp** layers, i.e. $1/\sqrt{24}$ of the distance between atoms in the close-packed layers.

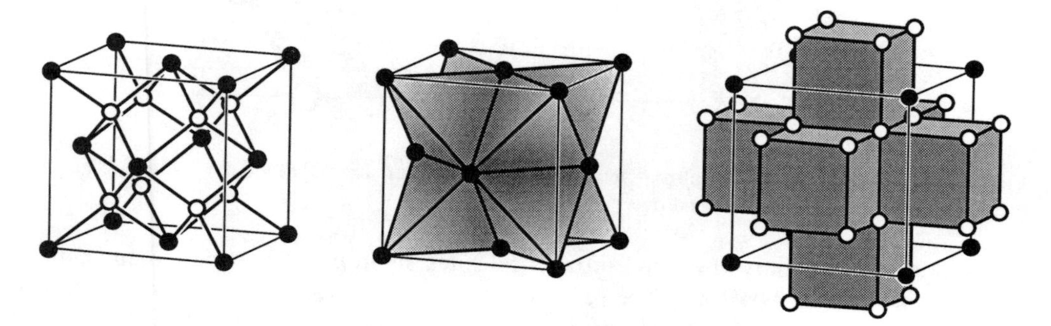

**Fig. 6.11.** Left: a unit cell of **ccp** (filled circles) with tetrahedral sites shown as open circles. If the open and filled circles are Ca and F respectively, we have a representation of the structure of $CaF_2$. Center: the same structure shown as $\{F\}Ca_4$ tetrahedra (shaded). Right: the same structure but now some $\{Ca\}F_8$ cubes are outlined.

Notice that because the F atoms are in 4-coordination, the Ca atoms must be in 8-coordination (there is only one Ca for two F atoms), and in light of the above discussion, it is not surprising to find that the coordination is $\{Ca\}F_8$ cubes.

If the roles of "cation" and "anion" are reversed in **fluorite** (as in, for example, $Li_2O$ with **ccp** O) we get the *antistructure*, called in this instance **antifluorite**.

In the (idealized) structure of **NiAs**, **Ni** atoms occupy the octahedral sites of an **hcp** array of **As**. The sequence is $A{\cdot}\gamma B{\cdot}\gamma A{\cdot}\gamma B\ldots$ The pattern of octahedral sites is $\gamma{\cdot\cdot}\gamma{\cdot\cdot}\gamma\ldots$ corresponding to points of a primitive hexagonal lattice [ideally with $c/a = \sqrt{(2/3)}$]. Interchanging **Ni** and **As** will produce the antistructure (as in PtB with **hcp** Pt).

In the **NaCl** structure, the octahedral sites of **ccp** are filled. The structure is $A{\cdot}\gamma B{\cdot}\alpha C{\cdot}\beta A{\cdot}\gamma B{\cdot}\alpha C\ldots$The octahedral sites alone are $\gamma{\cdots}\alpha{\cdots}\beta{\cdots}\gamma{\cdots}\alpha{\cdots}\beta\ldots$ i.e. again **ccp** (as observed in § 6.1.2). Interchanging **Na** and **Cl** produces the same structure (so it is its own antistructure).

One often sees structures such as **NaCl** and **NiAs** projected down an axis contained in the **cp** layers (normal to [111] for **NaCl** and normal to **c** for **NiAs**)—see for example Fig. 4.9, parts of which are repeated for convenience here as Fig. 6.12. The nature of the packing can be recognized quickly if it is realized that in **ccp** all the octahedral sites are related by primitive lattice translation vectors so all the octahedra have the same orientation; but in **hcp**, octahedra in alternate layers are related by reflection and two different orientations occur.

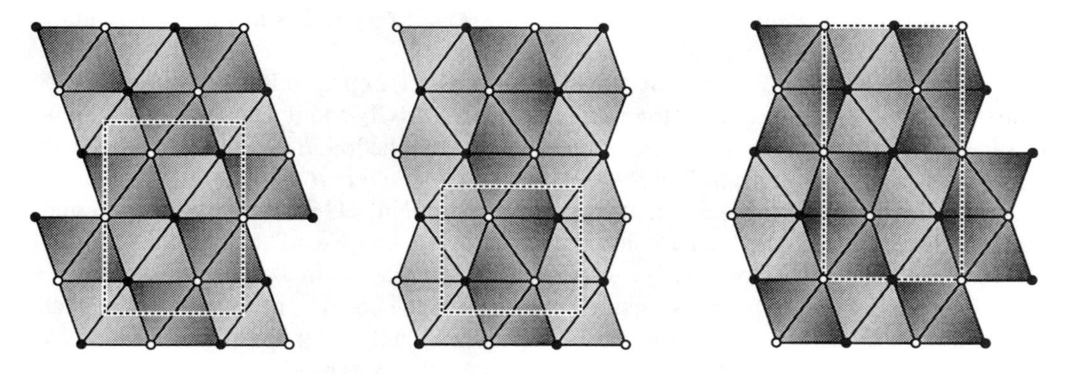

**Fig. 6.12.** From left to right **NaCl, NiAs** and **TiP** as cation-centered octahedra. Light and darker shaded polyhedra have elevations that differ by half the repeat distance in the direction out of the page.

In **TiP** there is *hc* packing of **P** and the {**Ti**}$\mathbf{P_6}$ octahedra again occur in two orientations. As now only every second layer is *h*, double layers of octahedra in one orientation alternate with double layers in the other orientation (see Fig. 6.12 again).

**NaCl** is the only one of these octahedral structures that is its own antistructure. As noted above, in **NiAs** the **Ni** are at the points of a primitive hexagonal lattice and the **As** are in {**As**}$\mathbf{Ni_6}$ trigonal prisms. In **TiP** the **P** are half in {**P**)$\mathbf{Ti_6}$ octahedra and half in {**P**)$\mathbf{Ti_6}$ trigonal prisms as illustrated in Fig. 6.13. Notice that crystallographically the structure is ternary, $\mathbf{Ti_2P(1)P(2)}$, and that there are antistructure compounds such as $RbScO_2$ with {Rb}$O_6$ trigonal prisms and {Sc}$O_6$ octahedra. The anion packing (now not **cp**) in such structures is discussed in § 6.4.1.

There are many patterns of partly filling octahedral sites in **cp** structures and we mention here just one or two of the simpler.

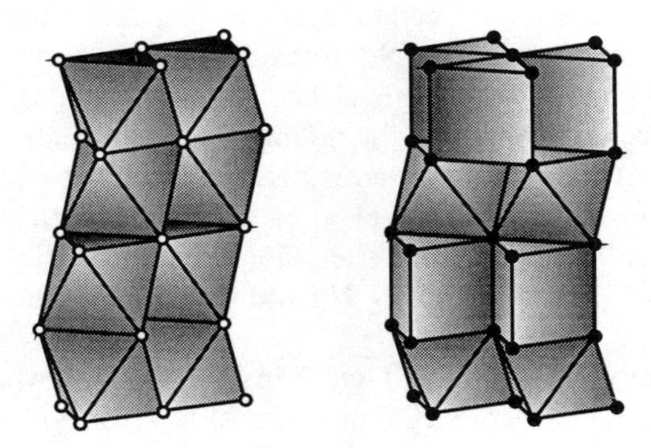

**Fig. 6.13**. Clinographic projections of **TiP** with **c** (normal to the **cp** layers of **P**) vertical. Left: as {**Ti**}$P_6$ octahedra. Right: as {**P**}$Ti_6$ octahedra and trigonal prisms.

Filling only alternate layers of octahedral sites in **hcp** gives the sequence $A{\cdot}\gamma{\cdot}B{\cdots}A\ldots$ The structure is generally known as **$CdI_2$**. There is a repeat every two **cp** layers and the symmetry is trigonal, $P\bar{3}m1$, with $c/a$ ideally = $\sqrt{(8/3)}$ = 1.633. **Cd** is in 1 $a$: (0,0,0) and **I** is in 2 $d$: ±(1/3,2/3,$z$) with $z$ ideally = 1/4.

Filling only alternate layers of octahedral sites in **ccp** produces the sequence $A{\cdot}\gamma{\cdot}B{\cdots}C{\cdot}\beta{\cdot}A{\cdots}B{\cdot}\alpha{\cdot}C{\cdots}A\ldots$. The structure is known as **$CdCl_2$** and it can be seen that now the repeat is every sixth **cp** layer. The symmetry is rhombohedral, $R\bar{3}m$, with $c/a$ ideally = $3\sqrt{(8/3)} = \sqrt{24} = 4.90$. **Cd** is in 3 $a$: $R$ + (0,0,0) and **Cl** is in 6 $c$: $R$ ± (0,0,$z$) with $z$ ideally = 1/4. A related structure type is that usually called **$\alpha$-$NaFeO_2$** in which **O** is **ccp** and alternate layers of octahedra sites are filled with **Na** and **Fe**. The crystallographic description is the same as for **$CdCl_2$** except there is also a cation in 3 $c$: $R$ + (0,0,1/2).

There is a number of different structures known in which one half of the octahedral sites are filled in every **cp** layer. An example with (approximately) **hcp** anions is provided by **$CaCl_2$** and the closely-related **rutile** ($TiO_2$) structure (see Exercise 7).

In considering possible patterns of filling interstitial sites in **cp** arrays, perhaps the most important consideration is that in "ionic" crystals short distances between cations (or anions) are generally unfavorable. This means that face sharing between coordination polyhedra (especially tetrahedra) is avoided if possible. A good example of this principle at work is provided by structures in which there are slabs in which all octahedral sites are filled and slabs in which all tetrahedral sites are filled. To avoid face sharing between polyhedra the **cp** layers between like slabs must be $c$, and between unlike slabs they must be $h$.[1] Thus in **$CaF_2$** (all tetrahedral slabs) and **NaCl** (all octahedral slabs) every layer is $c$. In **$La_2O_3$** with **cp La**, slabs (**$LaO_2$**) with **O** in all the tetrahedral sites alternate with slabs (**LaO**) with **O** in all the octahedral sites and the **La** layers are all $h$. In $Th_3N_4$ with **cp** Th (see Exercise 6) double slabs (2ThN) with N in all octahedral sites have Th in a $c$ layer at the center and these double slabs are separated by $h$ layers of Th from slabs with N in

[1]The reader will find that such statements are very readily verified if a few polyhedra are at hand.

tetrahedral sites ($ThN_2$). There are exceptions to these rules of course; in **NiAs** there is face sharing between octahedral slabs, and it is argued that, in some instances at least, this is due to metal-metal bonding (between **Ni** atoms across the shared face).

### *6.1.6 Stacking incomplete* **cp** *layers (honeycomb and kagome)*

The notation of 6.1.4 for stacking complete $3^6$ layers of atoms is in wide-spread use. It is useful to extend it to more complicated packings of hexagonal layers derived from $3^6$. There is no generally accepted notation—we use one that seems to us useful and is kept as simple as possible.[1]

Consider first the honeycomb pattern $6^3$. The unit cell for this is $\sqrt{3} \times \sqrt{3} = 3$ times as large as that for $3^6$, as illustrated in Fig. 6.14. The position of a honeycomb layer (symbol $G$ for graphite) is conveniently specified by the position of the center of a hexagon in the pattern. In stacking honeycomb layers, we recognize three *relative* positions (1, 2 and 3) indicated by small filled circles in the figure (upper left) at $0,0,z$ ; $1/3,2/3,z$ and $2/3,1/3,z$ in the 3× cell. This is particularly useful for describing partial filling of octahedral sites in **hcp** structures as illustrated next (note that it is less readily adaptible to **ccp**).

**Fig. 6.14.** Left: top a $\sqrt{3} \times \sqrt{3}$ cell of $3^6$ with below a $3^6$ net subdivided into a $G$ ($6^3$) layer (shaded circles) and a $g$ layer (open circles). Right: top a $2 \times 2$ cell of $3^6$ with below a $3^6$ net subdivided into an $N$ (3.6.3.6) layer (shaded circles) and an $n$ layer (open circles). The small filled circles mark the three choices of origin (1, 2 and 3).

In the (idealized) **corundum** structure of $Al_2O_3$, Al atoms are located in 2/3 of the octahedral sites of an **hcp** array of O, each Al layer being a honeycomb. We can code this structure as follows: $A{\cdot}G_1{\cdot}B{\cdot}G_2{\cdot}A{\cdot}G_3{\cdot}B{\cdot}G_1{\cdot}A{\cdot}G_2{\cdot}B{\cdot}G_3...$ By using dots as place markers we have automatically imparted the information that the Al atoms on the $G$ layers are in

[1]For another notation see particularly W. B. Pearson (Book List). Pearson describes over 100 structure types using a related system.

octahedral coordination. Note that the sequence requires six O layers to repeat. For ideal **hcp** of O atoms a distance $d$ apart, the $a$ axis would be $\surd 3d$ (see Fig. 6.14) and the $c$ axis would be $6\surd(2/3)d$ so that $c/a = \surd 8 = 2.83$ ; in the real structure $c/a = 2.73$.

Figure 6.15 shows the structure as $\{Al\}O_6$ octahedra. Notice that pairs of octahedra share faces; as also shown in the figure the Al atoms move away from the centers of the octahedra to avoid short Al...Al distances. $Al_2O_3$ with this structure is also known as $\alpha$-$Al_2O_3$. For crystallographic data see Appendix 5.

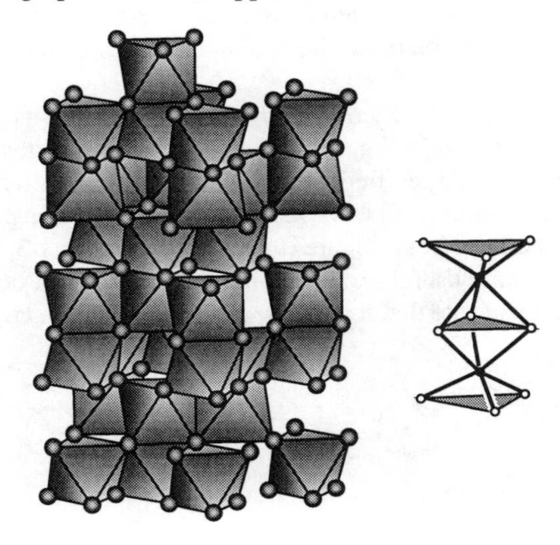

**Fig. 6.15**. The structure of $\alpha$-$Al_2O_3$ as articulated $\{Al\}O_6$ octahedra. **c** runs up the page. On the right the Al-O bonds are shown (on a larger scale) in a pair of face-sharing octahedra.

The centers of the hexagons of $6^3$ fall on a $3^6$ net with $\surd 3$ times the spacing of a **cp** layer (see Fig. 6.14 again) and we label such nets $g$ (note that combining $g$ and $G$ in one layer returns a **cp** layer). In the structure of $PdF_3$, the Pd atoms fill one-third of the octahedral sites of an **hcp** F array. The structure is $A{\cdot}g_1{\cdot}B{\cdot}g_2{\cdot}A{\cdot}g_3{\cdot}B{\cdot}g_1{\cdot}A{\cdot}g_2{\cdot}B{\cdot}g_3$... (see Fig. 6.28, p. 236). Note that the hexagonal cell is the same as for **corundum**; for $PdF_3$, $c/a = 2.82$.

Often layers in structures are kagome (3.6.3.6) nets which are symbolized $N$ (for net). The unit cell is now $2 \times 2 = 4$ times as large as that for $3^6$, as illustrated in Fig. 6.14, and again three relative positions (1, 2 and 3) of the net specified by the centers of the hexagons are recognized. These are now particularly useful for describing partial filling of octahedral sites in **ccp** structures. The centers of the hexagons fall on a $3^6$ net of twice the spacing of a **cp** layer and we label such $3^6$ nets $n$ (combining $n$ with $N$ produces a **cp** layer).

The three positions of the $N$ and $n$ nets are shown in Fig. 6.16. Some intermetallic compounds are variants of **cp** ordered in this way. In **$Cu_3Au$** the **Cu** atoms are on the kagome nets ($N$) and **Au** centers the hexagons of each layer ($n$). If atoms on the same layer are enclosed in parentheses, the code becomes: $(n_1N_1)\cdots(n_2N_2)\cdots(n_3N_3)$... or (omitting place markers) $(n_1N_1)$ $(n_2N_2)$ $(n_3N_3)$, i.e. a superstructure of $ABC$. We meet the related structure $N_1N_2N_3$ below as an 8-coordinated sphere packing (lattice complex $J$, § 6.3.5).

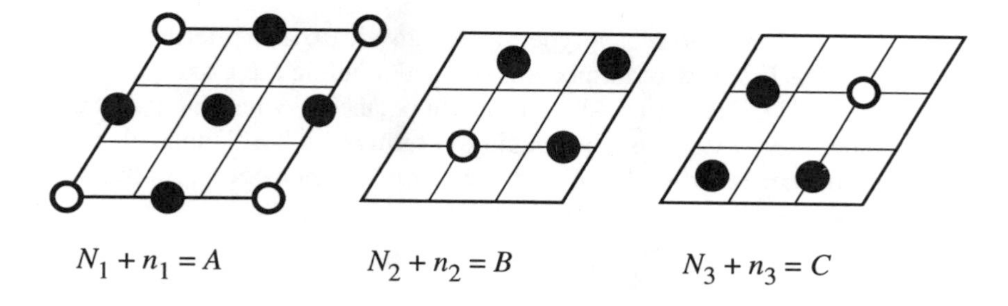


**Fig. 6.16**. The three positions of kagome ($N$) nets with respect to a 2 × 2 cell of $3^6$. We have (arbitrarily) labeled them so that $N_1$ combined with $n_1$ produce a **cp** layer with symbol $A$ etc.

In the **spinel** structure typified by $MgAl_2O_4$, O is on a **ccp** array with Mg in 1/8 of the tetrahedral sites and Al on 1/2 of the octahedral sites. The structure is cubic but for some purposes (such as describing related structures) is conveniently considered as stacking of layers normal to [111]: $N_1 \cdot Bn_1n_3n_2A \cdot N_2 \cdot Cn_2n_1n_3B \cdot N_3 \cdot An_3n_2n_1C\ldots$ Another notation which has been used refers to the $N$ layers as $O$ ("octahedral") layers, and layers of the type $n_1n_3n_2$ (which contain one octahedral and two tetrahedral sites) as $T_2$ layers.

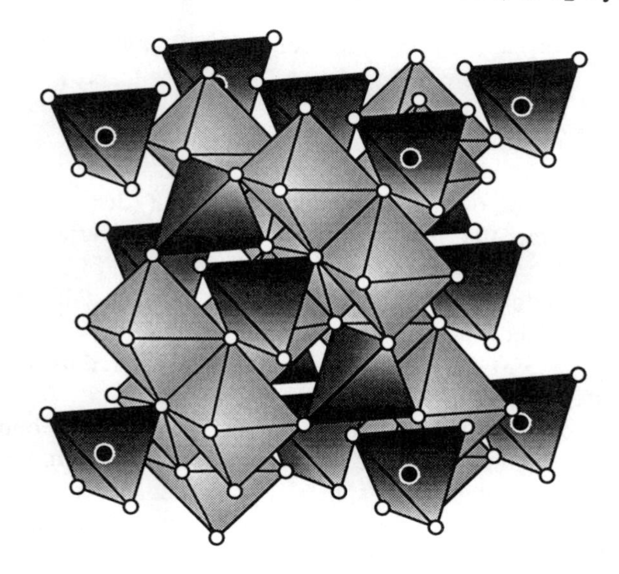

**Fig. 6.17**. $MgAl_2O_4$ as $\{Mg\}O_4$ tetrahedra and $\{Al\}O_6$ octahedra. Filled circles are Mg atoms at the corners of a cubic unit cell. For clarity some of the octahedra at the "back" of the unit cell have been omitted.

We cannot begin to do justice to **spinel** here.[1] Fig. 6.16 illustrates just one aspect of

[1]In Volume II of this series we devote more than a dozen pages to describing **spinel** and its close relatives. We find a notation based on that given here to be invaluable.

the structure—as cation centered polyhedra. As an aid to deciphering the code, it is expanded below. The first row of numbers is the height along the **c** axis of the hexagonal cell in multiples of $c/24$. Recall that Al atoms are in octahedral sites and that Mg atoms are in tetrahedral sites. As we describe the packing in terms of a $2 \times 2$ supercell of a **cp** layer there are 4 O atoms per layer $A$, $B$ or $C$ and 3 Al atoms corresponding to the $N$ layers. In the table we show these numbers as subscripts.

| 0 | 1 | 2 | 3 | 4 | 5 | 6 | 7 | 8 | 9 | 10 | 11 | 12 |
|---|---|---|---|---|---|---|---|---|---|---|---|---|
| $N_1$ | · | $B$ | $n_1$ | $n_3$ | $n_2$ | $A$ | · | $N_2$ | · | $C$ | $n_2$ | $n_1$ |
| $Al_3$ | · | $O_4$ | Mg | Al | Mg | $O_4$ | · | $Al_3$ | · | $O_4$ | Mg | Al |

| 12 | 13 | 14 | 15 | 16 | 17 | 18 | 19 | 20 | 21 | 22 | 23 | 24 |
|---|---|---|---|---|---|---|---|---|---|---|---|---|
| $n_1$ | $n_3$ | $B$ | · | $N_3$ | · | $A$ | $n_3$ | $n_2$ | $n_1$ | $C$ | · | $N_1$ |
| Al | Mg | $O_4$ | · | $Al_3$ | · | $O_4$ | Mg | Al | Mg | $O_4$ | · | $Al_3$ |

The Al array $N_1 \cdots n_3 \cdots N_2 \cdots n_1 \cdots N_3 \cdots n_2 \ldots$ also will be met again as a 6-coordinated sphere packing (lattice complex $T$, § 6.3.9).

Very many other compounds can be described in terms of partial filling of octahedral and/or tetrahedral sites of **cp** (Exercises 6 and 13 give examples) and it is virtually impossible to master systematic crystal chemistry without some appreciation of the principles involved.

### *6.1.7 The "size" of interstitial sites.*

For a regular tetrahedron of unit edge length the height (distance from a vertex to the center of an opposite face) is $\sqrt{(2/3)}$. The distance from a vertex to the center of the tetrahedron is 3/4 of the height (see Fig. 6.10) $= \sqrt{(3/8)} = 0.612$. Conversely, the edge length of an $\{A\}B_4$ coordination tetrahedron with unit length $A$-$B$ bonds is $\sqrt{(8/3)} = 1.633$. The radius of a sphere that exactly fits inside a tetrahedron of touching spheres of radius 1/2 (unit diameter) is $\sqrt{(3/8)} - 1/2 = 0.1124$. The ratio of the radius of the inner sphere to the radius of the outer spheres is $[\sqrt{(3/8)} - 1/2]/(1/2) = \sqrt{(3/2)} - 1 = 0.2247$.

It should be obvious that the perpendicular distance between opposite faces of an octahedron is the same as the height of a tetrahedron (see Fig. 6.10). For an octahedron of unit edge (formed by the centers of six spheres of unit diameter in contact) the distance to the center is $1/\sqrt{2}$. The ratio of the radius of the largest sphere that will fit inside an octahedral site to the radius of the close-packed spheres is $\sqrt{2} - 1 = 0.4142$.

It is sometimes stated that these "radius ratios" determine the coordination numbers of atoms in ionic crystals. The idea is that an atom (ion) that is too small for, say, an octahedral site (cation radius / anion radius less than 0.414) will instead go into a smaller (e.g. tetrahedral) site. Unfortunately to apply such considerations, radii must first be assigned to ions. Even when this is done, it will be found that the facts are not generally in accord with predictions even for the alkali halides (presumably the most "ionic" of crystals). Stated more bluntly the "rules" are generally not obeyed! The reader interested in this topic is referred also to § 6.8.3.

## 6.2 Body-centered cubic (bcc)

There are two interesting problems whose solutions are the same, and which lead to the body-centered cubic (**bcc**) array. The first problem is that of covering space completely with (partly overlapping) spheres of a given size such that their density is a minimum. The second concerns the filling of space with congruent tetrahedra. We have seen (Exercise 3, Chapter 5) that regular tetrahedra alone will not fill space, but a number of structures of metallic compounds are found in which space is divided into irregular tetrahedra.[1] The body-centered cubic array is the simplest such structure, and the only one in which all the tetrahedra have congruent faces and equivalent vertices. We refer to these tetrahedra as Sommerville tetrahedra.[2]

The Sommerville tetrahedron has faces that are isosceles triangles with one edge of length $a$ and two edges of $\sqrt{3}a/2$, where $a$ is the cubic unit cell edge for **bcc**. The angles of the triangles are $\cos^{-1}(1/3) = 70.53°$ and $\cos^{-1}(1/\sqrt{3}) = 54.74°$ (2×). The dihedral angles are 45° (4×) and 90° (2×). Fig. 6.18 shows how these tetrahedra are related to a body-centered cubic lattice. The figure also shows that four tetrahedra combine to form a space-filling octahedron with equivalent vertices and congruent faces so that the body-centered cubic array can be considered as arising from a packing of these (irregular) octahedra also. It is sometimes found stated (erroneously) that the body-centered cubic array divides space into tetrahedra *and* octahedra, but the octahedra are in fact clusters of four tetrahedra and the centers of the octahedra are the midpoints of the long edges of the tetrahedra so the term "octahedral site" is something of a misnomer (see below).[3] Contrast eutactic (**cp**) arrays in which space is divided into *separate* regions which are regular tetrahedra and regular octahedra.

The figure also shows that six octahedra (= 24 tetrahedra) of the **bcc** packing combine to form a rhombic dodecahedron with lattice points at the vertices *and* at the center.

Some facts (that will be useful later) about the body-centered array are included here. The symmetry is $Im\bar{3}m$ and the lattice points are in 2 $a$: $I$ + (0,0,0). The centers of the tetrahedra are in the faces of the cube at 12 $d$: $I$ + (±1/4,0,1/2)κ. The site symmetry at these points is $\bar{4}2m$ (tetragonal). Note that there are six tetrahedra for every lattice point. The tetrahedral sites in one unit cell are at the vertices of a truncated octahedron. We meet the pattern of tetrahedral sites as the "sodalite net" in the next chapter (§ 7.3.10). The tetrahedron around a tetrahedral site encloses all the space that is nearer to that tetrahedral site than to any other (it is the Voronoi polyhedron of the pattern of tetrahedral sites).

[1]These are sometimes referred to as "topologically close packed." The **β-W** ($A$15) structure (§ 6.6.4) is a well-known example.

[2]After D. M. Y. Sommerville who discussed this problem at length [*Proc. Edin. Math. Soc.* **41**, 49 (1923)]. Sommerville also discusses three other space-filling tetrahedra derived by dissecting the basic tetrahedron, but these are of little interest in the present connection.

[3]To a mathematician a *hole* in a lattice is a site where the distance to the nearest lattice point is a local maximum. Points where the distance to the nearest lattice point is a global maximum are known as *deep* holes, other holes are *shallow* holes. In **ccp** the octahedral sites are deep holes and the tetrahedral sites are shallow holes. In **bcc** there is only one kind of hole (the tetrahedral sites). The positions of the "octahedral" sites correspond, not to local maxima in distance from the nearest lattice point, but to saddle points.

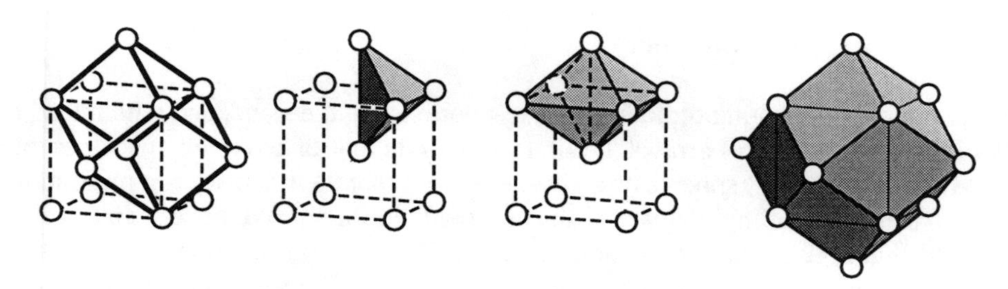

**Fig. 6.18**. Various aspects of the body-centered cubic lattice. At the left a cubic (centered) cell outlined with broken lines and a primitive rhombohedral cell ($\alpha$ = 109.47°) in full lines. Second from left: a Sommerville tetrahedron defined by four lattice points. Second from right: an octahedron composed of four Sommerville tetrahedra. Right: a rhombic dodecahedron composed of six octahedra = 24 tetrahedra.

The centers of the "octahedral sites" (better the midpoints of the long edges of the tetrahedra) are at 6 *b*: $I + (0,1/2,1/2)\kappa$. These correspond to the centers of the faces and edges of the unit cell. The site symmetry at 6 *b* is again tetragonal: 4/*mmm*. We meet the pattern of the octahedral sites as the NbO net in the next chapter.

There is an apparent paradox that has lead to confusion. There are six tetrahedral sites per **bcc** atom and three octahedral sites per **bcc** atom yet the octahedron is comprised of four tetrahedra. The resolution is to be found in the observation that the centers of long edges of a given octahedron are also octahedral sites. Thus if we placed (correctly oriented) octahedra with centers at *every* octahedral site we would cover space *twice*. In § 5.1.10 we called attention to a tetragonal tetrahedron with curved faces that filled space (Fig. 5.17). The six vertices of that polyhedron are arranged in space as the six vertices of the **bcc** octahedron, but because the surfaces are curved inward the volume of the tetragonal tetrahedron is only half as great as that of the **bcc** octahedron. In the space filling by these tetragonal tetrahedra, their centers are at the **bcc** octahedral sites.

The Voronoi polyhedron around an octahedral site is actually a polyhedron with 12 faces obtained by truncating four edges of the **bcc** octahedron. Three such polyhedra are shown in Fig. 6.19 in different orientations to suggest how they pack to fill space. The arrangement of the acute vertices in the packing is **bcc**.

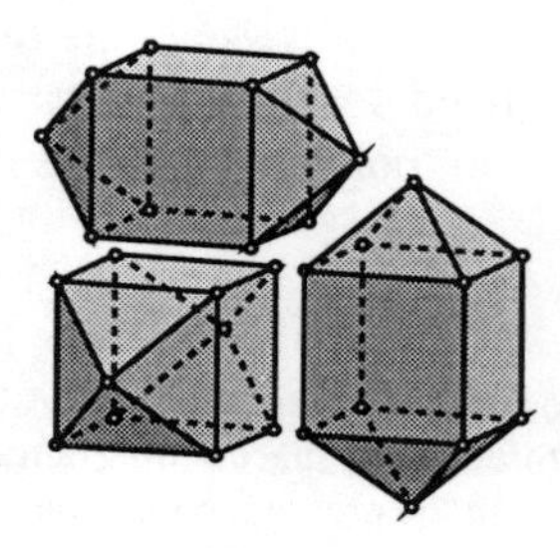

**Fig. 6.19**. The Voronoi polyhedron around "octahedral" sites in **bcc** shown in three different orientations to suggest how it fills space.

A nice example of filling all the tetrahedral sites in **bcc** is to be found in the structure of $M_6C_{60}$ ($M$ is an alkali atom such as K); in these compounds the centers of the approximately-spherical $C_{60}$ molecules are **bcc**.

Structures based on **bcc** and on **cp** occur with comparable frequency in intermetallic compounds. On the other hand, although "ionic" crystal structures can be described in terms of **cp** arrays with partial or complete filling of interstitial sites, ionic structures similarly based on **bcc** arrays are rather rare.[1] In the **cp** case the coordination polyhedra around the interstitial sites are regular whereas in the **bcc** case they are not.[2]

Note that we use **bcc** to refer to a geometrical arrangement, not to a symmetry (which would be written *bcc*). In $Cu_2O$ (Exercise 8, Chapter 3) the Cu atoms are **ccp** and the O atoms are **bcc** but the structure is primitive cubic (space group $Pn\bar{3}m$).

## 6.3 Sphere packings and relationships between them

Crystallographers have long been interested in the general problem of packings of spheres. *Stable* sphere packings are those in which each sphere is in contact with at least four others not all on the same hemisphere. *Equivalent* spheres are those related by symmetry operations (rotations, translations, etc.). We discuss here some interesting stable packings of equivalent spheres, referred to in this section just as "sphere packings" for brevity. Many packings with a given number of neighbors can be distorted smoothly to a higher density approaching arbitrarily close to that of closest packing so the interest is in finding low-density (rare) sphere packings which often correspond to a high-symmetry structures.[3] Although the problem is interesting, it is one of mathematics rather than chemistry. Our *real* interest is in describing structures (and their inter-relationships) that are of importance in crystal chemistry.

The two arrangements (**ccp** and **hcp**) we have identified of densest packing of *equivalent* spheres are the only such arrangements in which each sphere has twelve neighbors and, as they both have the same density, they are both the densest and the rarest packings of spheres with twelve neighbors.

Sphere packings are often characterized by the fraction of space occupied by the spheres (or density $\rho$). This is determined as the ratio of the volume of spheres in a unit cell to the volume of the unit cell. To illustrate: the **ccp** arrangement of spheres of unit diameter has a unit cell containing four spheres and cell edge $a = \sqrt{2}$. The volume of the spheres is $4 \times 4\pi r^3/3 = 2\pi/3$ and the cell volume is $a^3 = 2\sqrt{2}$. The ratio is $\rho = \pi/\sqrt{18} = 0.740\ldots$. We generally give coordinates for packings of unit diameter spheres, the density is then given

[1] If one considers *ordered* bcc arrays such as **CuZn** (§ 6.6.2) then one finds that **perovskite** $ABX_3$ and many derived structures are based on a **CuZn** array of cations $AB$ with anions approximately in the center of some of the octahedra sites. What is rare is the case of ionic crystals in which just one kind of atom is **bcc**.

[2] One can extend this remark by observing that less dense arrays, such as primitive cubic and primitive hexagonal, that do provide regular coordination sites are indeed more common than **bcc** (see § 6.4).

[3] As we remarked in Chapter 5, the problem is often stated (incorrectly) as that of finding the *densest* sphere packing with a given number of neighbors, *i.e.* the opposite of what we have stated.

by $\rho = z\pi/(6V)$, where $z$ is the number of spheres per cell and $V$ is the cell volume.

A topic of interest in connection with the description of structural relationships and possible transformation mechanisms is the description of paths between packings. We will focus on paths that preserve as much symmetry as possible (the space group of the intermediate structure will either be the same as, or a subgroup of, those of the two end structures) and which are completely specified by one free parameter. Often this parameter will correspond to the angle of rotation of a group of points. We have met such transformations already for finite objects such as that of a cuboctahedron to an icosahedron (§ 2.5.7) and a trigonal bipyramid to a square prism (§ 5.6.1). For plane patterns, important transformations are from 3.6.3.6 to $3^6$ and from $3^2$.4.3.4 to $4^4$ (§ 5.3.3).

### *6.3.1 11- coordination*

11-coordinated sphere packings are rather rare. Here we describe a simple and well known 11-coordinated structure that is important in crystal chemistry. A formal description for spheres of unit diameter is:

$P4_2/mnm$, $a = 1 + 1/\sqrt{2} = 1.707$, $c = 1$, $\rho = 0.7187$
Centers in 4 $f$: $\pm(x,x,0\ ;\ 1/2+x,1/2-x,1/2)$, $x = 1/(2 + \sqrt{2}) = 0.292$

The arrangement can be seen from Fig. 6.20 in which sphere centers are taken as defining the vertices of regular octahedra, and which also illustrates the relationship of the structure to **hcp**. In fact the structure is a special (minimum density) arrangement of an orthorhombic structure with space group *Pnnm* and centers of spheres in 4 $g$: $\pm(x,y,0;\ 1/2+x,1/2-y,1/2)$ with:

$$a = \sqrt{3}\cos\phi,\ b = \sqrt{(8/3)}\cos\phi + \sqrt{(1/3)}\sin\phi,\ c = 1,\ \rho = 2\pi/(3ab)$$
$$x = (\sqrt{2} - \tan\phi)/\sqrt{18},\ y = (1 + \sqrt{2}\tan\phi)/(4 + \sqrt{2}\tan\phi)$$

Eleven spheres are in contact for $0 < \phi < \sin^{-1}(1/3)$: the minimum density occurs at $\phi = (1/2)\sin^{-1}(1/3) = 9.74°$ at which point the structure is the tetragonal one given above. When $\phi = 0°$ or $\sin^{-1}(1/3) = 19.47°$ the structure is that of hexagonal eutaxy (**hcp**) with symmetry $P6_3/mmc$ (and 12-coordination). The tetragonal structure is close to the anion arrangement in the rutile form of $TiO_2$ (see Exercise 7) so we call it the **rutile** packing. The density is $4\pi/(9 + \sqrt{72}) = 0.7187$; no rarer 11-coordinated sphere packing seems to be known (for a second 11-coordinated sphere packing of the same density see § 6.4.1).

The transformation from **hcp** to the **rutile** packing and *vice versa* is accomplished by concerted rotations (by $\phi$) of columns of octahedra of atoms as suggested by Fig. 6.20 in which the rotation axis is normal to the plane of the figure. It is interesting that rotation of the columns of octahedra by 19.47° has the same effect as rotation of the whole pattern by 90°. (i.e. is a pseudorotation). It may be seen that *in projection* the transformation corresponds (approximately—not all edges are equal) to a transformation between $3^6$ and $3^2$.4.3.4.

We will encounter other such transformations involving concerted rotations of polyhedral groups subsequently (in particular the **rutile** packing can be transformed to a low-density 6-coordinated sphere packing by rotation of tetrahedral groups).

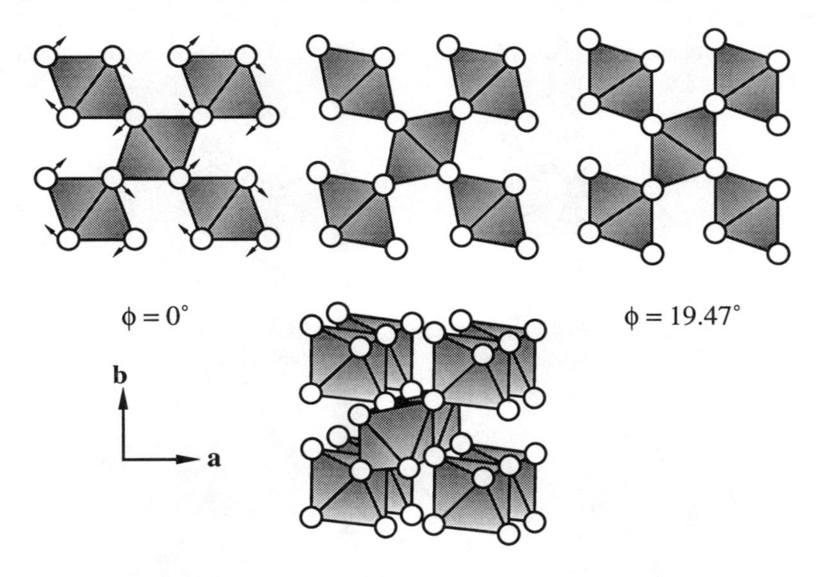


**Fig. 6.20**. Top: The transformation from **hcp** (left) to the minimum density tetragonal 11-coordinated sphere packing (center) and then to **hcp** rotated by 90° (right). The projection is along [001] of the orthorhombic cell. Bottom: A clinographic projection of the tetragonal structure.

### *6.3.2 10-coordination* (**bct**) *and a relationship between* **ccp** *and* **bcc**

A well-known 10-coordinated sphere packing is a lattice packing:

**bct** $I4/mmm$, $a = \sqrt{(3/2)} = 1.225$, $c = 1$, $\rho = 2\pi/9 = 0.698$. Centers in 2 $a$: $I$ + (0,0,0)

We refer to this structure as **bct** (short for body-centered tetragonal). Fig. 6.21 illustrates the 10-fold coordination. The primitive cell has $a' = b' = c' = 1$, $\alpha' = \beta' = \cos^{-1}(-1/4) = 104.48°$, $\gamma = \cos^{-1}(-1/2) = 120°$.

The reader may wish to verify that the points on (110) planes form regular $3^6$ nets as illustrated on the right in Fig. 6.21. A description of this structure as a non-close-packed stacking of $3^6$ nets is given in § 6.4.1 below.

Other special cases of the body-centered tetragonal lattice are of interest. If $a = c = 2/\sqrt{3} = 1.155$, the structure is the body-centered cubic (**bcc**) arrangement of unit spheres (symmetry $Im\bar{3}m$); and if $a = 1$, $c = \sqrt{2} = 1.414$, it is the face-centered cubic (**ccp**) arrangement (with symmetry $Fm\bar{3}m$) described in terms of a body-centered cell. It may be seen then that **bcc** and **ccp** are simply related to each other by a tetragonal compression or extension (Fig. 6.22). This relationship is known to metallurgists as the Bain relationship (or correspondence) and is of interest in connection with the transformation of iron from the $\gamma$ form (**ccp**) to the $\alpha$ form (**bcc**) on cooling. The high-temperature form of iron containing carbon is called austenite, on cooling it transforms rapidly with change of shape to a body-centered tetragonal (nearly cubic) form called martensite. Transformations of this type, which do not require diffusion of atoms, are called *martensitic* and have been extensively studied by metallurgists. [A. Martens (1850-1914) was a German metallurgist.]

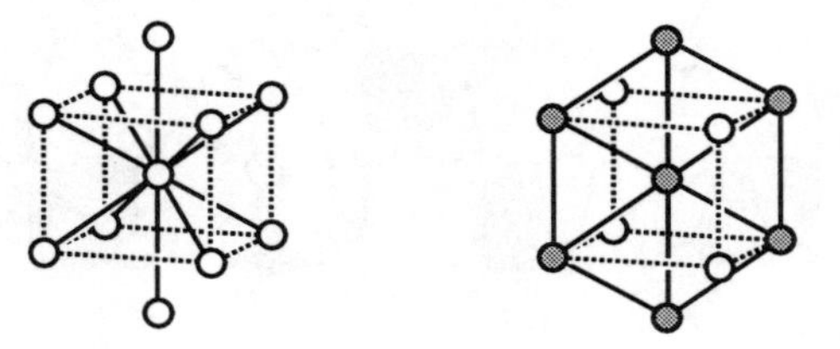

**Fig. 6.21**. Left: a fragment of the **bct** lattice showing 10-fold coordination of a central atom. Right: atoms on a $3^6$ net on a (110) plane shown shaded and connected by full lines.

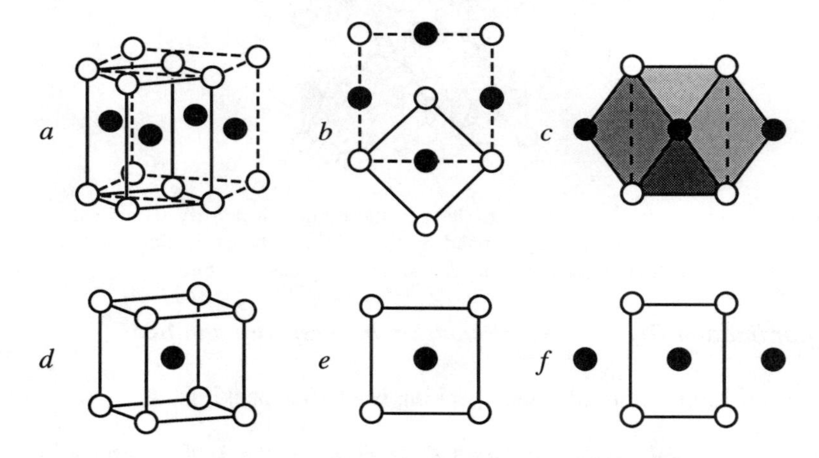


**Fig. 6.22**. (*a*) Face-centered cubic with the cubic cell outlined with broken lines and a body-centered tetragonal cell outlined with full lines. (*b*) *a* projected down the vertical axis. (*c*) *a* projected with the long axis of the tetragonal cell vertical on the paper. A cuboctahedron of atoms seen along a two-fold axis is depicted. (*d*) A body-centered cubic cell. Nearest neighbors are the same distance apart as they are in *a*. (*e*) *d* in projection. (*f*) The 10-coordinated tetragonal packing with its *c* axis *horizontal* on the paper. Its relationship to body-centered cubic can be seen by comparison with *e* and to face-centered cubic by comparison with *c*.

It should be obvious that the **bcc** structure can be tetragonally deformed until it is arbitrarily close to **ccp**, all the time keeping each sphere in contact with eight neighbors. The **bcc** structure is thus one at a local minimum (in coordinate space) of density.[1] It is less obvious that the 10-coordinate structure (**bct**) can be deformed to the **ccp** structure keeping

[1]We don't preclude the possibility of some of these structures being at global density minima for a given coordination number. The question has not been much discussed except for the case of four coordination which gives the rarest (?) stable sphere packing (§ 7.5.2).

each sphere in contact with ten neighbors and that it is at a local density minimum. However, this is simply demonstrated analytically and is similar to the 11- to 12-coordination transformation discussed earlier (§ 6.3.1, p. 228) in that again we need a lower symmetry cell for the general case. The symmetry is orthorhombic:

$$Immm,\ a^2 + b^2 = 3,\ 1 \le a,b \le \sqrt{2},\ c = 1.\ \text{Centers at } 2a\text{: } I + (0,0,0)$$

The special case $a = 1$, $b = \sqrt{2}$ (or $a = \sqrt{2}$, $b = 1$) corresponds to **ccp**. The case $a = b = \sqrt{(3/2)}$ corresponds to the tetragonal minimum density (**bct**).

The body-centered tetragonal arrangement is not common in elemental structures—examples are $\beta$-Hg ($c/a = 0.707$)[1] and Pa ($c/a = 0.825$)—but it is very often encountered in the structures of intermetallic compounds [e.g. **CuAu** and **$MoSi_2$** (§ 6.6.2)].

### 6.3.3 *Another 10-coordination: C-centered orthorhombic* (**cco**) *and further relationships between* **bcc**, **ccp** *and* **hcp**

Another 10-coordinated sphere packing with the same density as **bct** is derived by periodic twinning of **bct** on (101) planes. Fig. 6.23 shows, on the left, one such twin plane in **bct**. On the right in the figure is the new structure with the positions of twin planes indicated by arrows.

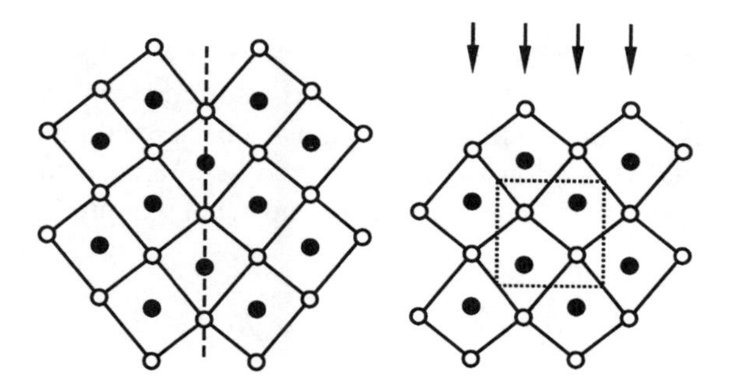

**Fig. 6.23**. Twinning of body-centered tetragonal (see text), on the right the broken lines indicate a unit cell (**b** vertical and **c** horizontal).

The crystallographic description of the new structure, which we call **cco** (for *C*-centered orthorhombic) is:

**cco** $Cmcm,\ a = \sqrt{(3/2)} = 1.225,\ b = \sqrt{(5/2)} = 1.581,\ c = \sqrt{(12/5)} = 1.549,\ \rho = 0.698$
Centers in 4 $c$: $C \pm (0,y,1/4)$, $y = 3/10$

The structure is a maximum volume form of a 10-coordinated sphere packing with the same symmetry and atoms in the same positions and with parameters given by:

[1]For mercury $c/a$ is sufficiently small that Hg atoms have only two nearest neighbors so that strings of Hg atoms run parallel to **c** (compare Fig. 6.16 *f*).

$$a = \sqrt{[(3 - 8y)/(1 - 2y)]},\ b = \sqrt{[1/(1 - 2y)]},\ c = \sqrt{(8y)},\ 1/4 < y < 1/3$$

When $y = 1/4$, the structure is **ccp**, with $a = b = c = \sqrt{2}$; when $y = 1/3$, the structure is **hcp**, with $a = 1$, $b = \sqrt{3}$, $c = \sqrt{(8/3)}$. For intermediate values of $y$ each sphere has 10 neighbors at a unit distance away. The position of maximum volume is for $y = 3/10$, when the unit cell volume is 3 (compare $\sqrt{8} = 2.828$ for **cp**). This is the same volume per sphere as in the 10-coordinated body-centered tetragonal lattice packing (**bct**) described in § 6.3.2.

Fig. 6.24 shows the structure again projected on (100), but now with **b** horizontal and **c** vertical. In going from **ccp** (on the left) to **hcp** (on the right), $a$ is decreased by $\sqrt{2}$, $b$ is increased by $\sqrt{(3/2)}$ and $c$ is increased by $\sqrt{(4/3)}$.

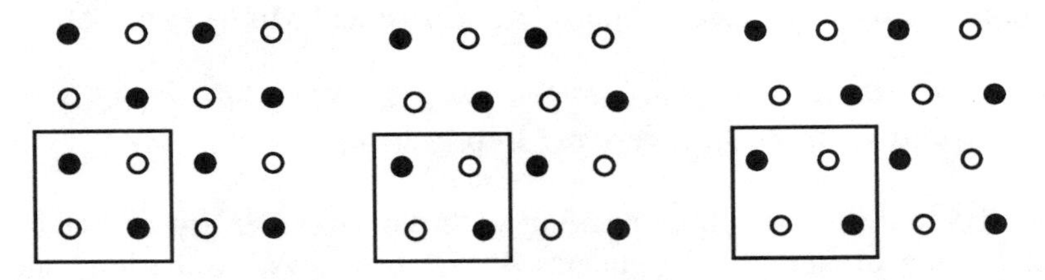

**Fig. 6.24**. Illustrating the relationship between **ccp** (left), $y = 1/4$; 10-coordinated packing (center) with $y = 3/10$; and **hcp** (right), $y = 1/3$.

If $a/b = a/c$ for **ccp** were reduced from 1 to $1/\sqrt{2} = 0.707$ the structure would be **bcc** (the Bain relationship again), so the figure would also represent the transformation from **bcc** to **hcp**. (So be careful in "reading" projections—check axial ratios.)

In general, periodic reflection twinning of body-centered orthorhombic cells (with parameters $a_o$, $b_o$ and $c_o$) on (011) in the manner indicated will produce the *Cmcm* structure with:

$$a = a_o,\ b = \sqrt{(b_o^2 + c_o^2)},\ c = 2b_oc_o/b,\ y = 1/2 - c_o^2/2b^2$$

For **bct**, $a_o = b_o = \sqrt{(3/2)}$, $c_o = 1$. Twinning this cell produces **cco** sphere packing.

For **ccp**, the "orthorhombic" (now actually tetragonal) cell with two lattice points has $a_o = c_o = 1$, $b_o = \sqrt{2}$. Twinning this cell produces **hcp** (the {101} planes of the tetragonal cell are {111} planes of the conventional cubic cell).

For **bcc**, $a_o = b_o = c_o = \sqrt{(4/3)}$. Twinning this cell produces **bcc** again which is not surprising as (101) planes of **bcc** are already mirror planes.

To summarize, here are the parameters for important sphere packings in terms of *Cmcm* with points in 4 *c*: $C \pm (0,y,1/4)$:

| structure | $a$ | $b$ | $c$ | $y$ |
|---|---|---|---|---|
| **ccp** | $\sqrt{2} = 1.414$ | $\sqrt{2} = 1.414$ | $\sqrt{2} = 1.414$ | 1/4 |
| **cco** | $\sqrt{(3/2)} = 1.225$ | $\sqrt{(5/2)} = 1.581$ | $\sqrt{(12/5)} = 1.549$ | 3/10 |
| **hcp** | 1 | $\sqrt{3} = 1.732$ | $\sqrt{(8/3)} = 1.633$ | 1/3 |
| **bcc** | $2/\sqrt{3} = 1.155$ | $\sqrt{(8/3)} = 1.633$ | $\sqrt{(8/3)} = 1.633$ | 1/4 |

### *6.3.4 8-coordination: packing of trigonal prisms*

By now the **bcc** arrangement should be familiar as an example of an 8-coordinated sphere packing.

Another simple arrangement is provided by the points of a primitive hexagonal lattice with $a = 1$, $c = 1$. This arrangement occurs as one of the high-pressure polymorphs of elemental Si ($a = 2.53$, $c = 2.37$ Å at 20 GPa). It also occurs in several compound structure types in which there is trigonal prismatic coordination of atoms. The vertices are those of a space filling by trigonal prisms of which there are two per lattice point. (Two prisms sharing a square face form a unit cell.) The density is $\rho = \pi/\sqrt{27} = 0.604$.

The relationship between these two packings (*cI* and *hP*) is very simple, and interesting in several contexts, although the general intermediate structure contains two kinds of sphere so it is not a packing of *equivalent* spheres. Fig. 6.25 compares a primitive hexagonal lattice, described using a super-cell with three points per cell, with **bcc** described using a similar cell.

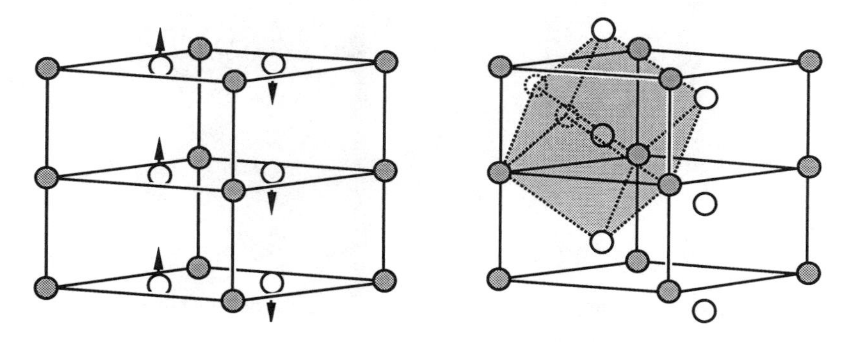

**Fig. 6.25.** The relationship between primitive hexagonal (= *hP*, left) and **bcc** (described using a hexagonal cell, right). The distance between neighboring lattice points is the same in each case. The conventional cubic cell for **bcc** is lightly shaded.

The intermediate structure is trigonal:

$P\bar{3}m1$, $a = \sqrt{[3(1 - z^2)]}$, $c = 1$
One sphere at 1 $a$: 0,0,0; two spheres at 2 $d$: $\pm(1/3,2/3,z)$

When $z = 0$ we have the primitive hexagonal structure (symmetry $P6/mmm$) and when $z = 1/3$ we have the **bcc** structure described with the hexagonal cell. Note that $c/a$ changes by only about 6% in the transformation. It should be noted also that in **bcc** we have $3^6$ layers in *A*, *B* and *C* positions separated by **c**/3. The primitive hexagonal structure has these three layers collapsed to one layer of three times the density.

Note that combining any *two* of *A*, *B* or *C* will give a honeycomb ($6^3$) layer. This occurs if $z = 1/2$; at which point the structure is formally that of $AlB_2$ (§ 5.3.5) with symmetry again $P6/mmm$ (the honeycomb layers are B). The symmetry-breaking transition from $z = 1/3$ (bcc) towards $z = 1/2$ is important in metallurgy and is known as the *bcc* $\rightarrow \omega$ transition.

There is a second way of filling space with trigonal prisms such that all vertices are equivalent (Fig. 6.26). Make a slab of face-sharing prisms with prism 3-fold axes all collinear. Now put such slabs together with the prism axes alternating. The prism vertices correspond to the centers of an 8-coordinated sphere packing:

$I4_1/amd$, $a = 1$, $c = \sqrt{12}$, $\rho = 0.6046$. Centers in 4 *a*: $I \pm (0,3/4,1/8)$

This is the Th arrangement of $ThSi_2$ (Si centers the trigonal prisms). It is also the structure of a high-pressure polymorph of Cs.

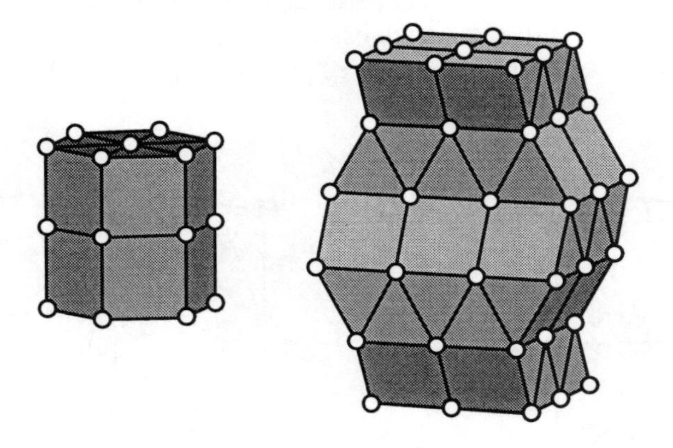

**Fig. 6.26**. Space fillings by trigonal prisms that correspond to 8-coordinated sphere packings. Left: primitive hexagonal. Right a tetragonal sphere packing.

### *6.3.5 Another 8-coordination: the J lattice complex*

Another 8-coordinated sphere packing is of very frequent occurrence in crystal structures. The arrangement is an example of an *invariant lattice complex*—an array of symmetry-related points on fixed positions. The more-common of these are sometimes described by symbols (see § 6.8.7) and the symbol for this one is *J* .[1]

***J*** $Pm\bar{3}m$, $a = \sqrt{2}$, $\rho = \pi/\sqrt{32} = 0.5554$. Centers in 3 *c*: $(0,1/2,1/2)\kappa$

It can be seen that this structure may be considered as derived from cubic eutaxy (**ccp**)

[1]The reason for this particular symbol is apparently an association with the "jack" of the common game of that name [W. Fischer *et al.*, *Space Groups and Lattice Complexes*, National Bureau of Standards Monograph 134 (1973)]. However the jacks familiar to us have six points and $m\bar{3}m$ symmetry whereas the points in the *J* complex have *4/mmm* symmetry and eight neighbors. Nevertheless we like the name because of the association with K. H. Jack who appears to have been the first to point out the important relationship of this array to **hcp** [K. H. Jack & V. Guttmann, *Acta Crystallogr.* **4**, 246 (1951)].

by removal of 1/4 of the spheres (those at 0,0,0) and thus it has 3/4 of the density of eutactic arrangements.

The $J$ complex occurs notably as the anion array in the cubic perovskite ($ABX_3$) structure (met earlier in § 5.3.4) which can be thought of as a three-dimensional array of corner-connected octahedra of anions $X$ centered by cations $B$. The $BX_3$ arrangement alone is known as **$ReO_3$**. The faces of the octahedra divide space into an equal number of regular octahedra and cuboctahedra (centered by $A$ in **perovskite**). The packing could therefore also be considered as arising from a space-filling packing of regular octahedra and cuboctahedra (Fig. 6.27). The $A$ atoms center the cuboctahedra, and we note, in passing, that the *combination* of $A$ and $X$ ($AX_3$) is **ccp** (see § 6.6.1).

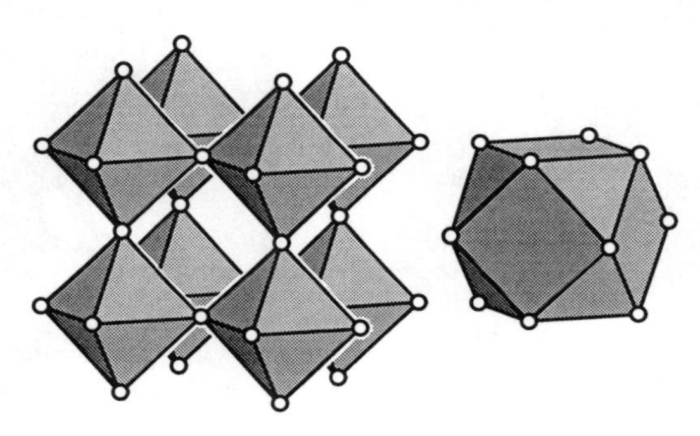

**Fig. 6.27**. Left: The $J$ lattice complex shown as an array of corner-connected regular octahedra. Right: The cuboctahedron "hole" in the packing.

In the cubic structure, the bond angle $\theta = \angle B\text{-}X\text{-}B = 180°$. If the octahedra are maintained rigid but tilted (thus "crumpling" the array and reducing the bond angle $\theta$) so that $\cos^{-1}(-2/3) = 131.81...° < \theta \leq 180°$ each vertex will still have eight nearest neighbors.[1]

If all the octahedra (which remain undistorted) are rotated by an angle $\pm\phi$ about a set parallel <111> axes the arrangement can be converted from the cubic $J$ structure ($\phi = 0$) to hexagonal eutaxy (**hcp**) ($\phi = \pm 30°$, symmetry $P6_3/mmc$). If the octahedra are filled the final $BX_3$ arrangement is **$PdF_3$** (§ 6.1.5). The intermediate and final structures have symmetry $R\bar{3}c$. Referred to a centered hexagonal cell and unit-diameter spheres:

$$a = \sqrt{8}\cos\phi,\ c = \sqrt{48}.\ \text{Centers in } 18\ e\text{: } R \pm (x,0,1/4;\ 0,x,1/4;\ \bar{x},\bar{x},1/4)],\ x = (\sqrt{3} - \tan\phi)/\sqrt{12}$$

It is interesting that the $J$ structure can be considered to be composed of intersecting kagome layers parallel to all {111}. The points at any one elevation in the hexagonal structure such as $z$ = 1/4 (with $x,y = x,0$ ; $0,x$ and $\bar{x},\bar{x}$) correspond to the points in a hexagonal cell relating the kagome $\rightarrow 3^6$ illustrated in Fig. 5.40 (see also Exercise 10 in Chapter 5). Hexagonal eutaxy corresponds to a stacking of the $3^6$ layers. In Fig. 6.28 try to see how collapsing the kagome nets to $3^6$ corresponds to rotations of the octahedra alternately clockwise and anti-clockwise about [111].

[1]For $\theta < \cos^{-1}(-2/3)$ inter-octahedral $X...X$ distances will be less than the intra-octahedral distances.

Using the symbols of § 6.1.6, the stacking of kagome layers along [111] is $N_1N_2N_3$.

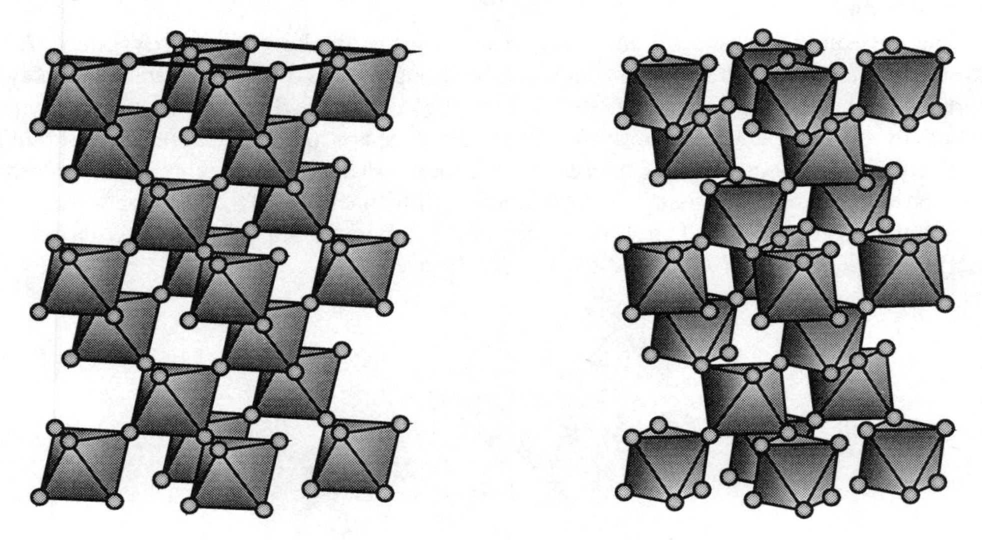

**Fig. 6.28**. Left: the octahedra of the *J* arrangement shown with [111] vertical on the page. Part of a kagome (3.6.3.6) net in a (111) plane is heavily outlined at the top. Right: the same set of octahedra after rotation as described in the text so that the vertices are **hcp**. The figure illustrates the transformation $\mathbf{ReO_3} \rightarrow \mathbf{PdF_3}$.

### 6.3.6 *Another 8-coordination: the* **pyrochlore** *packing*

Another interesting 8-coordinated arrangement can also be considered as arising from a different array of corner-connected octahedra. We call it the pyrochlore packing as it is an idealization of the octahedral framework in compounds with the pyrochlore structure[1]:

**pyrochlore** $Fd\bar{3}m$, $a = 16/\sqrt{18}$, $\rho = 27\sqrt{2}\pi/256 = 0.468$
centers in 48 *f*: $F \pm (x,1/8,1/8\ ;\ 1/4{-}x,1/8,1/8\ ;\ \bar{x},3/8,3/8\ ;\ 3/4{+}x,3/8,3/8)\kappa$, $x = 7/16$

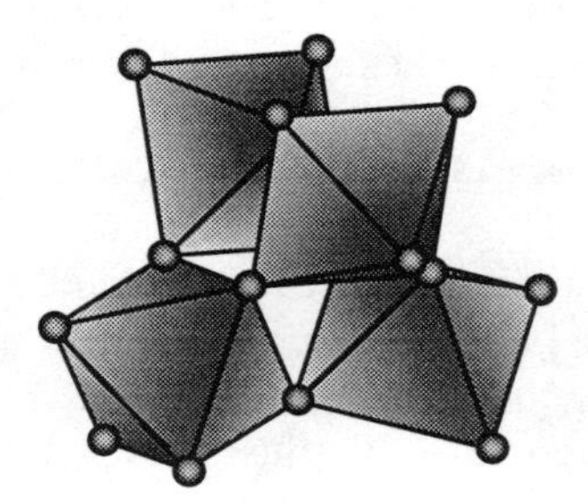

**Fig 6.29***a*. A "pyrochlore unit" of four octahedra sharing vertices. The space at the center is an "empty" octahedron.

[1]Pyrochlore is a mineral of variable composition: $(Ca,Na)_2(Nb,Ta)_2O_6(O,OH,F)$.

The structure may be thought of as composed of "pyrochlore units" of corner-sharing octahedra such as illustrated in Fig. 6.29*a*. Further corner sharing between units produces the structure shown in Fig. 6.29*b*.

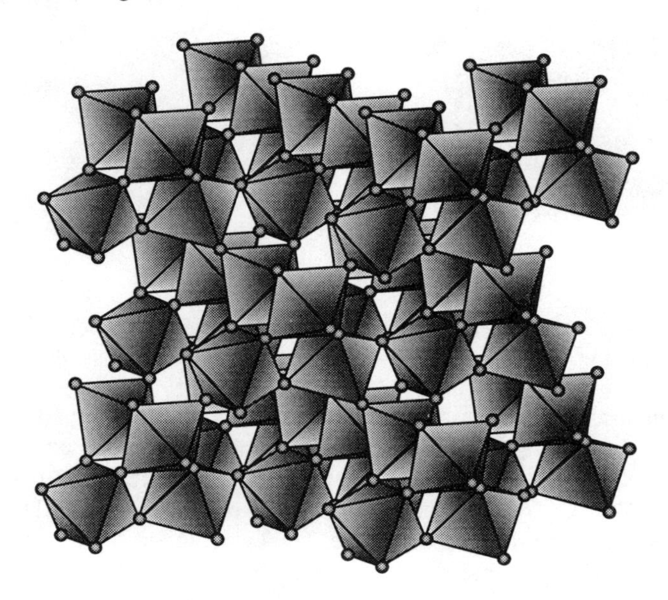

**Fig. 6.29***b*. The pyrochlore packing as corner-connected octahedra. The drawing consists of a face-centered cubic array of the "pyrochlore units" of Fig. 6.29*a*.

The same arrangement occurs in intermetallic compounds: for example as the W arrangement in $Fe_3W_3C$ (with C in the octahedra).

### *6.3.7 Another 8-coordination: the S lattice complex*

The final 8-coordinated sphere packing that we mention is another invariant cubic lattice complex (symbol *S*). A formal description for unit diameter spheres is:

*S* $\quad I\bar{4}3d$, $a = 8/\sqrt{14} = 2.138$ , $\rho = 2\pi/a^3 = 0.643$
Sphere centers in 12 *a*: $I$ + (3/8,0,1/4 ; 1/8,0,3/4)κ

The points of this arrangement divide space up into irregular octahedra (actually metaprisms) and irregular tetrahedra. The metaprisms form non-intersecting rods along <111> by sharing triangular faces. In Fig. 6.30 these rods are shown in a projection down [111]. The packing of rods in four different directions is the same as in the **garnet** cylinder packing to be described in § 6.7.3. Each metaprism also shares a face with a metaprism in each of three rods not parallel to it, so each metaprism shares a face with five adjacent metaprisms. A high-pressure form of elemental Ga has this structure. It also corresponds to the Th positions in the important $Th_3P_4$ structure type.

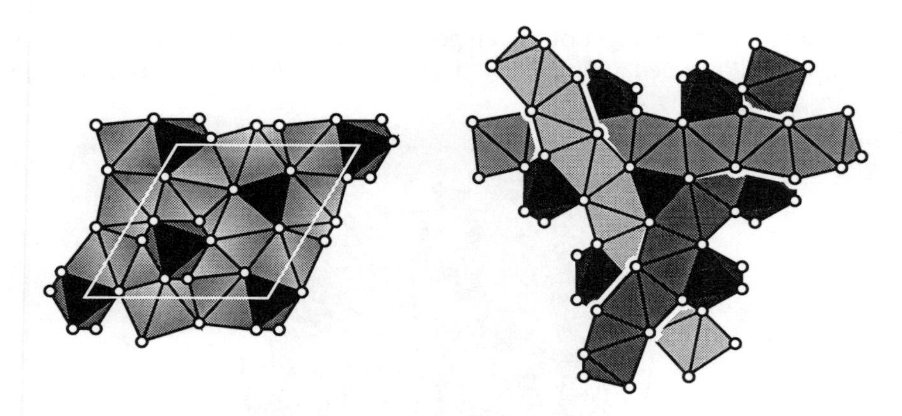

**Fig 6.30**. Left: The *S* structure projected down [111]. Dark shading indicates [111] rods of face-sharing metaprisms. Right: parts of three rods parallel to the other three <$\bar{1}$11> directions are emphasized.

There are four-thirds of an octahedron (metaprism) and one tetrahedron for each sphere in the *S* packing. In $\mathbf{Th_3P_4}$, **P** atoms center the metaprisms. The centroids (equidistant from all six vertices) of the metaprisms are in 16 *c*: $I$ + ($x,x,x$ ; $1/4+x,1/4+x,1/4+x$ ; $1/2-x,\bar{x},1/2+x$ ; $1/4-x,3/4-x,3/4+x$)κ, $x$ = 1/12. The real structure of $Th_3P_4$ (for data see Appendix 5) has $x$ close to this "ideal" value. The centroids of the tetrahedra are in 12 *b*: $I$ + (7/8,0,1/4 ; 5/8,0,3/4)κ. Note that the tetrahedral sites (12 *b*) also form an *S* lattice complex displaced from the 12 *a* positions by 1/2,0,0. Filling 12 *a* by *A* and 12 *b* by *B* would give a (hypothetical) 4-coordinated structure *AB* that is its own antistructure. The tetrahedral sites are far from regular; in the *AB* structure just described the *ABA* angles are 99.6° (4×) and 131.8 ° (2×). Taken together the *A* and *B* positions form a 4-connected net known as $S^*$ which is described in § 7.3.12.

There is a sense in which the *S* structure is intermediate between the primitive hexagonal and **cp** arrays. Thus consider the division of space into tetrahedra and/or 6-coordinated figures (prisms, metaprisms or antiprisms): (Here $N_6$ and $N_4$ are the numbers of six- and 4-coordinated sites per packing atom.)

| *structure* | $N_6$ | $N_4$ |
|---|---|---|
| primitive hexagonal | 2 | 0 |
| *S* ($Th_3P_4$) | 4/3 | 1 |
| eutactic (**cp**) | 1 | 2 |
| **bcc** | 0 | 6 |

### *6.3.8 7-coordination and a relation between* **FeSi** *and* **NaCl**

7-coordinated sphere packings can be obtained in a rather obvious way by prismatic stacking of 5-coordinated layers. We describe such a stacking of $3^3.4^2$ nets in § 6.4.2 and a stacking of $3^2.4.3.4$ nets in § 6.4.3.

Another, less obvious, 7-coordinated packing, again for spheres of unit diameter is:

$Pa\bar{3}$, $a = 4/(\sqrt{15} - \sqrt{3}) = 1.868$, $\rho = 0.6802$
Sphere centers in 8 $c$: $\pm(x,x,x\ ;\ (1/2+x,1/2-x,\bar{x})\kappa)$, $x = (\sqrt{5} - 1)/8 = 0.155$

Note that for $x = 1/4$ we have a primitive cubic structure with a doubled cell edge (8 lattice points). The transformation from the primitive cubic structure to the $Pa\bar{3}$ structure (and *vice versa*) involves displacements along all four <111> directions and is rather difficult to illustrate; the neighbors of the points in one unit cell are shown in the top part of Fig. 6.31.

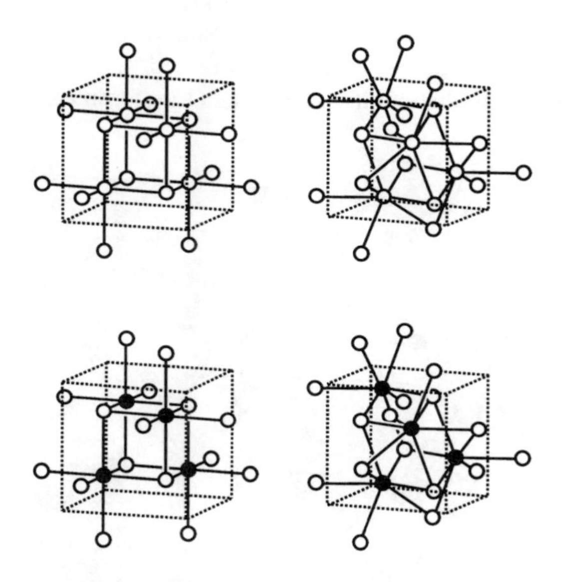

**Fig. 6.31**. Top: the $Pa\bar{3}$ 7-coordinated structure (right) compared with the primitive cubic structure (left). Bottom: alternately coloring the points in these structures produces **NaCl** (left) and **FeSi** (right).

We have seen that the anions in **fluorite** ($CaF_2$) are in a primitive cubic array. In $PdF_2$ recovered from high pressure (under which conditions it probably is **fluorite**) the F array is the $Pa\bar{3}$ structure. For crystallographic data for $PdF_2$ see Appendix 5.[1]

If the eight points in 8 $c$ of $Pa\bar{3}$ are alternately colored black and white in such a way that (e.g.) those at $x,y,z$ are black and those at $1-x,1-y,1-z$ are white, the symmetry is lowered to $P2_13$ and the two sets of points are at 4 $a$: $x,x,x\ ;\ (1/2+x,1/2-x,\bar{x})\kappa$. this corresponds to the situation in idealized **FeSi** with $x_{Fe} = 1-x_{Si}$. (In the real structure of FeSi, $x_{Fe} = 0.136$, $x_{Si} = 0.844$—see Appendix 5). Similarly coloring the points of the primitive cubic array produces **NaCl**. We have therefore derived a simple relationship between these last two structures as also shown in Fig. 6.31.

[1]The reported symmetry of this form of $PdF_2$ is actually $P2_13$, but the parameters are very close to what they would be for $Pa\bar{3}$ symmetry [A. Tressaud & G. Demazeau, *High Temp. High Press.* **16**, 303 (1984)] and these are given in Appendix 5.

### *6.3.9 6-coordination: the T lattice complex and* **cristobalite**

Structures with lower coordination numbers are better described as nets [systems of points (atoms) connected by edges (bonds)]. However we will consider here a few 6-coordinated sphere packings. The first is with spheres at the points of a simple cubic lattice. Note that a cube is a special case of a rhombohedron (with $\alpha = 90°$). It should be clear that a cube of unit edge can be smoothly distorted to a rhombohedron with $\alpha = 60°$ (a primitive cell of *fcc*) so we have a simple way of transforming from simple cubic to face-centered cubic. The intermediate symmetry is $R\bar{3}m$ and as long as $\alpha > 60°$ there are only six nearest neighbors. Therefore, this 6-coordinated sphere packing can approach arbitrarily close in density to that of closest packing (12-coordinated).

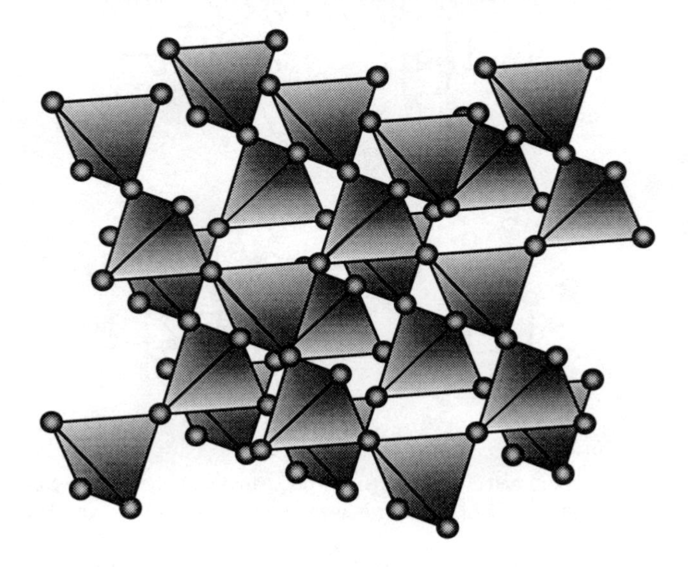

**Fig. 6.32**. The *T* structure shown as corner-connected tetrahedra.

A second 6-coordinated packing we discuss here has centers of spheres at the points of another invariant cubic lattice complex (symbol *T*).

***T*** $Fd\bar{3}m$, $a = \sqrt{8}$, $\rho = 0.370$. Sphere centers in 16 *c*: $F$ + (0,0,0; (0,1/4,1/4)κ)

This is another pattern that occurs in a wide variety of crystal structures. It is the oxygen positions of an *idealized* high-cristobalite ($\beta$-cristobalite) form of $SiO_2$, so it is sometimes (a little misleadingly) called the cristobalite arrangement. Other notable occurrences are as the sites of the octahedrally-coordinated atoms in the spinel structure and the pyrochlore structure, and the Cu atoms in the $MgCu_2$ structure (§ 6.6.3). The structure can be considered as made up of corner-connected tetrahedra as shown in Fig. 6.32. Examination of the figure will show that the points fall on kagome (3.6.3.6) nets parallel to {111} as they do in the *J* structure. However now the kagome nets normal to any one <111>

direction alternate with $3^6$ nets with twice the spacing between points. The empty space (not in the tetrahedra) in the structure is made up of a packing of truncated tetrahedra, accordingly the structure can be considered as arising from a packing of regular tetrahedra and truncated tetrahedra.[1]

Using the symbols of § 6.1.6, the stacking of atom layers normal to [111] in the $T$ structure is $N_1n_3N_2n_1N_3n_2$...—the same as that of the Al atoms in the spinel structure.

The $T$ structure can be collapsed by concerted rotations of the tetrahedra about $\bar{4}$ axes to produce denser arrangements (in much the same way as the $J$ structure is converted to **hcp** by rotations of octahedra) as illustrated in Fig. 6.33. Let $B$ represent the centers of the tetrahedra and $X$ the vertices. Analytical descriptions of these transformations are, in terms of $u = \cos\phi$, $s = \sin\phi$ and $t = \tan\phi$ ($\phi$ is a rotation angle discussed below):

(i) $T \rightarrow$ cubic eutaxy:

$I\bar{4}2d$, $a = 2u$, $c = \sqrt{8}$
$X$ in 8 $d$: $I$ + ($x$,1/4,1/8 ; $\bar{x}$,3/8,1/8 ; 3/4,$x$,7/8 ; 1/4,$\bar{x}$,7/8), $x = t/4$
$B$ in 4 $a$: $I$ + (0,0,0 ; 1/2,0,3/4)

When $\phi = 0°$ the structure is the cubic $T$ pattern described with a body-centered tetragonal cell. When $\phi = 45°$ it is **ccp** described with a doubled cell ($c = 2a$). Note that the density increases by a factor of two in going from $T$ to **ccp**.

If atoms $A$ are in positions 4 $b$: $I$ + (0,0,1/2 ; 1/2,0,1/4) we have stoichiometry $ABX_2$. If $\phi = 45°$, both $A$ and $B$ are tetrahedrally coordinated by $X$ and we have a superstructure of **sphalerite** (ZnS); an example is chalcopyrite, $CuFeS_2$. If $A$ is larger than $B$, then $\phi$ [$= \tan^{-1}(4x)$] is smaller, allowing the $\{A\}X_4$ tetrahedron to expand, thus for $CdSiP_2$, $x$ = 0.21 ($\phi = 40°$).

(ii) $T \rightarrow$ 11-coordinated.

$P4_12_12$, $a = 1 + u$, $c = \sqrt{8}u$
$X$ in 8 $b$ : ($x$,$y$,$z$ ; $y$,$x$,$\bar{z}$ ; $\bar{x}$,$\bar{y}$,1/2+$z$ ; $\bar{y}$,$\bar{x}$,1/2−$z$ ;1/2−$x$,1/2+$y$,1/4−$z$ ; 1/2−$y$,1/2+$x$,1/4−$z$ ; 1/2+$x$,1/2−$y$,3/4−$z$ ; 1/2+$y$,1/2−$x$,3/4−$z$),
$x = u/(2 + 2u)$, $y = s/(2 + 2u)$, $z = (1 + t)/8$
$B$ in 4 $a$: ($x$,$x$,0 ; $\bar{x}$,$\bar{x}$,1/2 ; 1/2−$x$,1/2+$x$,1/4 ; 1/2+$x$,1/2−$x$,3/4),
$x = (1 + u + s)/(4 + 4u)$

This $P4_12_12$ structure is actually an idealization (regular tetrahedra) of the anion positions in the low-cristobalite form of $SiO_2$. $\phi = 0°$ corresponds to $T$ and $\phi = 45°$ corresponds to the 11-coordinated packing described in § 6.3.1. Just as there are compounds $ABX_2$ with $I\bar{4}2d$ symmetry, there are also ternary compounds with $P4_12_12$ symmetry. Examples include $LiAlO_2$ and $NaAlO_2$ with different values of $\phi$. If $A = B$ we have the structure of $\beta$-BeO, which we meet again in § 7.3.3. Cristobalite is intermediate,

[1]This structure is just one of an infinite number of ways of packing these polyhedra (see the discussion of the $MgCu_2$ and the $MgZn_2$ structures in § 6.6.3).

with $\phi \approx 20°$ at room temperature.[1] Note that, like the quartz form of $SiO_2$, cristobalite exists in enantiomorphic forms with symmetry either $P4_12_12$ or $P4_32_12$.

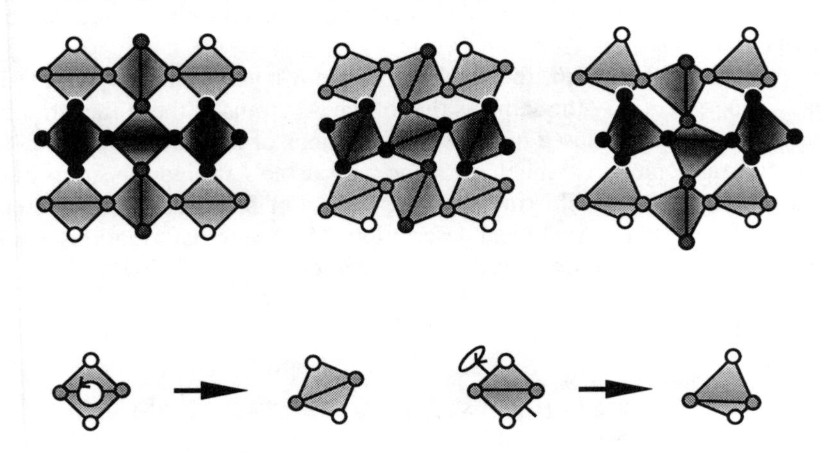

**Fig. 6.33**. Partial "collapse" of the ideal cristobalite (top left) to the $I\bar{4}2d$ structure (top middle) and to the $P4_12_12$ structure (top right). The structures are projected down **c** and the centers of the tetrahedra are at heights differing by $c/4$. $\phi = 22.5°$ in the two derived structures. The lower part of the figure shows how a tetrahedron appears in projection after rotation about a $\bar{4}$ axis.

There is a third way (with symmetry $Pna2_1$) to collapse the *T* structure by concerted rotations of tetrahedra, this time to give **hcp**, but in the intermediate arrangement the structure has two crystallographically-distinct spheres.[2]

### *6.3.10 Another 6-coordination: the Y lattice complex*

Our last sphere packing with 6-coordination is also cubic:

$$P2_13,\ a = \sqrt{[1/(8x^2 - 2x + 1/2)]},\ 0 \le x \le 1/4$$
Sphere centers in 4 *a*: $(x,x,x\ ;\ (1/2+x,1/2-x,\bar{x})\kappa)$

If $x = 0$ (or 1/4 or 1/2 or 3/4) the arrangement is face-centered cubic (cubic eutaxy) with 12-coordination. If $x$ is changed from 0 then there are six nearest and six next-nearest neighbors. Increasing $|x|$ from 0 corresponds to displacing the centers of spheres along all four <111> directions (cf. the $Pa\bar{3}$ 7-coordinated packing described in § 6.3.8 above). The density is a minimum for $|x| = 1/8$ at which point the unit cell parameter (for equal spheres)

[1]Note that if we consider the $\{Si\}O_4$ tetrahedra in cristobalite to be rigid, compression along **c** will require $\phi$ to increase (to decrease $c$), and hence $a$ will decrease also. Materials that behave in this way (contracting in a direction normal to an applied compressive stress) are rather rare, but cristobalite is one. In terms of elasticity theory, it is said that it has a negative Poisson's ratio.

[2]For a detailed account of these transformations, and their relevance to crystal chemistry, see M. O'Keeffe & B. G. Hyde, *Acta Crystallogr.* B**32**, 2923 (1976).

is √(8/3) and the density is $\sqrt{(3/2)}\pi/8 = 0.481$. The symmetry is now $P4_332$ and the sphere centers are on the invariant cubic lattice complex $^+Y$ with coordinates (1/8,1/8,1/8 ; 5/8,3/8,7/8)κ. Fig. 6.34 (which might be compared with Fig. 6.31) shows unit cells of **ccp** (using $x = 1/4$) and $^+Y$ for comparison. The six neighbors of $^+Y$ are at the vertices of a trigonal metaprism (symmetry 32) as indicated in the figure.

The **Fe** (or **Si**) atoms in **FeSi** are close to the $^+Y$ packing. As we saw in § 6.3.8 converting the **Fe** and **Si** arrays to two inter-penetrating **ccp** arrays produced **NaCl**. But two inter-penetrating **ccp** arrays also describes **sphalerite** (ZnS, see § 6.1.2), the difference is that in **NaCl** the two arrays are displaced by 1/2,1/2,1/2; in **sphalerite** they are displaced by 1/4,1/4,1/4 (or 3/4,3/4,3/4). Thus the transformation $^+Y \rightarrow$ **ccp** can also be used to describe **FeSi** $\rightarrow$ **sphalerite**. A description of the three structures in terms of a $P2_13$ cell is:

| | | |
|---|---|---|
| **NaCl** | $x_{Na} = 0.25$ | $x_{Cl} = 0.75$ |
| **FeSi** | $x_{Fe} = 0.15$ | $x_{Si} = 0.84$ |
| **ZnS** | $x_{Zn} = 0.0$ | $x_S = 0.75$ |

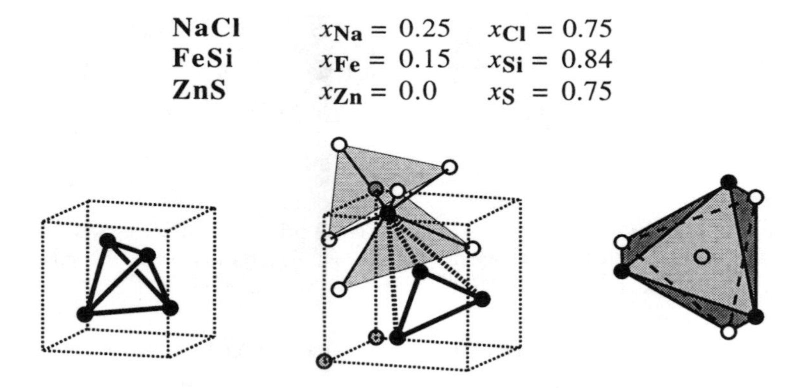

**Fig. 6.34**. Left: a unit cell of **ccp**. Center: a unit cell (filled circles) of $^+Y$; the neighbors of one point are shown as open circles and the two equilateral faces of the trigonal metaprism formed by those points are lightly shaded. Right: the same coordination metaprism viewed down a 3-fold axis.

## 6.4 Sphere packings with cube and trigonal prism sites as stackings of two-dimensional nets

Many of the sphere packings in crystal structures are simply described as a stacking of planar nets, particularly $3^6$ and $4^4$. If we remove the restriction to closest packings of such layers, structures can be generated with sites that are trigonal prisms or cubes. Here we extend our notation for describing structures based on closest packing to include these other packings also. We give many examples of real structures based on the idealized sphere packings; crystallographic data for these compounds are collected in Appendix 5.

### *6.4.1 Non-close-packed stackings of* $3^6$ *nets*

We have already discussed closest packings as a stacking of $3^6$ nets in positions labeled

*A*, *B* or *C*. We did not allow sequences involving two layers in the same position. In a sequence *AB* the spacing between layers of unit-diameter spheres is assumed to be (at least close to) $\sqrt{(2/3)} = 0.816$. A sequence *AA* requires layers to be (close to) unit distance apart and indeed the symbol *A* (= *AAA*...) could be taken to stand for a primitive hexagonal lattice sphere packing ($c/a = 1$). Allowing like letters to repeat generates trigonal prismatic sites between the layers. For *AA* these sites are at positions $\beta$ *and* $\gamma$ at the same level and together they form a $6^3$ net within the layer as shown in Fig. 6.35. For **$AlB_2$** (Greek letters now for **B**, and parentheses enclose sites at the same level) the sequence is $A(\beta\gamma)A(\beta\gamma)A$... showing that **B** atoms fill *all* the trigonal prism sites of a primitive hexagonal **Al** structure and form a $6^3$ (honeycomb) net, as described in § 5.3.5.

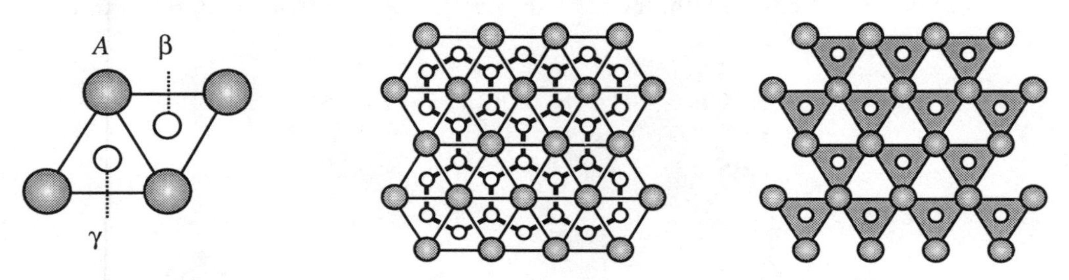


**Fig. 6.35**. Left: The position *A* of a **cp** layer and two trigonal prism sites $\alpha$ and $\beta$ between two *A* layers. Center: illustrating that the $\beta$ and $\gamma$ sites together for a $6^3$ net. Right: the pattern of trigonal prisms when just the $\beta$ sites are occupied.

Fig. 6.35 also shows the pattern of trigonal prisms when one half of the trigonal prism sites are occupied. A simple structure of this type is **WC** which may be written (with Greek letters for **C**) $A\beta A\beta A$.... As shown in Fig. 6.36, the **{C}$W_6$** trigonal prisms form infinite columns by sharing their triangular faces.[1] It might be noted that the structure is its own antistructure so there are also **{W}$C_6$** trigonal prisms.

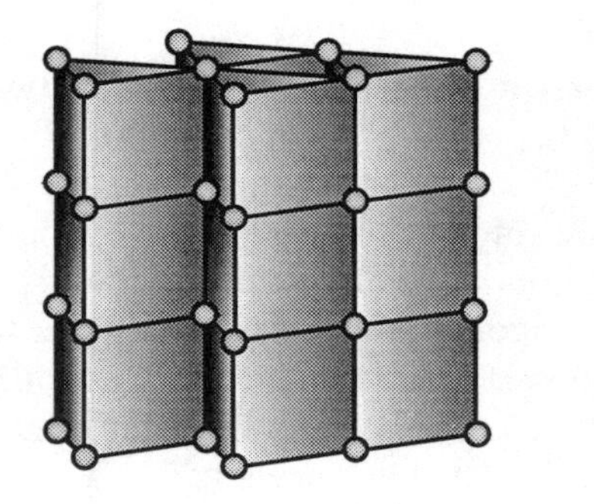

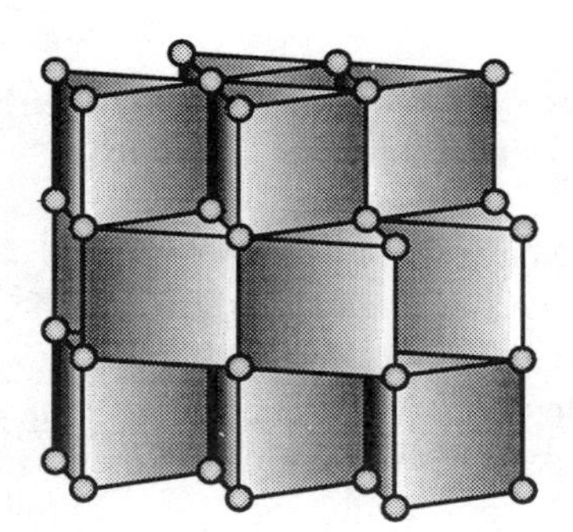

**Fig. 6.36**. Left: **WC** as **{C}$W_6$** trigonal prisms. Right: **NiAs** as **{As)$Ni_6$** trigonal prisms.

[1]Note that *B* and *C* are notations for layer positions, but B and C are symbols for chemical elements!

**NiAs** was described earlier (§ 6.1.5) in terms of **hcp** As and can be coded (with Greek letters for **Ni**) as $B\alpha C\alpha B$; the structure with symbols for cations and anions reversed is $A\beta A\gamma A$.... The layers of trigonal prisms are the same as in WC, but now adjacent layers are staggered so that trigonal prisms share only edges between the layers as illustrated in Fig. 6.36.

We next consider some patterns in which stackings of the *AA* type containing trigonal prism sites are mixed with those of the *AB* type that contain tetrahedral and octahedral holes. Special interest attaches to slabs *AA* with half the trigonal prism holes filled, and to slabs *AB* with either all the octahedral holes or all the tetrahedra holes filled. Fig. 6.37 shows these three kinds of slab. Notice that the faces of the polyhedra occupy one half of the triangles in a **cp** layer (cf. Fig. 6.35, right); a consequence is that slabs of these sorts can be joined without polyhedra sharing faces, as in the structures to be described next. Notice that an isolated octahedral or trigonal prism slab of *X* between two **cp** *Y* layers has stoichiometry $XY_2$; when the *Y* are shared between two slabs the contribution to the overall stoichiometry is $XY_{2/2} = XY$. An isolated tetrahedral layer has stoichiometry $X_2Y_2$ and a slab sharing *Y* with adjacent slabs contributes $X_2Y_{2/2} = X_2Y$ to the overall stoichiometry. We have already discussed (§ 6.1.5) $\mathbf{La_2O_3}$ with **hcp La** and alternating octahedral slabs (**OLa**) and tetrahedral slabs ($\mathbf{O_2La}$).

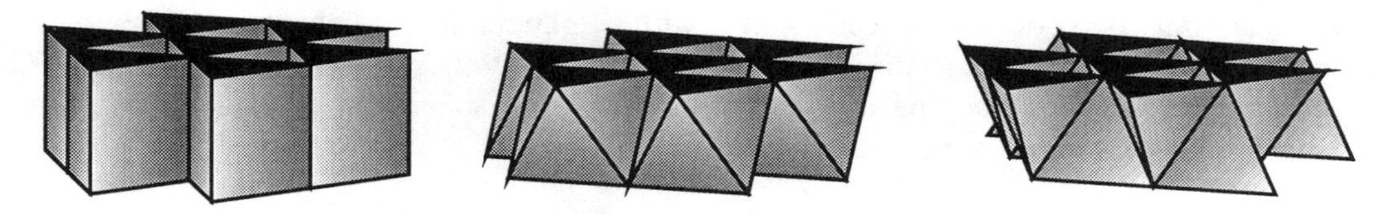

**Fig. 6.37**. From left to right: parts of layers of trigonal prisms, octahedra and tetrahedra.

Two 10-coordinated packings with the same density are: *AABB*... and *AABBCC*...

| | |
|---|---|
| *AABB* | $P6_3/mmc$, $a = 1$, $c = 2 + \sqrt{(8/3)} = 3.63$, $\rho = 2\pi/(\sqrt{18} + \sqrt{27}) = 0.6657$ |
| | Centers in 4*f*: $\pm(1/3,2/3,z\ ;\ 1/3,2/3,1/2-z)$, $z = 1/(4 + \sqrt{24}) = 0.112$ |

We have in fact met this structure as the Ti packing in TiP (see § 6.1.5, especially Fig. 6.13, p. 220). With Greek letters for P the packing is $A\gamma B\alpha B\gamma A\beta A$.... P(1) is in octahedral holes ($\gamma$) between *A* and *B* layers. P(2) is in trigonal prismatic holes: $\beta$ between *AA* layers and $\gamma$ between *BB* layers. The combined P packing $\gamma\alpha\gamma\beta$... corresponds to *hc* close packing (but note that the packings cannot correspond simultaneously to *ideal AABB* cation packing and *ideal hc* anion packing as the two packings have incompatible spacing requirements).

The same cation packing is found in $Sc_2O_2S$ which may be coded with Greek letters for O,S as $A\beta\alpha B\gamma B\alpha\beta A\gamma$... This shows that the tetrahedral sites between *AB* layers are filled (these are O atoms) and one-half the trigonal-prismatic sites between *AA* and *BB* layers are filled (these are S atoms).[1]

[1]If one wants to be completely explicit, the elemental symbols could be used as subscripts.

In the TiP and $Sc_2O_2S$ structures the filling of sites in the trigonal prism layers is dictated by the requirement that the trigonal prisms do not share triangular faces with occupied octahedra (in TiP) or occupied tetrahedra (in $Sc_2O_2S$) in adjacent layers. In the hexagonal structures of $MoS_2$ and $NbS_2$ the anion packing is *AABB* but the interstices in the *AB* slab are empty so there are isolated trigonal prism layers and it does not matter very much therefore which set of sites is filled in these layers. In the $2H_a$ structure of $NbSe_2$ the sequence is $A\gamma AB\gamma B$... and in the $2H_b$ structure of $MoS_2$ the sequence is $A\beta AB\alpha B$... .

In the structure of BaCu, Ba has the same (*AABB*) packing, with Cu in *all* the trigonal prism sites: with Greek letters for Cu, the stacking sequence is $A(\beta\gamma)AB(\alpha\gamma)B$. Notice that the Cu atoms are on $6^3$ nets and the structure could be described as an intergrowth of $\mathbf{AlB_2}$ slabs $BaCu_2$ with **cp** slabs Ba.

The second 10-coordinated sphere packing (*AABBCC*) is rhombohedral:

*AABBCC* $\quad R\bar{3}m$, $a = 1$, $c = 3 + \sqrt{6}$. Centers in 6 $c$: $R \pm (0,0,z)$, $z = 1/(6 + \sqrt{24}) = 0.092$

The S atoms in $3R$ $MoS_2$ have this arrangement; with Greek letters for Mo the sequence is $A\beta AB\gamma BC\alpha C$. The pattern of filling of trigonal prisms destroys the center of symmetry and there are two kinds of S atom in the structure (for data see Appendix 5).

There are other ways of stacking parallel $3^6$ nets.[1] Adjacent layers can be stacked so there are only two contacts for each sphere with neighbors in an adjacent layer. Fig. 6.38 indicates the appropriate positions, labeled *D*, *E* and *F*. Notice that positions *D*, *E* and *F* are only relevant for stacking over an *A* layer. For unit spheres in contact the spacing between layers such as *AD* is $\sqrt{3}/2 = 0.866$.

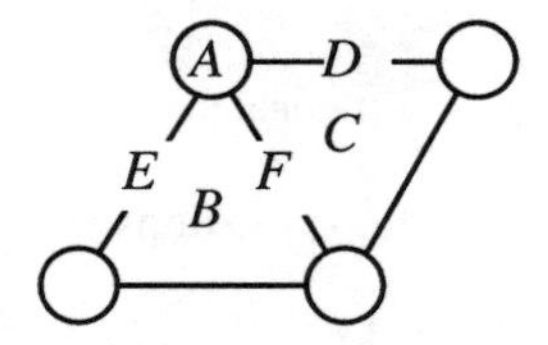


**Fig. 6.38**. Illustrating positions for stacking $3^6$ nets. *A*, *B* and *C* are the same as shown in Fig. 6.1.

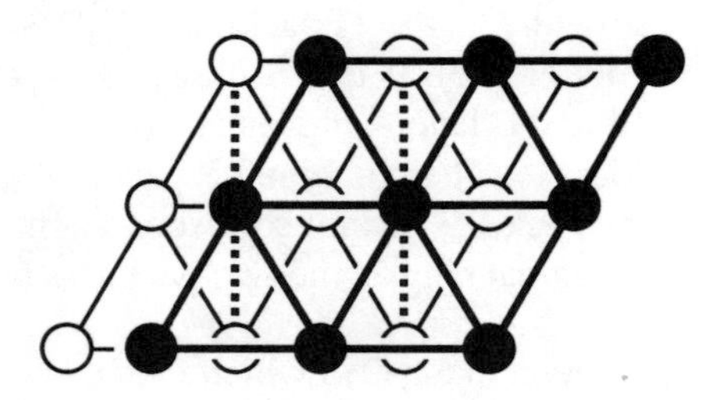

**Fig. 6.39**. **bct** projected on (110) showing the *AD* stacking of $3^6$ nets (filled and open circles respectively). The cell outlined by dotted lines is a face centered tetragonal cell $\sqrt{2}a \times \sqrt{2}a \times c$ ($a$ is the edge of the body-centered tetragonal cell). **c** is horizontal on the page.

[1]See also A. F. Wells *Structural Inorganic Chemistry* [5*th* Ed. Oxford (1984), Chapter 4].

A sequence such as *ADAD*... , illustrated in Fig. 6.39, corresponds to a 10-coordinated sphere packing (6 neighbors in each layer and 2 above and 2 below) and is, in fact, our old friend **bct** (§ 6.3.2). The nets are (110) layers of the body-centered cell (shown in Fig. 6.21, p. 230) or (100) layers of a face-centered cell (indicated in Fig. 6.39).

The 10-coordinated sequence *ADEF* is closely related to **bct** (which is *AD*) and has the same density. The structure is orthorhombic:

*ADEF* $Fddd$, $a = 1$, $b = \sqrt{3}$, $c = \sqrt{12}$. Sphere centers in 8 $a$: $F \pm (1/8,1/8,1/8)$

$\gamma$-Pu has this structure (see Exercise 9). Fig 6.40 (which should be compared with 6.39) illustrates the structure projected on (001).

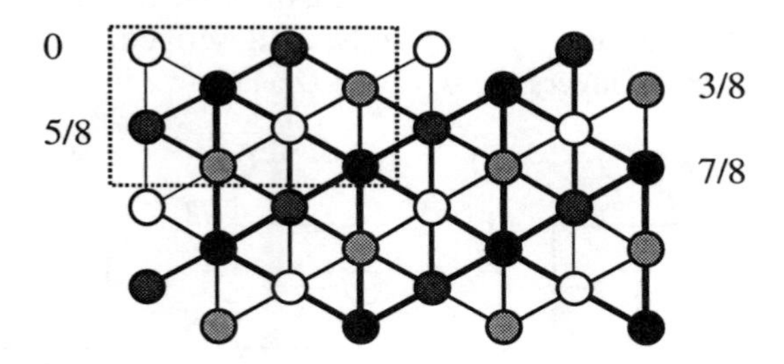


**Fig. 6.40**. The 4-layer sequence *ADEF* of stacked $3^6$ layers with *Fddd* symmetry projected down **c**. (**b** is horizontal on the page). Numbers are elevations in units of *c*.

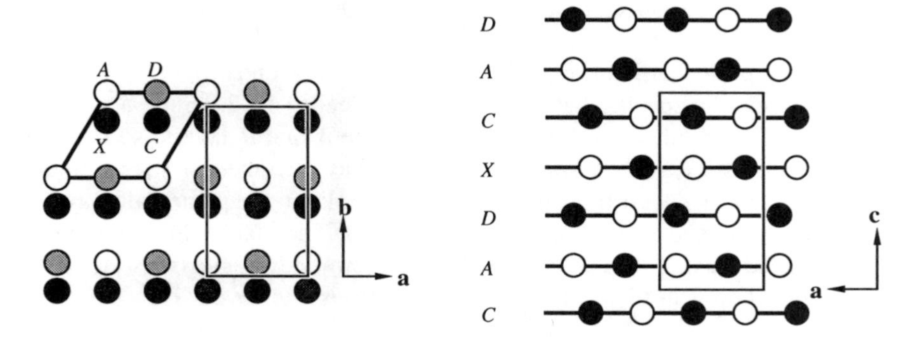


**Fig 6.41**. An 11-coordinated sphere packing obtained by stacking $3^6$ nets as described in the text. Left: projected on (001) (**b** vertical on the page). Depth of shading indicates increasing elevation. Right: projected on (100). Open and filled circles represent sphere centers with $x = 0$ and 1/2. The orthorhombic unit cell is outlined (**c** vertical on the page).

There are many other possibilities, but they are not very common in crystal chemistry. One of the simpler is the sequence of two close-packed layers, say *BA* followed by *D* which, in turn, is followed by another close-packed layer. This fourth layer is not in a position to which we have yet applied a label; it is shown in Fig. 6.41 where it is labeled *X*. This is an 11-coordinated sphere packing; for example, a sphere in an *A* layer has 6 neighbors in the layer, 3 in the $B$ layer and 2 in the $D$ layer. The crystallographic description of this packing is:

$Cmca$, $a = 1$, $b = \sqrt{3}$, $c = \sqrt{(8/3)} + \sqrt{3}$, $\rho = 4\pi/(9 + \sqrt{72}) = 0.719$
Sphere centers in 8 $f$: $C \pm (0,y,z\ ;\ 0,y,1/2-z)$, $y = 1/6$, $z = 3/\sqrt{2} - 2$

The density is the same as that of the tetragonal 11-coordinated sphere packing of § 6.3.1; it would be nice to know if a less dense 11-coordinated packing exists.

### *6.4.2 Non-close-packed stackings of* $4^4$ *nets*

Recall that **ccp** can also be described as a stacking of $4^4$ nets. A number of other sphere packings (not *closest* sphere packings) can also be so described. It is convenient now to recognize four positions for the layers: $A$, $B$, $C$ and $D$ as shown in Fig. 6.42.[1]

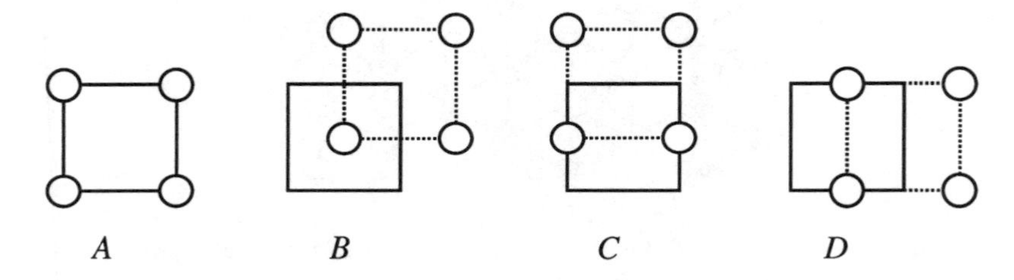


**Fig. 6.42**. The four relative positions for stacked $4^4$ nets. The full lines outline a square cell.

A pair of layers packed $AB$ (or, equivalently, $CD$) corresponds to a (100) slice of **ccp** and for unit spheres, the spacing between layers is (ideally) $1/\sqrt{2}$.[2] Such a slab contains tetrahedral sites. The octahedral sites of **ccp** lie in the $A$ and $B$ layers, so the slab also contains half octahedra (square pyramids). Thus to have octahedral sites a three-layer sequence such as $ABA$ is necessary. The tetrahedral sites lie in the plane half way between $A$ and $B$ in both the $C$ and $D$ positions (see Fig. 6.43). The slab with filled sites could therefore be coded (using Greek letters for the tetrahedral sites) as $A(\gamma\delta)B$ with parentheses again being used to enclose atoms at the same level. This unit is a very common element of crystal structures and we refer to it as a *tetragonal tetrahedral layer* (to distinguish it from the hexagonal slab of tetrahedral sites between two $3^6$ layers). It is a (100) layer of **fluorite**. Notice that the tetrahedral sites are on a $4^4$ net of twice the density (dotted lines on the left in Fig. 6.43).

Such layers occur in the litharge form of PbO with O in the tetrahedral sites and the mackinawite form of FeS (indeed they are sometimes called mackinawite layers) with Fe in

[1]There should be no risk of confusion with the symbols for stacking $4^4$ nets with those for stacking $3^6$ nets as long as it is made clear to which one refers. *Parallel* $3^6$ and $4^4$ nets of equal-sized spheres are incommensurate so mixed stackings do not normally occur.

[2]Beware! If the spacing between the layers is 1/2 one has an element of **bcc**.

tetrahedral sites. The overall cation packing in PbO is approximately **ccp** as the sequence is $A(\gamma\delta)BA(\gamma\delta)B\ldots$ but in the real material the inter-layer spacings are not exactly the ideal values.

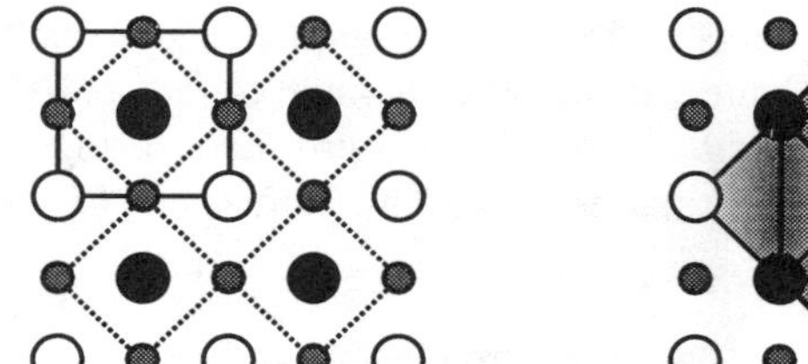

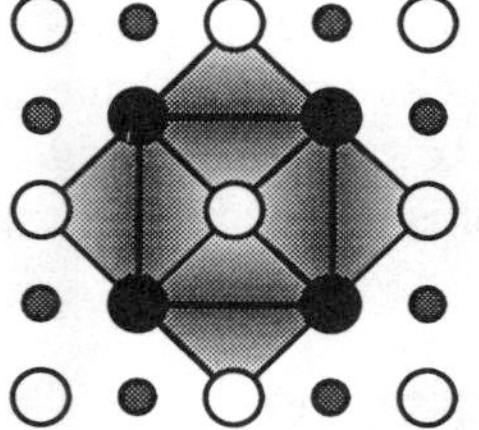

**Fig 6.43**. Left: plan view of two $4^4$ layers of spheres (open and filled large circles) stacked *AB*; the smaller circles indicate the location of tetrahedral sites between the layers. Center: some of the tetrahedra are shown. Right: the same in elevation.

**PbFCl (matlockite)** is an important example of a structure type with tetragonal tetrahedral layers, and as it is one of the most commonly-occurring of all ternary structures we digress a little to describe it here. The structure goes under several names and we also use **BaMgSi** for anti-structure compounds in which **Ba + Mg** replace **Cl + F**; in *Pearson's Handbook* (Book List) it is called **$Cu_2Sb$**. It is related to **PbO** in the sense that there are (**PbF**) tetragonal tetrahedral layers with stoichiometry **$Pb_2F_2$** but they are now interwoven with two layers of **Cl** on $4^4$ nets. Now the tetrahedral layers are further apart, and in isostructural compounds like LaO*X* (*X* = Cl, Br, I) with layers of $\{O\}La_4$ tetrahedra of almost constant size, the interlayer spacing is determined by the size of the interlayer atoms *X*.

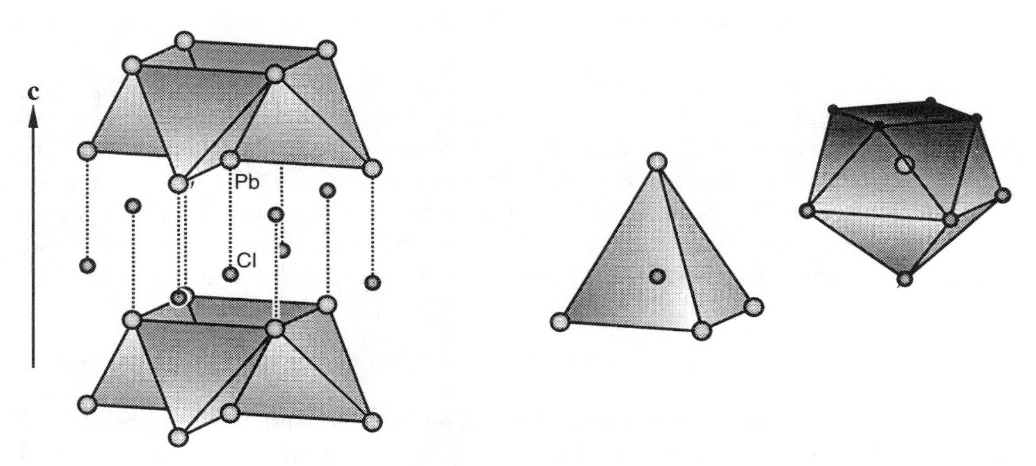


**Fig. 6.44**. The structure of PbFCl. Left: showing the layers of edge-sharing $\{F\}Pb_4$ tetrahedra with Cl atoms in intervening layers. Right: showing the $\{Cl\}Pb_5$ and $\{Pb\}Cl_5F_4$ coordination polyhedra around the two atoms identified with labels on the left. F atoms (not shown on the left) are small filled circles.

Figure 6.44 illustrates the structure of PbFCl. F is in a {F}$Pb_4$ tetrahedron and Cl is in a {Cl}$Pb_5$ square pyramid. The Pb coordination is an irregular square antiprism with a small $F_4$ square face and a larger $Cl_4$ square face; this larger face is capped by a fifth Cl to give Pb a 9-coordination, {Pb}$Cl_5F_4$.

We return now to a discussion of further packings of $4^4$ layers. A sequence such as *AC* with spacing $\sqrt{3}/2$ will generate an element of primitive hexagonal packing, in fact a $(10\bar{1}0)$ layer, and a continuing sequence *AC*... (or *AD*... or *BC*... or *BD*...) *is* primitive hexagonal with the $\bar{6}$ axes of the trigonal prisms perpendicular to the stacking direction as can be seen from Fig. 6.45.

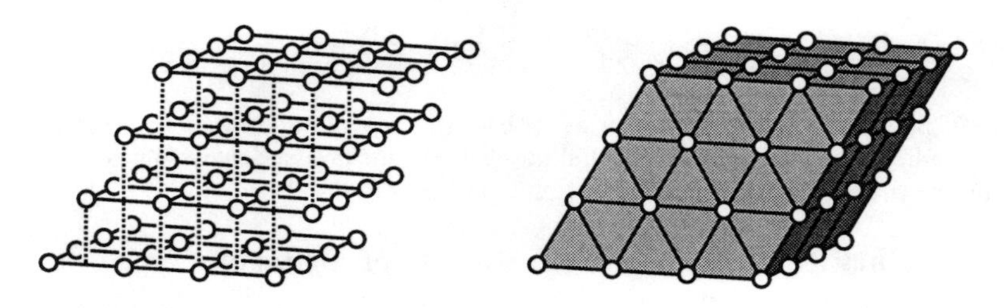

**Fig. 6.45**. Illustrating that non-close-packed stacking of $4^4$ layers can generate a primitive hexagonal array. Left: a sequence of $4^4$ layers. Right: the trigonal prismatic sites are shaded.

A sequence such as *AA*, with unit spacing between the layers, corresponds to a slice of primitive cubic in what should be an obvious way.

We now describe some other stackings of $4^4$ layers which correspond to packings of equivalent spheres and which are encountered in crystal chemistry. As discussed above they can contain trigonal prism and cube sites as well as tetrahedral sites between two layers that are in closest packing. We assume the interlayer spacings to be those for spheres in contact.

***ACBD***: This corresponds to a packing of trigonal prisms with their axes pointing in two orthogonal directions (and normal to the stacking directions). The symmetry is $I4_1/amd$ (with the 4-fold axis parallel to the stacking direction) so the prisms do not have 6-fold axes in the structure. We have described this 8-coordinated packing in § 6.3.5 (see Fig. 6.26).

***ABCD***. This is a 10-coordinated sphere packing (it has the same density as the 10-coordinated stackings of $3^6$ nets described above) and might be considered an intergrowth of **ccp** (the *AB* and *CD* parts) with primitive hexagonal (the *BC* and *DA* parts). The symmetry is now orthorhombic:

***ABCD*** *Cmcm*, $a = c = 1$, $b = \sqrt{2} + \sqrt{3}$. Centers in 4 *c*: $C \pm (0,y,1/4)$, $y = (3 - \sqrt{6})/4$

This packing is found as the structure of one of the high-pressure polymorphs of Ga. If the trigonal prism sites (between the *BC* and *DA* layers) are filled, the important **CrB** structure type (**B** in the trigonal prisms) is obtained. In **UBC**, in addition to **B** in the hexagonal prisms (of **U**), there are **C** atoms in the square pyramids.

The structure is shown in three ways in Fig. 6.46, first as a stacking of $4^4$ nets, secondly with the trigonal prisms shaded in, and finally as a projection down **a**. In the last case the trigonal prism centers are indicated; notice that they form zig-zag rods parallel to **c**.

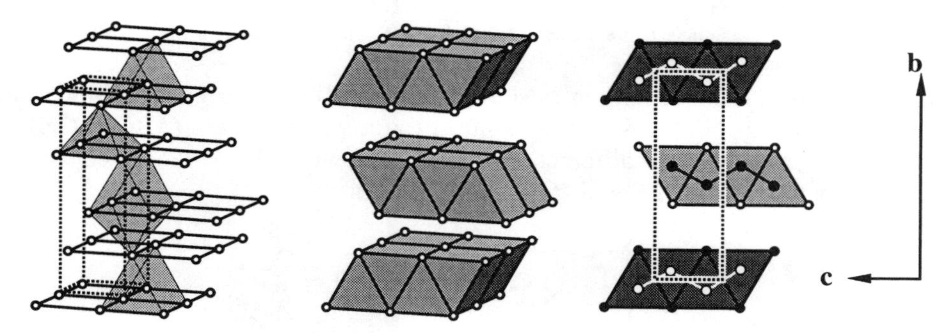


**Fig. 6.46**. Illustrating an *ABCD* stacking of $4^4$ layers (**CrB** packing). In the middle the trigonal prism layers are shaded. On the right the structure is shown in projection on (100) (light and dark shading is used to differentiate prisms with centers at $x = 0$ and $x = 1/2$). The trigonal prism sites are shown as larger circles connected by lines to their nearest neighbors.

We digress to describe a related 10-coordinated sphere packing of the same density that is not however usefully described as a stacking of layers. In this structure we again have trigonal prisms, but now in columns (rather than slabs) and the sphere packing corresponds to the **Fe** atom positions in (idealized) **FeB** (**B** is in the trigonal prisms). Fig. 6.47 compares the two packings. In both structures the trigonal prism centers form zig-zag rods which are normal to the plane of projection in Fig. 6.47.

**FeB packing** *Pnma*, $a = 1.9062$, $b = 1$, $c = 1.6506$
Sphere enters in 4 *c*: $\pm(x,1/4,z\ ;\ 1/2+x,1/4,1/2-z)$, $x = 0.1811$, $z = 0.1583$

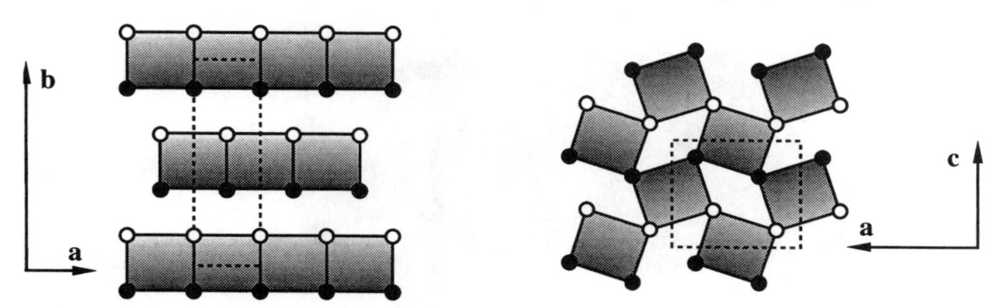


**Fig. 6.47**. Left: the **CrB** sphere packing (of **Cr**) projected on (001) (compare fig. 6.46). Open and filled circles are at $z = 1/4$ and 3/4 respectively. Right: the **FeB** sphere packing (of **Fe**)projected on (010). Open and filled circles are at $y = 1/4$ and 3/4 respectively. In both cases shaded rectangles are columns of trigonal prisms sharing square faces.

***ACDABDCB***: This is another 10-coordinated sphere packing of the same density and can similarly be considered as an inter-growth of alternate layers of **ccp** and primitive hexagonal. It is tetragonal:

***ACDABDCB*** $I4_1/amd$, $a = 1$, $c = \sqrt{8} + \sqrt{12} = 6.293$
Sphere centers in 8 *e*: $I \pm (0,1/4,z\ ;\ 0,3/4,1/4+z)$, $z = \sqrt{3}/[8(\sqrt{2} + \sqrt{3})] = 0.0688$

This occurs as the Mo packing in $\alpha$-MoB (with again B in the trigonal prismatic sites, now between *AC*, *DA*, *BD* and *CB*). In contrast to the previous case (*ABCD*, **CrB**), the trigonal prisms "point" in two different directions as shown in Fig. 6.48 (which should be compared with the middle of Fig. 6.46). $\beta$-MoB is **CrB**.

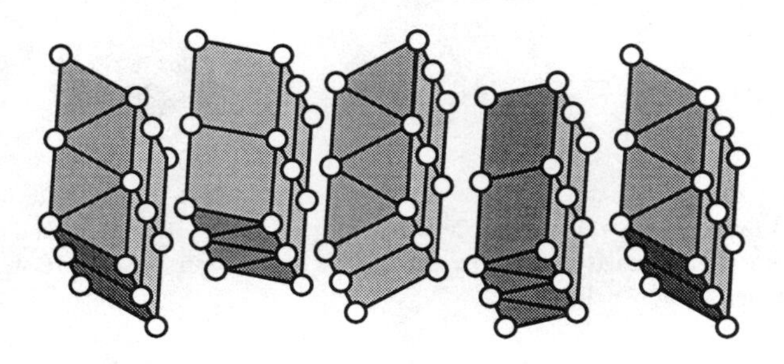

**Fig. 6.48**. Illustrating an *ACDABDCB* stacking of $4^4$ layers (**c** is horizontal).

***AABB***: This is a 9-coordinated sphere packing and might be considered an intergrowth of **ccp** (the *AB* and *BA* parts) with primitive cubic (the *AA* and *BB* parts). The symmetry is tetragonal:

***AABB*** $I4/mmm$, $a = 1$, $c = 2 + \sqrt{2}$, $\rho = 2\pi/(6 + \sqrt{18}) = 0.6134$
Sphere centers in 4 *e*: $I \pm (0,0,z)$, $z = 1/\sqrt{8} = 0.3536$

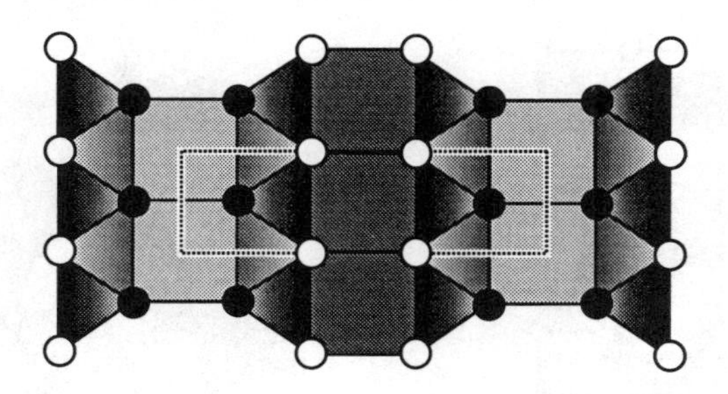

**Fig. 6.49**. A projection of the *AABB* stacking of $4^4$ layers of Te atoms in $Th_2TeN_2$ showing the layers of tetrahedra (centered by N) and cube sites (centered by Te). The projection is down **a** with **c** running across the page. The cube centers are at $x = 0$ and $x = 1/2$ (differentiated by light and darker shading).

In $Th_2TeN_2$ N atoms are between the *AB* and *BA* slabs of Th, forming tetragonal tetrahedral layers (cf. Fig. 6.43) with stoichiometry $N_2Th_2$. The Te atoms are between the *AA* and *BB* layers of Th. Using Greek letters for Te and N, the structure is $A\beta A(\gamma\delta)B\alpha B(\gamma\delta)$. An *idealized* version (with regular tetrahedra and cubes) is shown in

Fig. 6.49. $Th_2TeN_2$ is a member of the largest of all groups of ternary structure types. It is usually named for the antistructure compound $ThCr_2Si_2$ (in which Th and Cr play the role of Te and N in $Th_2TeN_2$) but in *Pearson's Handbook* (Book List) it is called **$BaAl_4$**. The structure type is similar to **PbFCl** in having tetragonal tetrahedral layers and a wide range of $c/a$ according to the size of the atom between the layers (in the "cube" sites).

*ACCA*: This is a 7-coordinated sphere packing and is an intergrowth of primitive cubic and primitive hexagonal. The symmetry is orthorhombic:

*ACCA*     *Cmmm*, $a = c = 1$, $b = 2 + \sqrt{3} = 2.7321$
Sphere centers in 4 *j*: $C \pm (0,y,1/2)$, $y = 1/(4 + \sqrt{12}) = 0.1340$

The *ACCA* packing is that of Fe in the $Fe_2AlB_2$ structure. Al fills the cube sites (between *AA* and *CC* layers of Fe) and B fills the trigonal prism sites (between *AC* and *CA* layers). The structure may be also considered as a prismatic stacking of $3^34^2$ nets as shown in Fig. 6.50.

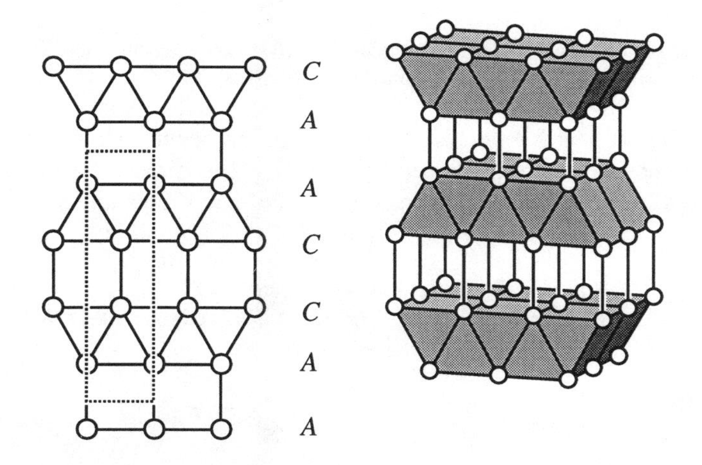


**Fig. 6.50.** *ACCA* stacking of $4^4$ nets of Fe in $Fe_2AlB_2$. Left: viewed normal to the stacking direction (vertical on the page). Right: as a clinographic projection. The trigonal prism sites (filled by B) are shaded, and the cubes (occupied by Al, not shown) are outlined.

### *6.4.3 Stacked* $3^2.4.3.4$ *nets*

Stacking of $3^2.4.3.4$ layers produces 7-coordinated sphere packings with a variety of interstitial sites—cubes, square antiprisms (better *metaprisms*), trigonal prisms and tetrahedra—and many crystal structures are based on filling some or all of the interstitial sites in these packings. In this section we briefly describe some of the simpler such structures (crystallographic data for real materials are in Appendix 5).

For a prismatic stacking (i.e. one directly above the other) of $3^2.4.3.4$ layers the formal description for spheres of unit diameter is:

$P4/mbm$, $a = \sqrt{(2 + \sqrt{3})} = 1.932$, $c = 1$, $\rho = 0.5612$
Sphere centers in $4g$: $\pm(x,1/2+x,0;1/2-x,x,0)$, $x = 1/\sqrt{(16 + \sqrt{192})} = 0.1830$

The interstices in the sphere packing are cubes and trigonal prisms. In $LiY_2Si_2$, Li is in all the cube sites and Si in all the trigonal prism sites of the Y packing as shown in Fig. 6.51. Notice that the atoms in the trigonal prism sites come together in pairs ($Si_2$ in $LiY_2Si_2$). The structure is usually called $\mathbf{U_3Si_2}$.

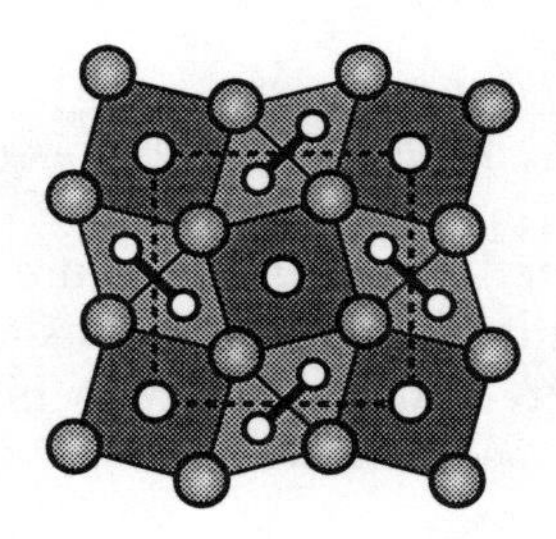

**Fig. 6.51**. Prismatic stacking of $3^2.4.3.4$ nets projected on (001) showing the cube sites (darker shaded) and trigonal prismatic sites that occur in pairs (joined by heavy lines).

Another stacking of $3^2.4.3.4$ nets is of rather common occurrence in crystal structures. Now the nets in each layer are displaced by 1/2,1/2,1/2 to make a two-layer stacking as shown in Fig. 6.52. The structure is tetragonal ($a$ and $x$ are the same as for the previous structure):

$I4/mcm$, $a = 1.932$, $c = \sqrt{(2\sqrt{3})} = 1.861$
Sphere centers in 8 $h$: $I \pm (x,1/2+x,0\ ;\ 1/2-x,x,0)$, $x = 0.1830$

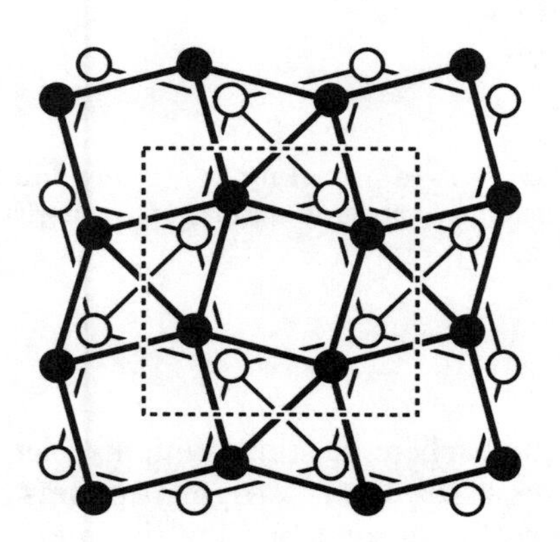

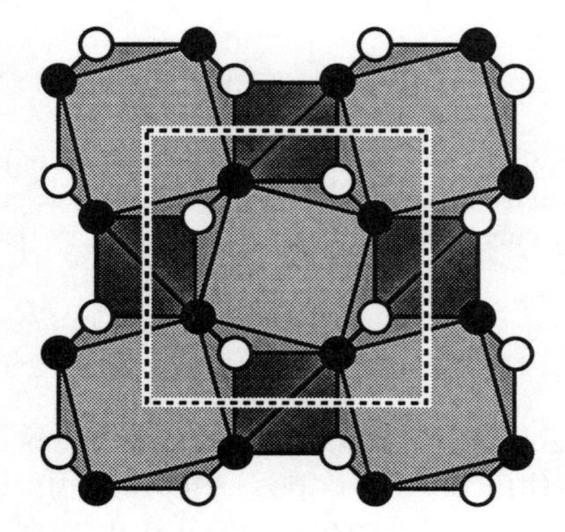

**Fig. 6.52**. A sphere packing generated by stacking alternating layers of $3^2.4.3.4$. Left: showing the nets. Right: emphasizing the square metaprisms and some of the tetrahedra (darker shaded). The projection is on (001) of the tetragonal cell. Empty and filled circles are at $z = 0$ and 1/2 respectively.

A notable feature of the structure is the columns of face-sharing square metaprisms. Note also the tetrahedra centered at 1/2,0,1/2 and 0,1/2,1/2. An example of the occurrence of this packing is the **Al** arrangement in $\mathbf{CuAl_2}$ in which **Cu** atoms center the $\mathbf{Al_8}$ metaprisms so that linear **-Cu-Cu-** rods run parallel to **c**. In $PtPb_4$ only half the $Pb_8$ metaprisms are filled by Pt (those in alternate layers perpendicular to **c**). In contrast, in $TaTe_4$, Ta fills only the $Te_8$ metaprisms centered at 0,0,*z* forming isolated rods of face-sharing {Ta}$Te_8$ metaprisms with their axes parallel to **c**. In $KInTe_2$, in addition to {K}$Te_8$ metaprisms, there are {In}$Te_4$ tetrahedra (the tetrahedral sites are also shown in Fig. 6.52).

A third stacking of $3^2.4.3.4$ nets is also important in crystal chemistry. *Pairs* of prismatically-stacked nets are displaced by 1/2,1/2,1/2 to produce a four-layer sequence, and the structure may be thought of as an intergrowth of the previous two. The symmetry is the same as in the previous sequence:

*I4/mcm*, $a = 1.9318$, $c = 3.8612$
Centers in 16 *l*: $I \pm (x,1/2+x,z\ ;\ 1/2-x,x,z\ ;\ x,1/2-x,1/2-z\ ;\ 1/2+x,x,1/2-z)$, $x = 0.1830$, $z = 0.1295$

Layers of square metaprism and tetrahedral sites alternate with layers containing trigonal prism and cube sites. In $Cr_5B_3$, Cr(2) atoms are in this sphere packing with Cr(1) in the cubes, and B atoms fill the square metaprisms and the trigonal prisms. In $PdGa_5$ the same packing of Ga(2) atoms occurs, now Ga(1) are in the cube sites, Pd in the metaprisms and the trigonal prisms are empty.

## 6.5 A summary of sphere packings

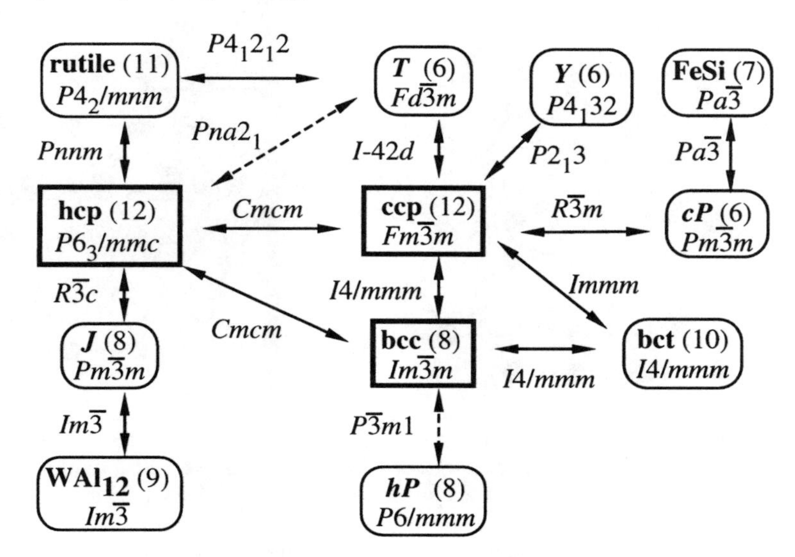


**Fig. 6.53**. Some of the structural relationships discussed in this chapter (arrows). Numbers in parentheses are coordination numbers. Notice the central positions of **cp** and **bcc**.

The diagram (Fig. 6.53) indicates some of the sphere packings and transformations between them that we have discussed. In the diagram the lines indicate a transformation path of the symmetry indicated; a broken line indicates that in the intermediate structure there is more than one kind of sphere (i.e. they are not all equivalent to each other). Attention is drawn to the central position occupied by **ccp**. Note that **cco** (symmetry *Cmcm*, see § 6.3.3) is not shown but is intermediate between **hcp** and **ccp**.

## 6.6 Some packings of two kinds of spheres

Here we describe some structures of simple binary intermetallic compounds. They are of interest both as the structures of large groups of compounds and as components of more complex structures. As an example we cite the fact that **$MgCu_2$** (§ 6.6.3) is the structure of many intermetallic compounds and also of the cation array (**$MgAl_2$**) in **spinel, $MgAl_2O_4$**.

### *6.6.1* $Cu_3Au$ *and* $Ni_3Sn$

**$Cu_3Au$** is a simple ordered derivative of **ccp**. The original *fcc* structure is replaced by a primitive cubic one (symmetry $Pm\bar{3}m$) with **Au** at 0,0,0 and **Cu** at (0,1/2,1/2)κ. The structure should be familiar: the **Cu** arrangement is the *J* structure described in § 6.3.5 and **Au** is in the cuboctahedral holes of this structure. (see Fig. 6.27).

(111) planes of **Cu** are kagome (3.6.3.6) nets and **Au** centers the hexagons of the nets so that the combined (111) nets are $3^6$. Fig 6.54 shows how the nets are stacked; using the symbolism of § 6.1.6, the stacking of kagome nets is $N_1N_2N_3$....

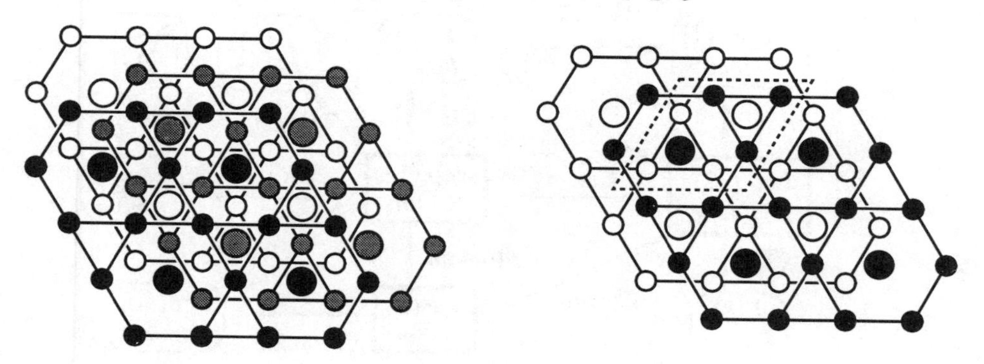

**Fig. 6.54**. Left: the cubic stacking of **$Cu_3Au$** nets projected down [111]. Larger circles are **Au**. Open, shaded and filled circles are at different levels: 0, 1/3 and 2/3 respectively in units of |**a** + **b** + **c**|. Right: the stacking of nets at $z$ = 1/4 and 3/4 (open and filled circles) in **$Ni_3Sn$**. Larger circles are **Sn**.

**$Ni_3Sn$** is the analogous structure derived by ordering of **hcp**. **Ni** atoms form kagome (3.6.3.6) layers with **Sn** centering the hexagons. The layers are therefore like (111) layers in **$Cu_3Au$** but now the stacking of kagome nets is hexagonal: $N_1N_2$... as also shown in

Fig. 6.54. For the ideal structure with atoms *d* apart:

**$Ni_3Sn$** — *$P6_3/mmc$*, $a = 2d$, $c = \sqrt{(8/3)}d$, $c/a = \sqrt{(2/3)} = 0.816$
**Ni** in 6 *h*: $\pm(x,2x,1/4\ ;\ x,\bar{x},1/4\ ;\ 2\bar{x},\bar{x},1/4)$, $x = -1/6$
**Sn** in 2 *c*: $\pm(1/3,2/3,1/4)$

Note that empty $\mathbf{Ni_6}$ octahedra (with centers at 0,0,0 and 0,0,1/2) share faces and form isolated rods parallel to **c**. Contrast $\mathbf{Cu_3Au}$ in which the empty $\mathbf{Cu_6}$ octahedra share vertices only (Fig. 6.27 again).

In the cubic high-temperature form of $BaTiO_3$ the $BaO_3$ arrangement is $\mathbf{AuCu_3}$ (Ti in the $O_6$ octahedra). In $BaNiO_3$ the $BaO_3$ arrangement is $\mathbf{SnNi_3}$ (Ni in the $O_6$ octahedra). These oxide structures are often referred to as cubic and hexagonal **perovskite** respectively. For crystallographic data see Appendix 5; cubic **perovskite** was described in § 5.3.4.

### *6.6.2* CuZn (*β*-brass), CuAu *and* $MoSi_2$

The simplest ordering of **bcc** is **CuZn** in which **Cu** is at the origin (0,0,0) and **Zn** at the body center (1/2,1/2,1/2) of a cubic cell. Notice that the structure is primitive cubic (symmetry $Pm\bar{3}m$). The structure is also that of CsCl which is often used as the eponymous compound, but as it is much more common for intermetallic compounds we prefer the name **CuZn** or ***β*-brass** (CuZn is *β*-brass). Notice that the structure is its own antistructure and each atom is coordinated by eight of the other kind at the corners of a cube (Fig. 6.55).

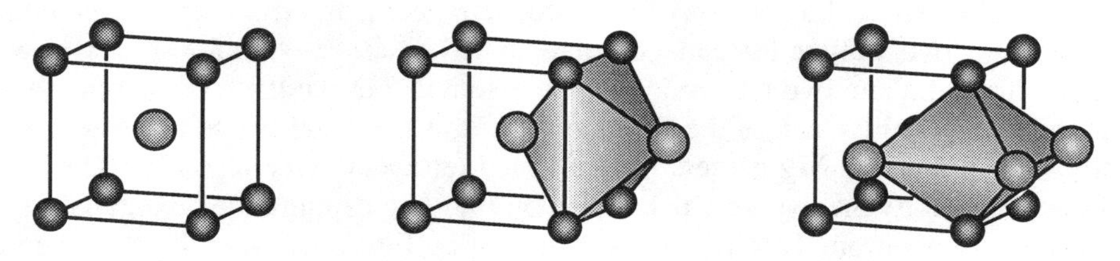

**Fig. 6.55.** Left a unit cell of **CuZn** (smaller, darker-shaded circles are **Cu**). Middle: a $\mathbf{Cu_4Zn_2}$ octahedron. Right a $\mathbf{Cu_2Zn_4}$ octahedron.

It may be recalled that in **bcc** the octahedra around octahedral sites fill space twice over (see § 6.2)—in **CuZn** there are two sets of octahedra: $\mathbf{Cu_4Zn_2}$ [at (0,1/2,1/2)κ] and $\mathbf{Cu_2Zn_4}$ [at (0,0,1/2)κ] as shown in Fig. 6.55. Each set of octahedra exactly fills space. In cubic **perovskite** $BaTiO_3$ the cation arrangement (BaTi) is **CuZn** and the anions are in $\{O\}Ba_4Ti_2$ octahedra.

Alternate {100} layers of **CuZn** are $4^4$ nets of **Cu** and **Zn** which are stacked *AB* (cf. § 6.4.2). For the structure to have cubic symmetry the layers must be *a*/2 apart, where *a* is the unit cell edge of the $4^4$ layers. If the layers are instead $\sqrt{2}a/2$ ($\approx 0.71a$) apart the structure is an ordering of **ccp** and the structure is referred to as **CuAu**. The symmetry is tetragonal (*P4/mmm*) and **Cu** is at 0,0,0 and **Au** at 1/2,1/2,1/2 so that **CuZn** is the special case of **CuAu** with $c/a = 1.0$; $c/a = \sqrt{2} = 1.414$ corresponds to an ideal ordering of **ccp**. The general case is called **CuAu**; examples of actual compounds are (with $c/a$ in parentheses): FeNi (1.41), CuAu (1.31), PtZn (1.22), NiZn (1.08) and MnHg (1.01).

In **$MoSi_2$** the atoms are again on $4^4$ nets in the sequence **MoSiSi** along **c**. The stacking alternates $AB$ so the structure has the six-layer repeat $A_{Mo}B_{Si}A_{Si}B_{Mo}A_{Si}B_{Si}$.

**$MoSi_2$**    $I4/mmm$
**Mo** in 2 $a$: $I + (0,0,0)$
**Si** in 4 $e$: $I \pm (0,0,z)$ with $z \approx 1/3$

$a$ is the spacing of the $4^4$ nets. If $c/a = 3\sqrt{2} = 4.23$ the structure is a superstructure of **ccp** and if $c/a = 3$ it is a superstructure of **bcc**. Many compounds have axial ratios between these two values. A third possibility is a superstructure of **bct** for which $c/a = 3\sqrt{(2/3)} = \sqrt{6} = 2.449$. For $MoSi_2$ $c/a = 2.45$ so clearly in this case the structure should be considered as a superstructure of the 10-coordinated **bct** packing. In Volume II of this series we adduce several examples of ionic compounds with **$MoSi_2$** cation packing; $K_2MgF_4$ is a good example (for data see Appendix 5).

### *6.6.3* **$MgCu_2$**

**$MgCu_2$** is often cited as an example of a structure that is based on an efficient packing of spheres of two sizes. The **Cu** arrangement is the $T$ sphere packing of § 6.3.9 (see Fig. 6.32) which can be described as an array of vertex sharing **$Cu_4$** tetrahedra. The space not occupied by **$Cu_4$** tetrahedra consists of truncated tetrahedra, and it is useful at the outset to see how these two polyhedra can combine to fill space. Fig. 6.56 shows how two truncated tetrahedra and two tetrahedra can be assembled into a large 60° rhombohedron. This is in fact a primitive cell of the $T$ structure. It is also a primitive cell of **$MgCu_2$** if **Cu** is in the $T$ positions and **Mg** centers the truncated tetrahedra (see also Fig. 6.58 below). **Mg** is coordinated by the twelve **Cu** and also by four **Mg** capping the hexagonal faces of the truncated tetrahedron. This 16-vertex coordination figure was identified as the Friauf polyhedron in § 5.1.7 (see Fig. 5.12). As shown below, **Cu** is 12-coordinated in a {**Cu**}**$Cu_6Mg_6$** icosahedron (Fig. 6.57).

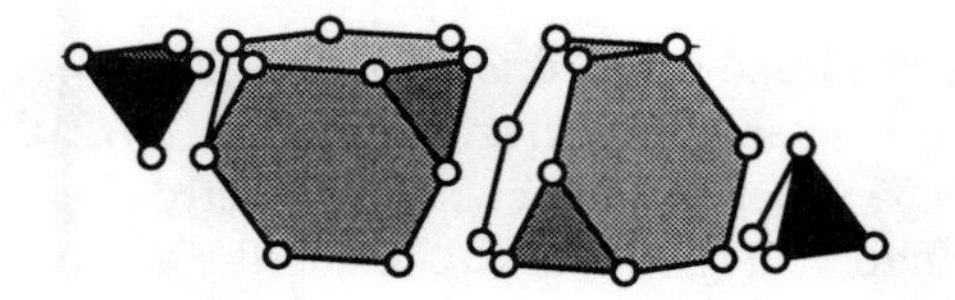

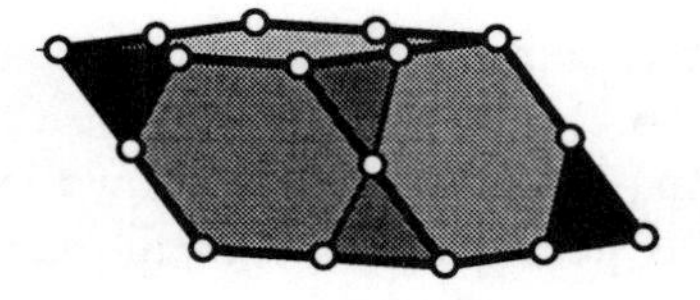

**Fig. 6.56**. Two tetrahedra and two truncated tetrahedra combining to make a 60° rhombohedron (right).

The formal description of the structure is:

**$MgCu_2$**    $Fd\bar{3}m$. **Mg** in 8 $a$: $F \pm (1/8,1/8,1,8)$ (lattice complex $D$ = **diamond**)
**Cu** in 16 $d$: $F \pm (1/2,1/2,1/2\ ;\ 1/2,1/4,1/4)\kappa$ (lattice complex $T$)

Fig. 6.57 shows the structure projected on (100). On the left, just the atom positions are shown, and most people will find that figure somewhat uninformative. However, in the center a $\mathbf{Cu_{12}}$ truncated tetrahedron around one **Mg** is picked out and on the right a $\mathbf{Cu_6Mg_6}$ icosahedron around one **Cu** is similarly depicted.

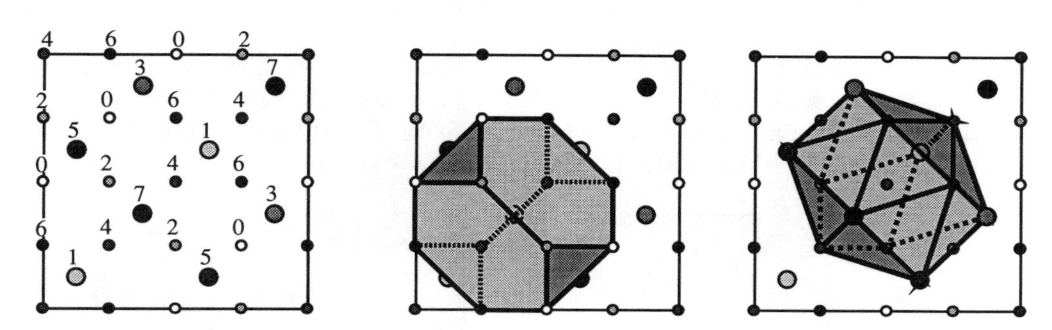


**Fig. 6.57**. Left: $\mathbf{MgCu_2}$ projected on (100). Larger circles are **Mg** and numbers are elevations in multiples of $a/8$. The intensity of shading is proportional to elevation. Center, an $\{\mathbf{Mg}\}\mathbf{Cu_{12}}$ truncated tetrahedron and right, a $\{\mathbf{Cu}\}\mathbf{Cu_6Mg_6}$ icosahedron picked out. Polyhedron edges obscured by front faces are shown as broken lines.

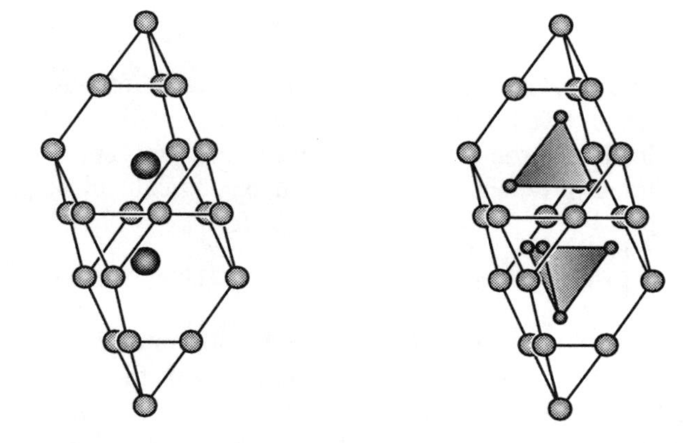

**Fig. 6.58**. Left: a primitive cell of $\mathbf{MgCu_2}$ (larger circles are **Mg**). Right: a primitive cell of $\mathbf{MgAl_2O_4}$ shown as $\{\mathbf{Mg}\}\mathbf{O_4}$ tetrahedra and **Al** atoms.

The structure can also be considered as a space-filling by tetrahedra (not all regular). These are of three sorts: per $\mathbf{MgCu_2}$ unit there are 4 $\mathbf{MgCu_3}$ tetrahedra (these are made up of an **Mg** and a triangular $\mathbf{Cu_3}$ face of the surrounding truncated tetrahedron), 12 $\mathbf{Mg_2Cu_2}$ tetrahedra and one regular $\mathbf{Cu_4}$ tetrahedron. Structures which are space fillings of tetrahedra are sometimes referred to as "topologically close packed."

We mentioned earlier (§ 6.1.6) that in **spinel** ($\mathbf{MgAl_2O_4}$) the cation array is $\mathbf{MgCu_2}$ and indeed to get **spinel** from $\mathbf{MgCu_2}$ all that has to be done is to fill all the $\mathbf{MgCu_3}$ tetrahedra with anions so that in $\mathbf{MgAl_2O_4}$ there are $\{\mathbf{O}\}\mathbf{MgAl_3}$ tetrahedra. It is not immediately obvious that this results in {Mg}$O_4$ tetrahedra, {Al}$O_6$ octahedra and (approximately) **ccp** O! For this aspect of the structure refer back to § 6.1.6, especially Fig. 6.17 (p. 223). Fig. 6.58 compares primitive cells of **$MgCu_2$** and **$MgAl_2O_4$**.

Close packed structures may be described as built up of rhombohedral units consisting of an octahedron and two tetrahedra (Fig. 6.4). Similarly, there is a family of structures built up of different stackings of the rhombohedral unit consisting of two centered truncated tetrahedra and two tetrahedra (Fig. 6.56). Fig. 6.59 illustrates the simpler possibilities. By analogy with close packing, the stacking in **$MgCu_2$** can be described as *c*; the stacking in **$MgZn_2$** is then *h* and in **$MgNi_2$** it is *hc* (see Fig. 6.59). The structures collectively are known variously as as "Friauf-Laves phases" or just "Laves phases."[1]

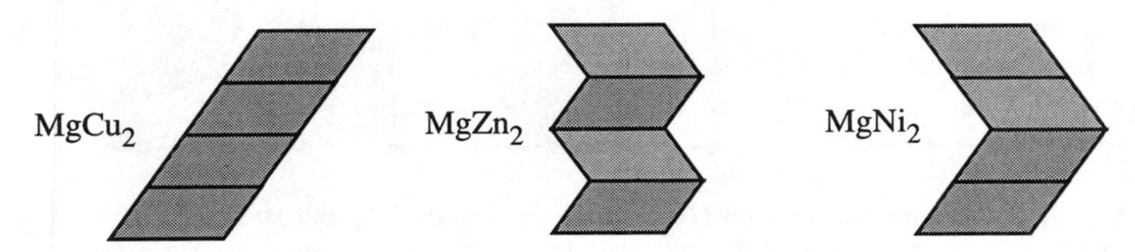


**Fig. 6.59**. The stacking of rhombohedral units in **$MgCu_2$**, **$MgZn_2$** and **$MgNi_2$**.

### *6.6.4* **$Cr_3Si$** (*A*15)

The structure known as *A*15 [2] or **$Cr_3Si$** has a number of features in common with **$MgCu_2$**. The atoms are in fixed positions (on the sites of invariant lattice complexes) so the structure is completely determined by the cubic cell constant. Space is again divided up into irregular tetrahedra, so it is another example of a topologically close-packed structure.

| | |
|---|---|
| **$Cr_3Si$** | $Pm\bar{3}n$. **Si** in 2 *a*: 0,0,0 ; 1/2,1/2,1/2 (**bcc**) |
| | **Cr** in 6 *c*: ±(1/4,0,1/2)κ [or 6 *d*: ±(1/4,1/2,0)κ] (lattice complex *W*) |

The structure is illustrated in Fig. 6.60. Notice the non-intersecting rods of Cr atoms along <100>; we describe a related cylinder packing with the same symmetry below [§ 6.7.3 (*a*)]. The **Cr** atoms form icosahedra with symmetry $m\bar{3}$ about the **Si**; this icosahedron was described in § 2.5.7 (see Fig. 2.25, p. 54). It is a good exercise to identify the icosahedron in projection as shown in Fig. 6.61.

Figure 6.61 also shows the coordination figure about **Cr**; it should be identified as the 14-vertex polyhedron obtained by capping the hexagonal faces of a hexagonal antiprism. The coordination of **Cr** is {**Cr**}**$Si_4Cr_{10}$**; the closest neighbors of **Cr** are two other **Cr** at a

[1]After the German crystallographer F. Laves, who contributed significantly to the understanding of intermetallic structures of this and related types. "Laves" is pronounced with two syllables: "La-ves."

[2]In *Structurbericht* (the predecessor of *Structure Reports*) simple structures were assigned symbols. *An* represents elemental structures such as *A*1 = **ccp**, *A*2 = **bcc**, *A*3 = **hcp**, *etc.* A form of W metal (or $W_3O$? = "*β*-tungsten") was early reported to be **$Cr_3Si$** and although this is no longer believed to be correct, the designation *A*15 is still used for this structure type. The designations *B*1 = **NaCl** and *B*2 = **CsCl** (**CuZn**) are also still used on occasion.

distance of $a/2$.

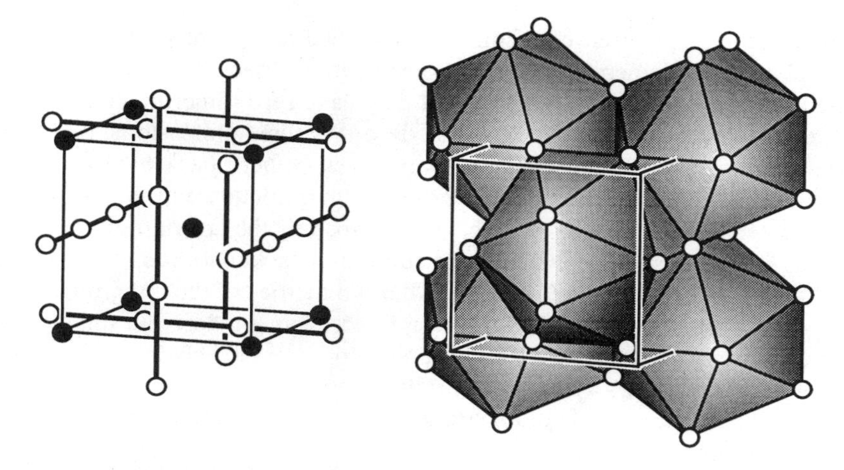

**Fig. 6.60.** $\mathbf{Cr_3Si}$ (**Cr** are open circles, **Si** are filled circles). Left: emphasizing the **Cr** rod packing. Right: showing some of the $\{\mathbf{Si}\}\mathbf{Cr_{12}}$ icosahedra (note the two orientations of the icosahedra).

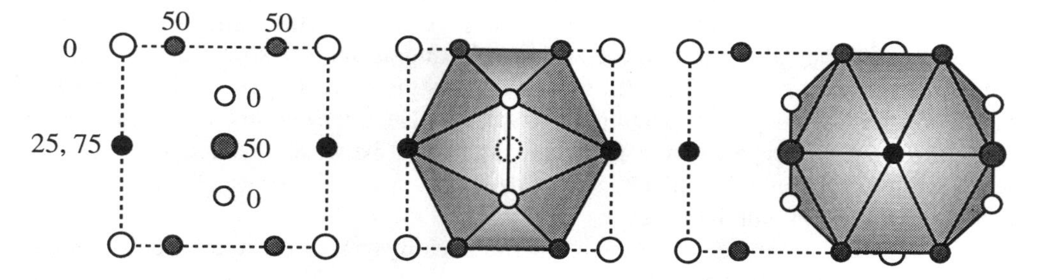


**Fig. 6.61**. Left: a unit cell of $\mathbf{Cr_3Si}$ projected on (001), numbers are heights in multiples of $c/100$. Larger circles represent **Si**. The shading scheme now indicates elevations. Center: The icosahedron centered at 1/2,1/2,1/2 is picked out. Right: The polyhedron around a **Cr** at $z = 1/4$.

Superconducting compounds with this structure are of considerable interest as many have high critical temperatures. $Nb_3Ge$ ($T_c$ = 23.2 K) has the highest superconducting transition temperature of any known material other than the copper-oxide superconductors.

The cation array in **garnet** oxides such as $Ca_3Al_2Si_3O_{12}$ is an ordered derivative of $\mathbf{Cr_3Si}$ with $(Ca_{3/2}Si_{3/2})$ in the **Cr** positions and Al in the **Si** positions. Indeed the complicated garnet structure is completely generated by putting O atoms in all the $Ca_2AlSi$ tetrahedra of the cation array. In Volume II of this series we will show that many other complex oxide structures are conveniently described as intermetallic structures "stuffed" with O atoms.[1]

[1]For an earlier account of this topic see M. O'Keeffe & B. G. Hyde, *Structure and Bonding* **61**, 77 (1985).

## 6.7 Cylinder (rod) packings[1]

Crystal structures are often conveniently described as sphere packings, with atoms instead of spheres, but sometimes we want to consider the packing of larger units. Atoms (or spheres) are *point* objects in the sense that they have zero-dimensional periodicity and crystals are three dimensional in the sense that they have three-dimensional periodicity. We have already described three-dimensional structures (sphere packings) as packings of *layers*, which are objects with two-dimensional periodicity, and we do so again in following chapters. The symmetry of layers is described by the layer groups (Appendix 1).

Sometimes it is convenient to consider structures as a packing of rods, which are objects with one-dimensional periodicity. These have the symmetries of the rod groups (Appendix 1).[2] The highest symmetry rod is the infinite cylinder; here we describe some packings of equivalent (symmetry-related) cylinders. In application to crystal chemistry, we replace the cylinders with *rods* of atoms. Examples of rods are strings of atoms, atoms at the vertices of rods of polyhedra (e.g. octahedra sharing opposite faces), atoms forming helices (as commonly found for rods of S atoms).

In the symmetrical packings we describe, symmetry axes coincide with the cylinder axes and are therefore non-intersecting. A line (such as a cylinder axis) corresponds to the locus of all points of a univariant lattice complex. For example $0,0,z$ or $\pm(0,0,z)$ corresponds to a line along **c** if $z$ is allowed to take all possible values. In the same way $\pm(0,0,z)\kappa$ corresponds to lines along **a**, **b** and **c** and intersecting at 0,0,0 (also at 1,0,0 ; etc.); this cannot correspond to a cylinder packing as the lines intersect. In general the location of axes is given as the line of intersection of two planes. The intersection of planes $x = x_0$ and $y = y_0$ is written as $x_0,y_0,z$ where $z$ can have any value. Likewise $1/3+u,2/3+u,u$ indicates the line of intersection of the planes $x = z + 1/3$ and $y = z + 2/3$. Unit cell parameters are given for cylinders of unit diameter.

In descriptive crystal chemistry the cubic rod packings are of most importance and are met repeatedly in that connection.

### *6.7.1 Cylinders with parallel axes*

If cylinders are packed with axes parallel to **c**, a cross-section $z$ = constant will just be a circle packing. In particular the closest packing of equal cylinders will be a hexagonal packing (i.e. based on a $3^6$ net):

*P6/mmm*, $a = 1$. Cylinder axes along 2 *e*: $\pm(0,0,z)$. Rod symmetry **p**6/*mmm*

The fraction of space occupied by the cylinders (the density) is the same as for closest

[1]For more on this topic see M. O'Keeffe & S. Andersson, *Acta Crystallogr.* A**33**, 914 (1977) and M. O'Keeffe, *Acta Crystallogr.* A**48**, 879 (1992).

[2]The reader anxious to learn about rod symmetries will find some good examples in this section. The less ambitious can skip the parts dealing with this topic. After all we got through § 6.4 (on the stacking of layers) without discussing layer groups (although the temptation to do so was strong). The reason was in part due to the fact that most of our "layers" were only one atom thick.

circle packing $\rho = \pi/\sqrt{12} = 0.9070$. The rod symmetry **p**6/*mmm* is that of an infinite column of hexagonal prisms stacked along the hexagonal axis. (**p** stands for the one-dimensional lattice).

An example of this packing is provided by the helical rods of P in NaP (compounds such as KP, NaAs and RbSb are isostructural) illustrated in Fig. 6.62. The rods have a 4-fold repeat so the axis is approximately a $4_1$ axis and all the helices are of the same hand. In this structure, the P-P bond length is 2.24 Å, and the P-P-P angles are 112° and 115°. LiP, LiAs, NaSb and KSb have a related structure containing helices of P of both hands. For crystallographic data see Appendix 5. Elemental Se has $3_1$ helices in the same rod packing.

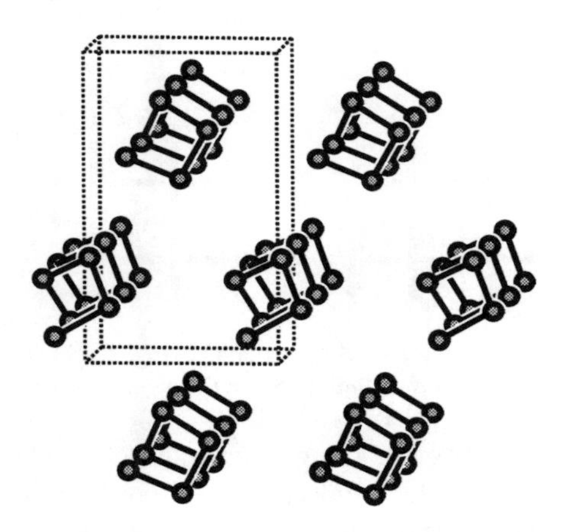

**Fig. 6.62.** The helical rods of P in NaP viewed at an angle slightly tilted from a $2_1$ axis.

Further cylinder packings can be derived from other regular and Archimedean tessellations. The least dense packing, based on $3.12^2$, has density $\rho = \sqrt{3}\pi/(7 + 4\sqrt{3}) = 0.3907$.

### *6.7.2 Cylinders with axes in parallel planes*

Here we describe packings of layers of cylinders in contact. They will all have the same density, $\rho = \pi/4 = 0.7854$. Cylinder axes lie along non-intersecting 2-fold rotation axes.

(a) Two-layer tetragonal. Here cylinder axes lie in layers perpendicular to **c**; for $z = 0$ the axes run in the [100] direction, for $z = 1/2$, they lie in the [010] direction. Thus the rods run along **a** through 0,0,0 and along **b** through 0,0,1/2; see Figs. 6.63 and 6.64. Note that we give unit cell parameters for a packing of cylinders of unit diameter; in a crystal structure the rod symmetry is **p***mmm* (which is the symmetry of, for example, an infinite stack of bricks) and the axial ratio *c*/*a* will not, in general, be equal to 2.

$P4_2/mmc$, $a = 1$, $c = 2$
Cylinder axes along 4 *l*: $\pm(x,0,1/2\ ;\ 0,x,0)$. Rod symmetry **p***mmm*

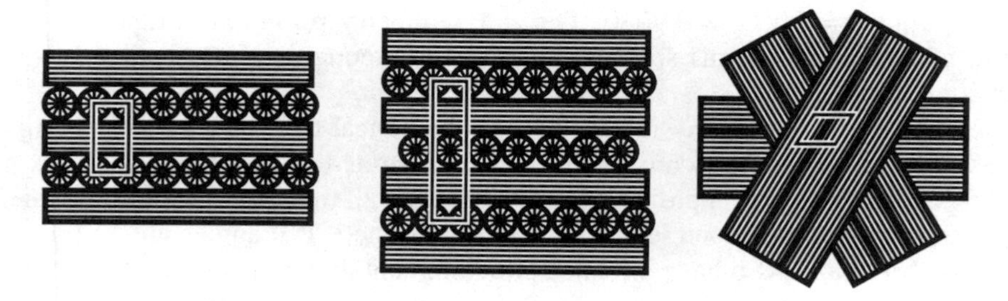

**Fig. 6.63**. Left: a two-layer tetragonal cylinder packing. Center: a four-layer tetragonal cylinder packing. **c** is up the page in both cases. Right: a three-layer hexagonal cylinder packing viewed down **c**. (See also Fig. 6.64 for the first two cases).

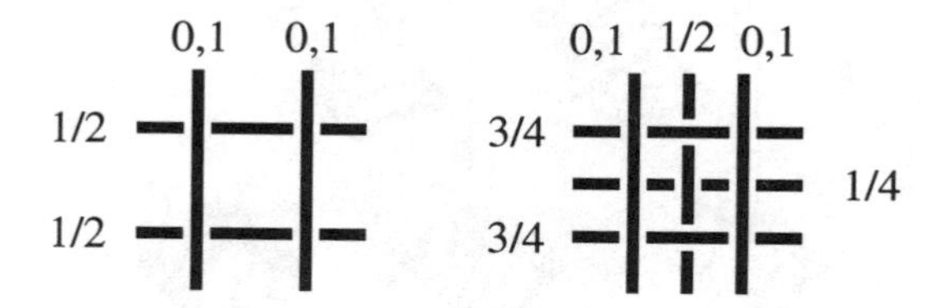


**Fig. 6.64**. Left: axes of a two-layer tetragonal cylinder packing. Right: axes of a four-layer tetragonal cylinder packing. The view is down **c** and the numbers are the elevations of the axes in units of $c$.

(b) Four-layer tetragonal. Rods run along **a** through 0,3/4,1/4 and 0,1/4,3/4 and along **b** through 0,0,0 and 1/2,0,1/2 (this description assures that there is a center of symmetry at 0,0,0). The arrangement should be apparent from Figs. 6.63 and 6.64.

$I4_1/amd$, $a = 1$, $c = 4$
Axes along $0,x,0$ ; $x,1/4,3/4$ ; $1/2,x,1/2$ ; $x,3/4,1/4$ derived from
16 $h$: $I \pm (0,x,z$ ; $0,1/2-x,z$ ; $1/4+x,1/4,3/4+z$, $3/4-x,1/4,3/4+z)$, $z = 0$
Rod symmetry **p***mcm*

A digression on rod symmetry: The rods in the four-layer structure (b) run along $2_1$ axes parallel to **a** and **b** of $I4_1/amd$. Consider the rod along **a**: it lies in the mirror plane normal to **b** and is normal to the mirror plane normal to **a**. The rod also lies in the $a$ glide plane normal to **c**.[1] The symbol for the rod symmetry group is accordingly **p***mma*. However in the rod groups it is conventional to take the translation direction as (with subscript "$r$" for the rod) $\mathbf{c}_r$. Making the substitution $\mathbf{a}_r = \mathbf{b}$, $\mathbf{b}_r = \mathbf{c}$, $\mathbf{c}_r = \mathbf{a}$, the rod symmetry group with these new axes becomes **p***mcm* as given above. The full symbol is **p** $2/m$ $2/c$ $2_1/m$. An infinite crankshaft or zig-zag (see § 7.3.5) has this symmetry. End digression.

A beautiful example of this cylinder packing is found in the structures of $Hg_{3-x}MF_6$ ($M$ = As,Sb,Nb,Ta and $x$ is typically about 0.12) The As compound is known as "alchemist's gold." Isolated $\{M\}F_6$ octahedra are surrounded by disordered rods of Hg atoms in positions 16 $h$ of $I4_1/amd$ with $z \approx 0$ as shown schematically in Fig. 6.65. Remarkably,

[1]It is a good exercise to get a copy of the *International Tables* and verify these statements.

the stoichiometric compound $Hg_3NbF_6$ has a simple layer structure with $3^6$ layers of Hg alternating with layers of isolated $\{Nb\}F_6$ octahedra.[1]

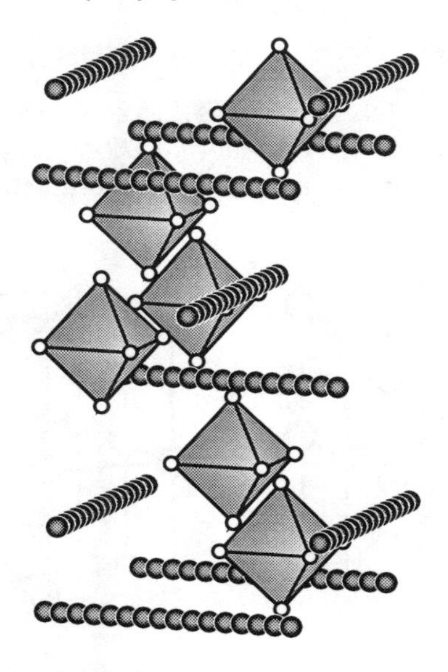

**Fig. 6.65**. The structure of $Hg_{2.88}SbF_6$ illustrated as $\{Sb\}F_6$ octahedra and rods of Hg atoms. The Hg atoms are disordered along the rods shown.

(c) Three-layer hexagonal. Rods run along **a** through 0,0,0 along **a** + **b** through 0,0,1/3 and along **b** through 0,0,2/3. The arrangement of cylinders should be apparent from Fig. 6.63.

$P6_222$, $a = 1$, $c = 3$
Axes along 6 $g$: ($\pm x$,0,0 ; $x,x$,1/3 ; $\bar{x},\bar{x}$,1/3 ; 0,$\pm x$,2/3). Rod symmetry **p**222

### *6.7.3 Cubic cylinder packings*

These are rather important as they sometimes form the basis for a description of cubic crystal structures that are otherwise difficult to describe. They are named for structures in which they are conspicuous features. We describe four packings and an intergrowth structure. The cylinder axes correspond to non-intersecting symmetry axes in cubic space groups (see § 3.3.6, p. 74) which are either (i) 2-fold or 4-fold axes parallel to <100> or (ii) 3-fold axes parallel to <111>. In the second case, the fact that the cylinder axes are inclined to each other makes structures based on this structural principle difficult to illustrate satisfactorily in projection. The most important packings are those named ***β*-W** and **garnet** [(a) and (b) below]. Cubic space groups with non-intersecting 3-fold axes are listed on p. 83.

[1]See I. D. Brown *et al., Inorg. Chem.* **23**, 405 (1984) and references therein.

(a) **β-W** (Figs. 6.66 and 6.67)

$Pm\bar{3}n$, $a = 2$, $\rho = 3\pi/16 = 0.5890$
Axes along 12 $g$: $\pm(x,0,1/2\ ;\ 1/2+x,0,1/2)\kappa$. Rod symmetry **p4$_2$/*mmc***

In the **β-W** structure of $A_3B$ (also known as $A15$ or **$Cr_3Si$**—see § 6.6.4), strings of $A$ atoms lie on the cylinder axes along <100> at $(u,0,1/2)\kappa$. (The $B$ atoms in $A_3B$ are at cell corners and at the body center and are in icosahedral coordination by $A$). The cylinder axes correspond to the $4_2$ axes of $Pm\bar{3}n$.

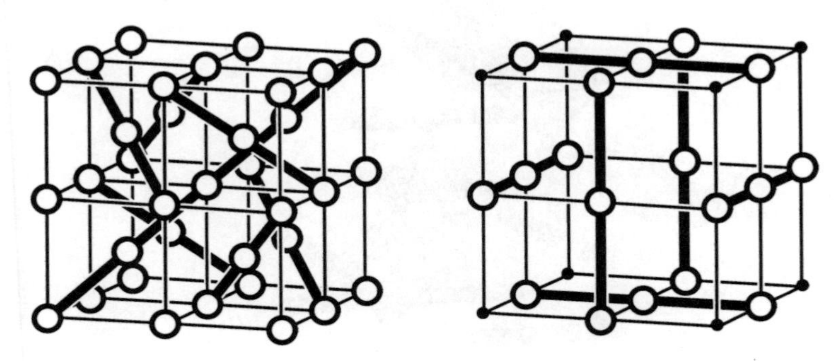

**Fig. 6.66**. Two cubic cylinder packings. Left: **garnet**. Right: **β-W**. The heavy lines indicate the locations of the cylinder axes. The open circles serve only to indicate high symmetry points on the rods.

(b) **garnet** (Fig. 6.66)

$Ia\bar{3}d$, $a = \sqrt{8}$, $\rho = \sqrt{3}\pi/8 = 0.6802$
Axes along 32 $e$: $I\pm(x,x,x\ ;\ 1/4+x,1/4+x,1/4+x\ ;\ 1/2-x,\ 1/2+x,x\ ;\ 3/4-x,1/4+x,3/4+x$
$x,1/2-x,1/2+x\ ;\ 3/4+x,3/4-x,\ 1/4+x\ ;\ 1/2+x,x,1/2-x\ ;\ 1/4+x,3/4+x,3/4-x)$
Rod symmetry **p$\bar{3}$*c*1**

In this cubic cylinder packing, cylinders are parallel to body diagonals of a cubic cell. Now they lie on non-intersecting $\bar{3}$ axes, i.e. along [111], [11$\bar{1}$], [1$\bar{1}$1] and [$\bar{1}$11].

In the **garnet** structure of $Ca_3Al_2Si_3O_{12}$ the cylinders are to be replaced by rods of alternating $\{Al\}O_6$ octahedra and empty $O_6$ trigonal prisms sharing opposite triangular faces. It might be noted that the centers of the octahedra are at $\bar{3}$ sites so they are trigonal antiprisms, but the absence of $\bar{6}$ axes means that the trigonal "prisms" are not strictly regular prisms (they are slightly twisted towards being metaprisms in the garnet structure). As discussed in Appendix 2, a rod of alternating *regular* octahedra and trigonal prisms has symmetry **p6$_3$/*mmc***; but the $6_3$ axis is incompatible with cubic symmetry.

(c) **β-Mn** (Fig. 6.67):

$I4_132$, $a = 4$, $\rho = 3\pi/32 = 0.2945$
Axes: $1/4,0,u\ ;\ 3/4,1/2,u\ ;\ u,1/4,0\ ;\ u,3/4,1/2\ ;\ 0,u,1/4\ ;\ 1/2,u,3/4$.
Rod symmetry **p4$_1$22**

A third cylinder packing is derived from **β-W** by removal of one-half of the cylinders. We call this the **β-Mn** cylinder packing as that otherwise enigmatic structure is simply described in terms of rods of face-sharing $Mn_4$ tetrahedra with this cylinder packing. Fig. 6.67 shows the **β-W** and the **β-Mn** packings, the former with a cell with doubled edge. The symmetry of the **β-Mn** packing is $I4_132$ and the axis equations are obtained by substituting $1/4,0,u$ in the general positions 48 *i*. The cylinder axes lie along the $4_1$ axes of this space group; the substitution $3/4,u,0$ puts the cylinder axes on the $4_3$ axes. Combination of the two enantiomorphs of the **β-Mn** packing recovers the **β-W** packing.

**Fig. 6.67**. The **β-W** (left) and **β-Mn** (right) cylinder packings. The true cell edge for **β-W** has half the edge of that shown.

(*d*) **$SrSi_2$** (Fig. 6.68):

> $I4_132$, $a = 6\sqrt{2}$, $\rho = \sqrt{3}\pi/72 = 0.0756$
> Axes: $1/3+u,2/3+u,u$ ; $1/6+u,2/3-u,u$ ; $2/3+u,5/6+u,-u$ ; $5/6-u,5/6+u,u$.
> Rod symmetry **p3$_1$2**

A fourth cylinder packing is obtained by removing eight-ninths of the cylinders of the **garnet** packing. We call it the **$SrSi_2$** packing because the Si atoms in that structure form $3_1$ (or $3_2$) helices with axes corresponding to those of the cylinder packing (see § 7.2). The positions of the cylinder axes ($3_1$) are obtained by substituting $1/3+u,2/3+u,u$ in the general positions 48 *i* of $I4_132$. The enantiomorphous structure with cylinders on $3_2$ axes is obtained by the substitution $2/3+u,1/3+u,u$. A sketch of the packing viewed down one of the $3_1$ axes is shown in Fig. 6.68.

(*e*) **γ-Si**:

> $Ia\bar{3}d$, $a = 6\sqrt{2}$
> $3_1$ axes $1/3+u,2/3+u,u$ ; $1/6+u,2/3-u,u$ ; $2/3+u,5/6+u,-u$ ; $5/6-u,5/6+u,u$
> $3_2$ axes $2/3+u,1/3+u,u$ ; $5/6+u,1/3-u,u$ ; $1/3+u,1/6+u,-u$ ; $1/6-u,1/6+u,u$.
> Rod symmetry **p3$_1$2** and **p3$_2$2**

The **$SrSi_2$** packing is very open, and the two enantiomorphs can intergrow without contact (so the intergrowth structure is not a *stable* cylinder packing). In the **γ-Si** polymorph of silicon (§ 7.3.12), the Si atoms fall on $3_1$ and $3_2$ helices with axes corresponding to this intergrowth structure. The cylinder axes are now obtained by the substitution of $1/3+u,2/3+u,u$ in the general positions of $Ia\bar{3}d$.

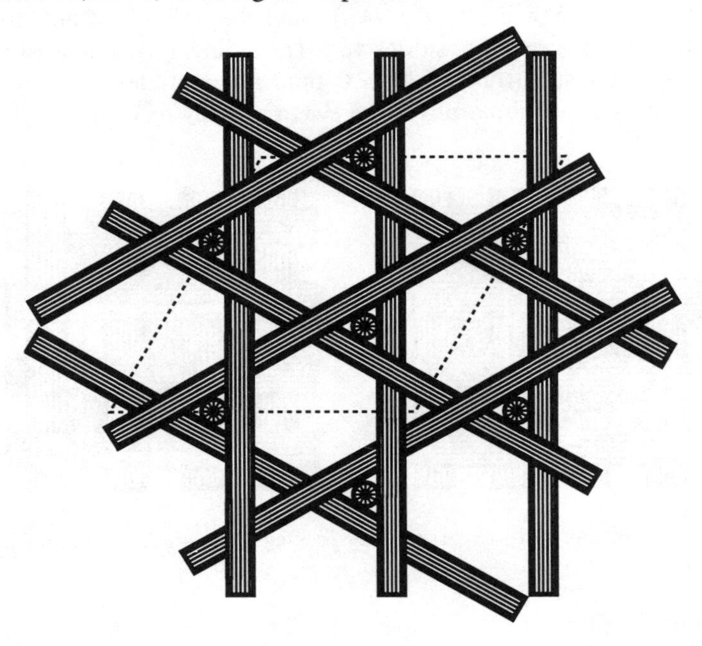

**Fig. 6.68**. The **$SrSi_2$** cylinder packing viewed down [111]. Note that for clarity only a few cylinders not parallel to the projection axis are shown.

## 6.8 Notes

### *6.8.1 Symmetries of arrays of closest packed spheres*

The space groups for arrays corresponding to closest sphere packings are $Fm\bar{3}m$ (only for **ccp**), $P6_3/mmc$, $P6_3mc$, $P\bar{6}m2$, $R\bar{3}m$, $R3m$, $P\bar{3}m1$ and $P3m1$. A useful discussion and table has been published [A. L. Patterson & J. S. Kasper, *International Tables*, vol. II] that allows the symmetry of complex sequences to be determined. In many crystal structures one or more sets of atoms are only approximately in closest packing and the symmetry may be lower.

The ideal symmetry is readily determined from the Zhdanov symbol (§ 6.1.3) of the packing from the rules given here, which should be applied in the order given until the space group is determined. In using the rules be sure to note that a sequence such as 2211 refers to ...221122112211... and could equally be written 1221 or 2112.

(a) Determine if there is a center of symmetry: This is revealed by symmetry of certain *numbers* in the succession of numbers. Thus 2111 can be written:

...(2)1(1)1(2)1(1)1(2)...

where the numbers in parentheses are located symmetrically. There will always be zero (no center of symmetry) or two such numbers (centrosymmetrical) in the symbol. Note that if the symbol contains just two numbers (as in 11 or 41) each number is symmetrically surrounded, and that in e.g. 2112 no *number* is symmetrically surrounded.

(b) If the first half of the symbol is the same as the second half the symmetry is either $P6_3/mmc$ (centrosymmetric) or $P6_3mc$ (non-centrosymmetric). Thus 11 (**hcp**) and 121121 have symmetry $P6_3/mmc$ and 123123 has symmetry $P6_3mc$.[1]

(c) If the symbol can be written so that the second half of the symbol is the reverse of the first half, but there is no center of symmetry, the symmetry is $P\bar{6}m2$. An example is 2112.

(d) The Zhdanov symbol always has an even number of terms: $N_1N_2N_3N_4....N_n$ ($n$ even). If the symmetry has not yet been determined [in (b) or (c)], subtract the sum of the even terms from the sum of the odd terms i.e.: $N_1 + N_3 +...+ N_{n-1} - (N_2 + N_4 +...+ N_n)$.

If the result is either zero or a multiple of 3 the symmetry is trigonal (but not rhombohedral): either $P\bar{3}m1$ (centrosymmetric) or $P3m1$ (non-centrosymmetric). Thus 41 has symmetry $P\bar{3}m1$ and 5211 has symmetry $P3m1$.

If the result is neither zero nor a multiple of three the symmetry is rhombohedral: either $R\bar{3}m$ (centrosymmetrical) or $R3m$ (non-centrosymmetrical). Thus 21 (*hhc*) has symmetry $R\bar{3}m$ and 3211 has symmetry $R3m$.

### 6.8.2 Neighbors, coordination sequences, and identifying packings

It should be noted that closest sphere packings differ in numbers of $n$th geometrical neighbors. For unit diameter spheres the number of neighbors at a given distance are listed for the first few shells of **ccp** and **hcp** below. Generally such numbers cannot be used to distinguish packings in crystal structures, as the arrangement often only approximates an ideal sphere packing, and the numbers of geometrical neighbors rapidly lose any relation to those in the ideal packing.

| distance | 1 | √2 | √(8/3) | √3 | √(11/3) | 2 |
|---|---|---|---|---|---|---|
| **ccp** | 12 | 6 | 0 | 24 | 0 | 12 |
| **hcp** | 12 | 6 | 2 | 18 | 12 | 6 |

In Chapter 7 we discuss *coordination sequences* which represent the numbers of *topological* neighbors in shells. In the context of sphere packings, a second topological neighbor of a sphere is one (other than the reference sphere) in contact with first neighbors; third neighbors are those (other than first neighbors) that are in contact with second neighbors; and so on. The number of $k$th neighbors in *this* sense is $n_k$. It is interesting that $n_k$ for $k > 1$ is greater for **hcp** than for **ccp**. For **hcp** the sequence is 12, 44, 96... for **ccp**

[1]The symbols for these packings could be abbreviated <1>, <121> and <123> respectively.

it is 12, 42, 92.... Be sure to distinguish *topological* neighbors (discussed here) with *geometrical* neighbors (discussed in the previous paragraph).

For *lattice* sphere packings [$cP$, $cI$, $cF$, $tI$ (**bct**) and $hP$ ($c/a$ = 1)] with $N$ first neighbors there is a simple expression for the numbers of topological neighbors:

$$n_k = (N - 2)k^2 + 2 \tag{6.1}$$

Some other equations are (brackets indicate rounding down to an integer):

**hcp** $$n_k = [21k^2/2 + 2] \tag{6.2}$$

**cco** $$n_k = [17k^2/2 + 2] \tag{6.3}$$

Once the nearest neighbors of atoms in a structure have been identified (for example, on the basis of interatomic distance) the coordination sequence for each atom is uniquely defined and the packing can often be identified from the coordination sequences even when the arrangement departs significantly from the ideal geometry. In particular each kind of atom in a **cp** structure (*h, c, hc,* etc.) has a unique coordination sequence and this fact may be exploited to determine the nature of the packing.[1]

### *6.8.3 Close packing or polyhedron packing? An unsolved problem*

Many "ionic" crystal structures are based on approximately **cp** arrangements of cations and/or anions (and just as importantly, many are not). The well known structures of spinel ($MgAl_2O_4$) and olivine ($Mg_2SiO_4$) are examples in which the anion arrangement is approximately **ccp** and **hcp** respectively. A popular view (to which we do not subscribe) is that the reason for such structures occurring is that "large" anions are close packed (why, for heaven's sake?) and the "small" cations fit more-or-less snugly in the tetrahedral and/or octahedral interstices. One objection to this proposal is that many (e.g. oxide) structures are *not* based on **cp** arrays, or if they are, they are often **cp** *cation* arrays; but nevertheless one is lead to ask why so many structures based on **cp** (or better eutaxy) do occur.

A possible answer is as follows. The most common coordination figures found in oxides and related materials are $\{M\}X_4$ tetrahedra and $\{M\}X_6$ octahedra [even in compounds which are not based on **cp** such as enstatite ($MgSiO_3$)]. In order to make a crystal of the appropriate stoichiometry, the individual polyhedra must be condensed together by sharing corners and/or edges and/or faces. To take a concrete example: $MgAl_2O_4$ is constructed of $\{Mg\}O_4$ tetrahedra and $\{Al\}O_6$ octahedra combined in the ratio 1:2 and sharing O atoms so that there are four O atoms per $MgAl_2$. With regular polyhedra of equal edges it is conjectured that *any* periodic way of combining them subject to the foregoing constraints will result in a **cp** array of O atoms.

The unsolved (we think) problem which we offer the reader is this: What stoichiometries

[1]The program EUTAX does this for a number of simple structures. Users of this program might like to find the **cp** array of the I atoms in Exercise 11. Note that the numbers in the coordination sequence are largest for **hcp** and smallest for **ccp** (*i.e.* all other **cp** structures have intermediate values).

and combinations of octahedra and tetrahedra will lead inexorably to the polyhedron vertices being on a **cp** (i.e. eutactic) array?

In applications to crystal chemistry additional constraints might be added, such as not allowing tetrahedra to share faces (which would allow their central atoms to come rather close together), and to eliminate configurations that result in very asymmetric coordinations around the polyhedron vertices [see E. W. Gorter, *J. Solid State Chem.* **1**, 279 (1970).]

The term "close packing" is sometimes used rather loosely. For example in a discussion of the stability of the feldspar structure (specifically sanidine = $KAlSi_3O_8$) in a well known text it is stated that the oxygen atoms "approximate rather crudely to cubic close packing...perhaps this relative compactness contributes to the stability." The feldspar structure is based on a framework of corner-sharing $O_4$ tetrahedra and it would therefore be expected (see § 6.8.5) that each O atom will have six near neighbors and indeed this is the case. The six nearest neighbors of O atoms in sanidine are in the distance range 2.60-2.74 Å (corresponding to tetrahedron edges) and the next six neighbors are in a distance range of 3.30-4.48 Å.

### *6.8.4 More on the relationship between* **bcc** *and* **hcp***:* **AuCd**

In § 6.3.3 we described a relationship between **bcc** and **hcp**. Some metallic elements such as Ti and Zr have both structures (the high temperature or $\beta$ form is **bcc**) and the transition occurs very nearly at constant volume. The orthorhombic cells given in § 6.3.3 become in units of $V^{1/3}$ for **bcc**: $a, b, c$ = 0.794, 1.122, 1.122 and for **hcp**: $a, b, c$ = 0.707, 1.225, 1.154, so that the transformation from **bcc** on cooling requires about a 10% decrease in $a$ and a 10% increase in $b$. In **CuZn** compounds alternate {100} layers of **bcc** are **Cu** and **Zn**; some of these transform at low temperatures to a superstructure of **hcp** by the mechanism described. Fig 6.69 shows the resulting structure which is called **AuCd**, as the martensitic transformation has been well studied in that compound. In the binary compound the symmetry is *Pmma* and the unit cell is derived from the one described here by (0 0 1 / 1 0 0 / 0 1 0). Data for two forms of AuCd are given in Appendix 5. The low temperature form is close to ideal **hcp**; the 12 shortest distances are 2.89-3.16 Å.

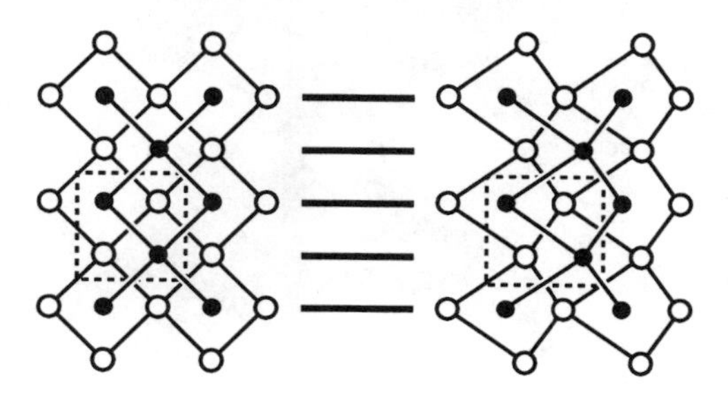

**Fig 6.69**. Left: the **CuZn** structure projected on (001). Right: the **AuCd** structure plotted on (010) of the *Pmma* cell with **c** horizontal on the paper. Heavy lines mark the position of (110) planes of **CuZn** that become **cp** planes in **AuCd**. If all the atoms were the same this figure would also illustrate the relationship of **bcc** (left) to **hcp** (right). Compare Fig. 6.24 (p. 232).

### *6.8.5 More sphere packings*

A complete enumeration of sphere packings would be a big task and has not been done to our knowledge. Cubic sphere packings have been enumerated by W. Fischer [*Zeits. Kristallogr.* **138**, 129 (1973); **140**, 50 (1974)] who found no examples with 10- or 11-coordination. An article on sphere packings is in the *International Tables C*. The study of lattice sphere packings in *N*-dimensional space is an active area of research in mathematics—see Appendix 2.

Here we describe some additional structures confined (so far!) to fewer crystal structure types than most of the structures described above.

#### *Another 10-coordinated sphere packing, and the* $Ti_5Te_4$ *structure*

Cubes with four faces capped by square pyramids (half octahedra) can be packed as shown in Fig. 6.70. The cubes share their uncapped faces to form rods parallel to **c** of a tetragonal cell, and the pyramids, which cap the other four faces of the cubes, share edges. The structure is less dense than the other 10-coordinated sphere packings described in this chapter:

$I4/m$, $a = \sqrt{[17/(14 - \sqrt{128})]} = 2.5156$, $c = 1$, $\rho = (56 - 32\sqrt{2})\pi/51 = 0.6619$
Centers in 8 $h$: $I \pm (x,y,0\ ;\bar{y},x,0)$, $x = (6 - \sqrt{2})/17 = 0.2697$, $y = (7 - \sqrt{32})/17 = 0.0790$

The Te arrangement in $Ti_5Te_4$ is quite close to this packing. The Ti atoms center all the cube faces, so the Ti arrangement consists of rods of octahedra sharing opposite vertices. About a dozen compounds (e.g. $V_5S_4$, $Ta_5Sb_4$) have the same structure.

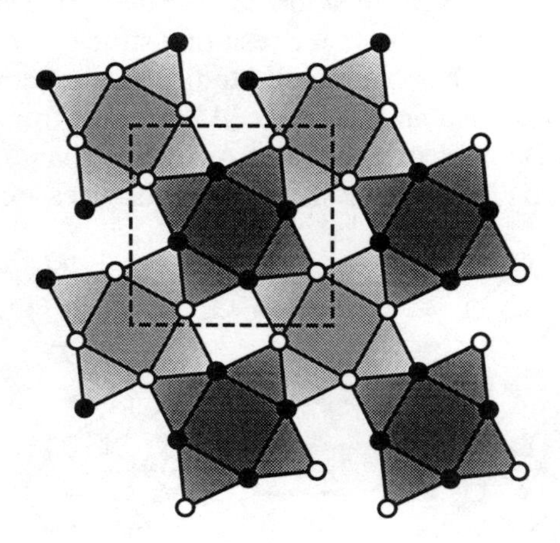

**Fig. 6.70**. A 10-coordinated sphere packing projected down the **c** axis. Open circles are at $z = 0$ and filled circles are at $z = 1/2$.

If the cubes with capped faces are deformed into cuboctahedra the result is a

9-coordinated packing corresponding to **ccp** with 1/5 of the atoms removed (compare with another 9-coordinated packing described below). The parameters would now be $a = \sqrt{5}$, $c = \sqrt{2}$, $a/c = 1.58$ $x = 3/10$, $y = 1/10$ and the sites at 4 $d$: $I + (0,1/2,1/4\ ;\ 1/2,0,1/4)$ are in regular tetrahedra. In $\beta$-$BaFe_2S_4$, $a/c = 1.45$, S atoms are in 8 $h$ with $x = 0.301$, $y = 0.120$, Ba centers the cuboctahedra of S and Fe is in the $S_4$ tetrahedra (for data see Appendix 5). $Ga_2Te_5 = Te(1)Ga_2Te(2)_4$ is iso-structural [with Te(1) playing the role of Ba and Te(2) playing the role of S].

*Some 9-coordinated sphere packings*

One symmetrical arrangement is:

$I\bar{4}3m$, $a = \sqrt{8}$. Centers in 24 $g$: $I + (x,x,z\ ;\ \bar{x},x,\bar{z}\ ;\ x,\bar{x},\bar{z}\ ;\ \bar{x},\bar{x},z)\kappa$, $x = 3/8$, $z = 1/8$

This arrangement represents a way of removing 1/4 of the spheres of cubic eutaxy so that each sphere has nine neighbors (compare the 8-coordinated $J$ arrangement, which can also be described as cubic eutaxy with 1/4 of the spheres removed). The density is accordingly $\rho = \pi/\sqrt{32} = 0.5554$. This packing is a special case of the anion packing in the mineral sodalite discussed under 6-coordinated sphere packings below (p. 274).

A second 9-coordinated sphere packing that occurs in a variety of contexts (for example as the Al arrangement in $WAl_{12}$) is also discussed below (p. 278) as an example of a sphere packing with icosahedral interstices. Fischer's compilation, referred to above, includes two other examples of 9-coordination with cubic symmetry.

*Another 8-coordinated sphere packing: the* $NaZn_{13}$ *structure*

About 50 compounds, mostly $MZn_{13}$ and $MBe_{13}$ (here $M$ is a "big" atom from the first three columns of the periodic table) have the $NaZn_{13}$ structure (for crystallographic data see Appendix 5). We describe it here, as it is an elegant example of how an apparently complex structure is built up from very simple principles.

We start by assembling an infinite structure by joining together snub cubes ($3^4.4$) sharing square faces in every possible direction; every polyhedron of one hand (recall the symmetry of a snub cube is 432) is joined in this way to six polyhedra of the other hand. Each vertex of this assembly will have eight nearest neighbors, so it may be considered an 8-coordinated sphere packing. A formal description is:

$Fm\bar{3}c$, $a = 4.5704$, $\rho = 0.5265$
Sphere centers in 96 $i$: $F \pm (0,\pm y,z\ ;\ 1/2,\pm z,y)\kappa$, $y = 0.1761$, $z = 0.1141$
Snub cube centers in 8 $a$: $F \pm (1/4,1/4,1/4)$
Icosahedron centers in 8 $b$: $F + (0,0,0\ ;\ 1/2,1/2,1/2)$

As well as the large holes at the centers of the snub cubes (filled by Na in $NaZn_{13}$) with symmetry 432, there are holes surrounded by 12 equidistant spheres forming almost regular icosahedra with symmetry $m\bar{3}$. In $NaZn_{13}$, Zn atoms center these icosahedra forming $Zn_{13}$ groups. (The unit cell therefore contains 96 + 8 = 104 Zn atoms and 8 Na atoms, i.e. eight $NaZn_{13}$ units.) In the real structure the free parameters for Zn are quite close to the ideal ones given above.

Figure 6.71 shows a beginning of the packing of snub cubes. The place where an icosahedron can nestle between the snub cubes should be identified. The figure also shows a snub cube sharing triangular faces with eight icosahedra (note that the latter occur in two different orientations). The triangular faces of the snub cube and the darker-shaded faces of the icosahedra are equilateral triangles. The icosahedron edges parallel to the cubic axes are about 4% longer than the others.

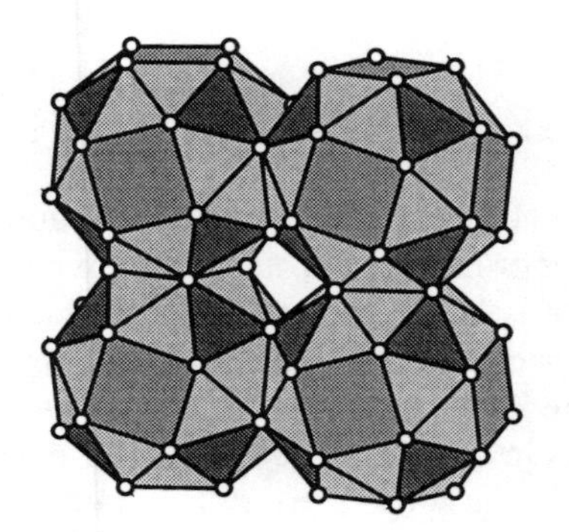
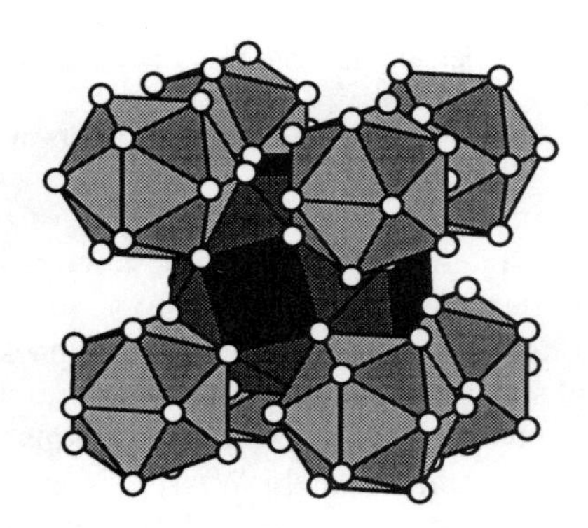

**Fig. 6.71**. Left: snub cubes (two of each hand) sharing square faces. Right: a snub cube (darker shaded and mostly obscured) sharing triangular faces with icosahedra as in the $NaZn_{13}$ structure.

A remarkable example of this structure is in a rare form of opal from Brazil. Two different-sized spheres of silica pack as Na and Zn. (Common opals have one size of sphere in **cp**). The spheres are now much bigger than atoms: about 0.5 μm [see J. V. Sanders & M. J. Murray, *Nature* **275**, 201 (1975) and *Phil. Mag.* **42**, 721 (1980)].

*More 6-coordinated sphere packings*

Frameworks of corner-connected (regular) tetrahedra such as the *T* structure are 6-coordinated sphere packings. The O atoms in quartz (§ 3.6) form another such framework. The O atoms in the sodalite structure are also an example. For reference we give coordinates for regular tetrahedra of unit edge length. The centers of tetrahedra (here labeled Si for convenience) are on a $W^*$ net:

$Im\bar{3}m$, $a = 2 + \sqrt{2} = 3.4142$, $\rho = 0.3157$
Sphere centers (O) in 24 *h*: $I + (0,\pm x,\pm x)\kappa$, $x = 1/\sqrt{8} = 0.3536$
Tetrahedron centers (Si) in 12 *d*: $I \pm (1/4,0,1/2)\kappa$

This is a special high-symmetry, low-density case of a more general 6-coordinated packing with symmetry $I\bar{4}3m$, in which the tetrahedron centers remain in the same positions (also labeled 12 *d*) but the vertices are in positions 24 *g*: $I + (x,x,z\ ;\ \bar{x},x,\bar{z}\ ;\ \bar{x},\bar{x},z$

$x,\bar{x},\bar{z})\kappa$. For a 6-coordinated sphere packing (regular tetrahedra sharing vertices), $x = \sqrt{(z^2 + 1/8)}$ with $0 \le z < 1/8$. When $z = 0$ we regain the $Im\bar{3}m$ structure with an Si-O-Si angle of 160.57°. In real sodalites (alumino-silicates) the bond angle is typically 140° corresponding to $z \approx 0.06$. When $z = 1/8$ the structure is the 9-coordinated structure referred to above, and the Si-O-Si angle would then be 109.48°.

Figure 6.72 shows a fragment of the structure in its minimum and maximum density forms. For unit tetrahedron edge length, the unit cell parameter is $a = 3.142$ in the minimum density form and $a = 2.818$ in the maximum density form and the density has increased by about 39%.[1]

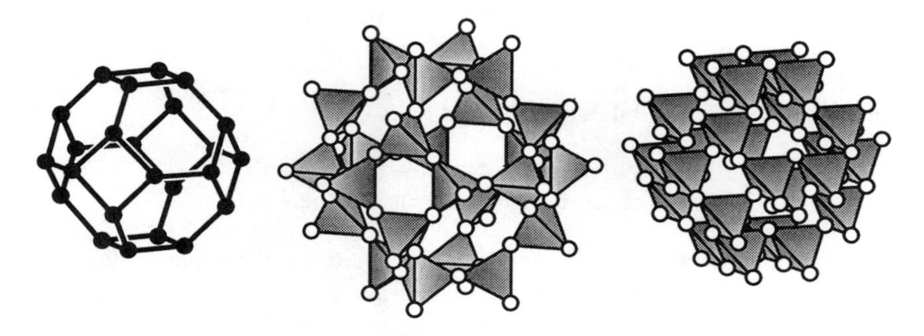

**Fig. 6.72.** Left: a truncated octahedron. Middle: A corner-connected array of tetrahedra (centered at the vertices of the truncated octahedron) as in the low-density, high-symmetry version of the sodalite anion structure. Right: the collapsed, high-density version of the same structure.

Yet another framework of regular tetrahedra has tetrahedron centers, $T$, on lattice complex $S^*$ (3/8,0,1/4 ; etc. of $Ia\bar{3}d$). The vertices, $X$ (sphere centers) are given by:

$Ia\bar{3}d$, $a = 2/\sqrt{(2-\sqrt{3})} = 3.8637$, $\rho = 0.4357$
$X$ in 48 $g$: $(1/8,x,1/4-x$ ; etc.), $x = 1/2 - \sqrt{3}/8 = 0.2835$ ; $T$-$X$-$T$ = 150°

An important elemental structure type is that of $\beta$-Sn (white tin). Here we give parameters for a slight idealization with six equal distances:

$I4_1/amd$, $a = \sqrt{15}/2$, $c = 1$, $\rho = 0.5585$. Sphere centers in 4 $a$: $I \pm (0,3/4,1/8)$

Except for the unit cell parameters this structure has the same description as the 8-coordinated structure described in § 6.3.4 (p. 234). The two structures are very different though; in **$\beta$-Sn** $c/a = 0.516$, in the 8-coordinated structure $c/a = 3.46$. This again emphasizes the fact that in non-cubic structures axial ratios should be carefully considered before concluding that two structures are the same (or related). In fact the same positions of $I4_1/amd$ with $c/a = 1.414$ corresponds to the diamond structure.

We could have devoted a section of this chapter to space-filling packings of polyhedra.

[1]The relevance of this structure and its transformations to crystal chemistry has been the subject of much discussion. Three papers on the topic appeared in *Acta Crystallogr.* A**37**, 1-17 (1981).

We have seen, for example, how space-filling packings of tetrahedra and octahedra give rise to eutactic arrangements; and the *T* structure may be considered a packing of tetrahedra and truncated tetrahedra. In the next chapter we will meet 4- and 5-connected structures that arise from other packings of regular and/or Archimedean polyhedra. Here are two 6-coordinated examples involving rhombicuboctahedra ($3.4^3$, Fig. 5.5) which we call *rco*'s for short.

An *rco* has two kinds of square faces: There are six having edges in common only with other squares; these are parallel to the faces of a cube. Joining *rco*'s through these faces will result in a structure in which the centers of the *rco*'s are on a primitive cubic lattice and the remaining space is a labyrinth of face-sharing cubes and cuboctahedra. Crystallographic data are:

(a) Cubes, cuboctahedra (3.4.3.4) and *rco*'s ($3^3.4$):

$Pm\bar{3}m$, $a = 1 + \sqrt{2} = 2.4142$, $\rho = 0.4465$
Sphere centers in 12 *i*: $(0,\pm x,\pm x)\kappa$, $x = 1/(2 + \sqrt{2}) = 0.2929$

We illustrate the structure in three ways in Fig. 6.73. Note particularly that if we consider just the packing of cubes and cuboctahedra (so that the "empty" space consists of *rco*'s) we have a continuous three-dimensional surface tiled with polygons. We will discuss such *infinite polyhedra* in the next chapter (see also Appendix 3). In this example all the vertices are equivalent and are $3.4^2.3.4^2$.

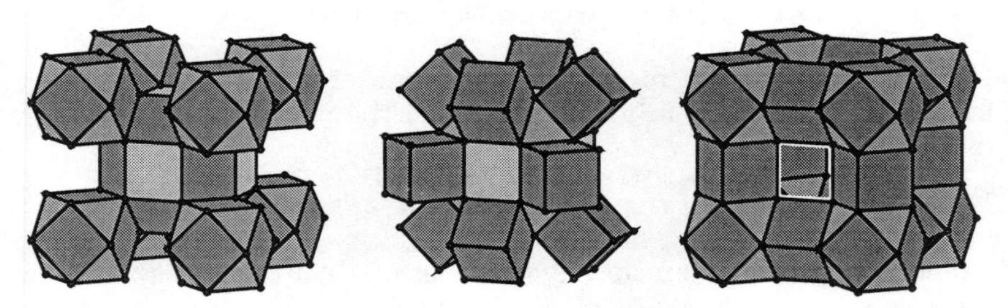

**Fig. 6.73.** Space filling by cubes, cuboctahedra and *rco*'s. Left: the combination of cuboctahedra and *rco*'s. Center: the combination of cubes and *rco*'s. Right: the combination of cubes and cuboctahedra.

(b) Tetrahedra, cubes and *rco*'s:

$Fm\bar{3}m$, $a = 2 + \sqrt{2} = 3.4142$, $\rho = 0.4210$
Sphere centers in 32 *f*: $F \pm (x,x,\pm x)\kappa$, $x = 1/(4 + \sqrt{8}) = 0.1464$

An *rco* has also twelve square faces with two edges in common with triangles. Joining them by these faces produces a structure in which the centers of the *rco*'s are on a face-centered cubic lattice. The remaining space consists of cubes and tetrahedra (1 and 2 respectively per *rco*) sharing vertices. Fig. 6.74 should provide sufficient information for model builders to proceed.

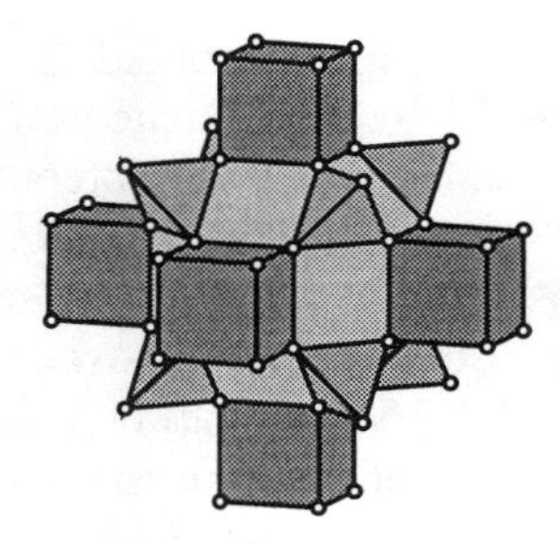

**Fig. 6.74.** Part of a space filling by cubes, tetrahedra and *rco*'s.

This is the O arrangement in compounds $Ag_7O_8X$ ($X$ is a monovalent anion such as $F^-$ or $NO_3^-$ which is in the center of the *rco*). Ag atoms are of two kinds: those centering the cubes (8-coordinated by O) and those centering the square faces (4-coordinated) of the *rco* that are not shared with cubes.

Another 6-coordinated structure is obtained by joining together *rco*'s and octahedra. Each triangular face of each *rco* is shared with an octahedron, and each octahedron shares a pair of opposite faces with *rco*'s (this makes a rather elegant model). We discuss this structure as an example of an infinite polyhedron $3^3.4^3$ in Appendix 3. Data are:

$Im\bar{3}m$, $a = 3.8857$, $\rho = 0.4284$
Sphere centers in 48 $k$: $I + (\pm x, \pm x, \pm z)\kappa$, $x = 0.3713$, $z = 0.1893$

Finally, we consider a fascinating 6-coordinated structure with only two parameters:

$P4_132$, $a = 8/[3\sqrt{(7 - \sqrt{33})}] = 2.3800$, $\rho = 0.4661$
Sphere centers in 12 $d$: $(1/8, x, 1/4+x$ ; etc.), $x = (9 - \sqrt{33})/16 = 0.2035$

In this structure three equilateral triangles twisted as in a three-bladed propeller meet at a point. There are two next-nearest neighbors at a distance 1.23 times the shortest distance. If all eight neighbors are counted, the structure may be described as a three-dimensional framework of corner-connected metaprisms. Three fifths of the Mn atoms in $\beta$-Mn have this arrangement; the remaining two fifths cap the equilateral triangular faces of the metaprisms, forming almost regular tetrahedra.

### *6.8.6 Sphere packings with icosahedra:* $WAl_{12}$ *and* $AuZn_3$

Strictly regular icosahedra are incompatible with crystallographic symmetry (which precludes the presence of 5-fold axes); but nature is very clever at designing periodic structures that feature *almost* regular icosahedra. Here we discuss two structures that may be considered as derived from the $J$ structure (§ 6.3.6, Fig. 6.27) and which arise in a variety of contexts.

The $J$ structure, considered as a packing of octahedra (sharing vertices with each other) and cuboctahedra (sharing faces), contains one of each polyhedron and a total of three vertices per unit cell. If the cubic cell edge is doubled there will be 8 (= 2 × 2 × 2) of each polyhedron in the larger cell. We can take the octahedron centers to be at ± (1/4,1/4,1/4 ; 1/4,3/4,3/4)κ in this cell and the cuboctahedron centers to be at 0,0,0 ; 1/2,1/2,1/2 and (1/2,0,0 ; 1/2,1/2,0)κ. The first two cuboctahedra (centered at the cell origin and body center each with 12 vertices) are isolated from each other and account for the 8 × 3 = 24 vertices in the doubled cell of the *J* structure. Let's now convert these two cuboctahedra to icosahedra as indicated in Fig. 2.25 (p. 54), and arrange them so that the shortest distance between vertices of neighboring icosahedra are the same as their edge lengths. It is remarkable that the transformation can be much more elegantly, and informatively, described in terms of concerted rotations of the corner-connected octahedra of the structure (which remain regular) as illustrated in Fig. 6.75. The resulting structure has symmetry $Im\bar{3}$.

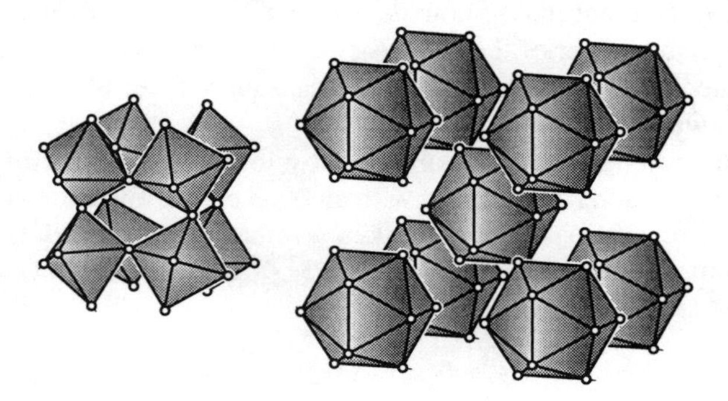

**Fig. 6.75**. The Al array in $WAl_{12}$. Left: as vertex-sharing octahedra. Right: as a body-centered array of icosahedra.

The octahedra are rotated about axes parallel to <111>. Let the angle of rotation (the same for every octahedron) be $\phi$. The structure is described as follows:[1]

$$Im\bar{3},\ a = (8\cos\phi + 4)/\sqrt{18}.\ \text{Vertices in } 24\ g\text{: } I \pm (0,y,\pm z)\kappa,$$
$$y = (3\cos\phi - \sqrt{3}\sin\phi)/(8\cos\phi + 4),\ z = (3\cos\phi + \sqrt{3}\sin\phi)/(8\cos\phi + 4)$$

For regular icosahedra the rotation angle is given by $\tan\phi = \sqrt{3}(\tau - 1)/(\tau + 1) = 22.2°$ and $y = 3/[4(1 + \tau + \sqrt{2})] = 0.1860$, $z = \tau y = 0.3010$. [Here, as usual, $\tau = (\sqrt{5} + 1)/2$.]

Centering the octahedra with atoms *B* produces stoichiometry $BX_3$. It is interesting that at ordinary pressure $ReO_3$ has the simple cubic structure with $\phi = 0$, but under pressure it suddenly crumples to produce the body-centered structure with $\phi > 0$.[2]

Centering the icosahedra of *X* with atoms *A* produces stoichiometry $AX_{12}$ and we have

[1]Readers who are anxious to derive this and related results for themselves should apply the rotation matrix of Eq. 2.3 (§ 2.5.1) to a point originally at 0,1/4,1/4 ($\phi = 0$).

[2]As the mechanism of compression changes from bond compression (initially), to buckling (angle bending), the bulk modulus (inverse of compressibility) of $ReO_3$ *decreases* with pressure [see B. Batlogg *et al.*, *Phys. Rev. B* **29**, 3762 (1984)]. For materials with just one mode of compression, the bulk modulus invariably increases with pressure.

the structure of $WAl_{12}$ which has coordinates very close to the ideal ones given here (see Appendix 5 for data).

What happened to the other six cuboctahedra originally ($\phi = 0$) in the unit cell? Four of their vertices have moved close to the center to produce rectangles with edges in the ratio 1:1.07 (i.e. almost square). A large family of oxides, typified by $CaCu_3Ti_4O_{12}$, is known in which Ca centers the icosahedra of the collapsed structure, Cu centers the rectangles and Ti centers the octahedra. Centering just the icosahedra and octahedra produces stoichiometry $AB_4X_{12}$; many compounds of this type are known, examples are $LaFe_4P_{12}$ and $LaFe_4Sb_{12}$.

Considered as a sphere packing, the ideal $X$ arrangement, with regular icosahedra, is 9-coordinated, although as mentioned above, each point has a tenth neighbor just 1.07 times as far away. Also to be noted is that although regular icosahedra are possible, icosahedral *symmetry* is not; in fact the symmetry at the center of the icosahedra is $m\bar{3}$.

The structure we have described represents just the most symmetrical way of collapsing the $J$ structure by rotations (or tilts) of the octahedra. In Volume II of this series we show how other important structures are derived by different patterns of tilts.

If the icosahedron at the body center of the cell in the above structure is rotated 90° about an axis parallel to a cube edge the symmetry is changed to $Pm\bar{3}n$ and, as before, there are 24 vertices in the unit cell A new sphere packing with only seven neighbors results if the shortest distance between vertices of neighboring icosahedra is the same as the edge length. This is illustrated in Fig. 6.76. Crystallographic data are:

$Pm\bar{3}n$, $a = 1/(2y) = 2.8859$, $\rho = 0.5232$
vertices in 24 $k$: $\pm(0,\pm y,z\ ;\ 1/2,1/2\pm z,1/2+y)\kappa$,
$y = 1/2 - \sqrt{(2+6\tau)/(4+4\tau)} = 0.1733$, $z = \tau y = 0.2803$

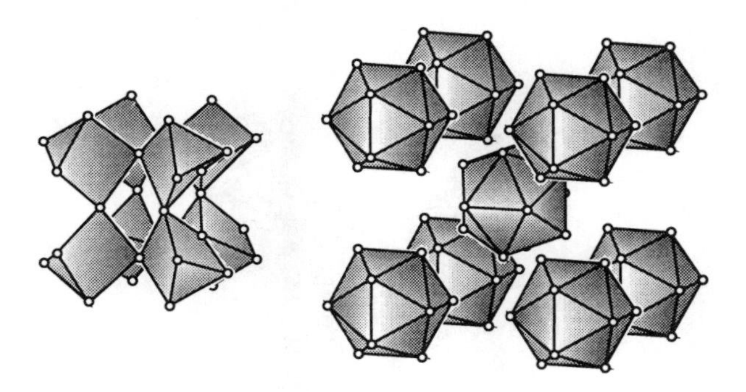

**Fig. 6.76**. The $Pm\bar{3}n$ arrangement of icosahedra. Left: as an array of corner-connected prisms. Right: the array of icosahedra. Compare with Fig. 6.75.

The array of corner-sharing octahedra in the $WAl_{12}$ structure has now become an array of corner-sharing (somewhat distorted) trigonal prisms. This structure has a nice surprise in store. The unit cell also contains six (again distorted) cuboctahedra with centers at 6 $c$: $\pm(1/4,0,1/2)\kappa$. If atoms $A$ (at 0,0,0 and 1/2,1/2,1/2) center the $X_{12}$ icosahedra and $B$ center the cuboctahedra, the stoichiometry is $AB_3X_{12}$. This is in fact the structure of $AuZn_3$ [=

$Au(1)Au(2)_3Zn_{12}$] in which the Zn parameters are quite close to the "ideal" ones given above (see Appendix 5 for crystallographic data).

The $AB_3$ arrangement in $AB_3X_{12}$ is the very common structure type $\mathbf{Cr_3Si}$ or **β-W** (§ 6.6.4). One form of $UH_3$, which we write as $U(1)U(2)_3H_{12}$ has the $AuZn_3$ structure. We might think of this as $\mathbf{Cr_3Si}$ $U(1)U(2)_3$ with H in tetrahedral interstices.

There are also some germanides and stannides, e.g. $Pr_3Rh_4Sn_{13}$, in which there are $Ge_{13}$ or $Sn_{13}$ groups obtained by centering the icosahedra. Pr atoms are in the cub-octahedra of Sn, and Rh atoms are in the trigonal prisms of Sn (for data see Appendix 5).

The symmetry of the $AuZn_3$ structure should lead us to expect (see § 6.8.10) to find a rod packing based on the **β-W** cubic cylinder packing. Fig. 6.77 illustrates the arrangement of cuboctahedra which consists of rods of face-sharing cuboctahedra packed in this way.

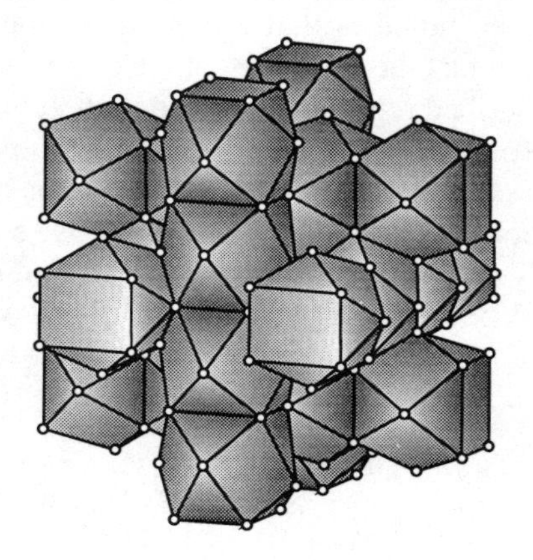

**Fig. 6.77**. The packing of rods of face-sharing cuboctahedra in the $AuZn_3$ structure. The rods run parallel to the three cube axes.

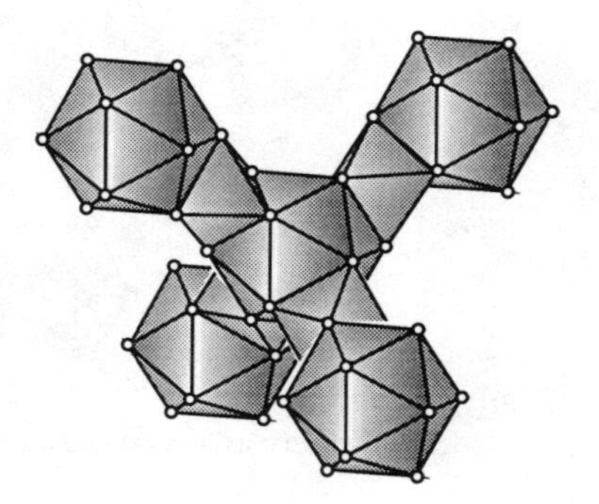

**Fig. 6.78**. A low density packing of icosahedra and octahedra; an infinite polyhedron $3^7$. Cf. Fig. 6.75.

Finally, we observe that half of the icosahedra of the $WAl_{12}$ structure (Fig. 6.75) could be omitted, leaving each icosahedron with just four neighboring icosahedra and with the centers of the icosahedra arranged as in the **diamond** net (§ 7.3.1). The resulting structure is illustrated in Fig. 6.78. Note that (excluding the shared face) seven equilateral triangles meet at each vertex so we have an infinite polyhedron $3^7$ as discussed in Appendix 3. The structure is a low-density 7-coordinated sphere packing, indeed the least dense that we know of. Crystallographic data for unit edge length are:

$Fd\bar{3}$, $a = 5.376$, $\rho = 0.3235$
Sphere centers in 96 *g*: (*x*,*y*,*z*, etc.), $x = 0.0320$, $y = 1/8$, $z = 0.2755$

### *6.8.7 Cubic invariant lattice complexes*

The symbols for some of these complexes (as given in the *International Tables A*) have been given already in this chapter but are summarized here for convenient reference, together with their highest symmetry occurrences. A prefixed "+" or "-" is used for enantiomorphous pairs and an affixed "*" indicates that the lattice complex is derived by combining two lattice complexes (one displaced from the other) with the same symbol but without the affix. In the table below *N* is the coordination number. The atoms in bold in chemical formulas lie on the lattice complex; the symmetry may be lower in the actual compound (cf. NbO). $V^*$ and $Y^{**}$ (a combination of two $Y^*$) correspond to two intergrown (but not inter-connected) nets.

| complex | space group | position | *N* | remarks |
|---|---|---|---|---|
| $F$ | $Fm\bar{3}m$ | 4 *a* | 12 | face-centered cubic |
| $I$ | $Im\bar{3}m$ | 2 *a* | 8 | body-centered cubic |
| $J$ | $Pm\bar{3}m$ | 3 *c* | 8 | Fig. 6.27 |
| $S$ | $I\bar{4}3d$ | 12 *a* or *b* | 8 | **$Th_3$**$P_4$, Fig. 6.30 |
| $P$ | $Pm\bar{3}m$ | 1 *a* | 6 | primitive cubic |
| $T$ | $Fd\bar{3}m$ | 16 *c* or *d* | 6 | Fig. 6.32 |
| $^+Y$ | $P4_332$ | 4 *a* | 6 | **Fe**Si, Fig. 6.34 |
| $^-Y$ | $P4_132$ | 4 *a* | 6 | enantiomorph of above |
| $D$ | $Fd\bar{3}m$ | 8 *a* or *b* | 4 | diamond, Fig. 7.9 |
| $^+V$ | $I4_132$ | 12 *c* | 4 | Fig. 7.36 |
| $^-V$ | $I4_132$ | 12 *d* | 4 | enantiomorph of above |
| $J^*$ | $Im\bar{3}m$ | 6 *b* | 4 | **NbO**, Fig. 7.31 |
| $W^*$ | $Im\bar{3}m$ | 12 *d* | 4 | sodalite, Fig. 7.30 |
| $S^*$ | $Ia\bar{3}d$ | 24 *d* | 4 | Fig. 7.35 |
| $V^*$ | $Ia\bar{3}d$ | 24 *c* | 4 | two *V* nets |
| $^+Y^*$ | $I4_132$ | 8 *a* | 3 | Sr**$Si_2$**, Fig. 7.6 |
| $^-Y^*$ | $I4_132$ | 8 *b* | 3 | enantiomorph of above |
| $Y^{**}$ | $Ia\bar{3}d$ | 16 *b* | 3 | two $Y^*$ nets, Fig.7.34 |
| $W$ | $Pm\bar{3}n$ | 6 *c* or *d* | 2 | **$Cr_3$**Si, Fig. 6.60 |

An invariant lattice complex may occur in more than one space group. Thus $J$ also occurs as positions $c$ and $d$ of $P\bar{4}3m$ and (with doubled cell) as positions $c$ and $d$ of $Fm\bar{3}c$.

Other lattice complexes with symbols include $^{+}Q$ for the Si atom positions in $\beta$-quartz (§ 3.6) with symmetry $P6_222$ and $^{-}Q$ for the enantiomorph ($P6_422$), $G$ for a $6^3$ layer and $N$ for a 3.6.3.6 (kagome) layer. We used these last two symbols in § 6.1.6.

### *6.8.8 Common cubic unit cells for arrays*

In § 5.6.12 we discussed common unit cells for plane patterns. There is an analogous problem in three dimensions involving cubic patterns. The smallest cubic supercell of a primitive cell has edges $2a$, i.e. 8 times the volume. The next will have 27 times the volume. The number of symmetry-related points in a cubic cell is a divisor of 192 so in practice, to investigate relationships between cubic structures of symmetry-related points in which a cubic cell is maintained, it is only necessary to consider relationships of this kind between structures with either the same number of points or differing by a factor of eight. There are then six sets of compatible numbers which are:

(i) 1, 8, 64
(ii) 2, 16
(iii) 3, 24, 192
(iv) 4, 32
(v) 6, 48
(vi) 12, 96

We now give some examples of the use of these numbers. The fact that the anions in the spinel structure (§ 3.4) are in 32 $e$ of $Fd\bar{3}m$: $F \pm (x,x,x\ ;\ (x,1/4-x,1/4-x)\kappa)$ (with $x$ typically = 0.26) suggests a possible relationship to **ccp** with four points per cubic cell [case (iv) above]. The reader is invited to confirm that if $x = 1/4$ the structure is indeed a face-centered cubic lattice described with a $2 \times 2 \times 2$ cell.[1]

Another example is provided by the structure of $Th_3P_4$ which has symmetry $I\bar{4}3d$ with P atoms in 16 $c$: $I + (x,x,x\ ;\ 1/4+x,1/4+x,1/4+x\ ;\ (1/2+x,1/2-x,\bar{x}\ ;\ 3/4+x,1/4-x,3/4-x)\kappa)$ with $x = 0.08$. In this case the reader may confirm that for $x = 0$, the arrangement is body-centered cubic [2 atoms per cell; case (ii) above] described with a $2 \times 2 \times 2$ cell.

Note also the relationship of the cubic 7-coordinated sphere packing of § 6.3.8 (p. 238) with symmetry $Pa\bar{3}$ and eight points per cell to the primitive cubic structure with one per cell [case (i)].

A formal description of the garnet structure of $Ca_3Al_2Si_3O_{12}$ was given in § 3.4. The cubic cell contains 96 O atoms. A well known reference work states that this structure has "oxygen ions in cubic close packing." Reference to case (vi) above shows that this

[1]It is amusing that a *rare* sphere packing (described in the next chapter) also has the same formal description (points in 32 $e$ of $Fd\bar{3}m$) but with $x = 1/(8 + \sqrt{96}) = 0.056\ldots$ so one must be a little cautious. Structures with the same formal description, but differing in the values of one or more parameters, may be very different.

arrangement with 96 atoms per cubic cell can only be compatible with other cubic structures with 12 atoms per unit cell. In particular it is incompatible with **ccp** (four per cell). Actually in $Ca_3Al_2Si_3O_{12}$ the first twelve O neighbors of an O atom range in distance from 2.57 - 3.85 Å and the thirteenth neighbor is at 3.89 Å so we can, in any case, discount the claim on those grounds. We can also use the coordination sequence (as described in § 6.8.2) based on the first twelve neighbors to show that the arrangement does not correspond topologically to *any* **cp** arrangement.

The Al positions in garnet are in 16 *a* of $Ia\overline{3}d$ and correspond to **bcc** (2 per cell) described by a $2 \times 2 \times 2$ cell [case(ii)] and the *combination* of the Ca and Si positions (24 *c* and 24 *d* respectively) corresponds to lattice complex *W* (6 points per cell) described by a $2 \times 2 \times 2$ cell [case (v)].

Cubic supercells of cubic cells with axes not parallel to the original one may occur, but they are probably not of much interest because of the size of the new cell. A simple example is that obtained by the transformation (3 4 0 / –4 3 0 / 0 0 5) with edge length of $5a$. A cubic supercell with edge an irrational multiple of $a$ (in the same way as a square cell has supercells $\sqrt{2}a \times \sqrt{2}a$ or $\sqrt{5}a \times \sqrt{5}a$) cannot occur however.

### *6.8.9 Packing of two sizes of sphere: "kissing" numbers*

In this chapter, the emphasis has been on sphere packings with one kind of sphere, because these commonly occur in simple crystal structures. However, some structures of intermetallic compounds can be considered as efficient (dense) packings of two or more kinds of sphere of different sizes. The structure of $MgCu_2$ (§ 6.6.3) is an often cited example of an efficient packing of two kinds of sphere and that of $NaZn_{13}$ (§ 6.8.5) is another. Very little systematic research has been done on the problem of classifying packings of spheres of two sizes, but it has an obvious relevance to crystal chemistry and some results would be expected to lead to useful insights into intermetallic structures. For packings of two sizes of circles see L. Fejes Tóth, *Regular Figures* {Pergamon Press, Oxford (1964)]. For the packing of two sizes of sphere see M. J. Murray & J. V. Sanders, *Phil. Mag.* A**42**, 721 (1980). These last authors were interested also in the structures of opals which are packings of (typically micron-sized) silica spheres. In contrast to crystals, opals really are packings of hard spheres.[1]

The maximum number of equal spheres that can touch a similar central one is known as the kissing number. It seems astonishing that the answer in three dimensions was once controversial and involved Newton (who correctly said twelve) and Gregory (who thought the answer might be thirteen). However, such questions are difficult to settle to the satisfaction of mathematicians, who are uncommonly hard to please in such matters.

In intermetallic structures, higher kissing numbers are commonly found; for example, in $NaZn_{13}$, Na has 24 equidistant Zn neighbors. In general, in such structures, nature contrives to design an arrangement in which every atom is highly coordinated. At the same time she is tolerant of small variations in interatomic distance—it is this aspect of the topic

[1]The "fire" in opal comes from Bragg diffraction of light from the planes in the periodic packing of spheres, in a similar way as crystals diffract X-rays.

which makes it virtually intractable from the point of view of formal geometry.

A related question which is of some importance to crystal chemistry concerns kissing numbers in binary (ternary etc.) compounds (extended structures) in which spheres (atoms) of one kind kiss only spheres of another kind. We call these numbers *heterosphere* kissing numbers. Typically in "ionic" crystals the "cations" have only "anions" as nearest neighbors and *vice versa*, and the question of heterosphere kissing numbers is particularly relevant for this class of compounds.

To take a specific example: in a binary compound $A_mB_n$ (in the strict sense in which all $A$ are equivalent as are all $B$) with $d(A\text{-}B) < d(A\text{-}A)$ and $d(B\text{-}B)$, what is the maximum possible coordination of $A$ by $B$?

We pause first to remark that if the coordination of $A$ by $B$ is $p$ and the coordination of $B$ by $A$ is $q$, then $pm = qn$. This almost trivial observation turns out to be of some consequence in determining possible coordination numbers in "ionic" crystals.

Our guess, based on observed crystal structures, for the answer to the above question for compounds $AB$, is that the maximum coordination number (heterosphere kissing number) is 8, and for compounds $A_mB_n$ that the average coordination number (i.e. averaged over all the atoms) can never exceed 8 as long as $d(A\text{-}B) < d(A\text{-}A)$ and $d(B\text{-}B)$.

*6.8.10 The occurrence of cubic cylinder packings*

| s.g. | **garnet** | **γ-Si** | **β-W** (×2) | |
|---|---|---|---|---|
| | $u,u,u$ | $\frac{1}{3}+u,\frac{2}{3}+u,u$ | $0,\frac{1}{4},u$ | $\frac{1}{4},0,u$ |
| $Ia\bar{3}d$ | $e$, $\mathbf{p\bar{3}c1}$ | $h$, $\mathbf{p3_12}$, $\mathbf{p3_22}$ | $f$, $\mathbf{p\bar{4}c2}$ | $h$, $\mathbf{p4_122}$, $\mathbf{p4_322}$ |
| $I\bar{4}3d$ | $c$, $\mathbf{p3c1}$ | $e$, $\mathbf{p3_1}$, $\mathbf{p3_2}$ | $d$, $\mathbf{p\bar{4}}$ | $e$, $\mathbf{p222_1}$ |
| $Ia\bar{3}$ | $c$, $\mathbf{p\bar{3}}$ | $e$, $\mathbf{p3_1}$, $\mathbf{p3_2}$ | $d$, $\mathbf{pcc2}$ | $e$, $\mathbf{p222_1}$ |
| $Pa\bar{3}$ | $c$, $\mathbf{p\bar{3}}$ | $d$, $\mathbf{p3_1}$, $\mathbf{p3_2}$ | $d$, $\mathbf{p1c1}$ | $d$, $\mathbf{p112_1}$ |

| s.g. | **garnet** | **SrSi$_2$** | | **β-W** (×2) | **β-Mn** | |
|---|---|---|---|---|---|---|
| | $u,u,u$ | $\frac{1}{3}+u,\frac{2}{3}+u,u$ | $\frac{2}{3}+u,\frac{1}{3}+u,u$ | $0,\frac{1}{4},u$ | $\frac{1}{4},0,u$ | $\frac{3}{4},0,u$ |
| $I4_132$ | $e$, $\mathbf{p3}$ | $i$, $\mathbf{p3_12}$ | $i$, $\mathbf{p3_22}$ | $f$, $\mathbf{p222}$ | $i$, $\mathbf{p4_122}$ | $i$, $\mathbf{p4_322}$ |
| $P4_132$ | $c$, $\mathbf{p3}$ | $e$, $\mathbf{p3_22}$ | $e$, $\mathbf{p3_12}$ | $e$, $\mathbf{p211}$ | $e$, $\mathbf{p4_1}$ | $e$, $\mathbf{p222_1}$ |
| $P4_332$ | $c$, $\mathbf{p3}$ | $e$, $\mathbf{p3_22}$ | $e$, $\mathbf{p3_12}$ | $e$, $\mathbf{p211}$ | $e$, $\mathbf{p222_1}$ | $e$, $\mathbf{p4_3}$ |
| $I2_13$ | $a$, $\mathbf{p3}$ | $c$, $\mathbf{p3_1}$ | $c$, $\mathbf{p3_2}$ | $b$, $\mathbf{p112}$ | $c$, $\mathbf{p222_1}$ | $c$, $\mathbf{p222_1}$ |
| $P2_13$ | $a$, $\mathbf{p3}$ | $b$, $\mathbf{p3_2}$ | $b$, $\mathbf{p3_1}$ | ($b$, $\mathbf{p1}$) | $b$, $\mathbf{p112_1}$ | $b$, $\mathbf{p112_1}$ |

| s.g. | **β-W** ($0,\frac{1}{2},u$ or $\frac{1}{2},0,u$) |
|---|---|
| $Pm\bar{3}n$ | $g$ or $h$, $\mathbf{p4_2/mmc}$ |
| $P\bar{4}3n$ | $h$ or $g$, $\mathbf{p\bar{4}2c}$ |
| $P4_232$ | $i$ or $j$, $\mathbf{p4_222}$ |
| $Pm\bar{3}$ | $f$ or $g$, $\mathbf{pmmm}$ |
| $P23$ | $g$ or $h$, $\mathbf{p222}$ |

Cubic cylinder packings are of particular interest in the description of complex crystal structures that otherwise resist description. Here (on the previous page) we list the cubic space groups with non-intersecting symmetry axes and the axis locations. The entries in the table are the Wyckoff positions and compatible rod symmetry (Appendix 1). For example we see that the substitution *u,u,u* in $Ia\bar{3}d$ corresponds to positions *e* and represents the axes of the **garnet** packing. Rods along these axes have symmetry $\mathbf{p}\bar{3}c1$.

Non-intersecting 4-fold or 2-fold axes also occur in the space groups with non-intersecting 3-fold axes. They fall into two sets, each corresponding to the axes of **β-W** described by a doubled cell. We label this **β-W** (×2). Thus, again for $Ia\bar{3}d$, the substitution 0,1/4,*u* will produce[1] the set (0,1/4,*u* ; 0,3/4,*u* ; 1/2,3/4,*u* ; 1/2,1/4,*u*)κ corresponding to positions 48 *f* and rod symmetries will be $\mathbf{p}\bar{4}c2$. This set is illustrated as a cylinder packing in Fig. 6.67 with the unit cell (**a** down the page, **b** horizontal and **c** up out of the page) origin appropriately located. The other set of non-intersecting 4-fold axes are $4_1$ and $4_3$ axes along the lines generated by the substitution 1/4,0,*u*. Please note that the table is appropriate only for the choices of origin made in the *International Tables*.

The **β-W** structure without a doubled cell occurs in a separate set of space groups as shown. In these space groups the 3-fold axes intersect.

## 6.9 Exercises

1. Americium is hexagonal:

| | |
|---|---|
| Am | $P6_3/mmc$, $a = 3.474$, $c = 11.25$ Å<br>Am(1) in 2*a* (0,0,0 ; 0,0,1/2) ; Am(2) in 2*d*: ±(1/3,2/3,3/4) |

Describe the structure in terms of stacking $3^6$ nets (*A*, *B* and *C*) and in terms of *h* and *c*.

2. Other **cp** arrays with just two kinds of sphere are:

| | |
|---|---|
| *hcc* (6*H*) | $P6_3/mmc$, $c/a = 6\sqrt{(2/3)}$<br>*h* in 2 *b*: ±(0,0,1/4) ; *c* in 4 *f*: ±(1/3,2/3,*z* ; 1/3,2/3,1/2–*z*), *z* = 1/12 |
| *hhc* (9*R*) | $R\bar{3}m$, $c/a = 9\sqrt{(2/3)}$<br>*c* in 3 *a*: *R* + (0,0,0) ; *h* in 6 *c*: *R* ± (0,0,*z*), *z* = 2/9 |
| *hhcc* (12*R*) | $R\bar{3}m$, $c/a = 12\sqrt{(2/3)}$<br>*c* in 6 *c*: *R* ± (0,0,*z*), *z* = 5/24 ; *h* in 6 *c*: *R* ± (0,0,*z*), *z* = 3/8 |

3. Mercury has a rhombohedral structure (space group $R\bar{3}m$) with one atom in the unit cell. $a = 2.993$ Å, $\alpha = 70.74°$. What are the distances to the 12 nearest neighbors of a Hg atom? Transform to a face-centered rhombohedral cell (four atoms per cell) for a comparison with face centered cubic. What is $\alpha$ for this cell? [Hint see § 4.4.]

[1]The reader who wishes to verify this statement should note, for example, that the line 0,1/4,*u* is the same *line* as 0,1/4,1/4–*u*; it is a line parallel to **c** passing through 0,1/4,0.

4. What is the arrangement obtained from two interpenetrating *fcc* lattices, one at 0,0,0 and one at 1/2,1/2,/12. (i.e. a combination of positions 4 *a* and 4 *b* of $Fm\overline{3}m$)? What is the arrangement of points in 8 *c* of $Fm\overline{3}m$ [$F \pm$ (1/4,1/4,1/4)] ? If we combine all these positions (4 *a*, 4 *b* and 8 *c*) we will have combined four *fcc* lattices (16 points). What is the arrangement now?

5. Positions 16*a* of $Ia\overline{3}d$ are $I$ + (0,0,0 ; 1/4,1/4,1/4 ; 0,1/2,1/2 ; 1/4,1/4,3/4)κ. What simple arrangement is this?

6. Here are two examples of structures based on **cp**. Note that although they are quite different, they differ only in the numerical values (i.e. they have the same space group and atoms in the same sets of positions—they are *isopuntal*).

(a) $Th_3N_4$ — $R\overline{3}m$, $a$ = 3.875 Å, $c$ = 27.39 Å, $c/a$ = 7.07
Th(1) in 3 *a*: $R$ + (0,0,0) ; Th(2) in 6 *c*: $R \pm (0,0,z)$, $z$ = 0.222
N(1) in 6 *c*, $z$ = 0.132 ; N(2) in 6 *c*, $z$ = 0.377

Th is approximately **cp**; what is the stacking sequence? N atoms fill tetrahedral and/or octahedral interstices. Answer (with *A*,*B*,*C* for cations and $\alpha,\beta,\gamma$ for anions):

$$hhc \text{ Th; sequence} = A{\cdot}\gamma{\cdot}B\alpha{\cdot}\beta A{\cdot}\gamma{\cdot}B{\cdot}\alpha{\cdot}C\beta{\cdot}\gamma B{\cdot}\alpha{\cdot}C{\cdot}\beta{\cdot}A\gamma{\cdot}\alpha C{\cdot}\beta\ldots$$

(b) $Fe_3S_4$ (smythite) $R\overline{3}m$, $a$ = 3.47 Å, $c$ = 34.5 Å, $c/a$ = 9.94
Fe(1) in 3 *a*: $R$ + (0,0,0) ; Fe(2) in 6 *c*: $R \pm (0,0,z)$, $z$ = 0.9171
S(1) in 6 *c*, $z$ = 0.7898 ; S(2) in 6 *c*, $z$ = 0.6270

S is approximately **cp**; what is the stacking sequence? Fe atoms fill octahedral interstices. Answer (with *A*,*B*,*C* for anions and $\alpha,\beta,\gamma$ for cations):

$$hhcc \text{ S; sequence} = A{\cdot}\gamma{\cdot}B{\cdot}\gamma{\cdot}A{\cdot}\gamma{\cdot}B\cdots C{\cdot}\beta{\cdot}A{\cdot}\beta{\cdot}C{\cdot}\beta{\cdot}A\cdots B{\cdot}\alpha{\cdot}C{\cdot}\alpha{\cdot}B{\cdot}\alpha{\cdot}C\ldots$$

7. The structures of the rutile form of $TiO_2$ and of $CaCl_2$ are closely related:

$TiO_2$ — $P4_2/mnm$, $a$ = 4.594 Å, $c$ = 2.958 Å
Ti in 2 *a*: (0,0,0 ; 1/2,1/2,1/2)
O in 4 *f*: ±($x$,$x$,0 ; 1/2+$x$,1/2−$x$,1/2), $x$ = 0.305

$CaCl_2$ — $Pnnm$, $a$ = 6.241 Å, $b$ = 6.432 Å, $c$ = 4.340 Å
Ca in 2 *a*: (0,0,0 ; 1/2,1/2,1/2)
Cl in 4 *g*: ±($x$,$y$,0 ; $x$+1/2,1/2−$y$,1/2), $x$ = 0.275, $y$ = 0.325

Plot both structures in projection down the short axis and compare with Fig. 6.20 (p. 229). What (approximately) is the anion packing?

8. The structure of $Mo_2BC$ is:

| | |
|---|---|
| $Mo_2BC$ | *Cmcm*, $a$ = 3.086, $b$ = 17.35, $c$ = 3.047 Å<br>Mo(1) in 4 *c*: *C* ± (0,*y*,1/4), *y* = 0.3139 ; Mo(2) in 4 *c*, *y* = 0.0721<br>B in 4 *c*, *y* = 0.4731 ; C in 4 *c*, *y* = 0.1920 |

Describe the Mo structure as a stacking of $4^4$ nets along **b** using the notation *A*, *B*, *C*, *D* and describe the type of site (octahedron, trigonal prism etc.) occupied by B and C. Compare the B-B and C-C distances in the structure.

9. Pa and γ-Pu are both 10-coordinated:

| | |
|---|---|
| Pa | *I4/mmm*, $a$ = 3.932, $c$ = 3.238 Å. Pa in 2 *a*: *I* + (0,0,0) |
| γ-Pu | *Fddd*, $a$ = 3.159, $b$ = 5.768, $c$ = 10.162 Å. Pu in 8 *a*: *F* ± (1/8,1/8,1/8) |

Plot the Pa structure projected on (110). How do the interatomic distances compare with those in the 10-coordinated **bct**? The γ-Pu structure is based on a stacking of $3^6$ nets along **c** in the sequence *ADEF* (see Fig. 6.40, p. 247). Calculate the ten shortest interatomic distances in γ-Pu. What is the next shortest distance?

10. $Ta_3B_4$ is orthorhombic:

| | |
|---|---|
| $Ta_3B_4$ | *Immm*, $a$ = 3.29, $b$ = 14.0, $c$ = 3.13 Å<br>Ta(1) in 2 *c*: *I* + (0,0,1/2); Ta(2) in 4 *g*: *I* ± (0,*y*,0), *y* = 0.180<br>B(1) in 4 *g*, *y* = 0.375; B(2) in 4 *h*: *I* ± (0,*y*,1/2), *y* = 0.444 |

Describe the Ta arrangement as a stacking of $4^4$ nets along **b** (as in Exercise 8). What is the coordination of the B atoms? Check your answers by drawing the structure in projection down **a**.

$V_3B_4$ is isostructural with $Ta_3B_4$ and VB is isostructural with CrB (p. 250). There is also reported a composition $V_5B_6$ (= 2VB + $V_3B_4$). Can you guess a possible structure for this last composition? [See Hyde & Andersson (Book List) p. 227.]

11. Many halides have structures with **cp** anions. Some are rather simple (e.g. $\mathbf{CdCl_2}$, and $\mathbf{CdI_2}$, § 6.1.5), but others have rather complicated low-symmetry structures. Data in abbreviated form for some iodides are given below.

| | |
|---|---|
| $ZnI_2$ | $I4_1/acd$, $a$ = 12.284, $c$ = 23.583 Å. Zn 32*g*, 0.3749,0.3625,0.0627<br>I(1) 16*d*, 0,1/4,0.0047 ; I(2) 16*e*, 0.262,0,1/4 ; I(3) 32*g*, 0.0113,0.9993,0.1267 |
| $HgI_2$ | $P4_2/nmc$, $a$ = 4.370, $c$ = 12.443 Å. Hg 2*a*, 1/4,3/4,1/4 ; I 4*d*, 1/4,1/4,0.3891 |
| $BiI_3$ | $R\bar{3}$, $a$ = 7.498, $c$ = 20.68 Å. Bi 6*c*, 0,0,0.167 ; I 18*f*, 0.342,0.340,0.0805 |
| $SnI_4$ | $Pa\bar{3}$, $a$ = 12.273 Å. Sn 8*c*, 0.125,0.125,0.125<br>I(1) 8*c*, 0.252,0.252,0.252 ; I(2) 24*d*, –0.002,–0.002,0.253 |

| | |
|---|---|
| $UI_4$ | $C2/c$, $a$ = 13.967, $b$ = 8.472, $c$ = 7.510 Å, $\beta$ = 90.54°. U 4$f$, 0,0.152,1/4<br>I(1) 8$f$, 0.123,0.118,–0.086 ; I(2) 8$f$, –0.134,0.382,0.100 |
| $HfI_4$ | $C2/c$, $a$ = 11.787, $b$ = 11.801, $c$ = 12.905 Å, $\beta$ = 116.3°. All atoms 8$f$.<br>Hf 0.4244,0.3610,0.3753 ; I(1) 0.3270,0.3830,0.1309<br>I(2) 0.4470,0.1351,0.3866 ; I(3) 0.1898,0.3761,0.3632<br>I(4) 0.4369,0.6154,0.3808 |
| $PrI_2$ | $F\bar{4}3m$, $a$ = 12.360 Å. All atoms 16 $e$, $x,x,x$ etc.<br>Pr, $x$ = 0.3606 ; I(1), $x$ = 0.1115 ; I(2(, $x$ = 0.6257 |

Identify the anion packings (all are **cp**) and the way the cations occupy the interstices.

12. A tetragonal tetrahedral layer (§ 6.4.2) consists of two $4^4$ nets of $Y$ stacked $AB$, with a $4^4$ net of $X$ of twice the density in between [so there are $\{Y\}X_4$ tetrahedra]. The stoichiometry is $X_2Y_2$. If $n$ such layers are joined together the stoichiometry is $X_{2n}Y_{n+1}$ (the case $n = \infty$ corresponds to **fluorite** structure $X_2Y$). In compounds with the $KCu_4S_3$ structure there are double tetrahedral layers ($n$ = 2) of $\{Cu\}S_4$ tetrahedra interwoven with layers of K in $\{K\}S_8$ cubes:

| | |
|---|---|
| $KCu_4S_3$ | $P4/mmm$, $a$ = 3.899, $c$ = 9.262 Å, $c/a$ = 2.38. K in 1 $b$: 0,0,1/2<br>Cu in 4 $i$: ±(0,1/2,$z$ ; 1/2,0,$z$), $z$ = 0.1603<br>S(1) in 1 $a$: 0,0,0 ; S(2) in 2 $h$: ±(1/2,1/2,$z$), $z$ = 0.2944 |

Verify that the S packing consists of $4^4$ nets stacked $ABB$... and that Cu atoms are in $\{Cu\}S_4$ tetrahedra . Compare the Cu...Cu distance in the layers with the Cu...Cu distance in elemental Cu (**ccp** with $a$ = 3.615 Å). Speculate on the oxidation states of the atoms.

13. $AlCr_2C$ is one of the so-called $H$ phases found for aluminum-transition metal carbides and nitrides. Many other isopuntal compounds are known including examples with two non-metallic components (which are still called $H$ phases):

| | |
|---|---|
| $AlCr_2C$ | $P6_3/mmc$, $a$ = 2.860, $c$ = 12.82 Å, $c/a$ = 4.48. Al in 2 $d$: ±(1/3,2/3,3/4)<br>Cr in 4 $f$: ±(1/3,2/3,$z$ ; 1/3,2/3,1/2–$z$), $z$ = 0.086 ; C in 2 $a$: (0,0,0 ; 0,0,1/2) |
| $Ti_2SC$ | $P6_3/mmc$, $a$ = 3.210, $c$ = 11.20 Å, $c/a$ = 3.49. S in 2 $d$: ±(1/3,2/3,3/4)<br>Ti in 4 $f$: ±(1/3,2/3,$z$ ; 1/3,2/3,1/2–$z$), $z$ = 0.099 ; C in 2 $a$: (0,0,0 ; 0,0,1/2) |

Describe the two compounds above in terms of stacking of $3^6$ nets (using $ABC,\alpha\beta\gamma$). What (if any) sets of atoms approximate closest packing in each case? What are the coordination polyhedra around S and C? [Compare $Ti_2SC$ with TiP (§ 4.6.3).]

14. Here is a simple packing of unit diameter spheres:

$I4/mmm$, $a = 1 + \sqrt{2}$, $c = \sqrt{2}$. Centers in 8 $i$: $I$ ± ($x$,0,0 ; 0,$x$,0), $x = 1/(2 + \sqrt{2})$

Identify the net in the layers normal to **c**, and the coordination number of the packing.

# CHAPTER 7

# NETS AND INFINITE POLYHEDRA

## 7.1 Introduction

In this chapter, in contrast to the last, we discuss some arrays of points with low coordination number, particularly 3- or 4-coordination. These are less usefully considered as sphere packings, and are more commonly described as *nets*. In some cases it is useful to consider the nets as the edges and vertices of packings of polyhedra. As in the previous chapter, the emphasis is mainly on the simpler high-symmetry patterns that occur in a variety of structural contexts. Now a systematic organization is more difficult as nets may be derived and described in more than one way. To improve continuity we have included in the main body of the text some material that might otherwise have been relegated to the Notes. In particular section numbers in this chapter that are marked with an asterisk may be of lesser interest to some readers and may be omitted in a first reading.

The diamond structure is a familiar example of a 4-coordinated (or 4-connected) net and many other 4-connected nets arise as structures of alumino-silicates (including the two most common crystalline materials in the earth's crust: quartz and feldspar). In the latter case the Si (or Al) atoms are the nodes (or vertices) of the net and the -O- bonds are to be considered the links (or edges). The frameworks of zeolites (mainly alumino-silicates and alumino-phosphates) are currently of great interest as their catalytic and other properties are largely determined by their structures. Other important 4-connected nets occur in covalent solids and as the hydrogen-bonded networks in polymorphs of ice and in hydrates.

Nets can also have mixed coordination; thus the net describing the atoms in $Si_3N_4$ (= ${}^{iv}Si_3{}^{iii}N_4$), in which Si is connected to four N, and N is connected to three Si, is referred to as (3,4)-connected. An important class of nets with mixed coordination is that corresponding to frameworks of corner-connected octahedra and tetrahedra. For example, in $Fe_2(SO_4)_3$, $\{Fe\}O_6$ octahedra share corners with $\{S\}O_4$ tetrahedra and *vice versa*. The Fe and S atoms are at vertices and the -O- links correspond to edges of a (4,6)-connected net.

We usually describe nets in crystallographic terms. We generally give unit cell parameters and coordinates of vertices that correspond to an idealized conformation in which the edges are of equal length, and in which the volume, subject to this constraint, is a maximum. This conformation is also one of maximum symmetry. Some nets occur in a variety of crystal structures and often then have lower symmetry.

There appears to be no *simple* method of giving a purely topological definition of nets, but a partial topological characterization of 3- and 4-connected nets is nevertheless useful, so we discuss this topic first. A *systematic* description of nets is difficult and efforts to enumerate possibilities have not succeeded in any rigorous manner (many hundreds of 4-connected nets have been described in the literature).

The topology of nets is a source of some fascinating, and mostly unsolved, problems. For comments on these aspects see Appendix 3.

We caution the reader that it is often *very* difficult to appreciate the structure of a three-dimensional net and virtually impossible to do such things as enumerate rings from a drawing. On the other hand models can be made simply and inexpensively (see Notes) and it will be found that these are invaluable (and sometimes essential) aids to understanding.

### *7.1.1 Circuits, rings and Schläfli symbols*

Three-dimensional nets can be considered as infinite periodic graphs; we then tend to talk of *vertices* (atoms) and *edges* (bonds)—a common practice in graph theory. A *path* is a continuous sequence of edges, and a *circuit* is a closed path beginning and ending at the same vertex. The term *ring* is used in a special sense, described below, that is consonant with chemical usage. Any two edges with a common vertex define an *angle* at that vertex.

Recall that in Chapter 5 we often characterized finite polyhedra and two-dimensional nets by Schläfli symbols, which gave in cyclic order the size of the polygons common to a vertex. For 3- and 4-connected three-dimensional nets it is a common practice to extend the idea of a Schläfli symbol to include these cases also. Now, instead of polygons, either *shortest circuits* or *shortest rings* are used and we must first make clear the definition of these terms and be careful to distinguish between them.

For each angle at a vertex we can find a circuit which is a path that starts out at the vertex in question (the *home* vertex), goes out along one edge, and returns home along the second edge of the angle. The shortest such path (one that traverses the least number of edges) is the shortest circuit associated with that angle and is signified by the number of edges it contains. Some authors characterize three-dimensional nets by giving a "Schläfli symbol" that indicates the size of the shortest circuit at each angle; we give an example of this procedure below.

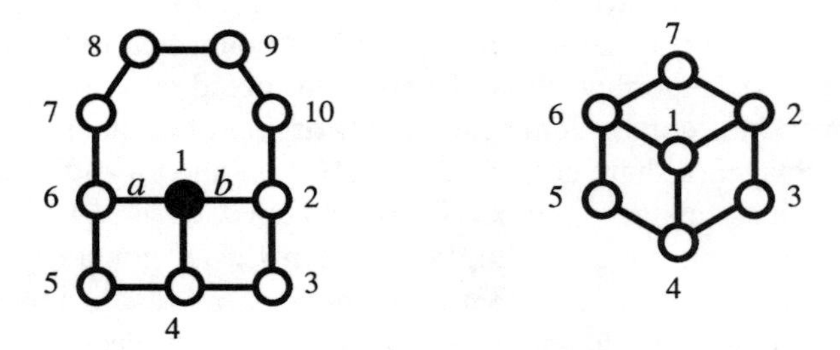


**Fig. 7.1**. Two fragments of nets discussed in the text.

The use of shortest circuits is not always consistent with our earlier treatment of polyhedra and two-dimensional nets as we now explain. In Fig. 7.1 (left) let the circle labeled "1" be the home vertex, and let *a* and *b* be two edges defining the angle *ab* at vertex "1." The rest of the numbered vertices represent a fragment of a net, which may be finite (i.e. the net of a polyhedron) or an infinite two- or three-dimensional net. There is a 6-circuit 1,2,3,4,5,6 containing this angle, but there is a "short cut" back to vertex 1 from vertex 4. To be consistent with earlier usage, we should not consider circuits that have such short cuts and count only those without them. Another circuit containing the angle *ab* is 1,6,7,8,9,10,2; this circuit does not contain short cuts as the path along the circuit between any two vertices on the circuit is a shortest path between them. Such circuits are variously called "fundamental circuits," "primitive rings," or just "rings;" here we will use the simplest term "ring" and reject the 6-circuit 1,2,3,4,5,6 as not being a ring but accept the 7-circuit 1,6,7,8,9,10,2 as a ring.

It is not hard to see that an infinite net will have only a finite number of rings for each vertex, whereas there is an infinite number of circuits. Interesting unsolved problems are how the number and sizes of rings affect properties such as density, and what constraints there are on ring size.

Referring to Fig. 7.1 (right) we can see that the circuit 1,2,3,4,5,6 is not counted as a ring, but 2,3,4,5,6,7 is. The latter ring is however made up of smaller ones (1,2,3,4 and 1,4,5,6 and 1,2,6,7) in the sense that traversing all the edges of the smaller rings will result in traversing all the edges of the larger one. Rings that cannot be decomposed in such a manner have been called "strong rings."

It is useful to recognize that the graphs of finite polyhedra usually contain rings that do not enter into the Schläfli symbol. Figure 7.2 (left) is a conventional representation of a cube: on the right is a Schlegel diagram. The circuit 1,4,8,7,6,2 (shown as heavier lines) is a 6-ring. The presence of 6-rings is not reflected in the Schläfli symbol ($4^3$) for the cube.

The reader might like to verify that there are also 6-rings (hexagons) in the cuboctahedron, 3.4.3.4 (see e.g. Fig. 6.8, p. 215).

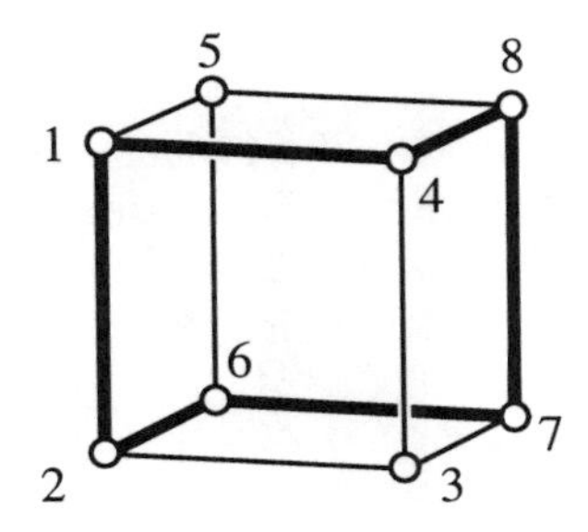


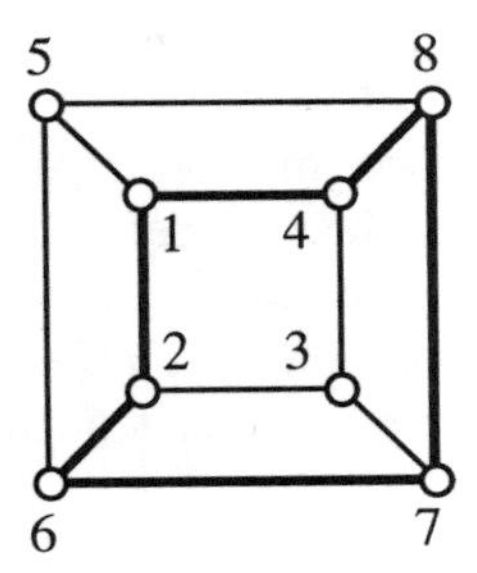


**Fig. 7.2**. Left: a conventional representation of a cube. Right: a corresponding Schlegel diagram. A 6-ring is shown as heavier lines.

Three angles meet at each node of a three-connected net. In contrast to plane nets, in three-dimensional nets more than one ring may be included in an angle (see also the next section), so we modify the Schläfli symbol to read $X_x{\cdot}Y_y{\cdot}Z_z$ where $X$, $Y$, $Z$ are numbers that represent the ring size and $x$,$y$,$z$ are numbers that indicate the numbers of rings meeting at that angle; subscript "1" is omitted. Thus $8{\cdot}8{\cdot}8_2$ indicates that at two of the angles there is an 8-ring and at the third angle there are two 8-rings. Note that many authors omit the subscripts, and the symbol for the vertex in this example is then written $8^3$.

Just as polyhedra often contain larger rings than those used to specify the Schläfli symbol, three-dimensional nets often contain larger rings in an analogous manner. (Remember that we use only the *shortest* rings at a vertex to construct the vertex symbol.)

To make clear the distinction between the use of circuits and rings, we give another example. Fig. 7.3 shows a fragment of a net. The 3-connected vertex shown as a filled circle has a 4-ring, a 6-ring and a 10-ring at the three angles, so using rings the symbol is 4.6.10 (the fragment shown might represent part of the net of the truncated icosidodecahedron). The 10-ring is at the angle *ab*. There is also an 8-circuit (1,2,3,4,5,6,7,8) at the same angle, but we do not count it as a ring because there is a short cut between vertices 1 and 4.

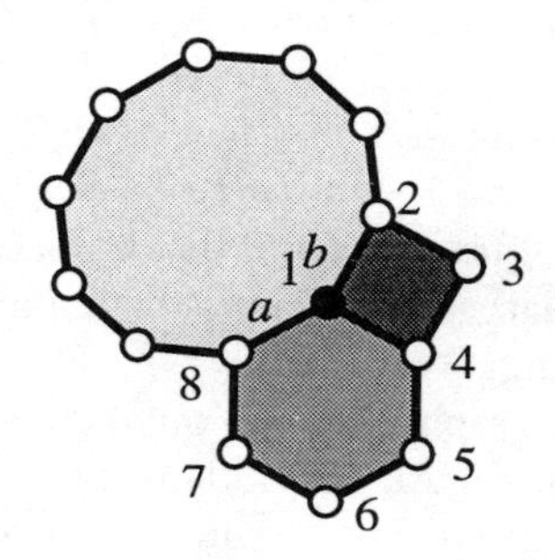


**Fig. 7.3.** Rings surrounding a 4.6.10 vertex (filled circle). The angle *ab* is contained in the 10-ring (lightly shaded) and also in the 8-circuit (not a ring) 1,2,3,4,5,6,7,8.

### *7.1.2 Schläfli symbols for 4-connected nets*

Let us arbitrarily label the four edges meeting at a vertex of a 4-connected net *a*, *b*, *c* and *d*. A pair of edges, such as *ab* define an angle at that vertex. There are six angles at each vertex defined by pairs of edges: (*ab*, *cd*), (*ac*, *bd*), (*ad*, *bc*). Pairs of angles in parentheses have no common edge and are referred to as *opposite* angles. Pairs of angles with a common edges are referred to as *adjacent*.

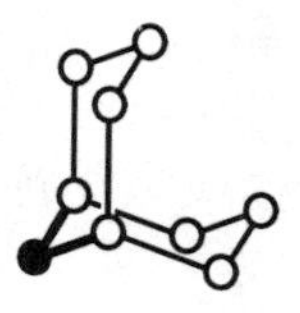

**Fig. 7.4.** Illustrating two 6-rings containing the same angle (heavy lines) at a vertex (filled circle) in the diamond net. This figure is a fragment of Fig. 7.10 (left).

For a given angle there may well be several distinct shortest rings. For example, in the diamond structure (§ 7.3.1) two 6-rings are contained in each angle as sketched for one angle in Fig. 7.4.

In order to facilitate comparison with the (rather large) literature on 4-connected nets we sometimes use two kinds of "Schläfli" symbol. A "short" one that specifies just the shortest circuit contained in each angle (most commonly found in the literature) and a "long" one that recognizes only rings and is described next by example.

In the structure of a feldspar (§ 7.3.9) such as $CaAl_2Si_2O_8$, the net of the Al and Si atoms has two kinds of vertex.[1] If we use just the shortest *circuits* at each angle, the symbols for both vertices are $4^2.6^3.8$; however, using *rings* we can distinguish the two vertices. The smallest rings and circuits associated with each angle are given below for each vertex. Also listed are the numbers of such rings and circuits for each angle.

| | vertex 1 | | | | vertex 2 | | | |
|---|---|---|---|---|---|---|---|---|
| angle | ring | number | circuit | number | ring | number | circuit | number |
| *ab* | 4 | 1 | 4 | 1 | 4 | 1 | 4 | 1 |
| *cd* | 6 | 1 | 6 | 1 | 6 | 2 | 6 | 2 |
| *ac* | 4 | 1 | 4 | 1 | 4 | 1 | 4 | 1 |
| *bd* | 6 | 1 | 6 | 1 | 8 | 1 | 8 | 3 |
| *ad* | 8 | 2 | 8 | 6 | 6 | 1 | 6 | 1 |
| *bc* | 10 | 10 | 6 | 1 | 6 | 2 | 6 | 2 |

We now write a *long Schläfli symbol* for each vertex as follows. We write, in order, the symbols of the rings with a subscript for the number of rings (omitting the subscript "1"). Note that we pair circuits by opposite angles and, subject to that constraint, write the smallest numbers first. The symbols for the two vertices are therefore:

| | |
|---|---|
| vertex 1 | $4{\cdot}6{\cdot}4{\cdot}6{\cdot}8_2{\cdot}10_{10}$ |
| vertex 2 | $4{\cdot}6_2{\cdot}4{\cdot}8{\cdot}6{\cdot}6_2$ |

Not all nets have distinctive vertex symbols even using long symbols: the pair **diamond** and **lonsdaleite** (see below) is a conspicuous example; for both nets the vertex symbol is $6_2{\cdot}6_2{\cdot}6_2{\cdot}6_2{\cdot}6_2{\cdot}6_2$.

Note that in our usage short symbols (using shortest circuits) employ *super*scripts which (including the implied "1"s) add up to three for 3-connected nets and to six (for the six angles) for 4-connected nets. Long symbols contain three entries for a 3-connected net and six entries for a 4-connected net and these are separated by a "·" and may (often do) employ *sub*scripts in some of the entries. Although the procedure may appear somewhat complicated, it is in fact very readily implemented using a computer, and is of considerable help in identifying nets of a given topology in structures.

### *7.1.3 Coordination sequences*

We briefly mentioned *coordination sequences* for structures in § 6.8.2. Each different kind of vertex in a net has associated with it a coordination sequence (*CS*) which is the

[1]But not (as one might first expect) with Si on one kind of vertex and Al on the other. The edges of the nets correspond to the -O- bonds to (Si,Al) atoms. Ca is accommodated in cavities in the net and is ignored in the present context.

sequence $n_1, n_2, \ldots n_k, \ldots$ of numbers of $k$th topological neighbors. In the language appropriate for discussion of nets we simply define a $k$th neighbor of a vertex to be one for which the *shortest* path to that vertex consists of $k$ edges.

Figure 7.5 illustrates the concepts of topological neighbors and *CS* for the two-dimensional net $4^4$. The reference vertex (filled circle) has four neighbors (dark shading) connected to it by an edge, so $n_1 = 4$. The second neighbors (light shaded) of the reference vertex are first neighbors of the first neighbors (other than the reference vertex) and clearly $n_2 = 8$. Similarly the third neighbors (open circles) of the reference vertex are the first neighbors of the second neighbors of the reference vertex that are not first neighbors of the reference vertex (or, more simply, the third neighbors are those for which the shortest path to the reference vertex consists of three edges). It should be clear that $n_3 = 12$. In fact it should be clear (drawing a few more shells may help) that $n_k = 4k$ in this instance and that the *CS* is 4, 8, 12, 16,....

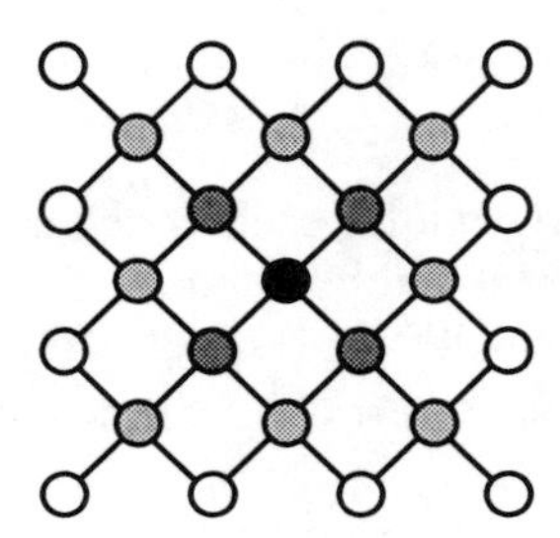

**Fig. 7.5.** Illustrating the topological neighbors of a vertex in the two-dimensional net $4^4$ (see text).

Recall that the *CS* is concerned only with the topology of the net. In Fig. 7.5 the $4^4$ net is illustrated in its most symmetrical form and it should be obvious that the eight second topological neighbors are not all the same geometrical distance from the reference vertex.

We give examples of *CS*'s for some nets in the sections below. One of their main uses is in computer recognition of nets, but it should be emphasized that occasionally two different nets have the same *CS* so that strictly speaking it can only be proved that two structures have different topologies.[1] The *CS* for the two vertices of the feldspar net are 4, 10, 20, 38, 58,... and 4, 10, 22, 38, 56,... respectively. We know of no example of two different nets in which the vertices have simultaneously the same *CS* and Schläfli symbol.

### *7.1.4 Further definitions*

Other terms that have obtained some currency are now defined.

A *uninodal* net is one in which all vertices are congruent. In its maximum symmetry form all vertices will be related by symmetry operations (be equivalent).

A *uniform* net is one in which the shortest rings at each angle are equal in size. A familiar two-dimensional example is $6^3$.

[1]The reader can readily verify that the *CS* for the two-dimensional net 3.4.6.4 is the same as that for $4^4$.

A *quasi-regular* net is one in which all vertices and all edges are equivalent.[1]

A *regular* net has all vertices, all edges and all angles equivalent. A regular net is necessarily uniform. The net of the diamond structure is the only regular three-dimensional 4-connected net; the corresponding regular 3-connected net is described next.

## *7.2. 3-connected nets

We have met some plane 3-connected nets before, notably the honeycomb ($6^3$). This is very frequently found as a layer in a crystal structure (e.g. graphite). A number of three-dimensional nets which can be realized with equal edges and angles of 120° between the edges, are known, and some are of sufficient interest to describe here.

The first three-dimensional net is an invariant cubic lattice complex:

| | |
|---|---|
| $Y^*$ | $I4_132$, $a = \sqrt{8}$ |
| | $10_5{\cdot}10_5{\cdot}10_5$ in 8 *a*: $I$ + (1/8,1/8,1/8 ; 3/8,7/8,5/8)κ |
| | or 8 *b*: $I$ + (7/8,7/8,7/8 ; 5/8,1/8,3/8)κ |

These two sets of positions produce structures that are enantiomers of each other and they are symbolized $^+Y^*$ and $^-Y^*$ respectively (for reasons which will be apparent later).

Although there are only four vertices in the repeat unit (the primitive cell), the structure is difficult to illustrate. Fig 7.6 shows two projections of $^+Y^*$. Note (on the left) that there are four-fold helices along [001] of one hand (anticlockwise along $+z$, i.e. $4_1$) in the structure, and that (right) there are likewise three-fold helices along [111] that are all of the same hand ($3_2$). The helices are of opposite hand in $^-Y^*$. The Schläfli symbol is $10_5{\cdot}10_5{\cdot}10_5$ (five 10-rings meet at each angle). This net occurs as the Si arrangement in $SrSi_2$ and isostructural compounds. It is the only regular three-dimensional 3-connected net. For a stereo picture of this net see § 7.11.8.

The axes of the $4_1$ helices are arranged as in the **β-Mn** cylinder packing and the axes of the $3_2$ helices are arranged as in the **$SrSi_2$** cylinder packing (hence the name of the latter).

It is interesting that the same structure can be derived in several ways. Positions 8 *c* of $P4_132$ are ($x,x,x$ ; $3/4-x,3/4-x,3/4-x$ ; $\bar{x},1/2+x,1/2-x$ ; $1/4-x,3/4+x,1/4+x$)κ and 8 *c* of the enantiomorphic group $P4_332$ are ($x,x,x$, ; $1/4-x,1/4-x,1/4-x$ ; $\bar{x},1/2+x,1/2-x$ ; $3/4-x,1/4+x,3/4+x$)κ. We have the special cases:

| | |
|---|---|
| $P4_132$ | 8 *c*: $x$ = 1/8 or 5/8 → 8 *a* of $I4_132$ |
| $P4_332$ | 8 *c*: $x$ = 3/8 or 7/8 → 8 *b* of $I4_132$ |

In the crystal structure of $SrSi_2$ (for data see Appendix 5), the Si atoms are at positions 8 *c* of $P4_332$ with $x = 0.423$ so they constitute a slightly distorted version of $^-Y^*$ (the Si-Si-Si angles are 118°). We may write the compound as $Sr^{2+}(Si^-)_2$ and then recognize that $Si^-$ is

[1]One sometimes sees the term "regular" for what we term "quasi-regular" but this conflicts with established usage for plane nets and polyhedra. The polyhedra 3.4.3.4 and 3.5.3.5 are quasiregular.

isoelectronic with P. The formation of three non-coplanar P-P bonds in elemental P is ascribed to the presence of a non-bonding pair of valence electrons, and it is tempting to suppose that similar considerations apply to the Si-Si bonding in $SrSi_2$.

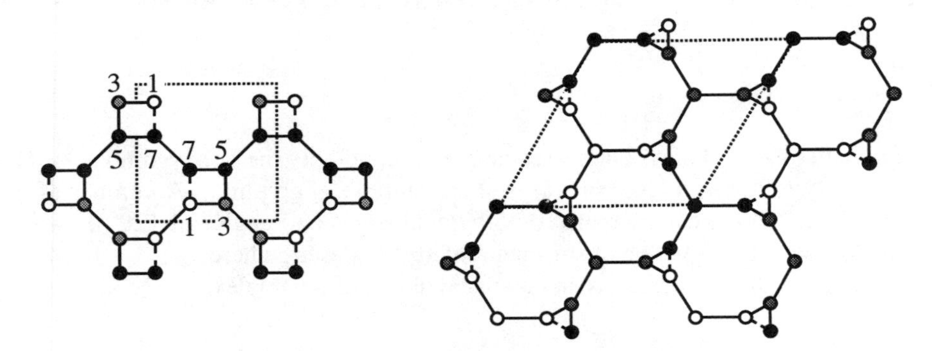


**Fig. 7.6.** Left: The lattice complex $^{+}Y^{*}$ projected on (001). The unit cell is shown by broken lines and numbers are heights in multiples of $c/8$. Right: The same structure projected on (111). Open, shaded and filled circles are at 0, 1/3 and 2/3 of a primitive translation vector {(**a**+**b**+**c**)/2 along [111]}.

The invariant positions of $P4_332$ are 4 $a$: (1/8,1/8,1/8 ; 3/8,7/8,5/8)κ and 4 $b$: (5/8,5/8,5/8 ; 1/8,7/8,3/8)κ and the invariant positions of $P4_132$ are 4 $a$: (3/8,3/8,3/8 ; 1/8,5/8,7/8)κ and 4 $b$: (7/8,7/8,7/8 ; 1/8,3/8,5/8)κ. The symbols for these lattice complexes are $^{+}Y$ and $^{-}Y$ respectively (see § 6.3.10). We can combine these as follows:

| | |
|---|---|
| $P4_132$ | 4 $a$ + 4 $b \rightarrow$ 8 $b$ of $I4_132$ ($2^{-}Y \rightarrow {}^{-}Y^{*}$) |
| $P4_332$ | 4 $a$ + 4 $b \rightarrow$ 8 $a$ of $I4_132$ ($2^{+}Y \rightarrow {}^{+}Y^{*}$) |

This shows incidentally that $I4_132$ has both $4_1$ and $4_3$ axes. If we had atoms $A$ on positions 4 $a$ and $X$ on 4 $b$ (of $P4_132$ or $P4_332$) we would get a simple (unknown) structure with $A$ surrounded by an equilateral triangle of $X$ (and *vice versa*). Note that it is its own "antistructure" (interchanging $A$ and $X$ produces the same structure).

A second cubic 3-connected net, called $6.8^2$ $D$, can be constructed with angles of 120°:

| | |
|---|---|
| $6.8^2$ $D$ | $Pn\bar{3}m$, $a = \sqrt{18}$ |
| | 6·8·8 in 24 $i$: $\pm(1/2,x,\bar{x}$ ; $0,1/2-x,\bar{x}$ ; $0,x,1/2+x$ ; $1/2,1/2-x,1/2+x)$κ, $x = 1/3$ |

A fragment of the structure is shown in Fig. 7.7. A notable feature is the groups (joined by edges) of 12 coplanar vertices parallel to {111}. The Schläfli symbol is 6·8·8. There are two kinds of edge: those on 6-rings, and those not on six-rings. The 6-rings and 8-rings of this structure can be considered as covering an infinite surface (named $D$ in Appendix 4) and hence form an infinite polyhedron as discussed later for some 4-coordinated structures.

A related net, called $6.8^2$ $P$, with the same Schläfli symbol (6·8·8) is also shown in Fig. 7.7. Parameters for unit edge length and regular hexagons are:

| | |
|---|---|
| $6.8^2$ *P* | $Im\bar{3}m$, $a = 5.266$ |
| | 6·8·8 in 48 *k*: $I + (\pm x,\pm x,\pm z)\kappa$, $x = 0.3275$ and $z = 0.0949$ |

The angles are 120° and 114.1° (2×). For slightly different parameters, the bond angles can be made all equal to 118.5°. Like the previous net this can be considered as an infinite polyhedron 6·8·8.[1]

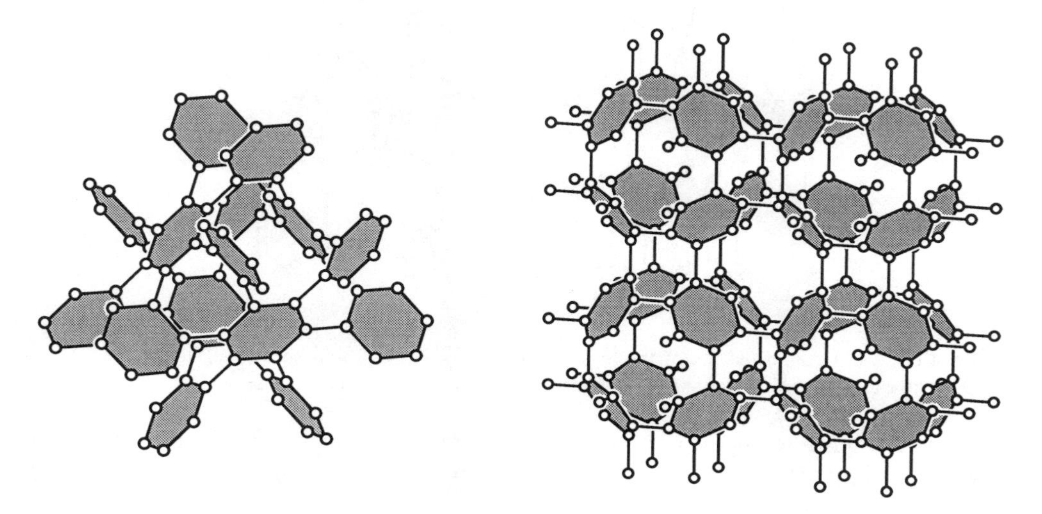

**Fig. 7.7**. Two cubic 3-connected nets, 6·8·8. Left: $6.8^2$ *D*. Right: $6.8^2$ *P*.

Another 3-connected net occurs in silicides such as $ThSi_2$ (for crystallographic data for the compound see Appendix 5). It can also be constructed with 120° bond angles:[2]

| | |
|---|---|
| $\mathbf{ThSi_2}$ net | $I4_1/amd$, $a = \sqrt{3}$, $c = 6$ |
| | vertices $(10_2{\cdot}10_4{\cdot}10_4)$ in 8 *e*: $I \pm (0,1/4,z\ ;\ 0,3/4,1/4+z)$, $z = -1/24$ |

This net is sketched in Fig. 7.8. The Schläfli symbol is $10_2{\cdot}10_4{\cdot}10_4$.

Three-connected nets can also occur in framework oxides with -O- links serving as edges as described for the structure of $B_2O_3$ below. In $P_2O_5$ $\{P\}O_4$ tetrahedra share *three* corners with neighbors (so the stoichiometry is $PO_{3/2}O = 1/2\ P_2O_5$). In one polymorph, isolated $P_4O_{10}$ molecules are formed in which the P atoms are at the vertices of a tetrahedron (see Fig. 5.18, p. 150). In a second form the P-O-P links form a honeycomb

[1]These two $6.8^2$ nets have been considered as possible structures for three-coordinated carbon; see M. O'Keeffe *et al.*, *Phys. Rev. Letts.* **68**, 2325 (1992); this reference explains the origin of the names. The $Pn\bar{3}m$ structure is a particularly favorable candidate; it is known to organic chemists as the "Riley structure" as it was apparently first suggested by H. L. Riley [see *e.g.* J. Gibson, *et al.*, *J. Chem. Soc.* 456 (1946)]. The (hypothetical) carbon is called "polybenzene."

[2]This is another net that has been considered as a possible carbon structure; interestingly carbon with this structure is predicted to be denser than graphite and metallic. [R. Hoffmann *et al.*, *J. Amer. Chem. Soc.* **103**, 4831 (1983)].

($6^3$) layer. In the third, and apparently most stable, form the P-O-P links have the topology of the $ThSi_2$ net (see Exercise 11).

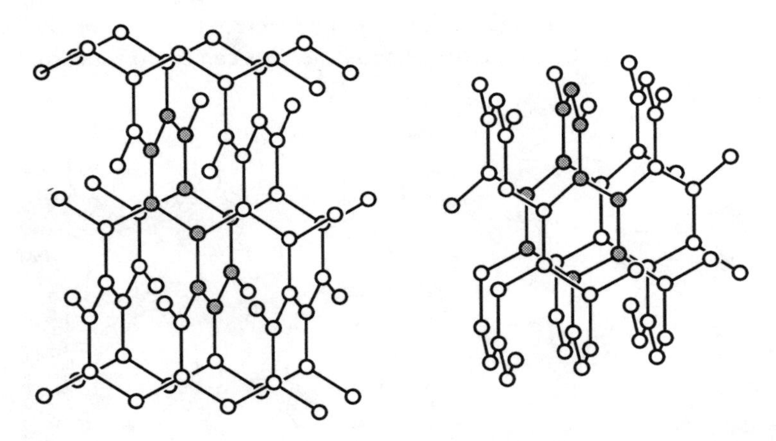

**Fig. 7.8**. Left: the 3-connected net of the Si atoms in $ThSi_2$. Right: the 3-connected net of the B atoms in $B_2O_3$. The **c** axis is vertical and the shaded vertices fall on a 10-ring in each case.

Another 3-connected net (Fig 7.8) that can be constructed with 120° bond angles again contains 10-rings. It is the net of the B atoms in $B_2O_3$ (for crystallographic data see Appendix 5) and is trigonal:

**$B_2O_3$** net $P3_112$, $a = \sqrt{3}$, $c = 9/2$
vertices ($10 \cdot 10_2 \cdot 10_2$) in 6 *c*: $(x,y,z\ ;\ \bar{y},x-y,1/3+z\ ;\ y-x,\bar{x},2/3+z\ ;\ x,x-y,\bar{z}\ ;$
$y-x,y,1/3-z\ ;\ \bar{y},\bar{x},2/3-z\ )$, $x = 1/3$, $y = 1/6$, $z = 1/9$

## 7.3. 4-Connected nets

The number of 4-connected nets found in crystal structures is very large. Well over 100 different topologies are known for framework silicates, particularly natural and synthetic zeolites. Some of these have a large number of topologically-distinct vertices and resist simple classification. In this chapter we confine ourselves mainly to relatively simple examples of nets with particular emphasis on those, such as that of the diamond structure, which arise in a number of different contexts. Some of the structures we describe do not appear to have been recognized yet in crystal structures. These have been assigned an arbitrary number for identification.[1]

Recall that unless explicitly stated otherwise, crystallographic parameters refer to unit edge (bond) length. For some nets (e.g. **diamond**) this is sufficient to completely

[1]These index numbers are known to the computer program EUTAX. [See also M. O'Keeffe & N. E. Brese, *Acta Crystallogr.* A**48**, 663 (1992).] With the recent flurry of activity in synthesis of new zeolites and related materials we find that we are continually replacing index numbers with names of known materials.

determine the structure; for other structures we give parameters for maximum volume subject to the constraint of equal edge length. Densities are expressed as $r$ = number of vertices per unit volume (for nets of unit edge). In the context of framework alumino-silicates, there is considerable interest in the framework density ($FD$), usually expressed as the number of tetrahedral atoms (Al,Si) per 1000 Å$^3$. As the Si...Si distance is typically 3.06 Å in framework silicates, $FD \approx 1000r/3.06^3 = 34.9r$ Å$^{-3}$.

### *7.3.1* **Diamond, lonsdaleite** *and their polytypes*

The diamond net is of course that of the diamond form of carbon and is also found as the structure of the stable forms of Si, Ge and (at low temperatures) Sn. As it occurs in many structure types it will prove profitable to become familiar with it. In the structure every point is connected to four neighbors at the vertices of a regular tetrahedron as shown in Fig. 7.9. A formal description is as follows:

**diamond** $Fd\bar{3}m$, $a = 4/\sqrt{3}$, $r = 0.650$
vertices in 8 $a$: $F \pm (1/8,1/8,1/8)$

The positions of the vertices correspond to the lattice complex $D$.

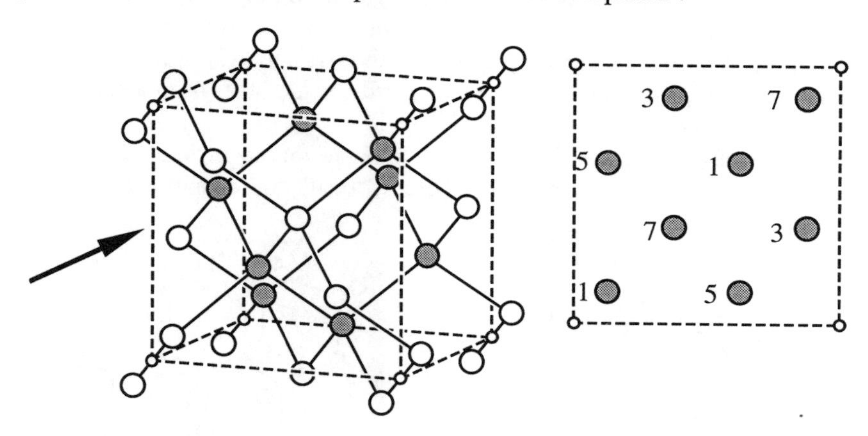


**Fig 7.9**. The diamond net. On the left in clinographic projection. The unit cell is outlined with dashed lines and points within the cell are shaded. On the right is shown a projection down the direction shown as an arrow (which corresponds to a unit cell edge) of the atoms shaded in the drawing on the left. Numbers are atom heights in multiples of 1/8. Small circles are centers of symmetry at the unit cell origin.

The long Schläfli symbol for this structure is $6_2 \cdot 6_2 \cdot 6_2 \cdot 6_2 \cdot 6_2 \cdot 6_2$. As we will see, some other nets have the same symbol.

It should be obvious from the formal description above, that the diamond array consists of two **ccp** arrays (origins at 1/8,1/8,1/8 and –1/8,–1/8,–1/8), each array occupying half of the tetrahedral sites of the other. If the two arrays are occupied by different kinds of atom, we have the **sphalerite** structure of ZnS. Recall that in § 6.1.5 we described **sphalerite** as a "close packing" of S with Zn in tetrahedral sites (or *vice versa*). Now we see that the structure is equally well (perhaps better) described as a bond network.

A related 4-connected net occurs as the structure of lonsdaleite, a rare allotrope of carbon.[1] A formal description is:

**lonsdaleite** $P6_3/mmc$, $a = \sqrt{(8/3)}$, $c = 8/3$, $r = 0.650$
vertices in 4 *f*: $\pm(1/3,2/3,z\ ;\ 2/3,1/3,1/2+z)$, $z = 1/16$

The parameters are for the idealized structure with unit edge length and angles equal to the "tetrahedral" angle $\cos^{-1}(-1/3) = 109.47°$.

It should be obvious from the above description [notice that $c/a = \sqrt{(8/3)}$] that, as in **diamond**, the vertices fall on two **cp** arrays (but now **hcp**) separated by $c/8$. The vertices of one array are in one half the tetrahedral sites of the second one, and *vice versa*. If the two arrays consist of different kinds of atom we have the **wurtzite** form of ZnS.

Fig. 7.10 shows the diamond and lonsdaleite structures side by side for comparison. In the figure the direction up the page is [111] for the diamond structure and [001] for lonsdaleite. To facilitate the comparison, note that the diamond structure can be described using a hexagonal cell with vertices in $R \pm (0,0,1/8)$, $a = \sqrt{(8/3)}$, $c = 4$. It should also be noted that the nets could also be described as a stacking of puckered $6^3$ nets (3-connected) with additional bonds between the layers. In terms of the symbols introduced in the previous chapter (§ 6.1.6) the stacking of (puckered) $6^3$ nets ($G$) in lonsdaleite is $G_1G_1$... and in diamond it is $G_1G_2G_3$.... Later we will describe some other 4-connected nets derived from $6^3$ in related ways.

In the diamond structure all the 6-circuits are skew hexagons in "chair" conformation. In lonsdaleite the hexagons not normal to [001] are in "boat" conformation (see Fig. 7.10).

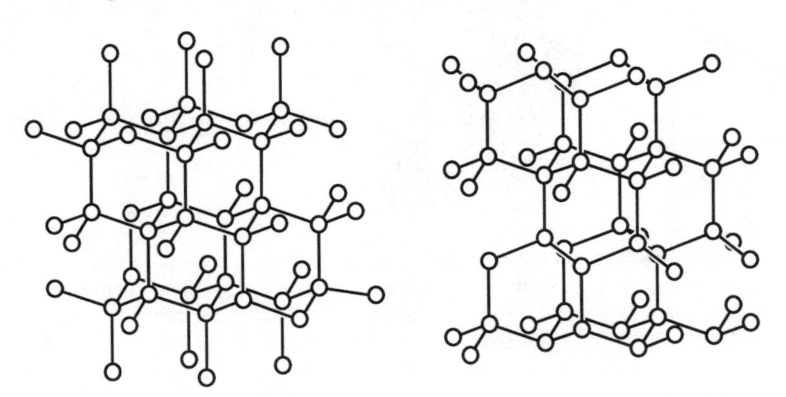

**Fig 7.10.** Left: the diamond structure in clinographic projection with [111] vertical on the page. Right: the lonsdaleite structure with [001] vertical on the page.

[1]Named for Kathleen Lonsdale who made many contributions to carbon chemistry, notably the first demonstration (by X-ray diffraction) of the planarity of the benzene ring. She was the first woman to be elected to fellowship in the Royal Society of London.

The Schläfli symbol of the lonsdaleite net is the same as that of diamond, *viz.* $6_2{\cdot}6_2{\cdot}6_2{\cdot}6_2{\cdot}6_2{\cdot}6_2$. Accordingly to distinguish between these nets topologically we need to consider numbers of *k*th neighbors. These are:

| $k$ | 1 | 2 | 3 | 4 | 5 | 6 | 7 | 8 | 9 | 10 |
|---|---|---|---|---|---|---|---|---|---|---|
| diamond | 4 | 12 | 24 | 42 | 64 | 92 | 124 | 162 | 204 | 252 |
| lonsdaleite | 4 | 12 | 25 | 44 | 67 | 96 | 130 | 170 | 214 | 264 |
| *difference* | 0 | 0 | 1 | 2 | 3 | 4 | 6 | 8 | 10 | 12 |

It may be seen that lonsdaleite has more topological neighbors (i.e. is *topologically denser*[1]) than diamond for third and subsequent neighbors. In these cases simple expressions, in which brackets indicate rounding *down* to an integer, obtain for $n_k$, the number of *k*th neighbors:

**diamond**: $$n_k = [5k^2/2] + 2 \qquad (7.1)$$

**lonsdaleite**: $$n_k = [21k^2/8] + 2 \qquad (7.2)$$

Diamond and lonsdaleite structures may be derived from, respectively, cubic and hexagonal eutaxy by filling one-half of the tetrahedral sites (either all those pointing down, or all those pointing up). Related *polytypes* can be obtained in a similar way from more complicated closest packings. The simplest of these, derived from *hc* (4*H*) packing (i.e. two *hc* **cp** arrays with points of one array in tetrahedral interstices of the other), is:

**carbon 4*H*** $P6_3/mmc$, $a = \sqrt{(8/3)}$, $c = 16/3$, $r = 0.650$
*c* vertices in 4 *e*: $\pm(0,0,z\ ;\ 0,0,1/2+z)$, $z = 3/32$
*h* vertices in 4 *f*: $\pm(1/3,2/3,z\ ;\ 1/3,2/3,1/2-z)$, $z = 5/32$

A commonly encountered projection of these, and related structures, is shown in Fig. 7.11. For diamond, this is a projection on (110) of the cubic cell; for the hexagonal cell of the other two nets it is on $(11\bar{2}0)$. Single lines represent bonds in the plane of the projection; double lines represent bonds out of the plane but superimposed in projection. Readers interested in topics such as stacking faults in Si and similar defects will find it well worth the effort it takes to learn to interpret such diagrams. In particular notice that the double lines represent a zig-zag chain of vertices seen in projection; we encounter such a motif repeatedly in the next few sections.

"Diamond" (used as an abrasive) made by shock compression of graphite is usually a rather disordered mixture of polytypes. SiC also occurs as many polytypes; a cubic **sphalerite** form (known as $\beta$-SiC), a **wurtzite** form, and other hexagonal and rhombohedral forms. The term $\alpha$-SiC is sometimes used for the non-cubic forms. Under the trade name "carborundum" it is used on a large scale as an abrasive. The polytypes have been studied very extensively as SiC is also potentially a valuable material for

[1]In the *topological* sense. In geometric terms the two nets in their most regular forms with equal edges have the same density (vertices per unit volume).

microelectronic applications. Unfortunately the electronic properties are very sensitive to structure and it is still far from certain what factors determine which polytype will form, and most preparations consist of intergrowths of different polytypes. The structure and isotopic composition of meteoritic SiC (moissanite) is also currently of considerable interest.

In the Notes (§ 7.11.4) we describe some of these polytypes in more detail and give coordinates for idealized versions of some of the simpler structures.

Forms of silica ($SiO_2$) with vertex-sharing $\{Si\}O_4$ tetrahedra and with the Si-(O)-Si nets having the **diamond** and **lonsdaleite** topology are known as cristobalite and tridymite respectively.

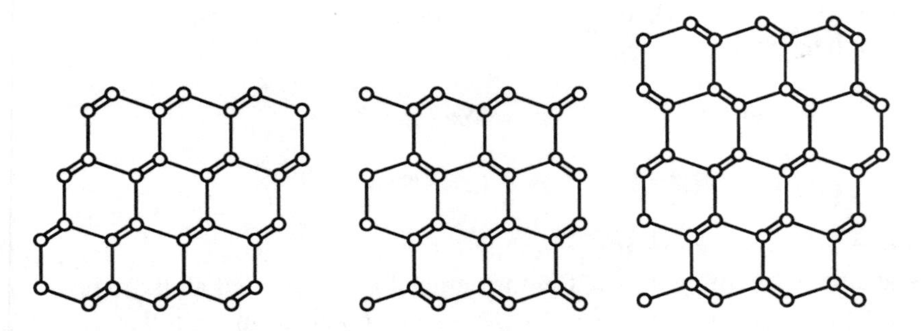

**Fig. 7.11**. Left: the diamond structure projected on (110). Middle: the lonsdaleite structure projected on $(11\bar{2}0)$. Right: the *hc* diamond polytype similarly projected. Double lines represent bonds up and down out of the plane of the paper.

### *7.3.2 Two more uniform nets, $6^6$

Fig. 7.12 shows a net that is a first cousin to the diamond-lonsdaleite family (compare Fig. 7.11). It is also uniform ($6^6$) but now the extended symbol is $6{\cdot}6{\cdot}6{\cdot}6_2{\cdot}6_2{\cdot}6_2$. There are also ten 10-rings meeting at each vertex in addition to the nine 6-rings. We have not found a good name for this net so it is arbitrarily named **net #9**. A crystallographic description is (note that it is denser than **diamond**):

**net #9** *Fddd*, $a = 4.644$, $b = 3.061$, $c = 1.532$, $r = 0.735$
$6{\cdot}6{\cdot}6{\cdot}6_2{\cdot}6_2{\cdot}6_2$ in 16 *e*: $F \pm (x,1/8,1/8\ ;\ 1/4{-}x,\ 1/8,1/8)$, $x = 0.3057$

Another simple uniform net (which we call **net #5**) is also shown in Fig. 7.12. Like the previous net (and **lonsdaleite**) there are only four vertices in the primitive cell. The extended symbol is $6{\cdot}6{\cdot}6_2{\cdot}6_2{\cdot}6_3{\cdot}6_3$. There are also eight 8-rings and ten 10-rings meeting at each vertex in addition to the nine 6-rings. A crystallographic description is:

**net #5** $P4_122$, $a = 2.030$, $c = 1.414$, $r = 0.686$
$6{\cdot}6{\cdot}6_2{\cdot}6_2{\cdot}6_3{\cdot}6_3$ in 4 *a*: $(0,x,0\ ;\ 0,\bar{x},1/2\ ;\ x,0,3/4\ ;\ \bar{x},0,1/4)$, $x = 0.3258$

Despite their simplicity, model builders will find constructing these last two nets a challenge, and counting rings by hand quite difficult.

It appears that all uniform 4-connected nets are $6^6$ (some more examples are given below).[1] In contrast uniform 3-connected nets ranging from $7^3$ to $12^3$ are known (see Notes).

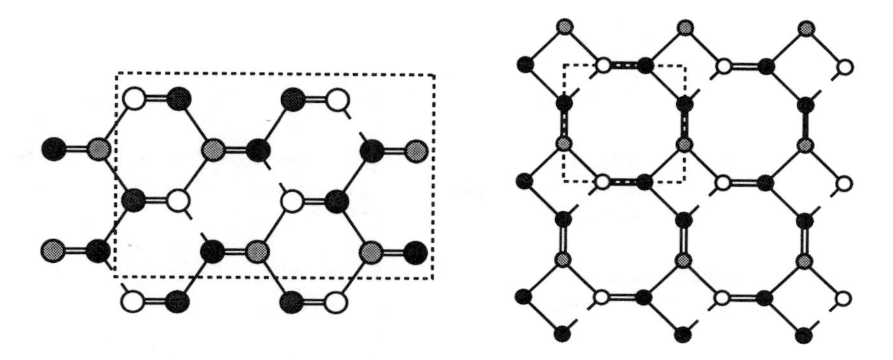

**Fig 7.12.** Left: **net #9** projected on (001) with **a** horizontal on the page. Open circles are at $z = 1/8$, lightly shaded circles at $z = 3/8$, darker shaded at $z = 5/8$ and filled circles at $z = 7/8$. Right: **net #5** projected on (001). Open circles are at $z = 0$, lightly shaded circles at $z = 1/4$, darker shaded at $z = 1/2$ and filled circles at $z = 3/4$. In both figures broken lines represent bonds to atoms with either $z < 0$ or $z > 1$.

### *7.3.3 Nets derived from* $6^3$: **$CrB_4$**

As mentioned above, it is possible to derive 4-connected nets in a formal way by linking stacked planar 3-connected nets. **Lonsdaleite** is a simple example of a 4-connected net derived by linking $6^3$ nets (Fig. 7.13, left). Two more ways of deriving 4-connected nets from $6^3$ are also shown in Fig. 7.13. In the figure, open and filled circles are to be interpreted as links up and down from a given layer and in opposite directions in adjacent layers so that each vertex is 4-connected. This description is sufficient to specify the topology (connectivity) of the net. Note that edges connecting pairs of open or filled circles correspond to 4-rings seen in projection. Except for **lonsdaleite**, all nets derived in this way contain 4-rings.

Each of the two new nets derived in the figure is uninodal and the long Schläfli symbol is also the same in both cases: $4{\cdot}6_2{\cdot}6{\cdot}6{\cdot}6{\cdot}6$ (short symbol $4.6^5$).

The first new net, in the middle in Fig. 7.13, is the net of the B atoms in $CrB_4$ (which has the orthorhombic structure described for $MnB_4$ in § 4.6.5—see also Appendix 5) so we call it the **$CrB_4$** net. The other net is found as the arrangement of Ga atoms in $CaGa_2O_4$ (edges correspond to Ga-O-Ga bonds). A formal description of the first net is:

**$CrB_4$**    *I4/mmm*, $a = 3$, $c = \sqrt{2}$, $r = 0.629$
$4{\cdot}6_2{\cdot}6{\cdot}6{\cdot}6{\cdot}6$ in 8 *h*: $I \pm (\pm x, x, 0)$, $x = 1/6$.

[1]An example of a net with only 7-rings is given below (§ 7.5.1), but one angle contains *no* rings.

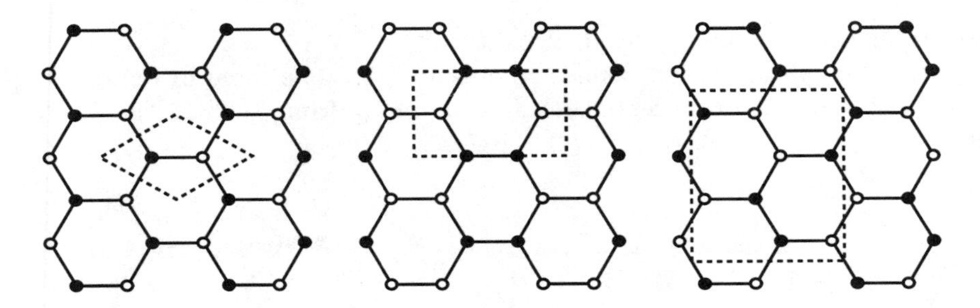

**Fig. 7.13**. Derivation of 4-connected nets from stacked $6^3$ nets (see text). On the left, **lonsdaleite** is derived, in the middle **$CrB_4$**, and on the right **$CaGa_2O_4$**. In each case additional bonds go up (down) from open (filled) circles to identical nets above (below) the ones shown.

The net is illustrated in several ways in Fig. 7.14. On the left it is projected on (001), a projection that suggests that it could also be derived from the planar $4.8^2$ net in an obvious way. This projection should also be compared with that of **net #109** on the left of Fig. 7.15. The projection in the middle of Fig 7.14 corresponds to that in the middle of Fig. 7.13 (but rotated by 90°).

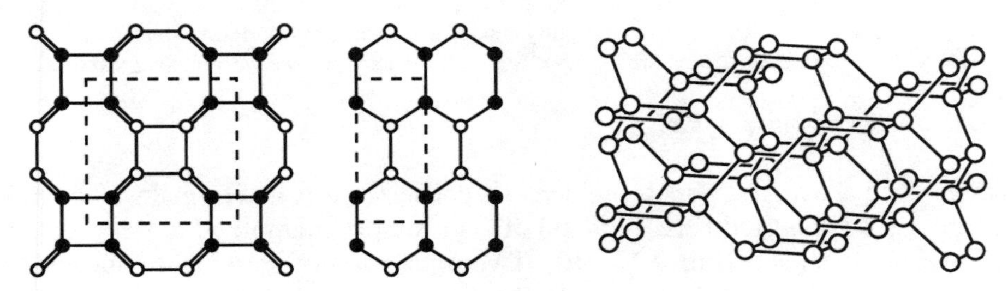

**Fig. 7.14**. The $CrB_4$ net. Left: a projection on (001); filled and open circles are at $z$ = 0 and 1/2 respectively. Middle: the structure projected on (100); filled circles are at $x = \pm 1/6$ and open circles at $x = \pm 1/3$. Right: as a clinographic projection (**c** vertical).

We meet the **$CrB_4$** net in several disguises. The 8 *h* positions of *I4/mmm* can split into two groups of 4 corresponding to the positions 4 *f* and 4 *g* of $P4_2/mnm$:

4 *f*: $\pm(x,x,0\ ;\ 1/2+x,1/2-x,1/2)$ 4 *g*: $\pm(x,\bar{x},0\ ;\ 1/2+x,1/2+x,1/2)$

If these positions are separately occupied by Be and O atoms (with $x \approx 1/6$) we obtain the structure of $\beta$-BeO (for the real $\beta$-BeO structure see Appendix 5).

The same net appears as the Al,Si net in one form of $CaAl_2Si_2O_8$ (cf. the **paracelsian** net in § 7.3.6 which is the net of another form of $CaAl_2Si_2O_8$).

The reader is encouraged to draw the second net (that of $CaGa_2O_4$). This net also occurs as the Al,P net in the mineral variscite, $AlPO_4{\cdot}2H_2O$ (the corresponding net in metavariscite, which has the same composition, has the **$CrB_4$** topology). The crystallographic description is:

**$CaGa_2O_4$** *Cmca*, $a = 2.823$, $b = 3.291$, $c = 2.794$, $r = 0.616$
$4{\cdot}6_2{\cdot}6{\cdot}6{\cdot}6{\cdot}6$ in 16 *g*: $C \pm (\pm x,y,z\ ;\ \pm x,1/2-y,1/2+z)$,
$x = 0.177$, $y = 0.133$, $z = 0.087$

### *7.3.4 Two nets related to **$CrB_4$** with zig-zag rods

Fig. 7.14 shows (left) how **$CrB_4$** may be derived from the two-dimensional net $4.8^2$ by replacing some of the edges (those not on squares) by zig-zag lines (which are shown as double lines in the projection) representing edges connecting layers above and below. A related net can be derived from $4.8^2$ by an analogous procedure if the squares are slightly tilted out of the plane as shown in Fig. 7.15 (left). We label this **net #109**.

Another net that occurs as the (Al,P) framework in a form of $AlPO_4$ known as $AlPO_4$-31 is derived similarly from the two-dimensional net 4.6.12 [see Fig. 7.15 (right)].

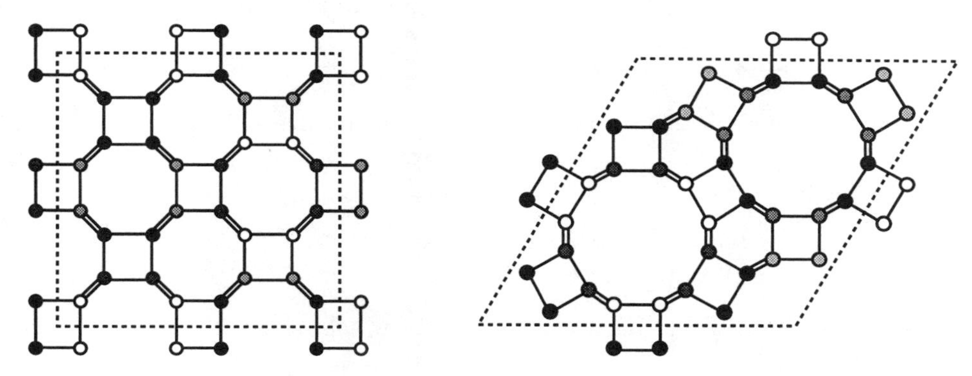

**Fig. 7.15**. Left: **net #109** projected on (001). Increasing depth of shading indicates elevations of $z/8$, $3z/8$, $5z/8$ and $7z/8$. Right: **$AlPO_4$-31** projected on (001). Increasing depth of shading indicates elevations of $z/12$, $3z/12$, $5z/12$, $7z/12$, $9z/12$ and $11z/12$.

Crystallographic data for these structures are:

**net #109** $I4_1/amd$, $a = 5.856$, $c = 1.423$, $r = 0.656$
$4{\cdot}6_2{\cdot}6{\cdot}6{\cdot}6{\cdot}6$ in 32 *i*: $I \pm (x,y,z$ ; etc.), $x = 0.085$, $y = 0.080$, $z = 0.125$

**$AlPO_4$-31** $R\bar{3}m$, $a = 6.800$, $c = 1.578$, $r = 0.570$
$4{\cdot}6_2{\cdot}6{\cdot}6_2{\cdot}6{\cdot}6_3$ in 36 *i*: $R \pm (x,y,z$ ; etc.), $x = 0.1992$, $y = 0.251$, $z = 0.25$

The nets of this and the previous section can be distinguished by numbers of topological neighbors $n_k$ (the coordination sequence). Values for the first three are rather close to each other as might be expected for nets with the same Schläfli symbols.

| *k* | *1* | *2* | *3* | *4* | *5* | *6* | *7* |
|---|---|---|---|---|---|---|---|
| **$CrB_4$** | 4 | 11 | 24 | 41 | 62 | 90 | 122 |
| **$CaGa_2O_4$** | 4 | 11 | 24 | 41 | 63 | 91 | 123 |
| **net # 109** | 4 | 11 | 24 | 42 | 65 | 95 | 131 |
| **$AlPO_4$-31** | 4 | 11 | 22 | 37 | 59 | 85 | 114 |

### *7.3.5* **$SrAl_2$, cancrinite,** *and related nets with double zig-zags*

Two more ways of connecting $6^3$ nets are shown in Fig. 7.16. These both give rise to uninodal nets that are of interest in crystal chemistry. The net on the left is found as the Al arrangement in $SrAl_2$, so we name it after that compound.[1] The same topology also occurs (considerably distorted from the geometry given above) as the network of bonds in $\alpha$-Np.

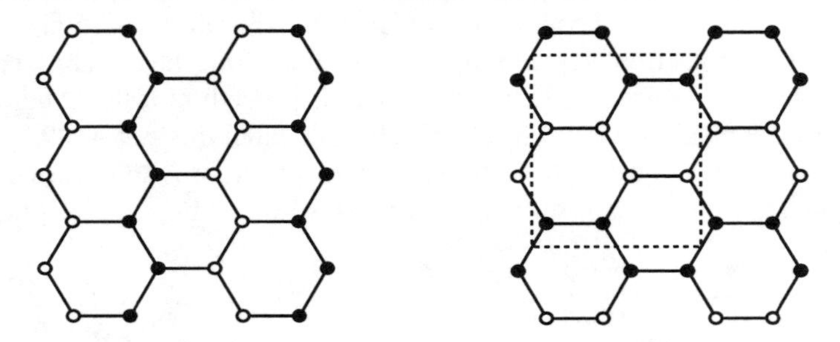

**Fig. 7.16**. Derivation of the **$SrAl_2$** (left) and **paracelsian** (right) nets from $6^3$ (compare Fig. 7.13).

A description of the **$SrAl_2$** net is:

**$SrAl_2$** *Imma*, $a = 3.268$, $b = 1.681$, $c = 2.631$, $r = 0.554$
$4{\cdot}6{\cdot}4{\cdot}6{\cdot}6{\cdot}8_2$ in 8 *i*: $I \pm (\pm x, 1/4, z)$, $x = 0.153$, $z = 0.103$

The structure is simply illustrated as a projection down the short axis (**b**); it is shown in this way as Fig. 7.17 (left).

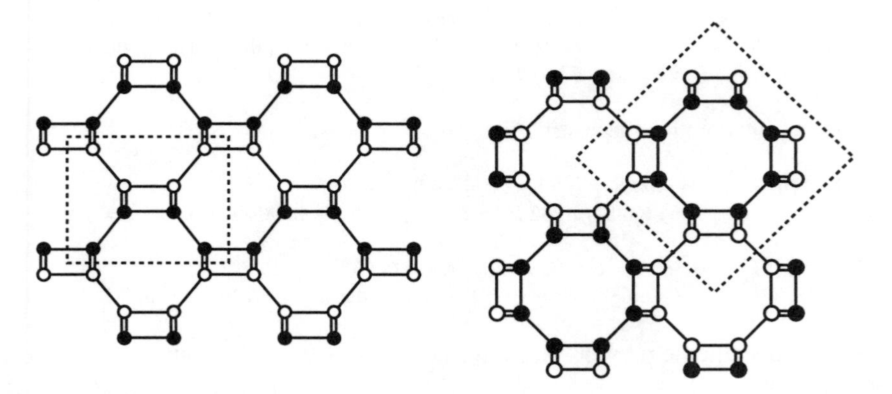

**Fig 7.17**. Left the $SrAl_2$ net projected on (010). **a** is horizontal on the page. Filled and empty circles are at $y = 1/4$ and 3/4 respectively. Right: **MAPO-39** projected on (001). Filled and empty circles are at $z = 0$ and 1/2 respectively. In both cases single lines are edges in the plane of the projection; double lines represent edges up and down out of the plane.

[1]The "type" compound is often taken as $CeCu_2$, structural data for which are to be found in Exercise 3.11. If you did that exercise, compare your drawing with Fig. 7.17.

Other occurrences of this net are as the (Al,Si) framework in $RbAlSiO_4$ and in the synthetic zeolite known as Li-A with stoichiometry $LiAlSiO_4 \cdot H_2O$.

A closely related net occurs in the synthetic zeolite with (Mg,Al,P) as framework atoms known as MAPO-39. The formal description of this net is:

**MAPO-39** $I4/mmm$, $a = 4.150$, $c = 1.708$, $r = 0.544$
4·6·4·6·6·8 in 16 *l*: $I \pm (\pm x,y,0\ ;\ \pm y,x,0)$, $x = 0.121$, $y = 0.291$

The relationship of this net to **$SrAl_2$** should be apparent from Fig. 7.17. Both structures feature double zig-zags of vertices which make up a puckered ladder as shown schematically on the left in Fig. 7.18. The two nets differ only in the way the double zig-zags are interconnected. Note that in Fig. 7.17 the projection is along the axis of the double zig-zag (which projects as a rectangle).

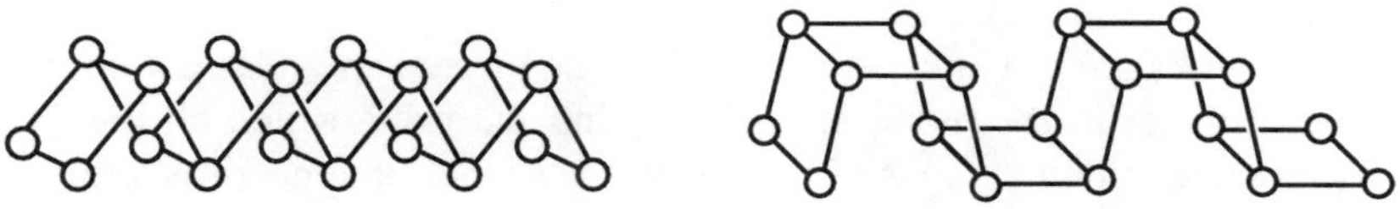

**Fig. 7.18**. Left: part of a double zig-zag. Right: part of a double crankshaft.

The net on the right in Fig. 7.16 occurs as the (Al,Si) framework in paracelsian and is discussed in § 7.3.6.

Another simple net containing double zig-zags occurs in the structure of cancrinite, which has ideal formula $CaNa_6Al_6Si_6O_{24}CO_3 \cdot 2H_2O$. This net is shown in projection in Fig. 7.19. Crystallographic data are:

**cancrinite** $P6_3/mmc$, $a = 4.000$, $c = 1.633$, $r = 0.530$
4·6·4·6·6·6 in 12 *j*: $\pm(x,y,1/4\ ;\ y,x,3/4\ ;\ \bar{y},x-y,1/4\ ;\ x-y,\bar{y},3/4\ ;$
$y-x,\bar{x},1/4\ ;\ \bar{x},y-x,3/4)$, $x = 1/12$, $y = 5/12$

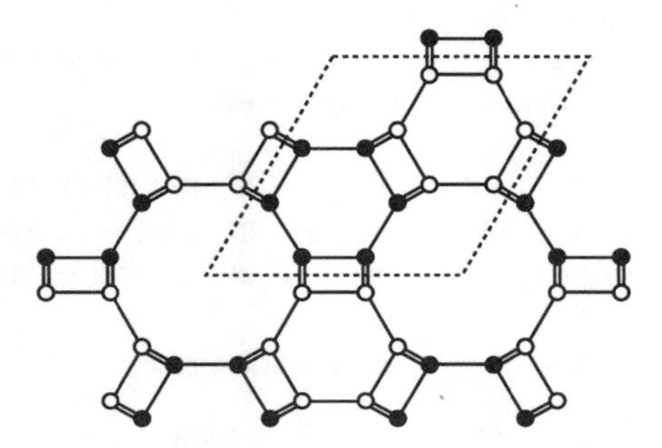

**Fig. 7.19**. The **cancrinite** net projected on (001). Open circles are at $z = 1/4$ and filled circles at $z = 3/4$. Double lines represent "zig-zags" viewed in projection.

Note that the unit cell consists of hexagons centered at 1/3,2/3,1/4 and 2/3,1/3,3/4 so the centers of the hexagons are stacked in a sequence *AB*... as in hexagonal close packing. We will meet a number of related nets based on stackings of hexagons in § 7.8.5.

### **7.3.6. Some nets derived from* $4.8^2$ *with double crankshafts*

The derivation in Fig. 7.13 of some 4-connected nets from $6^3$ by addition of a fourth link (either "up" to a net above or "down" to a net below), suggests that nets could be similarly derived from stacked $4.8^2$ nets. Four uninodal examples, one of which we have already met, are given here. Fig. 7.20, which is to be interpreted in the same way as Figs. 7.13 and 7.16, provides a definition of their topologies. Another way of deriving nets from $4.8^2$ is suggested by Fig. 7.17; we meet yet another way in our discussion of the feldspar net.[1]

A feature of the structures in this and the next section is the occurrence of rods of atoms arranged in what is known as a "double crankshaft" configuration as shown in the right-hand part of Fig. 7.18 (above).

On a double crankshaft rod, all vertices are three-connected and related 4-connected nets (discussed here) differ in the mode of cross-linking the rods, which all have their axes parallel. In projection along the axis of the double crankshaft it appears as a rectangle with two links "up" on one long edge and two links "down" on the opposite edge. (In Fig. 7.20 we have used an idealized representation in which the rectangles appear as squares—contrast Figs. 7.17 and 7.19.)

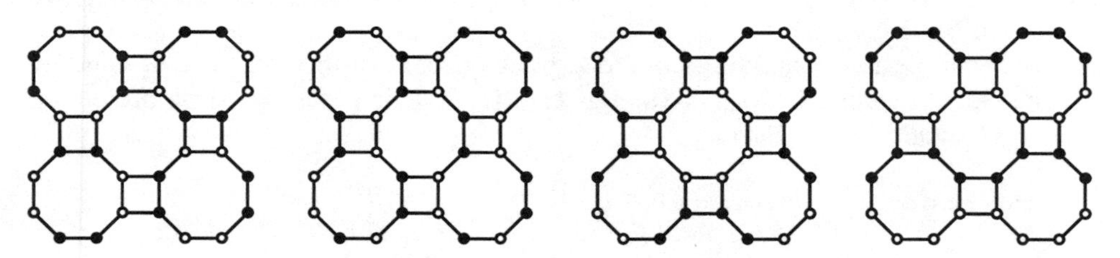

**Fig 7.20**. An idealized representation of 4-connected nets derived from stacked $4.8^2$. From the left: **net #75**, the **paracelsian** net [compare Fig. 7.16 (right)], the **merlinoite** net, and the **gismondine** net. Fourth edges go up from open circles and down from filled circles.

The net derived second from the left in Fig. 7.20 is in fact the same as that shown on the right of Fig. 7.16; (see below). It occurs as the (Al,Si) framework in paracelsian, $BaAl_2Si_2O_8$.[2] Other compounds with the same framework are danburite, $CaB_2Si_2O_8$ and hurlbutite, $CaBe_2P_2O_8$.

A crystallographic description of the **paracelsian** net is:

**paracelsian** *Cmcm*, $a = 3.252$, $b = 3.118$, $c = 2.850$, $r = 0.554$
$4{\cdot}6{\cdot}4{\cdot}6{\cdot}6{\cdot}8_3$ in 16 *h*: $C \pm (\pm x,y,z\ ;\ \pm x,y,1/2-z)$, $x = 0.154$, $y = 0.355$, $z = 0.075$

[1]For a systematic account of the derivation of 4-connected nets from $4.8^2$ see J. V. Smith, *Amer. Mineral.* **63**, 960 (1978).

[2]The celsian modification of $BaAl_2Si_2O_8$ has the feldspar structure (§ 7.3.9).

The net derived on the left in Fig. 7.20 has the same Schläfli symbol as **paracelsian**. We have not identified it in a known material so it is arbitrarily labeled **net #75**. The description is:

**net #75** $I4/mcm$, $a = 4.509$, $c = 2.844$, $r = 0.553$
$4{\cdot}6{\cdot}4{\cdot}6{\cdot}6{\cdot}8_3$ in 32 $m$: $I \pm (x,y,\pm z\ ;\ \bar{y},x,\pm z\ ;\ x,\bar{y},1/2\pm z\ ;\ y,x,1/2\pm z)$,
$x = 0.101$, $y = 0.243$, $z = 0.176$

As it is very useful to be able to interpret projections of nets, in Fig. 7.21 we repeat the two projections of **paracelsian** using the coordinates given above.[1] In the first (left) the projection [on (001)] is along the axes of the double crankshafts and the second (second left) the projection [on (100)] is normal to the double crankshafts.

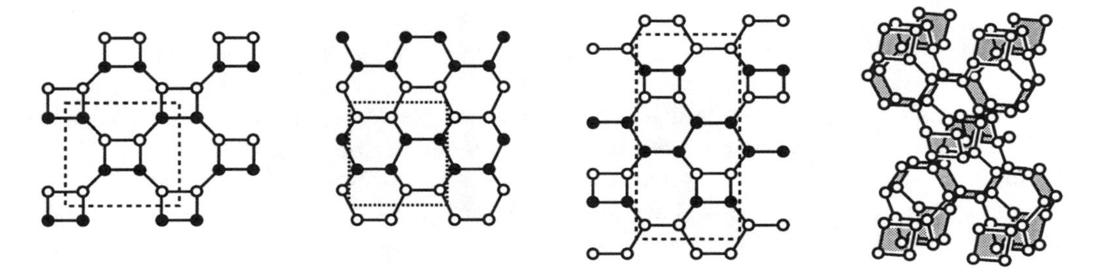

**Fig. 7.21.** Left: **paracelsian** projected on (001) with **b** up the page. Open circles are vertices at $z = 0.07$ and 0.43; filled circles are vertices at $z = 0.57$ and 0.93. Double crankshafts project as rectangles. Second left: the same net projected on (100) with **b** up the page. Open circles are vertices at $x = 0.15$ and 0.85; filled circles are vertices at $z = 0.35$ and 0.65. Double crankshafts project as single crankshafts of all filled or all open circles. Second right: **net #78** projected on (100) with **c** up the page. Open circles are vertices at $x = 0.17$ and 0.83; filled circles are vertices at $z = 0.33$ and 0.67. Right: the same net in clinographic projection with double crankshafts shaded.

These two projections suggest the possibility of a family of nets with double crankshafts running in two perpendicular directions. The simplest such net (and possibly the only one with one kind of vertex) is shown in the two right-hand drawings of Fig. 7.21. Data for this net (#78) are:

**net #78** $I4_1/amd$, $a = 3.020$, $c = 6.210$, $r = 0.565$
$4{\cdot}6{\cdot}4{\cdot}8_2{\cdot}6{\cdot}6$ in 32 $i$: $I \pm (x,y,z$ ; etc.), $x = 0.165$, $y = 0.085$, $z = 0.069$

The other two nets of Fig. 7.20 are found in the structures of merlinoite and in gismondine; they both have Schläfli symbols $4{\cdot}4{\cdot}4{\cdot}8_2{\cdot}8{\cdot}8$ (note the short symbol is now $4^3.6^2.8$ in both cases) and are illustrated in Fig. 7.57 (p. 342). These nets are less dense than the other two, and alumino-silicates based on these frameworks accommodate a substantial amount of water (as typical for zeolites). Ideal formulas are merlinoite, $K_5Ca_2Al_9Si_{23}O_{64}{\cdot}24H_2O$, and gismondine, $CaAl_2Si_2O_8{\cdot}4H_2O$.

[1] Once one learns to "read" such diagrams, it will be found very easy to construct "spaghetti" models.

**merlinoite** $I4/mmm$, $a = 4.482$, $c = 3.312$, $r = 0.481$
$4{\cdot}4{\cdot}4{\cdot}8_2{\cdot}8{\cdot}8$ in 32 $o$: $I \pm (x,\pm y,\pm z\ ;\ y,\pm x,\pm z)$, $x = 0.112$, $y = 0.269$, $z = 0.151$.

**gismondine** $I4_1/amd$, $a = 10/3$, $c = \sqrt{80}/3 = 2.981$, $r = 0.483$
$4{\cdot}4{\cdot}4{\cdot}8_2{\cdot}8{\cdot}8$ in 16 $g$: $I \pm (\pm x, 1/4 \pm x, 7/8)$, $x = 3/20$

Coordination sequences for nets of this and the last section, $n_k$, are:

| $k$ | *1* | *2* | *3* | *4* | *5* | *6* | *7* |
|---|---|---|---|---|---|---|---|
| **$SrAl_2$, MAPO-39** | 4 | 10 | 21 | 36 | 54 | 78 | 106 |
| **paracelsian** | 4 | 10 | 21 | 37 | 57 | 81 | 109 |
| **#75** | 4 | 10 | 21 | 37 | 57 | 81 | 110 |
| **#78** | 4 | 10 | 21 | 37 | 58 | 83 | 111 |
| **cancrinite** | 4 | 10 | 20 | 34 | 54 | 78 | 104 |
| **merlinoite** | 4 | 9 | 18 | 32 | 49 | 69 | 93 |
| **gismondine** | 4 | 9 | 18 | 32 | 48 | 67 | 92 |

The nets are generally distinguished by numbers of neighbors, $n_k$, although **$SrAl_2$** and **MAPO-39** have the same sequence (but these nets are distinguished by their Schläfli symbols). Notice that **$SrAl_2$, paracelsian, net #75** and **net #78** have very similar geometrical densities ($r$) and also have very similar numbers of topological neighbors ($n_k$). The same is true for **merlinoite** and **gismondine**. Generally it is found (see § 7.5) that the geometrical density, $r$, and the topological density as measured by sum of $n_k$ (over, say ten coordination shells), are rather well correlated.

### *7.3.7 *Another net with double crankshafts:* **gmelinite**

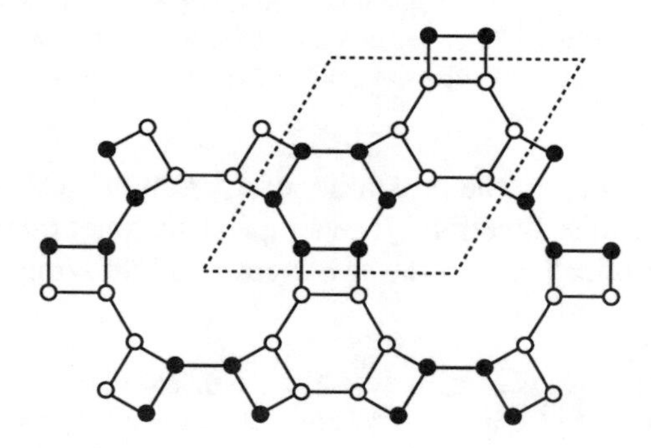

**Fig. 7.22**. The **gmelinite** net projected on (001). Open circles are vertices at $z = 0.09$ and 0.41 and filled circles are vertices at $z = 0.59$ and 0.91. Lines joining filled circles and lines joining open circles represent edges normal to **c** (i.e. parallel to the plane of the paper). The other lines represent edges going up and down so the rectangles represent double crankshafts in projection.

Another simple net with double crankshafts is found as the (Al,Si) framework of the natural zeolite gmelinite, which has approximate formula $Na_{2-x}Ca_xAl_2Si_4O_{12}{\cdot}6H_2O$. As shown in Fig. 7.22, the structure is now derived from the two-dimensional net 4.6.12.

Note that the structure contains hexagonal prisms centered at 1/3,2/3,1/4 and 2/3,1/3,1/4. The hexagon stacking could therefore be symbolized *AABB*... Contrast **cancrinite** (Fig. 7.19, p. 307) in which hexagons are stacked *AB*... Crystallographic data are:

**gmelinite** $P6_3/mmc$, $a = 4.418$, $c = 3.149$, $r = 0.451$
4·4·4·8·6·8 in 24 *l*: ±(*x*,*y*,*z* ; etc.), $x = 1/3$, $y = 0.440$, $z = 0.091$

### *7.3.8 Alternating "up-down" nets*

There is a large class of nets derived from a stacking of 3-connected two-dimensional nets with fourth links from each vertex in the layer alternating up to a layer above or down to a layer below. For such alternation to be possible, the rings in the two-dimensional net must all be even, and for a given two-dimensional net there is only one distinct three-dimensional "up-down" net. Accordingly there are only three nets in this class with one kind of vertex; they are derived from $6^3$, $4.8^2$ and 4.6.12 respectively. The net derived from $6^3$ is in fact **lonsdaleite** (§ 7.3.1). The net derived from $4.8^2$ is found as the Zn and Sb net of $TlZn_2Sb_2$ so we name it after that compound (recall that bold face names refer to structures—in this case a net). The net derived from 4.6.12 is found in the alumino-phosphate zeolite known as **$AlPO_4$-5**. Data for the last two (illustrated in Fig. 7.23) are:

**$TlZn_2Sb_2$** $I4/mcm$, $a = 3.235$, $c = 2.639$, $r = 0.580$
$4{\cdot}6_2{\cdot}6{\cdot}6_3{\cdot}6{\cdot}6_3$ in 16 *l*: *I* ± (*x*,*x*+1/2,*z* ; etc.), $x = 0.147$, $z = 0.190$

**$AlPO_4$-5** $P6/mcc$, $a = 4.515$, $c = 2.599$, $r = 0.523$
$4{\cdot}6_2{\cdot}6{\cdot}6_3{\cdot}6_2{\cdot}6_3$ in 24 *m*: ±(*x*,*y*,*z* ; etc.), $x = 0.455$, $y = 0.122$, $z = 0.192$

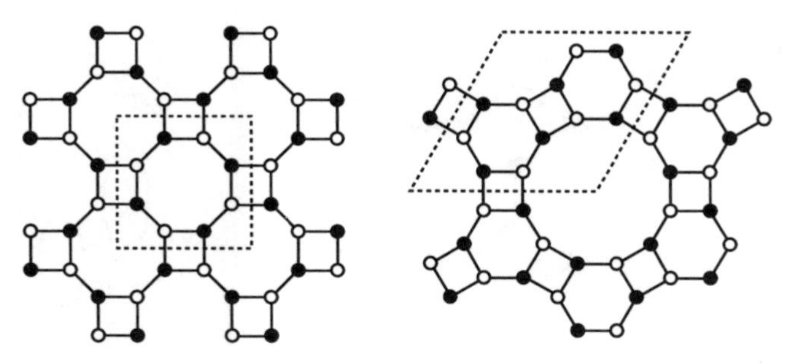

**Fig 7.23**. Left: projection of **$TlZn_2Sb_2$** on (001). Right: projection of **$AlPO_4$-5** on (001). In both cases open circles are vertices at $z = 0.39$ and 0.61, and filled circles are vertices at ±0.19. Edges go down from filled circles to a layer below, and up from empty circles to one above.

A conspicuous feature of these nets is the rods of atoms that project as a square with connections up-down-up-down (UDUD).[1] This may be contrasted with the double crankshaft which projects as a rectangle with connections UUDD. A fragment of an UDUD rod is shown in Fig. 7.24.

[1]For the case of of two or more rods fused together (when the "squares" become rectangles) see § 7.11.8

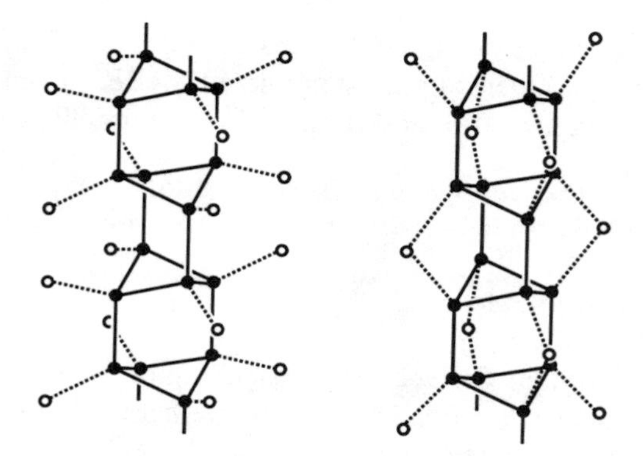

**Fig. 7.24**. Left: a fragment of $\mathbf{TlZn_2Sb_2}$ (Fig. 7.23) illustrating an UDUD rod (full lines and filled circles). Right: the same rod with connecting vertices as in **scapolite**.

Several other "up-down" nets found in zeolites are described in § 7.8.4. A simple net with UDUD rods occurs in minerals of the scapolite family.[1] We include it here as it illustrates several useful points. Crystallographic data are:

**scapolite** $I4/mmm$, $a = 4.338$, $c = 2.294$, $r = 0.556$
$4{\cdot}5_2{\cdot}8_2{\cdot}8_2{\cdot}8_2{\cdot}8_2$ in 8 $i$: $I \pm (x,0,0 \,;\, 0,x,0)$, $x = 0.163$
$4{\cdot}8_2{\cdot}5{\cdot}5{\cdot}5{\cdot}5$ in 16 $n$: $I \pm (0,y,\pm z \,;\, y,0,\pm z)$, $y = 0.339$, $z = 0.280$

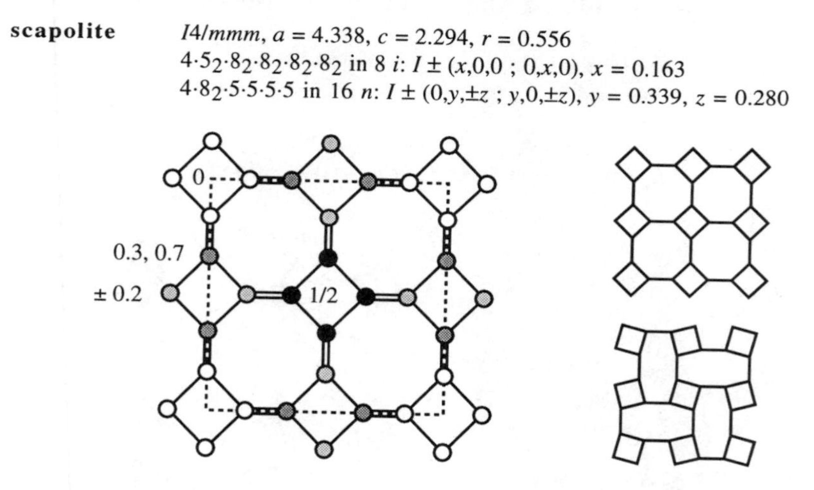


**Fig. 7.25**. Left: the **scapolite** net projected on (001). Numbers are elevations in units of $c$. Right: the $4.8^2$ net in its most-symmetrical, minimum-density form (top) and partly collapsed (bottom).

The net consists of UDUD rods linked by single squares of vertices. In Fig 7.24 (right) one vertex of each such square is shown as an open circle. Fig. 7.25 shows the structure in projection on (001) and it is worth the effort it takes to read the projection. Note in particular that squares at $z = 0$ and $z = 1/2$ represent planar square groups, but that the other squares represent UDUD rods. As usual we show pairs of edges inclined up and down as

[1]Scapolite has approximate formula $Na_4Al_3Si_9O_{24}Cl$. Remember that the net is the structure of only the tetrahedrally-coordinated atoms (in this case Al and Si).

double lines.

The crystallographic data given above for **scapolite** are for the highest symmetry form and the most open structure. In fact it is common for nets of this sort to "collapse" to a denser structure, with the extent of collapse determined by the nature (and size) of the material in the cavities of the net (in scapolite this is Na and Cl, other members of the family contain $CO_3$ and $SO_4$ groups). Fig. 7.25 also shows schematically a common mode of collapse of a rod structure (such as **scapolite**) based on $4.8^2$ nets.

An isolated UDUD rod of $\{T\}O_4$ tetrahedra has stoichiometry $T_2O_5$. In narsarsukite such rods are joined by columns of vertex-sharing $\{Ti\}O_6$ octahedra to produce a structure with composition $Na_2TiO(Si_2O_5)_2$. (For more on narsarsukite, see Exercise 7.12.14.)

### *7.3.9* **Feldspar** *and* **coesite**

The feldspars are a large and complex group of minerals with general formula $A(Al,Si)_4O_8$. The (Al,Si) atoms are on a 4-connected net linked by O atoms. In the structure of coesite (a high-pressure polymorph of $SiO_2$) the Si atoms are on a different, but closely related net. Derivation of the feldspar net from $4.8^2$ is shown in Fig. 7.26. To interpret this figure, it should be noted that distorted $4.8^2$ nets are packed in a two layer repeat; the net in solid lines alternating with that shown by broken lines. Filled and open circles represent respectively edges going (almost) vertically up and down from the solid-line net to the broken-line net above or below.

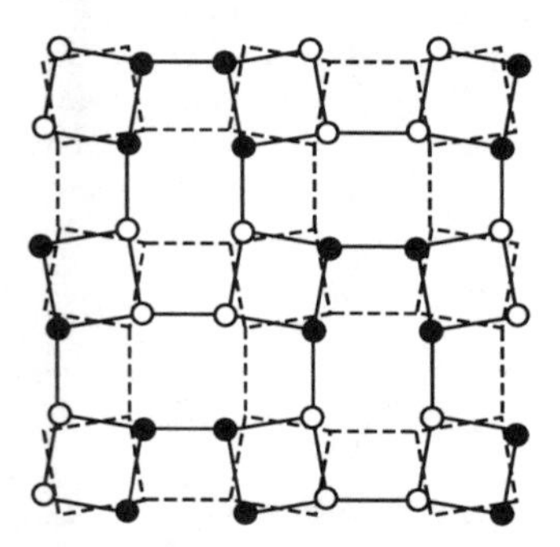

**Fig. 7.26**. Derivation of the feldspar net from $4.8^2$. See text.

A crystallographic description of this net (as usual with unit edges and in its highest-symmetry, maximum-volume form) is:

**feldspar** $C2/m$, $a = 3.189$, $b = 3.951$, $c = 2.346$, $\beta = 115.4°$, $r = 0.599$
vertices in 8 $j$: $C \pm (x,\pm y,z)$
$4{\cdot}6_2{\cdot}4{\cdot}8{\cdot}6{\cdot}6_2$: $x = 0.287$, $y = 0.373$, $z = 0.376$
$4{\cdot}6{\cdot}4{\cdot}6{\cdot}8_2{\cdot}10_{10}$: $x = 0.000$, $y = 0.231$, $z = 0.213$

Note that the net of the real mineral is somewhat collapsed from the maximum volume form. Parameters more representative of real minerals (again for unit edge) are $a = 2.780$, $b$

= 4.202, $c$ = 2.314, $\beta$ = 116.0°, $r$ = 0.659, vertices in 0.215, 0.381, 0.337 and 0.011, 0.176, 0.222. Ordering of silicon and aluminum in real materials also lowers the symmetry and leads in some instances to larger unit cells; the topology of the net stays the same, of course.

Why, with simple, symmetrical nets such as those of the previous sections available, did nature choose this more-complex, lower-symmetry net as the basis of the structure of the most common of all minerals on the face of the earth?

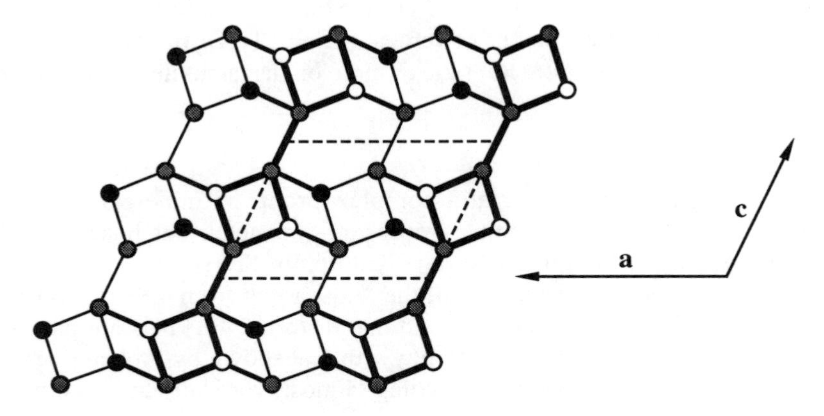


**Fig 7.27**. Projection of the feldspar net on (010). The points shown are a slab with $1/2 < y < 1$. A second slab lies beneath and related to the first by a mirror plane at $y = 1/2$. Shaded circles represent one type of vertex, open and filled circles the other. Vertices connected by heaviest lines are approximately in a plane and above the plane of those connected by the lightest lines. Additional edges go vertically up from vertices shown by open circles and down from vertices shown as filled circles.

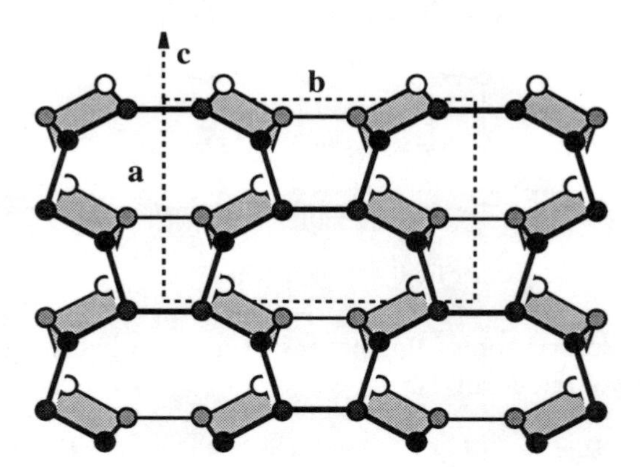


**Fig. 7.28. Feldspar** projected on (001) showing the twisted double crankshafts (shaded). An **ab** face of the unit cell is shown and the arrow marked **c** is the projection of that axis on the plane. Open, lightly shaded, darker shaded and filled circles are vertices at elevations approximately 0.45, 0.80, 1.32 and 1.67 above the plane. As **c** is not normal to the page successive layers are displaced up the page by the projection of **c**. The coordinates used in the drawing are for the denser of the two sets given above.

As an aid to constructing a model a projection of the **feldspar** net is shown in Fig. 7.27. The twisted double crankshafts now run horizontally across the page (parallel to **a**). In making a "spaghetti" model the best strategy is to first construct the double crankshafts, secondly link them to make the layer shown in the figure (the double crankshafts will now twist) and finally connect layers to their mirror images using the remaining unused links as shown in Fig. 7.28 in which the twisted crankshafts are seen in a projection on (001).

A related net, that is also made up of twisted double crankshafts, is that of the Si atoms in the coesite form of $SiO_2$ (a high-pressure polymorph; for data see Appendix 5). The net is made up of layers similar (topologically identical) to those shown for feldspar in Fig. 7.27. The linkage between layers makes the topology difficult to describe and results in rings of edges being looped as in a chain. Fig 7.29 (which should be contrasted with Fig. 7.28) shows how the layers of crankshafts are linked. Model builders should first construct double-crankshaft layers as for **feldspar**, then link them using Fig. 7.29 as a guide. Note that the middles of these linking edges are at centers of symmetry.

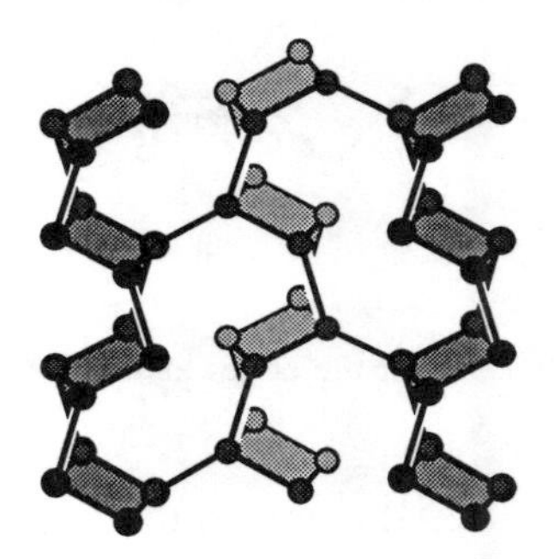
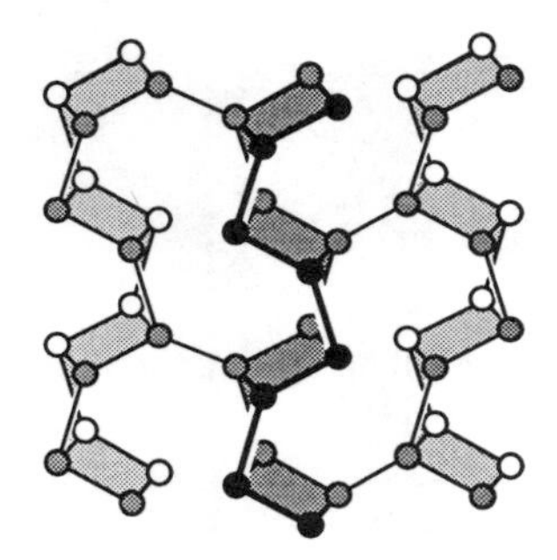

**Fig. 7.29**. The linkage of double-crankshaft layers in coesite shown in projection on (001). Only one double crankshaft of each layer is shown. On the left, the middle double crankshaft is connected to two higher in elevation, on the right the same (middle) double crankshaft is connected to two others of lower elevation. Vertices with the same shading have approximately the same elevation.

The coesite net contains odd (9-) rings. This means that if there are two kinds of atoms ($A$ and $B$) at the vertices they cannot be arranged so that $A$ has only $B$ neighbors and *vice versa*. Data for a form of the coesite net are given below. Note the high density ($r$). Coesite is the densest known form of silica with Si in 4-coordination by oxygen.

**coesite** $C2/c$, $a = 2.327$, $b = 4.152$, $c = 2.281$, $\beta = 120.8°$, $r = 0.845$
vertices in 8 $f$: $C \pm (x,y,z\ ;\ x,\bar{y},1/2+z)$
$4{\cdot}8{\cdot}4{\cdot}9_7{\cdot}6{\cdot}8$: $x = 0.157$, $y = 0.094$, $z = 0.088$
$4{\cdot}6{\cdot}4{\cdot}6{\cdot}8{\cdot}9_2$: $x = 0.019$, $y = 0.325$, $z = 0.041$

### *7.3.10* Sodalite

A simple quasi-regular four-connected net that arises in many contexts is the net we call the **sodalite** net. A formal description is, for unit edge length:

**sodalite** $Im\bar{3}m$, $a = \sqrt{8}$, $r = 0.530$
4·4·6·6·6·6 in 12 $d$: $I \pm (1/4,0,1/2)\kappa$

We met this pattern earlier (§ 6.2) as that of the tetrahedral sites of **bcc** packing. It may also be recalled that this arrangement is that of the vertices of a space-filling by truncated octahedra (illustrated in Fig. 7.30). The centers of the truncated octahedra are on the nodes of a *bcc* lattice.

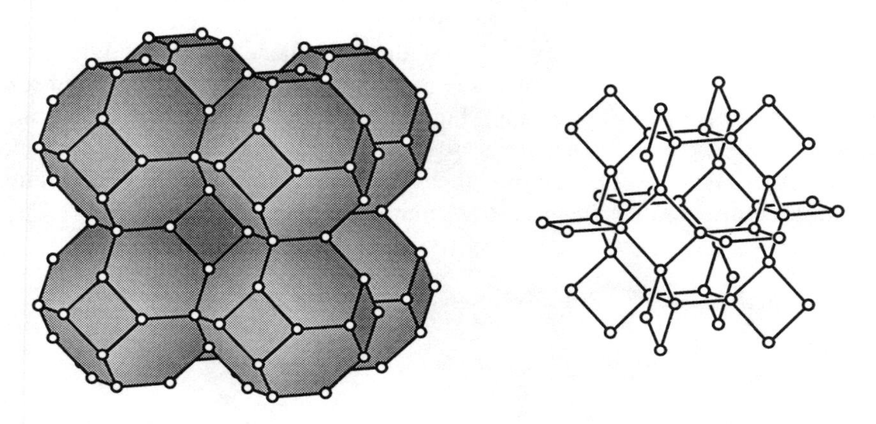

**Fig. 7.30**. Left: space-filling by truncated octahedra ($4.6^2$). Right: the edges and vertices shown as the four-connected **sodalite** net.

The net corresponds to the arrangement of the (Si,Al) atoms in the mineral sodalite[1]: $Na_4Si_3Al_3O_{12}Cl$, although in the real structure the O arrangement lowers the symmetry to $I\bar{4}3m$ (see Fig. 6.72, p. 275) and Si,Al ordering further lowers the symmetry to $P\bar{4}3n$. The positions of the vertices of the net are also those of the lattice complex $W^*$. The same pattern is shown as a 4-connected net in Fig. 7.30. The Schläfli symbol of the vertices is 4·4·6·6·6·6. Note that for a net derived from a packing of polyhedra each angle contains just one ring (the Schläfli symbol has no subscripts).

The number of topological neighbors is given by the very simple formula:

$$n_k = 2k^2 + 2 \tag{7.3}$$

### *7.3.11* **NbO** *and* **quartz**

The positions of the Nb and O atoms in the simple cubic structure of NbO (for crystallographic data see Appendix 5) taken together are at the nodes of another quasi-regular net (Fig. 7.31). For unit edge length, the crystallographic description is:

[1]Minerals of this group are often called ultramarines. Ordering and the occurrence of incommensurate phases in sodalites such as $Ca_4Al_6O_{12}WO_4$ (CAW) and $Sr_4Al_6O_{12}MoO_4$ (SAM) are currently lively topics of investigation. Yet another sodalite composition is $Ca_4Al_6O_{13}$ (with one O atom in the cage).

**NbO** $Im\bar{3}m$, $a = 2$, $r = 0.75$
$6_2 \cdot 6_2 \cdot 6_2 \cdot 6_2 \cdot 8_2 \cdot 8_2$ in 6 *b*: $I + (0,1/2,1/2)\kappa$

The vertices of the net correspond to the invariant lattice complex $J^*$.[1] They also represent the distribution of "octahedral sites" in the *bcc* structure and so the vertices are at the centers of the squares in the **sodalite** net. The Schläfli symbol of the vertices is $6_2 \cdot 6_2 \cdot 6_2 \cdot 6_2 \cdot 8_2 \cdot 8_2$. Note that the edge angles are four of 90° and two of 180° and that the vertices fall on three mutually-perpendicular strings that intersect in pairs.

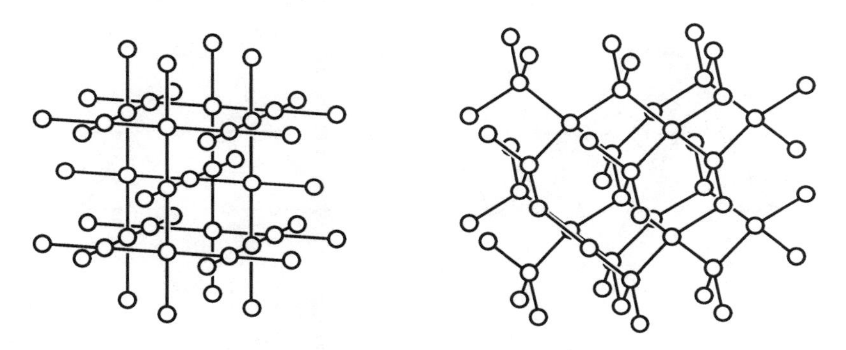

**Fig. 7.31**. Left: the **NbO** net. Right: the **quartz** net (**c** is vertical).

Another quasi-regular net, that we will see is related to the **NbO** net, is the **quartz** net. It describes the positions of the Si atoms in the quartz form of $SiO_2$ (the most stable form at room temperature and pressure). In the maximum volume configuration the structure is:

**quartz** $P6_222$, $a = \sqrt{(8/3)}$, $c = \sqrt{3}$, $r = 0.75$
$6 \cdot 6 \cdot 6_2 \cdot 6_2 \cdot 8_7 \cdot 8_7$ in 3 *c*: 1/2,0,0 ; 0,1/2,2/3 ; 1/2,1/2,1/3

The net is enantiomorphic; its mirror image has symmetry $P6_422$ with coordinates 1/2,0,0 ; 0,1/2,1/3 ; 1/2,1/2,2/3. The structure (also illustrated in Fig. 7.31) again corresponds to an invariant lattice complex; the two enantiomers are labeled $^+Q$ and $^-Q$ respectively. The edges are all equivalent, so the net is quasiregular. The angles are two of 90°, two of $\cos^{-1}(-1/3) = 109.47°$ and two of $\cos^{-1}(-2/3) = 131.81°$.

The **NbO** and **quartz** nets have the same short Schläfli symbol: $6^4.8^2$ and the same density ($r = 0.75$). The relationship between them is shown in Fig. 7.32 in which a projection of the **NbO** net on (111) is compared with a projection of the **quartz** net on (001). It is to be noted that the repeat distance normal to the plane of projection is the same in the two cases, and that both nets contain the same number of vertices per unit volume (i.e. they have the same geometrical density). The main difference is that **NbO** contains three-fold spirals of two hands whereas **quartz** contains spirals of only one hand.

[1]Note that in NbO the Nb and O atoms (each 4-coordinated in a square by the other) are each on a *J* lattice complex; two such complexes separated by 1/2,1/2,1/2 combine to give $J^*$.

A third net is related to these two. It is of some interest because, like them, it has only three vertices in the primitive cell. The projection in Fig. 7.33 should make the family relationship clear (compare Fig. 7.32). Wells (see the Notes at the end of this chapter) calls it "net 2" so we label it **W2**. It is found as the arrangement of Ni atoms in heazlewoodite, $Ni_3S_2$ (for crystallographic data see Appendix 5), and is a rare example of a net containing *only* odd rings. A formal description is:

**W2** $R32$, $a = 5/\sqrt{3}$, $c = \sqrt{5}$, $r = 0.558$
$3{\cdot}7{\cdot}7{\cdot}7{\cdot}7_2{\cdot}7_2$ in 9 *d*: $R + (x,0,0\ ;\ 0,x,0\ ;\ \bar{x},\bar{x},0)$, $x = 1/5$

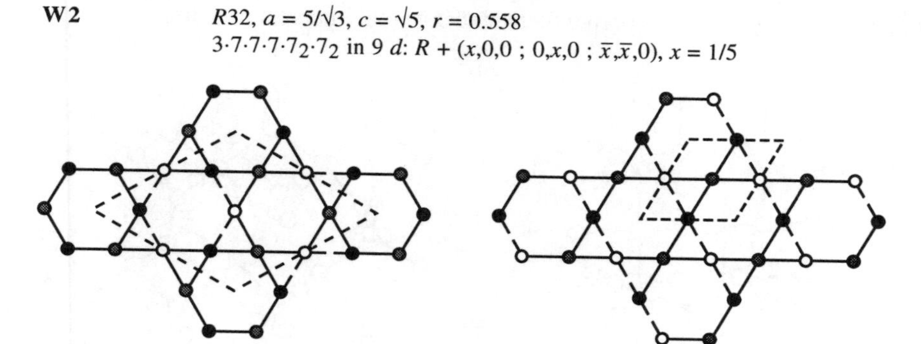

**Fig. 7.32**. Left: the **NbO** net projected on (111). Right: the **quartz** net projected on (001). Open, shaded and filled circles are at 0, 1/3 and 2/3 of the repeat distance normal to the plane of projection.

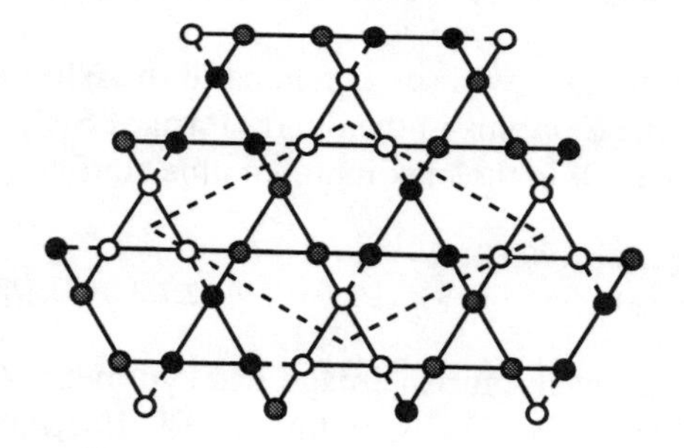

**Fig. 7.33**. The net **W2** drawn for comparison with Fig. 7.32

The nets differ in topological density: The values of $n_k$ are given by:

**NbO** $$n_k = 3k^2 + 2 - (k \bmod 2) \qquad (7.4)$$

**quartz** $$n_k = 19k^2/6 + 2 \qquad (k = 6i)$$
$$n_k = [(19k^2 + 10)/6] \qquad (2 < k \neq 6i) \qquad (7.5)$$

**W2** $$n_k = 5k^2/2 \qquad (k = 2i)$$
$$n_k = 5k^2/2 + 7/2 \qquad (k = 2i+1) \qquad (7.6)$$

Here $i$ is a positive integer, and brackets indicate rounding down to an integer. $n_1 = 4$ for **W2** (as, of course, for all 4-connected nets). Topologically, **quartz** is the densest of these nets (has most topological neighbors) and **W2** (which contains 3-rings) is topologically less dense than the other two. (See § 7.5 for a discussion of density.)

### **7.3.12 More quasi-regular and/or uniform nets:* **γ-Si**

Three other nets of interest are described briefly here. They are all cubic and although there is only a small number of vertices in the repeat unit, they provide an interesting challenge to the model builder.

The first is found as the structure of a high-pressure polymorph of elemental silicon (for data see Appendix 5), so we call it the **γ-Si** net. Data for the idealized net are:

**γ-Si** $Ia\bar{3}$, $a = 2/(\sqrt{6} - \sqrt{3})$, $r = 0.739$
$6{\cdot}6_2{\cdot}6{\cdot}6_2{\cdot}6{\cdot}6_2$ in 16 *c*: $I \pm (x,x,x\ ;\ x,1/2-x,1/2+x)\kappa$, $x = (\sqrt{2} - 1)/4 = 0.1035$

The vertices in **γ-Si** have long Schläfli symbol $6{\cdot}6_2{\cdot}6{\cdot}6_2{\cdot}6{\cdot}6_2$ so that this is another uniform net. Note that there are two different angles: three of 97.94° and three of 118.13°.

There are several interesting features of the structure. If the value of $x$ is increased to 1/8 = 0.125 the vertices are on the lattice complex $Y^{**}$ which is the positions 16 *b* of $Ia\bar{3}d$ and corresponds to an intergrowth of the $^{+}Y^{*}$ and $^{-}Y^{*}$ lattice complex. We saw (§ 7.2) that these latter represent the two enantiomers of the Si net in $SrSi_2$. Thus **γ-Si** can be considered as derived from two inter-grown $\mathbf{SrSi_2}$ (3-connected) nets. The **γ-Si** and $Y^{**}$ structures are compared in Fig. 7.34.

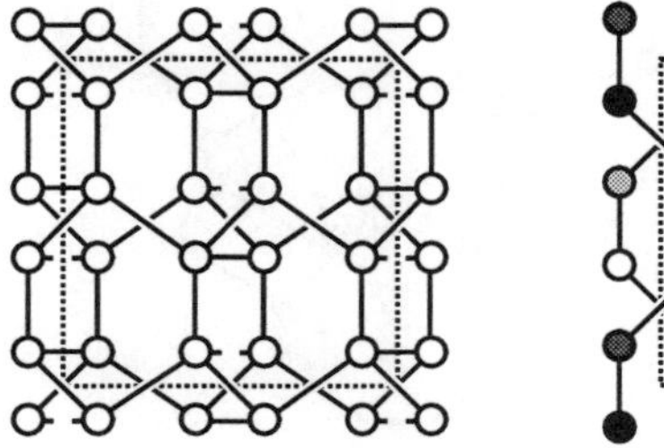
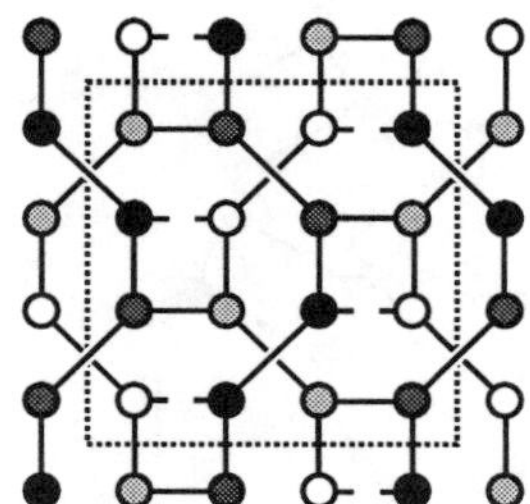

**Fig. 7.34**. Left: the γ-Si net projected on (001); vertices in the top layer are at approximately $z = 1.1c$ and $0.9c$ and those in the bottom layer at $0.4c$ and $0.6c$. Broken bonds go to layers above or below those shown in the figure. Right: The $Y^{**}$ lattice complex shown as two intergrown three-connected nets ($^{\pm}Y^{*}$) projected on (001). Open circles at $c/8$, light shaded at $3c/8$, darker shaded at $5c/8$ and filled at $7c/8$.

The positions 16 *c* of $Ia\bar{3}$ can also be considered to be a combination of two sets of 8 *a* of $I2_13$: $I + (x,x,x\ ;\ x,1/2-x,1/2+x)\kappa$, with $x = u_1$ and $x = u_2$. If $u_1 + u_2 = 1/2$ the symmetry is actually still $Ia\bar{3}$. In γ-Si $u_1 = 0.104$ and $u_2 = 0.386$. If these are changed to $u_1 = 0.0$ and $u_2 = 0.25$ the structure is transformed to the body-centered cubic array (described with a cell of twice the edge length). This suggests that the material found after application of pressure may have transformed from a body-centered cubic lattice at high pressure as the pressure was released. It has been suggested that the diamond form of carbon will transform to **γ-Si** at very high pressure (about 800 GPa).[1]

As **γ-Si** approximates the $Y^{**}$ structure (two intergrown $Y^*$ nets) it contains intergrown $3_1$ and $3_2$ helices of Si atoms each arranged as in **$SrSi_2$**. For this reason the cylinder packing consisting of two interlaced **$SrSi_2$** packings was named the **γ-Si** packing in § 6.7.3.

Two other invariant lattice complexes are 4-connected nets. They are of less importance in crystal chemistry and are frustratingly difficult to illustrate (as our figures below attest), but we include them here for completeness. Model makers will find that they are challenging to construct but very beautiful.

| | |
|---|---|
| $S^*$ | $Ia\bar{3}d$, $a = \sqrt{(32/3)}$, $r = 0.689$ |
| | $6{\cdot}6{\cdot}6_2{\cdot}6_2{\cdot}6_2{\cdot}6_2$ in 24 $d$: $I \pm$ (3/8,0,1/4 ; 1/8,0,3/4)κ |

This is another uniform net. It is also quasi-regular. A feature of the structure is that there are large non-intersecting tunnels parallel to <111> arranged as in the **garnet** cylinder packing (§ 6.7.3). The projection on (111) shown in Fig. 7.35 reveals one such set of tunnels.

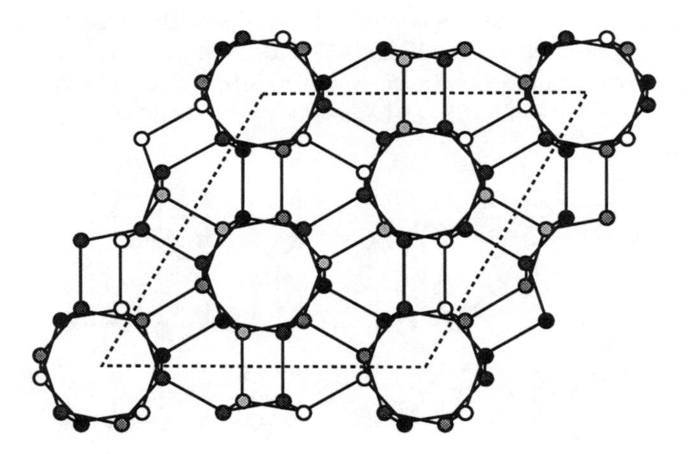

**Fig. 7.35**. The lattice complex $S^*$ projected on (111). Points in order of increasing depth of shading are at heights (in units of |**a**+**b**+**c**|/2) of 1/12, 3/12, 5/12, 7/12, 9/12 and 11/12.

The second invariant lattice complex is an enantiomorphous pair:

| | | |
|---|---|---|
| **$HL4_3$** | | $I4_132$, $a = \sqrt{(32/3)}$, $r = 0.344$ |
| | $^+V$ | $3{\cdot}3{\cdot}10_2{\cdot}10_2{\cdot}10_3{\cdot}10_3$ in 12 $c$: $I$ + (1/8,0,1/4 ; 3/8,0,3/4)κ |
| | $^-V$ | $3{\cdot}3{\cdot}10_2{\cdot}10_2{\cdot}10_3{\cdot}10_3$ in 12 $d$: $I$ + (5/8,0,1/4 ; 7/8,0,3/4)κ |

This is also a quasi-regular net. The net is very open; now two 3-rings meet at each vertex (Fig. 7.36 emphasizes this aspect of the structure) and the other rings in the structure

[1]R. Biswas *et al., Phys. Rev.* B**35**, 9559 (1988).

are ten 10-rings meeting at each vertex. The structure is the third of four 4-coordinated rare sphere packings discussed by Heesch and Laves (see § 7.5.2) so we label it $\mathbf{HL4_3}$.

In much the same way as $^+Y^*$ and $^-Y^*$ can interpenetrate (Fig. 7.34) so can $^+V$ and $^-V$ to produce the lattice complex $V^*$ which corresponds to the positions 24 *c* of $Ia\bar{3}d$.

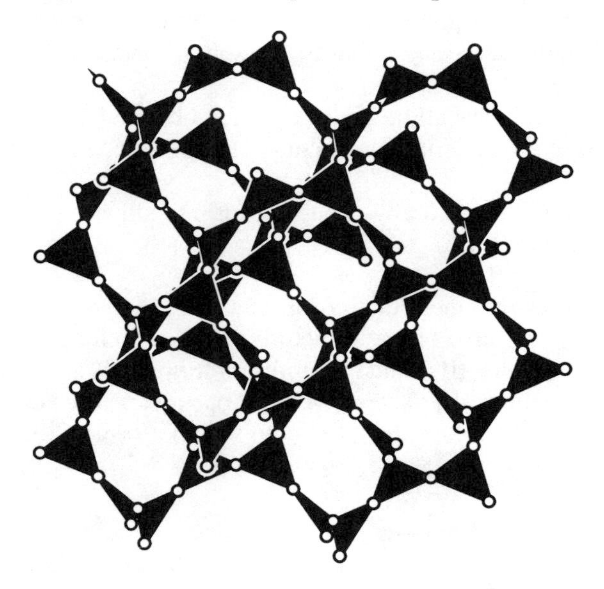

**Fig. 7.36**. A clinographic projection of the lattice complex $^+V$ ($\mathbf{HL4_3}$).

### *7.3.13 Silica* ($SiO_2$) *and water nets:* **keatite** *and* **moganite**

The planet we inhabit is made largely of silicates, and its surface consists largely of water (solid and liquid) and framework silicates. Silica ($SiO_2$) itself is of importance in a variety of contexts, and at least twelve polymorphs have been described. Low pressure forms of silica consist of framework structures of $\{Si\}O_4$ tetrahedra sharing vertices and the 4-connected nets corresponding to some of these structures have been met already: here we discuss several others. Silica is also found as a very-high-pressure **rutile** form (with six-coordinated Si) known as stishovite. $BeF_2$ and $GeO_2$ and ternary derivatives such as $AlPO_4$ also adopt at least some of the silica structures. Note that most of the silica polymorphs have lower symmetry than the idealized net on which the structure is based.

Solid water (ice) in its low pressure forms is also based on 4-connected nets of O atoms joined by -H... bonds and the nets are the same as in some of the silica polymorphs. In higher pressure forms the structures are based on two inter-penetrating 4-connected nets.

(i) Quartz, the stable form of silica at ordinary temperature and pressure, was described in § 3.6 and the 4-connected net discussed in § 7.3.11.

(ii) At high temperature, quartz transforms to cristobalite which is based on the

**diamond** net (§ 7.3.1). This is the net of ice $I_c$ which is stable at very low temperatures. Ice VII (formed at pressures above about 2 GPa = 20 kbar) consists of two interpenetrating cristobalite nets.

(iii) Tridymite is a (possibly metastable) form of silica based on the **lonsdaleite** net. This is the net of the familiar ice ($I_h$) stable at atmospheric pressure below 0 °C.

(iv) Coesite is the first crystalline phase of silica obtained when quartz is subject to pressure (about 2 GPa). The net of this structure was discussed in § 7.3.9.

(v) Melanophlogite is rare naturally-occurring form of silica that is based on the net of the Type I gas hydrate net (§ 7.6).

(vi) Keatite is another rare metastable form of silica. The net (Fig. 7.37) contains two kinds of node and occurs also as the structure of $\gamma$-Ge (a form recovered from high pressure) and as the net of ice III (which is produced from ice $I_h$ at a pressure of about 200 MPa). Data for keatite $SiO_2$ and $\gamma$-Ge are given in Appendix 5. The keatite cell is tetragonal ($P4_32_12$, $a$ = 7.46, $c$ = 8.61 Å); in the maximum volume form of the net, $r$ = 0.668. The Schläfli symbols are Si(1): $5{\cdot}5{\cdot}5_2{\cdot}7_2{\cdot}8_2{\cdot}8_2$ and Si(2): $5{\cdot}7{\cdot}5{\cdot}7{\cdot}5{\cdot}7_2$.

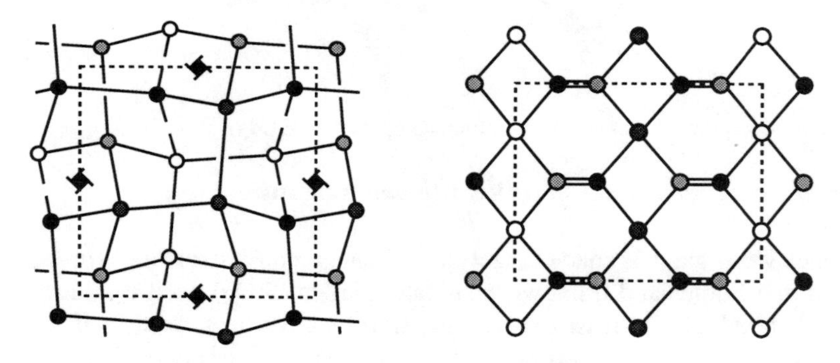

**Fig. 7.37**. Left: the Si atoms of keatite projected on (001). Increasing depth of shading indicated vertices at approximately $z$ = 0, 1/4, 1/2, and 3/4. The positions of the $4_3$ axes in the unit cell (broken lines) is indicated. Right: the **moganite** net projected on (010). **a** is horizontal across the page. Progressively darker shaded circles represent vertices at $y$ = 0, 1/4, 1/2 and 3/4.

(vii) Moganite is another polymorph of $SiO_2$ which is reported[1] to be monoclinic ($I2/a$). Recent results suggest that moganite occurs more commonly than once supposed in fibrous forms of "quartz" known as chalcedony, agate, chert, flint, etc.[2] The idealized 4-connected net of the Si atoms is illustrated in Fig. 7.37. For unit edge crystallographic data are:

[1]G. Miehe & H. Graetsch, *Eur. J. Mineral.* **4**, 693 (1992).
[2]P. J. Heaney & J. E. Post, *Science* **255**, 441 (1992).

**moganite** $Ibam$, $a = 3.53$, $b = 1.61$, $c = 2.89$, $r = 0.731$
$4{\cdot}4{\cdot}6_2{\cdot}6_2{\cdot}8_2{\cdot}8_2$ in 4 $a$: $I \pm (0,0,1/4)$
$4{\cdot}8_6{\cdot}6{\cdot}6{\cdot}6{\cdot}6$ in 8 $j$: $I \pm (x,y,0\ ;\ 1/2+x,1/2-y,0)$, $x = 0.167$, $y = 0.250$

For more on **moganite** and its relationship to **quartz** see § 7.11.7. Crystallographic data for the real material are given in Appendix 5.

(viii) When molten silica is cooled, it forms a glass (amorphous silica) which is a random network of vertex-sharing $\{Si\}O_4$ tetrahedra. This is often referred to as *quartz glass* but the term "quartz" should be restricted to the crystalline polymorph. An amorphous silica is also obtained when quartz is compressed at low temperatures; amorphous ice can similarly be obtained from crystalline ice.

(ix) The structure of ice VI (stable between about 0.6 and 2 GPa at room temperature) is based on two inter-penetrating **edingtonite** nets (see § 7.8.7)

## 7.4 Nets and infinite polyhedra

We now expand our consideration of nets constructed from polyhedra sharing faces. They may be derived as a space filling (tiling) by finite polyhedra or considered as an infinite surface tiled with polygonal faces (infinite polyhedra). The most important of these nets are those of zeolites. The simplest such structure, the **sodalite** net, was described in § 7.3.10.

### *7.4.1* **Linde A**: *an infinite polyhedron* $4^2.6^2$

The first such new net is that of the zeolite known as Linde $A$:

**Linde** $A$ $Pm\bar{3}m$, $a = 1 + \sqrt{8} = 3.828$, $r = 0.428$
$4{\cdot}6{\cdot}4{\cdot}6{\cdot}4{\cdot}8$ in 24 $k$: $(0,\pm y,\pm z\ ;\ 0,\pm z,\pm y)\kappa$, $y = 1/(4 + \sqrt{2}) = 0.1847$, $z = 2y$

In Fig. 7.38 (left) the structure is illustrated as an assembly of cubes and truncated octahedra ($4.6^2$) sharing square faces. Considered as a 4-connected *net* it has the Schläfli symbol $4{\cdot}6{\cdot}4{\cdot}6{\cdot}4{\cdot}8$ given above. However we can also consider this structure as an *infinite polyhedron*; at each vertex two squares and two hexagons meet, and the interior of this polyhedron is the space occupied by the cubes and truncated octahedra. Considered as a *polyhedron* the Schläfli symbol is $4^2.6^2$. Note the distinction between the net description and the polyhedron description—in the former we count rings at six angles at the vertices, in the latter we count (in cyclic order) only polygons on the surface of the polyhedron.

The empty space in the above structure consists of truncated cuboctahedra (4.6.8) sharing octagonal faces as shown in Fig. 7.38 (right). This is likewise an infinite polyhedron $4^2.6^2$ and the "interior" of the polyhedron is the space occupied by the truncated cuboctahedra.

It may be seen that we have two infinite polyhedra, each of which fills the empty space of the other. Such pairs of infinite polyhedra are termed *complementary*. Taken together they represent an example of space filling (tiling) by regular and/or Archimedean polyhedra. In each case the same polygons are on the surfaces of both polyhedra.

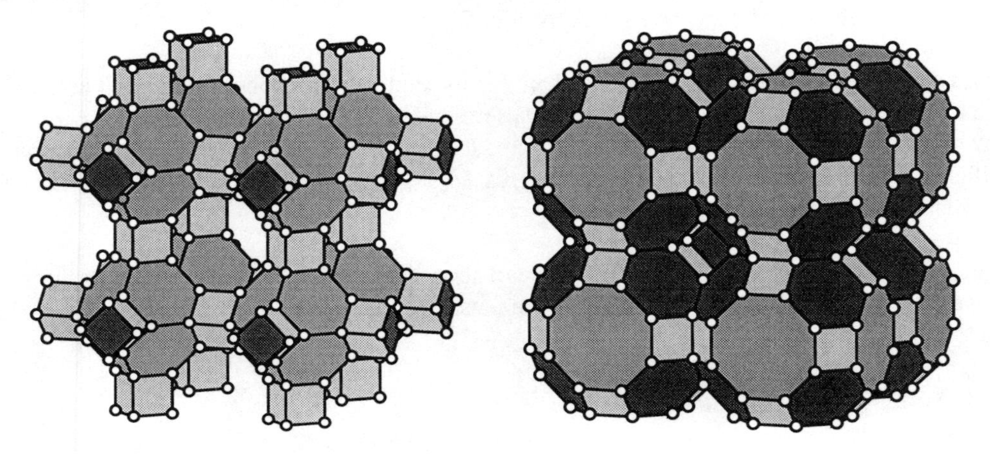

**Fig. 7.38**. The Linde *A* structure. Left: as an assembly of cubes and truncated octahedra. Right: as an assembly of truncated cuboctahedra.

For such an apparently complicated structure the number of topological neighbors is given by a very simple formula (brackets indicate rounding down to an integer):

$$n_k = [(8k^2 + 13)/5] \tag{7.7a}$$

*7.4.2 Zeolite* **rho**: *infinite polyhedra* 4³.6 *and* 4.8.4.8

The structure of the zeolite known as **rho** gives rise to another 4-connected net that can also be described as a space-filling by Archimedean polyhedra as shown in Fig. 7.39. Data for unit edge are:

**rho** $Im\bar{3}m$, $a = 2 + \sqrt{8} = 4.828$, $r = 0.426$
4·4·4·6·8·8 in 48 *i*: $I + (1/4,\pm x,1/2\pm x\ ;\ 1/4,1/2\pm x,\pm x)\kappa$, $x = (\sqrt{2} - 1)/4 = 0.1035$

This structure may be considered as constructed of truncated cuboctahedra (4.6.8) and octagonal prisms ($4^2$.8) sharing octagonal faces.[1] From this point of view it is an infinite polyhedron $4^3$.6. The empty space is an identical infinite polyhedron, so it is its own complement. The combination of the infinite polyhedron and its complement corresponds a space-filling by truncated cuboctahedra and octagonal prisms.

[1]Compare Fig. 7.38 in which truncated cuboctahedra share octagonal faces without the intervening octagonal prisms.

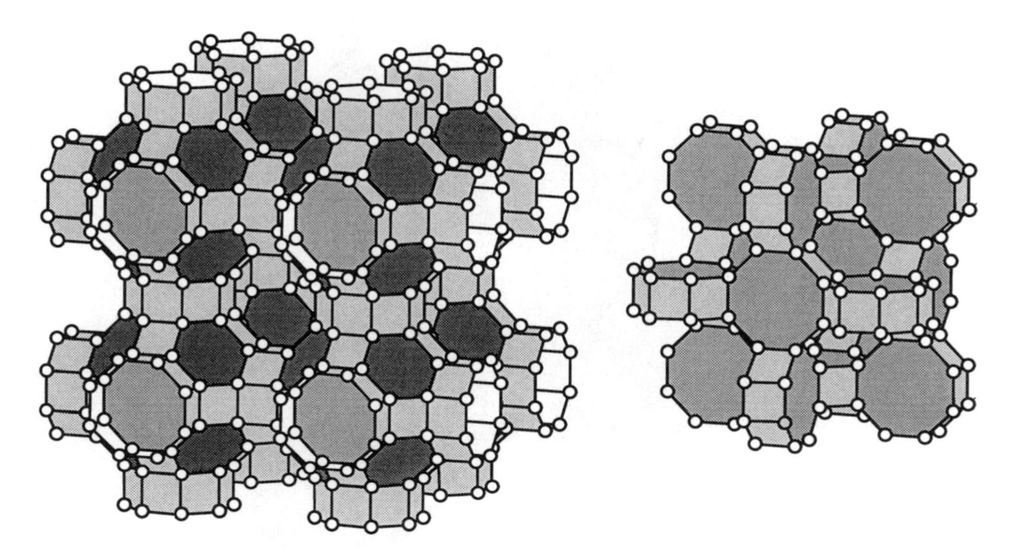

**Fig. 7.39**. The structure of zeolite **rho**. Left: as a packing of truncated cuboctahedra (4.6.8) and octagonal prisms ($4^2.8$) forming the polyhedron $4^3.6$. Right: as the polyhedron 4.8.4.8 formed from fused octagonal prisms.

Alternatively the same set of vertices may be described as derived from an assembly of octagonal prisms sharing square faces as also shown in Fig. 7.39. In this description, the structure is an infinite polyhedron 4.8.4.8. The complement of this infinite polyhedron is the one (not shown) derived from truncated cuboctahedra sharing their hexagonal faces.

**Rho** and **Linde A** have very similar densities, *r*. Remarkably, equation 7.7a (p. 324) holds for the coordination sequences of both structures.

### *7.4.3 Zeolite* **ZK-5** *and an infinite polyhedron* $4^3.8$

We have already met two structures involving the truncated cuboctahedron (4.6.8). One description of the Linde *A* structure involved linking them by sharing octagonal faces (note that they also are linked by cubes). Similarly the zeolite rho structure was obtained by linking truncated cuboctahedra by octagonal prisms attached to the octagonal faces. A third possibility involves linkage by hexagonal prisms attached to the hexagonal faces. This produces a structure (Fig. 7.40) that is the framework of the zeolite known as ZK-5 and is an infinite polyhedron $4^3.8$. Considered as a 4-connected net the Schläfli symbol is 4·4·4·8·6·8. The crystallographic description is:

**ZK-5** $Im\bar{3}m$, $a = 2/\sqrt{3} + \sqrt{8} + 2 = 5.983$, $r = 0.448$
4·4·4·8·6·8 in 96 *l*: $I \pm (x,\pm y,\pm z\ ;\ y,\pm x,\pm z)\kappa$,
$x = 1/2a = 0.0836$, $y = 1/4 - x/\sqrt{3} = 0.2018$, $z = 2y - x = 0.3199$

The *CS* is given by (cf. Eq. 7.7a):

$$n_k = [(12k^2 + 16)/7] \qquad (7.7b)$$

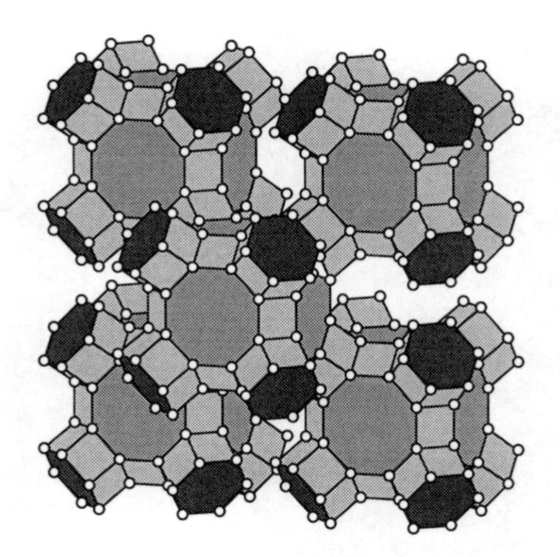

**Fig. 7.40**. Part of the zeolite **ZK-5** structure.

### *7.4.4* **Faujasite**: *a second infinite polyhedron* $4^3.6$

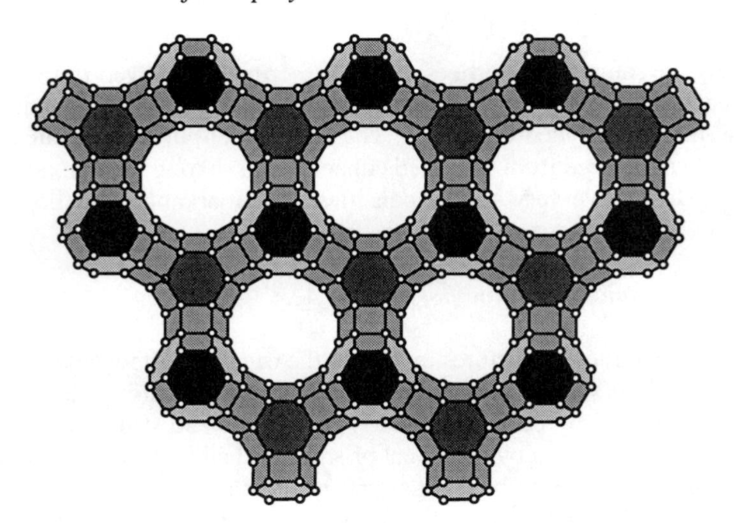

**Fig. 7.41**. A fragment of the faujasite structure projected on (111). The black hexagons are hexagonal prisms seen in projection and sharing a hexagonal face with a truncated octahedron underneath. The shaded regular hexagons are top faces of truncated octahedra connected to hexagonal prisms on the bottom face.

Another net that may be derived from a packing of polyhedra is that of the natural zeolite faujasite. This is obtained by fusing hexagonal prisms on four non-adjacent faces of a truncated octahedron ($4.6^2$). Adding four more truncated octahedra to the other hexagonal faces of the prisms results in a tetrahedral arrangement of the four truncated octahedra about the first one. Continuing so that a **diamond** array of truncated octahedra is obtained produces the **faujasite** net, which is an infinite polyhedron $4^3.6$. A polytype is obtained if the truncated cuboctahedra are connected as in **lonsdaleite** (rather than as in **diamond**); it should be clear that an infinite number of other polytypes is possible. Fig. 7.41 shows part of one layer of the structure. Considered as 4-connected nets all vertices in these structures have symbol 4·4·4·6·6·12.

Crystallographic data for the cubic (**c-**) and hexagonal (**h-**) **faujasite** nets are:

**c-faujasite** $Fd\bar{3}m$, $a = 20/(\sqrt{18} - \sqrt{3}) = 7.966$, $r = 0.380$
4·4·4·6·6·12 in 192 $i$: $F \pm (x,y,z\ ;\ y,x,z\ ;\ x,1/4-y,1/4-z\ ;\ x,1/4-z,1/4-y\ ;$
$y,1/4-z,1/4-x\ ;\ y,1/4-x,1/4-z\ ;\ z,1/4-x,1/4-y\ ;\ z,1/4-y,1/4-x)\kappa$,
$x = (\sqrt{6} - 1)/40 = 0.0362$, $y = 1/8$, $z = 3/8 - 2x = 0.3025$

**h-faujasite** $P6_3/mmc$, $a = 5.633$, $c = 9.199$, $r = 0.380$
all vertices 4·4·4·6·6·12 in 24 $l$: $\pm(x,y,z$ ; etc.)
vertex 1: $x,y,z =$ 0.371, 0.097, 0.0181 ; vertex 2: $x,y,z =$ 0.156, 0.489, 0.0706
vertex 3: $x,y,z =$ 0.430, 0.037, 0.1069 ; vertex 4: $x,y,z =$ 0.489, 0.156, 0.1957

### **7.4.5 An open structure, W*8 and a related zeolite*

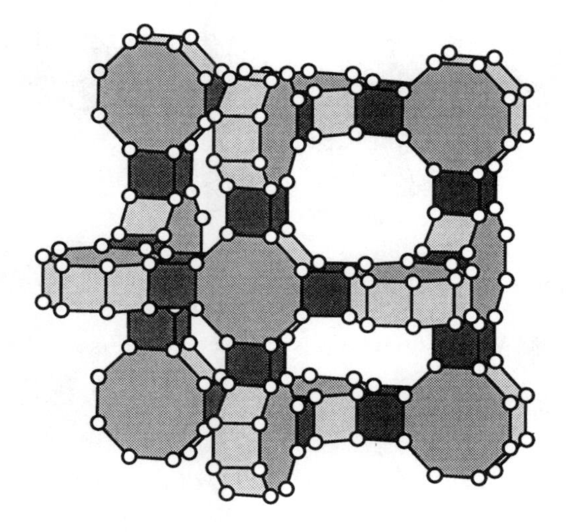

**Fig. 7.42**. Part of a rare net 4·4·4·8·4·12 ($W^*8$) shown as an infinite polyhedron $4^3.8$.

If cubes are inserted between the square faces common to the octagonal prisms in Fig. 7.39 (right), a very open 4-connected net 4·4·4·8·4·12 results (Fig. 7.42). An alternative description is as an infinite polyhedron $4^3.8$ (cf. the ZK-5 polyhedron, p. 325). For reasons to become apparent (§ 7.5.2) we label this net $W^*8$.

Crystallographic data for the $W^*8$ structure are (see also p. 373):

$W^*8$ $Im\bar{3}m$, $a = 4+\sqrt{8} = 6.828$, $r = 0.302$
4·4·4·8·4·12 in 96 $l$: $I \pm (x,\pm y,\pm z\ ;\ y,\pm x,\pm z)\kappa$, $2ax = 1$, $2ay = 3 + \sqrt{2}$, $2az = 3 + \sqrt{8}$

The density corresponds to a silicate with framework density $FD$ = 10.5 tetrahedral atoms per 1000 Å$^3$. We do not know of a zeolite based on this framework; the most open known structures based on 4-connected nets have $FD \approx 12.5$. The 12-rings in $W^*8$ are not planar; they have angles of 135° (compare 150° for plane dodecagons). The structure also contains 24-rings.

A closely related structure occurs in the synthetic zeolite known as CoAPO-50 which has a framework with composition $Co_3Al_5P_8O_{32}$. The structure (Fig. 7.43) contains cubes connected by squares forming hexagonal layers containing 12-rings. The layers are joined by edges connecting opposite vertices of the cubes. Crystallographic data for the ideal form of this structure (which has a density close to that of **faujasite**) are:

**CoAPO-50** $P\bar{3}1m$, $a$ = 4.1815, $c$ = 2.7321 $r$ = 0.387
4·8·4·8·4·8 in 4 $h$: ±(1/3,2/3,±$z$), $z$ = 0.1830
4·4·4·8·4·12 in 12 $l$: ±($x,y,z$ ; $\bar{y},x-y,z$ ; $y-x,\bar{x},z$ ; $y,x,z$ ; $x-y,\bar{y},z$ ; $\bar{x},y-x,z$), $x$ = 0.1381, $y$ = 2/3, $z$ = 0.3943

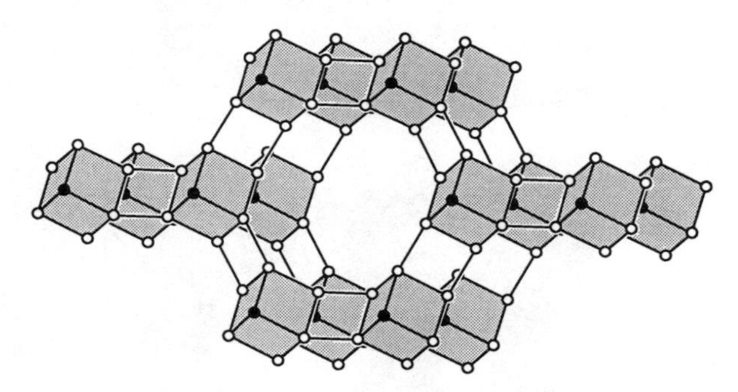

**Fig. 7.43**. The **CoAPO-50** net viewed almost down **c**.

A related net is that of **MAPSO-46** which is left as an exercise (7.12.12).

## *7.5 Rare and dense 4-connected nets

In the Chapter 6 we discussed sphere packings with particular attention to the densest packings of equivalent spheres. It is natural to ask also what is the rarest (least dense) packing of equivalent spheres. If we require the sphere packing to be *stable*, i.e. each sphere to be in contact with at least four others with points of contact not all on the same hemisphere, we must consider 4-coordinated sphere packings, or what is equivalent, 4-connected nets with equal edges.

We remind the reader that the density, expressed as vertices per unit volume, we call the *geometric* density. In the context of nets, we consider one to be *topologically* dense if it has a large number of $k$th neighbors (obviously, all 4-connected nets have four first neighbors). As a measure of density in this respect, we arbitrarily use the cumulative sum of the numbers of topological neighbors, $n_k$, for the first ten coordination shells (i.e. $k$ = 1 to 10) as a measure of the *topological* density. Appendix 3 elaborates on this topic.

The catalytic activity of zeolites is intimately bound up with the sizes of the rings in the structure, and considerable discussion has focused on topics such as the connection between ring size and density (in both senses). The same topic is also of interest in connection with glasses; for these a direct measure of ring sizes is generally unavailable, but can possibly be inferred from the density.

### **7.5.1 Two dense nets*

We consider here a second net (the first is **diamond**) with only two vertices in the repeat unit and a fourth net (the other three were described in § 7.3.11) with only three vertices in the repeat unit.

To derive the first new net we systematically remove one third of the edges of the 6-connected net of the primitive cubic lattice. The way that it is done is illustrated on the left in Fig. 7. 44. The arrangement of the vertices is cubic, but if the edges are considered, the symmetry is tetragonal: $P4_2/mmc$ $a = 1$, $c = 2$ with vertices in 2 $a$: 0,0,0 ; 0,0,1/2. The net can be distorted so that each vertex has only four (instead of six) geometrical nearest neighbors as suggested in the center of the figure. The symmetry is now $I4_1/acd$ and vertices are in 16 $e$: $I \pm (x,0,1/4\ ;\ x,1/2,3/4\ ;\ 1/4,1/4-x,0\ ;\ 1/4,3/4-x,0)$. For unit edge length and the next nearest distance as large as possible ($\sqrt{5}/2$), $a = 2$, $c = \sqrt{14}$ and $x = 1/8$ ($r = 1.069$). Note that distortion slightly increases the geometrical density.

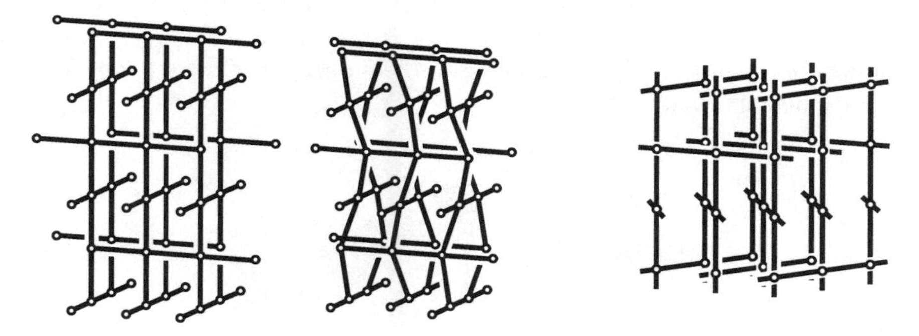

**Fig 7.44**. Left: the **$CdSO_4$** net derived from a primitive cubic array. Center: the same net distorted so that each vertex has only four geometric nearest neighbors. Right: a dense net with three vertices in the repeat unit.

This net is found as that of Cd,S (joined by -O-) in $CdSO_4$ ($HgSO_4$ is isostructural) hence the name, **$CdSO_4$** net. The short symbol for the vertices is $6^5.8$. It is interesting that one of the angles is not contained on any ring, as all circuits containing that angle have shortcuts (cf. § 7.1.1). We use $\infty$ to symbolize such an angle and the long Schläfli symbol becomes $6{\cdot}6{\cdot}6{\cdot}6{\cdot}6_2{\cdot}\infty$. It is very dense in the topological sense; the numbers of neighbors are:

$$n_1 = 4,\ n_2 = 12,\ n_k = 4k^2 - 6\ (k > 2) \tag{7.8}$$

Our second dense net (Fig. 7.44, right) is derived analogously by deleting one half of the edges corresponding to nearest neighbor distances in a primitive hexagonal lattice. Taking into account the edges the symmetry is $P6_222$. Unlike the previous one, this net does not appear to be realizable with shortest distances corresponding only to equal edges. It does, however, have some interesting properties that merit mention. Like the previous net, one angle is not contained in a ring and the long symbol is $7_2{\cdot}\infty{\cdot}7_3{\cdot}7_3{\cdot}7_3{\cdot}7_3$ (short symbol $7^5.9$). It is the only 4-connected net that has been described that does not have at least one 6- or smaller ring. It also has the largest number of topological neighbors of any known 4-connected net, so we call it **dense** net; the numbers of neighbors are given by:

| $k$ | 1 | 2 | 3 | 4 | 5 | 6 | >6 | |
|---|---|---|---|---|---|---|---|---|
| $n_k$ | 4 | 12 | 36 | 72 | 122 | 188 | $6k^2 - 30$ | (7.9) |

**7.5.2 Rare sphere packings*

This topic was considered many years ago by Heesch and Laves who found what was long considered to be the rarest (least dense) stable sphere packing. This structure is derived by replacing the vertices of the diamond net by groups of four spheres in contact (so that their centers form a regular tetrahedron). The tetrahedral groups are arranged so that they make contact along the diamond structure edges. We name this structure **HL4$_4$** or *D*4 (because the vertices of the *D* lattice complex are replaced by groups of four); see below for a crystallographic description.

A fragment of the structure is shown in Fig. 7.45. In the figure shaded tetrahedra replace vertices of the diamond net.

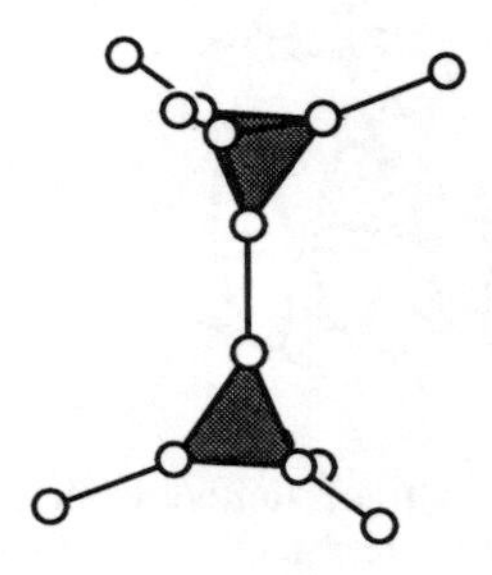

**Fig. 7.45**. Illustrating how **HL4$_4$** is obtained by decorating the vertices of **diamond**.

We call this process of replacing a vertex of a 4-connected net with a tetrahedral group of vertices *decorating*. It turns out that at least four other nets can be decorated in this way to give new uninodal nets. They are derived from the lattice complexes $^+Q$ (quartz), $S^*$, $W^*$ (sodalite) and $^+V$. We label them $^+Q4$, $S^*4$, $W^*4$ and $^+V4$. Here are their crystallographic data in abbreviated form:

**HL4$_4$** (*D*4) $\quad Fd\bar{3}m$, $a = \sqrt{8} + 4/\sqrt{3} = 5.1378$, $r = 0.236$ ($\rho = 0.1235$)
$3{\cdot}12_2{\cdot}3{\cdot}12_2{\cdot}3{\cdot}12_2$ in 32 $e$: $F \pm (x,x,x\ ;\ x,1/4{-}x,1/4{-}x)\kappa$, $x = 1/(8 + \sqrt{96}) = 0.0562$

| | |
|---|---|
| ${}^{+}Q4$ | $P6_222$, $a = 3.550$, $c = 3.885$, $r = 0.283$<br>$3{\cdot}12{\cdot}3{\cdot}12_2{\cdot}3{\cdot}16_7$ in 12 *k*, $x = 0.458$, $y = 0.115$, $z = 0.091$ |
| $S^{*}4$ | $Ia\bar{3}d$, $a = 7.168$, $r = 0.261$<br>$3{\cdot}12{\cdot}3{\cdot}12_2{\cdot}3{\cdot}12_2$ in 96 *h*, $x = 0.065$, $y = 0.224$, $z = 0.424$ |
| $W^{*}4$ | $Im\bar{3}m$, $a = 2 + 3\sqrt{2} = 6.2426$, $r = 0.197$ ($\rho = 0.1033$)<br>$3{\cdot}8{\cdot}3{\cdot}12{\cdot}3{\cdot}12$ in 48 *j*: $I \pm (0,y,\pm z\ ;\ 0,z,\pm y)\kappa$,<br>$y = (1/2 + \sqrt{2})/a = 0.3066$, $z = (1/2 + 3/\sqrt{2})/a = 0.4199$ |
| ${}^{+}V4$ | $I4_132$, $a = 4 + \sqrt{8} = 6.8284$, $r = 0.151$ ($\rho = 0.0789$)<br>$3{\cdot}6{\cdot}3{\cdot}20_2{\cdot}3{\cdot}20_3$ in 48 *i*, $x = y = \sqrt{2}/8 = 0.1768$, $z = 0$ |

The last of these is the rarest, but does not correspond to a stable sphere packing as the four contacts of spheres are all on the same hemisphere. The others, and *D*4, are stable sphere packings. $W^{*}4$ is illustrated in Fig. 7.46 which shows how it is derived by decorating the **sodalite** net. It is possibly the rarest stable sphere packing. The density (fraction of space filled by spheres in contact is $8\pi/(2 + 3\sqrt{2})^3 = 0.1033$ (compare with the density of $\pi/\sqrt{18} = 0.7404$ for closest packing).

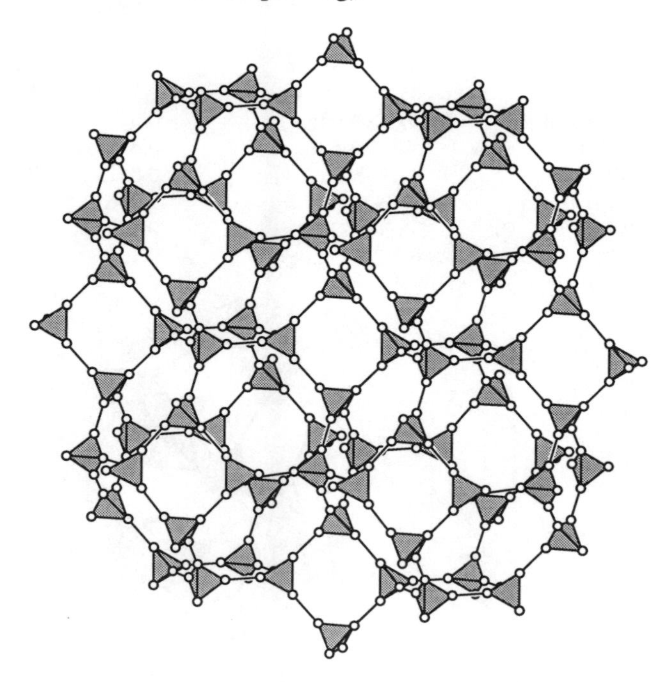

**Fig. 7.46.** The rare (low density) sphere-packing net $W^{*}4$ obtained by decorating the sodalite net with tetrahedral groups (shaded).

It is worth noting the apparently paradoxical fact that the most open nets are characterized by having a large number of small rings. The rare nets listed above all have three 3-rings or four 4-rings in their symbols (they also have large rings as a consequence). On the other hand dense nets generally have a small number of 3- or 4-rings, often shortest rings of 6 (7 in the case of the **dense** net).

It is interesting that density (or rarity) in the geometrical sense is correlated with density in the topological sense. In Table 7.1 we list (for maximum volume form) the number of vertices per unit volume, *r*, for some mostly dense or rare nets. Also listed is $c_{10}$, the cumulative number of topological neighbors out to tenth neighbors. The net 4·4·4·8·4·12 ($W^*8$) was described in § 7.4.5. It can be derived by replacing the vertices of the sodalite ($W^*$) net by cubes of vertices (the centers of the cubes are on a **sodalite** net). It is possibly the rarest uninodal net that does not contain 3-rings.

Note that if the restriction to uninodal nets is lifted, nets (but not *stable* sphere packings) can be constructed of arbitrarily low density by repeating the process of decoration.[1]

**Table 7.1.** Some dense and rare nets compared

| *net* | *Schläfli symbol* | *r* | $c_{10}$ |
|---|---|---|---|
| **dense** | $7_2{\cdot}\infty{\cdot}7_3{\cdot}7_3{\cdot}7_3{\cdot}7_3$ | 1.155 | 2078 |
| **CdSO$_4$** | $6{\cdot}6{\cdot}6{\cdot}6{\cdot}6_2{\cdot}\infty$ | 1.000 | 1488 |
| **coesite** | $4{\cdot}6{\cdot}4{\cdot}6{\cdot}8{\cdot}9_2$ | 0.845 | 1324 |
| | $4{\cdot}8{\cdot}4{\cdot}9_7{\cdot}6{\cdot}8$ | 0.845 | 1321 |
| **quartz** | $6{\cdot}6{\cdot}6_2{\cdot}6_2{\cdot}8_7{\cdot}8_7$ | 0.750 | 1230 |
| **NbO** | $6_2{\cdot}6_2{\cdot}6_2{\cdot}6_2{\cdot}8_2{\cdot}8_2$ | 0.750 | 1186 |
| **diamond** | $6_2{\cdot}6_2{\cdot}6_2{\cdot}6_2{\cdot}6_2{\cdot}6_2$ | 0.650 | 980 |
| **sodalite** | $4{\cdot}4{\cdot}6{\cdot}6{\cdot}6{\cdot}6$ | 0.530 | 790 |
| $W^*8$ | $4{\cdot}4{\cdot}4{\cdot}8{\cdot}4{\cdot}12$ | 0.302 | 453 |
| $D4$ = **HL4$_4$** | $3{\cdot}12_2{\cdot}3{\cdot}12_2{\cdot}3{\cdot}12_2$ | 0.236 | 496 |
| $W^*4$ | $3{\cdot}8{\cdot}3{\cdot}12{\cdot}3{\cdot}12$ | 0.197 | 409 |
| $^+V4$ | $3{\cdot}6{\cdot}3{\cdot}20_2{\cdot}3{\cdot}20_2$ | 0.151 | 350 |

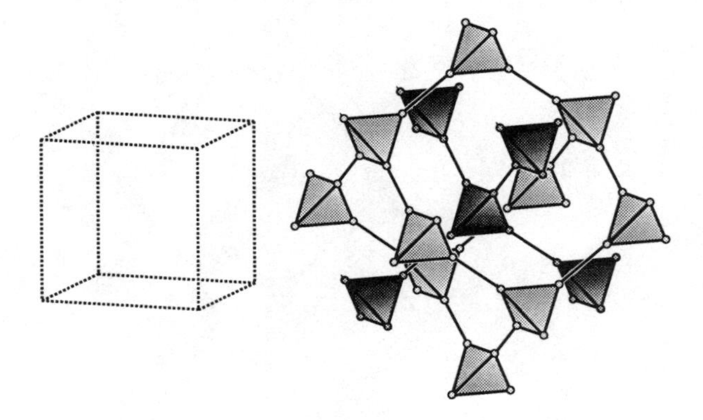

**Fig. 7.47.** The two inter-penetrating *D*4 nets (light and darker shaded) in $D4^*$.

The *D*4 (**HL4$_4$**) net (among others) can be intergrown with itself in a way such that the

[1]M. O'Keeffe & S. T. Hyde, *Zeits. Kristallogr.* (1996).

shortest distance between vertices of the two nets is greater than the edge length (in much the same way as $^+Y^*$ and $^-Y^*$ intergrow to produce $Y^{**}$, see § 7.3.12). The two intergrown *D*4 nets are called *D*4*. The unit cell edge is now only half that the original *fcc* cell and a lattice vector translates from a vertex on one *D*4 net to an identical vertex on the other net (see Fig. 7.47).[1] The structure of $LiCo(CO)_4$ is based on this principle with $\{Li\}O_4$ and $\{Co\}C_4$ tetrahedra joined by C-O bonds. $Zn(CN)_2 = Zn(1)Zn(2)(CN)_4$ is isostructural. In the structure of $ZnI_2$ there are tetrahedral $ZnI_4$ groups joined by common corrners to form a supertetrahedron (Fig. 5.18, p. 150) and the Zn arrangement is topologically the same. The Pb arrangement in NaPb is similar. (Data for these compounds are given in Appendix 5.)

*D*4* $\quad Pn\bar{3}m$, $a = \sqrt{2} + 2/\sqrt{3} = 2.5689$
vertice in 8 *e*: $\pm(x,x,x\ ;\ 1/2{+}x,1/2{+}x,\bar{x})\kappa$, $x = 1/(4 + \sqrt{24}) = 0.1124$

## 7.6 Clathrate hydrates, foam, and grains

Imagine a foam of equal-sized bubbles. The surface of the bubbles will form space-filling polyhedra with edges and faces curved so as to minimize their surface area. Three faces meet at an edge with dihedral angle 120° and four edges meet at a vertex with angles of 109.47°, so the vertices and edges will form a 4-connected net. The structures can also be considered as packings of polyhedra with 4-, 5- and 6-gon faces. As discussed by Kelvin over a hundred years ago, the simplest such net will be **sodalite** which, as we have seen (p. 315), is based on a space-filling by truncated octahedra.

Similar arrangements are found in a number of different contexts such as the crystallites of a fine-grained metal or ceramic and aggregates of biological cells. The crystal structures of clathrate hydrates are also based on these principles, for example the hydrogen-bonded framework of O atoms in $HPF_6{\cdot}6H_2O$ is **sodalite**. Framework silicates with structures based on these nets are known as *clathrasils*.

It is common in this context to use symbols for polyhedra that specify the number and types of faces. Specifically a symbol $[M^m.N^n....]$ refers to a polyhedron with *m* faces that are *M*-gons, *n* faces that are *N* gons, etc. Thus the space-filling truncated octahedron is a 14-hedron with six square faces and eight hexagonal and has symbol $[4^6.6^8]$. For polyhedra with three edges at every vertex the number of vertices is $(mM + nN + ...)/3$.

Interesting related space-filling polyhedra were discovered by Williams.[2] Converting two square faces and two hexagonal faces of $[4^6.6^8]$ into four pentagons will produce a polyhedron $[4^4.5^4.6^6]$ with symmetry *mm*2 that has four quadrangular, four pentagonal and six hexagonal faces as shown in Fig. 7.48. This polyhedron (with slightly curved edges) will fill space to produce the first Williams structure. Although there is just one kind of polyhedron, there are now four kinds of vertex.

Crystallographic data for the structure are:

[1]Note that two intergrown *D* (**diamond**) nets are just **bcc** and again the unit cell for the intergrowth has half the edge of the original.

[2]R. E. Williams, *Science* **161**, 276 (1968).

**Williams 1** $P4_2/ncm$, $a = 2.302$, $c = 7.909$, $r = 0.573$
4·4·6·6·6·6 in 4 *a*: ±(3/4,1/4,0 ; 3/4,1/4,1/2)
5·5·5·5·6·6 in 4*b*: ±(3/4,1/4,3/4 ; 1/4,3/4,3/4)
vertices in 8 *i*: ±(*x*,*x*,*z* ; 1/2–*x*,*x*,1/2+*z* ; *x*,1/2–*x*,1/2+*z* ; 1/2–*x*,1/2–*x*,*z*)
4·4·6·6·6·6, $x = 0.096$, $z = 0.0618$ ; 4·5·5·6·5·6, $x = 0.096$, $z = 0.1882$

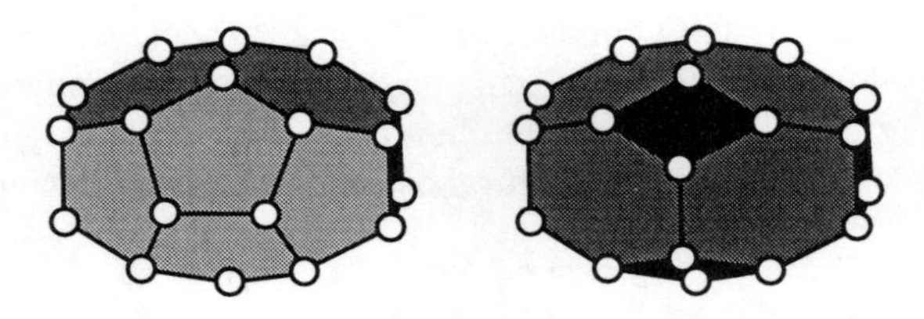

**Fig. 7.48**. The *mm*2 14-hedron in a space-filling configuration. The two aspects shown are related by rotation by 180° about a horizontal axis.

There is a second polyhedron with the same symbol $[4^4.5^4.6^6]$, now with symmetry 222, that also fills space. It is found in the structure of $BaCu_2P_4$ in which the Cu and P atoms form a 4-connected net (the coordinations are $\{Cu\}P_4$ and $\{P\}Cu_2P_2$). Crystallographic data for the net with unit edge are given below.

**$BaCu_2P_4$** *Fddd*, $a = 2.334$, $b = 8.207$, $c = 4.488$, $r = 0.558$
4·4·5·6·6·6 in 16 *f*: *F* ± (1/8,*y*,1/8 ; 1/8,1/4–*y*,5/8), $y = 0.5065$
4·5·5·6·6·6 in 32*h*: *F* ± (*x*,*y*,*z* ; 3/4–*x*,3/4–*y*,*z* ; 3/4–*x*,*y*,3/4–*z* ; *x*,3/4–*y*,3/4–*z*),
$x = 0.195$, $y = 0.1826$, $z = 0.8145$

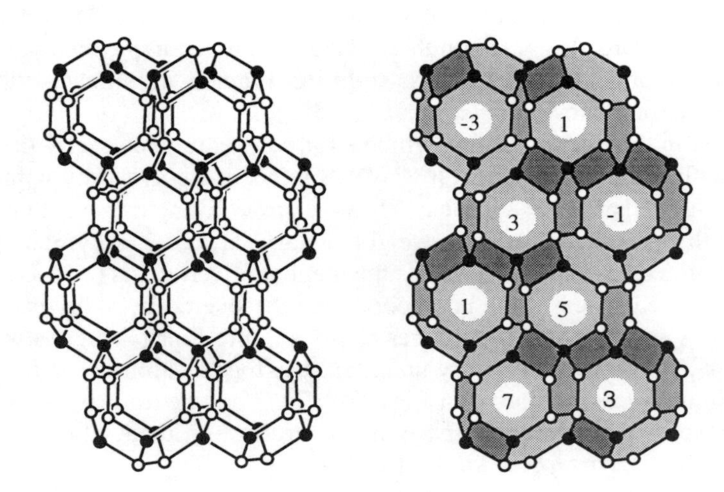


**Fig. 7.49**. The net of Cu (filled circles) and P in $BaCu_2P_4$ projected on (100). On the left as a 4-connected net and on the right as a packing of polyhedra. In the latter, the numbers are the *x* coordinates of the centroids of the polyhedra (Ba positions) in multiples of 1/8.

The structure (Fig. 7.49) contains equal numbers of both enantiomorphs of the polyhedron. Another feature of the structure is that it contains rods of {Cu}$P_4$ tetrahedra sharing opposite edges that run alternately along [101] and [10$\bar{1}$] and connected by P-P bonds. Ba atoms are at the centroids of the polyhedra (for crystallographic data see Appendix 5).

Continuing the process of converting squares + hexagons to pentagons will produce the second Williams space filling polyhedron [$4^2.5^8.6^4$] with two square, eight pentagonal and four hexagonal faces. The structure of the polyhedron packing is now rather simple (Fig. 7.50):

**Williams 2** $P4_2/mnm$, $a = 2.325$, $c = 3.880$, $r = 0.572$
5·5·5·5·6·6 in 4 *d*: ±(0,1/2,1/4 ; 1/2,0,1/4)
4·5·5·6·5·6 in 8 *j*: ±($x$,$x$,±$z$ ; 1/2+$x$,1/2−$x$,1/2±$z$), $x = 0.152$, $z = 0.129$

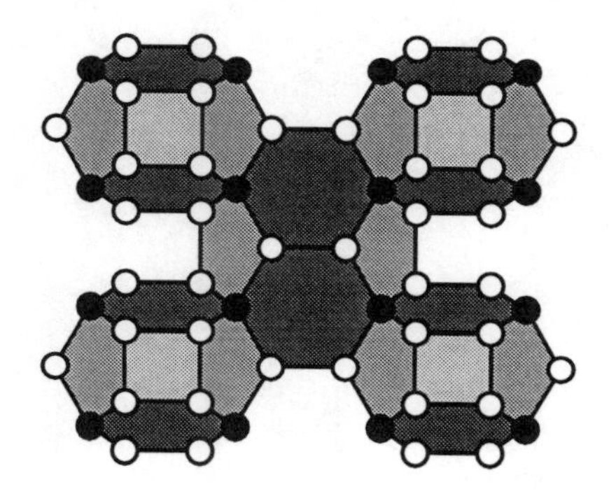

**Fig 7.50**. The $P4_2/mnm$ space filling by polyhedra shown projected on (110). Filled circles are the 5·5·5·5·6·6 vertices (in 4 *d*).

Finishing the process of eliminating squares produces a polyhedron [$5^{12}.6^2$] with twelve pentagonal faces and two hexagonal faces.[1] This polyhedron does not fill space but the structure of the hydrogen-bonded framework of the cubic chlorine hydrate (of approximate composition $2Cl_2.15H_2O$) is made of a packing of pentagonal dodecahedra [$5^{12}$] and these 14-hedra [$5^{12}.6^2$] in the ratio 1:3 (Fig. 7.51). This is sometimes known as the type I hydrate structure. The same framework occurs in the naturally-occurring (impure) form of silica known as melanophlogite. The same structure is also found in alkali silicides and germanides typified by $Na_4Si_{23}$ in which Si atoms are at the vertices and Na atoms center the larger polyhedra. A stereo view of the net is in Fig. 7.89 (§ 7.11.8).

The structure of the hydrates of a number of molecules such as $CHCl_3$ contains dodecahedra again and also 16-hedra [$5^{12}.6^4$] packed in the ratio 2:1. This is known as the type II hydrate structure. The 16-hedron is shown in Fig. 7.52; it has symmetry $\bar{4}3m$.[2]

Data for the nets with unit edge length (this condition is sufficient to fix all the coordinates) are:

[1]This polyhedron is the dual of the bicapped hexagonal antiprism (Fig. 5.12, p. 143).
[2]This polyhedron is the dual of the Friauf polyhedron (Fig. 5.12, p. 143).

**Type I** $Pm\bar{3}n$, $a = 4.3021$, $r = 0.578$
5·5·5·5·6·6 in 6 *c*: ±(0,1/2,1/4)κ
5·5·5·5·5·5 in 16 *i*: ±(*x*,*x*,±*x* ; 1/2+*x*,1/2+*x*,1/2±*x*)κ, $x = 0.1829$
5·5·5·5·5·6 in 24 *k*: ±(0,*y*,±*z* ; 1/2,1/2+*z*,1/2±*y* )κ, $y = 0.3099$, $z = 0.1162$

**Type II** $Fd\bar{3}m$, $a = 6.2054$, $r = 0.552$
5·5·5·5·5·5 in 8 *a*: *F* ± (1/8,1/8,1/8)
5·5·5·5·5·5 in 32 *e*: *F* ± (*x*,*x*,*x* ; *x*,1/4–*x*,1/4–*x*)κ, $x = 0.2180$
5·5·5·5·5·6 in 96 *g*: *F* ± (*x*,*x*,*z* ; *x*,1/4–*x*,1/4–*z* ; 1/4–*x*,*x*,1/4–*z* ; 1/4–*x*,1/4–*x*,*z*)κ, $x = 0.1820$, $z = 0.3709$

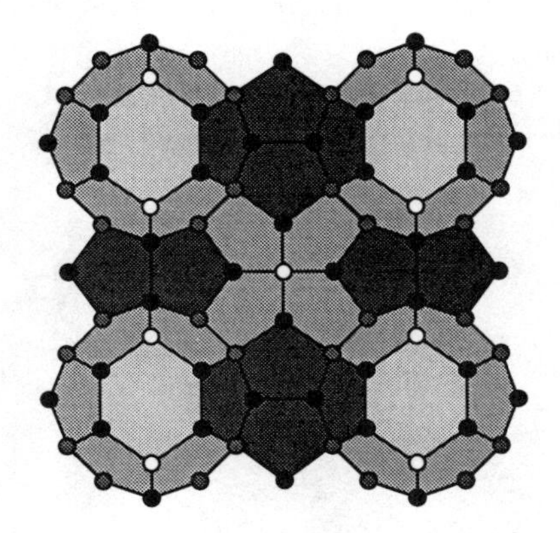

**Fig. 7.51**. A fragment of the Type I hydrate (clathrasil) structure viewed down [001]. Dodecahedra are shown with heavily shaded faces. Tetrakaidecahedra share hexagonal faces to form rods along <100>. The rods are packed (by sharing pentagonal faces) as in the **β-W** cylinder packing. Dodecahedra fill interstices in this rod packing.

The last structure (Type II) comes close to having all vertices $5^6$. The unit cell contains sixteen 12-hedra, eight 16-hedra, 144 5-gons and sixteen 6-gons. It does not appear possible to make 4-connected nets with *all* 5-rings, although model builders (see Notes at the end of this chapter) will find that remarkably large clusters of packed pentagonal dodecahedra can be made before strain becomes too severe to continue. To construct a "spaghetti" model of the type II net it is best to make one 16-hedron and then to construct dodecahedra on each of its pentagonal faces (there is only one way to do this); it should be obvious how to proceed thereafter. A stereo view of the net is in Fig. 7.90 (§ 7.11.8).

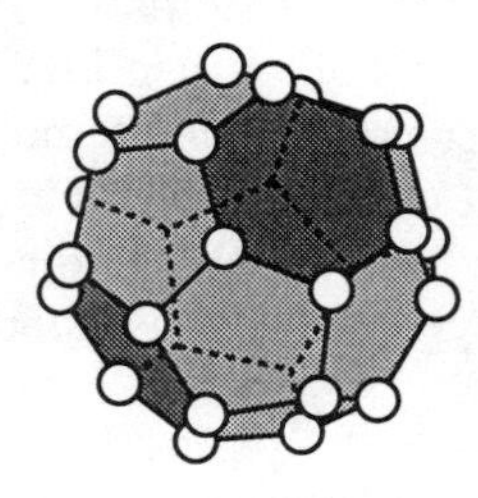

**Fig 7.52**. The hexakaidecahedron appearing in the Type II hydrate net.

The Type II net also occurs as Si or Ge frameworks in compounds $M_x$Si or $M_x$Ge formed by decomposition (loss of $M$ = alkali metal) of $M$Si or $M$Ge at high temperatures under vacuum. The synthetic zeolite dodecasil 3C is also based on this net.

The polyhedra in this section have three polygons (not necessarily regular or even planar) meeting at each vertex. For such a polyhedron (see the exercises in Chapter 5) the number of faces, $F$ and vertices, $V$ are related by $2F = V + 4$, and the number of edges, $E = 3V/2$. If there are $F_4$ faces with four edges and $F_5$ faces with five edges then $2F_4 + F_5 = 12$. There is no constraint on the number of faces with six edges.

We revisit the clathrate hydrate structures in Appendix 4 where two further structures are mentioned. Attention is also directed to the structures of the zeolite clathrasils (§ 7.8.6).

## 7.7 A summary of the simpler 4-connected nets

Here (Table 7.2) is a list of the simpler uninodal 4-connected nets either with less than 4 vertices in the topological repeat unit or quasi-regular. $Z$ is the number of vertices in the topological repeat unit (primitive unit cell). "l.c." refers to the symbol for an invariant lattice complex. It would be of interest to know if this list is complete. The list for $Z = 4$ would be quite long.[1] The **dense** net is uniform in that it contains only 7-rings.

**Table 7.2.** Names and properties of some simple 4-connected nets

| *net* | *Z* | *symbol* | l.c. | *regular* | *uniform* |
|---|---|---|---|---|---|
| **diamond** | 2 | $6_2\cdot6_2\cdot6_2\cdot6_2\cdot6_2\cdot6_2$ | $D$ | yes | yes |
| **$CdSO_4$** | 2 | $6\cdot6\cdot6\cdot6\cdot6_2\cdot\infty$ | | no | yes |
| **quartz** | 3 | $6\cdot6\cdot6_2\cdot6_2\cdot8_7\cdot8_7$ | $^+Q$ | quasi | no |
| **NbO** | 3 | $6_2\cdot6_2\cdot6_2\cdot6_2\cdot8_2\cdot8_2$ | $J^*$ | quasi | no |
| ***W*2** | 3 | $3\cdot7\cdot7\cdot7\cdot7_2\cdot7_2$ | | no | no |
| **dense** | 3 | $7_2\cdot\infty\cdot7_3\cdot7_3\cdot7_3\cdot7_3$ | | no | yes |
| **sodalite** | 6 | $4\cdot4\cdot6\cdot6\cdot6\cdot6$ | $W^*$ | quasi | no |
| **HL$4_3$** | 6 | $3\cdot3\cdot10_2\cdot10_2\cdot10_2\cdot10_2$ | $^+V$ | quasi | no |
| $S^* = Ia\bar{3}d$ 24$d$ | 12 | $6\cdot6\cdot6_2\cdot6_2\cdot6_2\cdot6_2$ | $S^*$ | quasi | yes |

## 7.8 Zeolite nets

The current considerable interest in zeolites stems from their value as catalysts and "molecular sieves" and each year sees a number of new structures discovered. Their

[1]Nets not realizable with shortest distances between vertices corresponding to edges probably should be excluded. There are five *uninodal* nets with Z = 4 in this chapter, we know of only one other. See M. O'Keeffe, *Phys. Chem. Minerals* **22**, 504 (1995).

properties are, to a large extent, determined by their structures, so we devote some space to this topic (but by no means exhaust it).[1]

An invaluable guide to zeolite nets is the *Atlas of Zeolite Structure Types*[2] which includes eighty 4-connected nets of natural and synthetic zeolites. The *Atlas* contains stereo diagrams of each net, coordination sequences for each vertex, references and synonyms. Some structures appear dauntingly complex, but many can be described rather simply as they contain a short axis suitable for projection. Once one learns to "read" the projection, it will be found that the three-dimensional structure may readily be reconstructed (model making is highly recommended). As in many instances we provide coordinates for idealized nets (not given in the *Atlas*), they can be readily studied by computer.[3]

The term *zeolite* is not rigorously defined: it is used loosely to refer to any oxide with an open structure (say $r < 0.6$) based on a framework of corner connected $\{T\}O_4$ tetrahedra.[4] Some authors use the term *clathrasil* to refer to those structures without large channels (shortest ring at each angle a 6-ring or smaller). From this point of view **sodalite** is a clathrasil. *Pentasils* are open silica-rich alumino-silicate structures in which the smallest rings are 5-rings. Some of these are referred to as *silicalites*.

A number of simpler zeolite nets have already been described (an index to zeolite nets in this chapter is given in § 7.8.8, p. 353). Here we describe some more, using easily-recognized structural units (such as "zig-zag" or "crankshaft" rods) as an organizing principle. The reader uninterested in zeolites is invited to scan through this section quickly, pausing perhaps to admire some of the more-beautiful structures that occur.

### **7.8.1 Zig-zag structures*

In § 7.3.3-7.3.5 we described some nets, including those of zeolites Li-A, MAPO-39, $AlPO_4$-31 and cancrinite, which contain parallel zig-zag rods of vertices. The repeat distance for a unit edge zig-zag is typically about 1.65 time the edge length (about $1.65 \times 3.05 \approx 5$ Å for zeolites) and many zig-zag structures have one short axis of about this size, and have all vertices lying on mirror planes, so that they are readily shown in projection. In such a projection the framework appears as a two-dimensional 3-connected net. **Cancrinite** for example (Fig. 7.19, p. 307), projects as the 4.6.12 net.

[1]A good introduction to the properties and uses of zeolites is the article by J. M. Newsam in *Solid State Chemistry: compounds* (A. K. Cheetham & P. Day, eds.) Oxford (1992). A good source of data concerning zeolites is *Handbook of Molecular Sieves* by R. Szostak [Van Nostrand, New York (1992)].

[2]W. M. Meier & D. H. Olsen, *Atlas of Zeolite Structure Types*, Third Ed. Butterworth-Heinemann (1992). This also appeared as issue 5 of the journal *Zeolites*, **12** (1992). *Natural Zeolites* by G. Gotardi & R. Galli [Springer, Berlin (1985)] has good drawings that will be appreciated by model builders.

[3]The symmetry of real materials is generally lower that the maximum symmetry of the net. For structures for which we do not give coordinates, see the references given in the *Atlas of Zeolite Structure Types*. Note also that our coordinates may, in some instances, be rather different from those in real structures; they do however serve to define the topology of the net.

[4]When heated, zeolite minerals give off water as steam, and the name comes from the Greek for *boiling stone*. Purists insist that the term "zeolite" should be restricted to alumino-silicate minerals, but the wider sense used in this section (and in the *Atlas*) now has general currency.

A particularly simple net with both double and single zig-zags occurs in the zeolite known as NaJ with ideal composition $Na_2Al_2Si_2O_8 \cdot H_2O$. The net is illustrated both as a projection down the zig-zags and in clinographic projection in Fig. 7.53. Crystallographic data are:

**NaJ** *Pmma*, $a$ = 1.576, $b$ = 2.525, $c$ = 2.525, $r$ = 0.597
$6 \cdot 6 \cdot 6 \cdot 6 \cdot 6_2 \cdot 6_2$ in 2 $g$: ±(1/4,1/2,$z$), $z$ = 0.378
$4 \cdot 6_2 \cdot 4 \cdot 6_2 \cdot 6 \cdot 8_2$ in 4 $k$: ±(1/4,±$y$,$z$), $y$ = 0.198, $z$ = 0.122

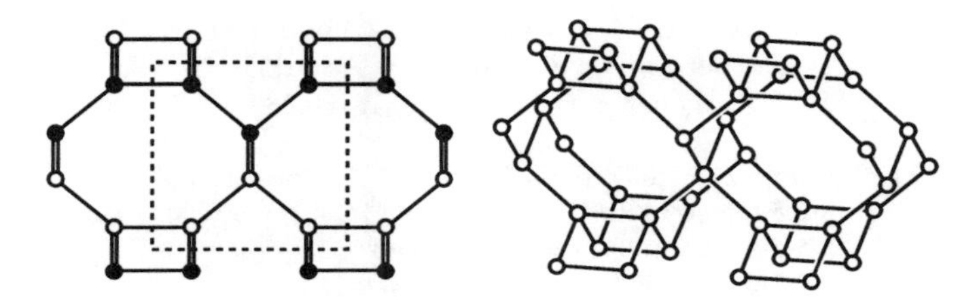

**Fig. 7.53**. The **NaJ** net. Left: projected on (100) with open and filled circles at $x$ = 1/4 and 3/4 respectively. Right: in clinographic projection. Compare with Fig. 7.62 (**$AlPO_4$-25**) p. 346.

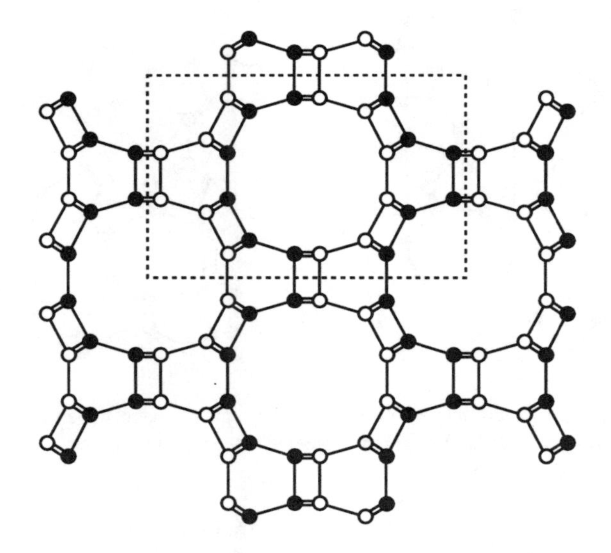

**Fig. 7.54**. The **MAPO-36** net projected on (001) with **b** horizontal on the page. Open and filled circles are at $z$ = 0 and 1/2 respectively. Note that zig-zags are shown as double lines and that double zig-zags project as rectangles.

The net of the zeolite MAPO-36 (with a $MgAl_{11}P_{12}O_{48}$ framework) projects as 4.6.12 with the squares changed to rectangles representing a double zig-zag in projection (and the hexagons and dodecagons also distorted) as shown in Fig. 7.54 (contrast with **cancrinite**, Fig. 7.19, p. 307). Data for this net are:

**MAPO-36** $Cmcm$, $a = 4.357$, $b = 6.687$, $c = 1.697$, $r = 0.485$
$4{\cdot}6{\cdot}4{\cdot}6{\cdot}6{\cdot}6_2$ in 8 $g$: $C \pm (\pm x,y,1/4)$, $x = 0.385$, $y = 0.040$
$4{\cdot}6_2{\cdot}4{\cdot}6_2{\cdot}6{\cdot}12_2$ in 8 $g$, $x = 0.319$, $y = 0.183$
$4{\cdot}6_2{\cdot}4{\cdot}6_2{\cdot}6{\cdot}12_2$ in 8 $g$, $x = 0.115$, $y = 0.251$

A number of zeolite nets with zig-zags project as two-dimensional nets containing pentagons. Two simple examples base on pentagon-octagon nets are those of bikitaite ($LiAlSi_2O_6.H_2O$) and $CsAlSi_5O_{12}$ (Fig. 7.55). Data are:

**bikitaite** $Cmcm$, $a = 2.365$, $b = 5.104$, $c = 1.656$, $r = 0.600$
$5_2{\cdot}6_2{\cdot}6{\cdot}6{\cdot}6{\cdot}6$ in 4 $c$: $C \pm (0,y,1/4)$, $y = 0.055$
$5{\cdot}5{\cdot}5{\cdot}5{\cdot}6{\cdot}8_2$ in 8 $g$: $C \pm (\pm x,y,1/4)$, $x = 0.289$, $y = 0.198$

**$CsAlSi_5O_{12}$** $Cmcm$, $a = 1.602$, $b = 4.713$, $c = 5.151$, $r = 0.617$
$5{\cdot}6{\cdot}5{\cdot}6{\cdot}5_2{\cdot}6$ in 8 $f$: $C \pm (0,y,z\ ;\ 0,y,1/2{-}z)$, $y = 0.045$, $z = 0.088$
$5{\cdot}5{\cdot}5{\cdot}5{\cdot}6{\cdot}8_2$ in 8 $f$, $y = 0.255$, $z = 0.058$
$5{\cdot}6{\cdot}5{\cdot}6{\cdot}6_2{\cdot}8_2$ in 8 $f$, $y = 0.440$, $z = 0.153$

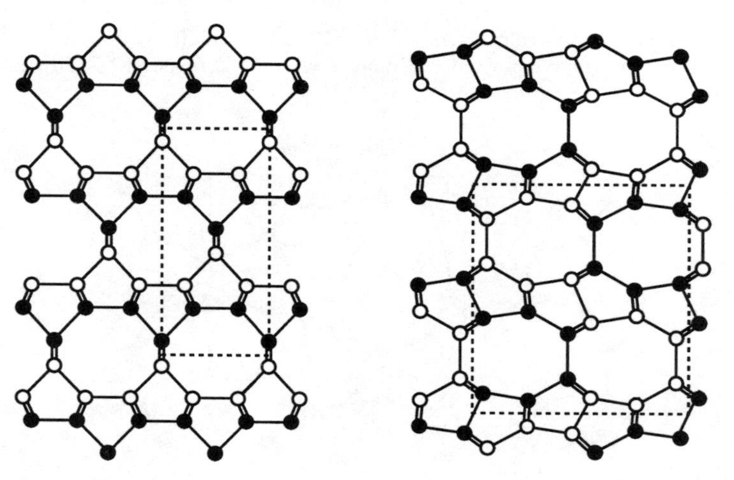

**Fig. 7.55**. Left **bikitaite** projected on (001) with **b** vertical on the page. Open and filled circles are at $z = 1/4$ and 3/4 respectively. Right: **$CsAlSi_5O_{12}$** projected on (100) with **c** vertical on the page. Open and filled circles are at $x = 0$ and 1/2 respectively.

The synthetic zeolites ZSM-12, ZSM-23 and theta-1 (essentially hydrous silica with small amounts of Na and Al) are also derived from two-dimensional nets, but now including either decagons or dodecagons (see Fig. 7.56). The nets of the first two contain seven different types of vertex, but **theta-1** has a simple description:

**theta-1** $Cmcm$, $a = 4.575$, $b = 5.638$, $c = 1.625$, $r = 0.573$
$5{\cdot}5{\cdot}5{\cdot}5{\cdot}6_2{\cdot}10_2$ in 4 $c$: $C \pm (0,y,1/4)$, $y = 0.262$
$5{\cdot}5{\cdot}5{\cdot}5{\cdot}{\cdot}6_2{\cdot}\infty$ in 4 $c$, $y = 0.635$ (note the absence of a ring at one angle)
$5{\cdot}5{\cdot}5{\cdot}5{\cdot}6{\cdot}10_2$ in 8 $g$: $C \pm (\pm x,y,1/4)$, $x = 0.209$, $y = 0.210$
$5_2{\cdot}6_2{\cdot}6{\cdot}6_2{\cdot}6{\cdot}6_2$ in 8 $g$, $x = 0.307$, $y = 0.052$

**ZSM-12** $C2/m$, $a = 8.03$, $b = 1.62$, $c = 3.92$, $\beta = 107.7$, $r = 0.58$

**ZSM-23** $Pmmn$, $a = 1.62$, $b = 6.94$, $c = 3.59$, $r = 0.59$

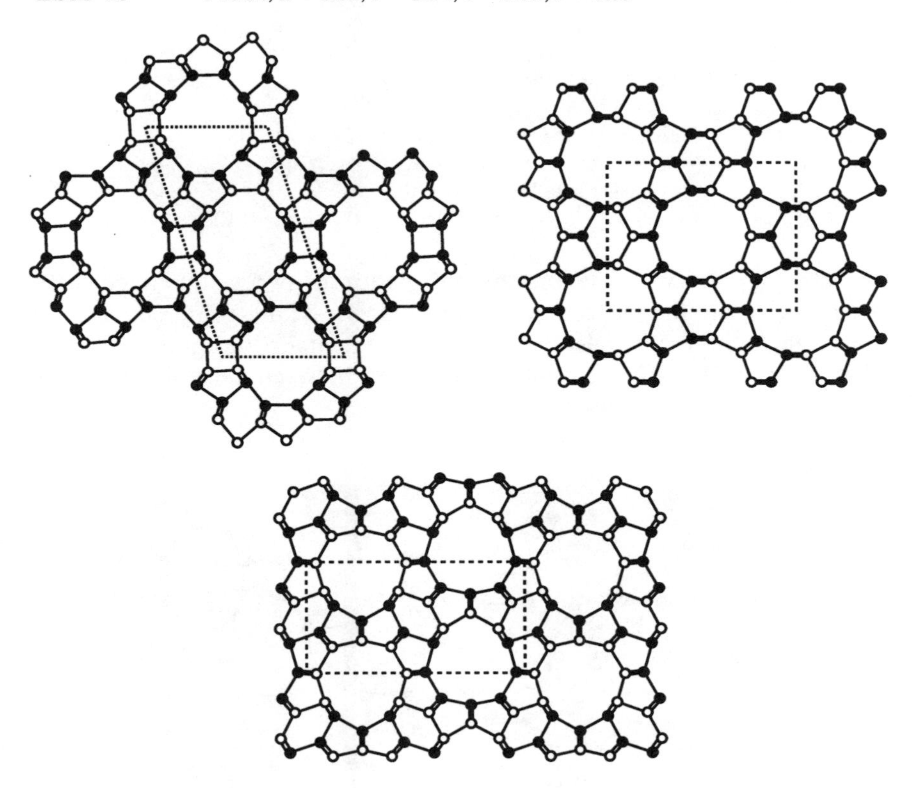

**Fig. 7.56.** Three zeolite nets with zig-zags (shown as double lines) projected down the short axis. Top left: **ZSM-12**. Top right: **theta-1**. Bottom: **ZSM-23**. Filled and open circles differ in elevation by 1/2 the vertical repeat distance.

### **7.8.2 Crankshaft structures*

In § 7.3.6 (p. 308) we discussed some nets derived from the two-dimensional net $4.8^2$ that contained vertices arranged on double crankshafts. In that section coordinates and a schematic illustration of **merlinoite** and **gismondine** were given. It might be noted that the repeat distance of a crankshaft is about 3.3 times an edge length: this translates into a repeat distance of about $3.3 \times 3$ Å = 10 Å for silicates and related materials (aluminophosphates etc.). In projections down the crankshaft axes, vertices are at elevations about $\pm 0.15$ (in units of the projection axis length) from mirror planes (which are either at 0 and 1/2 or at 1/4 and 3/4). Here we describe two more double crankshaft structures. The first is found in the aluminosilicate **phillipsite** and the second in a form of aluminum phosphate known as **$AlPO_4$-C**.

Crystallographic data for **phillipsite** and **AlPO4-C** are:

**phillipsite** *Cmcm*, $a = 3.446$, $b = 4.222$, $c = 4.359$, $r = 0.505$
both vertices $4{\cdot}4{\cdot}4{\cdot}8_2{\cdot}8{\cdot}8$ in 16 *h*: $C \pm (\pm x,y,z\ ;\ \pm x,y,1/2{-}z)$
vertex 1: $x = 0.145$, $y = 0.109$, $z = 0.044$
vertex 2: $x = 0.355$, $y = 0.243$, $z = 0.135$

**AlPO4-C** *Cmca*, $a = 2.988$, $b = 5.897$, $c = 3.334$, $r = 0.545$
$4{\cdot}4{\cdot}4{\cdot}8_2{\cdot}8{\cdot}8_2$ in 16 *g*: $C \pm (\pm x,y,z\ ;\ \pm x,1/2{+}y,1/2{-}z)$,
$x = 0.167$, $y = 0.051$, $z = 0.120$
$4{\cdot}6{\cdot}4{\cdot}6{\cdot}6{\cdot}8_2$ in 16 *g*, $x = 0.167$, $y = 0.221$, $z = 0.120$

These nets, together with **merlinoite** and **gismondine** are illustrated in Fig. 7.57. Note that in the figure the projection is down the axis of the double crankshaft (which projects as a rectangle). Note also that **merlinoite** contains octagonal prisms (shaded) centered at 0,0,0 and 1/2,1/2,1/2. Some relationships between the structures (all based on $4.8^2$) should be apparent from the figure.

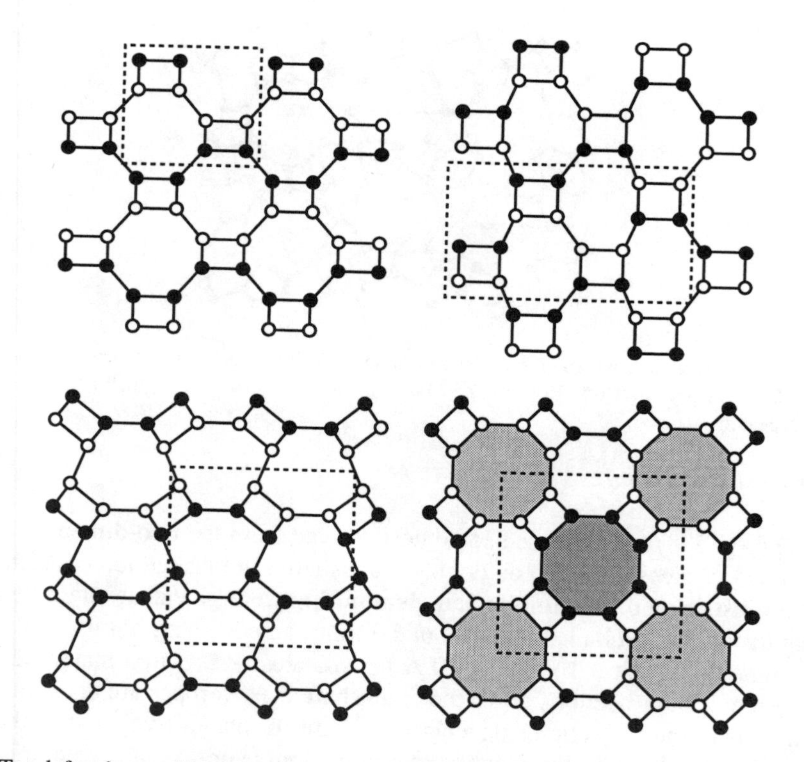

**Fig. 7.57.** Top left: **gismondine** projected on (100) with **c** vertical on the page. Top right: **AlPO4-C** projected on (100) with **c** vertical on the page. Bottom left: **phillipsite** (100) with **b** vertical on the page. Bottom right: **merlinoite** projected on (001). In each case open circles are vertices at about ±0.15 and filled circles are vertices at about 0.5±0.15. Double crankshafts project as rectangles.

Another zeolite framework with double crankshafts is **gmelinite** (§ 7.3.7, p. 310); this is based on the two-dimensional net 4.6.12.

A simple zeolite framework containing both *single* and *double* crankshafts is found in **$AlPO_4$-12** ($AlPO_4$-33 has the same framework). Data for this structure (illustrated in Fig. 7.58) are:

**$AlPO_4$-12**    *Pmma*, $a = 3.276$, $b = 2.498$, $c = 2.865$, $r = 0.512$
$4{\cdot}8_2{\cdot}4{\cdot}8_2{\cdot}6{\cdot}8_2$ in 4 *g*: $\pm(x,1/2,z\ ;\ 1/2-x,1/2,z)$, $x = 0.097$, $z = 0.356$
$4{\cdot}4{\cdot}4{\cdot}6{\cdot}8{\cdot}8$ in 8 *l*: $\pm(x,\pm y,z\ ;\ 1/2-x,\pm y,z)$, $x = 0.097$, $y = 0.200$, $z = 0.134$

The short axis of **$AlPO_4$-12** is **b**. It might be noted that along this direction (horizontal in Fig. 7.58) vertices form rods (either all filled or all empty circles in the figure) that are referred to as "saw-tooth." We use a projection down saw-tooth rods in § 7.8.3 (next).

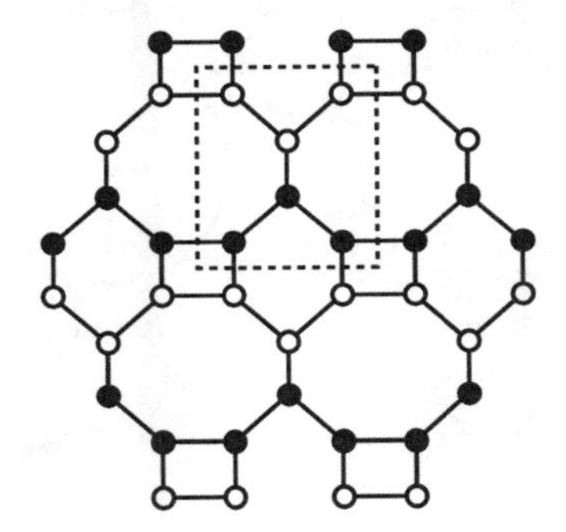

**Fig. 7.58. $AlPO_4$-12** projected on (100) with **b** horizontal on the page. Double crankshafts project as rectangles. Vertices not on a double crankshaft are on a single crankshaft. Open circles are vertices at $x = 0.1$ and 0.4; filled circles are vertices at $x = 0.6$ and 0.9. Compare with Fig. 7.53 (**NaJ**), p. 339.

### *7.8.3 Saw-tooth structures

A rod intermediate between a zig-zag and a crankshaft is known as a saw tooth. A double saw-tooth rod is illustrated in Fig. 7.59. Now there are two kinds of vertex—the "teeth" (T) and the "base" (B)—in the ratio 1:2 on the rod. The repeat distance of a saw-tooth rod is about 2.4-2.6 times the edge length (about 2.5 × 3.0 Å = 7.5 Å in zeolites). The T vertices are on mirror planes at elevation either 0 and 1/2 or at 1/4 and 3/4 in projection down the rod, and the B vertices are about ±0.2 from the mirror planes. Again some zeolite structures are conveniently shown as projections, but the figures must be interpreted with care (see, for example, the legend for Fig. 7.60).

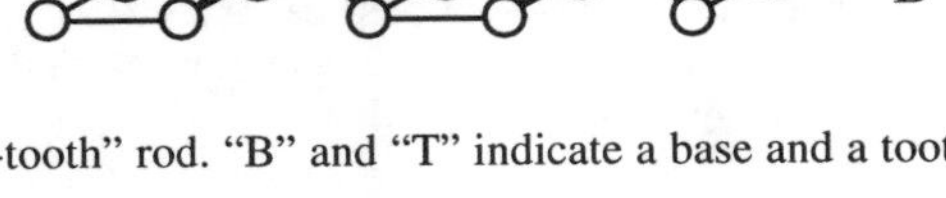

**Fig. 7.59**. A double "saw-tooth" rod. "B" and "T" indicate a base and a tooth vertex.

Two simple nets featuring double saw-tooth rods are found in the nets of the zeolites Linde-L and mazzite (both silica-rich aluminosilicates) shown in Fig. 7.60.

**Linde L** $P6/mmm$, $a = 6.007$, $c = 2.354$, $r = 0.489$
$4{\cdot}8_3{\cdot}4{\cdot}8_3{\cdot}6{\cdot}12$ in 12 $p$: $\pm(x,y,0$; etc.), $x = 0.096$, $y = 0.359$
$4{\cdot}4{\cdot}4{\cdot}6{\cdot}6{\cdot}8$ in 24 $r$: $\pm(x,y,z$ ; etc.), $x = 0.167$, $y = 0.500$, $z = 0.288$

**mazzite** $P6_3/mmc$, $a = 5.816$, $c = 2.505$, $r = 0.490$
$4{\cdot}8_2{\cdot}4{\cdot}8_2{\cdot}5{\cdot}6$ in 12 $j$: $\pm(x,y,1/4$; etc.), $x = 0.495$, $y = 0.161$
$4{\cdot}5{\cdot}4{\cdot}5{\cdot}8{\cdot}12$ in 24 $l$: $\pm(x,y,z$ ; etc.), $x = 0.096$, $y = 0.364$, $z = 0.050$

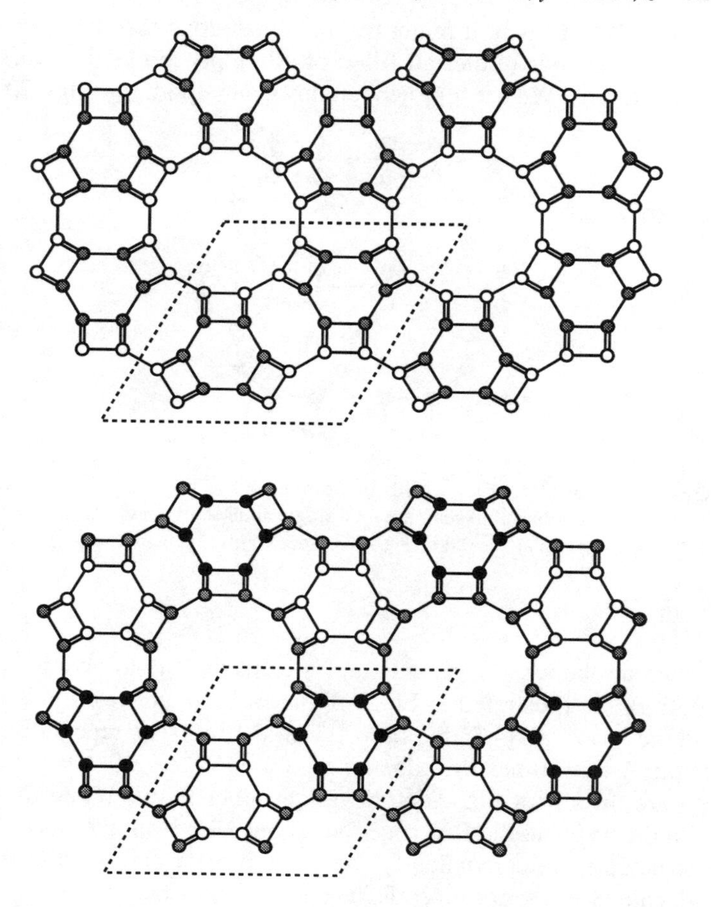

**Fig. 7.60**. Top: **Linde L** projected on (001). Open circles are T vertices at $z = 0$, shaded circles are B vertices at $z = \pm 0.29$. The latter form hexagonal prisms centered at 1/3,2/3,1/2 and 2/3,1/3,1/2. Bottom: **mazzite** projected on (001). Open and filled circles are T vertices at $z = 1/4$ and $z = 3/4$ respectively. Shaded circles on saw-tooth rods with teeth at $z = 1/4$ are B vertices at $z = -0.05$ and 0.55; shaded circles on saw-tooth rods with teeth at $z = 3/4$ are B vertices at $z = 0.45$ and 1.05. Note that here the rectangles are projections of the "double saw-tooth" rods of Fig. 7.59.

Some other saw-tooth nets found in silica-rich alumino-silicate nets are found in mordenite, dachiardite, ferrierite and ZSM-57. Symmetries and unit cell parameters (for unit edge) of these nets are:

| | |
|---|---|
| **mordenite** | *Cmcm*, $a = 5.60$, $b = 6.80$, $c = 2.42$, $r = 0.52$ |
| **dachiardite** | *C2/m*, $a = 5.84$, $b = 2.48$, $c = 3.30$, $\beta = 112.0°$, $r = 0.54$ |
| **ferrierite** | *Immm*, $a = 6.25$, $b = 4.34$, $c = 2.41$, $r = 0.55$ |
| **ZSM-57** | *Imm2*, $a = 2.44$, $b = 4.63$, $c = 6.13$, $r = 0.52$ |

The nets are shown in projection in Fig. 7.61. To interpret the diagrams note that (a) filled and open circles are T vertices on mirror planes separated by 1/2 the projection distance; (b) shaded circles are B vertices at elevations of approximately 0.3 above and below the T vertices to which they are connected, (c) saw-tooth rods are shown as double lines.

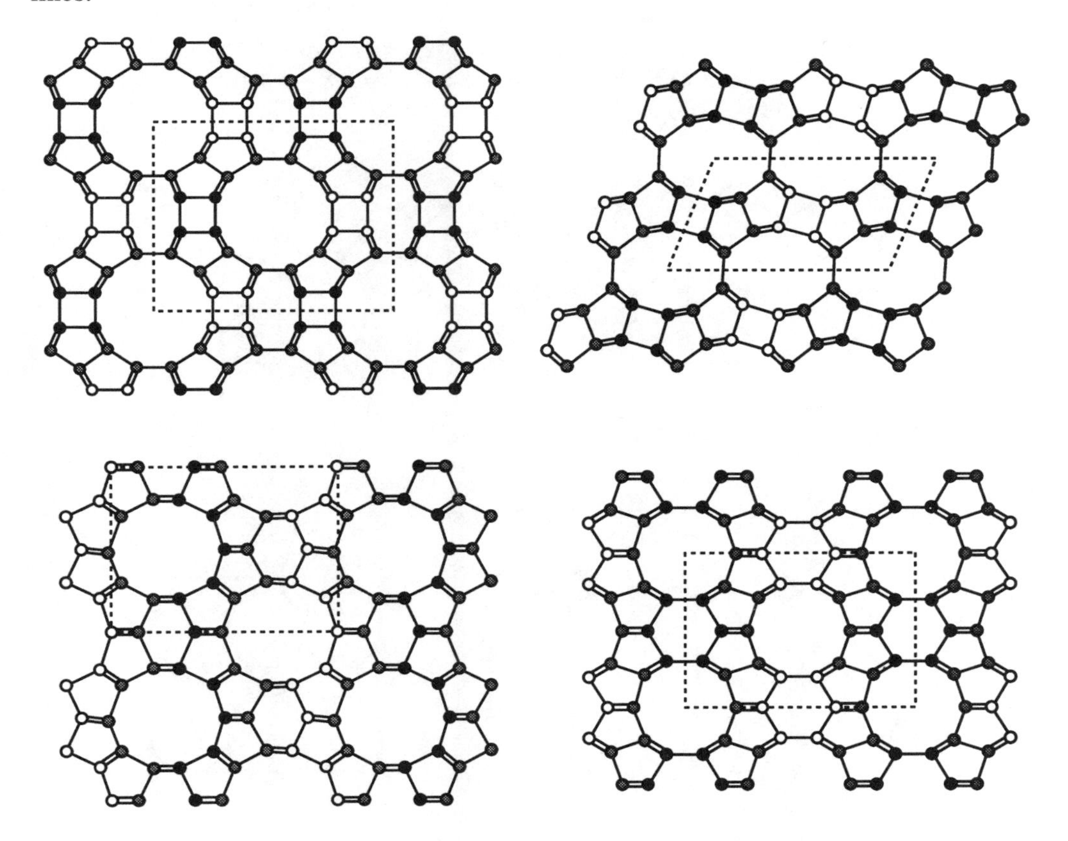

**Fig. 7.61**. Top left: **mordenite** projected on (001) with **a** horizontal on the page. Bottom left: **ZSM-57** projected on (100) with **c** horizontal on the page. Top right: **dachiardite** projected on (010) with **c** horizontal on the page. Bottom right: **ferrierite** projected on (001) with **a** horizontal on the page. See also the text.

### *7.8.4 More up-down structures

In § 7.3.8 we described two nets (one was of the zeolite **$AlPO_4$-5**) based on "up-down" connections of two-dimensional nets. Some rather complicated $AlPO_4$ structures that have recently been discovered are based on this principle, so again there is a short axis, and they are readily depicted as projections down that axis (Fig. 7.62). The repeat distance along the projection axis is about 2.7 the edge length (i.e. about 8.4 Å in alumino-phosphates). Note that these nets contain only even rings and that in the real materials Al and P alternate (lowering the symmetry).

Data for the two simpler of these nets with unit edge are:

**VPI-5** $P6_3/mcm$, $a = 6.086$, $c = 2.674$, $r = 0.406$
$4{\cdot}6_3{\cdot}4{\cdot}6_3{\cdot}6{\cdot}6_4$ in 12 $k$: $\pm(x,0,z$ ; etc.), $x = 0.4227$, $z = 0.063$
$4{\cdot}6_2{\cdot}6{\cdot}6_3{\cdot}6_2{\cdot}6_3$ in 24 $l$: $\pm(x,y,z$ ; etc.), $x = 0.1786$, $y = 0.5120$, $z = 0.563$

**$AlPO_4$-25** $Cmma$, $a = 2.603$, $b = 4.800$, $c = 3.149$, $r = 0.610$
$6{\cdot}6_2{\cdot}6{\cdot}6_2{\cdot}6_2{\cdot}6_2$ in 8 $n$: $C \pm (\pm x,1/4,z)$, $x = 0.308$, $z = 0.849$
$4{\cdot}6_2{\cdot}6{\cdot}6_3{\cdot}6_2{\cdot}6_3$ in 16 $o$: $C \pm (\pm x,y,z$ ; $\pm x,1/2-y,z)$,
$x = 0.192$, $y = 0.099$, $z = 0.651$

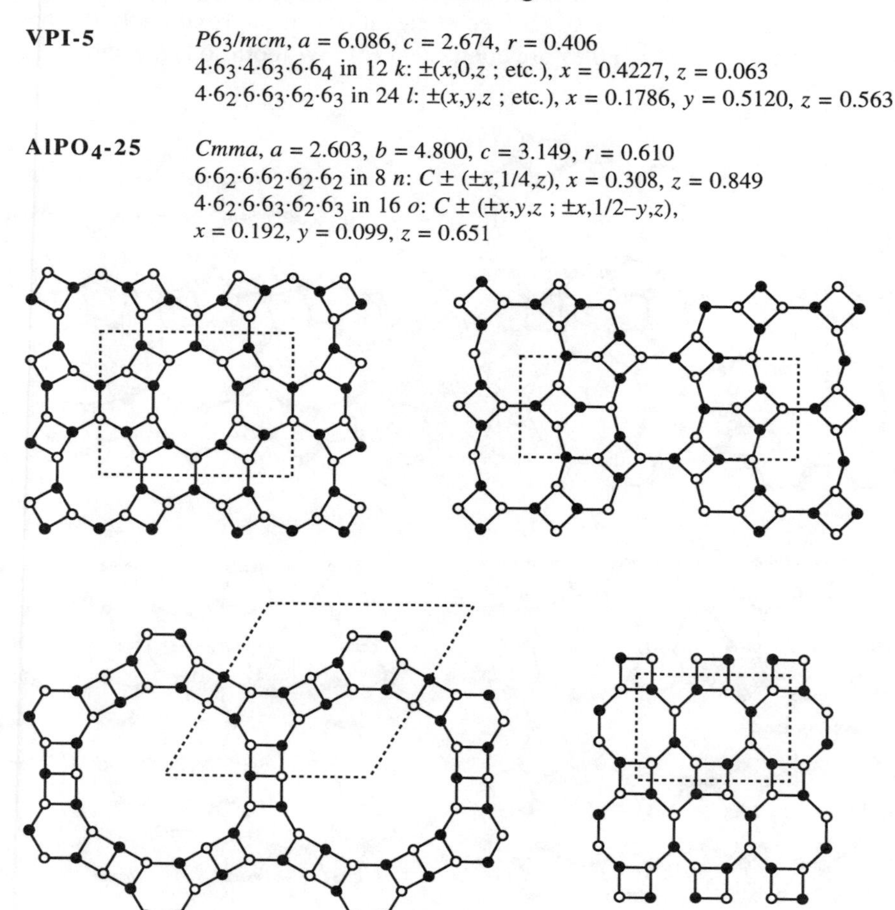

**Fig. 7.62.** Top left: **$AlPO_4$-11** projected on (100) with **b** horizontal on the page. Top right: **$AlPO_4$-41** projected on (001) with **b** horizontal on the page. Bottom left: **VPI-5** projected on (001). Bottom right: **$AlPO_4$-25** projected on (100) with **b** horizontal on the page. Additional edges go "up" from open circles and "down" from filled circles.

The **$AlPO_4$-41** net has symmetry *Cmcm* and four different kinds of vertex; The **$AlPO_4$-11** net has symmetry *Imma* and three different kinds of vertex.

Another member of this family that has recently been discovered is known as **$AlPO_4$-8**. This beautiful structure, shown (again slightly idealized) in Fig. 7.63, has five different kinds of vertex and contains 14-rings. The symmetry of the net is also *Cmcm*.

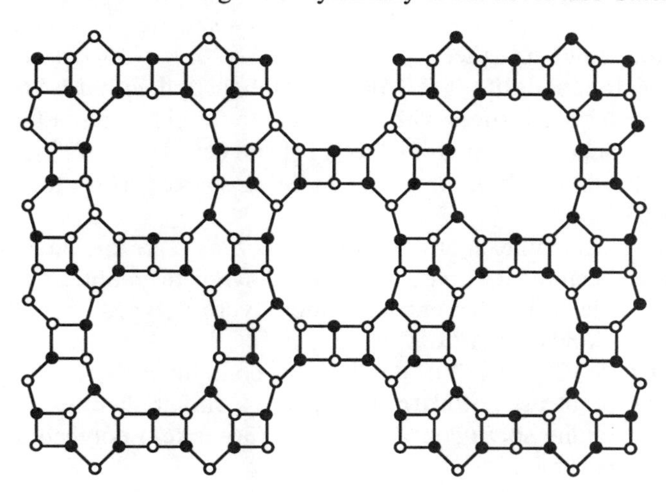

**Fig. 7.63**. **$AlPO_4$-8** projected on (001) with **a** horizontal on the page.

### *7.8.5 The "ABC-6" family

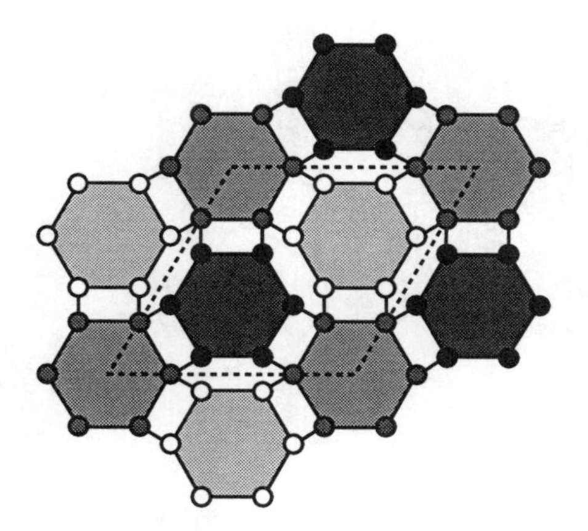

**Fig. 7.64**. **Sodalite** projected on (111) with a hexagonal cell outlined. Open circles at 1/6, shaded circles at 1/2 and filled circles at 5/6 of the repeat unit 1/2<111>. Edges joining vertices at 5/6 to those at 7/6 (i.e. 1/6) are not shown.

In § 7.3.5 (p. 306) we commented on **cancrinite** which we described as a stacking of plane hexagons centered at 1/3,2/3,1/4 and 2/3,1/3,3/4. By analogy with the nomenclature used for sphere packing (§ 6.1.1), we could describe that stacking sequence *AB*... The inclined edges linking the hexagons formed double zig-zag rods. Similarly in **gmelinite** (§ 7.3.7. p. 310) there are hexagonal prisms centered at 1/3,2/3,1/4 and 2/3,1/3,3/4 and the stacking of hexagons could be described as *AABB*.. and the edges not perpendicular to the stacking direction form a double crankshaft rod.

A sequence of hexagons stacked *AAB*... produces a structure based on saw-tooth rods. This is found in the zeolite offretite.

The primitive cell of **sodalite** (§ 7.3.10) is rhombohedral with $a = \sqrt{6}$, $\alpha = \cos^{-1}(-1/3) = 109.47°$ and contains six vertices with $x,y,z$ equal to the six permutations of 1/4,1/2,3/4. These comprise a planar hexagon normal to [111] and centered at 1/2,1/2,1/2. Packing the rhombohedral cells will result in the hexagons being stacked *ABC* along a 3-fold axis as shown in Fig. 7.65.[1]

Nets derived by stacking hexagons in positions *A*, *B* or *C* are known as ABC-6 nets and about a dozen have been recognized in natural and synthetic zeolites.[2] Some of these are summarized in Table 7.3. The entries under "vertex types" are the numbers of topologically-distinct kinds of vertex.

Chabazite is $Ca_3Al_6Si_{12}O_{36}{\cdot}20H_2O$; The net contains a rhombohedral stacking of hexagonal prisms (contrast **sodalite** which has a rhombohedral stacking of single hexagons). The remaining space consists of large (36-vertex) polyhedra as illustrated in Fig. 7.65. Data for the net are:

**chabazite** $R\bar{3}m$, $a = 4.404$, $c = 4.757$, $r = 0.451$
4·4·4·8·6·8 in 36 $i$: $R \pm (x,y,z$ ; etc.), $x = 0.106$, $y = 0.439$, $z = 0.062$

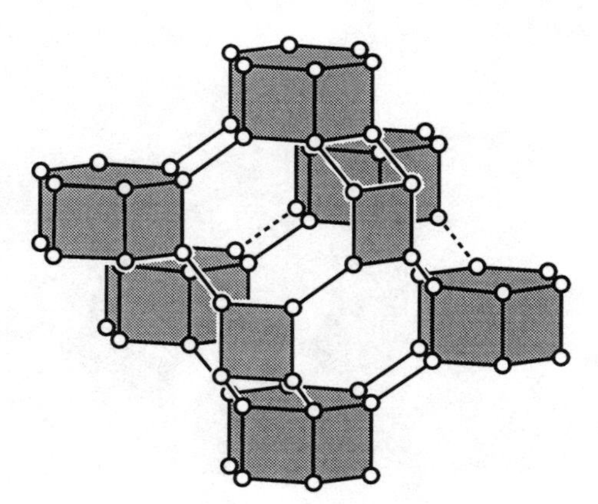

**Fig. 7.65**. The large polyhedron in **chabazite** formed by a linked stacking of hexagonal prisms. For two hexagonal prisms only one square face is shown (shaded).

[1]The skeptical reader may wish to transform **sodalite** to a hexagonal cell as outlined in § 4.4.2. The cell has $a = 4$ and $c = \sqrt{6}$ and contains 18 vertices. Compare with the 12-vertex cell of **cancrinite** given in § 7.3.5 (p. 306) which has the same $a$, and $c$ two-thirds as large.

[2]For a systematic discussion of possible ABC-6 structures and their symmetries, see J. V. Smith & J. M. Bennett, *Amer. Mineral.* **66**, 777 (1981).

**Table 7.3**. Some ABC-6 nets

| *sequence* | *net* | *vertex types* |
|---|---|---|
| *AB* | **cancrinite** | 1 |
| *ABC* | **sodalite** | 1 |
| *ABAC* | **losod** | 2 |
| *ABABAC* | **liottite** | 3 |
| *ABABACAC* | **afghanite** | 3 |
| *AABB* | **gmelinite** | 1 |
| *AABBCC* | **chabazite** | 1 |
| *AABBCCBBAACC* | **$AlPO_4$-52** | 3 |
| *AAB* | **offretite** | 2 |
| *ABBACC* | **TMA-E** | 2 |
| *AABAAC* | **erionite** | 2 |
| *AABCCABBC* | **levyne** | 2 |

### *7.8.6 Pentasils (silicalites), clathrasils and related structures*

We have already met the frameworks of **melanophlogite** ($SiO_2$) and **dodecasil-3C** in § 7.6 where they were identified as the frameworks of clathrate hydrates called **Type I** and **Type II** respectively. (A zeolite named ZSM-39 also has the latter structure). Another simple structure based on a space filling of polyhedra is **octadecasil** (the same framework has been found in $AlPO_4$-16). In this structure cubes pack with truncated rhombic dodecahedra in the ratio 1:1 to fill space as shown in Fig. 7.66. The large polyhedron with 18 faces (an octadecahedron = $[4^6.6^{12}]$) may be derived by truncating the acute vertices (where four edges meet) of a rhombic dodecahedron. The centers of each set of polyhedra fall on points of an *fcc* lattice; accordingly, taken together the centers have a **NaCl** arrangement. Data for this net are given on the next page.

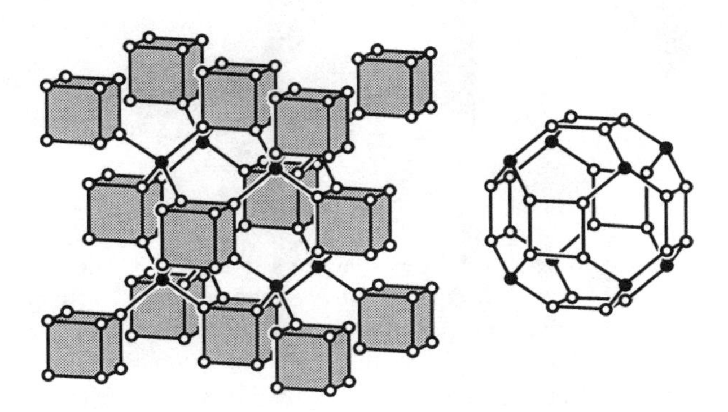

**Fig. 7.66**. The **octadecasil** net. Left: as cubes connected by isolated tetrahedral vertices (filled circles). Right: the octadecahedron.

| | |
|---|---|
| **octadecasil** | $Fm\overline{3}m$, $a$ = 4.309, $r$ = 0.500 |
| | 6·6·6·6·6·6 in 8 $c$: $F \pm$ (1/4,1/4,1/4) |
| | 4·6·4·6·4·6 in 32 $f$: $F \pm (\pm x,x,x)\kappa$, $x$ = 0.116 |

A particularly beautiful clathrasil known as sigma-2 has recently been described. This is based on a packing of 36-vertex icosahedra $[5^{12}.6^8]$ and enneahedra $(4^3.5^6]$. The large polyhedron, which has symmetry $\overline{4}2m$, is called the "tennis ball" because the pentagons form an endless strip rather like the seam of a tennis ball (see Fig. 7.67).[1] Like that of most clathrasils, the structure is difficult to illustrate satisfactorily, but it is easy to make a model. In **sigma-2** the "tennis balls" share opposite hexagonal faces to form rods along <100> as indicated in Fig. 7.67, and the smaller polyhedra (also shown in the figure) fill the interstices of the packing. The rods of face-sharing tennis balls are packed as in the 4-layer cylinder packing of § 6.7.2 (b) (p. 264).

The crystallographic description is fairly simple:

| | |
|---|---|
| **sigma-2** | $I4_1/amd$, $a$ = 3.28, $c$ = 11.17, $r$ = 0.53 |
| | 5·6·5·6·5·6 in 16 $h$: $I \pm$ (0,$y$,$z$ ; etc.), $y$ = 0.098, $z$ = 0.7133 |
| | 4·6·5·5·5·5 in 16 $h$: $I \pm$ (0,$y$,$z$ ; etc.), $y$ = 0.038, $z$ = 0.3669 |
| | 4·6·5·5·5·5 in 16 $h$: $I \pm$ (0,$y$,$z$ ; etc.), $y$ = 0.098, $z$ = 0.5529 |
| | 4·6·5·5·5·5 in 16 $f$: $I \pm$ ($x$,0,0 ; etc.), $x$ = 0.274 |

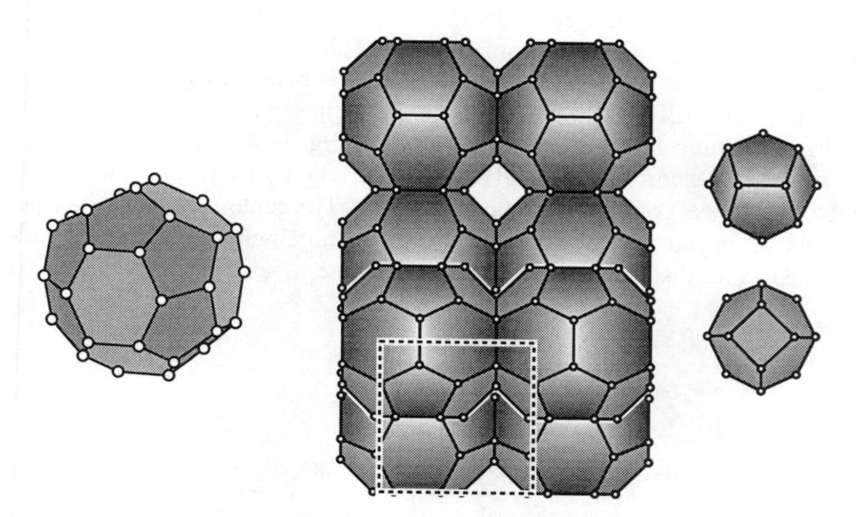

**Fig. 7.67**. Left: The "tennis ball" icosahedron as it appears in **sigma-2**. The $\overline{4}$ axis is vertical on the page. Middle: part of the icosahedron packing of **sigma-2** viewed down [001]. Right: the enneahedra that fill the interstices are shown in two different orientations.

A zeolite named **AlPO$_4$-22** is included here because it is another simple example of a 4-

[1]For more on this and related polyhedra see Appendix 4.

connected net derived from packing polyhedra. (As the net contains 8-rings, most authors would not classify this material as a clathrasil.) The net is shown in projection in Fig. 7.68. Model builders will discover that it is made up of equal numbers of two kinds of polyhedra; one with 10 faces $[4^6.6^4]$ and the other with 18 faces $[4^8.6^8.8^2]$ (Fig. 7.68). Each kind of polyhedron forms rods along **c** by sharing opposite faces. Crystallographic data for the ideal net are:

| | |
|---|---|
| **$AlPO_4$-22** | $P4/nmm$, $a = 4.324$, $c = 2.547$, $r = 0.504$ |
| | 4·4·4·6·6·6 in 8 $g$: $\pm(x,\bar{x},0$ etc.), $x = 0.134$ |
| | 4·4·6·6·6·8 in 16 $k$: $\pm(x,y,z$, etc.), $x = 0.029$, $y = 0.866$, $z = 0.345$ |

The 4·4·4·6·6·6 vertices form squares at $z = 0$ and the 4·4·6·6·6·8 vertices form octagons at $z = \pm 0.35$. How the polygons are connected should be evident from Fig. 7.68.

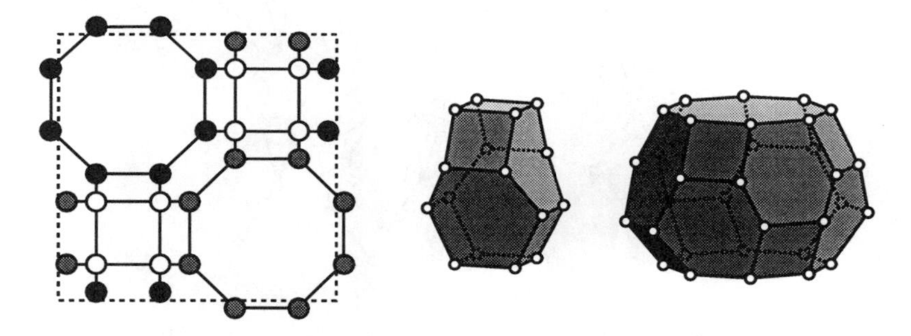

**Fig. 7.68**. Left: the **$ALPO_4$-22** net projected on (001). Open, shaded and filled circles are at $z = 0$, 0.35 and 0.65 respectively. Right: the two kinds of polyhedra in the structure (**c** is vertical). The top and bottom faces of the polyhedra are separated by $c$.

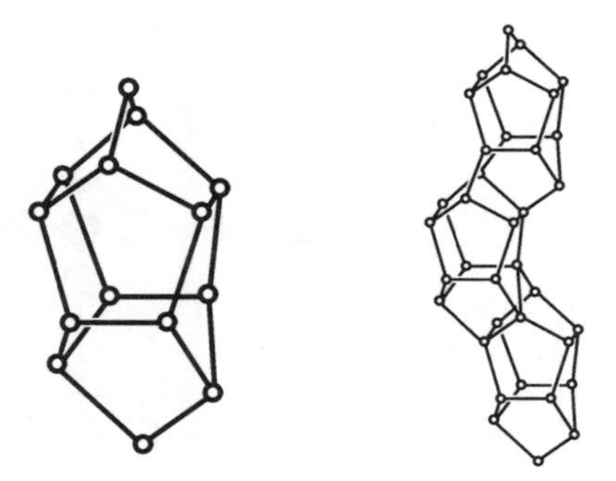

**Fig. 7.69**. Left: a pentasil unit. Right: three condensed pentasil units

The structures known as *silicalites* are rather more complicated. The net of silicalite 1 (ZSM-5) has twelve different kinds of vertex and that of silicalite 2 (ZSM-11) has eight different kinds. 5- and 6-rings dominate but there are also 4- and 10-rings in the structures. A basic building block in these structures is the "pentasil unit" (shown in Fig. 7.69) which is a polyhedron with eight pentagonal faces (a "pentagonal octahedron") and two divalent vertices. These can be condensed into corrugated slabs as hinted in the figure. The silicalite structures are then derived by joining the slabs either across mirror planes to form silicalite 2 or through inversion centers to form silicalite 1.[1]

### *7.8.7 Fibrous zeolites

The materials under this heading are a group of natural and synthetic zeolites with some fascinating crystal chemistry. The basic building unit is the rod of vertices shown on the left in Fig. 7.70. These rods can be linked in two directions perpendicular to the rod axis as shown on the right in the figure which illustrates the linkage in edingtonite.

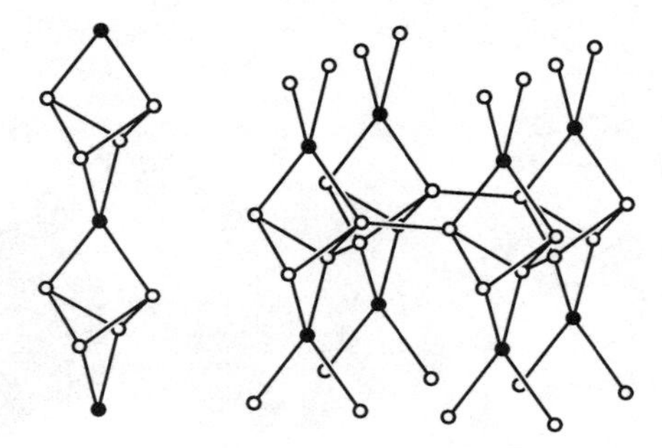

**Fig. 7.70**. Left the building unit of the fibrous zeolites. Right: the linkage of rods in **edingtonite**.

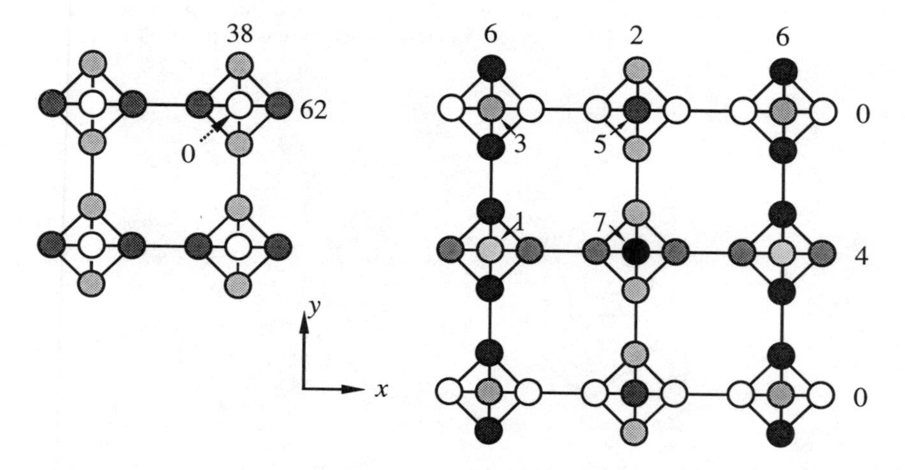


**Fig. 7.71**. Left: **edingtonite** projected on (001); numbers are elevations in multiples of $c/100$. Right: **natrolite** projected on (001); numbers are approximate elevations in multiples of $c/8$.

[1]Good illustrated accounts of these structures are given by C. A. Fyfe *et al., J. Amer. Chem. Soc.* **111**, 2470 (1989) and D. H. Olsen *et al., J. Phys. Chem.* **85**, 2238 (1981). Silicalites are important commercial catalysts.

The two simplest topologies are:

**edingtonite** $P\bar{4}m2$, $a = 2.204$, $c = 2.122$, $r = 0.485$
$4_2 \cdot 4_2 \cdot 8_4 \cdot 8_4 \cdot 8_4 \cdot 8_4$ in 1 *a*: 0,0,0
$4 \cdot 8_3 \cdot 4 \cdot 8_3 \cdot 4_2 \cdot 8_4$ in 4 *j*: ($\pm x$,0,$z$ ; 0,$\pm x$,$\bar{z}$), $x = 0.273$, $z = 0.376$

**natrolite** $I4_1/amd$, $a = 4.406$, $c = 2.124$, $r = 0.485$
$4_2 \cdot 4_2 \cdot 8_4 \cdot 8_4 \cdot 8_4 \cdot 8_4$ in 4*b*: $I \pm$ (0,1/4,3/8)
$4 \cdot 8_2 \cdot 4 \cdot 8_2 \cdot 4_2 \cdot 8_4$ in 16*h*: $I \pm$ (0,$y$,$z$ ; 1/2,$y$,1/2−$z$ ; 1/4+$y$,1/4,3/4+$z$ ; 1/4+$y$,3/4,1/4−$z$), $y = 0.387$, $z = -0.001$

Fig 7.71 shows, on the left, **edingtonite** in projection. The rods project as centered squares and together with the links between rods form a $4.8^2$ pattern. It is important to recognize that we have described the net in its most-symmetrical minimum-density form. In real material (edingtonite has the ideal formula $BaAl_2Si_3O_{10} \cdot H_2O$) the framework collapses as discussed for scapolite in § 7.3.8 (see Fig. 7.25, p. 312) and the unit cell is doubled. In the real material the symmetry is further lowered by Si,Al ordering and is in fact $P2_12_12$.

Fig. 7.71 also shows **natrolite** in projection. The links between rods are now at four different elevations and form rods arranged as in the four-layer cylinder packing of § 6.7.2 (b). Natrolite itself has the ideal composition $Na_2Al_2Si_3O_{10} \cdot 2H_2O$. The same framework is found in other minerals such as scolecite, $CaAl_2Si_3O_{10} \cdot 3H_2O$ (note the substitution of 2Na by Ca+$H_2O$) and also in anhydrous synthetic materials such as $Rb_2Ga_2Ge_3O_{10}$. Again the framework is partly collapsed and its symmetry is lowered from $I4_1/amd$ to $I\bar{4}2d$. Si,Al ordering in natrolite further lowers the symmetry to *Fdd*2 (with a doubled cell) and in scolecite Ca, $H_2O$ ordering reduces the symmetry further to *Cc*. It is common to use the same size of cell for these structures: this entails using a face centered cell for all four symmetries. The tetragonal space groups become $F4_1/ddm$ and $F\bar{4}d2$ (see § 3.3.4, p.73) and *Cc* becomes *Fd*. The descent in symmetry is in terms of full symbols:

| | | | | |
|---|---|---|---|---|
| parent structure | *F* | $4_1/d$ | $2/d$ | $2/m$ |
| collapsed | *F* | $\bar{4}$ | *d* | 2 |
| Si/Al order | *F* | *d* | *d* | 2 |
| Ca/$H_2O$ order | *F* | 1 | *d* | 1 |

The conventional *Cc* cell is obtained from *Fd* by (0 0 $\bar{1}$ / 0 1 0 / 1/2 0 1/2).

Thompsonite, which has approximate composition $NaCa_2Al_5Si_5O_{20} \cdot 6H_2O$, has a closely related structure with a different linkage of rods.[1]

### 7.8.8 Zeolite net nomenclature and index

The bewildering variety of names for natural and synthetic zeolites has lead the Structure Commission of the International Zeolite Association to establish three-letter symbols for

[1]For a systematic account of linkages and symmetries possible for fibrous silicates, see J. V. Smith, *Zeits. Kristallogr.* **165**, 191 (1983).

structure topologies.[1] These are given in Table 7.4 for the zeolite nets discussed in this chapter (three of these are in the Exercises § 7.12) together with the names that we have used, so this section can serve as an index to those structures. It is a pity that no generally agreed symbols are available for other common nets (such as **diamond** and **keatite**) which occur in many different contexts.

**Table 7.4**. Symbols, names and sections for some zeolite nets

| | | | | | |
|---|---|---|---|---|---|
| ABW | **$SrAl_2$** | § 7.3.5 | GIS | **gismondine** | § 7.3.6 |
| AEL | **$AlPO_4$-11** | § 7.8.4 | GME | **gmelinite** | § 7.3.7 |
| AET | **$AlPO_4$-8** | § 7.8.4 | JBW | **NaJ** | § 7.8.1 |
| AFG | **afghanite** | § 7.8.5 | KFI | **ZK-5** | § 7.4.3 |
| AFI | **$AlPO_4$-5** | § 7.3.8 | LEV | **levyne** | § 7.8.5 |
| AFO | **$AlPO_4$-41** | § 7.8.4 | LIO | **liottite** | § 7.8.5 |
| AFS | **MAPSO-46** | § 7.12.12 | LOS | **losod** | § 7.8.5 |
| AFT | **$AlPO_4$-52** | § 7.8.5 | LTA | **Linde A** | § 7.4.1 |
| AFY | **CoAPO-50** | § 7.4.5 | LTL | **Linde L** | § 7.8.3 |
| ANA | **analcime** | § 7.12.6 | MAZ | **mazzite** | § 7.8.3 |
| APC | **$AlPO_4$-C** | § 7.8.2 | MEL | **silicalite 2 (ZSM-11)** | § 7.8.6 |
| AST | **octadecasil** | § 7.8.6 | MEP | **melanophlogite** | § 7.6 |
| ATN | **MAPO-39** | § 7.3.5 | MER | **merlinoite** | § 7.3.6 |
| ATO | **$AlPO_4$-31** | § 7.3.4 | MFI | **silicalite 1 (ZSM-5)** | § 7.8.6 |
| ATS | **MAPO-36** | § 7.8.1 | MFS | **ZSM-57** | § 7.8.3 |
| ATT | **$AlPO_4$-12** | § 7.8.2 | MON | **montesommaite** | § 7.12.7 |
| ATV | **$AlPO_4$-25** | § 7.8.4 | MOR | **mordenite** | § 7.8.3 |
| AWW | **$AlPO_4$-22** | § 7.8.6 | MTN | **type II (ZSM-39)** | § 7.6 |
| BIK | **bikitaite** | § 7.8.1 | MTT | **ZSM-23** | § 7.8.1 |
| CAN | **cancrinite** | § 7.3.5 | MTW | **ZSM-12** | § 7.8.1 |
| CAS | **$CsAlSi_5O_{12}$** | § 7.8.1 | NAT | **natrolite** | § 7.8.7 |
| CHA | **chabazite** | § 7.8.5 | OFF | **offretite** | § 7.8.5 |
| DAK | **dachiardite** | § 7.8.3 | PHI | **phillipsite** | § 7.8.2 |
| EAB | **TMA-E** | § 7.8.5 | RHO | **rho** | § 7.4.2 |
| EDI | **edingtonite** | § 7.8.7 | SGT | **sigma-2** | § 7.8.6 |
| EMT | **hex. faujasite** | § 7.4.4 | SOD | **sodalite** | § 7.3.10 |
| ERI | **erionite** | § 7.8.5 | TON | **theta-1** | § 7.8.1 |
| FAU | **faujasite** | § 7.4.4 | VFI | **VPI-5** | § 7.8.4 |
| FER | **ferrierite** | § 7.8.3 | | | |

## 7.9 5-connected nets

Five-connected nets have received comparatively little attention. Describing them by Schläfli symbols gets a little cumbersome as there are now ten angles and they cannot be all equivalent.[2] Some examples are to be found in the structures of borides which often have extended B-B bonding with connectivity ranging from 2 (forming rods) through 3 (forming

[1]The index of the *Atlas of Zeolite Structure Types* has 332 entries.

[2]This follows from the fact that the complete graph with five points is not planar and therefore cannot represent the vertices and edges of a three-dimensional polyhedron.

layers) to 4, 5 or 6 (forming three-dimensional frameworks).

Our first three examples also represent space filling by regular and Archimedean polyhedra. The first of these is a space-filling by octahedra and truncated cubes ($3.8^2$) and is found in nature as the B structure of $CaB_6$ and similar borides such as $KB_6$ and $LaB_6$. It is illustrated in Fig.7.72. This is a simple cubic structure; data for unit edge length are:

**$CaB_6$** $\quad Pm\bar{3}m$, $a = 1 + \sqrt{2} = 2.4142$, $r = 0.426$
vertices in 6 $e$: $\pm (x,0,0)\kappa$, $x = 1/(2 + \sqrt{2}) = 0.2929$

In $CaB_6$ $a$ = 4.151 Å, Ca is at 1/2,1/2,1/2 in the center of the truncated cube (24-coordinated by B), and the $x$ parameter (0.302) for B is close to the ideal value given above.

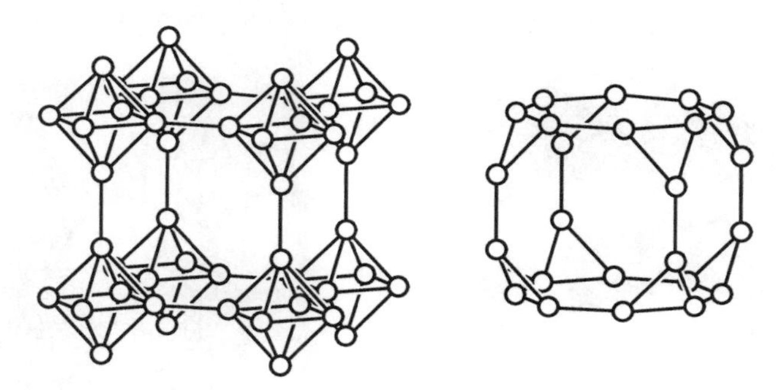

**Fig. 7.72**. Left: the net of B atoms in $CaB_6$. Right: a truncated cube formed by 24 vertices of the net.

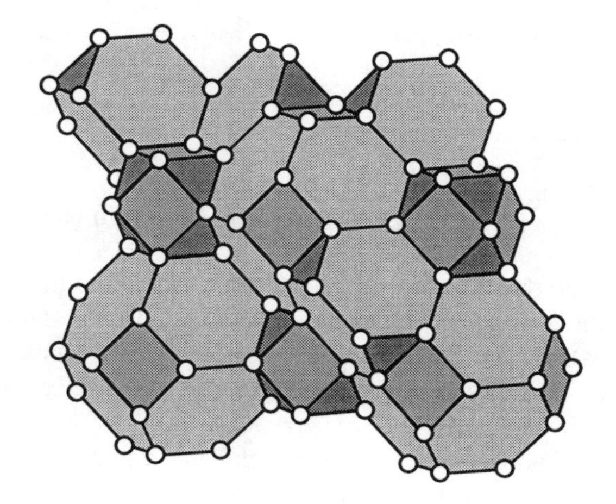

**Fig. 7.73**. The boron arrangement in $UB_{12}$.

The second five-connected structure is a space-filling by truncated octahedra, cuboctahedra and truncated tetrahedra that is the B structure of $UB_{12}$ and isostructural borides (e.g. $NiB_{12}$, $LuB_{12}$). The crystallographic description is again very simple:

**$UB_{12}$** $Fm\bar{3}m$, $a = \sqrt{18} = 4.2426$, $r = 0.629$ [$a = 7.473$ Å for $UB_{12}$]
vertices in 48 *i*: $F \pm (1/2,x,\pm x)\kappa$, $x = 1/3$

A sketch of part of the structure is shown in Fig. 7.73. In $UB_{12}$, U atoms in 4 *a*: $F$ + (0,0,0) center the $B_{24}$ truncated octahedra. The centers of the cuboctahedra are at 4 *b*: $F$ + (1/2,1/2,1/2) and the centers of the truncated tetrahedra are at 8 *c*: $F \pm$ (1/4,1/4,1/4).

If the truncated octahedra and half the truncated tetrahedra are omited, the remaining truncated tetrahedra and cuboctahedra form an infinite polyhedron $3.4.6^2.4$. The symmetry of the figure is now $F\bar{4}3m$, but the positions of the vertices are the same (see Fig. 7.74).

The third polyhedron packing is a space-filling by cubes, octagonal prisms, rhombicuboctahedra ($3.4^3$) and truncated cubes ($3.8^2$). In Fig. 7.74 the network of cubes and rhombicuboctahedra is shown as an infinite polyhedron $3.4^4$. The complementary polyhedron (also $3.4^4$) consists of truncated cubes ($3.8^2$) joined with octagonal prisms.

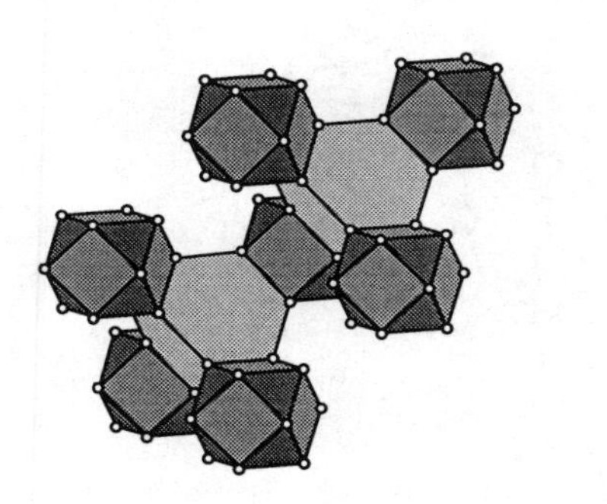
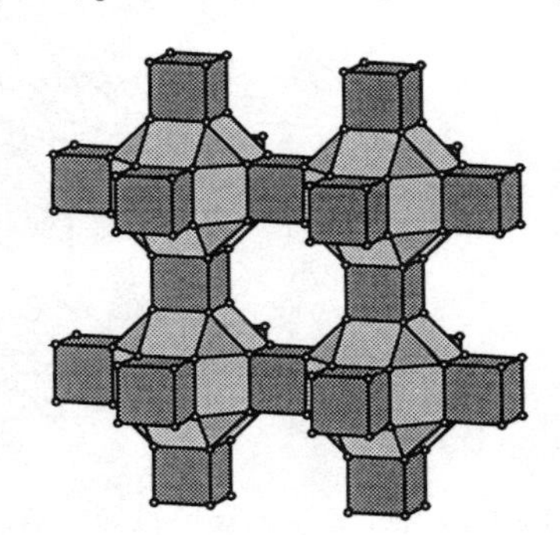

**Fig. 7.74**. Left: a fragment of an infinite polyhedron $3.4.6^2.4$. Right: Rhombicuboctahedra and cubes forming an infinite polyhedron $3.4^4$.

Crystallographic data for this structure are:

$3.4^4$ $Pm\bar{3}m$, $a = 2 + \sqrt{2} = 3.4142$, $r = 0.603$
vertices in 24 *m*: $(\pm x,\pm x,\pm z)\kappa$, $x = 1/(4 + \sqrt{8}) = 0.1465$, $z = 1/2-x$

Per unit cell there is one rhombicuboctahedron (center at cell corner), one truncated cube (center at cell center), three octagonal prisms (centers in cell faces) and three cubes (centers in middle of cell edges). As an example of the occurrence of this structure we cite the structure of $Pd_{17}Se_{15}$ (for crystallographic data see Appendix 5) which is truly a polyhedrist's delight. In the unit cell of this structure, 24 Pd atoms make up the 5-coordinated packing and additionally: {Pd}$Se_6$ octahedra center the rhombicuboctahedra, $Pd_6Se_{12}$ clusters consisting of $Pd_6$ octahedra edge-capped by Se forming a $Se_{12}$ cuboctahedron (cf. § 5.2.4, Fig. 5.32, p. 159) center the truncated cube and {Pd}$Se_4$ squares center the octagonal prisms. The unit cell content is accordingly $PdSe_6 \cdot Pd_6Se_{12} \cdot (PdSe_4)_3 \cdot Pd_{24} = Pd_{34}Se_{30}$. $Rh_{17}S_{15}$ is isostructural.

Two more five-connected infinite polyhedra follow. The first (Fig. 7.75) is $3^3.6^2$ and is made up of truncated tetrahedra sharing triangular faces with octahedra (so that two opposite faces of an octahedron are shared with truncated tetrahedra). Crystallographic data are :

| | |
|---|---|
| $3^3.6^2$ | $Fd\bar{3}m$, $a = 6.6024$, $r = 0.333$<br>vertices in 96 $g$: $F \pm (x,x,z$ etc.), $x = 0.0714$, $z = -0.0357$ |

In this structure, the connectivity of the truncated tetrahedra (by octahedra acting as links) has the **diamond** topology (the centers of the truncated tetrahedra form a **diamond** net). The structure can also be considered as octahedra joined to six neighboring octahedra by a fifth edge, as in **$CaB_6$**, but now the topology is different—the centers of the octahedra are at the points of the $T$ lattice complex.

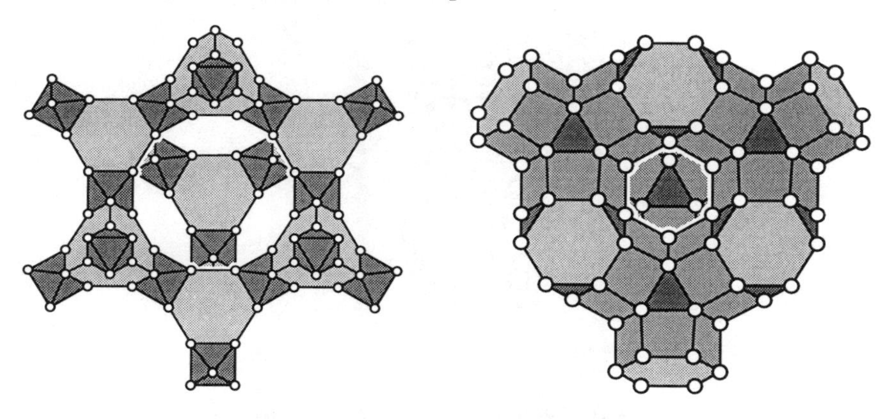

**Fig. 7.75**. Left: part of an infinite polyhedron $3^3.6^2$. Right: part of an infinite polyhedron $3.4^4$.

Another infinite polyhedron with vertices $3.4^4$ consists of truncated tetrahedra sharing hexagonal faces with hexagonal prisms (Fig. 7.75). This arrangement is a conspicuous part of the so-called $E$ structure which occurs for compositions such as $Mg_3Al_{18}Cr_2$, $ZrZn_{22}$ and $Al_{10}V$. Crystallographic data for the five-coordinated packing are:

| | |
|---|---|
| $3.4^4$ | $Fd\bar{3}m$, $a = 5.138$, $r = 0.709$<br>vertices in 96 $g$: $F \pm (x,x,z$ etc.), $x = 0.0562$, $z = 0.3314$ |

## 7.10 Nets with mixed connectivity

Nets with mixed connectivity inevitably involve more than one kind of vertex. Here we give some examples of such nets with just two kinds of vertex.

### **7.10.1 (3,4)-connected nets*

A very simple cubic structure (Fig. 7.76) has been proposed for $Pt_3O_4$ in which the atoms lie on invariant lattice complexes:

| | |
|---|---|
| $Pt_3O_4$ | $Pm\bar{3}n$, $a$ = 5.58 Å |
| | Pt in 6 $c$: ±(1/4,0,1/2)κ |
| | O in 8 $e$: ±(1/4,1/4,1/4 ; 1/4,3/4,3/4)κ |

In this structure there are {Pt}$O_4$ squares and {O}$Pt_3$ equilateral triangles with the Pt-O distance equal to $a/\sqrt{8}$. The O atoms are on the points of a primitive cubic lattice. The Pt atom positions correspond to lattice complex *W* and form non-intersecting rods of Pt atoms a distance $a/2$ apart. The rods, parallel to the cube axes, are packed as in the **β-W** cylinder packing (§ 6.7.3).

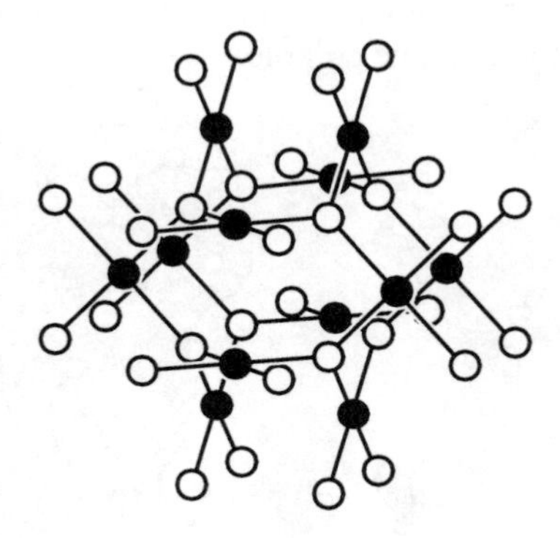

**Fig. 7.76**. The structure of $Pt_3O_4$ as a (3,4) connected net.

Boron in borates is commonly found as {B}$O_3$ triangles and {B}$O_4$ tetrahedra forming frameworks by sharing vertices. Exercise 2 gives an example of a 4-connected net derived from vertex sharing tetrahedra in $CaB_2O_4$ and the structure of $B_2O_3$ was cited (§ 7.2) as providing an example of a 3-connected net derived from vertex-sharing triangles. The structure of boracite, $Mg_3B_7O_{13}Cl$, which contains {B}$O_3$ triangles and {B}$O_4$ tetrahedra provides an elegant example of a (3,4) connected net of B atoms. Per formula unit there are $4BO_{3/2} + 3BO_{4/2} = B_7O_{12}$ in the B-O-B framework.

Here we describe just the idealized (3,4)-connected net. The basic unit consists of an octahedron of 4-connected vertices (*B*) with 3-connected vertices (*A*) centering four of the octahedron faces to form an $A_4B_6$ cluster (shown in Fig. 5.31, p. 159). Joining these octahedra by vertex sharing as shown in Fig. 7.77 [in the same way as in the *J* lattice complex (Fig. 6.27)] produces stoichiometry $A_4B_{6/2} = A_4B_3$. For unit edge the crystallographic description is:

| | |
|---|---|
| **boracite net** | $P\bar{4}3m$, $a = \sqrt{6}$ |
| | 3-connected in 4 $e$: ($x,x,x$ ; $\bar{x},\bar{x},x$)κ, $x$ = 1/6 |
| | 4-connected in 3 $d$: (1/2,0,0)κ |

In the real structure of boracite at high temperature, the unit cell edge is doubled to allow suitable B-O-B configurations, and the symmetry is $F\bar{4}3c$. Below 300 °C, the symmetry is lowered to $Pca2_1$ but the topology of the framework is unaltered.

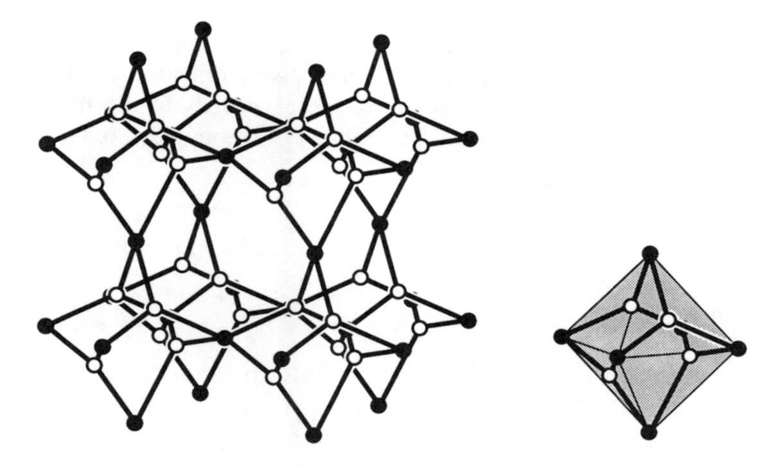

**Fig. 7.77**. The (3,4)-connected net of B atoms in boracite. On the right one $A_4B_6$ cluster is shown.

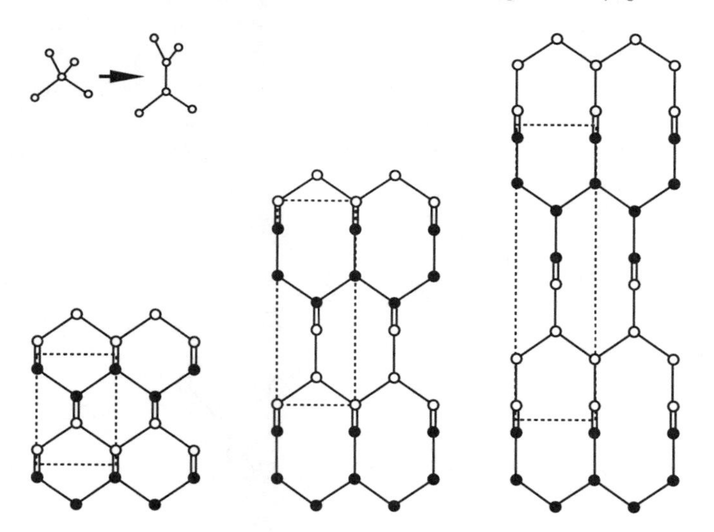

**Fig. 7.78**. Generation of 3- and (3,4)-connected nets from **diamond** (left). The (3,4)-connected net discussed in the text is shown in the middle and the **ThSi$_2$** net is shown on the right. In each case a body-centered tetragonal cell is shown and in the projection on (100), **c** is vertical, and points shown as filled and empty circles have elevations differing by $\delta x = 1/2$.

3-connected nets can simply be derived from 4-connected nets by replacing each 4-connected vertex by a pair of 3-connected vertices as shown in Fig. 7.78. The figure shows how to generate the **ThSi$_2$** net from **diamond** in this way. On the left **diamond** projected on (110) (cf. Fig. 7.11, p. 302) and on the right the **ThSi$_2$** net is projected on (100) of its tetragonal (*I*4$_1$/*amd*) cell (cf. Fig. 7.8, p. 298). Notice how each **diamond** vertex becomes a pair of 3-connected vertices with their edges in planes mutually at right angles. If only half the **diamond** vertices are so altered, a simple (3,4)-connected net is obtained. The ring size (all 8-rings) is intermediate between that in **diamond** (all 6-rings) and in **ThSi$_2$** (all 10-rings). The net has only three vertices in the repeat unit and is probably the simplest (3,4)-connected net. The cell is again body-centered tetragonal:

| | |
|---|---|
| (3,4)-connected net | $I\bar{4}m2$, $a$ = 1.800, $c$ = 3.744 |
| | 4-connected in 2 $a$: $I$ + (0,0,0) |
| | 3-connected in 4 $f$: $I$ + (0,1/2,$z$ ; 1/2,0,$\bar{z}$), $z$ = 0.3836 |

### *7.10.2 (4,6)-connected nets*

Simple examples of (4,6)-connected nets with two kinds of vertex occur as the structures of corundum ($^{vi}Al_2{}^{iv}O_3$) and $^{vi}Ni_2{}^{iv}S_3$.

In structures based on a framework of tetrahedra sharing corners with octahedra (and *vice versa*), the central tetrahedral and octahedral atoms form vertices of a (4,6)-connected net. Examples of such structures are those of the polymorphs of $Fe_2(SO_4)_3$ and $Al_2(WO_4)_3$; in these structures -O- links serve as the edges of the net. Just as for structures based on tetrahedral frameworks, "stuffed" variants are also found. In the following examples the atoms in bold face are on the net (different in every case and with -O- links again serving as edges) and the remaining metal atoms are in cavities in the framework: garnet, Ca$_3$**Al**$_2$(**Si**O$_4$)$_3$; langbeinite, K$_2$**Mg**$_2$(**S**O$_4$)$_3$; nasicon[1] = Na$_4$**Zr**$_2$(**Si**O$_4$)$_3$ and **Al**$_2$(**W**O$_4$)$_3$.

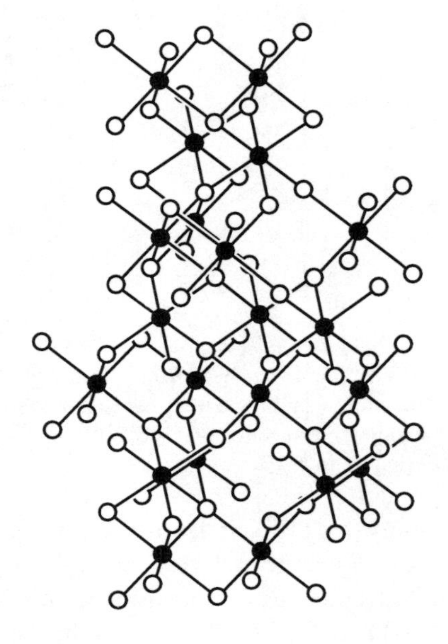

**Fig. 7.79. Corundum** as a (4,6)-connected net.

[1]Nasicon is a good Na-ion conducting silicate (**Na si**licate **con**ductor).

As for 4-connected nets, the same topology is often found in different contexts; for example, the same (4,6)-connected net is found in corundum, rhombohedral $Fe_2(SO_4)_3$ and in nasicon. We call it the **corundum** net. The corundum structure is possibly best appreciated as a packing of $\{Al\}O_6$ octahedra (see § 6.1.6), however it is shown as a bond (Al-O) network in Fig. 7.79.[1] Fig 7.80 shows the connection of octahedra and tetrahedra in rhombohedral $Fe_2(SO_4)_3$.

The occurrence of the **corundum** net in crystal structures illustrates the hierarchical way structures can develop. In $\mathbf{Al_2O_3}$, the edges of the net are Al-O bonds, in $\mathbf{Fe_2S_3}O_{12}$ the edges are Fe-O-S bonds. In $K_2\mathbf{Fe_2Zn_3}(CN)_{12}.xH_2O$, the edges are Fe-C-N-Zn bond groups (atoms in bold correspond to the vertices of the net). The molar volume of the last compound is over eight times that of the first one.

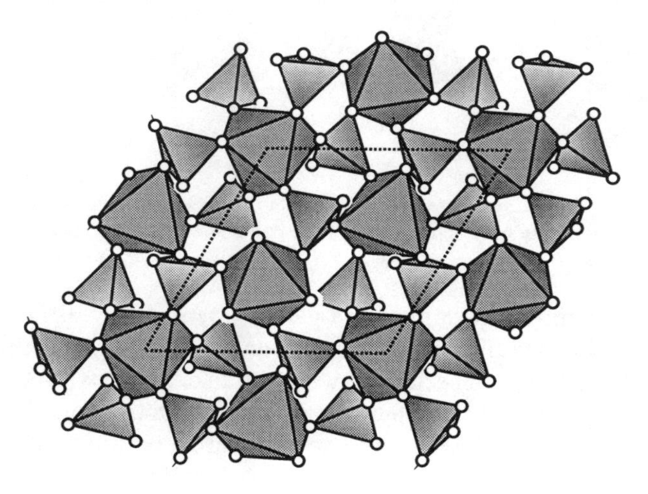

**Fig. 7.80**. The rhombohedral $Fe_2(SO_4)_3$ structure as a network of vertex-sharing $\{Fe\}O_6$ octahedra and $\{S\}O_4$ tetrahedra. The structure is projected down **c**. Although two kinds of octahedra are apparent in the drawing, they are topologically equivalent.

The different (4,6)-connected nets we have mentioned are readily distinguished by comparing their coordination sequences. The example below shows (a) that the (4,6)-nets of garnet and $Al_2(WO_4)_3$ are topologically distinct (despite a statement to the contrary sometimes encountered), and (b) that there are two topologically-distinct W atoms in $Al_2(WO_4)_3$.

| *net* | *atom* | $n_1$ | $n_2$ | $n_3$ | $n_4$ | $n_5$ | $n_6$ |
|---|---|---|---|---|---|---|---|
| garnet | Al | 6 | 12 | 42 | 50 | 114 | 110 |
| | Si | 4 | 16 | 28 | 74 | 76 | 162 |
| $Al_2(WO_4)_3$ | Al | 6 | 14 | 42 | 50 | 114 | 110 |
| | W(1) | 4 | 17 | 28 | 70 | 76 | 163 |
| | W(2) | 4 | 16 | 28 | 72 | 76 | 162 |

[1]For many people, Fig. 7.79 will merely illustrate the difficulty of interpreting "ball and stick" diagrams of structures!

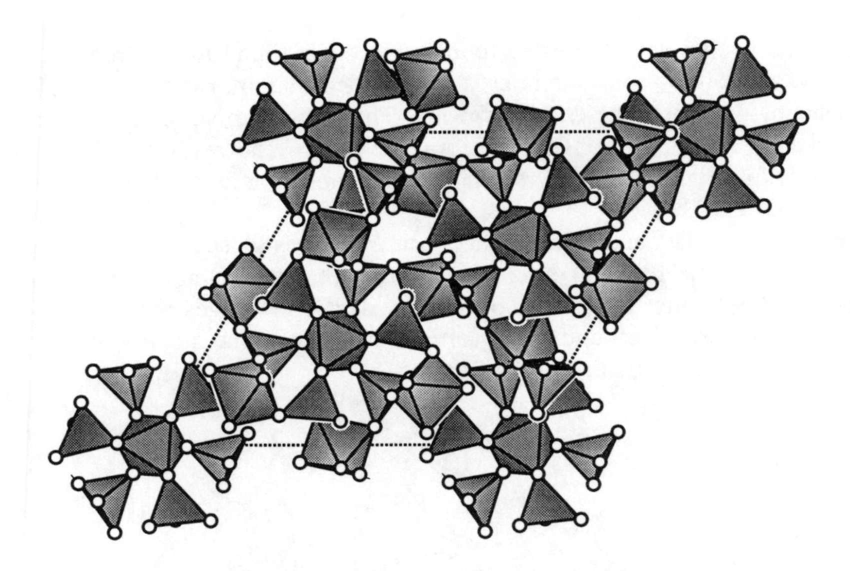

**Fig. 7.81**. A part of the garnet structure shown as linked {Si}$O_4$ tetrahedra and {Al}$O_6$ octahedra projected down [111]. For clarity only about a half of the polyhedra in the repeat unit [= (**a** + **b** + **c**)/2] out of the page are drawn.

The common, and important, structure of garnet is remarkably difficult to illustrate or describe. Here we illustrate (Fig. 7.81) just the connectivity of octahedra and tetrahedra which correspond to a (4,6)-connected net with -O- links as edges. Note that the vertices of the net (Al and Si positions) are at the sites of invariant lattice complexes (see § 3.4 for a list of coordinates); in § 6.6.4 we showed how the cation positions are related to $\mathbf{Cr_3Si}$.

For some low-density (4,6)-connected nets see Exercise 15.

## 7.11 Notes

### *7.11.1 More 3-connected nets*

Wells (reference in § 7.11.10) made a special study of uniform 3-connected nets (those in which the shortest rings at each angle are the same size). In contrast to uniform 4-connected nets, which are all $6^6$, Wells found nets $7^3$, $8^3$, $9^3$, $10^3$ (see § 7.2) and $12^3$. Here we list coordinates for a few of the simpler 3-connected nets with one kind of vertex. The reader may enjoy drawing them. They serve to illustrate the topologies that can occur.

A simple net $12_4{\cdot}12_7{\cdot}12_7$, which cannot be constructed with all angles equal to 120°, is for maximum volume:

$12^3$ $\quad$ $P6_222$, $a = 2.475$, $c = 2.026$
vertices in 6 $g$: $\pm(x,0,0\ ;\ x,x,1/3\ ;\ \bar{x},\bar{x},1/3\ ;\ 0,\pm x,2/3)$, $x = 0.298$

There are many $8^3$ nets. Known nets with one kind of vertex are all $8{\cdot}8{\cdot}8_2$. The first cannot have all angles equal to 120°, so we give coordinates for maximum volume; for the other two (Fig. 7.82), the coordinates are for bond angles of 120°:

I. $Im\bar{3}m$, $a = 6.352$
vertices in 96 $l$: $I + (\pm x,\pm y,\pm z\ ;\ \pm y,\pm x,\pm z)$, $x = 0.349$, $y = 0.190$, $z = 0.079$

II. $P6_222$, $a = 5/\sqrt{3}$, $c = \sqrt{6}$
vertices in 6 $i$: $(x,2x,0\ ;\ \bar{x},2\bar{x},0\ ;\ x,\bar{x},1/3\ ;\ \bar{x},x,1/3\ ;\ 2x,x,2/3\ ;\ \bar{x},2\bar{x},2/3)$, $x = 2/5$

III. $R\bar{3}m$, $a = 5$, $c = \sqrt{6}$
vertices in 18 $f$: $R \pm (x,0,0\ ;\ 0,x,0\ ;\ x,x,0)$, $x = 2/5$

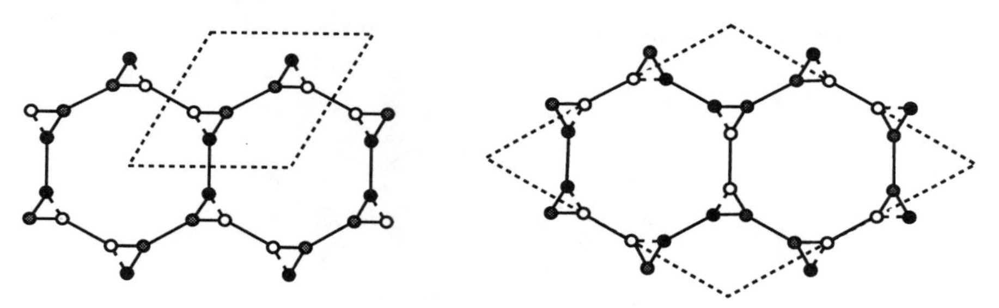

**Fig. 7.82.** The nets II (left) and III (right) projected down **c**. Open, shaded and filled circles are at elevations 0, 1/3 and 2/3, respectively.

Net II occurs as the (Cu,S) net in $BiCu_3S_3$ (Cu and S are 3-coordinated to each other).

As the $10^3$ $Y^*$ (§ 7.2) net is the only 3-dimensional 3-connected net with equivalent edges it is the only one that can be decorated with triangles to produce a uninodal 3-connected net (cf. § 7.5.2).[1] The 3-dimensional net is $3{\cdot}20_5{\cdot}20_5$ and is probably the least dense three-dimensional uninodal net ($c_{10} = 207$), we call it $Y^*3$. Data for this net are:

$Y^*3$ $I4_132$, $a = 4/(\sqrt{24} - \sqrt{18}) = 6.0944$, $r = 0.106$, $\rho = 0.0555$
vertices in 24 $g$: $I + (1/8,x,1/4-x\ ;\ 3/8,\bar{x},3/4-x\ ;\ 5/8,1/2-x,3/4+x\ ;$
$7/8,1/2+x,1/4+x)\kappa$, $x = (\sqrt{12} - 3)/8 = 0.0580$

The quasiregular 4-connected NbO ($J^*$) net could not be decorated with tetrahedra as the edges are coplanar; however if each vertex is replaced by a square, a 3-connected uninodal net is obtained[2] with symbol $4{\cdot}12_2{\cdot}12_2$:

$4.12^2$ $Im\bar{3}m$, $a = 2 + \sqrt{8}$
vertices in 24 $g$: $I \pm (x,0,1/2\ ;\ 0,x,1/2)\kappa$, $x = 1/(4 + \sqrt{8}) = 0.1464$

"Five-electron" compounds $AB$, where $A$ is an alkali metal and $B$ is from group 4A

[1]The analogous process in two dimensions produces $3.12^2$ from $6^3$.
[2]The analogy in two dimensions is the generation of $4.8^2$ from $4^4$.

(column 14) of the periodic table, have some fascinating structures, usually with *B* in three-coordination. The most common structures are **KGe** and **NaPb** in which *B* atoms form tetrahedral groups. In LiGe, however, a 3-dimensional net is formed in which the vertices are $8{\cdot}8{\cdot}10_3$ (LiSi and MgGa have the same structure). The angles are less than 120° as might be expected[1] and we give the actual coordinates for the structure.

| LiGe | $I4_1/a$, $a$ = 9.75 Å, $c$ = 5.78 Å<br>Li in 16 *f*: $I \pm (x,y,z\ ;\ 1/2-x,\bar{y},1/2+z\ ;\ 3/4-y,\ 1/4+x,1/4+z\ ;$<br>$3/4+y,\ 3/4-x,\ 3/4+z)$, with $x = 0.100$, $y = 0.100$, $z = -0.055$<br>Ge ($8{\cdot}8{\cdot}10_3$) in 16 *f*, with $x = 0.106$, $y = 0.051$, $z = 0.394$ |
|---|---|

With these parameters, the Ge-Ge distances are 2.55 (2×) and 2.60 Å. The next shortest Ge-Ge distances are 4.10 Å (2×). A sketch of the Ge structure is shown in Fig. 7.83.

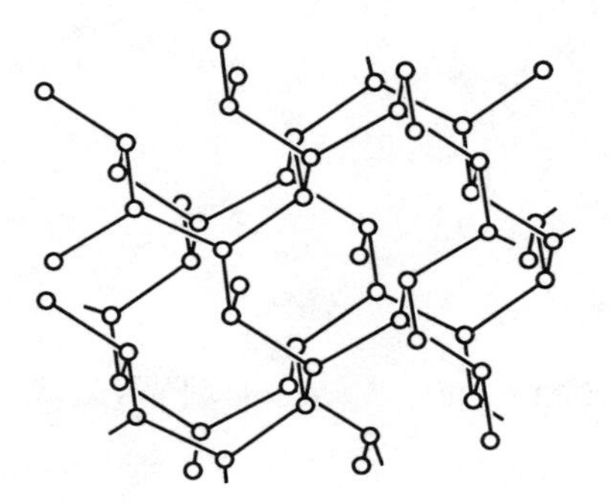

**Fig. 7.83**. The Ge net in LiGe.

Finally, a cubic net 6·9·9 which cannot be made with angles of 120°, so is given in its maximum volume form:

| $6.9^2$ | $Fd\bar{3}m$, $a$ = 6.521<br>vertices in 96 *g*: $F \pm (x,x,z\ ;\ x.1/4-x,1/4-z\ ;\ 1/4-x,x,1/4-z\ ;\ 1/4-x,1/4-x,z)\kappa$,<br>$x = 0.0708$, $z = 0.8875$ |
|---|---|

### *7.11.2 Model building*

Most three-dimensional nets are best appreciated by building models. Fortunately, "spaghetti" models of 3- and 4-connected nets are easily and cheaply built from readily available triangular and tetrahedral connectors and plastic tubing.[2] Connectors with 3 mm

[1] LiGe may be written as $Li^+Ge^-$; $Ge^-$ with five valence electrons is expected to have three 2-electron bonds pyramidally disposed and a non-bonding electron pair in an orbital at the apex of the pyramid. Compare $CaSi_2$ (§ 7.2)

[2] Available from many chemical supply houses. These are often sold as carbon, nitrogen, *etc.* atoms for building models of organic molecules, although the tetrahedral stars (also known as "caltrops" or "calthrops") probably more often end up as vertices in zeolite frameworks.

(1/8 inch) diameter spokes are suitable. As considerable bond strain occurs in small (3- or 4-) circuits, it is best to use tightly-fitting flexible plastic tubing and to make the edges just over twice the length of the spokes from the vertices.

### *7.11.3 Identifying nets*

It is sometimes quite difficult to identify nets in crystal structures, particularly when the structure is of low symmetry. A good way is to count numbers of (topological) neighbors using a computer (it is a good idea to count out to $n_k = 10$). *Usually* edges in nets correspond to shortest distances between vertices and it is very simple to count neighbors in this case. We have given the signature of a number of common nets in this chapter. It is also fairly simple also to get a computer to determine the Schläfli symbols of the vertices. Counting rings and numbers of neighbors by hand from a model can sometimes prove remarkably difficult (nets are known with more than 1000 shortest rings at an angle). If the net matches in both regards with a known net it is a *fairly* safe bet that they are the same.

The Exercises give some examples that are suitable for computer study; they are all done readily using EUTAX.

### *7.11.4 Diamond and* SiC *polytypes*

In § 7.3.1 we discussed polytypes of diamond. Polytypes of SiC are derived from these by replacing half the vertices by Si so that each Si is surrounded by four C and *vice versa*. In a formal sense the structure can be considered as **cp** Si with C in one half the tetrahedral sites (either all "up" or all "down"). The polytypes of SiC (a large number have been characterized) are usually named for the nature of the close packing, e.g. as *hcc* or 6*H* or, in Zhdanov notation 33 (see § 6.1.4).

There is an important distinction to be made between the description of SiC polytypes and the description of close packing. In the *hcc* sphere packing all the *c* spheres are related by symmetry (see Exercise 6.8.2), but in *hcc* SiC the *c* layers of Si are not so related (there are two distinct kinds) and to know which is which, it is necessary to know which of the two sets of tetrahedral sites is occupied by C. Usually the sequence of layers along the **c** axis is written on a line; here we use the convention that the direction from left to right is along the direction of a Si to C vector of a SiC bond parallel to **c**. Thus with Greek letters for carbon positions 6*H* SiC is coded (see § 6.1.4):

$$A\cdot\cdot\alpha B\cdot\cdot\beta A\cdot\cdot\alpha C\cdot\cdot\gamma B\cdot\cdot\beta C\ldots$$

In this sequence $A\cdot\cdot\alpha$, $B\cdot\cdot\beta$, $C\cdot\cdot\gamma$ correspond to Si-C bonds along **c**. The layer *B* is *h* and the layers *A* and *C* are *c*.

The possible symmetries for polytypes of diamond are the centro-symmetric groups: $Fd\bar{3}m$ (only for cubic diamond itself), $P6_3/mmc$, $R\bar{3}m$, and $P\bar{3}m1$. In each case the center of symmetry is in the midpoint of a C-C bond. In the polytypes of SiC the center of symmetry is destroyed and we have the possible symmetries: $F\bar{4}3m$ (only for the

**sphalerite** form, polytype symbol 3*C*), $P6_3mc$ (polytype symbol *NH*), *R*3*m* (polytype symbol *NR*), and *P*3*m*1 (polytype symbol *NT*). In giving coordinates for atoms in SiC polytypes it is convenient to take an origin midway along a Si-C bond so that for every Si at $x,y,z$ there is a C at $\bar{x},\bar{y},\bar{z}$. Thus we need only give explicitly coordinates for Si. It might be noted that in all polytypes except 3*C*, the symmetry at atom sites is 3*m*.

**Table 7.5**. Coordinates for idealized polytypes of SiC (see text)

| type | vertex | pos | $z_{Si}$ | type | vertex | pos | $z_{Si}$ |
|---|---|---|---|---|---|---|---|
| 3*C* | **c** | | | | | | |
| 2*H* | **h** | *b* | 1/16 | $10H_2$ | **c**chhh | *b* | 33/80 |
| 4*H* | **c**h | *a* | 13/32 | <311> | **c**hhhc | *b* | 1/80 |
| <2> | **h**c | *b* | 5/32 | | **h**hhcc | *a* | 9/80 |
| 6*H* | **c**ch | *b* | 17/48 | | **h**hcch | *b* | 17/80 |
| <3> | **c**hc | *b* | 1/48 | | **h**cchh | *a* | 5/16 |
| | **h**cc | *a* | 3/16 | $10H_3$ | **c**hchh | *b* | 49/80 |
| 9*R* | **c**hh | *a* | 69/72 | <221> | **h**chhc | *a* | 17/80 |
| <21> | **h**ch | *a* | 13/72 | | **c**hhch | *b* | 13/16 |
| | **h**hc | *a* | 53/72 | | **h**hchc | *b* | 33/80 |
| $8H_1$ | **c**cch | *b* | 21/64 | | **h**chch | *b* | 1/80 |
| <4> | **c**chc | *a* | 29/64 | $15R_1$ | **c**chch | *a* | 73/120 |
| | **c**hcc | *b* | 5/64 | <32> | **c**hchc | *a* | 41/120 |
| | **h**ccc | *b* | 45/64 | | **h**chcc | *a* | 3/40 |
| $8H_2$ | **c**hhh | *a* | 29/64 | | **c**hcch | *a* | 19/40 |
| <211> | **h**hhc | *b* | 5/64 | | **h**cchc | *a* | 7/8 |
| | **h**hch | *a* | 13/64 | $15R_2$ | **c**hhhh | *a* | 39/40 |
| | **h**chh | *b* | 21/64 | <2111> | **h**hhhc | *a* | 17/24 |
| 12*R* | **c**chh | *a* | 73/96 | | **h**hhch | *a* | 13/120 |
| <31> | **c**hhc | *a* | 17/96 | | **h**hchh | *a* | 101/120 |
| | **h**hcc | *a* | 19/32 | | **h**chhh | *a* | 29/120 |
| | **h**cch | *a* | 11/32 | 21*R* | **c**cchcch | *a* | 15/56 |
| 5*T* | **c**cchh | *c* | 9/40 | <43> | **c**chcchc | *a* | 55/56 |
| <41> | **c**chhc | *a* | 17/40 | | **c**hcchcc | *a* | 39/56 |
| | **c**hhcc | *b* | 5/8 | | **h**cchccc | *a* | 23/56 |
| | **h**hccc | *c* | 33/40 | | **c**chccch | *a* | 19/24 |
| | **h**ccch | *b* | 1/40 | | **c**hccchc | *a* | 29/168 |
| $10H_1$ | **c**ccch | *a* | 5/16 | | **h**ccchcc | *a* | 31/56 |
| <5> | **c**cchc | *b* | 33/80 | | | | |
| | **c**chcc | *b* | 1/80 | | | | |
| | **c**hccc | *a* | 9/80 | | | | |
| | **h**cccc | *b* | 57/80 | | | | |

In Table 7.5 we give a name (6*H* etc.) for all polytypes with five or fewer distinct kinds of Si (and hence C) atom and also for a 21*R* polytype. Underneath the name, the shortened Zhdanov symbol.[1] Next is given in bold the symbol (**h** or **c**) for the Si layer followed by

[1]If the number of symbols between angle brackets is odd the Zhdanov symbol is twice as long. Thus <221> refers to 221221.

the symbol for the following layers along the **c** axis in the direction specified by the sense of the Si to C bond. Next we give the Wyckoff symbol of the positions and finally the $z$ coordinate for Si in the ideal structure with regular tetrahedra. For this ideal structure $a = \sqrt{(8/3)}d$ where $d$ is the Si-C bond length. The coordinates of special positions ($z$ is given in the table) and the axial ratios are:

| | |
|---|---|
| *NH* | $P6_3mc$, $c/a = N\sqrt{(2/3)}$<br>2 $a$: (0,0,$z$ ; 0,0,1/2+$z$) ; 2 $b$: (1/3,2/3,$z$ ; 2/3,1/3,1/2+$z$) |
| *NR* | $R3m$, $c/a = N\sqrt{(2/3)}$<br>3 $a$ : $R$ + (0,0,$z$) |
| *NT* | $P3m1$, $c/a = N\sqrt{(2/3)}$<br>1 $a$: (0,0,$z$) ; 1 $b$: (1/3,2/3,$z$) ; 1 $c$: (2/3,1/3,$z$) |

Dozens of polytypes of SiC have been characterized. Most are intergrowths of *hc* (4*H*) and *hcc* (6*H*). Some polytypes have been assigned special names:

$$\beta = 3C,\ \ \mathrm{I} = 15R_1,\ \ \mathrm{II} = 6H,\ \ \mathrm{III} = 4H,\ \ \mathrm{IV} = 21R.$$

### *7.11.5 Two more nets derived from $6^3$: "C" and "D phases"*

Many compounds $MTT'X_4$ have structures with nets derived from $6^3$ nets (cf. § 7.3.3). These are often described as derived from tridymite but this is only correct if the net of the $T$ atoms is **lonsdaleite** (§ 7.3.1). Here we mention two binodal nets derived from $6^3$ (Fig. 7.84). The first is found as the (Be,P) net in the structure of **beryllonite**, $NaBePO_4$, and in related compounds, and is often called the "C phase" structure. The second occurs in compounds such as $KAlGeO_4$ [(Al,Ge) net] and is known as the "D phase structure."

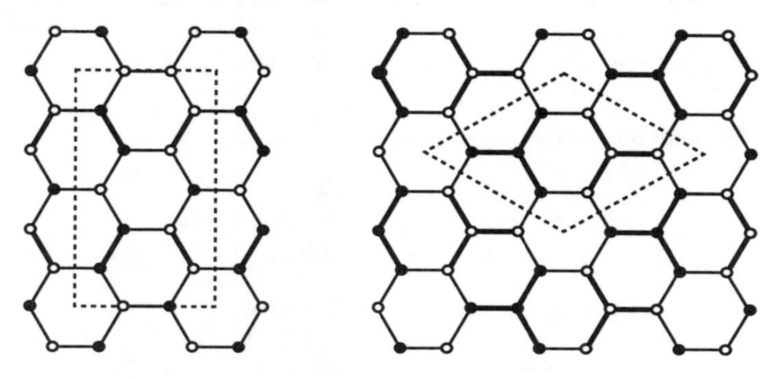

**Fig. 7.84**. Derivation of two 4-connected nets from $6^3$. Fourth bonds go up and down from open and filled circles respectively. Heavy lines are quadrangles seen in projection. Left: **beryllonite** (*C* phase), Right: *D* phase.

In cubanite, $CuFe_2S_3$, all the atoms are 4-connected, and the net of all the atoms is **beryllonite** (for crystallographic data for cubanite see Appendix 5).

### *7.11.6 Nets in* $CdP_2$ *and* $CdAs_2$

The nets in $CdP_2$ and $CdAs_2$ provide a treat. Cd is in tetrahedral coordination and each P (or As) atom is bonded to two other P (As) atoms so the net of all the atoms is 4-connected [compare the (Cu,P) net in $BaCu_2P_4$—see § 7.6]. In $CdAs_2$ (Fig 7.85, left), $4_3$ helices of As are cross-linked by Cd atoms. The net is remarkable for the large number of 11-rings (see the vertex symbols below). The net of the Cd atoms alone (with -As- acting as edges) is **diamond**, so the structure, considered as a framework of corner-connected $\{Cd\}As_4$ tetrahedra, is topologically the same as that of **cristobalite** $SiO_2$.

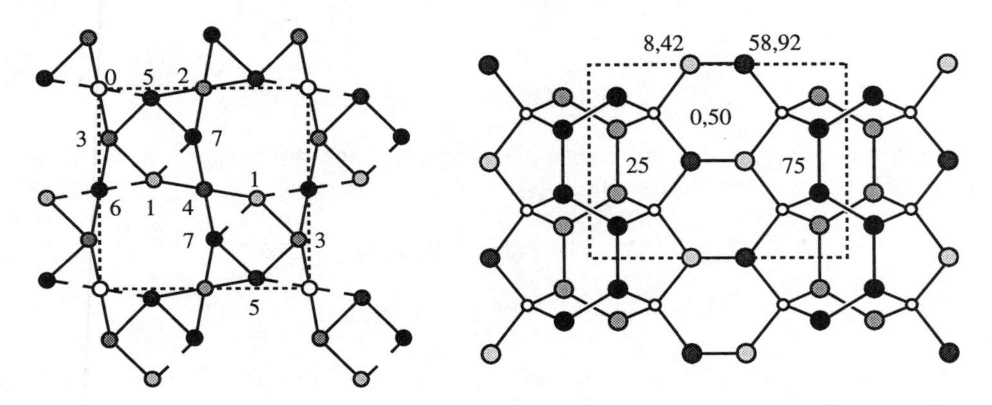


**Fig. 7.85**. Left: $\mathbf{CdAs_2}$ projected on (001). Numbers are elevations in multiples of $c/8$ with even numbers for **Cd**. Right: $\mathbf{CdP_2}$ projected on (110). **c** is horizontal on the page and numbers are elevations in units of $|(\mathbf{a}+\mathbf{b})|/100$. Smaller circles (at 0 and 50) are the $5{\cdot}5{\cdot}5{\cdot}5{\cdot}6{\cdot}6$ vertices.

$CdP_2$ occurs in two forms ($\alpha$ and $\beta$) that are topologically the same. The net of all the atoms has two kinds of vertex, but one kind is P and the second kind is alternating Cd and P, so there are three kinds of atom in the structure. The *net* is illustrated in Fig. 7.85 (right).

Crystallographic data for the nets (with unit edge) are:

| | |
|---|---|
| $\mathbf{CdAs_2}$ | $I4_122$, $a = 3.1706$, $c = 1.5804$, $r = 0.756$<br>$5{\cdot}5{\cdot}6{\cdot}6{\cdot}11_{26}{\cdot}11_{26}$ in 4 $a$: $I$ + (0,0,0 ; 0,1/2,1/4)<br>$5{\cdot}5{\cdot}5_2{\cdot}11_{20}{\cdot}6{\cdot}6$ in 8 $f$: $I$ + ($x$,1/4,1/8 ; etc.), $x = 0.4549$ |
| $\mathbf{CdP_2}$ | $P4_2/ncm$, $a = 2.1912$, $c = 3.9493$, $r = 0.633$<br>$5{\cdot}5{\cdot}5{\cdot}5{\cdot}6{\cdot}6$ in 4 $b$: ±(1/4,3/4,3/4 ; 1/4,3/4,1/4)<br>$5{\cdot}6{\cdot}5{\cdot}6{\cdot}5{\cdot}7_2$ in 8 $i$: ±($x,x,z$ ; etc.), $x = 0.0886$, $z = 0.3942$ |

$ZnP_2$ and $ZnAs_2$ are isostructural. The (Zn,P) or (Zn,As) nets have four kinds of vertex with 5-, 6- and 7-rings and are not discussed further. Notice that $BaCu_2P_4$ (p. 334) may be written as $Ba^{2+}[CuP_2^-]_2$ and that $CuP_2^-$ and $ZnP_2$ are isoelectronic.

### *7.11.7 More on the* **moganite**, **quartz** *and related nets*

On p. 322 we mentioned the net of the Si atoms in the **moganite** form of $SiO_2$. Here we describe a simple relationship between that net and the **quartz** net. In Fig 7.86 we show projections of the $P6_222$ ($^+Q$) and $P6_422$ ($^-Q$) enantiomorphs on (11$\bar{2}$0) of the conventional hexagonal cell (cf. § 7.3.11, p. 316). The two structures are related by reflection in a mirror plane at elevation 1/4. The bottom left shows the **moganite** net projected on (010) of *Ibam* as it appears in Fig. 7.37 (p. 322). Notice that the net consists of alternating bands of left- and right-hand **quartz** net. Interestingly real **moganite** $SiO_2$ can similarly decomposed into bands of left- and right-handed **quartz** $SiO_2$. In the amethyst form of quartz, Brazilian twins (intergrowths of left- and right-) are common and microscopic bands of moganite-structure material separate the two enantiomers [for details see B. G. Hyde & A. C. McLaren, *Aust. J. Chem.* (1996)].

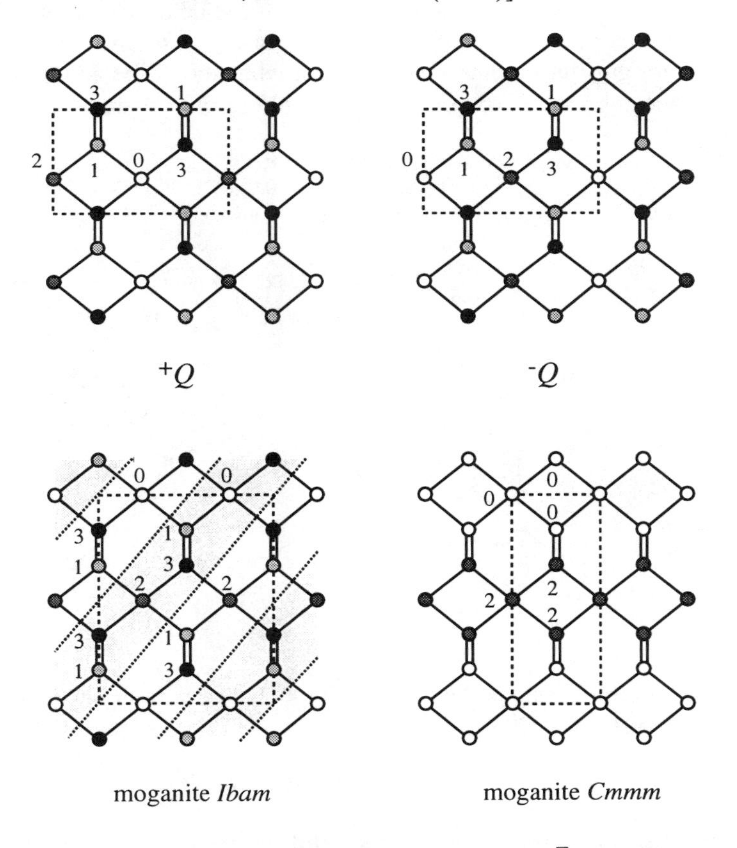


**Fig. 7.86**. Top: the two enantiomers of the **quartz** net projected on (11$\bar{2}$0) with the orthohexagonal cell outlined. Numbers are elevations in multiples of |**a**+**b**|/4. Bottom left: the **moganite** net projected on (010); numbers are now elevation in units of *b*/4. Note the alternating bands of $^+Q$ and $^-Q$ (shaded). Bottom right: an alternative conformation of **moganite** with 3 vertices in the repeat unit [projection on (010)].

It should be emphasized that the *Ibam* conformation of the moganite net is a close approximation to the Si structure reported for the mineral and for that reason was used to illustrate the structure.[1] However the figure shows that a simpler, more symmetrical (*Cmmm*) conformation of the net (with of course, the same *topology*) can be realized with only three vertices in the repeat unit. The nets of § 7.3.11 (**quartz, NbO** and **W2**) are the only others known with this property. Crystallographic data for this conformation are:

**moganite net** *Cmmm*, $a = 3.517$, $b = 1.786$, $c = 1.513$
$4\cdot4\cdot6_2\cdot6_2\cdot8_2\cdot8_2$ in 2 *a*: *C* + (0,0,0)
$4\cdot8_6\cdot6\cdot6\cdot6\cdot6$ in 4 *h*: $C \pm (x,0,1/2)$, $x = 0.186$

Fig. 7.86 should readily suggest ways of generating other nets in the **quartz-moganite** family. Two simple examples are shown in Fig. 7.87. That on the left (with 6- and 8-rings) is related to **quartz**—but notice that the vertices in 2 *a* (see below) are co-planar with their four neighbors (as in **NbO**). The net (which also has 4-rings) on the right of the figure is closely related to **moganite**. Parameters derived in what should be an obvious way from those for the orthohexagonal cell of **quartz** (with a shift of origin) are:

**Fig. 7.87 (left)** *Pmna*, $a = \sqrt{8}$, $b = \sqrt{3}$, $c = \sqrt{(8/3)}$
$6_2\cdot6_2\cdot6_2\cdot6_2\cdot8_4\cdot8_4$ in 2 *a*: (0,0,0 ; 1/2,0,1/2)
$6\cdot6\cdot6\cdot6\cdot6_2\cdot6_2$ in 4 *g*: $\pm(\pm1/4,y,1/4)$, $y = 1/3$

**Fig. 7.87 (right)** *Pccm*, $a = \sqrt{3}$, $b = \sqrt{(8/3)}$, $c = \sqrt{8}$
$4\cdot4\cdot6\cdot6\cdot8_2\cdot8_2$ in 2 *e*: $\pm(0,0,1/4)$
$4\cdot8_{10}\cdot6\cdot8_5\cdot6\cdot8_5$ in 4 *q*: $\pm(x,y,0 ; \bar{x},y,1/2)$, $x = 1/3$, $y = 1/4$

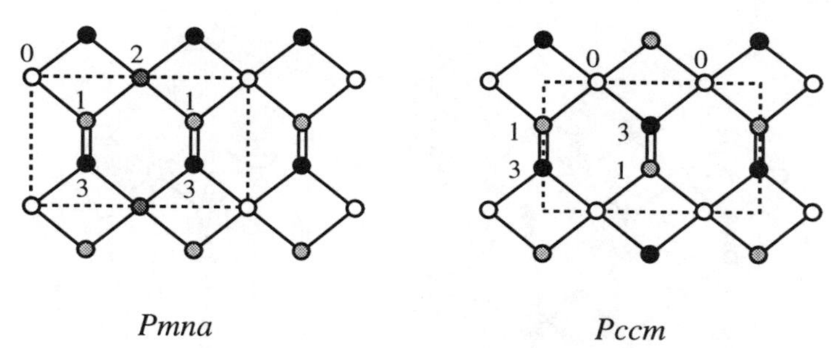


**Fig. 7.87**. Two nets related to **quartz**. Left: projected (001), elevations in multiples of $c/4$. Right: projected on (010), elevations in multiples of $b/4$.

## *7.11.8 Stereo picture of nets: Y*, clathrate hydrates I and II*

Many people find stereo pictures of nets helpful.[2] Here (Figs. 7.88-7.90) are such

[1] $BeH_2$ has the moganite structure, and the structure has symmetry *Ibam*. See Exercise 17.

[2] Stereo viewers (available in many bookstores) are helpful. Some people find the stereo perception easier when the picture is turned upside down.

pictures of three important nets that are difficult to illustrate.

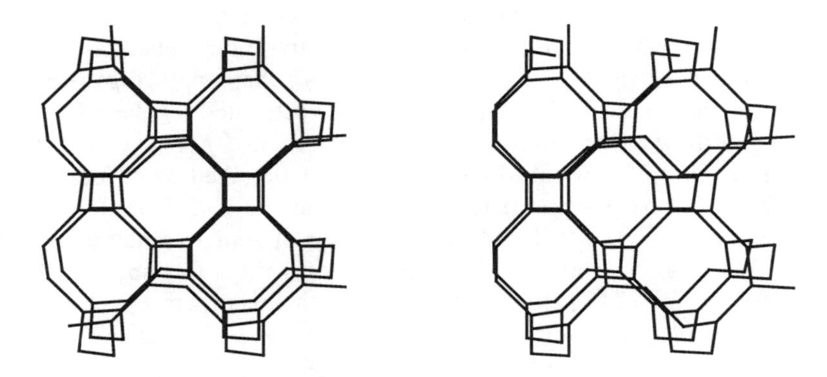

**Fig. 7.88**. A stereo view of the $Y^*$ net (compare Fig. 7.6).

**Fig. 7.89**. A stereo view of the Type I hydrate net (compare Fig. 7.51).

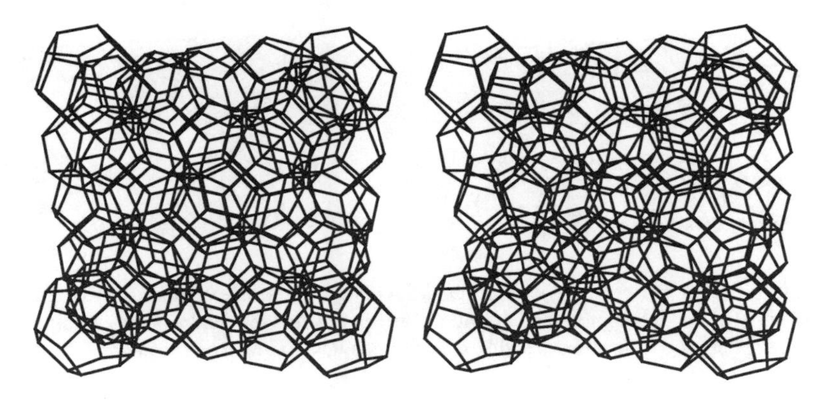

**Fig. 7.90**. A stereo view of the Type II hydrate net.

### *7.11.9 Anion positions and the possibility of open zeolite frameworks*

We have seen (§ 7.5.2, p. 330) that to make open (rare) nets with low density and large rings, configurations which also have a large concentration of small (3- or 4-) rings are generally needed. We have also seen (§ 5.2.1, p. 150) that small rings (<6-rings) put a constraint on the maximum *T-X-T* angle possible for catenated regular $\{T\}X_4$ tetrahedra. Thus for a ring of three tetrahedra (Fig. 5.19) the maximum *T-X-T* angle is 130.5°, and for clusters of 3-rings forming a tetrahedron (to give a supertetrahedron of four regular $\{T\}X_4$ tetrahedra), it should be obvious from Fig. 5.18 that the *T-X-T* angle is 109.47° [the tetrahedral angle, $\cos^{-1}(-1/3)$]. Similarly for a ring of four tetrahedra (Fig. 5.19) the maximum *T-X-T* angle is 160.5° and for clusters of 4-rings forming a cube (to give a $T_8X_{20}$ cluster of regular $\{T\}X_4$ tetrahedra, Fig. 5.20) the maximum *T-X-T* angle is 148.4°. To make a cluster of twelve tetrahedra with *T* atoms at the vertices of a pair of cubes sharing a face, the *T-X-T* angles are reduced to 109.47° (Fig. 7.91). The "cubes" are no longer cubes, but tetragonal prisms; the faces parallel to those shared are square, but the others have edges in the ratio 1:1.21.

A way of making nets of low density is to replace vertices in a net by tetrahedra of vertices as described in § 7.5.2 (this process can be repeated *ad nauseam* to produce nets of arbitrarily low density). Another way is to replace cubes in nets such as those of **Linde A** (Fig. 7.38) or $W^*8$ (Fig.7.42) by stacks of *N* cubes sharing faces; nets of arbitrarily low density can be made by increasing *N*. In both these cases however, some *T-X-T* angles must be as small as 109.47°. In silicas with framework structures the Si-O-Si angle is usually greater than about 140° (with similar values in related oxides), so these open structures cannot be formed.[1] In fact it appears that for alumino-silicates the **faujasite** structure is about the least dense that can be made. We give here coordinates for some tetrahedral framework structures based on simple low-density nets with cubic symmetry. The coordinates are for regular tetrahedra of unit *T-X* distance and are such as to maximize the minimum *T-X-T* angle. It may be seen that the tetrahedral structure based on $W^*8$ is not very likely to be formed for an alumino-silicate framework.

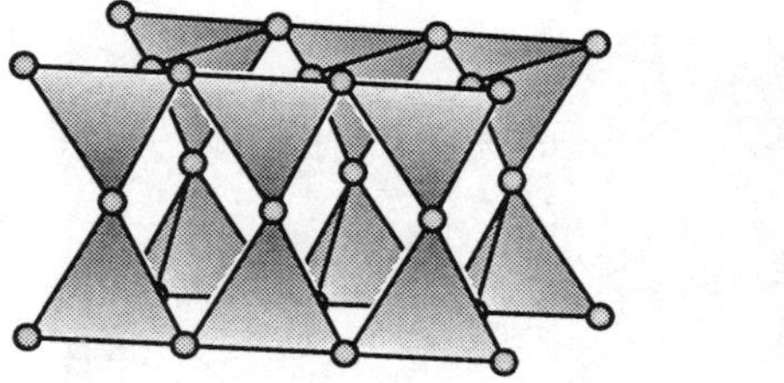

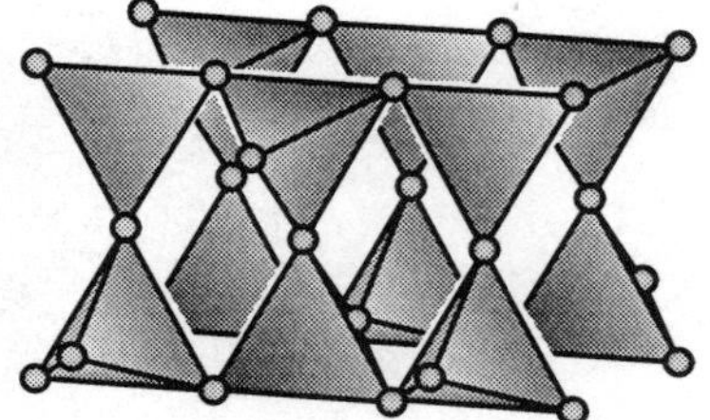

**Fig. 7.91.** Left: a cluster of twelve regular tetrahedra with centers at the vertices of two "cubes" (actually tetragonal prisms) sharing a common face. Right: a cluster of twelve tetrahedra corresponding to a fragment of a net containing "up-down" tetrahedra and based on rectangles sharing a common edge (see text).

[1]Minimum *T-X-T* angles are much smaller in sulfides than in oxides, so the former offer much greater promise for making open framework structures. See for example Exercise 16.

In the case of **sodalite** we give coordinates for the maximum *T-X-T* angle; decreasing the angle (as in the real mineral) will increase the density as described on p. 275. The transition from $\beta$-quartz to $\alpha$-quartz (§ 3.6) corresponds likewise to a decrease in density accomplished (for rigid tetrahedra) by decreasing the *T-X-T* angle.[1] Thus for a *given* topology, the density of tetrahedral framework structures can often be increased by decreasing the *T-X-T* angle; but to achieve frameworks of lower minimum density the minimum *T-X-T* angle has to be decreased (see e.g. the crystallographic data below).

**sodalite** — $Im\bar{3}m$, $a = 5.575$, density = 0.0692 *T* vertices per unit volume
*T* in 12 *d*: (1/4,0,1/2 ; etc.) ; *X* in 24 *h*: (0,*y*,*y* ; etc.), $y = 0.3536$, *T-X-T* = 160.6°

**rho** — $Im\bar{3}m$, $a = 9.2733$, density = 0.0602 *T* vertices per unit volume
*T* in 48 *i*: (1/4,*y*,1/2–*y* ; etc.), $y = 0.1036$
*X*(1) in 48 *j*: (0,*y*,*z* ; etc.), $y = 0.2242$, $z = 0.3809$, *T-X*(1)*-T* = 147.6°
*X*(1) in 48 *k*: (*x*,*x*,*z* ; etc.), $x = 0.1658$, $z = 0.3707$, *T-X*(1)*-T* = 147.6°

**Linde A** — $Pm\bar{3}m$, $a = 7.4339$, density = 0.0584 *T* vertices per unit volume
*T* in 24 *k*: (0,*y*,*z* ; etc.), $y = 0.1831$, $z = 0.3706$
*X*(1) in 24 *m*: (*x*,*x*,*z* ; etc.), $x = 0.1098$, $z = 0.3447$, *T-X*(1)*-T* = 148.4°
*X*(2) in 12 *h*: (*x*,0,1/2 ; etc.), $x = 0.2197$, *T-X*(2)*-T* = 148.4°
*X*(3) in 12 *i*: (0,*y*,*y* ; etc.), $y = 0.2929$, *T-X*(3)*-T* = 160.5°

**faujasite** — $Fd\bar{3}m$, $a = 15.1618$, density = 0.0551 *T* vertices per unit volume
*T* in 192 *i*: (*x*,*y*,*z* ; etc.), $x = 0.0361$ $y = 0.1240$, $z = 0.3045$
*X*(1) in 96 *h*: (0,*y*,$\bar{y}$ ; etc.), $y = 0.1059$, *T-X*(1)*-T* = 140.8°
*X*(2) in 96 *g*: (*x*,*x*,*z* ; etc.), $x = 0.0697$, $z = 0.3211$, *T-X*(2)*-T* = 140.8°
*X*(3) in 96 *g*, $x = 0.3284$, $z = 0.0374$, *T-X*(3)*-T* = 149.2°
*X*(4) in 96 *g*, $x = 0.2537$, $z = 0.1395$, *T-X*(4)*-T* = 152.7°

**w*8** — $Im\bar{3}m$, $a = 12.5567$, density = 0.0485 *T* vertices per unit volume
*T* in 96 *l*: (*x*,*y*,*z* ; etc.), $x = 0.0793$ $y = 0.3231$, $z = 0.4267$
*X*(1) in 48 *j*: (0,*y*,*z* ; etc.), $y = 0.1655$, $z = 0.3917$, *T-X*(1)*-T* = 133.8°
*X*(2) in 48 *i*: (1/4,*y*,1/2–*y* ; etc.), $y = 0.0983$, *T-X*(2)*-T* = 133.8°
*X*(3) in 48 *k*: (*x*,*x*,*z* ; etc.), $x = 0.1228$, $z = 0.3897$, *T-X*(3)*-T* = 133.8°
*X*(4) in 48 *j*, $y = 0.3309$, $z = 0.4281$, *T-X*(4)*-T* = 168.7°

An interesting way to obtain tetrahedral frameworks of low density has been described.[2] The nets are based on the "up-down" principle of coupling 3-connected two-dimensional nets. In § 7.3.8 (p. 311) examples are given in which "up-down" rods of vertices are derived from squares of the planar net. In **VPI-5** (§ 7.8.4) the net is derived from a two-dimensional net with pairs of squares sharing an edge (fusion of two up-down rods), and it should be obvious that **VPI-5** (Fig. 7.62) is simply derived from **$AlPO_4$-5** (Fig. 7.23) by replacing a square by two "squares" (actually now rectangles) sharing an edge. In Fig.

[1]Another tetrahedral framework than can have variable density is **cristobalite** (§ 6.3.9, p. 240). Note that in cristobalite (which has all 6-rings) the *T-X-T* angle can be as much as 180°. In the quartz structure, which has 6- and 8-rings, with regular tetrahedra the maximum angle is 155.6° (in the $\beta$- structure).

[2]J. V. Smith & W. J. Dytrych, *Nature* **309**, 607 (1984).

7.91, we show a fragment of the configuration of tetrahedra that results. In Fig. 7.92 we show a net derived analogously from **$TlZn_2Sb_2$** (Fig. 7.23). Clearly the strip of two rectangles could be replaced by strips of arbitrary length, so nets with large rings of any even size could be made, and with density approaching zero for very large strips. In nets of this type, the vertices on the octagon have Schläfli symbol $4{\cdot}6_2{\cdot}6{\cdot}6_3{\cdot}6{\cdot}6_3$; all the rest (on edges between two rectangles) have symbol $4{\cdot}6_3{\cdot}4{\cdot}6_3{\cdot}6{\cdot}6_4$. It may be seen that there at most two 4-rings meeting at a vertex, so that low (in the limit zero) density is achieved in this case with a relatively small number of small rings. For frameworks of regular tetrahedra, some of the *T-X-T* angles cannot exceed $\cos^{-1}(-5/9) = 123.75°$.

In the real $AlPO_4$ framework of VPI-5 [L. B. McCusker *et al.*, *Zeolites,* **11**, 308 (1991)] the tetrahedra are not very regular; in particular one $\{Al\}O_4$ "tetrahedron" is better thought of as part of an octahedron (with two water molecules completing the coordination sphere). The Al-O-P angles range from 137° to 162° (Al...P = 3.09 to 3.27 Å); compare berlinite $AlPO_4$ in which the values are Al-O-P = 142.5° and Al...P = 3.08 Å.

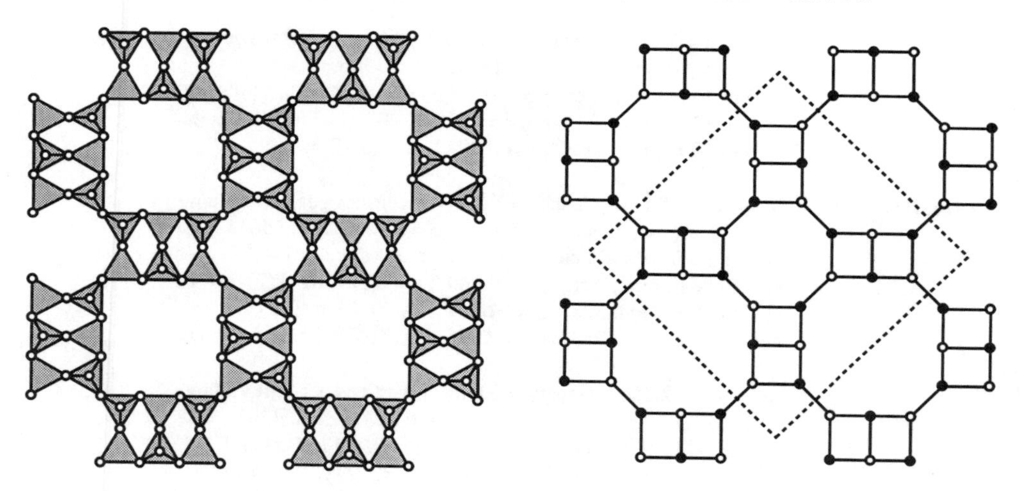

**Fig. 7.92.** An "up-down" net derived by replacing the squares of **$TlZn_2Sb_2$** (Fig. 7.23) by pairs of rectangles sharing an edge. Left: showing the $\{T\}X_4$ tetrahedra in projection down the 4-fold axis. Right: the *T* network in the same projection. Open (filled) circles are vertices with links to layers above (below).

In $TX_2$ frameworks of zeolite structures, the *T* atoms can be considered to lie on a surface with *X* atoms on either side. In a polyhedral cavity of *n T* atoms, there are $3n/2$ *X* atoms (one associated with each *T...T* "edge") generally slightly closer to the center of the cavity. Thus, for the framework of **Linde A** with the coordinates given above, the following atoms are at the given distances from the center of the cavity:

| cavity | *T* atoms | *X* atoms |
|---|---|---|
| 4.6.8 | 4.50 (48×) | 4.26 (48×), 4.31 (24×) |
| $4.6^2$ | 3.07 (24×) | 2.81 (24×), 3.08 (12×) |
| $4^3$ (cube) | 1.67 (8×) | 1.63 (8×), 1.63 (4×) |

Similarly the *TXTX*... rings on the surface of cages are usually puckered and, especially for larger rings, with *X* inside the ring. Fig. 7.93 shows the 8-ring in **Linde A**.

**Fig. 7.93.** The 8-ring in **Linde A** $TX_2$ with coordinates given on p. 373. Filled circles are *T*.

### *7.11.10 References*

The pioneering work on nets and infinite polyhedra by A. F. Wells is collected in his *Three-dimensional Nets and Polyhedra* [Wiley, New York (1977)]. A catalog of 4-connected nets (over 500) has been compiled by Smith and Bennett [see J. V. Smith, *Chem. Rev.* **88**, 149 (1988)]. References to zeolite nets are given in § 7.8 (p. 337) and some references to the topology of nets are given in Appendix 3 (§ A3.9). For a review of structures of clathrate hydrates and inclusion compounds see Chapter 7 of *Crystallography in Modern Chemistry* by T. C. W. Mak & G.-D. Zhou [Wiley, New York (1992)].

Those who are skeptical about the relevance of geometry to chemistry should know that some of the beautiful zeolite and other structures based on polyhedron packings that appear in this chapter were first described (and illustrated as models) by A. Andrieni in a classical paper *Sulle reti di poliedri regolari e semiregolari e sulle corrispondenti reti correlative* [*Mem. Soc. Ital. delle Scienze* Ser 3, **14**, 75 (1907)]. The Type I and Type II hydrate structures were predicted (and elegantly illustrated) by W. F. Clausen [*J. Chem. Phys.* **19**, 259 and 662 (1951)]. Many structures now known were predicted in advance, especially by A. F. Wells and by J. V. Smith and collaborators (*loc. cit. supra*). Indeed some zeolite structures could only be solved with the knowledge of possible structures, their symmetries and approximate coordinates. Some nets that were described as "unknown" in earlier drafts of this chapter, were subsequently found in recently determined crystal structures.

## 7.12 Exercises[1]

1. A body-centered array of cubes connected corner to corner by additional edges will produce a simple 4-connected net somewhat analogous to the connection of octahedra in the 5-connected $CaB_6$ net.[2]

[1]Most of the Exercises in this chapter will be tedious to do by hand, but are readily done using a computer program such as EUTAX (see the Note to the Reader).

[2]This structure ("*supercubane*") has been proposed as a possible form of carbon. See R. L. Johnston & R. Hoffmann, *J. Amer. Chem. Soc.* **111**, 810 (1989). Compare with the **octadecasil** net (§ 7.8.6).

$Im\bar{3}m$, $a = 2 + 2/\sqrt{3} = 3.1457$
$4{\cdot}8_2{\cdot}4{\cdot}8_2{\cdot}4{\cdot}8_2$ in 16 *f*: $I \pm (\pm x,x,x)\kappa$, $x = \sqrt{3}/(4 + \sqrt{48}) = 0.1585$

2. The 4-connected net of the B atoms in $CaB_2O_4$ may described as follows for unit edge (the same net appears in the structures of $SrB_2O_4$, $BaAl_2S_4$ and $BaGa_2S_4$):

$Pa\bar{3}$, $a = 3.5334$, $r = 0.544$.
$3{\cdot}6{\cdot}7{\cdot}7{\cdot}7_2{\cdot}7_2$ in 24 *d*: $\pm(x,y,z\ ;\ x,1/2-y,1/2-z\ ;\ 1/2+x,\ y,1/2-z\ ;$
$1/2-x,1/2+y,z)\kappa$, $x = 0.1046$, $y = 0.1709$, $z = 0.3295$.

3. $\gamma$-$LiAlO_2$ is tetragonal:

$\gamma$-$LiAlO_2$ $\quad$ $P4_12_12$, $a = 5.169$, $c = 6.268$ Å
Li and Al in 4 *a* : $(x,x,0\ ;\ \bar{x},\bar{x},1/2\ ;\ \pm(1/2-x,1/2+x,1/4))$
For Li, $x = 0.688$ ; for Al, $x = 0.324$
O in 8 *b*: $(x,y,z\ ;\ \bar{x},\bar{y},1/2+z\ ;\ 1/2-y,1/2+x,1/4+z\ ;\ 1/2+y,1/2-x,3/4+z\ ;\ y,x,\bar{z}\ ;$
$\bar{y},\bar{x},1/2-z\ ;\ 1/2-x,1/2+y,1/4-z\ ;\ 1/2+x,1/2-y,\ 3/4-z)$,
$x = 0.210$, $y = 0.164$, $z = 0.228$.

Li and Al are each on **diamond** nets and the atoms taken all together are on a $\mathbf{CrB_4}$ net (this structure is derived from that of $\beta$-BeO by the substitution $2Be \rightarrow Li + Al$).

4. $\beta$-$LiGaO_2$ is orthorhombic:

$\beta$-$LiGaO_2$ $\quad$ $Pna2_1$, $a = 5.402$, $b = 6.372$, $c = 5.007$ Å
all atoms in 4 *a*: $(x,y,z\ ;\ \bar{x},\bar{y},1/2+z\ ;\ 1/2+x,1/2-y,z\ ;\ 1/2-x,1/2+y,1/2+z)$
Li: $x = 0.421$, $y = 0.127$, $z = 0.494$ ; Ga: $x = 0.082$, $y = 0.126$, $z = 0.0$
O(1): $x = 0.070$, $y = 0.112$, $z = 0.371$ ; O(2): $x = 0.407$, $y = 0.139$, $z = 0.893$

Li and Ga are each on **diamond** nets and the atoms taken all together are on a **lonsdaleite** net (this structure is derived from that of ZnO by the substitution $2Zn \rightarrow Li + Ga$). The O atoms are **hcp**, as are the metal atoms (combined).

5. A 4-connected net considered by Heesch & Laves is derived from a space filling by truncated tetrahedra, truncated cubes and truncated cuboctahedra. We call this net $\mathbf{HL4_2}$.

$\mathbf{HL4_2}$ $\quad$ $Fm\bar{3}m$, $a = 2 + \sqrt{18} = 6.2426$, $r = 0.395$
$3{\cdot}4{\cdot}6{\cdot}8{\cdot}6{\cdot}8$ in 96 *k*: $F + (\pm x,\pm x,\pm z\,)\kappa$, $z = 1/(4 + \sqrt{72}) = 0.0801$, $x = (1 + \sqrt{2})z = 0.1934$

A "spaghetti" model is easily made if it is realized that each triangular face of the truncated cubes is shared with a similar face of a truncated tetrahedron and *vice versa*. The truncated cuboctahedra and truncated cubes likewise share octagonal faces.

6. Another uninodal 4-connected net occurs as the (Si,Al) framework of the natural zeolite analcime, $NaAlSi_2O_6{\cdot}H_2O$. A formal description of the net with unit edge length is:

**analcime** $\quad$ $Ia\bar{3}d$, $a = \sqrt{(96/5)} = 4.3818$, $r = 0.571$ ; $4{\cdot}4{\cdot}6{\cdot}6{\cdot}8_4{\cdot}8_4$ in 48 *g*: $(1/8,x,1/4-x$ ; etc.), $x = 1/3$

Try to make or find a good picture of this net (difficult!).

In the bixbyite structure of $Sc_2O_3$, the O atom positions are in positions 48 *e* of $Ia\overline{3}$: $x,y,z$ etc., with $x = 0.390$, $y = 0.155$, $z = 0.380$. Each O atom has four O nearest neighbors. Show that the net of the O atoms has the same topology as that of analcime.

7. The net of the recently described mineral montesommaite [approximate composition $(K,Na)_9Al_9Si_{23}O_{64}\cdot 10H_2O$] is very simple [hint: project on (100)]:

| | |
|---|---|
| **montesommaite** | $I4_1/amd$, $a = 2.258$, $c = 5.795$, $r = 0.542$ |
| | $4\cdot 5_2\cdot 5\cdot 8_2\cdot 5\cdot 8_2$ in 16 *h*: $I \pm (0,y,z$ ; etc.), $y = 0.028$, $z = 0.0856$ |

8. The diamond net is ubiquitous (see Exercises 3 and 4). Here are two more examples of its occurrence as the metal arrays in oxides. (Note that we are concerned with the *topology* of the structure, in this instance the topology of the net defined by edges joining the first four nearest neighbors.)

| | |
|---|---|
| **corundum** ($Al_2O_3$) | $R\overline{3}c$, $a = 4.759$, $c = 12.991$ Å |
| | Al in 12 *c*: $R \pm (0,0,z$ ; $0,0,1/2+z)$, $z = 0.3523$ ; O in 18 *e*: 0.3064,0,1/4 |
| **anatase** ($TiO_2$) | $I4_1/amd$, $a = 3.785$, $c = 9.514$ Å |
| | Ti in 4 *a*: $I \pm (0,3/4,1/8)$ ; O in 8 *e*: 0,1/4,0.0816 |

9. Another net that occurs in many different contexts is **$SrAl_2$** (the Al net in $SrAl_2$). Two examples of compounds iso-structural with $SrAl_2$ were given in Exercise 3.8.11. Here are three other examples of its occurrence as the net of (a) Mg and Si in SrMgSi, (b) Both atoms in $\alpha$-Np and (c) Al and Si in synthetic zeolite Li-A ($LiAlSiO_4\cdot 2H_2O$):

| | |
|---|---|
| SrMgSi | *Pnma*, $a = 7.78$, $b = 4.56$, $c = 8.49$ Å |
| | Sr in 4 *c*: $\pm(x,1/4,z$ ; $1/2+x,1/4,1/2-z)$, $x = 0.515$, $z = 0.683$ |
| | Mg in 4 *c*, $x = 0.640$, $z = 0.057$ ; Si in 4 *c*, $x = 0.276$, $z = 0.110$ |
| $\alpha$-Np | *Pnma*, $a = 6.661$, $b = 4.271$, $c = 4.888$ Å |
| | Np(1) in 4 *c*: $\pm(x,1/4,z$ ; $1/2+x,1/4,1/2-z)$, $x = 0.036$, $z = 0.208$ |
| | Np(2) in 4 *c*, $x = 0.319$, $z = 0.842$ |
| Li-A | $Pna2_1$, $a = 10.31$, $b = 8.18$, $c = 5.00$ Å |
| | Al in 4 *a*: $(x,y,z$ ; $\overline{x},\overline{y},1/2+z$ ; $1/2+x,1/2-y,z$ ; $1/2-x,1/2+y,1/2+z)$, |
| | $x = 0.136$, $y = 0.072$, $z = 0.25$ ; Si in 4 *a*, $x = 0.358$, $y = 0.378$, $z = 0.252$ |

10. The compound $NaGaSn_5$ can be considered in a formal sense as $Na^+(Ga^-Sn_5)$ with the $(Ga^-Sn_5)$ part having four valence electrons per atom, so it is not surprising to find these atoms forming a 4-connected net. The (Ga,Sn) structure was reported as:

| | |
|---|---|
| $NaGaSn_5$ | $P3_112$, $a = 6.328$, $c = 6.170$ Å |
| | M1 in 3 *a*: $(x,\overline{x},1/3$ ; $x,2x,2/3$ ; $2\overline{x},\overline{x},0)$, $x = 0.431$ |
| | M2 in 3 *b*: $(x,\overline{x},5/6$ ; $x,2x,1/6$ ; $2\overline{x},\overline{x},1/2)$, $x = 0.903$ |

with M1 and M2 being a disordered combination of Ga and Sn. Show that the net of the atoms is a uninodal net $5{\cdot}5{\cdot}5{\cdot}5{\cdot}5{\cdot}8_4$ which, in its most symmetrical form has symmetry $P6_122$. The net is illustrated in Fig. 7.94 below (note that $a$ and $c$ are approximately equal).

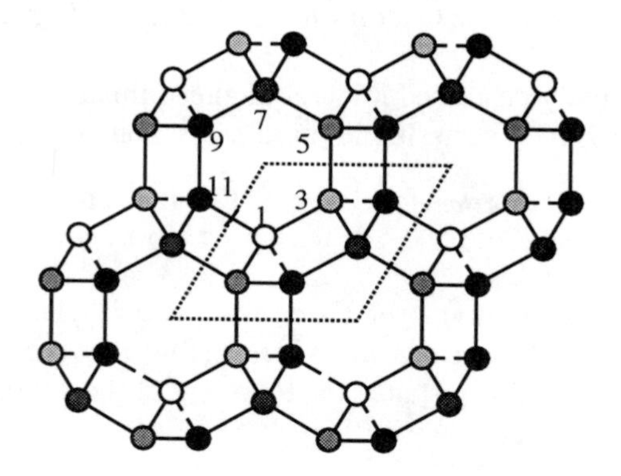


**Fig 7.94.** A net $5{\cdot}5{\cdot}5{\cdot}5{\cdot}5{\cdot}8_4$ projected down the **c** axis of $P6_122$. Elevations are in multiples of $c/12$.

11. The P atom positions in one form of $P_2O_5$ have been given as:

$Fdd2$, $a$ = 16.3, $b$ = 8.12, $c$ = 5.25 Å
P in 16 $b$: $F$ + ($x,y,z$ ; $\bar{x},\bar{y},z$ ; 1/4–$x$,1/4+$y$,1/4+$z$ ; 1/4+$x$,1/4–$y$;1/4+$z$),
$x$ = 0.075, $y$ = 0.083, $z$ = 0.153

Verify that the five shortest P-P distances are 2.79, 2.92 (2×) and 4.34 (2×) Å and that the three shortest P-P distances define a net with the topology of the Si net in $ThSi_2$.

12. A zeolite we haven't discussed, but which is nice to draw or to explore using computer graphics, is known as **MAPSO-46** (symbol AFS). Here are data for the maximum volume form of the net with unit edge:

**MAPSO-46** $P6_3/mcm$, $a$ = 4.363, $c$ = 8.034, $r$ = 0.366
$4{\cdot}8{\cdot}4{\cdot}8{\cdot}4{\cdot}8$ in 8 $h$: ±(1/3,2/3,±$z$ ; 1/3,2/3,1/2±$z$), $z$ = 0.438
$4{\cdot}4{\cdot}4{\cdot}8_2{\cdot}6_2{\cdot}8$ in 24 $l$: ±($x,y,z$ ; etc.), $x$ = 0.364, $y$ = 0.496, $z$ = 0.366
$4{\cdot}4{\cdot}4{\cdot}6{\cdot}6{\cdot}12$ in 24 $l$, $x$ = 0.570, $y$ = 0.703, $z$ = 0.312

The net is closely related to that of **CoAPO-50** (Fig. 7.43, p. 328), the main difference being that the cubes in that net are replaced by polyhedra with nine faces.

13. $BaCu_2S_2$ is polymorphic. One form has the $ThCr_2Si_2$ structure (§ 6.4.2) with Ba between tetragonal layers of {Cu}$S_4$ tetrahedra. A second form forms a three-dimensional 4-connected Cu,S net derived from $4.8^2$ by double zig-zag connections in the manner similar to that shown in Fig. 7.17 (p. 306). The vertices are all $4{\cdot}6{\cdot}4{\cdot}6{\cdot}6{\cdot}8_2$ but there are two topologically-different kinds. Crystallographic data for the compound are:

$BaCu_2S_2$ *Pnma*, $a = 9.308$, $b = 4.061$, $c = 10.408$ Å. Atoms in 4 *c*: ±(*x*,1/4,*z* ; 1/2+*x*,1/4,1/2–*z*)
Ba: $x = 0.240$, $z = 0.822$ ; Cu(1): $x = 0.056$, $z = 0.111$ ; Cu(2): $x = 0.083$, $z = 0.545$
S(1): $x = 0.483$, $z = 0.169$ ; S(2): $x = 0.339$, $z = 0.559$

Plot the Cu and S positions in projection down **b** to identify the net.
The net symmetry is *Cmcm*, but Cu,S ordering lowers the symmetry to *Pmcn* (*Pnma*).

14. A 4-connected net with "up-down" rods (§ 7.3.8) occurs as the Ga net in $Mg_2Ga_5$.

$Mg_2Ga_5$ *I4/mmm*, $a = 8.627$, $c = 7.111$ Å
Mg in 8 *h*: *I* ± (*x*,±*x*,0), $x = 0.300$; Ga(1) in 4 *e*: *I* ± (0,0,*z*), $z = 0.288$
Ga(2) in 16 *n*: *I* ± (0,*y*,±*z* ; *y*,0,±*z*), $y = 0.298$, $z = 0.181$

Ga(2) atoms form "up-down" rods (shaded in Fig. 7.95) and Ga(1) atoms link the rods.

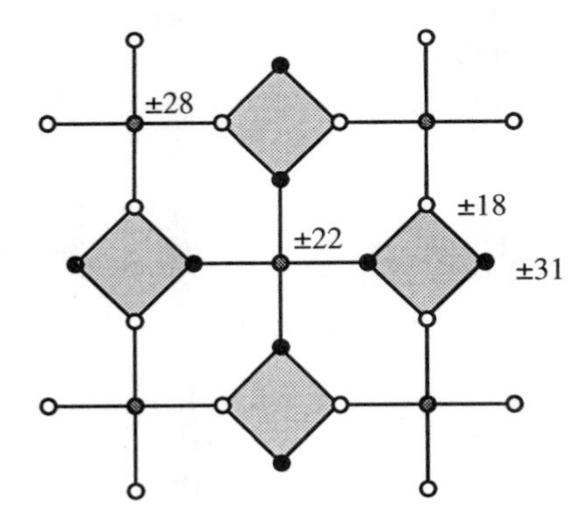


**Fig. 7.95**. The Ga arrangement in $Mg_2Ga_5$ projected on (001). Numbers are elevations in units of $c/100$.

The silicate, narsarsukite, was mentioned in § 7.3.8. In that compound Ti atoms link up-down rods of $\{Si\}O_4$ tetrahedra to produce the same 4-connected net with stoichiometry $(TiSi_4)O_{10}$. An additional O atom links the Ti atoms in the **c** direction producing distorted $\{Ti\}O_6$ octahedra and the composition is often written $Na_2TiOSi_4O_{10}$. Drawing the structure of narsasukite and identifying the net is a nice exercise [for data see D. R. Peacor & M. J. Buerger, *American Mineralogist*, **47**, 539 (1962)]. If you do this, you may notice that the metal atom arrangements are nearly identical (except for a change of scale) in $Mg_2Ga(1)Ga(2)_4$ and $Na_2TiSi_4O_{11}$.

15. Open (zeolite-like) (4,6)-connected networks are often produced by linking clusters of tetrahedra by octahedra and *vice versa* and indeed some authors include such structures under the heading of zeolites. The narsasukite framework (Exercise 14) is a (4,6)-connected net if *all* -O- links are counted. Here are some other examples for exploration:

(a) Part of the pharmacosiderite structure was illustrated as clusters of octahedra linked by tetrahedra in § 5.2.2 (see Fig. 5.23, p. 154). The coordinates of the metal atoms are: space group $P\bar{4}3m$: Fe (octahedral) in 4 *e*: *x*,*x*,*x* ; $(x,\bar{x},\bar{x})\kappa$ ; As (tetrahedral) in 3 *d*:

(1/2,0,0)κ. Fe has 3 Fe and 3 As neighbors, and As has 4 Fe neighbors. If all edges are equal, $x = 5/\sqrt{2}(\sqrt{6} - 1) = 0.145$ (quite close to the actual value). For the full structure see M. J. Buerger *et al.*, *Zeits. Kristallogr.* **125**, 93 (1967).

(b) An open structure consisting of rings of three tetrahedra (corner sharing) joined by further corner sharing with octahedra occurs in catapleite, $Na_2ZrSi_3O_9.2H_2O$ in which rings of three corner-sharing $\{Si\}O_4$ tetrahedra are joined by $\{Zr\}O_6$ tetrahedra. Data for the non-water atoms are:

catapleite

$P6_3/mmc$, $a = 7.40$, $c = 10.07$ Å, $V = 477$ Å$^3$.
Zr in 2 *a*: (0,0,0 ; 0,0,1/2) ; Na in 4 *f*: ± (1/3,2/3,*z* ; 1/3,2/3,1/2+*z*), $z = 0.08$
Si in 6 *h*: ± ($x,2x,1/4$ ; $x,\bar{x},1/4$ ; $2\bar{x},\bar{x},1/4$), $x = 0.20$ ; O(1) in 6 *h*, $x = 0.47$
O(2) in 12 *k*: ± ($x,2x,z$ ; $x,\bar{x},z$ ; $2\bar{x},\bar{x},z$ ; $x,2x,z+1/2$ ; $x,\bar{x},z+1/2$ ; $2\bar{x},\bar{x},z+1/2$),
$x = 0.135$, $z = 0.125$

In the Zr,Si net, Zr has 6 Si neighbors and Si has 2 Si and 2 Zr neighbors. Not unexpectedly, the Zr-Si distances are rather different (larger) than the Si-Si distances.

(c) A related, but different, structure is found in benitoite, $BaTiSi_3O_9$

benitoite

$P\bar{6}c2$, $a = 6.61$, $c = 9.72$ Å, $V = 368$ Å$^3$
Ba in 2 *f*: (2/3,1/3,0 ; 2/3,1/3,1/2) ; Ti in 2 *c*: (1/3,2/3,0 ; 1/3,2/3,1/2)
Si in 6 *k*: ($x,y,1/4$ ; $\bar{y},x-y,1/4$ ; $y-x,\bar{x},1/4$ ; $\bar{y},\bar{x},3/4$ ; $y-x,y,3/4$ ; $x,x-y,3/4$),
$x = 0.0711$, $y = 0.2894$ ; O(1) in 6 *k*, $x = 0.2535$, $y = 0.1972$
O(2) in 12 *l*: (as 6 *k* but 1/4 replaced by $z$ and $1/2-z$ and 3/4 replaced by $\bar{z}$, $1/2+z$)
$x = 0.0880$, $y = 0.4302$, $z = 0.1127$

Compare the Zr,Si net in catapleite with the Ti,Si net in benitoite. Note that the latter is much denser (compare the volumes of the unit cells which contain 6 Si atoms in each case).

16. A fascinating open structure based on a 4-connected net is that of the zeolite-like compound: $[N(CH_3)_4]_2MnGe_4S_{10}$ with a framework based on vertex sharing $\{Mn\}S_4$ and $\{Ge\}S_4$ tetrahedra. $Ge_4S_{10}$ "supertetrahedron" units (Fig. 5.18, § 5.2.1) and $MnS_4$ units are linked as in **diamond** (or better, as in **sphalerite**) so the net of the metal atoms is intermediate between **diamond** and ***D*4**. Here are data for the framework (explore!):

$I\bar{4}$, $a = 9.513$, $c = 14.281$ Å.
Mn in 2 *d*, *I* + (0,1/2,1/4)
Ge in 8 *g*: *I* + ($x,y,z$ ; $\bar{x},\bar{y},z$ ; $y,\bar{x},\bar{z}$ ; $\bar{y},x,\bar{z}$), $x = 0.570$, $y = 0.325$, $z = 0.089$

The vertex symbols are Mn: $9_2{\cdot}9_2{\cdot}9_2{\cdot}9_2{\cdot}9_2{\cdot}9_2$ and Ge: $3{\cdot}9_2{\cdot}3{\cdot}9_2{\cdot}3{\cdot}9_2$.

17. Draw the structure of $BeH_2$ and show that it is topologically the same as that of moganite ($SiO_2$).

$BeH_2$

*Ibam*, $a = 9.082$, $b = 4.160$, $c = 7.707$ Å ; Be(1) in 4 *a*: *I* ± (0,0,1/4)
Be(2) in 8 *j*: *I* ± ($x,y,0$ ; $1/2-x,1/2+y,0$), $x = 0.1699$, $y = 0.1253$
H(1) in 8 *j*, $x = 0.3055$, $y = 0.2823$
H(2) in 16 *k*: *I* ± ($x,y,\pm z$ ; $1/2-x,1/2+y,\pm z$), $x = 0.0895$, $y = 0.1949$, $z = 0.1515$

## APPENDIX 1

# MORE INFINITE SYMMETRY GROUPS

In this appendix we describe some infinite symmetry groups other than the space groups discussed in Chapters 1 and 3. Three-dimensional objects with translational symmetry in only two dimensions are *layers*. The symmetry groups of these objects are the 80 layer groups that are given below. Likewise three-dimensional objects with translational symmetry in only one dimension are *rods*. The 75 crystallographic rod groups are also listed.[1] Two-dimensional objects with one-dimensional translational symmetry are called *bands* or *friezes* and we describe the 7 band groups also.

A convenient way to consider these groups is as derived from space groups by removing translations in one or two dimensions. The reason for doing this is that the coordinates of general and special positions (and their site symmetries), and the nature and location of symmetry elements, can be obtained directly from the space group tables in the *International Tables* (abbreviated here to *IT*). As the coordinates of the general and special positions are the same as those of the space groups from which they are derived, the same labels (Wyckoff notation) are used for them here.

For completeness we also mention the *cylindrical* and *spherical* point groups that describe the symmetries of objects with ∞-fold rotation axes.

## A1.1 Layer groups

In the coordinate system used here it is assumed that the translations are along the $x$ and $y$ directions. The lattice can be oblique, either primitive ($p$) or centered ($c$) rectangular, hexagonal or square as for the two-dimensional space groups. The position in the plane group symbol has the same significance as for the three-dimensional space groups.

Once the space group from which the layer group is derived has been identified (and, if necessary, the axes relabeled as explained below) the symmetry elements and their locations and the coordinates of special and general positions are obtained directly from the *IT* (but of course there are no translations along $z$. In fact $z$ is now to be considered as the height above the $z = 0$ plane, and as such, has dimensions (e.g. $z$ may be measured in Å). The symmetry elements of the layer group are those of the space group which are contained in, or which intersect, the plane $z = 0$.

Comments and examples are taken in order of the system of the corresponding three-dimensional groups. For the full table of groups see § A1.6 (p. 389).

**Monoclinic**. The cases to be considered are classes 2, $m$ and $2/m$. The 2-fold axis of the layer group can be along $z$ in which case the lattice is oblique. The symbol for the layer

[1]With translations in only one direction, there is no restriction on the nature of rotation axes in rod groups. Here we restrict ourselves to those containing only 1-, 2-, 3-, 4- and 6-fold axes.

group is the same as that for the setting "unique axis *c*" in the *IT*. Only those monoclinic groups with primitive lattices and that do not have screw axes will have layer groups as subgroups in this instance.[1]

If the 2-fold axis of the layer group is parallel to one of the translations, it is taken as the **b** direction and the symbol for the layer group is the same as that for the three-dimensional group in the setting "unique axis *b*" except that (a) the lattice symbol is lower case and (b) the glide direction (if present) must be along **a** (so the glide planes are *a*). The lattice is now rectangular (either *p* or *c*).

It may be seen, for example, that information in the *IT* about *p*112/*b* (oblique) and *p*12/*a*1 (rectangular) are both contained under "nonstandard" settings of number 13, *P*12/*c*1 ("unique axis *c*, cell choice 3" and "unique axis *b*, cell choice 3" respectively).

**Orthorhombic.** The cases correspond to classes 222, *mm*2 and *mmm*. Layer glide planes can now be *a* or *b* or, for glide planes in the *xy* plane, *n*. Thus there is a layer group *pban* derived from *Pban*.

Another layer group is $p2_1am$ derived from $P2_1am$, which is a nonstandard setting of $Pmc2_1$ (number 26). Thus to get the information about $p2_1am$, one should first transform $Pmc2_1$ to $P2_1am$. This involves interchanging *x* and *z*. Thus from the *IT* for $Pmc2_1$ we find the general positions: ($\pm x,y,z$ ; $\pm x,\bar{y},1/2+z$). For $p2_1am$ the corresponding general positions are: ($x,y,\pm z$ ; $1/2+x,\bar{y},\pm z$) and the symmetry elements of the layer group are those of the space group intersecting, or contained in, the old $x = 0$ (new $z = 0$) plane.

A second layer group derived from $Pmc2_1$ is $p2_1ma$. We first get the general positions of $P2_1ma$ from those of $Pmc2_1$ by cyclic permutation $x \rightarrow y \rightarrow z \rightarrow x$ as: ($x,\pm y,z$ ; $1/2+x,\pm y,\bar{z}$). These are also the general positions of $p2_1ma$ and the symmetry elements of the layer group are those of the space group contained in the old $y = 0$ (new $z = 0$) plane.

At the time of writing, there does not appear to be a generally agreed "standard" setting for oblique and rectangular layer groups. Thus, to continue with the same examples, $p2_1am$ could be (and sometimes is) written $pb2_1m$, and $p2_1ma$ could be written $pm2_1b$. In both cases the labels of the *x* and *y* axes have been interchanged.

If in doubt the transformations of axes for different settings of the orthorhombic groups are given on p. 441-442. Note that the last position in the symbol given here for rectangular layer groups *always* refers to the unique direction (normal to the lattice). Layer groups such as *p*2*mm* and *pmm*2 are distinct groups although they are both derived from *Pmm*2 (number 25). *cmm*2 and *c*2*mm* are also distinct groups but are now subgroups of two different space groups: *Cmm*2 (number 35) and *C*2*mm* (a non-standard setting of *Amm*2, number 38). On the other hand *cm*2*m* is an alternative setting of *c*2*mm*.

**Tetragonal and Hexagonal.** The layer groups are derived from the space groups with a primitive lattice that do not have symmetry elements with translational components along *z*. The layer group symbols in these instances are simply derived from the space group symbols by the substitution of *p* for *P*. The positions of the individual entries of the layer group symbol have exactly the same significance as they do for the three-dimensional

[1]Clearly one cannot have symmetry elements that involve translations out of the plane of **a** and **b**.

tetragonal and hexagonal space groups. The symmetry elements of the layer group are now those of the space group contained in, or intersecting $z = 0$.

## A1.2 Rod groups

The only one dimensional lattice is primitive and we use the symbol **p** for it.

Many of the remarks of § A1.1 apply also to the rod group symbols. The major difference is that the translational symmetry is along the $z$ direction, so symmetry elements with translational components must be along that direction. In particular glide is always $c$.

Note that for tetragonal, trigonal and hexagonal rods the last two positions of the rod group symbol can be interchanged. This is because the orientation of the $x$ and $y$ axes is not determined by the directions of lattice translations. Thus although $P3m1$ and $P31m$ are different space groups and $p3m1$ and $p31m$ are different layer groups, **p**$3m1$ and **p**$31m$ are the same rod group with the orientation of $x$ and $y$ chosen differently (just as $31m$ and $3m1$ are also the same point group). Such redundant rod groups are in parentheses in the Table below (p. 389).

The coordinates of general and special positions are directly available from the *IT*: now $z$ is a dimensionless fraction of $c$, and $x$ and $y$ must be considered to have the dimensions of length. The symmetry elements of the rod group are those of the space group which are on or pass through the line $z = 0$.

## A1.3 Examples of layers and rods

There is a large group of layer compounds made up of $MX_2$ layers of either $\{M\}X_6$ octahedra or $\{M\}X_6$ trigonal prisms sharing edges. The lattices in both cases are hexagonal (Fig. A1.1).

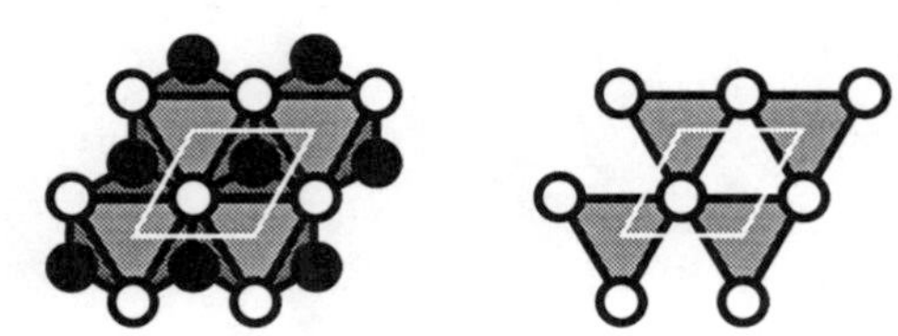

**Fig. A1.1.** Left part of a layer of $MX_6$ octahedra. Right part of a layer of $MX_6$ prisms.

In the octahedral layer $M$ is at 0,0,0 and $X$ is at $1/3,2/3,z$ and $2/3,1/3,\bar{z}$; these are positions 1 $a$ and 2 $d$ of $p\bar{3}m1$. The site symmetries are $\bar{3}m$ and $3m$ respectively. In the trigonal prism layer $M$ is again at 0,0,0 and $X$ is at $2/3,1/3,z$ and $2/3,1/3,\bar{z}$. These are positions 1 $a$ and 2 $i$ of $p\bar{6}m2$ and the site symmetries are $\bar{6}m2$ and $3m$ respectively.

Another common unit in crystal chemistry is a rod of alternating trigonal prisms and antiprisms (octahedra) sharing triangular faces normal to the rod axis (Fig. A1.2). The vertices are at $\pm(x,2x,z\ ;\ 2\bar{x},\bar{x},z\ ;\ x,\bar{x},z\ ;\ \bar{x},2\bar{x},1/2{+}z\ ;\ 2x,x,1/2{+}z\ ;\ \bar{x},x,1/2{+}z)$. These are positions 12 $k$ of **p**6$_3$/*mmc* with site symmetry $m$. The centers of the antiprisms are at 2 $a$: 0,0,0 ; 0,0,1/2 with site symmetry $\bar{3}m$ and the centers of the prisms are at 2 $b$: ± (0,0,1/4) with site symmetry $\bar{6}m2$.

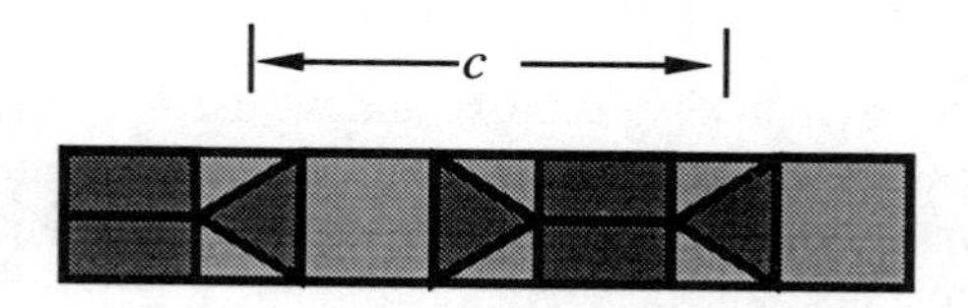


**Fig. A1.2**. Part of a rod of alternating trigonal prisms and octahedra.

Exercises:

(i) Find $a$, and $z$ for the layers of Fig. A1.1 made of regular polyhedra of edge 1 Å.

(ii) Find $x$, $z$ and $c$ for the rod of Fig. A1.2 made of regular polyhedra of edge 1 Å.

Answers:

(i) $a$ = 1 Å, octahedral layer $z = 1/\sqrt{6}$ Å, prism layer $z$ = 1/2 Å.

(ii) $x$ = 1/3 Å, $z = 1/(4 + \sqrt{24})$, $c = 2 + \sqrt{(8/3)}$ Å. [Note that $x$ has dimensions and $z$ is dimensionless in Exercise (ii)].

## A1.4 One- and two-dimensional "rods" (bands)

In one dimension there are but two point symmetry operations, reflection in a point which we represent by the symbol $m$, and the identity 1. There are therefore, just two point groups: 1 and $m$. Combined with the lattice **p** we get the two one-dimensional space groups **p**1 and **p***m* illustrated below (mirror points are shown as small circles, and the combination of long and short lines, — -, represents an asymmetric object). Note that in **p***m* there are two mirror points per unit cell.

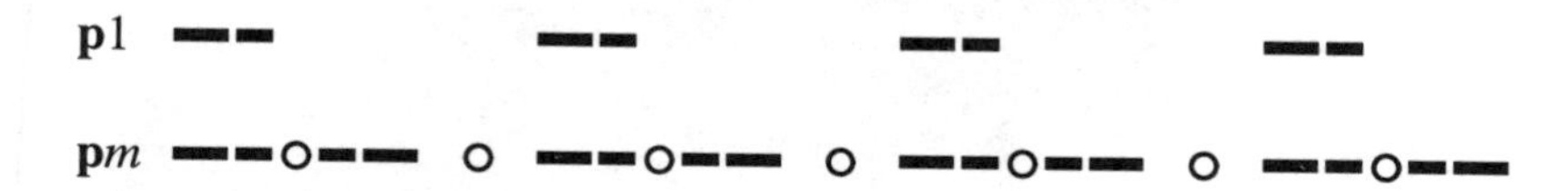


**Fig. A1.3**. One-dimensional space groups

A two-dimensional object with one-dimensional periodicity is variously referred to as a *band*, *frieze*, or *border*. The symmetry groups of such objects are readily enumerated, and can be related to the two-dimensional space groups just as done above for three-dimensional objects.

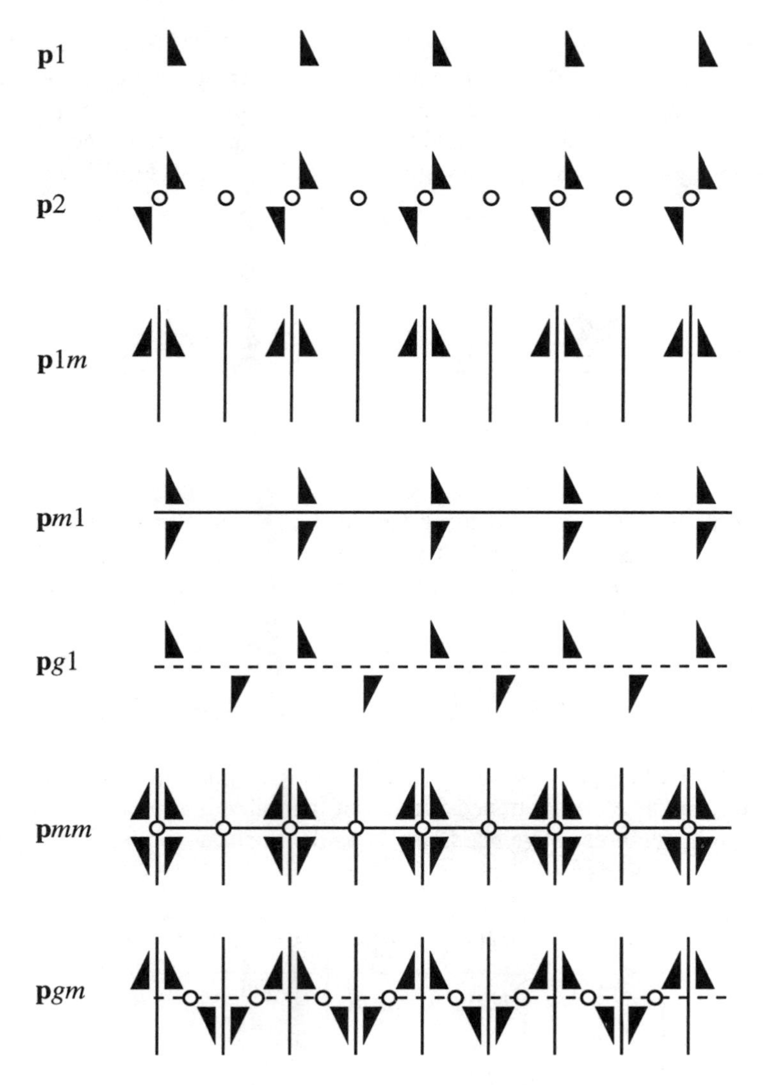


**Fig. A1.4.** The band groups. In each case the unit cell is the same size. Full lines represent mirror lines and broken lines represent glide lines. Small open circles represent 2-fold rotation points.

The permissible point groups are 1, 2, *m* and *mm*.[1] The first two simply give **p**1 and **p**2. For groups containing mirror lines we must specify their orientations, and we employ Cartesian axes $x$ and $y$ with the lattice translations along $y$. In accord with the conventions for three-dimensions we take the first position after **p** in the symmetry group symbol to refer to mirror (or glide) lines perpendicular to $x$ and the next symbol to refer to mirror lines

[1]It should be obvous that a one-dimensional lattice is not compatible with 3-, 4-, or 6-fold rotation points in the plane containing the lattice.

perpendicular to $y$. If there is no symmetry element we use "1" as a place marker. Note that a glide line must be along $y$ (perpendicular to $x$). Thus corresponding to point group $m$ we get band groups **p**$m$1, **p**1$m$ and **p**$g$1. From point group $mm$ we get **p**$mm$ and **p**$gm$. Patterns with these symmetries (generated by the symmetry elements acting on a scalene triangle) and the symmetry elements are illustrated in Fig. A1.4. Here they are tabulated as in § A1.6:

| $N$ | 2$D$ groups | band groups |
|---|---|---|
| 1 | $p1$ | **p**1 |
| 2 | $p2$ | **p**2 |
| 3 | $pm$ | **p**1$m$, **p**$m$1 |
| 4 | $pg$ | **p**$g$1 |
| 6 | $pmm$ | **p**$mm$ |
| 7 | $pmg$ | **p**$gm$ |

$N$ is the number of the space group given in the *IT*. Note that (*a*) in the *IT* the long symbols for the space groups are used. To be consistent with the usage in the *IT* we should write as "long" symbols **p**2$mm$ and **p**2$gm$ instead of **p**$mm$ and **p**$gm$. (*b*) for groups **p**$m$1 and **p**$gm$, $x$ and $y$ have to be interchanged from the setting used in the *IT*.

## A1.5 Point groups of infinite order and the symmetry of vectors

In Table 2.4 (§ 2.5.6, p.52) we listed non-crystallographic groups containing a single axis of arbitrary order $N$. A shortened version of that table is reproduced here using only Hermann-Maugin symbols so we need only consider the cases $N$ even and $N$ odd. We use short symbols for convenience.

| $N =$ | $N$ | $\bar{N}$ | $N2(2)$ | $Nm(m)$ | $N/m$ | $N/m2/m2/m$ | $\bar{N}+m$ |
|---|---|---|---|---|---|---|---|
| $2n$ | $N$ | $\bar{N}$ | $N22$ | $Nmm$ | $N/m$ | $N/mmm$ | $\bar{N}2m$ |
| $2n+1$ | $N$ | $\bar{N}$ | $N2$ | $Nm$ | | | $\bar{N}m$ |

There is an interesting relationship between these groups and the band groups. Imagine a finite fragment of the patterns of Fig. A1.4 containing $N$ translations with $N$ even. Now fold the fragment round to make a circle with beginning and end motifs superimposed.[1] If this is done with the pattern of **p**1, an object of symmetry $N$ is obtained. The same exercise repeated with the **p**2 pattern will produce symmetry $N22$ (note the two sets of 2-fold axes). Similarly **p**1$m$ will produce $Nmm$ (again note the two sets of mirrors), **p**$m$1 will produce

[1]The reader who finds the mental exercise difficult is invited to copy Fig. A1.4 and cut the different patterns (which show $N = 4$ translations) into bands, which can then be folded into a ring.

$N/m$, and **p***mm* will produce $N/mmm$. The patterns with glide will probably require a little more thought, but it will be found that $\bar{N}$ is obtained from a fragment of **p***g*1, and $\bar{N}m2$ is obtained from a fragment of **p***gm*. As $N$ gets larger the finite patterns approach the band patterns more and more closely and we can think of the band groups as limiting cases with $N = \infty$ of the point groups with $N$ even. Thus there is a one-to-one correspondence between the band groups and the entries for even $N$ in Table 2.3.

Let us consider the correspondence $Nmm \rightarrow$ **p**1*m* a little further. Fig. A1.5 shows two mirrors inclined at an angle $\alpha = 360°/2N$. Successive reflections in the mirrors generate an $N$-fold axis at their line of intersection. We can reduce the angle between the mirrors by increasing $r$ and at the same time keeping $d$ constant (see the figure). In the limit $r \rightarrow \infty$ we have parallel mirrors separated by $d$ and successive reflections in these generate a translation by $2d$. Thus the correspondence between point groups and band groups discussed in the previous paragraph is equivalent to considering the infinite set of translations as equivalent to rotations about an axis infinitely far away.

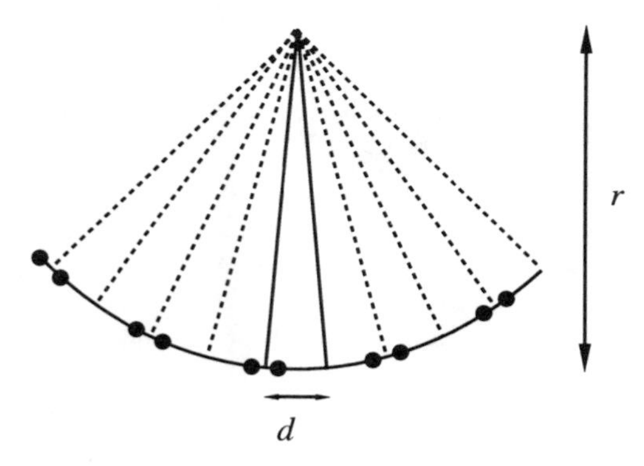


**Fig. A1.5**. Illustrating the effects of successive reflections in two inclined mirrors (shown as solid straight lines). See the text for details.

There is a second way of reducing the angle between the mirrors. This is by reducing $d$ while keeping $r$ constant; as $d$ goes to zero we will have a finite pattern with an $\infty$-fold symmetry axis. The symmetry group is now $\infty m$ (only one "$m$" as it is meaningless to talk of *sets* of mirror planes) and this is the symmetry of, for example, a cone. There are in fact five cylindrical symmetry groups. The reason that there are only five (rather than seven as in the case of the band groups) is that for a finite object with an $\infty$-fold axis, $\infty/m$ cannot be distinguished from $\bar{\infty}$ (recall that, for example, $\bar{6} \equiv 3/m$). Thus the cylindrical point groups are to be considered as limiting cases as $N \rightarrow \infty$ of the entries in Table 2.3 for $N$ odd. Accordingly the cylindrical groups are:

| | | |
|---|---|---|
| $\infty$ | $C_\infty$ | rotating cone |
| $\bar{\infty} \equiv \infty/m$ | $C_{\infty h} \equiv C_{\infty i}$ | rotating cylinder or rotating double cone |
| $\infty 2$ | $D_\infty$ | antirotating double cone |
| $\infty m$ | $C_{\infty v}$ | cone |
| $\bar{\infty} m \equiv \infty/m 2/m$ | $D_{\infty d} \equiv D_{\infty h}$ | cylinder or double cone |

In the table we give the point group symbols (first Hermann-Maugin, then Schoenflies) and then examples of objects with these symmetries. The last two symmetries in the table are realized in linear molecules either without a center of symmetry (as in CO = $C_{\infty v}$) or with one (as in $O_2 = D_{\infty d}$). To realize the first three symmetries we must consider rotating objects. Fig. A1.6 illustrates these. Note that the rotating cone and anti-rotating double cone exist in left- and right-handed (enantiomorphic) forms. In contrast a rotating double cone (or cylinder) has a center of symmetry. The reader who finds this puzzling is asked to look at Fig. A1.6 and then turn the book upside down, and to look at the figure again.[1]

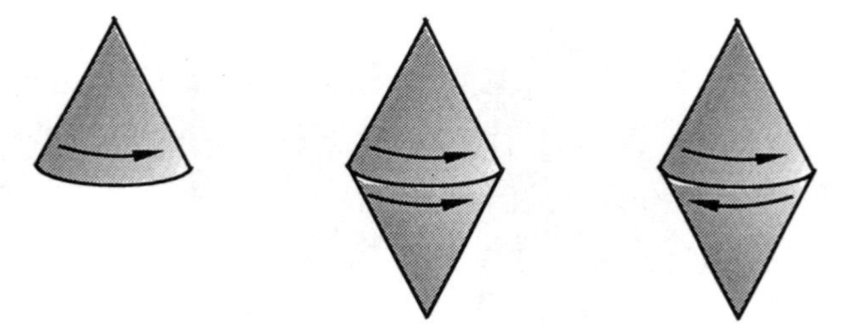

**Fig. A1.6**. From left to right: a rotating cone, a rotating double cone, and an antirotating double cone.

It is important to recognize that vectors can have one of two symmetries. *Polar* vectors, such as one corresponding to an electric dipole moment have the symmetry ($\infty m$) of a cone. Axial vectors, such as one representing a rotation axis[2] or a magnetic moment have the symmetry ($\bar{\infty}$) of a rotating cylinder. Confusion can arise because it is conventional to represent both kinds of vector by the same symbol—an arrow.[3] As shown in Fig. 1.7 improper operations such as reflection and inversion transform the two kinds of vector differently.

[1] It is useful to recall that the apparent sense of rotation depends on one's point of view. A person looking at a clock will see the hands rotating "clockwise." However from the point of view of the clock, the hands rotate "anticlockwise." Similarly the rotating double cone in the figure is rotating clockwise when viewed from the top but anti-clockwise when viewed from the bottom.

[2] A 2-fold rotation axis is a special case because $2^+ = 2^-$ so the symmetry is $\bar{\infty} m$ and a 2-fold axis does not have a sense (direction). Ironically the symbol for a 2-fold rotation axis in crystallography is an arrow.

[3] See Chapter 1 of *Icons and Symmetry* by S. L. Altmann [Oxford University Press (1992)]. Altmann gives an informative and entertaining account of the Ørsted experiment in which a magnetic needle is deflected by a current flowing in a parallel direction above it. If the magnet and current are both represented by arrows, it appears that the plane containing the magnet and wire is a plane of symmetry. In fact the magnet and current have different symmetries (of axial and polar vectors respectively), and as illustrated in Fig. A1.7, they behave differently on reflection in that plane, which is therefore not really a mirror plane.

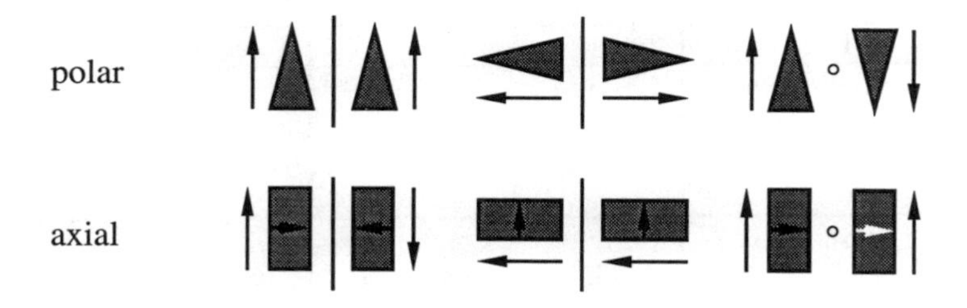


**Fig. A1.7**. Showing how polar and axial vectors (arrows) and cones and rotating cylinders are transformed by reflection in a mirror (shown as vertical line) and inversion in a center (open circle).

What about groups with more than one $N$-fold rotation axis? In Chapter 2 we enumerated all the possibilities for finite $N$. With one $\infty$-fold axis we get the cylindrical groups. With more than one $\infty$-fold axis we get the *spherical* groups. A sphere has in fact an infinity of $\infty$-fold axes and also a center of symmetry so they are actually $\bar{\infty}$ axes. We need at least two to generate spherical symmetry so the symmetry of a sphere is expressed as $\bar{\infty}\,\bar{\infty}$. If the center is removed we get the infinite pure rotation group $\infty\infty$. This second case is hard to imagine: think of a sphere rotating about an axis, and that axis rotating about a second axis at right angles to the first. Thus the spherical groups are ($K$ and $K_h$ are the Schoenflies symbols):

| | | |
|---|---|---|
| $\infty\infty$ | $K$ | rotating sphere |
| $\bar{\infty}\,\bar{\infty}$ | $K_h$ | stationary sphere |

In two dimensions the point symmetry groups are $N$ and $Nm$. If again we let $N$ go to infinity we get the two one-dimensional space groups **p**1 and **p***m* (§ A1.4), and the two infinite point groups (*circular* groups) which are the symmetries of a rotating circle ($\infty$) and of a stationary circle ($\infty m$).

## A1.6 Table of layer and rod group symbols

The table on the following pages is intended to be self-explanatory. $N$ is the sequence number in the *IT* of the three-dimensional space groups, which are given in their standard settings. The first page consists of groups with at most 2-fold axes. Then follow groups with 4-fold axes (tetragonal) and groups with 3-fold or 6-fold axes (trigonal and hexagonal)

| $N$ | 3-D groups | layer groups | | rod groups | |
|---|---|---|---|---|---|
| | triclinic | oblique | | | |
| 1 | $P1$ | $p1$ | | $\mathbf{p}1$ | |
| 2 | $P\bar{1}$ | $p\bar{1}$ | | $\mathbf{p}\bar{1}$ | |
| | monoclinic | oblique | rectangular | | |
| 3 | $P121$ | $p112$ | $p121$ | $\mathbf{p}112$ | $\mathbf{p}121$ |
| 4 | $P12_11$ | | $p12_11$ | $\mathbf{p}112_1$ | |
| 5 | $C121$ | | $c121$ | | |
| 6 | $P1m1$ | $p11m$ | $p1m1$ | $\mathbf{p}11m$ | $\mathbf{p}1m1$ |
| 7 | $P1c1$ | $p11b$ | $p1a1$ | | $\mathbf{p}1c1$ |
| 8 | $C1m1$ | | $c1m1$ | | |
| 10 | $P12/m1$ | $p112/m$ | $p12/m1$ | $\mathbf{p}112/m$ | $\mathbf{p}12/m1$ |
| 11 | $P12_1/m1$ | | $p12_1/m1$ | $\mathbf{p}112_1/m$ | |
| 12 | $C12/m1$ | | $c12/m1$ | | |
| 13 | $P12/c1$ | $p112/b$ | $p12/a1$ | | $\mathbf{p}12/c1$ |
| 14 | $P12_1/c1$ | | $p12_1/a1$ | | |
| | orthorhombic | rectangular | | | |
| 16 | $P222$ | $p222$ | | $\mathbf{p}222$ | |
| 17 | $P222_1$ | $p2_122$ | | $\mathbf{p}222_1$ | |
| 18 | $P2_12_12$ | $p2_12_12$ | | | |
| 21 | $C222$ | $c222$ | | | |
| 25 | $Pmm2$ | $pmm2$ | $p2mm$ | $\mathbf{p}mm2$ | $\mathbf{p}2mm$ |
| 26 | $Pmc2_1$ | $p2_1ma$ | $p2_1am$ | $\mathbf{p}mc2_1$ | |
| 27 | $Pcc2$ | $p2aa$ | | $\mathbf{p}cc2$ | |
| 28 | $Pma2$ | $pma2$ | $p2mb$ | $\mathbf{p}c2m$ | |
| 29 | $Pca2_1$ | $p2_1ab$ | | | |
| 30 | $Pnc2$ | $p2an$ | | | |
| 31 | $Pmn2_1$ | $p2_1mn$ | | | |
| 32 | $Pba2$ | $pba2$ | | | |
| 35 | $Cmm2$ | $cmm2$ | | | |
| 38 | $Amm2$ | $c2mm$ | | | |
| 39 | $Abm2$ | $c2mb$ | | | |
| 47 | $Pmmm$ | $pmmm$ | | $\mathbf{p}mmm$ | |
| 49 | $Pccm$ | $pmaa$ | | $\mathbf{p}ccm$ | |
| 50 | $Pban$ | $pban$ | | | |
| 51 | $Pmma$ | $pmma$ | $pmam$ | $\mathbf{p}mcm$ | |
| 53 | $Pmna$ | $pbmn$ | | | |
| 54 | $Pcca$ | $pbaa$ | | | |
| 55 | $Pbam$ | $pbam$ | | | |
| 57 | $Pbcm$ | $pmab$ | | | |
| 59 | $Pmmn$ | $pmmn$ | | | |
| 65 | $Cmmm$ | $cmmm$ | | | |
| 67 | $Cmma$ | $cmma$ | | | |

*Tetragonal (square) layer and rod groups*

| $N$ | 3-D group | layer group | rod group |
|---|---|---|---|
| 75 | $P4$ | $p4$ | $\mathbf{p}4$ |
| 76 | $P4_1$ | | $\mathbf{p}4_1$ |
| 77 | $P4_2$ | | $\mathbf{p}4_2$ |
| 78 | $P4_3$ | | $\mathbf{p}4_3$ |
| 81 | $P\bar{4}$ | $p\bar{4}$ | $\mathbf{p}\bar{4}$ |
| 83 | $P4/m$ | $p4/m$ | $\mathbf{p}4/m$ |
| 84 | $P4_2/m$ | | $\mathbf{p}4_2/m$ |
| 85 | $P4/n$ | $p4/n$ | |
| 89 | $P422$ | $p422$ | $\mathbf{p}422$ |
| 90 | $P42_12$ | $p42_12$ | |
| 91 | $P4_122$ | | $\mathbf{p}4_122$ |
| 93 | $P4_222$ | | $\mathbf{p}4_222$ |
| 95 | $P4_322$ | | $\mathbf{p}4_322$ |
| 99 | $P4mm$ | $p4mm$ | $\mathbf{p}4mm$ |
| 100 | $P4bm$ | $p4bm$ | |
| 101 | $P4_2cm$ | | $\mathbf{p}4_2cm$ |
| 103 | $P4cc$ | | $\mathbf{p}4cc$ |
| 111 | $P\bar{4}2m$ | $p\bar{4}2m$ | $\mathbf{p}\bar{4}2m$ |
| 112 | $P\bar{4}2c$ | | $\mathbf{p}\bar{4}2c$ |
| 113 | $P\bar{4}2_1m$ | $p\bar{4}2_1m$ | |
| 115 | $P\bar{4}m2$ | $p\bar{4}m2$ | [$\mathbf{p}\bar{4}m2 \equiv \mathbf{p}\bar{4}2m$] |
| 117 | $P\bar{4}b2$ | $p\bar{4}b2$ | |
| 123 | $P4/mmm$ | $p4/mmm$ | $\mathbf{p}4/mmm$ |
| 124 | $P4/mcc$ | | $\mathbf{p}4/mcc$ |
| 125 | $P4/nbm$ | $p4/nbm$ | |
| 127 | $P4/mbm$ | $p4/mbm$ | |
| 129 | $P4/nmm$ | $p4/nmm$ | |
| 131 | $P4_2/mmc$ | | $\mathbf{p}4_2/mmc$ |

*trigonal and hexagonal layer and rod groups*

| $N$ | 3-D group | layer group | rod group |
|---|---|---|---|
| 143 | $P3$ | $p3$ | $\mathbf{p}3$ |
| 144 | $P3_1$ | | $\mathbf{p}3_1$ |
| 145 | $P3_2$ | | $\mathbf{p}3_2$ |
| 147 | $P\bar{3}$ | $p\bar{3}$ | $\mathbf{p}\bar{3}$ |
| 149 | $P312$ | $p312$ | $[\mathbf{p}312 \equiv \mathbf{p}321]$ |
| 150 | $P321$ | $p321$ | $\mathbf{p}321$ |
| 152 | $P3_121$ | | $\mathbf{p}3_121$ |
| 154 | $P3_212$ | | $\mathbf{p}3_221$ |
| 156 | $P3m1$ | $p3m1$ | $\mathbf{p}3m1$ |
| 157 | $P31m$ | $p31m$ | $[\mathbf{p}31m \equiv \mathbf{p}3m1]$ |
| 158 | $P3c1$ | | $\mathbf{p}3c1$ |
| 162 | $P\bar{3}1m$ | $p\bar{3}1m$ | $[\mathbf{p}\bar{3}1m \equiv \mathbf{p}\bar{3}m1]$ |
| 164 | $P\bar{3}m1$ | $p\bar{3}m1$ | $\mathbf{p}\bar{3}m1$ |
| 165 | $P\bar{3}c1$ | | $\mathbf{p}\bar{3}c1$ |
| 168 | $P6$ | $p6$ | $\mathbf{p}6$ |
| 169 | $P6_1$ | | $\mathbf{p}6_1$ |
| 170 | $P6_5$ | | $\mathbf{p}6_5$ |
| 171 | $P6_2$ | | $\mathbf{p}6_2$ |
| 172 | $P6_4$ | | $\mathbf{p}6_4$ |
| 173 | $P6_3$ | | $\mathbf{p}6_3$ |
| 174 | $P\bar{6}$ | $p\bar{6}$ | $\mathbf{p}\bar{6}$ |
| 175 | $P6/m$ | $p6/m$ | $\mathbf{p}6/m$ |
| 176 | $P6_3/m$ | | $\mathbf{p}6_3/m$ |
| 177 | $P622$ | $p622$ | $\mathbf{p}622$ |
| 178 | $P6_122$ | | $\mathbf{p}6_122$ |
| 179 | $P6_522$ | | $\mathbf{p}6_522$ |
| 180 | $P6_222$ | | $\mathbf{p}6_222$ |
| 181 | $P6_422$ | | $\mathbf{p}6_422$ |
| 182 | $P6_322$ | | $\mathbf{p}6_322$ |
| 183 | $P6mm$ | $p6mm$ | $\mathbf{p}6mm$ |
| 184 | $P6cc$ | | $\mathbf{p}6cc$ |
| 185 | $P6_3cm$ | | $\mathbf{p}6_3cm$ |
| 187 | $P\bar{6}m2$ | $p\bar{6}m2$ | $\mathbf{p}\bar{6}m2$ |
| 188 | $P\bar{6}c2$ | | $\mathbf{p}\bar{6}c2$ |
| 189 | $P\bar{6}2m$ | $p\bar{6}2m$ | $[\mathbf{p}\bar{6}2m \equiv \mathbf{p}\bar{6}m2]$ |
| 191 | $P6/mmm$ | $p6/mmm$ | $\mathbf{p}6/mmm$ |
| 192 | $P6/mcc$ | | $\mathbf{p}6/mcc$ |
| 194 | $P6_3/mmc$ | | $\mathbf{p}6_3/mmc$ |

APPENDIX 2

# A GLIMPSE INTO HIGHER DIMENSIONS

## A2.1 Introduction: polytopes

There are many reasons for considering structures in higher dimensions—some of them practical, some of them aesthetic. Here we can only give a few teasing hints of the richness of the geometry of higher dimensions, but it is hoped they will remove some of the irrational fear of the subject and perhaps kindle a desire to delve more deeply into the subject. Our main purposes are to introduce terms that are gaining increasing currency in crystallography and to give some insight into how higher dimensional problems are handled.

We start by generalizing a three-dimensional Cartesian axis system to $n$ dimensions so that there are $n$ orthogonal axes $\mathbf{a}_1, \mathbf{a}_2, \ldots \mathbf{a}_n$. The $n$ coordinates of a point in this space are $x_1, x_2, \ldots x_n$. The Pythagoras theorem gives the distance between two points $x_{11}$, $x_{21}, \ldots x_{n1}$ and $x_{12}, x_{22}, \ldots x_{n2}$ as

$$d^2 = (x_{11} - x_{12})^2 + (x_{21} - x_{22})^2 + \ldots + (x_{n1} - x_{n2})^2 \quad \text{(A2.1)}$$

The generalization of a polygon (in two dimensions) and a polyhedron (in three dimensions) to $n$ dimensions is a *polytope*. A *hypercube* or *measure polytope* of edge $b$ with center at the origin has $2^n$ vertices at $(\pm b/2, \pm b/2, \ldots, \pm b/2)$ and has content ("volume") $b^n$. It is an example of a *regular* polytope. It is amusing that for a four-dimensional cube (known as a *tesseract*) the distance from the center to a vertex [which is $\surd(b^2/4 + b^2/4 + b^2/4 + b^2/4)$] is equal to the edge length ($b$).

Another regular polytope is the *cross polytope*. For edge length $b$ this has $2n$ vertices at $\pm(b/\surd 2, 0, \ldots 0)\kappa$. The three-dimensional version is the regular octahedron. The volume is $V = b^n 2^{n/2}/n!$.

A third polytope of importance is the *simplex* which has $n + 1$ vertices each of which is joined to all the others by edges, so that there are $n(n + 1)/2$ edges. A two-dimensional simplex is a triangle and a three-dimensional simplex is a tetrahedron. A regular simplex has all edges equal, say $b$, and volume $V = b^n \surd(n + 1)/[n! 2^{n/2}]$. The angle subtended by an edge at the center is $\cos^{-1}(-1/n)$ and the vertex to center distance is $b\surd[n/(2n + 2)]$. The order of the symmetry group of a regular simplex is $(n + 1)!$.

These are the only regular polytopes for $n \geq 5$, but four-dimensional space has some beautiful surprises (recall that in three dimensions we have two more regular polyhedra, the icosahedron and dodecahedron) that are mentioned below.

Points on the surface of a *hypersphere* (or just *sphere*) in $n$ dimensions with center at the origin $(0,0,\ldots,0)$ and with radius $r$ are given by:

$$r^2 = x_1^2 + x_2^2 + \ldots + x_n^2 \quad \text{(A2.2)}$$

The volume is given by (the second form avoids non-integral factorials for odd $n$):

$$V = r^n \pi^{n/2}/(n/2)! = (2r)^n \pi^{(n-1)/2}[(n-1)/2]!/n! \qquad \text{(A2.3)}$$

An often useful formula is for the volume of a general simplex with vertices (in Cartesian coordinates) $(x_{01},x_{02},\dots x_{0n})$, $(x_{11},x_{12},\dots x_{1n})$, …, $(x_{n1},x_{n2},\dots x_{nn})$ [compare Eq. 4.25]:

$$V = \frac{1}{n!}\begin{vmatrix} 1 & x_{01} & \dots & x_{0n} \\ 1 & x_{11} & \dots & x_{1n} \\ & \dots & & \dots \\ 1 & x_{n1} & \dots & x_{nn} \end{vmatrix} \qquad \text{(A2.4)}$$

The sign of $V$ depends on the order of numbering of the vertices.

## A2.2 Four-dimensional polytopes and honeycombs

It is convenient to rewrite the Schläfli symbols we have been using on one line. Thus the symbol for a cube $4^3$ becomes {4,3} and an octahedron $3^4$ becomes {3,4}. The symbol for a 4-dimensional hypercube becomes {4,3,3}. This is interpreted as a polytope whose hyperfaces (*cells*) are {4,3} (i.e. cubes) and three of these meet at an edge.

Two-dimensional tessellations can be considered as degenerate cases of three-dimensional polyhedra. For example {3,3}, {3,4} and {3,5} are the tetrahedron, octahedron and icosahedron respectively whereas {3,6} is a covering of the plane by (an infinite number of) triangles. Likewise {4,3} is a cube but {4,4} is a covering of the plane by squares. The other regular tessellation of the plane is {6,3}. In the same way a three-dimensional space filling by cubes (with four meeting at every edge) can be considered a degenerate case of a four-dimensional polytope and symbolized {4,3,4}. The way to interpret $\{p,q,r\}$ is that $r$ figures that are $\{p,q\}$'s meet at an edge. A space filling by polytopes is often called a *honeycomb* (elsewhere we reserve this term for {6,3})

The four-dimensional simplex is {3,3,3} and the four-dimensional cross polytope is {3,3,4}. There is also a polytope {3,3,5} (which might be thought of as the four-dimensional analog of the icosahedron) whose 600 cells are tetrahedra. Its dual is the regular polytope {5,3,3} (again we might consider this the analog of the dodecahedron) whose 120 cells are dodecahedra. It is noted in passing that the order of the symmetry group of these polyhedra is 14400—this is small for mathematicians but rather impressive compared to the order (48) of the largest three-dimensional crystallographic point group ($m\bar{3}m$) and the order (120) of the icosahedral group.

It transpires that there is a sixth polytope in four dimensions. This is {3,4,3} whose 24 cells are octahedra {3,4} with three meeting at every edge. {3,3,4} and {3,4,3} are space-filling polytopes so there are also four-dimensional honeycombs {3,3,4,3} and {3,4,3,3}.

Only in two and four dimensions are there regular honeycombs other than a space filling by hypercubes. The reader who wants to learn about these beautiful figures can do no better than read the book by Coxeter cited at the end of this appendix.

It is interesting that four dimensions is richest in *regular* polytopes. The geometry of higher dimensions is not dull however, there some remarkable lattices with special properties known. One teaser: the problem of the number of equal spheres that can contact a central one (this is sometimes called the maximum "kissing number") is remarkably difficult to solve. It is known to be 12 in three dimensions, but mathematicians don't know the answer in four dimensions.[1] The maximum kissing number *is* known in 24 dimensions and is 196560; a lattice with this coordination number is also known. The order of the point symmetry group of this lattice is 8315553613086720000 which is *still* small by mathematical standards (the order of the point symmetry group of the primitive hypercubic lattice is $2^n n!$—work *that* out for $n = 24$).

### A2.3 Four- and higher-dimensional lattices

In four-dimensions we specify a lattice by four vectors $\mathbf{a}_1$, $\mathbf{a}_2$, $\mathbf{a}_3$, $\mathbf{a}_4$ of length $a_1$, $a_2$, $a_3$, $a_4$ and with angles between them of $\alpha_{12}$, $\alpha_{13}$, $\alpha_{14}$, $\alpha_{23}$, $\alpha_{24}$, $\alpha_{34}$; i.e., in general, ten unit cell parameters. It transpires that there are 64 Bravais lattices. We look at one or two here. Conceptually the simplest of these is the primitive hypercubic with $a_1 = a_2 = a_3 = a_4 = a$ and all angles equal to 90°. The point symmetry of this lattice is the same as that of the hypercube and has order 384. Each lattice point has 8 neighbors. The primitive hypercubic lattice in $n$ dimensions may be symbolized $\mathbf{Z}^n$.

Recall that in four dimensions the distance from the center to a vertex of a hypercube is equal to an edge length. This means that a copy of the simple hypercubic lattice displaced by 1/2,1/2,1/2,1/2 can be fitted into the first one and generate a new lattice which is known to crystallographers as *Z*-centered hypercubic (mathematicians know it as $D_4$). Each point of the new lattice has 24 neighbors. The point at 0,0,0,0 has as neighbors, the original eight of the primitive lattice at $(\pm1,0,0,0)\kappa$ and 16 more from the second hypercubic lattice at the vertices of a hypercube with coordinates ±1/2,±1/2,±1/2,±1/2. If spheres of unit radius are put at the lattice points the arrangement with $a = 1$ corresponds to the densest (lattice) packing of spheres in four dimensions. Another way of looking at this lattice is as the vertices of the regular honeycomb {3,3,4,3}. The interstices of the lattice at (1/2,1/2,0,0 ; 1/2,0,1/2,0 ; 1/2,0,0,1/2 ; 0,1/2,1/2,0 ; 0,1/2,0,1/2 ; 0,0,1/2,1/2) are at the centers of the {3,4,3} polytopes.

Now for some vertigo. The positions of the interstices given above are the regular honeycomb {3,4,3,3} and the Voronoi polyhedron of the lattice points is a {3,4,3}. This figure is just the same as that of the 24 lattice points surrounding a given lattice point. In

[1]Mathematicians are hard lot to convince. It is a safe bet that the 24-coordinated lattice described below provides the answer to this problem and that of densest sphere packing in four dimensions. Only recently has it been *proved* that there is not a sphere packing in three dimensions that is denser than cubic closest packing.

fact the point symmetry of the lattice is the same as that of {3,4,3} and is *higher* than that of the primitive hypercubic lattice; the order of the point group is 1152. This means that in four dimensions there are distinct crystal systems corresponding to primitive hypercubic and centered hypercubic.

If we combine the lattice positions (0,0,0,0 ; 1/2,1/2,/12,1/2) with the positions of the interstices (given above) we have eight points which fall back at the points of the same lattice but with separation reduced by a factor of $\sqrt{2}$. In other words we can describe the same lattice using a non-primitive hypercubic unit cell with either 2 points per cell or 8 points per cell. Thus we have two descriptions of the same lattice: (*a*) 0,0,0,0 ; 1/2,1/2,1/2,1/2 ; (*b*) 0,0,0,0 ; 1/2,/12,1/2,1/2 and all six permutations of 1/2,1/2,0,0.

The matrix transforming the first description into the second is:

$$\mathbf{S} = \begin{pmatrix} 1 & 1 & 0 & 0 \\ 1 & -1 & 0 & 0 \\ 0 & 0 & 1 & 1 \\ 0 & 0 & 1 & -1 \end{pmatrix} \qquad \text{(A2.5)}$$

In more than four dimensions there are two distinct lattices, reciprocal to each other, that can be described using a centered hypercubic cell and that are generalizations of the one above. The first, called $D_n^*$, has lattice points at 0,0,...,0 and 1/2,1/2,...1/2. The second, called $D_n$, has lattice points at 0,0,...,0 and all combinations of 0 and 1/2 that add up to an integer. If the coordinates of the points in this second unit cell are doubled, it may be seen that they consist of combinations of integers that add up to an even number, and the lattice points consist of half of the lattice points of a primitive hypercubic lattice; for this reason $D_n$ is also called the checker-board lattice. The unit cell of $D_n$ contains $2^{n-1}$ points, and for unit distance between lattice points $a = \sqrt{2}$. We have seen that the reciprocal lattice of the *Z*-centered hypercubic lattice ($D_4^*$) is the same lattice ($D_4$). Only when $n = 4$ is the point symmetry of $D_n$ different from that of $\mathbf{Z}^n$.

We can't resist mentioning that in eight dimensions two copies of $D_8$ fit together (in the same way as two primitive hypercubic lattices gave $D_4$) to give a new lattice $E_8$. The second $D_8$ is displaced by 1/4,1/4,...,1/4 from the first. The reader will possibly find it a stimulating exercise to verify that $E_8$ represents a 240-coordinated sphere packing (the densest known in eight dimensions). For spheres of unit radius, the cell edge remains $\sqrt{2}$ and there are 256 points in the unit cell, i.e. 16 per unit volume. $E_8$ has attracted some attention in connection with quasicrystal structures; also of interest is that the points divide space into regular simplices and cross polytopes (how many and where?). Note that two $D_3$ ($\equiv A_3 \equiv fcc$) lattices displaced by 1/4,1/4,1/4 give the diamond structure (not a lattice).

Another family of lattices has non-orthogonal lattice vectors. In a plane we can have three vectors equally inclined to each other at an angle of $\cos^{-1}(-1/2) = 120°$. Two of these of equal length generate a hexagonal lattice. In three-dimensions four vectors all making an equal angle with the others are possible; the angle is the tetrahedral angle, $\cos^{-1}(-1/3) = 109.47°$. Three of these vectors of equal length generate a body-centered cubic lattice (recall that a primitive cell for body-centered cubic has $a = b = c$ and $\alpha = \beta = \gamma = 109.47°$). In general in $n$ dimensions $n+1$ vectors can be equally inclined at an angle of $\cos^{-1}(-1/n)$ and we can define a lattice using $n$ of them of equal length. Note that the $n + 1$ vectors are equivalent so there will be a symmetry operation of order $n + 1$ relating them as discussed below. The four-dimensional lattice with $a_1 = a_2 = a_3 = a_4 = a$ and $\alpha_{12} = \alpha_{13} = \alpha_{14} = \alpha_{23} = \alpha_{24} = \alpha_{34} = \cos^{-1}(-1/4) = 104.48°$ is known as the primitive icosahedral lattice. The order of the point symmetry group of the lattice is 240.

The lattice reciprocal to this 4-dimensional lattice is easily found to have $a_1 = a_2 = a_3 = a_4 = 1/a$ and $\alpha_{12} = \alpha_{13} = \alpha_{14} = \alpha_{23} = \alpha_{24} = \alpha_{34} = \cos^{-1}(1/2) = 60°$ . (Recall that in three dimensions, the lattice reciprocal to *bcc* is the *fcc* lattice which has primitive cell $a = b = c$ ; $\alpha = \beta = \gamma = 60°$). The simplest way to see this is to find the reciprocal of the **G** matrix:

$$\mathbf{G} = a^2 \begin{pmatrix} 1 & -1/4 & -1/4 & -1/4 \\ -1/4 & 1 & -1/4 & -1/4 \\ -1/4 & -1/4 & 1 & -1/4 \\ -1/4 & -1/4 & -1/4 & 1 \end{pmatrix}$$

$$\mathbf{G}^{-1} = \mathbf{G}^{*} = a^{-2} \begin{pmatrix} 1 & 1/2 & 1/2 & 1/2 \\ 1/2 & 1 & 1/2 & 1/2 \\ 1/2 & 1/2 & 1 & 1/2 \\ 1/2 & 1/2 & 1/2 & 1 \end{pmatrix} \qquad \text{(A2.6)}$$

The off-diagonal elements of the matrices are $\cos\alpha_{ij}$.

The new lattice can be described using a centered cell of the same shape as the primitive icosahedral lattice and it is then called the *SN*-centered icosahedral lattice. In this description there are 125 lattice points in the (non-primitive) unit cell.

We have remarked that these two lattices are four-dimensional analogs of the *bcc* and *fcc* lattices which, it so happens, are cubic. It is sometimes useful though to consider the *bcc* lattice in terms of its primitive cell. It is left as a non-trivial exercise for the reader to describe the *fcc* lattice in terms of a unit cell with the same shape (it needs 16 points per cell!). The general $n$-dimensional lattice with primitive cell edges all equal and all angles equal to 60° is known as $A_n$ and its reciprocal lattice with all angles equal to $\cos^{-1}(-1/n)$ is known as $A_n^*$. $A_n$ represents a lattice sphere packing with $n(n+1)$ neighbors and $A_n^*$ represents a lattice sphere packing with $n + 2$ neighbors.

## A2.4 Symmetry operations in four dimensions

It is instructive to see how symmetry operations arise as the dimensionality of space is increased. We restrict ourselves to point symmetry operations (those that leave *at least* one point invariant). In one dimension there is just one operation $m$ that takes a point to another place by reflection in a point. In two dimensions $m$ reflects in a line and in three dimensions

$m$ reflects in a plane. In general $m$ changes position in one direction (the normal to the mirror) and leaves an $n-1$ figure unchanged. Thus in four dimensions there are mirror hyperplanes.

In the transition from one to two dimensions we recognize a new operation that changes two coordinates (leaves a point invariant). The symmetry element is a rotation point. In three dimensions we have a rotation axis (a line is left invariant) and in four dimensions a rotation plane (a plane is left invariant).

Continuing from two to three dimensions, we identify a new operation that changes three coordinates (leaves just a point invariant). This is a (rotation-) inversion point. In four dimensions there are inversion axes[1] and we must seek new symmetry operations that leave only a point invariant. To see their nature it is easiest to appeal to the matrix representations of symmetry operations.

If we use a suitable basis, the matrices representing symmetry operations will always have elements ±1 and a determinant ±1 (such matrices are termed *unimodular*). For example an inversion point (at 0,0,0) in three dimensions is represented by the matrix shown below converting $x,y,z$ to $\bar{x},\bar{y},\bar{z}$:

$$\begin{pmatrix} -1 & 0 & 0 \\ 0 & -1 & 0 \\ 0 & 0 & -1 \end{pmatrix}\begin{pmatrix} x \\ y \\ z \end{pmatrix} = \begin{pmatrix} -x \\ -y \\ -z \end{pmatrix} \tag{A2.7}$$

The four-dimensional inversion *axis* (along the $w$ direction) is represented by:

$$\begin{pmatrix} -1 & 0 & 0 & 0 \\ 0 & -1 & 0 & 0 \\ 0 & 0 & -1 & 0 \\ 0 & 0 & 0 & 1 \end{pmatrix}\begin{pmatrix} x \\ y \\ z \\ w \end{pmatrix} = \begin{pmatrix} -x \\ -y \\ -z \\ w \end{pmatrix} \tag{A2.8}$$

An operation that leaves only a point (0,0,0,0) unchanged is:

$$\begin{pmatrix} -1 & 0 & 0 & 0 \\ 0 & -1 & 0 & 0 \\ 0 & 0 & -1 & 0 \\ 0 & 0 & 0 & -1 \end{pmatrix}\begin{pmatrix} x \\ y \\ z \\ w \end{pmatrix} = \begin{pmatrix} -x \\ -y \\ -z \\ -w \end{pmatrix} \tag{A2.9}$$

There are no generally agreed symbols for these symmetry operations, let's call the last

[1]There appears not to be an agreement on the meaning of the word "axis". We use it to mean a one-dimensional figure (straight line) left invariant by a symmetry operation. However one commonly sees the statement that one can have 5-fold symmetry axes in four dimensions; as discussed next the five-fold symmetry operation only leaves a point invariant. The axes in four dimensions are (roto-) inversion axes. The difficulty arises in part because (for example) in three dimensions one needs to specify the direction of the rotation component of a $\bar{4}$ symmetry operation as well as the location of the point of the inversion component even though neither the rotation axis or the inversion point exist separately in this example.

one Ξ.[1] It is instructive to break this operation into components.

$$\begin{pmatrix} -1 & 0 & 0 & 0 \\ 0 & -1 & 0 & 0 \\ 0 & 0 & -1 & 0 \\ 0 & 0 & 0 & -1 \end{pmatrix} = \begin{pmatrix} -1 & 0 & 0 & 0 \\ 0 & -1 & 0 & 0 \\ 0 & 0 & 1 & 0 \\ 0 & 0 & 0 & 1 \end{pmatrix} \begin{pmatrix} 1 & 0 & 0 & 0 \\ 0 & 1 & 0 & 0 \\ 0 & 0 & -1 & 0 \\ 0 & 0 & 0 & -1 \end{pmatrix} \quad \text{(A2.10)}$$

To identify the two matrices on the right we recall that in a three-dimensional Cartesian axis system a two-fold rotation about $z$ is represented as:

$$\begin{pmatrix} -1 & 0 & 0 \\ 0 & -1 & 0 \\ 0 & 0 & 1 \end{pmatrix} \quad \text{(A2.11)}$$

so that the first of the two matrices on the right of Eq. A2.10 can be identified as rotation about the $zw$ plane. The second is clearly a rotation about the $xy$ plane. Thus the operation Ξ corresponds to two 2-fold rotations about orthogonal planes (in four dimensions orthogonal planes have only a point on common). It should be obvious that repeating Ξ twice produces the identity, so the order of the operation is 2. It transpires that there are a number of other operations that may be considered as combinations of rotations and that have orders of 3, 4, 6, 8, or 12. The rotations are restricted to 2-, 3-, 4- or 6-fold rotations as in lower dimensions, but the new symmetry operations may have different order.

There is also a 5-fold symmetry operation compatible with translational symmetry in four dimensions. It is instructive to approach it by first considering lower dimensions. A three-fold rotation about a point in two dimensions can be represented by:

$$\mathbf{A} = \begin{pmatrix} 0 & -1 \\ 1 & -1 \end{pmatrix}, \quad \mathbf{A}^2 = \begin{pmatrix} -1 & 1 \\ -1 & 0 \end{pmatrix}, \quad \mathbf{A}^3 = \begin{pmatrix} 1 & 0 \\ 0 & 1 \end{pmatrix} \quad \text{(A2.12)}$$

In particular this operation takes the point 1,0 to 0,1 to $\bar{1},\bar{1}$ and back to 1,0; i.e. it interchanges the three equivalent axes of a hexagonal lattice.

In three dimensions there is a 4-fold operation that leaves only a point invariant; this is $\bar{4}$. We describe this operation using the basis of a primitive *bcc* cell [i.e. axes inclined at $\cos^{-1}(-1/3)$ to each other]. Successive powers of $\mathbf{A} = \bar{4}$ in this basis are represented by

$$\mathbf{A} = \begin{pmatrix} 0 & 0 & -1 \\ 1 & 0 & -1 \\ 0 & 1 & -1 \end{pmatrix}, \mathbf{A}^2 = \begin{pmatrix} 0 & -1 & 1 \\ 0 & -1 & 0 \\ 1 & -1 & 0 \end{pmatrix}, \mathbf{A}^3 = \begin{pmatrix} -1 & 1 & 0 \\ -1 & 0 & 1 \\ -1 & 0 & 0 \end{pmatrix}, \mathbf{A}^4 = \begin{pmatrix} 1 & 0 & 0 \\ 0 & 1 & 0 \\ 0 & 0 & 1 \end{pmatrix} \quad \text{(A2.13)}$$

The nature of **A** can be verified (see § 3.7.3) by checking that $\mathrm{Tr}(\mathbf{A}) = -1$ and $\det(\mathbf{A}) = -1$ as well as observing that it has an order of 4. Now the point 1,0,0 is successively taken to

[1]Symbols encountered include 22, 2222, I↑ and $\partial\varepsilon_{22}$.

0,1,0; 0,0,1; $\bar{1},\bar{1},\bar{1}$ and back to 1,0,0. Normally the $\bar{4}$ operation is described using an orthogonal basis so that the matrix representation is different; but of course the trace and determinant of the matrix are unchanged.

Finally coming to four dimensions, there is a five-fold operation that can be represented simply using a basis with axes at $\cos^{-1}(-1/4)$ to each other and which is represented by the matrix **V**:

$$\mathbf{V} = \begin{pmatrix} 0 & 0 & 0 & -1 \\ 1 & 0 & 0 & -1 \\ 0 & 1 & 0 & -1 \\ 0 & 0 & 1 & -1 \end{pmatrix} \tag{A2.14}$$

The reader should verify that $\mathbf{V}^5 = \mathbf{E}$ (the unit matrix) and that successive operations on e.g. 1,0,0,0 produce 0,1,0,0 ; 0,0,1,0 ; 0,0,0,1 ; $\bar{1},\bar{1},\bar{1},\bar{1}$ and back to 1,0,0,0. Note that $\det(\mathbf{V}) = 1$, so this is a proper operation. As it transforms a set of integers $p,q,r,s$ representing a lattice point into another set of integers representing another lattice point it is a symmetry element of a lattice and hence compatible with translational symmetry.

Some words of caution. The two-dimensional hexagonal lattice actually has a six-fold symmetry element (6-fold rotation) and the four-dimensional icosahedral lattice has a ten-fold symmetry element (which is a combination of **V** with $\Xi$ [1]) but the three-dimensional lattice does not have an eight-fold symmetry element.[2] The general four-dimensional lattice with a ten-fold symmetry element is referred to as *decagonal*. The unit cell parameter constraints are $a_1 = a_2 = a_3 = a_4$ and $\alpha_{12} = \alpha_{23} = \alpha_{34} = \alpha$, $\alpha_{13} = \alpha_{14} = \alpha_{23} = \beta$, with $\cos\alpha + \cos\beta = -1/2$ (i.e. there are two independent unit cell parameters), we have referred in the previous paragraph to the case $\alpha = \beta$.

To completely identify symmetry operations corresponding to matrices in four dimensions one needs as well as the trace and determinant, the *second invariant* which, for a matrix with elements $a_{ij}$, is:

$$\begin{aligned}&(a_{11}a_{22} - a_{12}a_{21}) + (a_{11}a_{33} - a_{13}a_{31}) + (a_{11}a_{44} - a_{14}a_{41}) + \\ &(a_{22}a_{33} - a_{23}a_{32}) + (a_{22}a_{44} - a_{24}a_{42}) + (a_{33}a_{44} - a_{34}a_{43})\end{aligned} \tag{A2.15}$$

Interest in quasicrystals, some of whose *diffraction patterns* have icosahedral symmetry, prompts the observation that in six dimensions, 5-fold "axes" [3] oriented as in icosahedral symmetry are possible (as well as 7-, 9-, 14- and 18-fold symmetry operations).

With orthogonal axes the **V** operation is represented by the matrix:

[1] It should be obvious that just as every three-dimensional *lattice* has $\bar{1}$ for a symmetry element, every four-dimensional lattice has $\Xi$ (which reverses the direction of lattice vectors) as a symmetry element.

[2] A combination of $\bar{4}$ with $\bar{1}$ (*existing separately*) is $4/m$.

[3] See footnote on p. 398.

$$\mathbf{V}' = \begin{pmatrix} \cos\rho_1 & \sin\rho_1 & 0 & 0 \\ -\sin\rho_1 & \cos\rho_1 & 0 & 0 \\ 0 & 0 & \cos\rho_2 & \sin\rho_2 \\ 0 & 0 & -\sin\rho_2 & \cos\rho_2 \end{pmatrix} \tag{A2.16}$$

which represents a double rotation first about the $zw$ plane by $\rho_1$ and then about the $xy$ plane by $\rho_2$ (compare the discussion of $\Xi$ given above). From the fact the trace must be $-1$ and the second invariant must be 1 (as for $\mathbf{V}$) it can easily be found that $\rho_1 = 2\pi/5$ and that $\rho_2 = 4\pi/5$. Thus the $\mathbf{V}$ operation is the *combination* of rotations by 1/5 and 2/5 of a circle about orthogonal planes.

## A2.5 Numbers of crystallographic symmetry groups

There are 4895 four-dimensional space groups and, as we have seen, the order of the point group can be as high as 1152 so that an *International Tables* for four dimensions with the same detail as Volume *A* (1983) would run to hundreds of volumes, and five dimensions would be out of the question. The table below lists the numbers of crystallographic groups in $n$ dimensions and with lattices of differing ($\leq n$) dimensions of periodicity.

| space | lattice dimension → | 0 | 1 | 2 | 3 | 4 | $n$ |
|---|---|---|---|---|---|---|---|
| 0 | | 1 | | | | | |
| 1 | | 2 | 2 | | | | |
| 2 | | 10 | 7 | 17 | | | |
| 3 | | 32 | 75 | 80 | 230 | | |
| 4 | | 227 | ? | ? | 1651 | 4895 | |
| $n$ | | ? | ? | ? | ? | ? | ? |

The entry under column 0 is the number of crystallographic point groups. It *is* known that the entry in every column is finite for finite $n$.

## A2.6 Generalization of Euler's formula for polyhedra

The generalization to $n$ dimensions of Euler's formula for polyhedra is:

$$\sum_{i=0}^{n-1} (-1)^i P_i = 1 - (-1)^n \tag{A2.17}$$

Where $P_i$ is the number of $i$-dimensional figures ($P_0$ is the number of vertices, $P_1$ is the number of edges etc.). For $n = 3$, this gives the familiar formula $P_0 - P_1 + P_2 = 2$ (i.e. $V - E + F = 2$). In four dimensions: $P_0 - P_1 + P_2 - P_3 = 0$. For a four-dimensional simplex $P_0 = 5$, $P_1 = 10$, $P_2 = 10$, $P_3 = 5$.

From Eq. 2.17 the analogous formula for a honeycomb (space filling by $n$-dimensional polytopes)can be derived[1]:

$$\sum_{i=0}^{n} (-1)^i P_i = 0 \tag{A2.18}$$

Where now the numbers should be interpreted as *relative* numbers of figures of the appropriate dimensionality. For a plane tessellation Eq. A2.18 gives the result: $V - E + F = 0$ which we have used many times. We use the analogous expression for three-dimensional honeycombs in Appendix 3.

### A2.7 References

One of the best introductions to elementary $n$-dimensional geometry is still *An Introduction to the Geometry of n Dimensions* by D. M. Y. Sommerville (Dover, New York, 1958). The definitive work on regular polytopes in $n$-dimensions is *Regular Polytopes* by H. S. M. Coxeter [third edition, Dover, New York, 1973] and on polytopes in general *Convex Polytopes* by B. Grünbaum [Interscience, New York, 1967]. Four-dimensional lattices are described fully by H. Wondratschek *et al.*, *Acta Crystallogr.* A**27**, 523 (1971) and an account of the four-dimensional space groups is in *Crystallographic Groups of Four-Dimensional Space* by H. Brown *et al.* [Wiley, New York, 1978]. A good introduction to four-dimensional crystallographic point groups and symmetry operations has been given by E. J. W. Whittaker [*An Atlas of Hyperstereograms of the Four-Dimensional Crystal Classes*, Clarendon Press, Oxford, 1985]. The four-dimensional hyperlayer groups (i.e. the symmetry groups of four-dimensional objects with translations in three dimensions) are described in *Colored Symmetry* by A. V. Shubnikov and N. V. Belov [Pergammon Press, Oxford, 1964]. For a wild ride into many-dimensional space (but with some excellent introductory material) *Sphere Packings, Lattices and Groups* [Springer-Verlag (1988)] by J. H. Conway and N. J. A. Sloan is highly recommended (this book also includes an enormous bibliography). There have been many recent examples of higher-dimensional crystallography applied to real world problems a good starting point is T. Janssen, *Acta Crystallogr.* A**42**, 261-271 (1986). Applications to quasicrystals will be found in the review by W. Steurer, *Zeits. Kristallogr.* **190**, 179 (1990). Two commonly used notations for point symmetry operations in four dimensions are those of C. Hermann [*Acta Crystallogr.* **2**, 139-145 (1949)] and A. C. Hurley [*Proc. Cambridge Philos. Soc.* **47**, 650-661 (1951)].

[1]For Eqs. A2.17 and A2.18, see the Coxeter reference cited in the next section.

## APPENDIX 3

# THE TOPOLOGY OF POLYHEDRA, NETS AND MINIMAL SURFACES

## A3.1 Introduction

Topological aspects of crystal chemistry are attracting increasing attention. One reason for the interest is that zeolite catalysts are of major economic importance and their properties are intimately related to structural features such as the size of the pores and cages. These in turn are related to topological properties such as the connectivity and the sizes and numbers of rings in the net of the framework atoms. This Appendix describes some topological aspects of structures, particularly of 3- and 4-connected nets and infinite polyhedra. It will be seen that the subject poses some interesting and challenging unsolved problems.

## A3.2 Finite polyhedra

Normally a polyhedron is thought of as a simple convex object topologically equivalent to a sphere: thus if the faces were deformable, it could be "blown up" so that it became a sphere in the same way as a truncated icosahedron ($5.6^2$) becomes a soccer ball. A finite polyhedron of this sort is topologically equivalent to a tiling of the surface of a sphere. The well-known Euler equation for the number of faces ($F$), edges ($E$), and vertices ($V$) of such a tiling is $F - E + V = 2$

A torus (an object shaped like a doughnut or the inner tube of a tire) has a hole through it, and is topologically different from a sphere. For a tiling of a torus, $F - E + V = 0$. This is the same as for a tiling of the plane (see § 5.6.11).

What about surfaces with more than one hole in them? A teacup with one handle is topologically the same as a torus and contains one hole. A soup bowl with two handles has two holes and is thus topologically distinct. The number of holes in a surface ($H$) is related to the *Euler-Poincaré characteristic* $\chi$ of a surface by:[1]

$$\chi = 2 - 2H \qquad \text{A3.1}$$

and for a surface with characteristic $\chi$:

$$V - E + F = \chi \qquad \text{A3.2}$$

Thus the torus (and the infinite plane) have $\chi = 0$ and a simple closed surface (such as that of a sphere) has $\chi = 2$ ($H = 0$).

[1]See H. S. M. Coxeter, *Introduction to Geometry* (Book List). A simple proof of Eq. A3.2 is given by R. Courant & H. Robbins, *What is Mathematics?* [4th Ed. Oxford (1947)].

A simple example of a polyhedron with $\chi = 0$ can be obtained from a ring of eight cubes fused together as shown on the left in Fig. A3.1. There are 32 faces, 64 edges and 32 vertices. Note that there are eight vertices of the sort $4^3$, sixteen $4^4$ and eight $4^5$.

Also shown in the figure is a polyhedron made by fusing together 20 cubes. It has five[1] holes ($\chi = -8$) and 72 faces, 64 vertices and 144 edges, so $V - E + F = -8$ confirming the formula above.

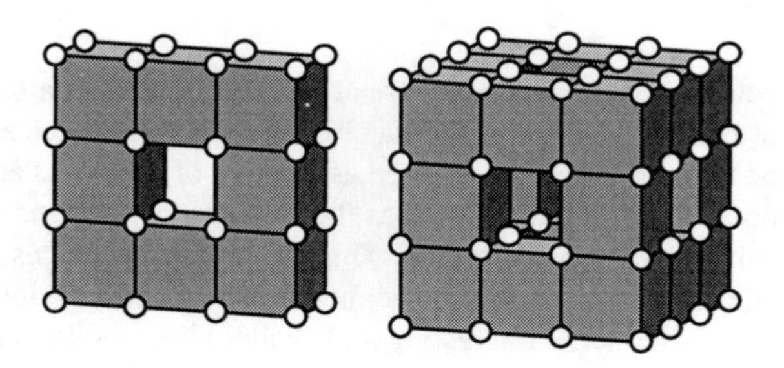

**Fig. A3.1**. Polyhedra with holes.

## A3.3 Infinite polyhedra

An obvious next step is to repeat the above process to make an infinite polyhedron with connecting holes. Consider a cubic unit cell with side 2 containing a cube of side 1 centered at 1/2,1/2,1/2 and fused to unit cubes centered at the face centers (0,1/2,1/2)κ. Repetition of the unit cell will produce a solid figure with 8 vertices, 12 faces, and 24 edges per unit cell. The empty space left behind will be exactly the same: empty cubes at 0,0,0 and at (0,0,1/2)κ (the same coordinates as before but displaced by 1/2,1/2,1/2). The structure shown on the right in Fig. A3.1 is an element of the infinite structure; it may be seen that square holes emerge through each face of the unit cell. As holes emerging from opposite faces are equivalent (really the same hole) the number of holes per unit cell is 6/2 = 3.

It is interesting to consider this figure as an infinite periodic polyhedron with $\chi < 0$. The numbers of vertices, edges, faces and holes per repeat unit are expressed as lower case letters ($v$, $e$, $f$ and $h$ respectively), we then have with $x = 2 - 2h$:

$$v - e + f = 2 - 2h = x \qquad \text{A3.3}$$

This formula applied to the infinite polyhedron constructed from fused cubes gives 8 - 24 + 12 = 2 – 2 × 3 = –4. Note that each vertex is $4^6$. This polyhedron is an example of a *skew polyhedron.*[2] The edges and vertices form a 6-connected net.

[1]Counting holes can be tricky. Start with a cube and make a hole between two opposite faces (one hole). Make a hole through a third face to join the first hole (the total is now two holes). Repeat for the fourth, fifth and sixth faces (three more holes for a total of five).

[2]H. S. M. Coxeter, *Proc. Lond. Math. Soc.* **43**, 33 (1937).

The polyhedron and the complementary empty space have the same connectivity. In this case the connectivity of the network of face-sharing cubes is just that of the primitive cubic lattice (six) and $h$ is one half this connectivity. As discussed particularly by Wells (reference in § A3.9) it is often simpler to consider the connectivity of the polyhedron than to count holes. In some cases (those in which the polyhedron in question and its complementary polyhedron are topologically identical) care must be exercised in deciding the repeat unit and the connectivity, and we now give some examples which also illustrate how to count vertices, edges and faces.

For a polyhedron with just one kind of vertex, once the number of vertices in the repeat unit is identified, counting edges and faces is easy. The number of edges is just one-half the number of vertices times the connectivity (i.e. $2v$ if four polygons meet at a vertex). To count faces note that each $n$-gon is shared by $n$ faces, so that if there are (for example) $v$ vertices of the type $n_1.n_2.n_3.n_4$, the number of faces is $v/n_1 + v/n_2 + v/n_3 + v/n_4$.

An infinite polyhedron can be obtained from the sodalite net (§ 7.3.10) by considering one-half the polyhedra of a space filling by truncated octahedra. Again the filled and empty space regions are identical. The cubic unit cell contains a truncated octahedron at 0,0,0 and it is fused together with its neighbors by sharing square faces, and has the same connectivity ($6 = 2h$) as the first infinite polyhedron. The reader should confirm that $v = 12$, $f = 8$ (the hexagonal faces) and $e = 24$, so that again $v - e + f = -4$ for a unit cell including three holes. This structure is also a regular (skew) polyhedron, in this case $6^4$.

In § 7.4.1 (p. 323) we described the two complementary Archimedean polyhedra $4^2.6^2$ of the Linde A structure. In this structure the number of vertices in the repeat unit (unit cell) is $v = 24$. Accordingly $e = 48$ and $f = 20$. The connectivity is again six ($h = 3$ and $x = -4$) as can be seen readily from Fig. 7.38 (p. 324) and again $v - e + f = x$.

Consider next the infinite polyhedron $4^3.6$ corresponding to the zeolite rho net (Fig. 7.39, p. 325). The body-centered cubic cell has again three holes and contains 48 vertices. Note that in considering this structure as an infinite polyhedron we consider points at the origin as *inside* the polyhedron and points at 1/2,1/2,1/2 as *outside* so in *this* sense the unit cell is primitive (so care must be taken in deciding what is the repeat unit). Thus the net of the tunnels is again the 6-connected simple cubic array and $h = 3$, $x = -4$.

We also described the same set of points as the polyhedron 4.8.4.8 which was shown (Fig. 7.39) as a packing of octagonal prisms; the structure is now connected as the 8-coordinated body-centered cubic array so in this case $h = 4$ and $x = -6$ for the primitive unit as shown in Fig. A3.2. The number of vertices in the primitive unit is 24 so the number of edges is 48, the number of faces is 18 and $v - e + f = 24 - 48 + 18 = -6 = x$.

**Fig. A3.2.** Left: a repeat unit of **bcc**. Right: the repeat unit of the diamond structure; each bond (except the one in the center) traverses a face of the primitive unit cell.

The infinite polyhedron corresponding to the faujasite net (Fig. 7.41, p. 326) is another $4^3.6$. There are 192 vertices in the face-centered cell, or 48 in the primitive cell. The network of tunnels in the faujasite structure has a diamond connectivity so now a hole enters (and leaves) through each face of the primitive cell (see Fig. A3.2) so there are now three holes per *primitive* cell. Enumeration shows that for the primitive cell $v = 48$, $e = 96$ and $f = 44$. Again $v - e + f = -4$ as expected for $h = 3$.

So far we have considered infinite polyhedra with four polygons meeting at each vertex. In § 7.2 (Fig. 7.7) we described two nets $6.8^2$. The one we called $6.8^2P$ is a tiling of a surface with primitive cubic connectivity ($h = 3$, $x = -4$). The repeat unit contains 48 vertices (eight hexagons) and as the net is 3-connected the number of edges is $48 \times 3/2 = 72$ and $f = 48/6 + 48/4 = 20$. Note that the repeat unit is taken as the body-centered cell; this because, just as for the polyhedron $4^3.6$ (zeolite rho) discussed above, the complementary polyhedron is the same as the original polyhedron.

The other polyhedron $6.8^2$ (also shown in Fig. 7.7) called $6.8^2D$ also requires some care, again because the complementary polyhedron is the same. The repeat unit contains 24 vertices. The net of the structure (the surface being tiled) is that of diamond but now the unit cell contains just one "diamond" unit and the connectivity is 4 ($h = 2$, $x = -2$) (contrast the discussion of the faujasite structure above). The number of edges is $24 \times 3/2 = 36$ and the number of faces is $24/6 + 24 \times 2/8 = 10$ and $v - e + f = 24 - 36 + 10 = -2 = x$ as expected.

Other infinite polyhedra with just one kind of vertex have also been described. In Chapter 6 on the right of Fig. 6.73 (p. 276) an infinite polyhedron $3.4^2.3.4^2$ was illustrated. The reader may wish to verify that for this structure $x = -4$, $v = 12$, $e = 36$, and $f = 20$ and that again $f - e + v = x$.

In § 6.8.5 we also mentioned (p. 277) the polyhedron $3^3.4^3$ made from rhombicuboctahedra and octahedra sharing faces (not illustrated). The net of the surface being tiled is eight connected (as in **bcc**) so $h = 4$.

Examples of infinite polyhedra with five faces meeting at a vertex were described in § 7.9. These are: $3.4^4$ and $3^3.6^2$ (Fig. 7.75, p. 357) and a second $3.4^4$ (Fig. 7.74, p. 356); all have $h = 3$ (for the primitive cell in the last two cases).

A polyhedron $3^7$ was described in § 6.8.6 (Fig. 6.78, p. 280). Again $h = 3$ for the primitive cell which contains 24 vertices, 84 edges (7-connected) and 56 faces. This polyhedron was derived from the packing corresponding to the **Al** vertices in $\mathbf{WAl_{12}}$. The reader might verify that in the latter structure nine triangles meet at every vertex and they tile a surface with the connectivity corresponding to **bcc** ($h = 4$) and can therefore be considered as an infinite polyhedron $3^9$. It should be clear that many of the sphere packings of Chapter 6 can also be described as infinite polyhedra. As examples we leave the reader to verify that the **pyrochlore** packing (p. 236) and that of the atoms on the snub cubes in $\mathbf{NaZn_{13}}$ (p. 273) are both polyhedra $3^8$.[1] These polyhedra are included in Table A3.1.

An interesting (and equivalent) way of considering the above polyhedra is in terms of the sums of angles of the polygons meeting at a vertex. The interior angle at the vertex of a

[1]The second of these requires some care. The primitive repeat unit is a pair of (opposite hand) snub cubes. The ten (so far) unshared square faces of this unit are each shared with other pairs so $h = 10/2 = 5$.

regular $N$-gon is $\alpha = \pi(1 - 2/N)$. Now for a polyhedron the angular defect at each vertex $k$ is defined as $\delta_k = 2\pi$ minus the sum of the angles $\alpha$. Thus consider a vertex common to an $n_1$-gon, $n_2$-gon, ..., $n_i$-gon:

$$\delta_k = 2\pi[1 - \sum_i (\tfrac{1}{2} - \tfrac{1}{n_i})] \tag{A3.4}$$

This formula defines $\delta_k$ for any vertex even if the individual polygons are not regular; all that matters is the $n_i$, the number of edges of the polygons. With $\delta_k$ so defined an alternative way of writing equation A3.3 is:

$$\sum_k \delta_k = 2\pi x \tag{A3.5}$$

Thus for a finite convex polyhedron ($x = 2$) we have Descartes' formula for the sum over the $k$ vertices:

$$\sum_k \delta_k = 4\pi \tag{A3.6}$$

For a plane tiling (net) with $x = 0$:

$$\sum_k \delta_k = 0 \tag{A3.7}$$

For infinite polyhedra with a topology characterized by $h$, the sum over the $v$ vertices in the primitive unit gives:

$$\sum_k \delta_k = 4\pi(1 - h) \tag{A3.8}$$

These formulas hold for all polyhedra of the types indicated even if there are many kinds of vertex and irregular polygons. They are quite useful for determining the numbers of different kinds of vertex that can be combined in a polyhedron (or a plane net).

To eliminate the factors of $\pi$ we define $\Delta = \delta/4\pi$, then for a polyhedron with one kind of vertex, the number of vertices in the repeat unit for a given connectivity are $v = (1 - h)/\Delta$. Table A3.1 below lists some of the infinite polyhedra we have mentioned above. It may be verified by referral to the crystallographic data given in the text that indeed $v$ is given correctly.

Yet another useful form of Eq. A3.3 can be obtained for infinite polyhedra with all vertices with the same connectivity $r$ (the number of polygons meeting at a point). Let there be $f_i$ polygons with $i$ sides per repeat unit so that $\Sigma f_i = f$. The number of edges is $e = rv/2$ and, as each polygon contributes $i/r$ vertices, $v = \Sigma i f_i / r$. Substituting these expressions in

Eq. A3.3:

$$\sum_i [2r + (2 - r) if_i] = 2rx \qquad \text{A3.9}$$

With the special cases:

$$r = 3 \text{ (three-connected)} \qquad \Sigma(6 - i)f_i = 6x \qquad \text{A3.9a}$$
$$r = 4 \text{ (four-connected)} \qquad \Sigma(4 - i)f_i = 4x \qquad \text{A3.9b}$$

As polygons with faces all having $r$ sides are dual to polyhedra with $r$-connected vertices, the number $p_n$ of $n$-connected vertices of polyhedra made of $r$-gons is:

$$\sum_n [2r + (2 - r) np_n] = 2rx \qquad \text{A3.10a}$$

In particular for simplicial polyhedra (all faces triangles), $r = 3$ and:

$$\Sigma(6 - n) = 6x \qquad \text{A.3.10b}$$

**Table A3.1.** Data for some infinite polyhedra (see text for symbols). Notice that a vertex symbol can refer to more than one polyhedron.

| polyhedron | name or comment | page | $-1/\Delta$ | $h$ | $v = (h - 1)(-1/\Delta)$ |
|---|---|---|---|---|---|
| $6.8^2$ | $6.8^2$ *D* | 296 | 24 | 2 | 24 |
| $6.8^2$ | $6.8^2$ *P* | 296 | 24 | 3 | 48 |
| $4^3.6$ | rho, faujasite | 324, 326 | 24 | 3 | 48 |
| $4^2.6^2$ | Linde A | 323 | 12 | 3 | 24 |
| 4.6.4.6 | analcime | 376 | 12 | 3 | 24 |
| $6^4$ | regular | 405 | 6 | 3 | 12 |
| $4^6$ | regular | 404 | 4 | 3 | 8 |
| $4^3.8$ | rho, $W^*8$ | 325, 327 | 16 | 4 | 48 |
| 4.8.4.8 | Fig. 7.39 | 325 | 8 | 4 | 24 |
| $3^3.6^2$ | Fig. 7.75 | 357 | 12 | 3 | 24 |
| $3.4^4$ | Figs. 7.74 and 7.75 | 356, 357 | 12 | 3 | 24 |
| $3.4.6^2.4$ | part of $UB_{12}$ packing | 356 | 6 | 3 | 12 |
| $3.4^2.3.4^2$ | Fig. 6.73 (right) | 276 | 6 | 3 | 12 |
| $3^3.4^3$ | not illustrated | 277 | 8 | 4 | 24 |
| $3^7$ | Fig. 6.78 | 280 | 12 | 3 | 24 |
| $3^8$ | pyrochlore | 236 | 6 | 3 | 12 |
| $3^8$ | $NaZn_{13}$ | 273 | 6 | 5 | 24 |
| $3^9$ | $WAl_{12}$ | 257 | 4 | 4 | 12 |

Note that a 4-connected net has six angles at a vertex and only four of these are considered when we consider the structure as a 4-connected infinite polyhedron. As the omitted angles are an opposite pair, there can be as many as three different descriptions of a given structure as an infinite polyhedron. Thus the net A·B·C·D·E·F could be described as polyhedra A.C.B.D, C.E.D.F or A.E.B.F.[1]

## A3.4 Space filling by polyhedra: nets and ring sizes

Instead of considering infinite polyhedra (tilings of surfaces with holes), we could consider the patterns arising from the vertices of a space *filling* (or tiling) by finite polyhedra. Let there be $P$ polyhedra, $F$ faces, $E$ edges and $V$ vertices. Euler's equation becomes now (see Eq. A2.18):

$$P - F + E - V = 0 \qquad \text{A3.11}$$

We will discuss infinite patterns with $p, f, e$ and $v$ polyhedra, faces, edges and vertices respectively per repeat unit. We start by verifying Eq. A3.11 for some simple cases.

In closest sphere packing there are two tetrahedra and one octahedron ($p = 3$) per vertex ($v = 1$). Each vertex is connected to 12 others, so (per vertex) $e = 6$. The octahedron has 8 faces and each of the tetrahedra has 4 faces, but as each face is shared with another the total number of faces is $f = (8 + 2 \times 4)/2 = 8$. In this case $p - f + e - v = 3 - 8 + 6 - 1 = 0$.

The body centered cubic structure divides space into six tetrahedra per atom so for $v = 1$, $p = 6, f = 6 \times 4/2 = 12$. To count edges we note that the edges of the tetrahedra are half cube body diagonals and cube edge lengths, so we consider the vertices as coordinated to the 8 nearest neighbors and the 6 next-nearest neighbors; accordingly $e = (8 + 6)/2 = 7$. Again $p - f + e - v = 0$.

Many intermetallic structures ("topologically close packed") are made up of a packing of tetrahedra (sharing faces) only. For these $f = 2p$ and so $p = e - v$. If further there are $N_n$ vertices that are $n$-coordinated, then $e$ is half the sum of $nN_n$, i.e.

$$p = \sum_n nN_n / 2 - v \qquad \text{A3.12}$$

We now apply this formula to the **$\beta$-W** structure of $A_3B$ (§ 6.6.4) in which $A$ is 14-coordinated and $B$ is 12-coordinated. Per unit $A_3B$, $p = (3 \times 14 + 12)/2 - 4 = 23$.

Note that if we divide Eq. A3.12 on both sides by $p$ we get the result that the number of tetrahedra per vertex is half the average coordination number minus one:

$$p / v = \sum_n nN_n / 2v - 1 \qquad \text{A3.12a}$$

An example of a 4-connected net derived from a polyhedral packing is that of sodalite (see Fig. 7.30, p. 316). Per primitive cell: $p = 1$, $v = 6$, $e = 12$ and $f = 7$ (half the number

[1]Note also that we group the angles by opposite pairs when determining the long Schläfli symbol for a 4-connected net; but, in accord with established usage, give the sizes of rings contained in angles in *cyclic* order when describing a polyhedron.

of faces of a truncated octahedron). Again $p - f + e - v = 0$. We use below the fact that there are 3 square faces and 4 hexagonal faces in the primitive cell.

For a 4-connected net $e = 2v$ and Eq. A3.11 becomes:

$$f - p = v \qquad \text{A3.13}$$

Let $f_n$ be the number of $n$-faces per repeat unit ($f_4 = 3$ and $f_6 = 4$ for sodalite) then as each vertex is shared with 6 faces (a 4-connected net has six angles), the number of vertices is:

$$v = \Sigma n f_n / 6 \qquad \text{A3.14}$$

The average ring size is $\langle n \rangle = \Sigma n f_n / f$ so that from Eqs. A3.12-14:

$$\langle n \rangle = 6 - 6p/f \qquad \text{A3.15}$$

It should be verified that for the sodalite net $p/f = 1/7$ and $\langle n \rangle = 36/7$.

The derivation of Eq. A3.15 shows that we only count the shortest rings at each angle (corresponding to polyhedron faces) and not the larger rings inside the polyhedra (for example the 12-rings around the truncated octahedra in the sodalite structure).

A number of nets derived from a space filling by polyhedra were described in Chapter 7. (See the exercises § 7.12.5 for $HL4_2$). Some of those with just one kind of vertex are listed in Table A3.2 below. $r$ is the density expressed as the number of vertices per unit volume (for unit edge length). Clearly there is not a strong correlation between average ring size and density. Generally the larger the *range* of ring size the smaller the density. The nets of § 7.6 (clathrate hydrates etc.) with 4-, 5- and 6- rings all have $r$ about 0.57. This suggests that the geometric mean ring size should be considered.[1] This is shown in the last column of the table as $\{n\}$ and clearly correlates better with $r$ than does $\langle n \rangle$.

Be sure to distinguish the two different descriptions of a structure (such as **rho**) as an infinite polyhedron ($4^3.6$) and as a four-connected net (4·4·4·6·8·8). Nets that can be described as polyhedron packings have no subscripts in the long Schläfli symbol.

Other 4-connected nets based on polyhedron packings were described in § 7.6. For the packings of 14-hedra, the average ring size is the same as for **sodalite** ($\langle n \rangle = 5.142$). For the packings of dodecahedra and larger polyhedra corresponding to the nets of the hydrates the average ring size is given in Appendix 4 (§ A4.5) and ranges from 5.06 to 5.11. The average ring size in all known zeolite structures (including those with more than one kind of vertex) that are derived from polyhedron packings falls in the narrow range $144/29 = 4.966 \leq \langle n \rangle \leq 36/7 = 5.142$. We conjecture that for any 4-connected net realizable with edges of equal length, and derived from packings of finite polyhedra, the average ring size is in the range $9/2 = 4.5 \leq \langle n \rangle \leq 36/7 = 5.14$.[2]

[1]Remember to count rings *per vertex*. An $n$-ring at a vertex is $1/n$ of an $n$-ring per vertex as an $n$-ring belongs to $n$ vertices.

[2]The restriction to *finite* polyhedra is necessary. A net like **MAPO-39** (Fig. 7.17, p. 306) is composed of finite and infinite (in one-dimension) polyhedra and the average ring size is 16/3 = 5.33.

**Table A3.2.** Average <*n*>, and geometric meanring size {*n*}, and density *r* of some polyhedron packings.

| net | symbol | <*n*> | *r* | {*n*} |
|---|---|---|---|---|
| $W^{*}4$ | 3·8·3·12·3·12 | 144/31 = 4.645 | 0.197 | 3.945 |
| $W^{*}8$ | 4·4·4·8·4·12 | 144/29 = 4.966 | 0.302 | 4.636 |
| Linde *A* | 4·6·4·6·4·8 | 144/29 = 4.966 | 0.428 | 4.806 |
| faujasite | 4·4·4·6·6·12 | 36/7 = 5.142 | 0.380 | 4.858 |
| $HL4_2$ | 3·4·6·8·6·8 | 36/7 = 5.142 | 0.395 | 4.799 |
| rho | 4·4·4·6·8·8 | 36/7 = 5.142 | 0.426 | 4.917 |
| ZK5 | 4·4·4·8·6·8 | 36/7 = 5.142 | 0.448 | 4.917 |
| sodalite | 4·4·6·6·6·6 | 36/7 = 5.142 | 0.530 | 5.043 |

In § 7.6 we mentioned the similarity between bubble packings and the hydrate structures. Experimental studies of froths of approximately equal-volume bubbles show that some of the bubble faces have four sides and that the average number of edges per face is 5.14.[1] It has been calculated[2] that for a *random* froth (i.e. with a random distribution of edge lengths) the average ring size is $6/(1 + 35/24\pi^2) = 5.23$.

Many of the 4-connected nets of interest in crystal chemistry have more than one ring of a given size at an angle. (**Diamond** has two 6-rings at each angle). They cannot be considered as packings of polyhedra with three edges meeting at each vertex and for this reason the above analysis does not apply directly to them. However we saw in § 5.1.10 (Fig. 5.16, p. 148) that **diamond** could be considered as a packing of hexagonal tetrahedra (containing divalent vertices) to produce a structure with 2 faces common to each angle (12 common to each vertex). Let there be $\mu$ faces meeting at a vertex in a 4-connected net derived from a space filling by polyhedra, then instead of A3.15 (which applies for $\mu = 6$):

$$\langle n\rangle = \mu(1 - p/f) \qquad \text{A3.16}$$

In the case of **diamond** $\mu = 12$, and for tetrahedra sharing faces $p/f = 1/2$, so Eq. A3.16 gives $\langle n\rangle = 6$ as is indeed the case.

Note that for **quartz**, which has Schläfli symbol $6{\cdot}6{\cdot}6_2{\cdot}6_2{\cdot}8_7{\cdot}8_7$, $\mu = 20$ and $\langle n\rangle = 80/11$. If we were to consider that net as derived from a packing of polyhedra, Eq. A3.16 shows that $p/f = 7/11$; so some, at least, of the "polyhedra" have less than four faces.

## A3.5 Coordination sequences and topological density

In Chapter 7 the number $n_k$ of $k$th topological neighbors of a vertex of a net was defined. The sequence of numbers $n_k$ is called the coordination sequence. A measure of the local topological density is defined as $\rho_k$ which is the sum of all the topological neighbors in the first $k$ shells divided by $k^3$.

[1]For illustrations of bubble shapes see E. B. Matzke, *Amer. J. Botany* **33**, 58 (1946).

[2]J. L. Meijering, *Philips Res. Rep.* **8**, 270 (1953).

$$\rho_k = \sum_{i=1}^{k} n_i \, / \, k^3 \qquad \text{A3.17}$$

As $n_k$ is often given by a quadratic in $k$, $n_k = ak^2 + bk + c$ (see Chapter 7 for examples), the limit of $\rho$ as $k \rightarrow \infty$ is the global topological density $\rho_\infty = a/3$. For uninodal 4-connected nets $\rho_\infty$ seems to be a rational number between 1/3 and 2 and correlates well with the geometrical density.[1]

## A3.6 Enumerating and identifying nets

It is difficult to enumerate nets in a systematic and comprehensive way. The problem can be considered purely topological, or it could be required that the nets be *realizable*. A realizable net is defined to be one that can be made in Euclidean space with equal edges and all shortest distances between vertices corresponding to edges. The problem then becomes one of enumerating sphere packings. But note that some zeolite nets have as many as a dozen topologically-distinct vertices.

The topology of a realizable net is very simply specified. Each vertex must be connected to others either in the same unit cell or one of the 26 contiguous ones (i.e. the 6 sharing a cell face, the 12 sharing a cell edge and the 8 sharing a cell corner). Fig A3.3 shows the repeat unit of **diamond** (compare Fig. A3.2). The vertices in the primitive cell are labeled "1" and "2" and are connected to each other; each "1" is also connected to a "2" in a neighboring cell related by a primitive lattice translation in the $x$, $y$ or $z$ direction, and similarly for each "2." We could code this information in what we call a *connectivity table* as follows:

| | | | | |
|---|---|---|---|---|
| 1 | 2 [000] | 2 [100] | 2 [010] | 2 [001] |
| 2 | 1 [000] | 1 [$\bar{1}$00] | 1 [0$\bar{1}$0] | 1 [00$\bar{1}$] |

Note that the second line is redundant in the sense that it follows immediately from the first. In general for an $n$-connected net with $v$ vertices in the repeat unit, we need only specify the connectivity of the $nv/2$ edges in the repeat unit. In what follows we omit redundancies in the connectivity tables.

The regular $Y^*$ (**$SrSi_2$**) net has four vertices in the primitive cell and the connectivity is also shown in Fig. A3.3. The connectivity table is:

| | | | |
|---|---|---|---|
| 1 | 2 [000] | 3 [000] | 4 [000] |
| 2 | | 3 [100] | 4 [010] |
| 3 | | | 4 [001] |

[1]For more on coordination sequences for 4-connected nets see M. O'Keeffe, *Zeits. Kristallogr.* **196**, 21 (1991) and M. O'Keeffe & S. T. Hyde, *Zeits. Kristallogr.* (1996).

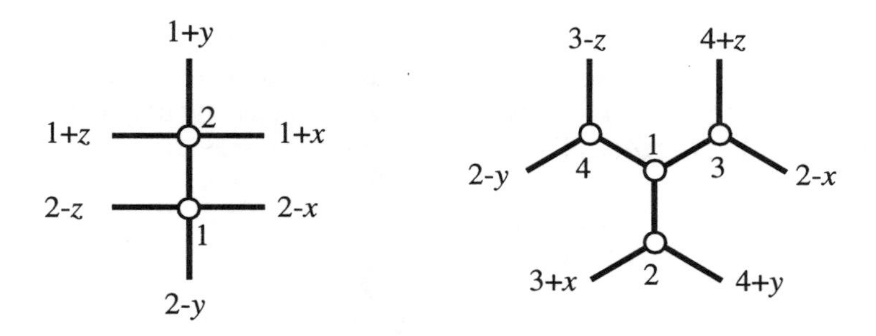


**Fig. A3.3.** The connectivity of **diamond** (left) and $Y^*$ (**$SrSi_2$**) nets (right).

The connectivity table of the **quartz** net is:

| | | | | |
|---|---|---|---|---|
| 1 | 2 [000] | 2 [100] | 3 [000] | 3 [010] |
| 2 | | | 3 [011] | 3 [$\bar{1}$01] |

It should be noted that the topology of the net is completely specified by the connectivity table and all topological properties such as numbers of rings and coordination sequences may be obtained from it. Unfortunately as both the numbering of the vertices ($v$! possibilities) and the labeling of directions are arbitrary, there are many apparently different connectivity tables that can describe the same net. A further difficulty in crystal structures is that frequently the crystallographic unit cell is larger than the topological repeat unit. In practice coordination sequences combined with Schläfli symbols (both of which are readily found by computer from the connectivity table) serve to identify a net with some reliability.

This method of describing nets allows ready generalization to higher dimensions (see references § A3.9). An example of the many interesting unsolved problems in this area is that of identifying the regular $n$-dimensional $m$-connected nets ($3 \leq m \leq n + 1$) corresponding to the three-dimensional nets $Y^*$ ($n = 3$, $m = 3$) and **diamond** ($n = 3$, $m = 4$).

## A3.7 Curvature and periodic minimal surfaces

There is currently great interest in periodic minimal surfaces (see § A3.9 for references) in a variety of contexts. We touch on some aspects of relevance to crystal structure here, starting with some elementary definitions and a discussion of *curvature*—a concept that has not entered into our discussion so far as we have been considering discrete, rather than continuous, structures.[1]

The curvature at a point $P$ on a two-dimensional curve can be defined informally as follows (see Fig. A3.4). Draw a circle through $P$ and two neighboring points $P_1$ and $P_2$ on

[1]We are content simply to state results that are derived in standard mathematics texts. Two excellent books (see Book List) are *Introduction to Geometry* by H. S. M. Coxeter and *Geometry and the Imagination* by D. Hilbert & S. Cohn-Vossen. We have borrowed heavily from the latter for this section.

either side of $P$ and on the curve (three points define a circle). As $P_1$ and $P_2$ approach closer and closer to $P$ we will obtain a limiting circle with radius $r$. $r$ is called the radius of curvature at $P$ and its reciprocal $k = 1/r$ is called the curvature. It should be clear that the curvature of a straight line is zero and that the curvature of a circle is constant.

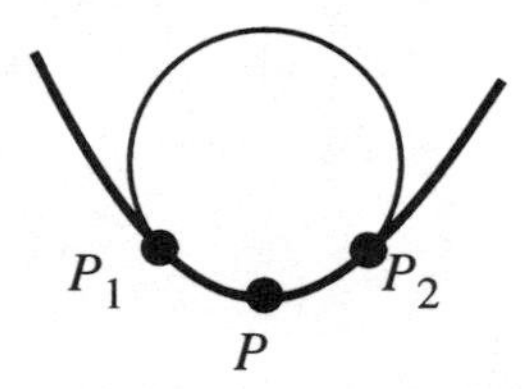


**Fig. A3.4**. Defining the curvature at a point on a curve (see text).

Now consider a three-dimensional surface such as an ellipsoid (Fig. A3.5) or a hill top. We can find two plane sections through the surface such that the lines of intersection are (a) of maximum curvature, $k_1$ and (b) of minimum curvature, $k_2$. These are known as the *principal curvatures*. Their product $K = k_1k_2$ is known as the *Gaussian curvature*. We adopt the convention that as the centers of curvature are "inside" the surface, the curvatures are considered positive.[1]

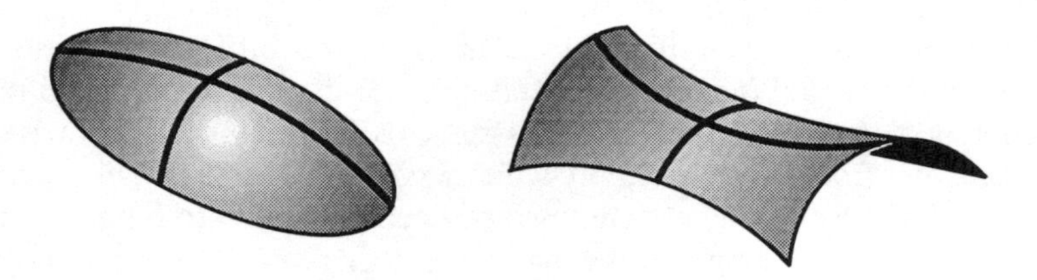

**Fig. A3.5**. Left: illustrating a point of positive Gaussian curvature and the directions of principal curvatures (heavy lines; the point in question is the intersection of these two lines). Right: similarly illustrating a point of negative Gaussian curvature.

An important property of the Gaussian curvature of a surface is that it remains invariant under bending; bending in this context referring to a deformation in which distances and angles on the surface remain invariant.[2]

The *mean curvature*, $H = (k_1 + k_2)/2$. A sphere has constant mean curvature and constant Gaussian curvature and is the only closed surface with these properties. The

[1]It may happen at a point on a surface that the curvature is the same in all directions, so that the principal directions are indeterminate. Such a point is called an *umbilical point*. All points on a sphere or a plane are of this type. More generally there are isolated umbilical points on a surface (there are four on an ellipsoid). If one of the principle curvatures is zero at a point, the point is referred to as *parabolic*.

[2]Thus imagine a plane sheet of paper (with zero Gaussian curvature) being rolled up into a cylinder, the minimum curvature (in a direction parallel to the axis of the cylinder) remains zero; so too does the Gaussian curvature. The maximum curvature becomes equal to the reciprocal of the radius of the cylinder.

sphere is also the solid with smallest total mean curvature for a given surface area; the total mean curvature being the mean curvature integrated over the surface.

We turn now to a hyperbolic surface (or a mountain pass or a saddle) as shown schematically in Fig. A3.5 (right). The two directions along which the principal curvatures are measured are indicated. The centers of the circles defining the radii of curvature are now on opposite sides of the surface so we adopt the convention that now one of the curvatures is negative. At such a point on such a surface the Gaussian curvature is negative and the mean curvature may be positive, negative *or zero*.

A *minimal* surface is one for which the principal curvatures are everywhere equal in magnitude but opposite in sign; i.e. the mean curvature is everywhere zero. A film of soap solution formed inside an arbitrarily-shaped loop of wire forms a bounded minimal surface.

The *integral curvature* of a surface is the integral of the Gaussian curvature over the surface. A remarkable result (the Gauss-Bonnet theorem) is that this surface integral is simply related to the characteristic of the surface (discussed in § A3.2):[1]

$$\int_S K \cdot d\sigma = 2\pi\chi \qquad \text{A3.18}$$

Thus the Gaussian curvature of a sphere ($\chi = 2$) is $1/r^2$ and the integral over the surface (area = $4\pi r^2$) = $4\pi$.

A *geodesic* line connecting two points on a surface represents the shortest line *on the surface* joining the two points.

The surfaces of infinite polyhedra are (faceted) infinite surfaces, and some, at least, are closely related to periodic minimal surfaces.

The cuprite ($Cu_2O$) structure can be described as two interpenetrating nets. Imagine the nets to be replaced by hollow tubes of elastic material and then that the tubes were inflated equally until they met at a surface. The surface (clearly periodic) divides space into two halves. It also has negative integral curvature and (less obviously) zero mean curvature, and is an example of a periodic minimal surface. As the topology of the labyrinth of each set of tunnels (inflated tubes) is the same as that of the diamond net, this is usually called the *D* surface.[2] The symmetry of the surface is the same as that of cuprite ($Pn\bar{3}m$).

A second minimal surface can be derived from two interpenetrating tubes with a six-connected simple cubic topology (Fig. A3.6). Again the surface separating the two sets of equally inflated tubes (in contact) is a periodic minimal surface, this time designated *P*. The symmetry is $Im\bar{3}m$. An alternative way to generate this surface is to arrange red and blue balloons as in **CsCl**; the *P* surface will *separate* the red and blue balloons after they are all equally inflated. Red balloons will touch red and blue will touch blue also; but the *P* surface corresponds to the boundary between red and blue. An approximation to the *P*

[1]An interesting property of a convex closed surface (such as an ellipsoid) is that it can not be bent (in the sense used above). Recall that a convex polyhedron is rigid (§ 5.6.3). When a football is deformed by kicking, the surface must be *stretched* as well as bent. Those who haven't tested it will be astonished by the rigidity of a plastic globe. Or, for that matter, of a ping pong ball.

[2]In the older literature this is also called the *F* surface.

surface can be obtained by packing truncated octahedra to fill space as illustrated in Fig. 7.30 (p. 316). The hexagonal faces approximate the minimal surface.

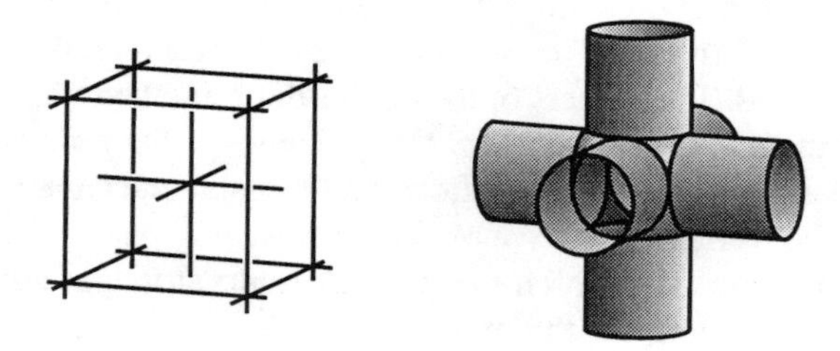

**Fig A3.6.** Left: part of two interpenetrating network of rods with cubic symmetry. Right: part of one set of rods shown partly "blown up" to form intersecting tubes.

A third minimal surface that is often discussed is known as Schoen's gyroid. The thought experiment suggested to the reader is now to inflate the two sets of cylinders in the intergrowth packing **γ-Si** (§ 6.7.3). The surface separating the two sets of equally inflated cylinders (no longer cylinders when "blown up") is the gyroid. The symmetry is $Ia\bar{3}d$; because of this, the surface is particularly difficult to illustrate satisfactorily. The references in § A3.9 may be consulted for help.

Much of the interest in crystal chemistry arises from the fact that some structures can be considered as tilings of periodic minimal surfaces. A good example of a 4- connected tiling of the *P* surface by hexagons and squares is provided by the net of the zeolite rho structure (see Fig. 7.39, p. 325—if the framework shown in the figure is mentally replaced by a curved surface one gets an idea of the appearance of the *P* surface). An example of a 3-connected tiling of the same surface by a hexagons and octagons is the 6.8.8 net of Fig. 7.7 (p. 297). A similar tiling of the *D* surface is the 6.8.8 net also shown in Fig. 7.7. The surface of the faujasite structure (Fig. 7.41, p. 326) is topologically the same as the *D* surface. Perhaps the most striking example is provided by the framework of the zeolite analcime (Exercise 7.12.6) which can be described as a tiling of the gyroid. This last structure is listed as an infinite polyhedron in Table A3.1.

## A3.8 Some conjectures about numbers and sizes of rings

The following observations (in this and the next paragraph) apply *only* to uninodal 4-connected nets and the statements should be construed as conjectures. It is possible that proving (or disproving!) them could lead to better constraints on the numbers and sizes of rings in 4-connected nets in general. For convenience the shortest ring at an angle is referred to as a SR. For 4-connected nets at least one of the SR's is a 6-ring or larger. For nets realizable with equal length edges corresponding to shortest distances between vertices at least one of the SR's is a 6-circuit or smaller. (Our "dense" net of § 7.5.1 contains only 7-rings but cannot be realized with shortest distances only corresponding to equal length edges). The largest SR is a 20-ring and the largest ring is a 24-ring.

In a net containing 5-rings there is always exactly one 5-ring per vertex.[1] In nets containing 10-rings, the total number of such rings meeting at a vertex is always a multiple of 5, and in a net containing 7-rings or 14-rings, the total number of such rings meeting at a vertex is always a multiple of 7. Uninodal 4-connected nets containing 11-, 13-, 15-, 17-, 19-, 22- or 23-rings have not been described.

Although the number of rings meeting at a vertex must be finite, the number can be quite large. Here for entertainment, and to test ring-counting skills of computer programs, are coordinates for a uninodal 4-connected net (Schläfli symbol $4\cdot6_2\cdot4\cdot6_3\cdot6\cdot18_{1422}$) with two 4-rings, six 6-rings and 3615 18-rings meeting at each vertex (1422 of the 18-rings are the shortest rings contained in one of the angles):

$R\bar{3}c$, $a = 6.700$, $c = 1.881$, $r = 0.984$
vertices in 36 *f*: (*x*,*y*,*z* etc.), $x = 0.029$, $y = 0.445$, $z = 0.0$

In § A3.6 we mentioned that it might be interesting to examine nets in *n*-dimensions. $n$+1-connected nets [with $n(n + 1)/2$ angles at each of the vertices] that are generalizations of **diamond** and **sodalite** have been described by M. O'Keeffe, *Acta Crystallogr.* A**47**, 748 (1991). The generalizations of diamond are all regular nets with vertex symbols $(6_{n-1})^{n(n+1)/2}$; that paper may be consulted for coordinates. Some connectivity tables are given below for other simple *n*-dimensional *m*-connected nets. The challenge is to derive coordinates for uniform edge lengths and to derive the systematics of the connection between ring size and dimensionality and connectivity. Note that $3 \leq m \leq n+1$. Is there a regular net for every *n* and *m*? How many?

The connectivity table for a regular 3-connected 4-dimensional net $12_4\cdot12_4\cdot12_4$ is:

| | | | |
|---|---|---|---|
| 1 | 2 [0000] | 4 [0000] | 6 [0000] |
| 2 | | 3 [0000] | 5 [0000] |
| 3 | | 4 [1000] | 6 [0100] |
| 4 | | | 5 [0010] |
| 5 | | | 6 [0001] |

This net, which also has fourteen 14-rings meeting at each angle (3 per vertex), might be compared with the 3-dimensional regular net ($Y^*$) $10_5\cdot10_5\cdot10_5$ (p. 295).

A very simple uniform (but not regular) 4-connected, 4-dimensional net with vertex symbol $8_6\cdot8_6\cdot8_7\cdot8_7\cdot8_7\cdot8_7$ has connectivity table:

| | | | | |
|---|---|---|---|---|
| 1 | 2 [0000] | 2 [1000] | 3 [0100] | 3 [0010] |
| 2 | | | 3 [0000] | 3 [0001] |

A regular 4-connected 6-dimensional net, $10_{10}\cdot10_{10}\cdot10_{10}\cdot10_{10}\cdot10_{10}\cdot10_{10}$, is:

[1]This means five 5-rings (each of which belongs to five vertices) meet at each vertex and the net cannot be composed entirely of 5-rings (even though the *average* ring size may be 5). In the nets of § 7.6, which have more than one kind of vertex there is more than one 5-ring per vertex; for example the type II hydrate net has 18/17 5-rings per vertex.

| | | | | |
|---|---|---|---|---|
| 1 | 2 [000000] | 3 [000000] | 4 [000000] | 5 [000000] |
| 2 | | 3 [100000] | 4 [010000] | 5 [001000] |
| 3 | | | 4 [000100] | 5 [000010] |
| 4 | | | | 5 [000001] |

## A3.9 References

There is a large literature on the topology of nets and polyhedra. Some references were given in § 7.11.10. The classic references are A. F. Wells' works: *Three-dimensional Nets and Polyhedra* [Wiley, New York (1977)] and *Further Studies of Three-Dimensional Nets* [American Crystallographic Association Monograph No. 8 (1979)].

The number and sizes of rings in 4-connected nets has been discussed by C. S. Marians & L. W. Hobbs, *J. Non-Crystalline Solids* **124**, 242 (1990); L. Stixrude & M. S. T. Bukowinski, *Amer. Mineral.* **75**, 1159 (1990); K. Goetzke & H.-J. Klien, *J. Non-Crystalline Solids* **127**, 215 (1991); M. O'Keeffe, *Zeits. Kristallogr.* **196**, 21 (1991). On the density of three-dimensional nets and its relationship to ring size, see S. T. Hyde, *Acta Crystallogr.* A**50**, 753 (1994).

The topological characterization of linkages of polyhedra has also given rise to quite a large literature. Some recent papers include E. Parthé, *Zeits. Kristallogr.* **189**, 101 (1989); E. Parthé & B. Chabot, *Acta Crystallogr.* B**46**, 7 (1990); N. Engel, *Acta Crystallogr.* B**47**, 217 (1991).

A topological topic, which we don't discuss, but which is nevertheless of considerable interest, is that of *percolation* in nets. A good introduction to this topic is D. Stauffer, *Introduction to Percolation Theory* [Taylor & Francis (1985)].

For applications of topology to molecular chemistry see *Chemical Applications of Topology and Graph Theory* [R. B. King (ed.) Elsevier, Amsterdam (1983)], and *Graph Theory and Topology in Chemistry* [R. B. King & D. Rouvray, (eds.) Elsevier, Amsterdam (1987)].

The literature on periodic minimal surfaces is rapidly expanding. A good introduction is *Crystalline frameworks as hyperbolic films* by S. T. Hyde [in *Defects and Processes in the Solid State: Geoscience Applications*, J. N. Boland & J. D. Fitz Gerald (eds.), Elsevier, (1993)]. Other references (which should be consulted for illustrations) that emphasize crystal-chemical applications are: S. T. Hyde & S. Andersson, *Zeits. Kristallogr.* **174**, 225 and 237 (1986); H. G. von Schnering & R. Nesper, *Angew. Chem. (Int. Ed.)* **26**, 1059 (1987), S. Andersson, S. T. Hyde, K. Larsson & S. Lidin, *Chem. Rev.* **88**, 221 (1988); W. Fischer & E. Koch, *Acta Crystallogr.* A**45**, 726 (1989). E. Koch & W. Fischer, *Acta Crystallogr.* A**46**, 33 (1990). The last two references describes a number of surfaces. A number of applications to chemistry, physics and biology are described in a collection of papers in *J. Phys.* C**7** (1990).

## APPENDIX 4

# LARGE POLYHEDRA

## A4.1 Introduction

In Chapter 5 we discussed some polyhedra with emphasis mainly on those polyhedra with a small number of vertices that commonly occur as coordination figures. Here we discuss some larger polyhedra (with more than 20 vertices) that are of increasing interest in several areas of chemistry and biochemistry.[1] First we review some basic material.

*Simple* polyhedra are those for which three edges meet at every vertex; clearly for $V$ vertices there are $3V/2$ edges, so the number of vertices is even. *Simplicial* polyhedra have only triangular faces. It should be obvious that they are the duals of simple polyhedra.

Polyhedra with $F$ faces, all of which are either $m$-gons or $(m+1)$-gons, where $m = [6 - 12/F]$,[2] are sometimes called *medial*. Their duals are simplicial polyhedra with $m$- and $(m+1)$-connected vertices. For $m \leq 4$ these latter are topologically equivalent to the deltahedra of § 5.1.6 (i.e. they, and only they, can be realized with equilateral triangles as faces). The interest in this appendix is mainly with the case of simple medial polyhedra with $m = 5$, i.e. those simple polyhedra with pentagon and hexagon faces (and their duals). For convenience we refer to these polyhedra as 5-6 polyhedra in what follows. In general, polyhedra cannot be realized with all faces as plane regular polygons although they are often "almost" regular polygons.[3]

Non-crystallographic symmetries are commonly encountered; as well as icosahedral symmetry, 5-fold and $\overline{10}$ and $\overline{12}$ axes often occur. Now the (probably more familiar) Schoenflies symmetry symbols are more appropriately used. Commonly encountered non-crystallographic symmetries in Hermann-Mauguin notation are $D_{6d} = \overline{12}m2$, $D_{5h} = \overline{10}m2$ and $D_{5d} = \overline{5}m$. We use the number of vertices to identify the polyhedron as this is generally more useful in chemistry (i.e. it is the number of atoms making up the polyhedron); the notation $V_N$ refers to a polyhedron with $N$ vertices.

## A4.2 5-6 Polyhedra

For 5-6 polyhedra there are exactly 12 pentagon faces and for $V$ vertices there are $E =$

[1] See *e.g.* T. G. Schmaltz *et al.*, *J. Amer. Chem. Soc.* **110**, 1113 (1988) for chemical applications and D. L. D. Caspar & A. Klug, *Cold Spring Harbor Symp. Quant. Biol.* **27**, 1 (1962) for biological applications.

[2] Here brackets indicate rounding down to the nearest integer.

[3] The dodecahedron and the truncated icosahedron are the only 5-6 polyhedra that can be constructed from regular plane polygons. See V. A. Zalgaller, *Convex Polyhedra with Regular Faces* [Consultants Bureau, New York (1969)].

$3V/2$ edges and $F = V/2 + 2$ faces.

The first member of this family is the pentagonal dodecahedron ($V_{20}$), already described in Chapter 5. This and some of the next members are encountered in the structures of the clathrate hydrates and clathrasils (see § 7.6 for $V_{24}$ and $V_{28}$ which are also illustrated in Figs. A4.1 and A4.2). Note that $V_{22}$ is topologically impossible. For $V > 26$ there is more than one isomer: two for $V_{28}$ (Fig. A4.2) and three for $V_{30}$ (Fig. A4.3) and the number of isomers grows very rapidly with $V$ so we only list some of the simpler cases in Table A4.1. The first four of the entries in the table are the duals of the Frank-Kasper polyhedra (§ 5.1.7, p. 143). It follows that these are the only 5-6 polyhedra in which hexagonal faces are completely surrounded by pentagons ("isolated" hexagons).

**Table A4.1** Some smaller 5-6 polyhedra

| $V$ | | symmetry |
|---|---|---|
| 20 | pentagonal dodecahedron | $I_h = m\overline{35}$ |
| 24 | 14-hedron of Type I hydrates | $D_{6d} = \overline{12}m2$ |
| 26 | 15-hedron | $D_{3h} = \overline{6}m2$ |
| 28 | 16-hedron of Type II hydrates | $T_d = \overline{4}3m$ |
| 28 | isomer of above | $D_2 = 222$ |
| 30 | pentagonal barrel | $D_{5h} = \overline{10}m2$ |
| 30 | isomer of above (9 kinds of vertex) | $C_{2v} = mm2$ |
| 30 | isomer of above (10 kinds of vertex) | $C_{2v} = mm2$ |
| 36 | hexagonal barrel | $D_{6h} = 6/mmm$ |
| 36 | tennis ball | $D_{2d} = \overline{4}m2$ |
| 36 | isomer of above | $D_{3h} = \overline{6}m2$ |

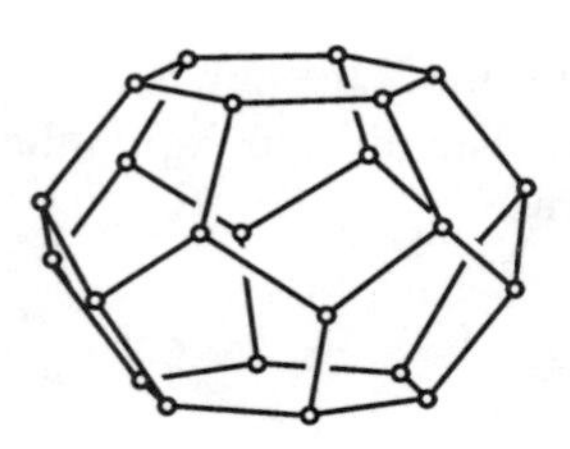

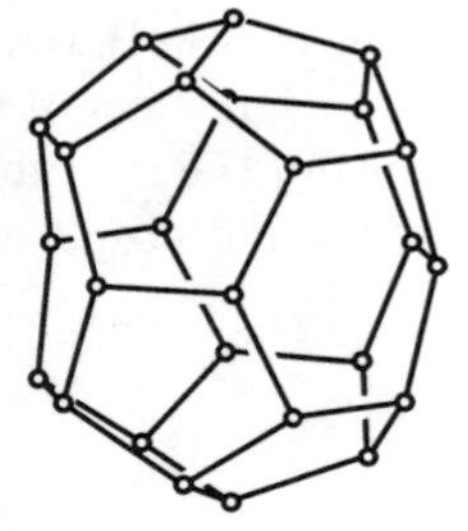

**Fig. A4.1**. The $V_{24}$ (left) and $V_{26}$ (right) polyhedra of Table A4.1.

The isomers of $V_{28}$ are illustrated in Fig. A4.2 and those of $V_{30}$ are illustrated in Fig. A4.3. It might be noted that the latter are easily inter-converted by rotation of two vertices and their connecting edge in the manner shown in Fig. 5.76 (p. 206).

Also listed in the Table are three symmetrical isomers of $V_{36}$ (see Fig. A4.3). The "tennis ball" is so named as the pentagons form a continuous edge-sharing strip that goes round the polyhedron in the same fashion as the seam goes round a tennis ball (which has the same symmetry). This is the smallest 5-6 polyhedron in which there are no vertices $5^3$ (i.e. vertices at which three pentagons meet).

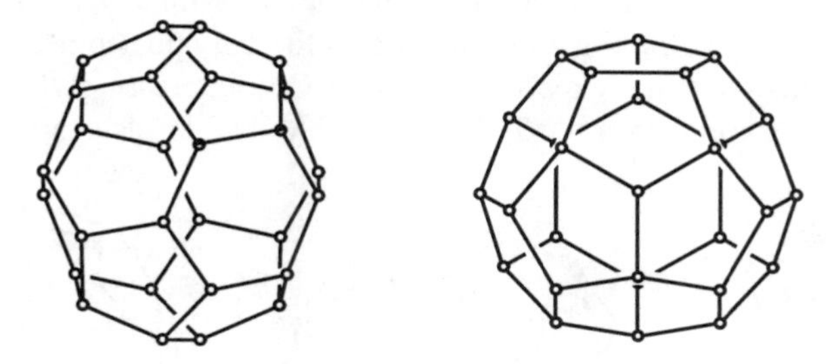

**Fig. A4.2**. The isomers of $V_{28}$ (Table A4.1). Left: $D_2$ viewed down a 2-fold axis. Right: $T_d$ viewed down a 3-fold axis.

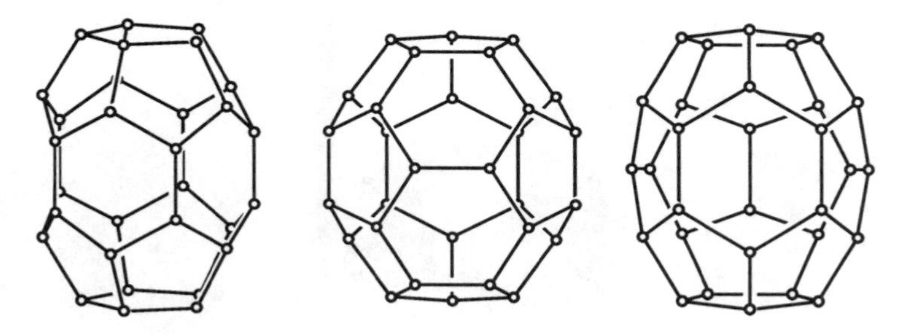

**Fig. A4.3**. The isomers of $V_{30}$ (Table A4.1). Left: $D_{5h}$ (pentagonal barrel). Middle and right: the two $C_{2v}$ isomers viewed down the 2-fold axis.

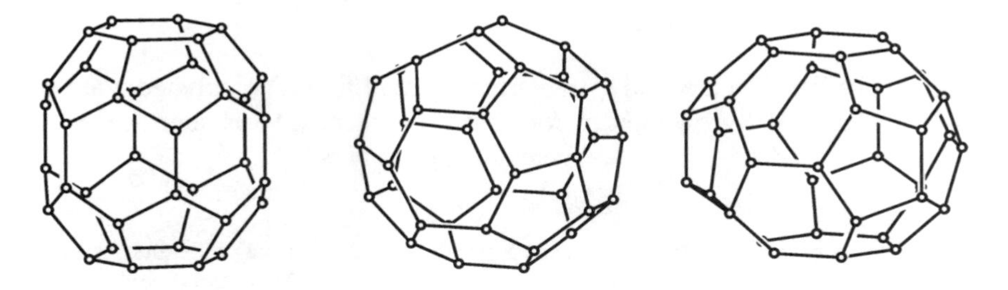

**Fig. A4.4**. Three isomers of $V_{36}$. Left: $D_{6h}$ (hexagonal barrel). Middle: $D_{2h}$ (tennis ball). Right: $D_{3h}$.

## A4.3 Fullerene polyhedra

Large polyhedral carbon molecules $C_n$ ($n \geq 60$) have come to be known collectively as *fullerenes*. These have structures based on 5-6 polyhedra in which pentagons are completely surrounded by hexagons. This restriction (known as the isolated pentagon rule or IPR), which can be justified by simple chemical arguments, greatly reduces the number of possible structures. The simplest possibility, and the easiest to prepare pure, is $C_{60}$ which has the structure of a truncated icosahedron ($5.6^2$). The next possibility is $C_{70}$ which is also fairly easy to prepare. Thereafter all stoichiometries with an even number of vertices are realizable as IPR polyhedra. Fig. A4.5 illustrates the only possibilities for $C_{70}$, $C_{72}$ and $C_{74}$ and Fig. A4.6 illustrates the two possibilities for $C_{76}$; interestingly, the $D_2$ isomer with 19 distinct vertices is favored over the $T_d$ isomer with only 5 kinds of vertex.[1]

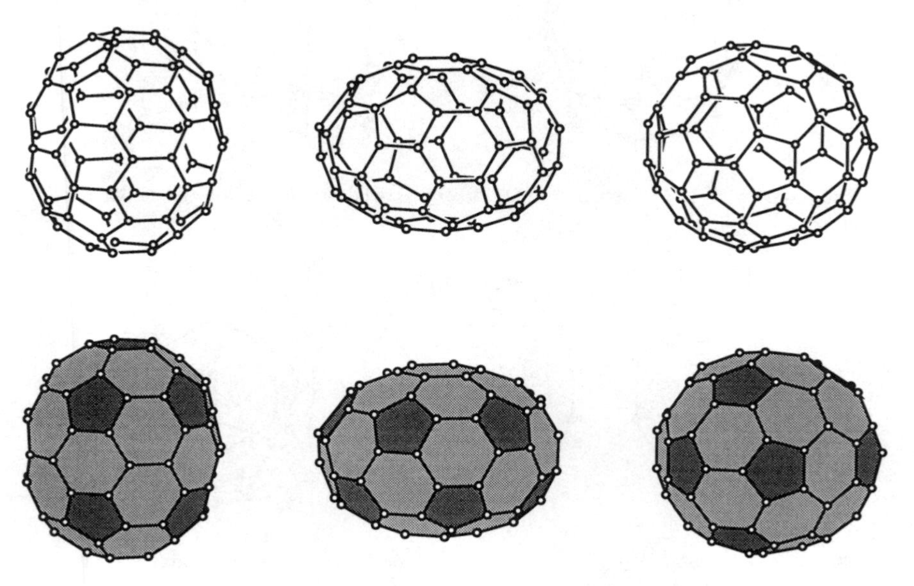

**Fig. A4.5**. From left to right: the structures of the fullerenes $C_{70}$, $C_{72}$ and $C_{74}$. On the top as "ball and stick" models, and below as opaque polyhedra.

The numbers of IPR isomers and symmetries of small fullerene polyhedra are given in Table A4.2.[2] The numbers grow rapidly; for 82 vertices there are 9 isomers, for 84 vertices there are 24 isomers and for 96 vertices there are 196 isomers.

[1]There have been very many reviews, conference proceedings, etc. describing fullerenes. A collection of articles appears in *Accounts of Chemical Research* **25**, No. 3 (1992). A good review with many illustrations and references is *Electronic Structure Calculations on Fullerenes and Their Derivatives* by J. Cioslowski [Oxford University Press (1995)]. Like other new materials that have brought so much excitement to solid state chemistry in recent years (quasicrystals, oxide superconductors) fullerenes are very easy and inexpensive to prepare. Indeed, it has been said, with some justification, that the greatest obstacle to fullerene synthesis was recognition of how remarkably simple it could be; and, as was the case for the other materials mentioned, it took non-chemists (W. Krätschmer and D. R. Huffman in this instance) to lead the way.

[2]We give the highest symmetry of polyhedra; molecules which are predicted to have degenerate ground states with that symmetry will probably distort to lower symmetry in accordance with the predictions of the Jahn-Teller theorem (this is the case for $T_d$ $C_{76}$ and $I_h$ $C_{80}$ for example).

Possible cubic symmetries for fullerene polyhedra are $T$ (23), $T_d$ ($\bar{4}3m$) and $T_h$ ($m\bar{3}$). Compositions for $T_d$ include $C_{76}$, and $C_{84}$. Fig. A4.7 illustrates $C_{116}$ with $T_h$ symmetry (the smallest IPR fullerene with this symmetry) and $C_{120}$ with $T_d$ symmetry. The smallest IPR polyhedron with $T$ symmetry is $C_{88}$.[1] Icosahedral polyhedra are discussed in the next section.

**Table A4.2** Symmetries of the isomers of the first few fullerenes.

| $V$ | isomers | symmetries |
|---|---|---|
| 60 | 1 | $I_h$ |
| 70 | 1 | $D_{5h}$ |
| 72 | 1 | $D_{6d}$ |
| 74 | 1 | $D_{3h}$ |
| 76 | 2 | $T_d$, $D_2$ |
| 78 | 5 | $D_{3h}$ (2), $D_3$, $C_{2v}$ (2) |
| 80 | 6 | $I_h$, $D_{5d}$, $D_{5h}$, $D_2$,$C_{2v}$ (2) |

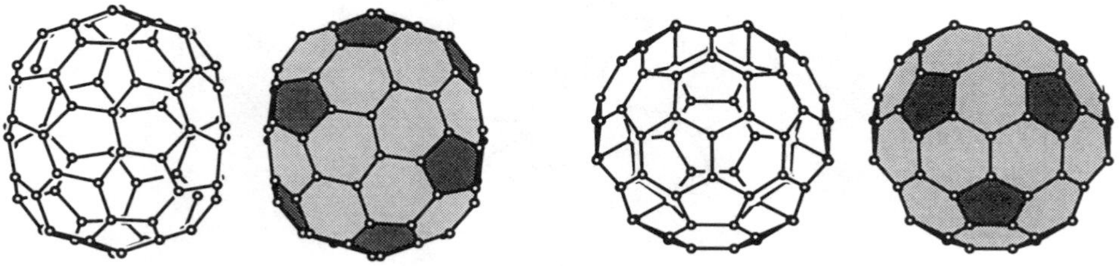

**Fig. A4.6**. The structures of the isomers of $C_{76}$. Left: $D_2$ projected down a 2-fold axis. Right: $T_d$ projected down a 3-fold axis (note that some vertices and edges are superimposed in the projection).

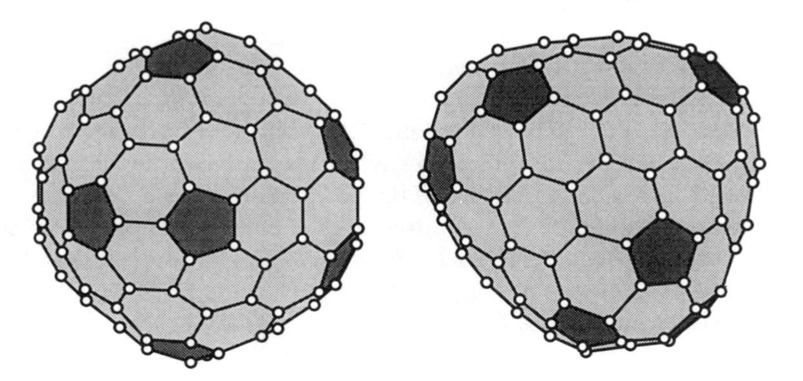

**Fig. A4.7.** The structures of the fullerene polyhedra $T_h$ $C_{116}$ (left) and $T_d$ $C_{120}$ (right).

Large fullerene (IPR) polyhedra can be generated from smaller 5-6 polyhedra by a

[1]The interested reader will find that knowing the number of vertices and the symmetry makes it easy to construct models of the more-symmetrical polyhedra that are not illustrated (see Notes).

process known as "leap-frogging." In this method, edges joining two new vertices are placed as a perpendicular bisector of each of the edges of the parent polyhedron. The new vertices form (smaller) hexagons inside the original hexagons and pentagons inside the original pentagons; additionally each of the old vertices is in the center of a new hexagon as shown in Fig. A4.8. It should be clear that even if the original polyhedron had adjoining pentagonal faces, the new polyhedron will have isolated pentagonal faces. Indeed the truncated icosahedron $5.6^2$ ($V_{60}$) is obtained by leap-frogging from the dodecahedron $5^3$ ($V_{20}$). A polyhedron that is obtained by leap-frogging must have three times the number of vertices of the original polyhedron and hence a multiple of six vertices. Note that only one of the five isomers of $C_{78}$ can be obtained by leap-frogging from $V_{26}$.

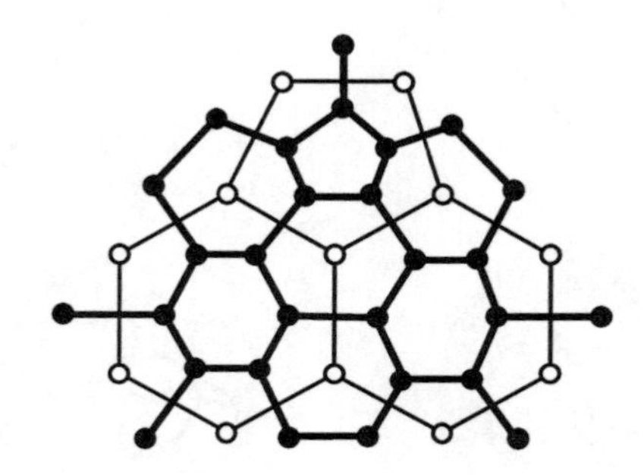

**Fig. A4.8.** Generating part of a larger polyhedron (heavy lines and filled circles) from a smaller polyhedron (lighter lines and open circles) by leap-frogging.

## A4.4. Icosahedral polyhedra

We discuss here only 5-6 polyhedra and their duals.[1] The icosahedral 5-6 polyhedra are of interest as symmetrical isomers of larger fullerenes and their duals are of considerable interest in connection with virus structures and have been extensively investigated in that connection.[2] To have icosahedral symmetry each of the twelve pentagons centers have to be on 5-fold axes. The remaining surface of the polyhedron is made up of hexagons. The arcs of great circles joining each pair of 5-fold axes of icosahedral symmetry enclose an equilateral spherical triangle covering 1/20 of a sphere as illustrated in Fig. A4.9.

To see how the polyhedron is developed, it is convenient to consider such a triangular patch which will be made up of a $6^3$ net with the hexagons at the corners replaced by

[1]These icosahedral polyhedra are sometimes called Goldberg polyhedra, as they appear to have first been described by M. Goldberg, *Tohoku Math. J.* **43**, 104 (1937). We must admit to not having read this paper. After discovery of the fullerenes, many authors (including us) independently "rediscovered" the icosahedral polyhedra. A treatment similar to the present one is to be found in T. G. Schmaltz *et al.*, *J. Amer. Chem. Soc.* **110**, 1113 (1988) who give further references and describe some other polyhedra.

[2]A beautiful picture of the polyoma virus which is based on the dual of 2,1 (see the discussion below) appeared on the cover of the February 13, 1992 issue of *Nature*. See J. P. Griffith *et al., Nature* **335**, 652 (1992).

pentagons.

Fig. A4.10 shows a fragment of $6^3$ with a triangular tile that is half a unit cell outlined. The tile contains one point. Suppose we take a larger hexagonal cell with new axes **a'** and **b'** related to the axes **a** and **b** of the elementary cell by:

$$\begin{pmatrix} \mathbf{a}' \\ \mathbf{b}' \end{pmatrix} = \begin{pmatrix} p+q & q \\ -q & p \end{pmatrix} \begin{pmatrix} \mathbf{a} \\ \mathbf{b} \end{pmatrix} . \qquad \text{A4.1}$$

The new cell is larger by a factor equal to the determinant of the matrix above, i.e. $p^2 + pq + q^2$ and a triangular tile (again one half of a unit cell) will contain $p^2 + pq + q^2$ points. If now the hexagons at the corners of the triangle are changed to pentagons we will have a patch corresponding to 1/20 of a polyhedron. The example of $p,q = 3,2$ is shown in Fig. A4.11 and basic units for some smaller polyhedra shown in Fig. A4.12.

There are distinct icosahedral polyhedra corresponding to each distinct pair $p,q$ with $p \geq q \geq 0$. The number of vertices is $V = 20(p^2 + pq + q^2)$. The number of edges is $3V/2$ and the number of faces (of which 12 are pentagons) is $V/2+2$. For the symmetry to be $I_h$, the edges of the triangles must lie on mirror planes; it should be evident that this is only the case if $p = q$ or $q = 0$, otherwise the symmetry is $I$. Note that there can be distinct icosahedral polyhedra with the same number of vertices; the first case is for $p,q = 7,0$ and 5,3 each of which have 980 vertices.

The first few simple icosahedral polyhedra with hexagon and pentagon faces are listed in Table A4.3 below. For their duals interchange $V$ and $F$. As the number of vertices increases the shape of the polyhedron tends to that of an icosahedron with pentagons at the twelve vertices. Fig. A4.13 illustrates $V_{240}$.

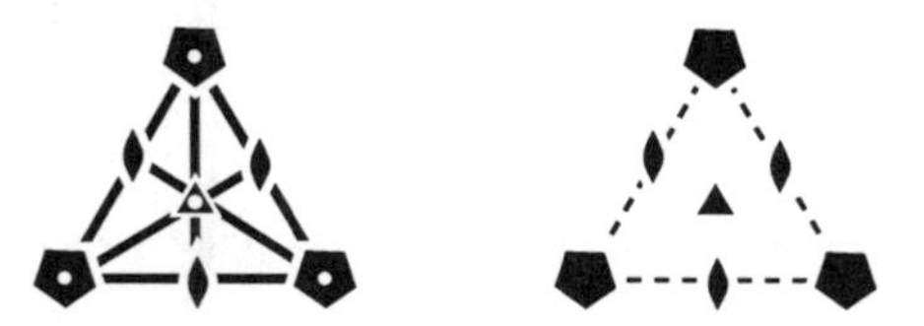

**Fig. A4.9**. Some of the icosahedral symmetry elements. On the left 2, $\bar{5}$ and $\bar{3}$ axes of $I_h$ are shown with heavy lines representing the traces of mirror planes. On the right are shown the corresponding 2, 3 and 5 axes of $I$.

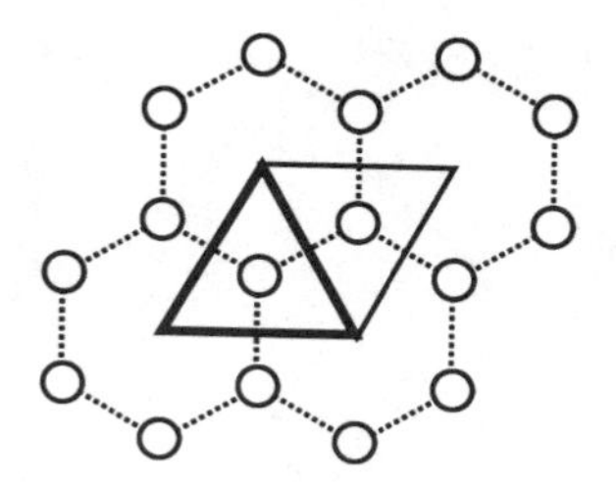

**Fig. A4.10.** A fragment of $6^3$ showing a unit cell and a triangular tile.

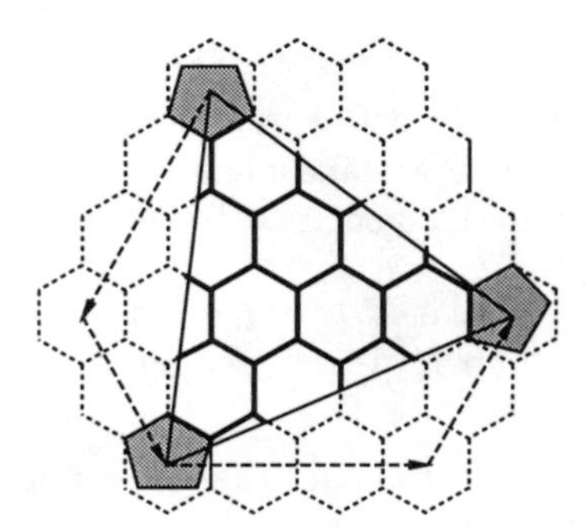

**Fig. A4.11**. A tile of the icosahedral polyhedron 3,2 (triangular outline). The arrows correspond to vectors of length $3a$ and $2a$ ($a$ is the distance between centers of hexagons). It might be verified that there are $3^2 + 3 \times 2 + 2^2 = 19$ vertices in the tile.

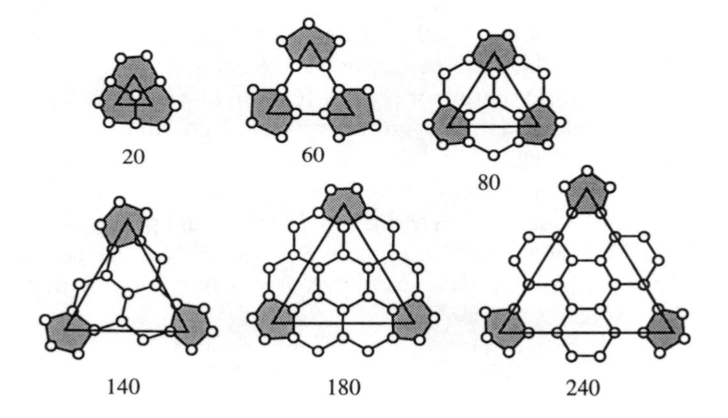


**Fig. A4.12**. Triangular patches of icosahedral polyhedra. The number under each diagram is the number of vertices in the full polyhedron.

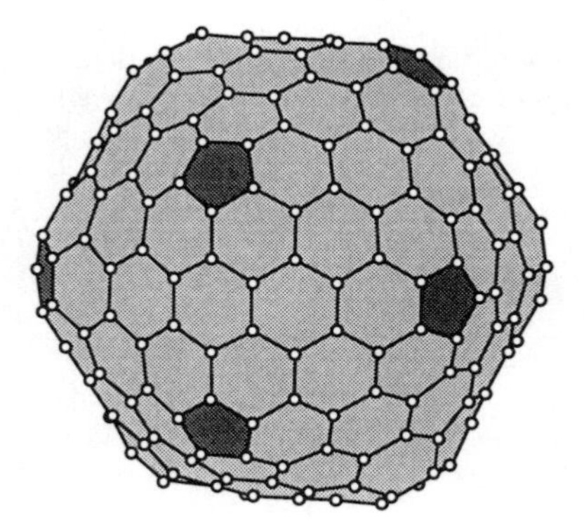

**Fig. A4.13.** The icosahedral polyhedron $V_{240}$.

**Table A4.3** Properties of the first few icosahedral 5-6 polyhedra. "pg" is the point group and $k_V$, $k_E$ and $k_F$ are respectively the numbers of topologically-distinct vertices, edges and faces.

| $p,q$ | $V$ | $E$ | $F$ | pg | $k_V$ | $k_E$ | $k_F$ |
|---|---|---|---|---|---|---|---|
| 1,0 | 20 | 30 | 12 | $I_h$ | 1 | 1 | 1 |
| 1,1 | 60 | 90 | 32 | $I_h$ | 1 | 2 | 2 |
| 2,0 | 80 | 120 | 42 | $I_h$ | 2 | 2 | 2 |
| 2,1 | 140 | 210 | 72 | $I$ | 3 | 4 | 2 |
| 3,0 | 180 | 270 | 92 | $I_h$ | 3 | 4 | 3 |
| 2,2 | 240 | 360 | 122 | $I_h$ | 3 | 3 | 4 |
| 3,1 | 260 | 390 | 132 | $I$ | 5 | 7 | 3 |
| 4,0 | 320 | 480 | 162 | $I_h$ | 5 | 6 | 4 |
| 3,2 | 380 | 570 | 192 | $I$ | 7 | 10 | 4 |
| 4,1 | 420 | 630 | 212 | $I$ | 7 | 11 | 5 |
| 5,0 | 500 | 750 | 252 | $I_h$ | 7 | 9 | 5 |
| 3,3 | 540 | 810 | 272 | $I_h$ | 6 | 9 | 5 |

## A4.5 Space filling packings of 5-6 polyhedra

The structures of two clathrate hydrates (Types I and II) were mentioned in § 7.6. These are based on space-filling packings of dodecahedra ($V_{20}$) and, respectively, 14-hedra ($V_{24}$) and 16-hedra ($V_{28}$). It is also possible to fill space with combinations of $V_{20}$, $V_{24}$ and $V_{26}$. These structures are difficult to illustrate clearly, but it easy to make models of them (see § A4.7). The simplest is a combination of $V_{20}$, $V_{24}$ and $V_{26}$ in the ratio 3:2:2, for want of a better name we call this structure III.[1] Crystallographic data for unit edge length are:

Packing III    $P6/mmm$, $a = 4.401$, $c = 4.399$
5.5.5.5.5.5 in 4 $h$, $x = 1/3$, $y = 2/3$, $z = 0.1137$
5.5.5.5.5.6 in 6 $l$, $x = 0.1312$, $y = 2x$, $z = 0.0$
5.5.5.5.6.6 in 6 $k$, $x = 0.2272$, $y = 0$, $z = 1/2$
5.5.5.5.5.5 in 12 $o$, $x = 0.2085$, $y = 2x$, $z = 0.1386$
5.5.5.5.5.6 in 12 $n$, $x = 0.3579$, $y = 0$, $z = 0.3140$

Another combination is known in the hydrate of tetra $n$-butyl ammonium benzoate which we call structure IV.[2] The relative proportions of polyhedra and the average ring size, $\langle n \rangle$, are listed in the table below.

The small range of average ring size is striking. Obviously $5 \leq \langle n \rangle < 6$ for any packing of 5-6 polyhedra, but it would be nice to have tighter bounds and to know the largest

[1]This is the framework of the hydrate of tetra *iso*-amyl ammonium fluoride. The real crystal structure is orthorhombic [D. Feil & G. A. Jeffrey, *J. Chem. Phys.* **35**, 1863 (1961)].

[2]This is a rather complex structure: there are 172 vertices of 17 crystallographic kinds in the tetragonal cell. See M.Bonamico, G. A. Jeffrey & R. K. McMillan, *J. Chem. Phys.* **37**, 2219 (1962). It should be remarked that much of our knowledge of hydrate structures is due to the work of Jeffrey and collaborators published in the early 1960s.

polyhedron which can participate in such packings. We are not aware of a *proof* that a space filling by polyhedra topologically equivalent to pentagonal dodecahedra is impossible. Notice that these packings only involve polyhedra with isolated hexagons.

| *structure* | $V_{20}$ | $V_{24}$ | $V_{26}$ | $V_{28}$ | $\langle n \rangle$ |
|---|---|---|---|---|---|
| I | 1 | 3 | | | 5.111 |
| II | 2 | | | 1 | 5.100 |
| III | 3 | 2 | 2 | | 5.106 |
| IV | 5 | 8 | 2 | | 5.109 |

## A4.6 Large coordination polyhedra

In § 5.1.7 (p. 143) we described the Frank-Kasper polyhedra which are simplicial polyhedra with $3^5$ and $3^6$ vertices and commonly found as coordination polyhedra in intermetallic compounds. The 16-vertex tetra-capped truncated tetrahedron (Friauf polyhedron) is the largest such polyhedron without adjacent $3^6$ vertices (i.e. $3^6$ vertices sharing an edge). To have larger coordination numbers without adjacent $3^6$ vertices, the coordination polyhedra must have $n$-gon faces with $n > 3$. An example is provided by the snub cube ($3^4.4$) which has triangular and quadrangular faces and occurs as the $\{Na\}Zn_{24}$ polyhedron in $NaZn_{13}$ (p. 273); it is shown again in Fig. A4.14.

There are polyhedra with $3^4.4$, $3^5$, and $3^6$ vertices in which the $3^6$ vertices are not adjacent. The largest is perhaps the polyhedron obtained by capping the eight hexagonal faces of the truncated octahedron ($4.6^2$). The resulting polyhedron (Fig. 14.14) has 32 vertices ($8 \times 3^6$ and $24 \times 3^4.5$).

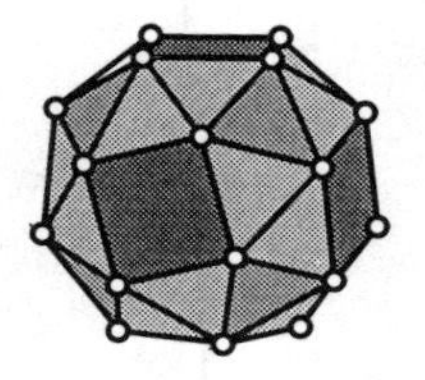

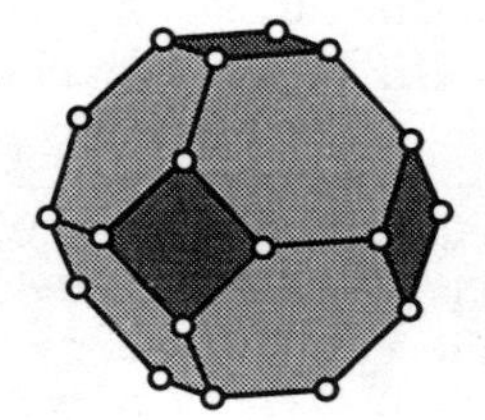

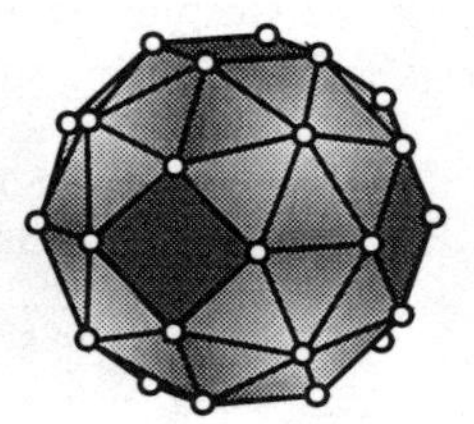

**Fig. A4.14.** Left: the snub cube ($3^4.4$). Right: the truncated octahedron ($4.6^2$) and the octacapped truncated octahedron with 32 vertices (in the conformation shown they are all equidistant from the center).

In the structures of $BaHg_{11}$ and $ThMn_{12}$ (these are the prototypes of fairly large families) there is a tetragonal (4/*mmm*) 20-vertex coordination polyhedron ($\{Ba\}Hg_{20}$ or $\{Th\}Mn_{20}$). This polyhedron can be considered to be derived by capping the hexagonal faces of a polyhedron [$3^8.4^2.6^4$] known as a "tetragonal hexagon prism"[1] to produce a polyhedron with four $3^6$ vertices, eight $3^5$ vertices and eight $3^5.4$ vertices as shown in Fig.

[1]The term "tetragonal hexagon prism" is used by E. Hellner & W. B. Pearson [*Physik Datan / Physics Data* **16**, Nr. 5 (1986)] who discuss its occurrence in intermetallic compounds. The polyhedron is not a *prism* in the usual sense.

A4.15.

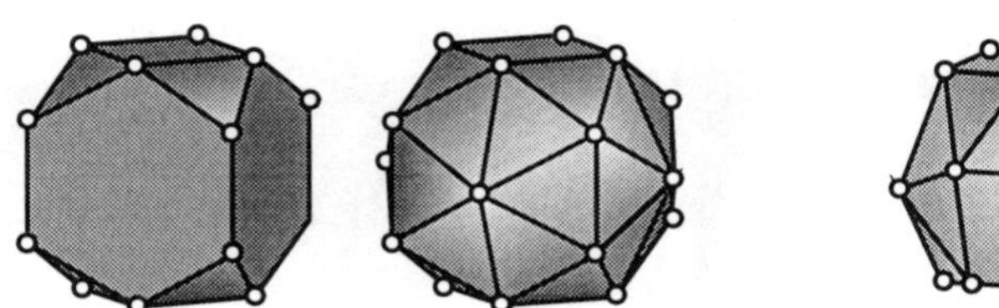

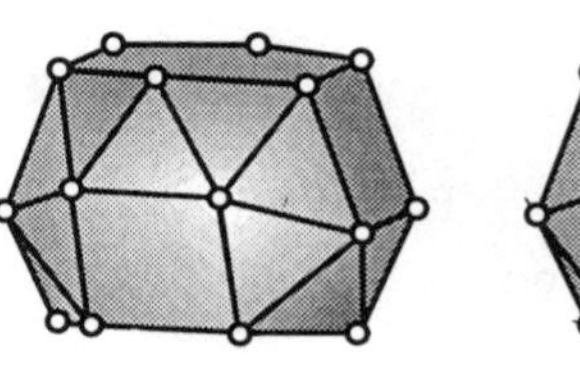

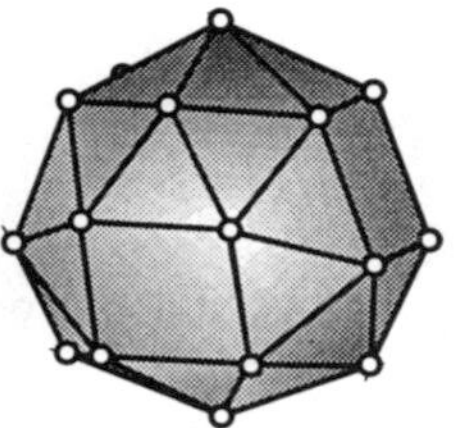

**Fig. A4.15.** Left: The "tetragonal hexagon prism" and the 20-vertex polyhedron obtained by capping its hexagonal faces. Right: a polyhedron with two hexagonal faces (top and bottom) and the polyhedron with 22 vertices obtained by capping the hexagonal faces.

Another family of structures is named for $BaCd_{11}$. In this structure there are $\{Ba\}Cd_{22}$ polyhedra. These are again tetragonal (now the symmetry is $\bar{4}m2$) and contain two $3^6$ vertices, four $3^5$ vertices and sixteen $3^4.5$ vertices. This is shown in Fig. A4.15 together with the polyhedron obtained by removing the two $3^6$ vertices. The following table, listing the number of vertices and faces in the four polyhedra with triangular and quadrangular faces, suggests that there are more polyhedra between $V_{24}$ and $V_{32}$.

| *polyhedron* | $3^5$ | $3^6$ | $3^4.4$ | *triangles* | *quadrangles* |
|---|---|---|---|---|---|
| $V_{20}$ | 8 | 4 | 8 | 32 | 2 |
| $V_{22}$ | 4 | 2 | 16 | 32 | 4 |
| $V_{24}$ | 0 | 0 | 24 | 32 | 6 |
| $V_{32}$ | 0 | 8 | 24 | 48 | 6 |

## A4.7 Models of large polyhedra

Most of the polyhedra describe here have irregular faces, so constructing models with (e.g.) cardboard faces is rather difficult. However satisfying "ball and stick" models can be made using tetrahedral or triangular stars. For smaller polyhedra, the tetrahedral star is more suitable (compare the angles of 108° in $V_{20}$ with the tetrahedral angle of 109.5°).

The net III of § A4.5 may be made the following way: (a) construct a column of $V_{24}$ polyhedra sharing hexagon faces, (b) surround the neck between alternate pairs of these polyhedra with a ring of six $V_{20}$ sharing the exposed pentagon faces (it will be found that there is only one way to do this). It will now be found that a ring of $V_{26}$'s fits snugly (again sharing pentagon faces) in the depressions of the new structure. Remember to keep the hexagon faces of $V_{26}$ parallel to the axis of the original column (which is the **c** axis).

To make large fullerene polyhedra 3-pointed "stars" are best. A good strategy is to construct the 12 isolated pentagons first, and then to explore ways in which they can be linked using the appropriate number of connectors of a different color (these are $6^3$).

## APPENDIX 5

# CRYSTAL STRUCTURE DATA

## A5.1 Introduction

In this volume we have been almost exclusively concerned with structure, and have paid little attention to the chemical compositions that have a given structure. There are several reasons for this approach. One is that some common structure types occur for a rather wide range of compounds; another is our belief that it is important to have some feeling for structures in general without being too weighed down with the baggage of theories that pretend to explain the occurrence of certain structure types for different compositions.

Nevertheless chemists, at least, should have some idea of the sorts of compounds that adopt the common structure types described in the text, so here we indicate some typical compositions for these. Of course many compounds are polymorphic, and in particular many transform under pressure. The general rule (there are some exceptions) is that increasing pressure causes a transformation to a structure with higher coordination numbers. Thus **sphalerite** compounds with 4-coordination generally transform to **NaCl** with 6-coordination; and **NaCl** compounds transform under pressure to **CsCl** with 8-coordination with, of course, an increase in density in each case.[1]

Crystal data for some compounds, referred to in the text, are given in the final section.

## A5.2 Elements

The metallic elements are nearly all either **cp** or **bcc** (Mn, and the early actinides are exceptions and they have generally rather complex structures). Periodic trends are fairly well developed; for example Ni, Pd, Pt and Cu, Ag, Au are all **ccp** and V, Nb, Ta and Cr, Mo, W are all **bcc**. However many metals are polymorphic; for example Fe is **bcc** at low temperature, **fcc** at higher temperature, and **hcp** under pressure. It should be noted that for most polymorphic metals the **bcc** and **cp** forms have very similar densities; in particular the **cp** forms are not always the densest modification.

C (at high pressure), Si, Ge and Sn (at low temperature) are **diamond** with 4-coordination. The remaining non-metallic elements have "covalent" structures with low coordination numbers (3 for P, As, Sb, Bi; 2 for S, Se, Te; 1 for N, O, F, Cl, Br, I).

[1]But note that although the density is greater in the high-pressure phase, the $A$-$X$ bond length is also greater; indeed the driving force for the transformation under pressure is to reduce the repulsion between atoms forced to be close together. Notice also in the examples we have given, that the structure of $AX$ *combined* is **diamond** for **sphalerite**, **pc** (primitive cubic) for **NaCl** and **bcc** for **CsCl** so the overall array is transforming to a more efficient packing under pressure.

## A5.3 Composition *AB* (*AX*)

There are hundreds of different *AB* (or *AX*) structure types. Some different structures we have mentioned briefly are those of AuCd, BaCu, FeSi, LiGe, LiP, MoB, NaP, NaPb, NbO, PbO and WC. These have either just one or at most a few representatives. Other structure types such as **ZnS, NaCl, NiAs, CuZn** (**CsCl**), **CuAu**, **CrB** and **FeB** have dozens or even hundreds of examples and we discuss them below.

### *A5.3.1* ***Sphalerite*** *and* ***wurtzite***

The polymorphs of ZnS (see § 4.6.4 and § 6.1.5) lend their names to the two common structure types with atoms in tetrahedral coordination.

**Wurtzite** is based on **hcp** arrays of both cations and anions (each in tetrahedral holes of the other array) and occurs mainly for oxides and nitrides such as BeO, ZnO, AlN and GaN. Very many ternary, quaternary, etc. compounds such as $\beta$-$NaFeO_2$, LiSiON, etc. have derived structures.

**Sphalerite** is more common and occurs for many binary compounds particularly of elements of columns 13-17 of the periodic table with eight valence electrons (excluding d electrons) per atom pair. Examples are "III-V" compounds *AX* with *A* = Al, Ga, In and *X* = P, As, Sb. and "II-VI" compounds *AX* with *A* = Be, Zn, Cd and *X* = S, Se and Te. Some compounds have both structures (ZnS!) and then it is common also to find polytypes based on more complex close packings. SiC is notable in this regard (see § 7.11.4). Interestingly, SiC is the only known "IV-IV" *compound*—GeC is unstable (has not yet been made) and GeSi is a disordered composition in the $Ge_xSi_{1-x}$ solid solution series. Just as for **wurtzites** there are many derived ternary, etc. structures; the most common type is **chalcopyrite** (**$CuFeS_2$**).

### *A5.3.2* **NaCl** *and* **NiAs**

**NaCl** and **NiAs** (see § 6.1.5) are the 6-coordinated analogs of **sphalerite** and **wurtzite** which are based on **ccp** and **hcp** respectively.

**NaCl** (also known as **rock salt**) is of course one of the more common structure types. It is often considered the prototypical "ionic" crystal structure and at normal pressures and temperatures it is that of all the alkali hydrides and halides other than CsCl, CsBr and CsI. Other **NaCl** compounds are the alkaline earth chalcogenides (MgO, SrS, BaTe etc.) other than MgTe and Be compounds (these have ZnS structures) Further examples are compounds *AX* with *A* = Sc, Y and lanthanide and *X* = N, P, As, Sb. Carbide examples include ThC and TiC. Hundreds of ternary compounds $ABX_2$ with the $\alpha$-$NaFeO_2$ structure (Na and Fe order) are also known. These are mainly oxides and sulfides, but also include antistructure compounds with ordered anions such as $Ba_2PBr$ and $Ca_2NCl$.

**NiAs** is also a very common structure type. For most of the compounds *AX*, *A* is a transition metal and *X* is S, Se, Te, V, As or Sb. Recall that the structure is hexagonal (see § 4.6.3) and it should be noted that many compounds have an axial ratio ($c/a$) rather different from the ideal value for **hcp As**. Many of the compounds assigned to this type also have a wide range of stoichiometry and some authors include compositions in the range $A_2X$ to $AX_2$ in the **NiAs** classification.

### *A5.3.3* **CuZn (CsCl)** *and* **CuAu**

**CsCl** ionic crystals are rare (CsCl, CsBr and CsI are the only alkali halides) at normal pressures, but many **NaCl** compounds transform to **CsCl** under pressure. The structure type, now often called **CuZn** is found for about a hundred intermetallic compounds such as LiAg, BeCu, CaTl, YIn, MnNi, etc. We saw in § 6.6.2 that the structure is in sense a special case of **AuCu** (which appears to be confined to intermetallic examples)

### *A5.3.4* **CrB** *and* **FeB**

**CrB** and **FeB** were described briefly in § 6.4.2. Together they comprise the structures of another large group (well over 100 binary examples) of intermetallic compounds. These orthorhombic structures have a number of free parameters (see data below) and the trigonal prisms (occupied by **B**) can vary significantly in shape from one compound to another and some authors recognize subgroups according to the shapes of the trigonal prisms (e.g. "short and fat" or "tall and skinny"). Some **CrB** compounds are CaAg, BaSi, ScGa, LaNi, VB and NiB; some **FeB** compounds are BaAg, LaSi, ZrGe, LuNi, TiB and MnB. Some have been found to transform to **CuZn** with increasing temperature and/or pressure. Although common as silicide and boride structures, no isostructural carbides are known.

## A5.4 Composition $AB_2$ ($AX_2$ and $A_2X$)

Just as for compounds *AB*, the composition $AB_2$ give rise to hundreds of structure types. Some like **cuprite** (known only for $Cu_2O$ and $Ag_2O$) and **quartz** (known only for forms of $SiO_2$, $GeO_2$ and $BeF_2$) have only a few examples. Notice that silica ($SiO_2$) and water ($H_2O$) are perhaps the most intensively studied of all binary compounds and the structures of their crystalline forms have much in common (§ 7.3.13).

We give examples of compositions belonging to some of the larger families below. We should point out here that **BaMgSi (PbFCl)** is also called **$Cu_2Sb$**, and **SrMgSi** is called either **$PbCl_2$** or **$Co_2Si$**. Both these very large groups are really ternary structure types. Another large group of compounds is known as **$Fe_2P$** and in fact this is really a quaternary structure type $A_3B_3X_2Y$ although most reported compositions are *ABX* (i.e. $X = Y$).

### *A5.4.1 $AB_2$ compounds with $\{A\}B_6$ octahedra or trigonal prisms*

An important structure type with $\{A\}B_6$ octahedra is **rutile** ($TiO_2$) which is found for oxides (e.g. of Ti, Ge and Sn) and also for fluorides of (e.g. of Mg, Mn, and Zn) and for $MgH_2$. It is closely related to **$CaCl_2$** (see Exercise 6.9.7) which is found only for a form of $PtO_2$ and for halides other than fluorides.

A second group of structures is that of $\mathbf{CdCl_2}$ and $\mathbf{CdI_2}$ with respectively **ccp** and **hcp** anions and all the octahedral sites in alternate layers filled with cations (§ 6.1.5). These are mostly halides (but not fluorides) but also include a few chalcogenides. Examples of $\mathbf{CdCl_2}$ compounds are $MgCl_2$, $NiCl_2$ and $ZnBr_2$; some anti-structure compositions are $Cs_2O$, $Sr_2N$, $Y_2C$ and $Ag_2F$. $\mathbf{CdI_2}$ compositions include $MgBr_2$, $MgI_2$, $TiS_2$, $PtSe_2$ and $IrTe_2$; anti-structure compositions are $W_2C$ and $Ti_2O$. If the H atoms are ignored, then hydroxides such as $Mg(OH)_2$ (brucite) and $Ca(OH)_2$ are $\mathbf{CdI_2}$.

A related series of structures has trigonal prism layers These are mostly compounds of the early transition elements (Nb, Ta, Mo and W) with S, Se and Te and most compounds are polymorphic—the simplest forms are $2H_a$, $2H_b$ and $3R$ described in § 6.4.1.

### *A5.4.2* ***Fluorite*** *and* ***antifluorite*** *compounds*

**Fluorite** (§ 6.1.5) compounds form a large group of mainly "ionic" crystals such as $CaF_2$, $SrCl_2$, $ThO_2$, $UO_2$ but the family also includes compounds such as $CoSi_2$ and $NiSi_2$. The group of **antifluorite** compounds is also large; typical "ionic" compositions are $Li_2O$, $Rb_2O$, $K_2S$, $Rb_2Se$. Some other isostructural phases are $Be_2C$, $Mg_2Si$, $Mg_2Sn$, $PtAl_2$ and $PtIn_2$.

### *A5.4.3 Intermetallic structures*: $\mathbf{AlB_2}$, $\mathbf{CuAl_2}$ *and* $\mathbf{MgCu_2}$

These three structure types are probably the largest groups of intermetallic $AB_2$ structures. There are hundreds of $\mathbf{AlB_2}$ compounds with the characteristic **graphite**-like honeycomb layers of B (§ 5.3.5), typical compositions are $ThAg_2$, $ThAl_2$, $CrB_2$, $MgB_2$, $CaGa_2$, $Li_2Pt$, $ScSi_2$, $YHg_2$ and $ThNi_2$. The largest group of compounds is that of borides and gallides, but as the examples given show, a rather wide range of intermetallic compositions occur.

Some $\mathbf{CuAl_2}$ compounds (§ 6.4.3) are $Th_2Ag$, $Th_2Al$, $Cr_2B$, $Zr_2Ni$, $Th_2Cu$ and $Ta_2Si$. Notice that the first three examples contain the same elements, but in different proportion, as the first three $\mathbf{AlB_2}$ compounds.

The $\mathbf{MgCu_2}$ (§ 6.6.3) group is the largest. *Pearson's Handbook* (Book List) has many hundreds of entries under this heading (not all binary compounds). Most compositions involve one or two transition elements, but there are also compounds such as $CaAl_2$ and $CsBi_2$ with the same structure; some other compositions are $PbAu_2$, $TaCo_2$, $DyPt_2$, $ZrW_2$, and $ZrZn_2$.

## A5.5 Other binary structure types

It would take a sizable book to do justice to binary structures in general.[1] Important

[1] The classic reference is *Kristallstrukturen zweikomponentiger Phasen* by K. Schubert [Springer-Verlag, Berlin (1964)].

structure types we have met include **corundum** (**$\alpha$-$Al_2O_3$**—§ 6.1.6) which is mainly an oxide structure (of Al, Cr, Fe, etc.) with $6^3$ layers of **Al** in octahedral sites of **hcp O**. A derivative structure is **ilmenite** (**$FeTiO_3$**) in which **Al** is replaced by alternate layers of **Fe** and **Ti**.

**$Th_3P_4$** (§ 6.3.7) is the structure of a large group of compounds $A_3X_4$ with $A$ = Th, U, Y, Ln and $X$ = P, As, Sb, S, Se and Te. Antistructure compounds $A_4X_3$ are also common, typical compositions are $La_4As_3$ (contrast $La_3S_4$) and $La_4Ge_3$. The structure type is notable for existing over a range of composition with (presumably) incomplete occupation of one set of atomic sites. Notable compounds of this sort are compositions in the range $Ln_2S_3$-$Ln_3S_4$.

## A5.6 Ternary structure types

With over $10^2$ elements, there are over $10^6/3!$ different combinations of three elements; in many of these cases compounds of several different stoichiometries are formed, and these in turn are often polymorphic. It may be seen therefore, that to give a *comprehensive* account of crystal chemistry would be a daunting task. Here we just mention typical compositions of some popular ternary structure types that have been met in the text.

### *A5.6.1 Oxide structures:* ***spinel*** *and* ***perovskite***

The prototypical **spinel** (§ 6.1.6) composition is $MgAl_2O_4$; other compositions $AB_2X_4$ (with $A$ in one eigth of the tetrahedral sites and $B$ in one half of the octahedral sites of **ccp** $X$) are $Mn_2GeO_4$, $Na_2WO_4$, $ZnAl_2S_4$ and $CdCr_2Se_4$. "Inverse" spinels are compounds $AB_2X_4$ with $B$ on tetrahedral and half the octahedral sites and $A$ on the other half of the octahedral sites; examples are $Li_2NiF_4$ and $ZnTi_2O_4$. Notice that there are now two kinds of $B$ atom (with different coordination) and the structure is really quaternary and compositions such as $LiZnNbO_4$ are also included as **spinels** (but in some, at least, of these last compositions, cation ordering occurs to produce a lower-symmetry structure). In magnetite, $Fe_3O_4$, the tetrahedral sites are occupied by $Fe^{3+}$ and the octahedral sites are a disordered combination of $Fe^{2+}$ and $Fe^{3+}$.

The mineral perovskite is $CaTiO_3$; its structure is a small orthorhombic distortion of the cubic $ABX_3$ structure (described in § 6.6.2 and § 5.3.4) with 12-coordinated $A$ and 6-coordinated $B$. Many perovskites have the orthorhombic structure, usually known as **$GdFeO_3$**. Examples are $NaMgF_3$, $KPdF_3$, $CaZrO_3$, $SmAlO_3$ and $NaUO_3$. $MgSiO_3$ at high-pressure has the same structure and is thought to be the major component of the earth's lower mantle, and thus to be the major phase in the planet. Other cubic **perovskite** compositions are compounds $AB_3X$ with $A$ and $B$ being metallic elements with **$Cu_3Au$** arrangement and $X$ = B, C or N in $\{X\}B_6$ octahedra. Examples are $ScIr_3B$, $ZnCo_3C$ and $SnMn_3N$. An anti-**perovskite** composition is $Na_3ClO$ (with $\{Cl\}Na_{12}$ and $\{O\}Na_6$).

### *A5.6.2 Intermetallic structures*: **BaMgSi (PbFCl), SrMgSi ($PbCl_2$)** *and* **$ThCr_2Si_2$**

**BaMgSi** or **PbFCl** (§ 6.4.1) is also known as **$Cu_2Sb$** or **$Fe_2As$**, but the first two designations are preferred as they emphasize the ternary nature of the structure. There are two main groups of compounds. In the first group there are two "anions" as in PbFCl, YOCl, BaHI and UAsSe with the smaller (given first) in tetragonal layers of (e.g.) {F}$Pb_4$ tetrahedra. The second group are antistructure compounds with two metal atoms such as BaMgSi, NaLiS and YFeSi; $Cu_2Sb$ and $Fe_2As$ are in this category also.

**SrMgSi** (§ 5.3.7) is also known as **$PbCl_2$** or **$Co_2Si$**, but again we prefer the first name which makes the ternary nature of the structure clear. In compounds with two crystallographically-distinct "anions," these are often the same element as in $PbCl_2$, $BaH_2$, $BaI_2$, $US_2$ and $ThP_2$. Compounds like SrMgSi, NbFeP and ReCoB are formally the antistructure, but often the composition consists of two or three metallic elements as in $Lu_2Au$ and LuCoSn.

**$ThCr_2Si_2$** (§ 6.4.2) is also named after a chemically binary composition viz. **$BaAl_4$**, but again we use the ternary designation. This has the most known examples of all structure types and we just mention a few typical compositions here. In **$ThCr_2Si_2$** there are layers of {**Cr**}**$Si_4$** tetrahedra. In the following formulas the tetrahedrally-coordinated atom is second: $BaAg_2Sn_2$, $UOs_2Si_2$, $CaCo_2As_2$, $LaPt_2Ge_2$, $YNi_2P_2$, $TlNi_2S_2$. Some compounds with two elements are $BaAl_4$, $CaGa_4$, $RbIn_4$ and $ThZn_4$. In the anti-structure type, **$Th_2TeN_2$**, there are {**N**}**$Th_4$** tetrahedral layers; another composition is $La_2TeO_2$.

## A5.7 Crystallographic data

Crystallographic data for some of the simpler structures discussed in Chapters 5-7 are given here in condensed form (the *International Tables* or some other source should be consulted for equivalent positions). The listing is in alphabetical order of the chemical formula as normally written. See the Book List part D for sources of data.

| | |
|---|---|
| $AlB_2$ | *P6/mmm*, *a* = 3.005, *c* = 3.245 Å. Al 1 *a*, 0,0,0 ; B 2 *d*, 1/3,2/3,1/2 |
| $Al_2O_3$ | $R\bar{3}c$, *a* = 4.759, *c* = 12.991 Å. Al 12 *c*, 0,0,0.3523 ; O 18 *e*, 0.3064,0,1/4 |
| AuCd (HT) | $Pm\bar{3}m$, *a* = 3.323 Å. *V* = 36.7 Å$^3$. Au 0,0,0 ; Cd 1/2,1/2,1/2 (**CuZn** structure) |
| AuCd (LT) | *Pmma*, *a* = 4.765, *b* = 3.154, *c* = 4.864 Å. *V* = 2 x 35.5 Å$^3$<br>Au 2 *f*: 1/4,0,0.312 ; Cd 2 *e*: 1/4,1/2,0.812 |
| $AuZn_3$ | $Pm\bar{3}n$, *a* = 7.903 Å. Au(1) 2 *a*, 0,0,0 ; Au(2) 6 *c*, 1/4,0,1/2 ; Zn 24 *k*, 0,0.165,0.300 |
| $B_2O_3$ | $P3_1$, *a* = 4.336, *c* = 8.340 Å, all atoms in 3 *a*<br>B(1), 0.223,0.393,0.980 ; B(2), 0.828,0.603,0.092<br>O(1), 0.547,0.397,0.0 ; O(2), 0.149,0.600,0.078 ; O(3), 0.005,0.161,0.871 |
| BaCu | $P6_3/mmc$, *a* = 4.499, *c* = 16.25 Å, *c*/*a* = 3.61<br>Ba 4 *f*, 1/3,2/3,0.1217 ; Cu(1) 2 *b*, 0,0,1/4 ; Cu(2) 2 *d*, 1/3,2/3,3/4 |

| | |
|---|---|
| $BaCu_2P_4$ | *Fddd*, $a$ = 5.345, $b$ = 18.973, $c$ = 10.244 Å.<br>Ba 8 *a*, 1/8,1/8,1/8 ; Cu 16 *f*, 1/8,0.5048,1/8 ; P 32 *h*, 0.195,0.1789,0.817 |
| β-$BaFe_2S_4$ | *I4/m*, $a$ = 7.678, $c$ = 5.292 Å<br>Ba 2 *a*, 0,0,0 ; Fe 4 *d*, 0,1/2,1/4 ; S 8 *h*, 0.6196, 0.1986,0 |
| BaMgSi | *P4/nmm*, $a$ = 4.610, $c$ = 7.870 Å<br>Ba 2 *c*, 1/4,1/4,0.339 ; Mg 2 *a*, 3/4,1/4,0 ; Si 2 *c*, 1/4,1/4,0.794 |
| $BaNiO_3$ | $P6_3/mmc$, $a$ = 5.629, $c$ = 4.811 Å<br>Ba 2 *c*, 1/3,2/3,1/4 ; Ni 2 *a*, 0,0,0 ; O 6 *h*, 0.1462,0.2924,1/4 |
| $BaTiO_3$ (200 °C) | $Pm\bar{3}m$, $a$ = 4.012. Ba 1 *a*, 0,0,0 ; Ti 1 *b*, 1/2,1/2,1/2 ; O 3 *c*, 0,1/2,1/2 |
| β-BeO | $P4_2/nmm$, $a$ = 4.75, $c$ = 2.74 Å. Be 4 *g*, 0.164,0.836,0 ; O 4 *f*, 0.190,0.190,0 |
| C graphite | $P6_3/mmc$, $a$ = 2.461, $c$ = 6.709 Å. C(1) 2 *b*, 0,0,1/4 ; C(2) 2 *c*, 1/3,2/3,1/4 |
| $CaB_6$ | $Pm\bar{3}m$, $a$ = 4.151 Å. Ca 1 *b*, 1/2,1/2,1/2 ; B 6 *e*, 0.302,0,0 |
| $CdCl_2$ | $R\bar{3}m$, $a$ = 3.846, $c$ = 17.493 Å. Cd 3 *a*, 0,0,0 ; Cl 6 *c*, 0,0,0.2519 |
| $CdI_2$ | $P\bar{3}m1$, $a$ = 4.224, $c$ = 6.859 Å. Cd 1 *a*, 0,0,0 ; I 2 *d*, 1/3,2/3,0.2492 |
| $Co_9S_8$ | $Fm\bar{3}m$, $a$ = 9.927 Å. Co(1) 4 *b*, 1/2,1/2,1/2 ; Co(2) 32 *f*, 0.1266, 0.1266,0.1266<br>S(1) 8 *c*, 1/4,1/4,1/4 ; S(2) 24 *e*, 0.2624,0,0 |
| CrB | *Cmcm*, $a$ = 2.978, $b$ = 7.870, $c$ = 2.935 Å<br>Cr 4 *c*, 0,0.1453,1/4 ; B 4 *c*, 0,0.4360,1/4 |
| $CrB_4$ | *Immm*, $a$ = 4.744 , $b$ = 5.477, $c$ = 2.866 Å. Cr 2 *a*, 0,0,0 ; B 8 *n*, 0.175,0.346,0 |
| $CuAl_2$ | *I4/mcm*, $a$ = 6.067, $c$ = 4.877 Å. Cu 4 *a*, 0,0,1/4 ; Al 8 *h*, 0.1581,0.6851,0 |
| $CuFeS_2$ | $I\bar{4}2d$, $a$ = 5.289, $c$ = 10.423 Å<br>Cu 4 *a*, 0,0,0 ; Fe 4 *b*, 0,0,1/2 ; S 8 *d*, 0.2574,1/4,1/8 |
| $CuFe_2S_3$ | *Pnma*, $a$ = 6.231, $b$ = 11.117, $c$ = 6.467 Å<br>Cu 4 *c*, 0.123,1/4,0.417 ; Fe 8 *d*, 0.137,0.0870,0.915<br>S(1) 4 *c*, 0.258,1/4,0.087 ; S(2) 8 *d*, 0.267,0.0846,0.588 |
| $Fe_2AlB_2$ | *Cmmm*, $a$ = 2.923, $b$ = 11.034, $c$ = 2.870 Å<br>Al 2 *a*, 0,0,0 ; Fe 4 *j*, 0,0.3540,1/2 ; B 4 *i*, 0,0.2071,0 |
| FeB | *Pnma*, $a$ = 5.495, $b$ = 2.946, $c$ = 4.053 Å<br>Fe 4 *c*, 0.180,1/4,0.125 ; B 4 *c*, 0.036,1/4,0.610 |
| $Fe_2P$ | $P\bar{6}2m$, $a$ = 5.868, $c$ = 3.465 Å. Fe(1) 3 *f*, 0.2568,0,0 ; Fe(2) 3 *g*, 0.5946,0,1/2<br>P(1) 1 *b*, 0,0,1/2 ; P(2) 2 *c*, 1/3,2/3,0 |
| $FeS_2$ (pyrite) | $Pa\bar{3}$, $a$ = 5.418 Å. Fe 4 *a*, 0,0,0 ; S 8 *c*, 0.384,0.384,0.384 |

| | |
|---|---|
| FeSi | $P2_13$, $a$ = 4.517 Å. Fe 4 $a$, 0.136,0.136,0.136 ; Si 4 $a$, 0.844,0.844,0.844 |
| $\gamma$- Ge | $P4_32_12$, $a$ = 5.93, $c$ = 6.98 Å. [origin chosen for comparison with $SiO_2$—keatite]<br>Ge(1) 4 $a$, 0.4088,0.4088,0 ; Ge(2) 8 $b$, 0.3270,0.1216,0.2486 |
| $Hg_3NbF_6$ | $P\bar{3}m1$, $a$ = 5.02, $c$ = 7.68 Å.<br>Hg(1) 1 $b$, 0,0,1/2 ; Hg(2) 2 $d$, 1/3,2/3,0.500 ; Nb 1 $a$, 0,0,0 ; F 6$h$, 0.309,0,0.143 |
| $Hg_{2.9}SbF_6$ | $I4_1/amd$, $a$ = 7.655, $c$ = 12.558 Å. Hg disordered in positions 16 $h$: 0,$y$,$z$ with $z \approx 0$<br>Sb: 4 $b$, 0,1/4,3/8 ; F(1) 8 $e$, 0,1/4,0.230 ; F(2) 16 $g$, 0.672,0.922,1/4 |
| $KInTe_2$ | $I4/mcm$, $a$ = 8.52, $c$ = 7.39 Å<br>K 4 $a$, 0,0,1/4 ; In 4 $b$, 0,1/2,1/4 ; Te 8 $h$, 0.177,0.678,0 |
| $K_2MgF_4$ | $I4/mmm$, $a$ 3.995, $c$ = 13.706 Å<br>K 4 $e$, 0,0,0.350 ; Mg 2 $a$, 0,0,0 ; F(1) 4 $c$, 0,1/2,0 ; F(2) 4 $e$, 0,0,0.150 |
| $K_3V_5O_{14}$ | $P31m$, $a$ = 8.680, $c$ = 4.991 Å. K 3 $c$, 0.605,0,0.000 ; V(1) 3 c, 0.231,0,0.472<br>V(2) 2 $b$, 1/3,2/3,0.472 ; O(1) 2 $b$, 1/3,2/3,0.796 ; O(2) 3 $c$, 0.240,0,0.782<br>O(3) 3 $c$, 0.838,0,0.367 ; O(4) 6 $d$, 0.469,0.177,0.366 |
| $LaB_2C_2$ | $P\bar{4}2c$, $a$ = 3.822, $c$ = 7.924 Å<br>La 2 $e$, 0,0,0 ; B 4 $h$, 1/2,0.226,1/4 ; C 4 $i$, 0.173,1/2,1/4 |
| $La_2O_3$ | $P\bar{3}m1$, $a$ = 3.938, $c$ = 6.136 Å. La 2 $d$, 1/3,2/3,0.2467<br>O(1) 1 $a$, 0,0,0 ; O(2) 2 $d$, 1/3,2/3,0.6470 |
| LiP | $P2_1/c$, $a$ =5.582, $b$ = 4.940, $c$ = 10.255 Å, $\beta$ = 118.15°. All atoms in 4 $e$<br>Li(1): 0.2151,0.3876,0.3299 ; Li(2): 0.2257,0.6597,0.0293<br>P(1): 0.3165,0.8952,0.2920 ; P(2): 0.3050,0.1565,0.1125 |
| $LiY_2Si_2$ | $P4/mbm$, $a$ = 7.105, $c$ = 4.144 Å.<br>Li 2 $a$, 0,0,0 ; Y 4 $h$, 0.181,0.681,1/2 ; Si 4 $g$, 0.383,0.883,0 |
| $Mn_2Hg_5$ | $P4/mbm$, $a$ = 9.758, $c$ = 2.998 Å<br>Mn 4 $h$, 0.180,0.680,1/2 ; Hg(1) 2 $d$, 0,1/2,0 ; Hg(2) 8 $i$, 0.063,0.204,0 |
| $\alpha$-MoB | $I4_1/amd$, $a$ = 3.105, $c$ = 16.97 Å. Mo 8 $e$, 0,1/4,0.071 ; B 8 $e$, 0,1/4,0.227 |
| $MoS_2$ ($2H_b$) | $P6_3/mmc$, $a$ = 3.161, $c$ = 12.295 Å. Mo 2 $d$, 1/3,2/3,3/4 ; S 4 $f$, 1/3,2/3,0.1275 |
| $MoS_2$ (3$R$) | $R3m$, $a$ = 3.166, $c$ = 18.41 Å<br>Mo 3 $c$, 0,0,0.0 ; S(1) 3 $c$, 0,0,0.2477 ; S(2) 3 $c$, 0,0,0.4190 |
| NaP | $P2_12_12_1$, $a$ = 6.038, $b$ = 5.643, $c$ = 10.142 Å. All atoms in 4 $a$<br>Na(1): 0.4174,0.9089,0.0318 ; Na(2): 0.1338,0.6367,0.3313<br>P(1): 0.3086,0.1404,0.2838 ; P(2): 0.4287,0.4020,0.1341 |
| NaPb | $I4_1/acd$, $a$ = 10.580, $c$ = 17.466 Å. Na(1) 16 $e$, 0.375,0,1/4<br>Na(2) 16 $f$, 0.375,0.625,1/8 ; Pb 32 $g$, 0.0694,0.3814,0.4383 |

| | |
|---|---|
| $Na_3Pt_4Ge_4$ | $I\bar{4}3m$, $a$ = 7.614 Å<br>Na 6 *b*, 0,1/2,1/2 ; Pt 8 *c*, 0.1366,0.1366,0.1366 ; Ge 8 *c*, 0.3136,0.3136,0.3136 |
| $NaZn_{13}$ | $Fm\bar{3}c$, $a$ = 12.284 Å<br>Na 8 *a*, 1/4,1/4,1/4 ; Zn(1) 8 *b*, 0,0,0 ; Zn(2) 96 *i*, 0,0.1806,0.1192 |
| $Nb_6F_{15}$ | $Im\bar{3}m$, $a$ = 8.19 Å. Nb 12 *e*, 0.242,0,0 ; F(1) 6 *b*, 0,1/2,1/2 ; F(2) 24 *h*, 0,0.25,0.25 |
| NbO | $Pm\bar{3}m$, $a$ = 4.21 Å. Nb 3 *c*, 0,1/2,1/2 ; O 3 *d*, 0,0,1/2 |
| $NbSe_2$ ($2H_a$) | $P6_3/mmc$, $a$ = 3.445, $c$ = 12.554 Å. Nb 2 *b*, 0,0,1/4 ; Se 4 *f*, 1/3,2/3,0.1172 |
| PbFCl | $P4/nmm$, $a$ = 4.106, $c$ = 7.230 Å<br>Pb 2 *c*, 1/4,1/4,0.800 ; F 2 *a*, 3/4,1/4,0 ; Cl 2 *c*, 1/4,1/4,0.350 |
| PbO | $P4/nmm$, $a$ = 3.972, $c$ = 5.018. Pb 2 *c*, 1/4,1/4,0.7615 ; O 2 *a*, 3/4,1/4,0 |
| $PdF_2$ | $Pa\bar{3}$, $a$ = 5.239 Å. Pd 4 *a*, 0,0,0 ; F 8 *c*, 0.343,0.343,0.342 |
| $Pd_{17}Se_{15}$ | $Pm\bar{3}m$, $a$ = 10.606 Å. Pd(1) 2 *b*, 1/2,1/2,1/2 ; Pd(2) 3 *d*, 1/2,0,0<br>Pd(3) 6 *e*, 0.238,0,0 ; Pd(4) 24 *m*, 0.352,0.352,0.150 ; Se(1) 6 *f*, 0.257,1/2,1/2<br>Se(2) 12 *i*, 0,0.230,0.230 ; Se(3) 12 *j*, 1/2,0.168,0.168 |
| $Pr_3Rh_4Sn_{13}$ | $Pm\bar{3}n$, $a$ = 9.698 Å. Pr 6 *d*, 1/4,1/2,0 ; Rh 8 *e*, 1/4,1/4,1/4<br>Sn(1) 2 *a*, 0,0,0 ; Sn(2) 24 *k*, 0,0.3073,0.1535 |
| $Re_3B$ | $Cmcm$, $a$ = 2.890, $b$ = 9.313, $c$ = 7.258 Å.<br>Re(1) 4 *c*, 0,0.426,1/4 ; Re(2) 8 *f*, 0.0,0.135,0.062 ; B 4 *c*, 0,0.744,1/4. |
| $Sc_2O_2S$ | $P6_3/mmc$, $a$ = 3.520, $c$ = 12.519 Å, $c/a$ = 3.56<br>Sc 4 *f*, 1/3,2/3,0.3930 ; O 4 *f*, 1/3,2/3,0.0661 ; S 2 *b*, 0,0,1/4 |
| γ-Si | $Ia\bar{3}$, $a$ = 6.636 Å. Si 16 *c*, 0.1003,0.1003,0.1003 |
| $SiO_2$ (coesite) | $C2/c$, $a$ = 7.135, $b$ = 12.372, $c$ = 7.173 Å, $\beta$ = 120.36°<br>Si(1) 8 *f*, 0.140,0.1084,0.072 ; Si(2) 8 *f*, 0.506,0.1590,0.540<br>O(1) 4 *a*, 0,0,0 ; O(2) 4 *e*, 0,0.3839,1/4 ; O(3) 8 *f*, 0.266,0.1233,0.940<br>O(4) 8 *f*, 0.311,0.1037,0.328 ; O(5) 8 *f*, 0.018,0.2119,0.478 |
| $SiO_2$ (keatite) | $P4_32_12$, $a$ = 7.464, $c$ = 8.620 Å. Si(1) 4 *a*, 0.410,0.410,0<br>Si(2) 8 *b*, 0.326,0.120,0.248 ; O(1) 8 *b*, 0.445,0.132,0.400<br>O(2) 8 *b*, 0.117,0.123,0.296 ; O(3) 8 *b*, 0.344,0.297,0.143 |
| $SiO_2$ (moganite) | I2/*a*, $a$ = 8.758, $b$ = 4.786, $c$ = 10.715 Å, $\beta$ = 90.08°. Si(1) 4 *c*, 1/4,0.9908,0<br>Si(2) 8 *f*, 0.0115,0.2533,0.1678 ; O(1) 8 *f*, 0.9686,0.0680,0.2860<br>O(2) 8 *f*, 0.1711,0.1770,0.1050 ; O(3) 8 *f*, 0.8657,0.2148,0.0739 |
| SrMgSi | $Pnma$, $a$ = 7.78, $b$ = 4.56, $c$ = 8.49 Å. Sr 4 *c*, 0.515,1/4,0.683<br>Mg 4 *c*, 0.640,1/4,0.057 ; Si 4 *c*, 0.276,1/4,0.110 |
| $SrSi_2$ | $P4_332$, $a$ = 6.540 Å. Sr 4 *a*, 1/8,1/8,1/8 ; Si 8 *c*, 0.428,0.428,0.428 |

$TbB_2C$ — $P4_2/mbc$, $a$ = 6.791, $c$ = 7.522 Å. Tb 8 $g$, 0.313,0.813,1/4
B(1) 8 $h$, 0.095,0.595,0 ; B(2) 8 $h$, 0.140,0.035,0 ; C 8 $h$, 0.456,0.322,0

$ThB_4$ — $P4/mbm$, $a$ = 7.256, $c$ = 4.113 Å. Th 4 $g$, 0.313,0.813,0
B(1) 4 $e$, 0,0,0.212 ; B(2) 4 $h$, 0.087,0.587,1/2 ; B(3) 8 $j$, 0.170,0.042,1/2

$ThB_2C$ — $R\bar{3}m$, $a$ = 6.700, $c$ = 14.467 Å
Th(1) 3 $a$, 0,0,0 ; Th(2) 6 $c$, 0,0,0.3156 ; B 18 $g$, 0.276,0,1/2 ; C 9 $d$, 1/2,0,1/2

$ThCr_2Si_2$ — $I4/mmm$, $a$ = 4.043, $c$ = 10.577 Å
Th 2 $a$, 0,0,0 ; Cr 4 $d$, 0,1/2,1/4 ; Si 4 $e$, 0,0,0.374

$ThMoB_4$ — $Cmmm$, $a$ = 7.481, $b$ = 9.658, $c$ = 3.771 Å. Mo 4 $g$, 0.171,0,0 ; Th 4 $i$, 0,0.302,0
B(1) 4 $h$, 0.379,0,1/2 ; B(2) 4 $j$, 0,0.093,1/2 ; B(3) 8 $g$, 0.234,0.155,1/2

$Th_3P_4$ — $I\bar{4}3d$, $a$ = 8.618 Å. Th 12 $a$, 3/8,0,1/4 ; P 16 $c$, 0.083,0.083,0.083

$Th_3Pd_5$ — $P\bar{6}2m$, $a$ = 7.149, $c$ = 3.899 Å
Th 3 $g$, 0.350,0,1/2 ; Pd(1) 2 $c$, 1/3,2/3,0 ; Pd(2) 3 $f$, 0.780,0,0

$ThSi_2$ — $I4_1/amd$, $a$ = 4.134, $c$ = 14.375 Å. Th 4 $a$, 0,3/4,1/8 ; Si 8 $e$, 0,1/4,0.2915

$Th_2TeN_2$ — $I4/mmm$, $a$ = 4.094, $c$ = 13.014 Å
Th 4 $c$, 0,0,0.344 ; Te 2 $a$, 0,0,0 ; N 4 $d$, 0,1/2,1/4

$Tl_3VS_4$ — $I\bar{4}3m$, $a$ = 7.51 Å. Tl 6 $b$, 0,1/2,1/2 ; V 2 $a$, 0,0,0 ; S 8 $c$, 0.175,0.175,0.175

$WAl_{12}$ — $Im\bar{3}$, $a$ = 7.580 Å. W 2 $a$, 0,0,0 ; Al 24 $g$, 0.0,0.184,0.309

$WC$ — $P\bar{6}m2$, $a$ = 2.906, $c$ = 2.837 Å. W 1 $a$, 0,0,0 ; C 1 $d$, 1/3,2/3,1/2

$YCrB_4$ — $Pbam$, $a$ = 5.972, $b$ = 11.46, $c$ = 3.461 Å.
Y 4 $g$, 0.125,0.150,0 ; Cr 4 $g$, 0.125,0.419,0 ; B(1) 4 $h$, 0.280,0.315,1/2
B(2) 4 $h$, 0.340,0.465,1/2 ; B(3) 4 $h$, 0.385,0.050,1/2 ; B(4) 4 $h$, 0.485,0.180,1/2

$Y_2ReB_6$ — $Pbam$, $a$ = 9.175, $b$ = 11.55, $c$ = 3.673 Å
Y(1) 4 $g$, 0.823,0.087,0 ; Y(2) 4 $g$, 0.445, 0.131,0 ; Re 4 $g$, 0.138, 0.178,0
B(1) 4 $h$, 0.050,0.060,1/2 ; B(2) 4 $h$, 0.250,0.075,1/2 ; B(3) 4 $h$, 0.300,0.240,1/2
B(4) 4 $h$, 0.140,0.310,1/2 ; B(5) 4 $h$, 0.480,0.290,1/2 ; B(6) 4 $h$, 0.110,0.470,1/2

$Zn(CN)_2$ — $P\bar{4}3m$, $a$ = 5.928 Å. Zn(1) 1 $a$, 0,0,0 ; Zn(2) 1 $b$, 1/2,1/2,1/2
C 4 $e$, 0.1938,0.1938,0.1938 ; N 4 $e$, 0.3092,0.3092,0.3092

$ZnO$ — $P6_3mc$, $a$ = 3.250, $c$ = 5.207 Å. Zn 2 $b$, 1/3,2/3,0.0 ; O 2 $b$, 1/3,2/3,0.3819

$ZrFe_4Si_2$ — $P4_2/mnm$, $a$ = 7.004, $c$ = 3.755 Å
Zr 2 $b$, 0,0,1/2 ; Fe 8 $i$, 0.0920,0.3468,0 ; Si 4 $g$, 0.2201,0.7799,0

## TABLES OF 3-DIMENSIONAL SYMMETRY GROUP SYMBOLS

### Crystallographic point groups

| system | point group | Schoenflies | spacegroups | center? |
|---|---|---|---|---|
| triclinic (anorthic) | $1$ | $C_1$ | 1 | no |
| | $\bar{1}$ | $C_i$ | 2 | yes |
| monoclinic | $2$ | $C_2$ | 3-5 | no |
| | $m$ | $C_s$ | 6-9 | no |
| | $2/m$ | $C_{2h}$ | 10-15 | yes |
| orthorhombic | $222$ | $D_2$ | 16-24 | no |
| | $mm2$ | $C_{2v}$ | 25-46 | no |
| | $mmm$ | $D_{2h}$ | 47-74 | yes |
| tetragonal | $4$ | $C_4$ | 75-80 | no |
| | $\bar{4}$ | $S_4$ | 81-82 | no |
| | $4/m$ | $C_{4h}$ | 83-88 | yes |
| | $422$ | $D_4$ | 89-98 | no |
| | $4mm$ | $C_{4v}$ | 99-110 | no |
| | $\bar{4}2m$ | $D_{2d}$ | 111-122 | no |
| | $4/mmm$ | $D_{4h}$ | 123-142 | yes |
| trigonal | $3$ | $C_3$ | 143-146 | no |
| | $\bar{3}$ | $C_{3i}$ | 147-148 | yes |
| | $32$ | $D_3$ | 149-155 | no |
| | $3m$ | $C_{3v}$ | 156-161 | no |
| | $\bar{3}m$ | $D_{3d}$ | 162-167 | yes |
| hexagonal | $6$ | $C_6$ | 168-173 | no |
| | $\bar{6}$ | $C_{3h}$ | 174 | no |
| | $6/m$ | $C_{6h}$ | 175-176 | yes |
| | $622$ | $D_6$ | 177-182 | no |
| | $6mm$ | $C_{6v}$ | 183-186 | no |
| | $\bar{6}m2$ | $D_{3h}$ | 187-190 | no |
| | $6/mmm$ | $D_{6h}$ | 191-194 | yes |
| cubic (isometric) | $23$ | $T$ | 195-199 | no |
| | $m\bar{3}$ | $T_h$ | 200-206 | yes |
| | $432$ | $O$ | 207-214 | no |
| | $\bar{4}3m$ | $T_d$ | 215-220 | no |
| | $m\bar{3}m$ | $O_h$ | 221-230 | yes |

## Monoclinic space group symbols for various cell choices and settings

In the following list the first symbol in each row is "standard" and those on the right of it are other possibilities. It is a common practice to omit the "1" space markers when **b** is the unique axis (so that, e.g. $C12/c1$ becomes $C2/c$). The number is the space group number in the *International Tables.* For numbers 9 and 15 interchanging the labels of the oblique axes results in additional "legal" symbols not used in the *Tables*.

| | **b** unique | | | **c** unique | | |
|---|---|---|---|---|---|---|
| 3 | $P121$ | | | $P112$ | | |
| 4 | $P12_11$ | | | $P112_1$ | | |
| 5 | $C121$ | $A121$ | $I121$ | $A112$ | $B112$ | $I112$ |
| 6 | $P1m1$ | | | $P11m$ | | |
| 7 | $P1c1$ | $P1n1$ | $P1a1$ | $P11a$ | $P11n$ | $P11b$ |
| 8 | $C1m1$ | $A1m1$ | $I1m1$ | $A11m$ | $B11m$ | $I11m$ |
| 9 | $C1c1$ | $A1n1$ | $I1a1$ | $A11a$ | $B11n$ | $I11b$ |
| | $A1a1$ | $C1n1$ | $I1c1$ | $B11b$ | $A11n$ | $I11a$ |
| 10 | $P12/m1$ | | | $P112/m$ | | |
| 11 | $P12_1/m1$ | | | $P112_1/m$ | | |
| 12 | $C12/m1$ | $A12/m1$ | $I12/m1$ | $A112/m$ | $B112/m$ | $I112/m$ |
| 13 | $P12/c1$ | $P12/n1$ | $P12/a1$ | $P112/a$ | $P112/n$ | $P112/b$ |
| 14 | $P12_1/c1$ | $P12_1/n1$ | $P12_1/a1$ | $P112_1/a$ | $P112_1/n$ | $P112_1/b$ |
| 15 | $C12/c1$ | $A12/n1$ | $I12/a1$ | $A112/a$ | $B112/n$ | $I112/b$ |
| | $A12/a1$ | $C12/n1$ | $I12/c1$ | $B112/b$ | $A112/n$ | $I112/a$ |

## Orthorhombic space group symbols for various settings

The following table gives orthorhombic space groups for various choices of axes.The second column headed **a b c** is the "standard" setting. The remaining columns are the symbols for different labeling of the axes. For example in the column headed **c a b**, the new **a** axis corresponds to the old **c** (in the standard setting), the new **b** is the old **a** and the new **c** is the old **b**.

| | **a b c** | **c a b** | **b c a** | **a -c b** | **b a -c** | **-c b a** |
|---|---|---|---|---|---|---|
| 16 | $P222$ | $P222$ | $P222$ | $P222$ | $P222$ | $P222$ |
| 17 | $P222_1$ | $P2_122$ | $P22_12$ | $P22_12$ | $P222_1$ | $P2_122$ |
| 18 | $P2_12_12$ | $P22_12_1$ | $P2_122_1$ | $P2_122_1$ | $P2_12_12$ | $P22_12_1$ |
| 19 | $P2_12_12_1$ | $P2_12_12_1$ | $P2_12_12_1$ | $P2_12_12_1$ | $P2_12_12_1$ | $P2_12_12_1$ |
| 20 | $C222_1$ | $A2_122$ | $B22_12$ | $B22_12$ | $C222_1$ | $A2_122$ |
| 21 | $C222$ | $A222$ | $B222$ | $B222$ | $C222$ | $A222$ |
| 22 | $F222$ | $F222$ | $F222$ | $F222$ | $F222$ | $F222$ |
| 23 | $I222$ | $I222$ | $I222$ | $I222$ | $I222$ | $I222$ |
| 24 | $I2_12_12_1$ | $I2_12_12_1$ | $I2_12_12_1$ | $I2_12_12_1$ | $I2_12_12_1$ | $I2_12_12_1$ |

| | **a b c** | **c a b** | **b c a** | **a -c b** | **b a -c** | **-c b a** |
|---|---|---|---|---|---|---|
| 25 | *Pmm2* | *P2mm* | *Pm2m* | *Pm2m* | *Pmm2* | *P2mm* |
| 26 | $Pmc2_1$ | $P2_1ma$ | $Pb2_1m$ | $Pm2_1b$ | $Pcm2_1$ | $P2_1am$ |
| 27 | *Pcc2* | *P2aa* | *Pb2b* | *Pb2b* | *Pcc2* | *P2aa* |
| 28 | *Pma2* | *P2mb* | *Pc2m* | *Pm2a* | *Pbm2* | *P2cm* |
| 29 | $Pca2_1$ | $P2_1ab$ | $Pc2_1b$ | $Pb2_1a$ | $Pbc2_1$ | $P2_1ca$ |
| 30 | *Pnc2* | *P2na* | *Pb2n* | *Pn2b* | *Pcn2* | *P2an* |
| 31 | $Pmn2_1$ | $P2_1mn$ | $Pn2_1m$ | $Pm2_1n$ | $Pnm2_1$ | $P2_1nm$ |
| 32 | *Pba2* | *P2cb* | *Pc2a* | *Pc2a* | *Pba2* | *P2cb* |
| 33 | $Pna2_1$ | $P2_1nb$ | $Pc2_1n$ | $Pn2_1a$ | $Pbn2_1$ | $P2_1cn$ |
| 34 | *Pnn2* | *P2nn* | *Pn2n* | *Pn2n* | *Pnn2* | *P2nn* |
| 35 | *Cmm2* | *A2mm* | *Bm2m* | *Bm2m* | *Cmm2* | *A2mm* |
| 36 | $Cmc2_1$ | $A2_1ma$ | $Bb2_1m$ | $Bm2_1b$ | $Ccm2_1$ | $A2_1am$ |
| 37 | *Ccc2* | *A2aa* | *Bb2b* | *Bb2b* | *Ccc2* | *A2aa* |
| 38 | *Amm2* | *B2mm* | *Cm2m* | *Am2m* | *Bmm2* | *C2mm* |
| 39 | *Abm2* | *B2cm* | *Cm2a* | *Ac2m* | *Bma2* | *C2mb* |
| 40 | *Ama2* | *B2mb* | *Cc2m* | *Am2a* | *Bmb2* | *C2cm* |
| 41 | *Aba2* | *B2cb* | *Cc2a* | *Ac2a* | *Bba2* | *C2cb* |
| 42 | *Fmm2* | *F2mm* | *Fm2m* | *Fm2m* | *Fmm2* | *F2mm* |
| 43 | *Fdd2* | *F2dd* | *Fd2d* | *Fd2d* | *Fdd2* | *F2dd* |
| 44 | *Imm2* | *I2mm* | *Im2m* | *Im2m* | *Imm2* | *I2mm* |
| 45 | *Iba2* | *I2cb* | *Ic2a* | *Ic2a* | *Iba2* | *I2cb* |
| 46 | *Ima2* | *I2mb* | *Ic2m* | *Im2a* | *Ibm2* | *I2cm* |
| 47 | *Pmmm* | *Pmmm* | *Pmmm* | *Pmmm* | *Pmmm* | *Pmmm* |
| 48 | *Pnnn* | *Pnnn* | *Pnnn* | *Pnnn* | *Pnnn* | *Pnnn* |
| 49 | *Pccm* | *Pmaa* | *Pbmb* | *Pbmb* | *Pccm* | *Pmaa* |
| 50 | *Pban* | *Pncb* | *Pcna* | *Pcna* | *Pban* | *Pncb* |
| 51 | *Pmma* | *Pbmm* | *Pmcm* | *Pmam* | *Pmmb* | *Pcmm* |
| 52 | *Pnna* | *Pbnn* | *Pncn* | *Pnan* | *Pnnb* | *Pcnn* |
| 53 | *Pmna* | *Pbmn* | *Pncm* | *Pman* | *Pnmb* | *Pcnm* |
| 54 | *Pcca* | *Pbaa* | *Pbcb* | *Pbab* | *Pccb* | *Pcaa* |
| 55 | *Pbam* | *Pmcb* | *Pcma* | *Pcma* | *Pbam* | *Pmcb* |
| 56 | *Pccn* | *Pnaa* | *Pbnb* | *Pbnb* | *Pccn* | *Pnaa* |
| 57 | *Pbcm* | *Pmca* | *Pbma* | *Pcmb* | *Pcam* | *Pmab* |
| 58 | *Pnnm* | *Pmnn* | *Pnmn* | *Pnmn* | *Pnnm* | *Pmnn* |
| 59 | *Pmmn* | *Pnmm* | *Pmnm* | *Pmnm* | *Pmmn* | *Pnmm* |
| 60 | *Pbcn* | *Pnca* | *Pbna* | *Pcnb* | *Pcan* | *Pnab* |
| 61 | *Pbca* | *Pbca* | *Pbca* | *Pcab* | *Pcab* | *Pcab* |
| 62 | *Pnma* | *Pbnm* | *Pmcn* | *Pnam* | *Pmnb* | *Pcmn* |
| 63 | *Cmcm* | *Amma* | *Bbmm* | *Bmmb* | *Ccmm* | *Amam* |
| 64 | *Cmca* | *Abma* | *Bbcm* | *Bmab* | *Ccmb* | *Acam* |
| 65 | *Cmmm* | *Ammm* | *Bmmm* | *Bmmm* | *Cmmm* | *Ammm* |
| 66 | *Cccm* | *Amaa* | *Bbmb* | *Bbmb* | *Cccm* | *Amaa* |
| 67 | *Cmma* | *Abmm* | *Bmcm* | *Bmam* | *Cmmb* | *Acmm* |
| 68 | *Ccca* | *Abaa* | *Bbcb* | *Bbab* | *Cccb* | *Acaa* |
| 69 | *Fmmm* | *Fmmm* | *Fmmm* | *Fmmm* | *Fmmm* | *Fmmm* |
| 70 | *Fddd* | *Fddd* | *Fddd* | *Fddd* | *Fddd* | *Fddd* |
| 71 | *Immm* | *Immm* | *Immm* | *Immm* | *Immm* | *Immm* |
| 72 | *Ibam* | *Imcb* | *Icma* | *Icma* | *Ibam* | *Imcb* |
| 73 | *Ibca* | *Ibca* | *Ibca* | *Icab* | *Icab* | *Icab* |
| 74 | *Imma* | *Ibmm* | *Imcm* | *Imam* | *Immb* | *Icmm* |

## Tetragonal space groups

| | | | |
|---|---|---|---|
| 75 $P4$<br>79 $I4$ | 76 $P4_1$<br>80 $I4_1$ | 77 $P4_2$ | 78 $P4_3$ |
| 81 $P\bar{4}$ | 82 $I\bar{4}$ | | |
| 83 $P4/m$<br>87 $I4/m$ | 84 $P4_2/m$<br>88 $I4_1/a$ | 85 $P4/n$ | 86 $P4_2/n$ |
| 89 $P422$<br>93 $P4_222$<br>97 $I422$ | 90 $P42_12$<br>94 $P4_22_12$<br>98 $I4_122$ | 91 $P4_122$<br>95 $P4_322$ | 92 $P4_12_12$<br>96 $P4_32_12$ |
| 99 $P4mm$<br>103 $P4cc$<br>107 $I4mm$ | 100 $P4bm$<br>104 $P4nc$<br>108 $I4cm$ | 101 $P4_2cm$<br>105 $P4_2mc$<br>109 $I4_1md$ | 102 $P4_2nm$<br>106 $P4_2bc$<br>110 $I4_1cd$ |
| 111 $P\bar{4}2m$<br>115 $P\bar{4}m2$<br>119 $I\bar{4}m2$ | 112 $P\bar{4}2c$<br>116 $P\bar{4}c2$<br>120 $I\bar{4}c2$ | 113 $P\bar{4}2_1m$<br>117 $P\bar{4}b2$<br>121 $I\bar{4}2m$ | 114 $P\bar{4}2_1c$<br>118 $P\bar{4}n2$<br>122 $I\bar{4}2d$ |
| 123 $P4/mmm$<br>127 $P4/mbm$<br>131 $P4_2/mmc$<br>135 $P4_2/mbc$<br>139 $I4/mmm$ | 124 $P4/mcc$<br>128 $P4/mnc$<br>132 $P4_2/mcm$<br>136 $P4_2/mnm$<br>140 $I4/mcm$ | 125 $P4/nbm$<br>129 $P4/nmm$<br>133 $P4_2/nbc$<br>137 $P4_2/nmc$<br>141 $I4_1/amd$ | 126 $P4/nnc$<br>130 $P4/ncc$<br>134 $P42/nnm$<br>138 $P4_2/ncm$<br>142 $I4_1/acd$ |

## Trigonal space groups

| | | | |
|---|---|---|---|
| 143 $P3$ | 144 $P3_1$ | 145 $P3_2$ | 146 $R3$ |
| 147 $P\bar{3}$ | 148 $R\bar{3}$ | | |
| 149 $P312$<br>153 $P3_212$ | 150 $P321$<br>154 $P3_221$ | 151 $P3_112$<br>155 $R32$ | 152 $P3_121$ |
| 156 $P3m1$<br>160 $R3m$ | 157 $P31m$<br>161 $R3c$ | 158 $P3c1$ | 159 $P31c$ |
| 162 $P\bar{3}1m$<br>166 $R\bar{3}m$ | 163 $P\bar{3}1c$<br>167 $R\bar{3}c$ | 164 $P\bar{3}m1$ | 165 $P\bar{3}c1$ |

## Hexagonal space groups

| | | | |
|---|---|---|---|
| 168 $P6$<br>172 $P6_4$ | 169 $P6_1$<br>173 $P6_3$ | 170 $P6_5$ | 171 $P6_2$ |
| 174 $P\bar{6}$ | | | |
| 175 $P6/m$ | 176 $P6_3/m$ | | |
| 177 $P622$<br>181 $P6_422$ | 178 $P6_122$<br>182 $P6_322$ | 179 $P6_522$ | 180 $P6_222$ |
| 183 $P6mm$ | 184 $P6cc$ | 185 $P6_3cm$ | 186 $P6_3mc$ |
| 187 $P\bar{6}m2$ | 188 $P\bar{6}c2$ | 189 $P\bar{6}2m$ | 190 $P\bar{6}2c$ |
| 191 $P6/mmm$ | 192 $P6/mcc$ | 193 $P6_3/mcm$ | 194 $P6_3/mmc$ |

## Cubic space groups

| | | | |
|---|---|---|---|
| 195 $P23$<br>199 $I2_13$ | 196 $F23$ | 197 $I23$ | 198 $P2_13$ |
| 200 $Pm\bar{3}$<br>204 $Im\bar{3}$ | 201 $Pn\bar{3}$<br>205 $Pa\bar{3}$ | 202 $Fm\bar{3}$<br>206 $Ia\bar{3}$ | 203 $Fd\bar{3}$ |
| 207 $P432$<br>211 $I432$ | 208 $P4_232$<br>212 $P4_332$ | 209 $F432$<br>213 $P4_132$ | 210 $F4_132$<br>214 $I4_132$ |
| 215 $P\bar{4}3m$<br>219 $F\bar{4}3c$ | 216 $F\bar{4}3m$<br>220 $I\bar{4}3d$ | 217 $I\bar{4}3m$ | 218 $P\bar{4}3n$ |
| 221 $Pm\bar{3}m$<br>225 $Fm\bar{3}m$<br>229 $Im\bar{3}m$ | 222 $Pn\bar{3}n$<br>226 $Fm\bar{3}c$<br>230 $Ia\bar{3}d$ | 223 $Pm\bar{3}n$<br>227 $Fd\bar{3}m$ | 224 $Pn\bar{3}m$<br>228 $Fd\bar{3}c$ |

# BOOK LIST

Here is a short list of books mostly in English that we have found particularly useful. Some more-specialized books we have referred to in the text. References to all the crystal structure data given in this book are to be found in the various compilations listed in D (this is where they came from).

A. INTERNATIONAL TABLES FOR X-RAY CRYSTALLOGRAPHY

These are: Volume A: Space-Group Symmetry, 3rd ed. 1992 [the indispensable reference]. Volume B: Reciprocal Space, 1993. Volume C: Mathematical, Physical and Chemical Tables, 1992. Kluwer Academic, Dordrecht.

B. Some books on crystallography and crystal chemistry

BLOSS, F. D.

*Crystallography and Crystal Chemistry*, reprinted by Mineral. Soc. Amer., Washington, D.C. (1994). Very clear exposition of the crystallographic point groups.

BURNS, G & GLAZER, A. M.

*Space Groups for Solid State Scientists* 2nd Ed., Academic Press, New York (1990). A good informal account of space groups with useful tables.

BOISEN, M. B. & GIBBS, G. V.

*Mathematical Crystallography*, Reviews in Mineralogy **15**, Mineral. Soc. of Amer., Washington, D.C. (Revised, 1990). A systematic account of how to do crystallographic calculations, and a derivation of the three-dimensional point and space groups.

DE JONG, W. F.

*General Crystallography*. Freeman, San Francisco (1959). Subtitled "A brief compendium" this useful little book contains a wide variety of information. Particularly useful for geometric aspects.

HYDE, B. G. & ANDERSSON, S.

*Inorganic Crystal Structures,* John Wiley & Sons, New York (1989). Systematic description of crystal structures with special emphasis on the development of "complex" structures from simpler ones using simple building principles. Numerous tables of data.

MEGAW, H. D.

*Crystal Structures: A Working Approach*, Saunders, Philadelphia (1973). Clear descriptions of symmetry and the crystallographic description of structures.

PEARSON, W. B.

*The Crystal Chemistry and Physics of Metals and Alloys*, John Wiley & Sons, New York (1972). A comprehensive account of the subject at the time and still very useful.

SMITH, J. V.

*Geometrical and Structural Crystallography*, John Wiley & Sons, New York (1982). A good introduction to formal crystallography. Intended for those who are prepared to *work* through a number of carefully considered examples.

WELLS, A. F.

*Structural Inorganic Chemistry,* 5th Ed., Clarendon Press, Oxford (1984). Contains a wealth of organized structural information with due attention to crystal structures. Introductory chapters discuss polyhedra, sphere packings *etc*. Every chemist should own a copy.

C. Some books on geometry

COXETER, H. S. M.
*Introduction to Geometry*, John Wiley & Sons, New York (1971). A classic that everyone should own and read. Includes a good account of two-dimensional symmetry groups.

CUNDY, H. M. & ROLLET, A. P.
*Mathematical Models*, 2nd Ed., Clarendon Press, Oxford (1961). Written for English sixth-formers, this book has, among other things, a lot of useful information on, and practical tips for making, polyhedra.

GRÜNBAUM, B. & SHEPHARD, G. C.
*Tilings and Patterns,* W. H. Freeman, New York (1986). An astonishing book, beautifully illustrated, that should dispel any illusions that two-dimensional patterns are boring. Heavy going in places.

HILBERT, D. & COHN-VOSSEN, S.
*Geometry and the Imagination,* Chelsea, New York (1952). Another classic, and one that is as fresh today as when it was written (originally published in 1932). 330 figures, some very beautiful. Every personal library should contain a copy.

SHUBNIKOV, A. V. & KOPSTIK, V. A.
*Symmetry in Science and Art,* Plenum Press, New York (1974). A readable guide to symmetry groups (including black-and-white and color groups) in different dimensions (*e.g.* rod and layer groups). Also material of general interest as suggested by the title.

WENNIGER, M. J.
*Polyhedron Models*, Cambridge University Press (1971). Illustrations of beautiful polyhedra and instructions for making them.

D. Reference books and data bases

STRUCTURE REPORTS
Originally *Strukturbericht* (Vols. 1-7), these are annual compendia of crystal structure data. The earlier volumes gave enough information to be a sufficient reference source, but later volumes have become mainly bibliographies (references without author's names!). The last volume was for 1990 and the series which started so auspiciously under P. P. Ewald and C. Hermann in 1931 seems to have died ignominiously without even a whimper. Hampered by a highly eccentric indexing system.

VILLARS, P. & CALVERT, L. D.
*Pearson's Handbook of Crystallographic Data for Intermetallic Phases.* 2nd edition, American Society of Metals, Metals Park, Ohio (1991). "Handbook" may not be the correct term for four volumes, each weighing more than 3 kg. Comprehensive and valuable reference source of structural data for compounds involving metallic elements, also includes data for pnictides and chalcogenides. Only *binary* oxides and no halides are included, and the price precludes personal ownership for most.
The second edition omits some compounds reported in Volume I of the first edition, so one needs to have both. The assignments to structure types are not always reliable and one should check the original papers if unit cells and/or origins are changed. We have found a number of errors and misprints.

WYCKOFF, R. W. G.
*Crystal Structures.* John Wiley & Sons, New York. Vol 1 (1963), Vol 2 (1964), Vol 3 (1965), Vol 4 (1968). A convenient reference organized by structure type if one does not want the latest information.

INORGANIC CRYSTAL STRUCTURE DATA BASE - Gmelin Institut, Karlsruhe.
Contains 39,000 entries and growing at a rate of ≈ 1500 per year.
Criteria for inclusion are (a) no C-C and/or C-H bonds (these are in the Cambridge Structure Data Base) (b) contains at least one of H, He, B, C, N, O, F, Ne, Si, P, S, Cl, Ar, Se, Br, Te, I, Xe, At and Rn (*i.e.* elements right of the "Zintl line" on the periodic table). Neither perfect (we find errors) or comprehensive (some surprising omissions) but excellent value for money.

# Formula Index

This index lists chemical compounds for which some structural information appears in Chapters 1-7 or Appendix 5. If the page number is in bold face full crystallographic data are given. Look also for trivial names in the subject index (especially for minerals and zeolites).

# Subject Index

The Table of Contents should be consulted for major topics discussed in this book (space groups, polyhedra, sphere packings, etc.). The subject index serves mainly to guide the reader to definitions of terms, mineral names, etc. so only the principal (defining) occurrences of a term are indicated. Note also we normally use the "natural" order of words as in "clathrate hydrate" not "hydrate, clathrate" (but try different combinations). For chemical compounds, also consult the formula index.